321

Trends in Banach Spaces and Operator Theory

A Conference on
Trends in Banach Spaces and Operator Theory
October 5–9, 2001
University of Memphis

Anna Kamińska
Editor

American Mathematical Society
Providence, Rhode Island

This volume contains the proceedings of a conference on Trends in Banach Spaces and Operator Theory which was held at the University of Memphis, October 5–9, 2001.

2000 *Mathematics Subject Classification.* Primary 22A22, 46Axx, 46Bxx, 46E30, 46Lxx, 47Axx, 47Bxx, 47Hxx, 47Lxx, 51F15.

Library of Congress Cataloging-in-Publication Data

Trends in Banach spaces and operator theory : a conference on trends in Banach spaces and operator theory, October 5–9, 2001, University of Memphis / Anna Kaminska, editor.

p. cm. — (Contemporary mathematics, ISSN 0271-4132 ; 321)

Includes bibliographical references.

ISBN 0-8218-3234-4 (softcover : acid-free paper)

1. Banach spaces—Congresses. 2. Operator theory—Congresses. I. Kaminska, Anna, 1950– II. Contemporary mathematics (American Mathematical Society) ; v. 321.

QA322.2.T69 2003
515′.732—dc21 2003041485

∞ The paper used in this book is acid-free and falls within the guidelines
established to ensure permanence and durability.
Visit the AMS home page at `http://www.ams.org/`

10 9 8 7 6 5 4 3 2 1 08 07 06 05 04 03

Contents

Preface

This volume contains the proceedings of the conference Trends in Banach Spaces and Operator Theory, held at the University of Memphis during the week of October 5–9, 2001. The conference was devoted to the recent advances in theories of Banach spaces and linear operators. These theories are intimately related, and they lie at the core of such diverse fields as harmonic analysis, partial differential equations, approximation theory, dynamical systems, probability, and financial mathematics. Stefan Banach, the inventor of the spaces that were later known by his name, wrote a monograph in 1932 under the very meaningful title *Thèorie des opérations linéaires*. The monograph presented the theory of normed and locally convex spaces and showed how the existing examples of function spaces and classical operators motivated the foundations of the new theory. It also described how this in turn influenced the development of the theory of linear operators. The constant interplay between the structure of spaces and properties of operators acting on them is crucial for both theories.

The conference brought together more than 130 participants from nineteen different countries. It gathered mathematicians who represented the entire spectrum of researchers and included both some of the most experienced specialists in the field as well as young mathematicians at the early stages of their research programs.

During the meeting the following thirteen principal speakers delivered one-hour plenary lectures.

Yuri Abramovich, Indiana University-Purdue University Indianapolis
Sheldon Axler, San Francisco State University
John B. Conway, University of Tennessee in Knoxville
Carl C. Cowen, Purdue University
Joe Diestel, Kent State University
Nigel Kalton, University of Missouri in Columbia
Barbara MacCluer, University of Virginia in Charlottesville
Edward W. Odell, University of Texas at Austin
Aleksander Pełczyński, Polish Academy of Sciences, Warsaw, Poland
Gilles Pisier, Texas A&M University in College Station
and Université de Paris VI
Haskell P. Rosenthal, University of Texas at Austin
Thomas B. Schlumprecht, Texas A&M University in College Station
Nicole Tomczak-Jaegermann, University of Alberta in Edmonton

In addition to plenary lectures, there were also eighty 20-minute talks organized in four parallel sessions. A panel discussion held during the meeting provided for a discussion of opinions, conclusions, and reflections on the current state of the field. The panel also identified multiple possibilities for future development of the theory and its applications. The discussion evolved into a lively and interesting event thanks to all participants, but especially because of the efforts of the principal speakers who volunteered to be members of the panel. Special recognition should also go out to Professor Sheldon Axler, who very skillfully moderated this rather unusual session.

Sadly, one of our principal speakers recently passed away. Yuri Abramovich died on February 5, 2003, after a four-year illness. In his memory, the Department of Mathematical Sciences at Indiana University-Purdue University Indianapolis has established the Yuri Abramovich Memorial Scholarship.

The proceedings presented here contain 25 papers; some of them are expository articles, while others are presentations of new results. Most of these results were presented during the conference. Some of the topics include Banach and operator space structure of C^*-algebras, Banach spaces with small spaces of operators or with abundance of non-trivial operators, non-commutative L^p-spaces, sectorial operators, composition operators, tensor products, isometric properties of the norm, bases in quasi-Banach spaces, Fourier algebras for locally compact groupoids, and geometry of Coxeter-invariant polyhedra.

The conference was sponsored by grants from the National Science Foundation, the University of Memphis Foundation, the Office of Academic Affairs and the Office of Research of the University of Memphis, and the University of Mississippi.

The members of the Mathematical Analysis seminar group from Memphis served as the organizational team for the conference. The group met over the past several years and included Jim Jamison, Anna Kamińska and Pei-Kee Lin from the University of Memphis, as well as Przemo Kranz from the University of Mississippi. Ray Clapsadle from the University of Memphis was also an invaluable addition to the organizational team. He created the web page for the conference, handled registration, and designed the logo and T-shirts for this event. Special acknowledgement should go to Jim Jamison, whose deep involvement and energy were crucial for the successful preparation and running of the meeting.

On behalf of the organizing committee I would like to take this opportunity to thank all our colleagues, secretarial staff members, and graduate students from the math department of the University of Memphis. Thanks to all for your efforts in assisting in running the meetings, chairing the sessions, helping with registration and doing several other necessary jobs.

Anna Kamińska

Contemporary Mathematics
Volume **321**, 2003

Characterizations of the reflexive spaces in the spirit of James' Theorem

María D. Acosta, Julio Becerra Guerrero, and Manuel Ruiz Galán

Since the famous result by James appeared, several authors have given characterizations of reflexivity in terms of the set of norm-attaining functionals. Here, we offer a survey of results along these lines. Section 1 contains a brief history of the classical results. In Section 2 we assume some topological properties on the size of the set of norm-attaining functionals in order to obtain reflexivity of the space or some weaker conditions. Finally, in Section 3, we consider other functions, such as polynomials or multilinear mappings instead of functionals, and state analogous versions of James' Theorem.

Hereinafter, we will denote by B_X and S_X the closed unit ball and the unit sphere, respectively, of a Banach space X. The stated results are valid in the complex case. However, for the sake of simplicity, we will only consider **real** normed spaces.

1. James' Theorem

In 1950 Klee proved that a Banach space X is reflexive provided that for every space isomorphic to X, each functional attains its norm [**Kl**]. James showed in 1957 that a separable Banach space allowing every functional to attain its norm has to be reflexive [**Ja1**]. This result was generalized to the non-separable case in [**Ja2**, Theorem 5]. After that, a general characterization of the bounded, closed and convex subsets of a Banach space that are weakly compact was obtained:

THEOREM 1.1. ([**Ja3**, Theorem 4]) *In order that a bounded, closed and convex subset K of a Banach space be weakly compact, it suffices that every functional attain its supremum on K.*

See also [**Ja3**, Theorem 6] for a version in locally convex spaces. James gave a wider list of characterizations of reflexivity in [**Ja4**, Theorems 1, 2 and 3]. A simpler proof of this result for the unit ball appeared in [**Ja5**, Theorem 2].

Since then, some authors have tried to obtain easier proofs for this result. For instance, Pryce [**Pr**] simplified in some aspects the proof of James' Theorem. In the

1991 *Mathematics Subject Classification.* Primary 46A25, 47A07 ; Secondary 47H60, 47A12.
Key words and phrases. Reflexivity, James' Theorem, norm attaining functional, numerical radius attaining operator, norm attaining polynomial.
The first and third author were supported in part by D.G.E.S., project no. BFM 2000-1467. The second author was partially supported by Junta de Andalucía Grant FQM0199.

text by Holmes [**Ho**, Theorem 19.A], one can find Pryce's proof in the case that the dual unit ball is weak-$*$ sequentially compact, which enables us to eliminate some steps. Another proof, owing to Simons [**Si1**], makes use of a minimax inequality. We will show a proof of Theorem 1.1 for those spaces that do not contain an isomorphic copy of ℓ_1 or are separable. The key result in this proof is another result by Simons, the so-called "Simons' inequality" [**Si2**]. We also refer to J. Diestel [**Di**] and K. Floret [**Fl**].

No proof of James' Theorem, in its general version, can be considered elementary. However, in the text by Deville, Godefroy and Zizler [**DGZ**] an easy proof is obtained from Simons' inequality in the separable case.

PROPOSITION 1.2. ([**Si2**]) *Suppose that X is a Banach space, $B \subset A$ are bounded subsets of X^*, $\{x_n\}$ is a bounded sequence in X, C is the closed convex hull of $\{x_n : n \in \mathbb{N}\}$ and it is satisfied that*

$$\forall x \in C \ \exists\, b^* \in B \ : \ b^*(x) = \sup_{a^* \in A} a^*(x).$$

Then

$$\sup_{b^* \in B} \limsup_n b^*(x_n) = \sup_{a^* \in A} \limsup_n a^*(x_n).$$

Let us recall that a Banach space has the*Grothendieck property* provided that each weak-$*$ null sequence in X^* is actually weak null. It is very easy to check that a separable Banach space satisfying the Grothendieck property is reflexive.

PROPOSITION 1.3. *Let X be a Banach space such that every functional attains its norm. Then X has the Grothendieck property.*

PROOF. Let us assume that X does not satisfy the Grothendieck property. Then there exist a weak-$*$ null sequence $\{x_n^*\}$ in S_{X^*} and a functional $\varphi \in B_{X^{***}} \backslash \{0\}$ such that φ is a cluster point of $\{x_n^*\}$ in the w^*-topology of X^{***} and

$$\forall x \in X, \quad \varphi(x) = 0. \tag{1}$$

Let us fix $x_0^{**} \in B_{X^{**}}$ with $\varphi(x_0^{**}) > 0$. If we take as $B := B_X$ and $A := B_{X^{**}}$ in Simons' inequality, we obtain

$$\sup_{x \in B_X} \limsup_n x_n^*(x) = \sup_{x^{**} \in B_{X^{**}}} \limsup_n x^{**}(x_n^*),$$

which is impossible, since

$$\sup_{x \in B_X} \limsup_n x_n^*(x) = 0$$

and

$$\sup_{x^{**} \in B_{X^{**}}} \limsup_n x^{**}(x_n^*) \geq \varphi(x_0^{**}) > 0.$$

■

It is also known that a Banach space is reflexive if it does not contain ℓ_1 and has the Grothendieck property [**Va**].

A different proof of James' Theorem under some restrictions can be found in [**FLP**, Theorem 5.9]. Here, the authors use the following key result:

If a convex and w^*-compact subset $K \subset X^*$ has a boundary B which is norm separable, then K is the closure (in the norm topology) of the convex hull of B [**FLP**, Theorem 5.7]. The above result can be easily deduced from Simons' inequality (see also [**Go**, Theorem I.2]).

Azagra and Deville proved that in any infinite-dimensional Banach space X, there is a bounded and starlike body $A \subset X$ (subset containing a ball centered at zero such that every ray from zero meets the boundary of A once at the most) such that every functional attains infinitely many local maxima on A [**AD**].

2. Other results relating the size of the set of norm attaining functionals with properties of the Banach space

In the following, we will write $A(X)$ for the set of norm attaining functionals on X. Let us recall that this set is always dense in the dual space (Bishop-Phelps Theorem [**BP1**]).

Apart from James' Theorem, some other results imply isomorphic properties of X by assuming a weaker condition on the size of $A(X)$. In order to state the first of these results, let us note that for a dual space X, say Y^*, then $Y \subset Y^{**} = X^*$ is a closed subspace of X^*, which is w^*-dense (Goldstine's Theorem) and such that $Y \subset A(Y^*) = A(X)$. Petunin and Plichko got the converse result under additional conditions.

THEOREM 2.1. ([**PPl**]) *Assume that X is separable and there is a Banach space $Y \subset X^*$ such that $Y \subset A(X)$ and Y is w^*-dense in X^*. Then X is isometric to a dual space.*

The previous result was also shown to hold in the case that X is weakly compactly generated [**Ef**].

In order to motivate the next result, let us begin with a simple example. For the space c_0, the set $A(c_0)$ is the subset of finite supported sequences, so $A(c_0) \subset \ell_1$ is of the first Baire category. Of course, the unit ball of c_0 has no slices of small diameter, since any slice of it has diameter equal to 2. That is, the unit ball of c_0 is not dentable. For separable spaces sharing with c_0 this isometric property of the unit ball, as a consequence of the techniques developed by Bourgain and Stegall, we obtain a similar result. Before stating it, let us introduce the following technical result:

LEMMA 2.2. ([**Bou**, Lemma 3.3.3]) *Let X be a Banach space, $x_0, y \in X$, $x^*, y^* \in S_{X^*}$ and $t > 0$. Suppose that $x^*(y) < x^*(x_0)$, $\frac{2}{t}\|x_0 - y\| \leq 1$ and $y^*(x_0) > \sup\{y^*(x) : x \in y + (\ker x^* \cap tB_X)\}$. Then*

$$\|x^* - y^*\| \leq \frac{2}{t}\|x_0 - y\|.$$

THEOREM 2.3. ([**Bou**, Theorem 3.5.5]) *Let X be a Banach space and $C \subset X$ be a closed, bounded and convex subset which is separable and non-dentable. Then*

$$A(C) := \{x^* \in X^* : x^* \text{ attains its maximum on } C\}$$

is of the first Baire category in X^*.

PROOF. Since C is not dentable, there is $\delta > 0$ such that any slice of C has diameter at least 3δ. Since C is separable, there is a dense sequence $\{x_n\}$ in C. We write, for all $n \geq 1$,

$$C_n = C \cap (x_n + \delta B_X)$$

and

$$O_n = \{x^* \in X^* : S(C, x^*, \eta) \cap C_n = \emptyset, \text{ for some } \eta\},$$

where $S(C, x^*, \eta) = \{x \in C : x^*(x) > \sup x^*(C) - \eta\}$. Since $\{x_n : n \in \mathbb{N}\}$ is dense in C, it holds that $C = \cup_n C_n$ and so it is immediate to check that $A(C) \subset X^* \backslash (\cap_n O_n)$.

It suffices to prove that O_n is an open and dense set, for every n. Assume that $x^* \in O_n$, therefore $S(C, x^*, \eta) \cap C_n = \emptyset$ for some $\eta > 0$. For $\varepsilon > 0$ such that $\varepsilon \sup_{c \in C} \|c\| < \frac{\eta}{4}$, it is satisfied that

$$y^* \in x^* + \varepsilon B_{X^*} \Rightarrow S\Big(C, y^*, \frac{\eta}{2}\Big) \subset S(C, x^*, \eta) \tag{1}$$

since

$$|x^*(x) - y^*(x)| \leq \varepsilon \sup_{c \in C} \|c\|, \quad \forall x \in C.$$

Therefore,

$$\begin{aligned} x \in S\Big(C, y^*, \frac{\eta}{2}\Big) &\Rightarrow x^*(x) \geq y^*(x) - \varepsilon \sup_{c \in C} \|c\| > \\ &> \sup y^*(C) - \frac{\eta}{2} - \varepsilon \sup_{c \in C} \|c\| \geq \\ &\geq \sup x^*(C) - \frac{\eta}{2} - 2\varepsilon \sup_{c \in C} \|c\| > \sup x^*(C) - \eta. \end{aligned}$$

Since $S(C, x^*, \eta) \cap C_n = \emptyset$, then by (1), we know that $S\Big(C, y^*, \frac{\eta}{2}\Big) \cap C_n = \emptyset$ and $y^* \in O_n$, so O_n is open.

Now we will check that O_n is dense in X^*. Let us fix $x^* \in X^*$ and $0 < \varepsilon < 1$. Since $S(C, x^*, \eta) = S(C, tx^*, t\eta)$ for any $\eta, t > 0$, then $\mathbb{R}^+ O_n \subset O_n$ and we can assume that $\|x^*\| = 1$.

C is bounded and $x^* \neq 0$; hence we can choose $y_0 \in X$ such that

$$x^*(x) > x^*(y_0), \quad \forall x \in C.$$

Let $M = \sup\{\|x - y_0\| : x \in C\}$, fix $t > \frac{2M}{\varepsilon}$, and write $V = \ker x^* \cap tB_X$, $A = V + y_0$. By the choice of y_0, it holds that $C_n \backslash A \neq \emptyset$, since $x^*(a) = x^*(y_0)$ for any $a \in A$ and $C_n \subset C$, and so $x^*(x) > x^*(y_0) = x^*(a)$ for any $x \in C$.

Since diam $C_n \leq 2\delta < 3\delta$ and $C_n \backslash A \neq \emptyset$, by the Hahn-Banach separation Theorem, there is a functional whose supremum on C_n is greater than the supremum on A. An appropriate slice of $\overline{\text{co}}(C_n \cup A)$ determined by this functional is essentially contained in C_n (see [**Bou**, Theorem 3.4.1]). Hence, there is a slice S of the set $\overline{\text{co}}(C_n \cup A)$ with diameter at most 3δ and such that $S \cap C_n \neq \emptyset$. Then in the case that $C \subset \overline{\text{co}}(C_n \cup A)$ since $C_n \subset C$, $S \cap C$ would be a slice of C with diameter less than 3δ, which is not possible because of the choice of δ. Thus, there is an element $c_0 \in C \backslash \overline{\text{co}}(C_n \cup A)$. Hence, there is a functional $y^* \in S_{X^*}$ such that

$$y^*(c_0) > \sup y^*(C_n \cup A).$$

Finally, in view of $\frac{2}{t}\|c_0 - y_0\| \leq \frac{2}{t} M \leq \varepsilon < 1$, from Lemma 2.2, it follows that $\|x^* - y^*\| \leq \frac{2}{t}\|c_0 - y\| \leq \frac{2}{t} M < \varepsilon$.

Since $\sup y^*(C) \geq y^*(c_0) > \sup y^*(C_n)$, then $S(C, y^*, \eta) \cap C_n = \emptyset$ for $\eta < \sup y^*(C) - \sup y^*(C_n)$, that is, $y^* \in O_n$ and O_n is dense. ∎

As far as the authors know, it remains open whether the previous result also holds by dropping the separability of the space. However, Kenderov, Moors and Sciffer proved the following result:

THEOREM 2.4. ([**KMS**]) *If K is any infinite compact and Hausdorff topological space, then $A(C(K))$ is of the first Baire category.*

By considering the weak-$*$ topology in X^*, instead of the norm topology, Debs, Godefroy and Saint Raymond proved that a separable and non reflexive Banach space X satisfies that $A(X)$ does not contain weak-$*$ open sets [**DGS**].

The analogous result for any Banach space was proven to hold by Jiménez-Sevilla and Moreno [**JiM**].

LEMMA 2.5. ([**JiM**, Lemma 3.1]) *If the dual unit ball of a Banach space X contains a slice of norm attaining functionals, then X is reflexive.*

PROOF. Assume that $x_0^{**} \in S_{X^{**}}$ determines a slice such that
$$S \equiv S(B_{X^*}, x_0^{**}, \eta) \subset A(X).$$
Let us fix $x_0^* \in B_{X^*}$ with $\|x_0^*\| < 1$ satisfying that $x_0^{**}(x_0^*) > 1 - \frac{\eta}{2}$. We define
$$B := (S - x_0^*) \cap (-S + x_0^*) \subset (B_{X^*} - x_0^*) \cap (B_{X^*} + x_0^*).$$
By the choice of x_0^*, we know that an element in $(B_{X^*} - x_0^*) \cap (B_{X^*} + x_0^*)$ is in fact in B, since $x^* - x_0^* = y^* + x_0^* \quad (\|x^*\|, \|y^*\| \leq 1)$ implies that $\frac{x^* - y^*}{2} = x_0^*$ and we know that
$$1 - \frac{\eta}{2} < x_0^{**}(x_0^*) = \frac{x_0^{**}(x^*) - x_0^{**}(y^*)}{2} \leq \frac{1 - x_0^{**}(y^*)}{2},$$
that is,
$$x_0^{**}(-y^*) > 1 - \eta \Rightarrow y^* \in -S,$$
and by a similar argument $x^* \in S$ also, so
$$x^* - x_0^* = y^* + x_0^* \in (S - x_0^*) \cap (-S + x_0^*),$$
as we wanted to check.

Hence the balanced and convex set $B = (B_{X^*} - x_0^*) \cap (B_{X^*} + x_0^*)$ is a w^*-closed set of X^* and it is immediate that $rB_{X^*} \subset B$ for $0 < r < 1 - \|x_0^*\|$. That is, B is the dual unit ball of an equivalent norm $\|\!|\ \|\!|$ on X.

By assumption, for any $x^* \in X^*$ with $\|\!| x^* \|\!| = 1$, then $x^* + x_0^* \in S$ and $-x^* + x_0^* \in S$. Also, $\|x^* + x_0^*\| = 1$ or $\|x^* - x_0^*\| = 1$. In the first case, there is $x_0 \in X \backslash \{0\}$ with
$$\begin{aligned} (x^* + x_0^*)(x_0) \geq y^*(x_0), &\qquad \forall y^* \in X^*, \|y^*\| \leq 1 \\ \Rightarrow x^*(x_0) \geq z^*(x_0), &\qquad \forall z^* \in B_{X^*} - x_0^* \\ \Rightarrow x^*(x_0) \geq z^*(x_0), &\qquad \forall z^* \in B, \end{aligned}$$
that is, x^* attains its (new) norm at the element $\frac{x_0}{\|\!| x_0 \|\!|}$. In the second case, we argue in a similar way to arrive at the same conclusion. We checked that for the norm $\|\!|\ \|\!|$ any functional in the unit sphere attains the norm, and so, by James' Theorem, the space is reflexive. ∎

THEOREM 2.6. ([**JiM**, Proposition 3.2]) *Let X be a Banach space; then either X is reflexive or $S_{X^*} \cap A(X)$ contains no (non empty) open set in (S_{X^*}, w^*). Also in the second case, the convex hull of $S_{X^*} \backslash A(X)$ is (norm) dense in the dual unit ball.*

PROOF. If $S_{X^*} \cap A(X)$ contains a (non-empty) w^*-open set in (S_{X^*}, w^*), we can assume without losing generality that there is a subset $W \subset X$ that can be expressed in the form

$$W = \{x^* \in X^* : x^*(x_i) \geq \delta_i, i = 1, \cdots, n\},$$

where $x_i \in S_X$, $\delta_i > 0 \quad (1 \leq i \leq n)$, such that $W \cap S_{X^*}$ has a non-empty interior in the relative w^*-topology and such that $W \cap S_{X^*} \subset A(X)$.

An element x^* in the border (norm topology) of $V := W \cap B_{X^*}$ has to satisfy that $\|x^*\| = 1$ or $x^*(x_i) = \delta_i$ for some $1 \leq i \leq n$. In the first case, $x^* \in S_{X^*} \cap W$, so x^* attains its norm. As a consequence, there is an element $x \in S_X$ such that

$$x^*(x) \geq y^*(x), \qquad \forall y^* \in V. \tag{1}$$

In the second case, the above inequality holds for $x = -x_i$. If y_0^* satisfies that $\|y_0^*\| < 1$ and $y_0^*(x_i) > \delta_i$ for every $1 \leq i \leq n$, then y_0^* is interior (norm topology) to V and so the subset

$$B := (V - y_0^*) \cap (y_0^* - V)$$

contains a ball centered at zero. It is clear that B is a bounded, convex, balanced and w^*-closed subset of X^*. Hence B is the dual unit ball for an equivalent norm on X^*.

By using that every element x^* in the border of V satisfies (1), it is easy to check that any element in the border of B that can be written as $w^* - y_0^* = y_0^* - z^*$, where w^* or z^* belongs to the border of V, attains its (new) norm. By James' Theorem, X is reflexive.

Finally, if the convex hull of $S_{X^*} \backslash A(X)$ were not dense in B_{X^*}, then by the Hahn-Banach separation Theorem, there is a slice S of B_{X^*} such that $S \cap S_{X^*} \subset A(X)$. It follows that $S \subset A(X)$, and by Lemma 2.5, X is reflexive. ∎

By using Lemma 2.5 again then, for spaces with special geometric properties of smoothness, it can be stated that the space is reflexive or $A(X)$ contains no open subset for the norm topology. Let us recall that a Banach space X has the *Mazur intersection property* if, and only if, every bounded closed convex subset of X is an intersection of closed balls [**Ma**]. For instance, spaces with a Fréchet differentiable norm satisfy the Mazur intersection property.

THEOREM 2.7. ([**JiM**, Proposition 3.3]) *A Banach space X satisfying the Mazur intersection property is either reflexive or $A(X)$ contains no (non-empty) open set.*

PROOF. If $A(X)$ contains an open ball, since $A(X)$ is a cone, then $A(X)$ contains an open ball centered at an element x_0^* in S_{X^*} of radius $r > 0$. Since X has the Mazur intersection property, then the subset of w^*-denting points of the unit sphere is dense in S_{X^*} [**GGS1**, Theorem 2.1]. Therefore, there is an element $y_0^* \in S_{X^*}$ with $\|y_0^* - x_0^*\| < \frac{r}{2}$ such that y_0^* has w^*-slices of small diameter.

Then, for every $0 < \varepsilon < \frac{r}{2}$, there is an element $y_0 \in X$ determining a slice on B_{X^*} with diameter less than ε such that the slice contains y_0^*. Hence for some $\eta > 0$ we have

$$S(B_{X^*}, y_0, \eta) \subset y_0^* + \varepsilon B_{X^*}.$$

Therefore, the above slice is contained in

$$y_0^* + \frac{r}{2} B_{X^*} \subset x_0^* + rB_{X^*} \subset A(X).$$

By Lemma 2.5, X is reflexive. ∎

By using a different idea of proof, we will show a new result along the same lines. Before that, let us note that every Banach space can be renormed such that the set of norm attaining functionals contains an open ball [**AR1**, Corollary 2]. Therefore, no trivial isomorphic assumption implies reflexivity in the case that the set of norm attaining functionals contains a ball. Let us recall that a space X is *weak Hahn-Banach smooth* if every $x^* \in A(X)$ has a unique Hahn-Banach extension to X^{**}.

THEOREM 2.8. ([**AR2**, Theorem 1]) *A weak Hahn-Banach space X, such that $A(X)$ has a non-empty interior, is reflexive.*

PROOF. Every weak Hahn-Banach space is Asplund [**GGS2**, Theorem 3.3], so X does not contain ℓ_1. We will assume that X is not reflexive, therefore by [**Va**, Theorem 5], X does not have the Grothendieck property. That is, there is a sequence $\{x_n^*\}$ in S_{X^*} which is w^*-convergent to zero, but not weakly convergent. Therefore, there is a non zero w^*-cluster point of $\{x_n^*\}$ in X^{***}, $x_0^{***} \in B_{X^{***}}$, also satisfying $x_0^{***}{}_{|X} \equiv 0$.

If we fix $x_0^{**} \in S_{X^{**}}$ and $\varepsilon > 0$, by passing to a subsequence we will assume that

$$x_0^{**}(x_n^*) > x_0^{***}(x_0^{**}) - \varepsilon, \qquad \forall n \in \mathbb{N}.$$

By assumption, there is $x_0^* \in S_{X^*}, r > 0$ with $x_0^* + rB_{X^*} \subset A(X)$. By using Simons' inequality (Proposition 1.2) for the sequence $x_0^* + rx_n^*$ and the subsets $B = B_X$ and $A = B_{X^{**}}$, we will get that

$$\|x_0^*\| \geq x_0^*(x_0^{**}) + r\Big(x_0^{***}(x_0^{**}) - \varepsilon\Big).$$

The previous inequality holds for any positive number ε, so

$$\|x_0^*\| \geq (x_0^* + rx_0^{***})(x_0^{**}).$$

Since x_0^{**} is any element in $S_{X^{**}}$, then

$$\|x_0^*\| \geq \|x_0^* + rx_0^{***}\|,$$

where $x_0^{***} \in X^{***}\backslash\{0\}$. Since $x_0^{***}(x) = 0$, for every $x \in X$, the previous inequality says that the functional $x_0^* + rx_0^{***}$ is a Hahn-Banach extension of $x_0^* \in A(X)$ to X^{**}, which is impossible since X is weak Hahn-Banach smooth. Therefore, X is reflexive. ∎

It is known that the previous result is not true for smooth spaces (spaces such that every element at S_{X^*} has a unique normalized support functional). In fact, every separable Banach space can be renormed in order to be smooth while preserving the subset of norm attaining functionals [**DGS**]. Therefore, by using that every Banach space is isomorphic to a new one satisfying that the set of norm attaining functionals has a non-empty interior, in the separable case we obtain a new norm which is smooth and satisfies that the set of norm attaining functionals contains a ball.

Let us observe that Theorem 2.8 was obtained by Jiménez Sevilla and Moreno for the weak topology [**JiM**, Corollary 3.7]. In relation to the weak topology we can state the following result:

THEOREM 2.9. *Let X be a separable Banach space such that $A(X) \cap S_{X^*}$ contains a (non-empty) open set in the topological space (S_{X^*}, w). Then X is quasi-reflexive.*

PROOF. Suppose that there is a (relative) weak open and convex subset $U \subseteq B_{X^*}$ such that its closure W satisfies $W \cap S_{X^*} \subseteq A(X)$. We may assume that $0 \in U$. If not, take $x^* \in U$, $\|x^*\| < 1$, and consider the new equivalent dual norm $|\!|\!| \, . \, |\!|\!|$ whose unit ball is $B = (-x^* + B_{X^*}) \cap (x^* + B_{X^*})$. Observe that $W_1 = (-x^* + W) \cap (x^* - W)$ satisfies $W_1 \cap S_{(X^*, |\!|\!| \ |\!|\!|)} \subseteq A(X, |\!|\!| \ |\!|\!|)$.

We will write $\| \ \|$ for the norm instead of $|\!|\!| \ |\!|\!|$ and from now on we assume that U is of the form

$$U = \{x^* \in B_{X^*} : |x_i^{**}(x^*)| < \delta \ , \ i = 1, ..., n\}$$

where $\{x_i^{**}\}_{i=1}^n \subseteq S_{X^{**}}$ and δ is a positive number. Suppose that the linear space generated by $X \cup \{x_i^{**}\}_{i=1}^n$ is a proper subspace of X^{**}. Then there is a functional $\phi \in S_{X^{***}}$ such that $\phi(z) = 0$ for all $z \in X \cup \{x_i^{**}\}_{i=1}^n$. Given $\varepsilon > 0$, there exists $y^{**} \in S_{X^{**}}$, such that $\phi(y^{**}) > 1 - \varepsilon$. We are assuming that X is separable, so the restriction of the $\sigma(X^{***}, X \cup \{x_i^{**}\}_{i=1}^n \cup \{y^{**}\})$-topology to bounded sets is metrizable. Hence, in view of the w^*-denseness of B_{X^*} in $B_{X^{***}}$, we can find a sequence $\{x_n^*\}$ in S_{X^*} that converges to ϕ in the $\sigma(X^{***}, X \cup \{x_i^{**}\}_{i=1}^n \cup \{y^{**}\})$-topology while satisfying

$$x_n^* \in U, \quad \forall n \geq 1$$

and

$$x_n^*(y^{**}) > 1 - \varepsilon \ , \ \forall n \in \mathbb{N} \ .$$

Now fix $n_0 \in \mathbb{N}$. Since U is open, $U \cap S_{X^*} \subset A(X)$ and $A(X)$ is a cone, we can consider $r > 0$, such that the ball of radius r centered at $x_{n_0}^*$ is contained in $A(X)$. Clearly we are in conditions to apply Proposition 1.2, taking $B = B_X$, $A = B_{X^{**}}$ and $\{x_{n_0}^* + r x_n^*\}$, so we get

$$\sup_{x \in B_X} \limsup_n (x_{n_0}^* + r x_n^*)(x) = \sup_{x^{**} \in B_{X^{**}}} \limsup_n (x_{n_0}^* + r x_n^*)(x^{**}) \quad (1)$$

But,

$$\sup_{x \in B_X} \limsup_n (x_{n_0}^* + r x_n^*)(x) = \|x_{n_0}^*\| = 1$$

and

$$\sup_{x^{**} \in B_{X^{**}}} \limsup_n (x_{n_0}^* + r x_n^*)(x^{**}) \geq (x_{n_0}^* + r x_n^*)(y^{**}) \geq (1 - \varepsilon)(1 + r).$$

Since $\varepsilon > 0$ is arbitrary, the above equality (1) implies $1 \geq 1 + r$, a contradiction. ∎

By using essentially the same idea of proof as in Theorem 2.8, but combining it with other results it was shown:

THEOREM 2.10. ([**ABR2**, Theorem 1]) *A Banach space X is reflexive if it does not contain ℓ_1 and for some $r > 0$,*

$$B_{X^*} = \overline{\text{co}}^{w^*}\{x^* \in S_{X^*} : x^* + rB_{X^*} \subset A(X)\},$$

where we denote by $\overline{\text{co}}^{w^*}$ the w^*-closure of the convex hull.

For the space ℓ_1 the dual unit ball satisfies the previous assumption for $r = \frac{1}{2}$. In fact, every Banach space can be renormed such that the convex hull of the interior of the norm attaining functionals in the unit sphere is (norm) dense in the dual unit ball [**ABR2**, Proposition 3].

Finally, let us observe that there is also a version of Theorem 2.10 for operators instead of functionals (see [**ABR1**]) and a result along the same lines for the space of operators (or compact operators) between two Banach spaces.

3. James-type results for polynomials, multilinear forms and numerical radius

This section is devoted to the study of some versions of James' Theorem in which polynomials and multilinear forms are considered instead of linear functionals. Moreover, we give a James-type result in terms of numerical radius attaining operators. In spite of the variety of problems, it is possible to offer a unifying treatment of them, thanks to a reformulation of James' Theorem stated below.

Let us recall that, given a Banach space X and a bounded subset C of X^*, a subset B of C is said to be a *boundary* for C if

$$\forall x \in X\ \exists\, b^* \in B : b^*(x) = \sup_{c^* \in C} c^*(x).$$

Bauer's optimization Principle provides some examples: the set of extreme points of the dual unit ball of a Banach space is a boundary for that ball.

The reformulation above mentioned is established in these terms:

THEOREM 3.1. ([**Ru**]) *Let X be a Banach space, let $C \subset X^{**}$ and let B be a boundary for C. If $B \subset X$, then $C \subset X$.*

Observe that James' Theorem follows easily from this result.

Let us recall that a continuous homogeneous polynomial P on a Banach space X *attains the norm* when

$$\exists\, x_0 \in B_E\ :\ |P(x_0)| = \|P\| := \sup_{x \in B_X} |P(x)|.$$

In order to pose the right version of James' Theorem for polynomials, let us note that even for reflexive spaces there are polynomials not attaining the norm. For instance, consider the continuous 2-homogeneous polynomial on ℓ_2 given by

$$P(x) := \sum_{n \geq 1} \Big(1 - \frac{1}{n}\Big) x(n)^2, \qquad (x \in \ell_2)$$

which does not attain the norm, since $\|P\| = 1$ and if $x \neq 0$ then $|P(x)| < \|x\|^2$.

On the other hand, notice that, given n functionals $x_i^* \in X^*$ $(1 \leq i \leq n)$, the polynomial $P_{x_1^* \cdots x_n^*}$ defined by

$$P_{x_1^* \cdots x_n^*}(x) := x_1^*(x) \cdots x_n^*(x), \quad (x \in X)$$

attains the norm provided that X is reflexive. These polynomials are precisely those that allow us to establish a polynomial James' Theorem:

THEOREM 3.2. ([**Ru**]) *A Banach space X is reflexive if (and only if) there exist $n \geq 1$ and $x_1^*, \ldots, x_n^* \in X^*\backslash\{0\}$ such that, for all $x^* \in X^*$, the polynomial $P_{x_1^* \cdots x_n^* x^*}$ attains the norm.*

Let us emphasize that Theorem 3.2 follows from Theorem 3.1 for suitable subsets B of X and C of X^{**}.

A different question from the one considered here, the reflexivity of the space of all continuous n-homogeneous polynomials on a Banach space, has been studied by several authors (see [**JM**] and the references therein).

Next we exhibit a James' Theorem by means of multilinear forms. A continuous n-linear form φ on a Banach space X *attains its norm* when there exist $x_1, \ldots, x_n \in B_X$ such that

$$|\varphi(x_1, \ldots, x_n)| = \|\varphi\| := \sup\{|\varphi(y_1, \ldots, y_n)| : y_1, \ldots, y_n \in B_X\}).$$

The multilinear version of Theorem 3.2 can be easily derived form James' Theorem, but this does not happen in the symmetric case. If $x_1^*, \ldots, x_n^* \in X^*$, we will write $S_{x_1^* \cdots x_n^*}$ for the symmetrization of the multilinear form

$$(x_1, \ldots, x_n) \mapsto x_1^*(x_1) \cdots x_n^*(x_n), \quad (x_1, \ldots, x_n \in X),$$

i.e.,

$$S_{x_1^* \cdots x_n^*}(x_1, \ldots, x_n) = \frac{1}{n!} \sum_{\sigma \in \Delta_n} x_{\sigma(1)}^*(x_1) \cdots x_{\sigma(n)}^*(x_n)$$

(Δ_n is the set of all permutations of n elements). For this kind of multilinear forms we have, making use of Theorem 3.1,

THEOREM 3.3. ([**Ru**]) *A Banach space X is reflexive if (and only if) there are $n \geq 1$ and $x_1^*, \ldots, x_n^* \in X^*\backslash\{0\}$ such that for all $x^* \in X^*$ the symmetric $(n+1)$-linear form*

$$S_{x_1^* \cdots x_n^* x^*}$$

attains the norm.

Let us observe that it is not possible to derive Theorem 3.2 or 3.3 directly from James' Theorem: we will give two simple examples of a polynomial and a symmetric bilinear form of the finite type attaining the norm and such that this does not happen for all the functionals involved.

Example. Let us fix an element $x^* \in B_{\ell_1}$ with $x^*(e_1) = 0$ and consider the polynomial P on c_0 given by

$$x \mapsto (e_1^* - x^*)(x)(e_1^* + x^*)(x).$$

If $x \in B_{c_0}$ is a fixed element and we write $\alpha := x^*(x)$, then

$$|P(x)| = |(x(1) - \alpha)(x(1) + \alpha)| = |x(1)^2 - \alpha^2| \leq \max\{1, \alpha^2\} = 1,$$

and $P(e_1) = 1$, so P attains its norm and the functionals $e_1^* + x^*, e_1^* + x^*$ do not attain the norm if the support of x^* is not finite.

For the symmetric bilinear form, we will take $X = \ell_1$ and a functional $x^* = \sum_{i=1}^{\infty} \alpha_n e_n^* \in \ell_\infty$ where $\alpha_1 = \frac{1}{2}, 0 < \alpha_n < 1$ and $\lim \{\alpha_n\} = 1$. The symmetric bilinear form S on ℓ_1 given by

$$(x,y) \mapsto \Big(\sum_{n=1}^{\infty} \alpha_n x(n)\Big) y(1) + \Big(\sum_{n=1}^{\infty} \alpha_n y(n)\Big) x(1)$$

attains the norm at (e_1, e_1) since for $x, y \in B_{\ell_1}$, if we write $a = x(1), b = y(1)$ and we assume that $x(n), y(n) \geq 0$, we get

$$|S(x,y)| \leq \Big(\frac{1}{2}a + (1-a)\Big)b + \Big(\frac{1}{2}b + (1-b)\Big)a =$$
$$= a + b - ab = a(1-b) + b \leq 1.$$

Since $\|x^*\| = 1$ and $\{\alpha_n\}$ satisfies $0 < \alpha_n < 1$ for all n, x^* does not attain the norm.

We now consider a version of James' Theorem for numerical radius. The *numerical range* of an operator $T \in L(X)$ (Banach algebra of all bounded and linear operators on X) is given by

$$V(T) := \{x^*(Tx) : (x, x^*) \in \Pi(X)\},$$

where

$$\Pi(X) := \{(x, x^*) \in X \times X^* : \|x\| = \|x^*\| = x^*(x) = 1\}$$

and its *numerical radius* by

$$v(T) := \sup\{|\mu| : \mu \in V(T)\}.$$

An operator $T \in L(X)$ is said to *attain its numerical radius* if

$$\exists\ (x_0, x_0^*) \in \Pi(X)\ :\ |x_0^*(Tx_0)| = v(T).$$

Theorem 3.1 makes it possible to obtain a James-type result in terms of numerical radius:

THEOREM 3.4. ([**AR3**]) *A Banach space is reflexive provided that each rank-one operator attains its numerical radius.*

The converse is false; in fact, in [**AR3**] it was proven that a separable Banach space is finite-dimensional if, and only if, for any equivalent norm, every rank-one operator attains the numerical radius. It is not known if every reflexive Banach space has an equivalent norm for which every rank-one operator attains its numerical radius. A partial answer was given in [**ABR3**].

THEOREM 3.5. *A Banach space X is reflexive if, and only if, there exist $x_0 \in S_X$ and an equivalent norm for which each rank-one operator of the form $x^* \otimes x_0$ ($x^* \in X^*$) attains its numerical radius.*

By using the refinement of James' Theorem proven by Jiménez Sevilla and Moreno (Theorem 2.6), some slight improvements to the results we mentioned can be given:

THEOREM 3.6. ([**ABR3**]) *Let X be a Banach space. Then the following conditions are equivalent:*

i) *X is reflexive.*

ii) *There is* $n \geq 1$ *and* $x_1^*, \ldots, x_n^* \in X^* \backslash \{0\}$ *such that the set*

$$\{x^* \in X^* : P_{x_1^* \cdots x_n^* x^*} \text{ attains its norm}\}$$

has a non-empty weak-$*$ interior.

iii) *For some* $n \geq 1$ *and* $x_1^*, \ldots, x_n^* \in X^* \backslash \{0\}$ *the set*

$$\{x^* \in X^* : S_{x_1^* \cdots x_n^* x^*} \text{ attains its norm}\}$$

has a non-empty weak-$*$ interior.

If there is an element $x_0 \in S_X$ *satisfying that the set*

$$\{x^* \in X^* : x^* \otimes x_0 \text{ attains its numerical radius}\}$$

has a non-empty weak-$*$ interior, the space X *is reflexive.*

4. Open questions

1. If X is a Banach space such that X^* contains a w^*-dense linear subspace which is norm closed and contained in $A(X)$, is X isometric to a dual space?
2. If B_X is non-dentable, is $A(X)$ of the first Baire category (norm topology)?
3. If for any equivalent norm on X, the subset $A(X)$ has a non-empty interior, is X reflexive?
4. If $A(X) \cap S_{X^*} \supset W \cap S_{X^*} \neq \emptyset$ for some w-open set W, is X reflexive?
5. Given a Banach space X and $\varepsilon > 0$, is there a Banach space Y with $d(Y, X) < 1 + \varepsilon$ and such that

$$B_{Y^*} = \overline{\text{co}}^{w^*}\{x^* \in S_{X^*} : x^* \text{ in the interior of } A(Y)\}?$$

6. Suppose that X has the Mazur intersection property and $NA(X, Y)$ has a non-empty interior ($Y \neq \{0\}$), is X reflexive?
7. If X is separable and the unit ball is non-dentable, is $NA(X, Y)$ of the first Baire category, for any Banach space Y?
8. Can we assure that any reflexive Banach space has an equivalent norm for which every rank-one operator attains its numerical radius?

References

[ABR1] M. D. Acosta, J. Becerra Guerrero and M. Ruiz Galán, *Norm attaining operators and James' Theorem,* Recent Progress in Functional Analysis (Valencia, 2000), (Bierstedt, Bonet, Maestre and Schmets, Eds.), North. Holland Math. Stud., 189, North-Holland, Amsterdam, 2001, pp. 215–224.

[ABR2] M. D. Acosta, J. Becerra Guerrero and M. Ruiz Galán. *Dual spaces generated by the interior of the set of norm attaining functionals,* Studia Math. **149** (2002), 175–183.

[ABR3] M. D. Acosta, J. Becerra Guerrero and M. Ruiz Galán, *James type results for polynomials and symmetric multilinear forms,* preprint.

[AR1] M.D. Acosta and M. Ruiz Galán, *New characterizations of the reflexivity in terms of the set of norm attaining functionals.* Canad. Math. Bull. **41** (1998), 279–289.

[AR2] M.D. Acosta and M. Ruiz Galán, *Norm attaining operators and reflexivity.* Rend. Circ. Mat. Palermo **56** (1998), 171–177.

[AR3] M.D. Acosta and M. Ruiz Galán, *A version of James' Theorem for numerical radius.* Bull. London Math. Soc. **31** (1999), 67–74.

[AD] D. Azagra and R. Deville, *James' Theorem for starlike bodies.* J. Funct. Anal. **180** (2001), 328–346.

[BP1] E. Bishop and R.R. Phelps, *A proof that every Banach space is subreflexive.* Bull. Amer. Math. Soc. **67** (1961), 97–98.

[Bo] J. Bourgain, *On dentability and the Bishop-Phelps property.* Israel J. Math. **28** (1977), 265–271.

[Bou] R.D. Bourgin, *Geometric aspects of convex sets with the Radon-Nikodym property,* Lecture Notes in Math. 993, Springer-Verlag, Berlin, 1983.

[DGS] G. Debs, G. Godefroy and J. Saint Raymond, *Topological properties of the set of norm-attaining linear functionals.* Canad. J. Math. **47** (1995), 318–329.

[DGZ] R. Deville, G. Godefroy and V. Zizler, *Smoothness and renormings in Banach spaces,* Pitman Monographs and Surveys Pure Appl. Math., vol. 64, Longman Sci. Tech., New York, 1993.

[Di] J. Diestel, *Geometry of Banach Spaces-Selected Topics,* Lecture Notes in Math., vol. 485, Springer-Verlag, Berlin, 1975.

[Ef] N.M. Efremov, *On an adjoint condition for Banach space.* Soviet Math. **28** (1984), 12–14.

[Fl] K. Floret, *Weakly Compact Sets,* Lecture Notes in Math., vol. 801, Springer-Verlag, Berlin, 1980.

[FLP] V.P. Fonf, J. Lindenstrauss and R.R. Phelps, *Infinite-dimensional convexity, In: Handbook of the geometry of Banach spaces,* vol. I (Johnson and Lindenstrauss, Eds.), North-Holland, Amsterdam, 2001, pp. 599–670.

[GGS1] J.R. Giles, D.A. Gregory and B. Sims: *Characterisation of normed linear spaces with Mazur's intersection property.* Bull. Austral. Math. Soc. **18** (1978), 105–123.

[GGS2] J.R. Giles, D.A. Gregory and B. Sims, *Geometrical implications of upper-semicontinuity of the duality mapping on a Banach space.* Pacific J. Math. **79** (1978), 99–109.

[Go] G. Godefroy, *Boundaries of a convex set and interpolation sets.* Math. Ann. **277** (1987), 173–184.

[Ho] R. Holmes, *Geometric Functional Analysis and its Applications,* Graduate Texts in Math. 24, Springer-Verlag, New York, 1975.

[Ja1] R.C. James, *Reflexivity and the supremum of linear functionals.* Ann. of Math. **66** (1957), 159–169.

[Ja2] R.C. James, *Characterizations of reflexivity.* Studia Math. **23** (1964), 205–216.

[Ja3] R.C. James, *Weakly compact sets.* Trans. Amer. Math. Soc. **113** (1964), 129–140.

[Ja4] R.C. James, *Weak compactness and reflexivity.* Israel. J. Math. **2** (1964), 101–119.

[Ja5] R.C. James, *Reflexivity and the sup of linear functionals.* Israel J. Math. **13** (1972), 289–300.

[JM] J.A. Jaramillo and L.A. Moraes, *Duality and reflexivity in spaces of polynomials.* Arch. Math. **74** (2000), 282–293.

[JiM] M. Jiménez Sevilla and J.P. Moreno, *A note on norm attaining functionals.* Proc. Amer. Math. Soc. **126** (1998), 1989–1997.

[KMS] P.S. Kenderov, W.B. Moors and S. Sciffer, *Norm attaining functionals on $C(T)$.* Proc. Amer. Math. Soc. **126** (1998), 153–157.

[Kl] V.L. Klee, *Some characterizations of reflexivity.* Revista de Ciencias **52** (1950), 15–23.

[Ma] S. Mazur, *Über schwache Konvergenz in den Raumen ℓ_p*. Studia Math. **4** (1933), 128–133.

[PPl] J.I. Petunin and A.N. Plichko, *Some properties of the set of functionals that attain a supremum on the unit sphere.* Ukrain. Math. Zh. **26** (1974), 102–106, MR 49:1075.

[Pr] J.D. Pryce, *Weak compactness in locally convex spaces.* Proc. Amer. Math. Soc. **17** (1966), 148–155.

[Ru] M. Ruiz Galán, *Boundaries and weak compactness*, preprint.

[Si1] S. Simons, *Maximinimax, minimax, and antiminimax theorems and a result of R.C. James.* Pacific J. Math. **40** (1972), 709–718.

[Si2] S. Simons, *An eigenvector proof of Fatou's lemma for continuous functions.* Math. Intelligencer, **17** (1995), 67–70.

[Va] M. Valdivia, *Fréchet spaces with no subspaces isomorphic to ℓ_1.* Math. Japon. **38** (1993), 397–411.

Departamento de Análisis Matemático, Universidad de Granada, 18071 Granada, Spain

E-mail address: dacosta@ugr.es

Departamento de Matemática Aplicada, Universidad de Granada, 18071 Granada, Spain

E-mail address: juliobg@ugr.es

Departamento de Matemática Aplicada, Universidad de Granada, 18071 Granada, Spain

E-mail address: mruizg@ugr.es

Contemporary Mathematics
Volume **321**, 2003

Uniqueness of Unconditional Bases in Quasi-Banach Spaces

Fernando Albiac, Nigel J. Kalton, and Camino Leránoz

ABSTRACT. We present an overview of the problem of uniqueness of unconditional basis up to permutation in quasi-Banach spaces, showing the latest results as well as the techniques (different from the locally convex case) used in the proofs.

1. Definitions and Notation

A (real) *quasi-Banach space* X is a complete metrizable vector space whose topology is given by a quasi-norm on X. That is, a map $\|\cdot\| : X \to \mathbb{R}$, $x \mapsto \|x\|$, satisfying

i) $\|x\| > 0 \quad (x \in X, x \neq 0)$
ii) $\|\alpha x\| = |\alpha|\|x\| \quad (\alpha \in \mathbb{R}, x \in X)$
iii) $\|x_1 + x_2\| \leq C(\|x_1\| + \|x_2\|) \quad (x_1, x_2 \in X)$,

where C is a constant independent of x_1 and x_2. If $C = 1$, X is a Banach space.

If, in addition, X is a vector lattice and $\|x\| \leq \|y\|$ whenever $|x| \leq |y|$ we say that X is a *quasi-Banach lattice.*

We recall that a quasi-Banach lattice X is said to be *p-convex*, where $0 < p < \infty$, if there is a constant $M > 0$ such that for any $n \in \mathbb{N}$ and any $x_1, \dots, x_n \in X$ we have

$$\Big\|(\sum_{i=1}^{n} |x_i|^p)^{1/p}\Big\| \leq M\Big(\sum_{i=1}^{n} \|x_i\|^p\Big)^{1/p}.$$

The procedure to define the element $(\sum_{i=1}^{n} |x_i|^p)^{1/p} \in X$ is exactly the same as in Banach lattices (see [**19**], pp. 40-41).

A quasi-Banach space X is called *natural* if it is isomorphic to a closed subspace of a quasi-Banach lattice which is p-convex for some $p > 0$ (see [**11**]).

When dealing with a quasi-Banach space X, it is often convenient to know which is the "smallest" Banach space containing X. To define it formally, if $(X, \|\cdot\|)$ is a quasi-Banach space whose dual separates the points of X, the *Banach envelope* of X, denoted by $\hat{X}$, is the completion of the normed space $(X, \|\cdot\|_c)$, where $\|\cdot\|_c$

2000 *Mathematics Subject Classification.* Primary 46A35, 46A45, 46A16; Secondary 46B15, 46B42, 46B45.

The second author was sopported by NSF grant DMS-9870027.

is the Minkowski functional of the convex hull of the unit ball of X. For instance, the Banach envelope of the sequence spaces ℓ_p $(0 < p < 1)$ is ℓ_1. It is easy to see that any unconditional basis of a quasi-Banach space X is also an unconditional basis of its Banach envelope $\hat{X}$, equivalent to its normalization and with the same unconditional constant. Our basic references for Banach envelopes are [**12**] and [**13**].

An unconditional basic sequence $(u_n)_{n=1}^{\infty}$ in X is *complemented* if there is a bounded projection $P : X \longrightarrow [u_n]$.

If $(e_n)_{n=1}^{\infty}$ is an unconditional basis of X and $(u_n)_{n=1}^{\infty}$ is an unconditional basic sequence of the form $u_n = \sum_{k\in\mathcal{A}_n} e_k^*(u_n)e_k$, where the sets $(\mathcal{A}_n)_n$ are disjoint subsets of $\mathbb{N}$, we say that $(u_n)_{n=1}^{\infty}$ is *disjoint* with respect to $(e_n)_{n=1}^{\infty}$.

Let $(u_n)_{n=1}^{\infty}$ be a set of disjointly supported vectors in the unconditional basis $(e_n)_{n\in\mathbb{N}}$ of X. Then, $(u_n)_{n=1}^{\infty}$ is an unconditional basic sequence which is complemented if and only if there exists a biorthogonal sequence $(u_n^*)_{n=1}^{\infty} \in X^*$, $u_n^*(u_m) = \delta_{nm}$, and such that the projection

$$P(x) = \sum_{n=1}^{\infty} u_n^*(x)u_n$$

is well defined and bounded.

2. The problem of uniqueness of unconditional basis in quasi-Banach spaces

If $(X, \|.\|)$ is a quasi-Banach space with a normalized unconditional basis $(e_n)_{n=1}^{\infty}$ (i.e. $\|e_n\| = 1$ for all $n \in \mathbb{N}$), we say that X has *unique unconditional basis* if whenever $(x_n)_{n=1}^{\infty}$ is another normalized unconditional basis of X, then $(x_n)_{n=1}^{\infty}$ is *equivalent* to $(e_n)_{n=1}^{\infty}$. That is, there is a constant D so that

$$D^{-1}\Big\|\sum_{i=1}^{n} a_i x_i\Big\| \le \Big\|\sum_{i=1}^{n} a_i e_i\Big\| \le D\Big\|\sum_{i=1}^{n} a_i x_i\Big\|,$$

for any choice of scalars $(a_i)_{i=1}^{n}$ and every $n \in \mathbb{N}$. For notation, we write $(x_n)_{n=1}^{\infty} \sim (e_n)_{n=1}^{\infty}$ when $(x_n)_{n=1}^{\infty}$ and $(e_n)_{n=1}^{\infty}$ are equivalent.

The problem of the uniqueness of unconditional basis is classical. The earliest results were obtained by Lindenstrauss and Pelczynski, who proved in 1968 that c_0 and ℓ_1 have a unique unconditional basis ([**18**]). On the other hand, it was well known that the Hilbert space ℓ_2 had a unique unconditional basis too ([**15**]). Lindenstrauss and Zippin proved in 1969 ([**20**]) that c_0, ℓ_1 and ℓ_2 are the only Banach spaces with this property. Therefore, the uniqueness of unconditional basis is a rare property for Banach spaces.

The situation for quasi-Banach spaces which are not Banach spaces is quite different. Indeed, since the structure of a locally bounded space which is not locally convex is more rigid, it is more difficult for such a space to have a basis. Because of that, when a space has an unconditional basis, it is more likely to be unique. For example, in 1977 it was shown in [**10**] that a wide class of non-locally convex Orlicz sequence spaces, including the spaces ℓ_p for $0 < p < 1$, have a unique unconditional basis. Nawrocki and Ortynski ([**21**]) investigated the Lorentz sequence spaces $d(w,p)$ for $0 < p < 1$, described their Banach envelopes and proved that if the sequence $\omega = (\omega_n)_{n=1}^{\infty}$ verifies

$$\inf_n \frac{(\omega_1 + \cdots + \omega_n)^{1/p}}{n} = 0$$

then $d(\omega, p)$ has uncountably many non-equivalent unconditional basis. Answering a question raised in [**21**], the authors proved in [**14**] that if $0 < p < 1$ and

$$\lim_{n\to\infty} \frac{(\omega_1 + \cdots + \omega_n)^{1/p}}{n} = \infty$$

then $d(w, p)$ has a unique unconditional basis.

If an unconditional basis is unique, then it must be equivalent to all its permutations and hence must be symmetric. Consequently, for spaces without symmetric basis it is more natural to consider the property of *uniqueness of unconditional basis up to permutation*: whenever $(x_n)_{n=1}^{\infty}$ and $(e_n)_{n=1}^{\infty}$ are two normalized unconditional bases of X, there is a permutation $\pi : \mathbb{N} \longrightarrow \mathbb{N}$ such that $(x_n)_{n\in\mathbb{N}} \sim (e_{\pi(n)})_{n\in\mathbb{N}}$.

In [**8**], Edelstein and Wojtaszczyk proved that any finite direct sum of the spaces c_0, ℓ_1 and ℓ_2 has a unique unconditional basis, up to permutation.

Attempting a complete classification of the Banach spaces with unique unconditional basis up to permutation, Bourgain, Casazza, Lindenstrauss and Tzafriri studied in [**4**] infinite direct sums of these spaces. They showed that $c_0(\ell_1)$, $c_0(\ell_2)$, $\ell_1(c_0)$ and $\ell_1(\ell_2)$ all have unique unconditional basis up to permutation, while the result is not true for $\ell_2(c_0)$ and $\ell_2(\ell_1)$. They also found an unexpected additional space, the 2-convexified Tsirelson space $T^{(2)}$ (see [**7**]) for the definition), with a unique unconditional basis up to permutation. Their results indicate that a complete classification of Banach spaces with this property is very unlikely to be achieved. More recently, further examples of "pathological" spaces with unique unconditional basis up to permutation have been provided in [**5**] and in [**9**].

It seemed only natural to translate the question of uniqueness of unconditional basis (up to permutation) into the setting of non-locally convex spaces that are infinite direct sums of the classical quasi-Banach spaces with a unique unconditional basis, which are the following:

$$\begin{aligned}
\ell_p(\ell_q) &= (\ell_q \oplus \ell_q \oplus \cdots \oplus \ell_q \oplus \cdots)_p\\
\ell_p(c_0) &= (c_0 \oplus c_0 \oplus \cdots \oplus c_0 \oplus \cdots)_p\\
\ell_p(\ell_1) &= (\ell_1 \oplus \ell_1 \oplus \cdots \oplus \ell_1 \oplus \cdots)_p\\
\ell_p(\ell_2) &= (\ell_2 \oplus \ell_2 \oplus \cdots \oplus \ell_2 \oplus \cdots)_p\\
c_0(\ell_p) &= (\ell_p \oplus \ell_p \oplus \cdots \oplus \ell_p \oplus \cdots)_0\\
\ell_1(\ell_p) &= (\ell_p \oplus \ell_p \oplus \cdots \oplus \ell_p \oplus \cdots)_1\\
\ell_2(\ell_p) &= (\ell_p \oplus \ell_p \oplus \cdots \oplus \ell_p \oplus \cdots)_2,
\end{aligned}$$

where $0 < p, q < 1$.

All the above can be seen as spaces of infinite matrices: if $(E, \|.\|_E)$ and $(F, \|.\|_F)$ are one of the quasi-Banach sequence spaces, $(\ell_p, \|.\|_p)$ $(p \in (0,1] \cup \{2\})$ or $(c_0, \|.\|_\infty)$, then the space $E(F)$ consists of all infinite matrices $(x_{lk})_{l,k=1}^{\infty}$ satisfying $\mathbf{x}_l = (x_{lk})_{k=1}^{\infty} \in F$ for all $l \in \mathbb{N}$ and $(\|\mathbf{x}_l\|_F)_{l=1}^{\infty} \in E$. The quasi-norm of an element

$(x_{lk})_{l,k=1}^{\infty} \in E(F)$ is defined as

$$\|(x_{lk})_{l,k}\|_{E(F)} = \|(\|(x_{lk})_{k=1}^{\infty}\|_F)_{l=1}^{\infty}\|_E.$$

The space $E(F)$ is a quasi-Banach space whose Banach envelope is $\hat{E}(\hat{F})$ and whose dual can be identified with $E^*(F^*)$.

These spaces have a canonical unconditional basis that will be denoted by $(e_{lk})_{l,k=1}^{\infty}$. The (l,k) coordinate of $e_{l_0 k_0}$ is 1 if $l = l_0$ and $k = k_0$, and 0 otherwise.

The lattice structure induced by the canonical basis on $E(F)$ is clearly p-convex if $(E, \|.\|_E)$ and $(F, \|.\|_F)$ are p-convex.

While the case of the space $\ell_2(\ell_p)$ $(0 < p < 1)$ remains open, the uniqueness of unconditional basis up to permutation has been proved for complemented subspaces with unconditional basis of all the other spaces. We can summarize the results in a theorem:

THEOREM 2.1. *Let X be one of the following quasi-Banach spaces: $\ell_p(\ell_q)$, $\ell_p(c_0)$, $\ell_p(\ell_1)$, $\ell_p(\ell_2)$, $c_0(\ell_p)$, and $\ell_1(\ell_p)$, $0 < p, q < 1$; and let Q be a bounded linear proyection from X onto a subspace Z which has a normalized unconditional basis $(x_n)_{n=1}^{\eta}$ with unconditional constant $K \geq 1$. Then, $(x_n)_{n=1}^{\eta}$ is equivalent to a permutation of a subbasis of the canonical basis of X. The equivalence constant only depends on K and $\|Q\|$. As a consequence, the following quasi-Banach spaces have unique unconditional basis up to permutation: $\ell_p \oplus \ell_q$, $\ell_p(\ell_q^n)_{n=1}^{\infty}$, $\ell_q \oplus \ell_p(\ell_q^n)_{n=1}^{\infty}$, $\ell_p(\ell_q)$, $\ell_p \oplus c_0$, $\ell_p(\ell_\infty^n)_{n=1}^{\infty}$, $c_0 \oplus \ell_p(\ell_\infty^n)_{n=1}^{\infty}$, $\ell_p(c_0)$, $\ell_1 \oplus \ell_p$, $\ell_1(\ell_p^n)_{n=1}^{\infty}$, $\ell_p \oplus \ell_1(\ell_p^n)_{n=1}^{\infty}$, $\ell_1(\ell_p)$, $c_0(\ell_p^n)_{n=1}^{\infty}$, $\ell_p \oplus c_0(\ell_p^n)_{n=1}^{\infty}$, $c_0(\ell_p)$, with $0 < p, q < 1$.*

The proof of the above theorem is quite different depending on the space X. Next, we sketch the proofs. For the sake of simplicity, we will take $Z = X$ and $(x_n)_{n=1}^{\infty}$ a normalized unconditional basis of X. The general case is analogous.

3. Uniqueness of the unconditional basis of $\ell_p(\ell_q)$ and $\ell_p(X)$, $0 < p, q < 1$.

The main references for this section are [**14**] and [**16**] .

For $0 < p, q \leq 1$ fixed, $\ell_p(\ell_q)$ is the space of infinite matrices $(x_{lk})_{l,k=1}^{\infty}$ satisfying

$$\|(x_{lk})_{l,k}\|_{p,q} = \Big(\sum_{l=1}^{\infty} \big(\sum_{k=1}^{\infty} |x_{lk}|^q\big)^{p/q}\Big)^{1/p} < \infty.$$

For each $0 < p, q < 1$, $(\ell_p(\ell_q), \|\cdot\|_{p,q})$ is a quasi-Banach space whose Banach envelope is $(\ell_1, \|\cdot\|_1)$ and whose dual space can be identified with ℓ_∞.

The lattice structure induced by the canonical basis on $\ell_p(\ell_q)$ is $\min\{p, q\}$-convex.

The proof of the uniqueness of unconditional basis in $\ell_p(\ell_q)$ relies on a technique, specific to the non-locally convex case, that has been used in all the proofs of uniqueness of unconditional basis so far. It was introduced in [**10**] to prove the uniqueness of unconditional basis in non-locally convex Orlicz sequence spaces (cf. Theorem 2.3 of [**14**]):

LEMMA 3.1 ("Large coefficients" technique). *Let X be a natural quasi-Banach space with normalized unconditional bases $(e_{lk})_{l,k=1}^{\infty}$ and $(x_n)_{n=1}^{\infty}$. Suppose there is a constant $\beta > 0$ (independent of n) and an injective map $\pi : S \subseteq \mathbb{N} \longrightarrow \mathbb{N} \times \mathbb{N}$ so that*

$$|e_{\pi(n)}^*(x_n)| \geq \beta \quad \text{and} \quad |x_n^*(e_{\pi(n)})| \geq \beta$$

for all $n \in S$. Then, there exist positive constants ρ, ρ' so that

$$\rho\Big\|\sum_{n\in S}\alpha_n e_{\pi(n)}\Big\| \leq \Big\|\sum_{n\in S}\alpha_n x_n\Big\| \leq \rho'\Big\|\sum_{n\in S}\alpha_n e_{\pi(n)}\Big\|$$

for any sequence of scalars (α_n) finitely non zero.

A normalized unconditional basis $(x_n)_{n=1}^{\infty}$ of a quasi-Banach space is called *strongly absolute* if for any $\varepsilon > 0$ there exist $C_\varepsilon > 0$ so that

$$\sum_{n=1}^{N}|\alpha_n| \leq C_\varepsilon \sup_n |\alpha_n| + \varepsilon\|\sum_{n=1}^{N}\alpha_n x_n\|$$

for any choice of scalars $(\alpha_n)_{n=1}^{N}$ and $N \in \mathbb{N}$.

Intuitively, if a quasi-Banach space has a strongly absolute unconditional basis, it is far from being a Banach space.

LEMMA 3.2. *For $0 < p < 1$, the canonical basis of ℓ_p is strongly absolute.*

Lemma 3.2 is fundamental when we want to apply Lemma 3.1 to those quasi-Banach spaces in which ℓ_p $(0 < p < 1)$ is involved. It means that, if for a sequence $(a_n)_{n\in\mathbb{N}} \in \ell_p$ $(0 < p < 1)$, its ℓ_1-norm and its ℓ_p-quasi-norm are comparable, then so is its ℓ_∞ norm i.e if for some constants C_1, C_2, we have

$$C_1 \leq \|(a_n)_n\|_1 \leq \|(a_n)_n\|_p \leq C_2,$$

then for $\varepsilon = \dfrac{C_1}{2C_2}$, we get

$$\|(a_n)_n\|_\infty \geq \frac{C_1}{2C_\varepsilon}.$$

So we can ensure the existence of a large coordinate in the sequence $(a_n)_n$.

Lemma 3.1 and Lemma 3.2 imply that any normalized unconditional basis $(x_n)_{n=1}^{\infty}$ of $\ell_p(\ell_q)$ is equivalent to a subset of the canonical basis $(e_{lk})_{l,k=1}^{\infty}$ and, therefore, equivalent to it (Theorem 2.7 of [**14**]).

This result is actually true for any space $\ell_p(X)$ $(0 < p < 1)$ of sequences of elements of a natural quasi-Banach space X such that

$$\|(y_n)_{n=1}^{\infty}\|_{\ell_p(X)} = \Big(\sum_{n=1}^{\infty}\|y_n\|_X^p\Big)^{1/p} < \infty,$$

whenever X has a strongly absolute normalized unconditional basis.

If $(e_n)_{n=1}^{\infty}$ is an unconditional basis of X then by repeating $(e_n)_{n=1}^{\infty}$ in each coordinate we get a strongly absolute unconditional basis of the natural quasi-Banach space $\ell_p(X)$ that we denote by $\ell_p(e_n)$. Using Lemma 3.1 and Lemma 3.2 the authors proved:

THEOREM 3.3. (Proposition 3.1 of [**14**]) *Let X be a natural quasi-Banach space with a strongly absolute normalized unconditional basis $(e_n)_{n=1}^{\infty}$, and assume $0 < p < 1$. Then, if $(x_n)_{n=1}^{\infty}$ is another normalized unconditional basis of X, the bases $\ell_p(e_n)$ and $\ell_p(x_n)$ of $\ell_p(X)$ are equivalent up to permutation.*

We improved the previous result by utilizing the following result of Wojtaszczyk:

LEMMA 3.4. (Theorem 2.12 of [**22**]) *Let Y be a natural quasi-Banach space with strongly absolute unconditional basis. Assume also that Y is isomorphic to some of its cartesian powers Y^s, $s = 2, 3, \ldots$ Then, all normalized unconditional bases in Y are permutatively equivalent.*

Now, as an easy consequence we obtain:

THEOREM 3.5. *Let X be a natural quasi-Banach space with a strongly absolute normalized unconditional basis $(e_n)_{n=1}^{\infty}$ (not necessarily unique up to permutation), and let $0 < p < 1$. Then, $\ell_p(X)$ has a unique unconditional basis up to permutation.*

4. Uniqueness of the unconditional basis of $c_0(\ell_p)$, $0 < p < 1$.

The main references for this section are [**16**] and [**17**].

For $0 < p \leq 1$ fixed, $c_0(\ell_p)$ is the space of infinite matrices $(x_{lk})_{l,k=1}^{\infty}$ such that $(x_{lk})_{k=1}^{\infty} \in \ell_p$ for all $l \in \mathbb{N}$, $\sum_{k=1}^{\infty} |x_{lk}|^p \overset{l\to\infty}{\longrightarrow} 0$, and

$$\|(x_{lk})_{l,k}\|_p = \sup_l \Big(\sum_{k=1}^{\infty} |x_{lk}|^p \Big)^{1/p} < \infty.$$

For each $0 < p < 1$, $(c_0(\ell_p), \|\cdot\|_p)$ is a quasi-Banach space whose Banach envelope is $(c_0(\ell_1), \|\cdot\|_1)$ and whose dual space can be identified with $\ell_1(\ell_\infty)$. That is, the Banach space of infinite matrices $a = (a_{lk})_{l,k=1}^{\infty}$ satisfying

$$\|a\| = \sum_{l=1}^{\infty} \sup_k |a_{lk}| < \infty.$$

The lattice structure induced by the canonical basis on $c_0(\ell_p)$ is p-convex.

Bourgain, Casazza, Lindenstrauss and Tzafriri proved in [**4**]:

THEOREM 4.1 (Theorem of uniqueness in $c_0(\ell_1)$). *Let $(x_n)_{n=1}^{\infty}$ be a normalized unconditional basis of $c_0(\ell_1)$. Then, there exist a constant Δ, and a partition of the integers into mutually disjoint subsets $(B_j)_{j=1}^{\infty}$, such that*

$$\Delta^{-1} \sup_j \sum_{n\in B_j} |a_n| \leq \Big\| \sum_{n=1}^{\infty} a_n x_n \Big\|_c \leq \Delta \sup_j \sum_{n\in B_j} |a_n| \tag{4.1}$$

for any finitely non-zero sequence of scalars $(a_n)_n$.

Our aim is to show:

THEOREM 4.2. (Theorem 3.2 of [**16**], Theorem 2.1 of [**17**]) *Let $(x_n)_{n=1}^{\infty}$ be a normalized unconditional basis of $c_0(\ell_p)$ $(0 < p < 1)$. Then, $(x_n)_{n=1}^{\infty}$ is equivalent to a permutation of a subbasis of the canonical basis of $c_0(\ell_p)$. That is, there exist constants Δ_1, Δ_2 and a partition of $\mathbb{N}$ into mutually disjoint subsets $(B_j)_{j=1}^{\infty}$, such that*

$$\Delta_1 \sup_j \Big(\sum_{n\in B_j} |a_n|^p \Big)^{1/p} \leq \Big\| \sum_{n=1}^{\infty} a_n x_n \Big\| \leq \Delta_2 \sup_j \Big(\sum_{n\in B_j} |a_n|^p \Big)^{1/p},$$

for any finitely non-zero sequence of scalars $(a_n)_n$.

Despite the fact that the duality techniques the authors used in Theorem 4.1 cannot be translated to the non-locally convex case, it is essential to our arguments that the property we want to prove for $c_0(\ell_p)$ is shared by its Banach envelope.

PROOF. From the fact that $(x_n)_{n=1}^{\infty}$ is a normalized unconditional basis of $c_0(\ell_p)$, it follows that $(x_n)_{n=1}^{\infty}$ is an unconditional basis of its Banach envelope $c_0(\ell_1)$, where Theorem 4.1 occurs.

First we see that for each j, $(x_n)_{n\in B_j}$ is equivalent to an ℓ_q-basis in the q-Banach envelope $c_0(\ell_q)$ for any $0 < p < q \leq 1$, where $(B_j)_{j=1}^{\infty}$ is the partition in (4.1). To prove that, we use "large coefficients" techniques.

Then, we study the behaviour of the sequence $(x_n^*)_{n=1}^{\infty} \subset \ell_1(\ell_\infty)$, and prove that, except for a uniformly finite number of n's in each B_j, the norm $\|x_n^*\| = \sum_{l=1}^{\infty} \sup_k |x_n^*(e_{lk})|$ is concentrated in a uniformly finite number of l's. Hence, there are large coefficients among the coordinates of each of those x_n^*, and we use Lemma 3.1 to prove that the corresponding x_n are equivalent to a subbasis of the canonical basis of $c_0(\ell_p)$. The subbasis of $(x_n)_{n\in\mathbb{N}}$ of elements for which the norm $\|x_n^*\|$ is not concentrated in a uniformly finite number of l's is equivalent in $c_0(\ell_p)$ to a c_0-basis. □

5. Uniqueness of the unconditional basis of $\ell_p(c_0)$ and $\ell_p(\ell_2)$, $0 < p < 1$.

The main references for this section are **[1]** and **[2]**.

For $0 < p \leq 1$ fixed, $\ell_p(c_0)$ is the space of infinite matrices $(x_{lk})_{l,k=1}^{\infty}$ satisfying $x_{lk} \overset{k\to\infty}{\longrightarrow} 0$ for all $l \in \mathbb{N}$ and

$$\|(x_{lk})_{l,k}\|_p = \Big(\sum_{l=1}^{\infty} \sup_k |x_{lk}|^p\Big)^{1/p} < \infty.$$

For each $0 < p < 1$, $(\ell_p(c_0), \|\cdot\|_p)$ is a quasi-Banach space whose Banach envelope is $(\ell_1(c_0), \|\cdot\|_1)$ and whose dual space can be identified with $\ell_\infty(\ell_1)$, that is, the Banach space of infinite matrices $a = (a_{lk})_{l,k=1}^{\infty}$ satisfying

$$\|a\| = \sup_l \sum_{k=1}^{\infty} |a_{lk}| < \infty.$$

The lattice structure induced by the canonical basis on $\ell_p(c_0)$ is clearly p-convex.

THEOREM 5.1. (Theorem 2.1 of **[1]**, Theorem 2.1 of **[2]**) *Let $(x_n)_{n=1}^{\infty}$ be a normalized unconditional basis for $\ell_p(c_0)$ $(0 < p < 1)$. Then, $(x_n)_{n=1}^{\infty}$ is equivalent to a permutation of a subbasis of the canonical basis of $\ell_p(c_0)$. That is, there exist constants Δ_1, Δ_2 and a partition of $\mathbb{N}$ into mutually disjoint subsets $(B_j)_{j=1}^{\infty}$, such that*

$$\Delta_1\Big(\sum_{j=1}^{\infty} \sup_{n\in B_j} |a_n|^p\Big)^{1/p} \overset{(2)}{\leq} \Big\|\sum_{n=1}^{\infty} a_n x_n\Big\| \overset{(1)}{\leq} \Delta_2\Big(\sum_{j=1}^{\infty} \sup_{n\in B_j} |a_n|^p\Big)^{1/p}, \tag{5.1}$$

for any finitely non-zero sequence of scalars $(a_n)_n$.

Bourgain, Casazza, Lindenstrauss and Tzafriri proved in **[4]**:

THEOREM 5.2 (Theorem of uniqueness in $\ell_1(c_0)$). *Let $(x_n)_{n=1}^{\infty}$ be a normalized unconditional basis for $\ell_1(c_0)$. Then, there exist a constant Δ, and a partition of the integers into mutually disjoint subsets $(B_j)_{j=1}^{\infty}$, such that*

$$\Delta^{-1}\sum_{j=1}^{\infty} \sup_{n\in B_j} |a_n| \leq \Big\|\sum_{n=1}^{\infty} a_n x_n\Big\|_c \leq \Delta\sum_{j=1}^{\infty} \sup_{n\in B_j} |a_n| \tag{5.2}$$

for any finitely non-zero sequence of scalars $(a_n)_n$.

As we saw in the $c_0(\ell_p)$-case, although the techniques used in Theorem 5.2 cannot be translated to the non-locally convex case, the result itself is essential to our arguments, and the partition given by that result is the one that works for our Theorem.

PROOF. We prove Theorem 5.1 in two parts corresponding to each of the inequalities (1) and (2) in the equation (5.1). Apart from Theorem 5.2, the proof of inequality (1) is based on the following result:

LEMMA 5.3. (Theorem 3.3 of [**13**]) *Let Y be a natural quasi-Banach space with unconditional basis such that Y^* has finite cotype. Then, there exists a constant A, depending only on p and the cotype constant of Y^*, such that*

$$\|y\|_Y \leq A\|y\|_c$$

for every $y \in Y$. (That is, Y is isomorphic to its Banach envelope.)

From the fact that $(x_n)_{n=1}^{\infty}$ is a normalized unconditional basis of $\ell_p(c_0)$, it follows that $(x_n)_{n=1}^{\infty}$ is an unconditional basis of its Banach envelope $\ell_1(c_0)$, where Theorem 5.2 applies.

For each j let us consider $X_j = \overline{\{x_n; n \in B_j\}}^{\ell_p(c_0)}$, where $(B_j)_{j=1}^{\infty}$ is the partition in (5.2). Then,

$$\hat{X}_j = \overline{\{x_n; n \in B_j\}}^{\ell_1(c_0)} \simeq c_0.$$

By Lemma 5.3, $X_j \simeq c_0$; or equivalently

$$\Big\|\sum_{n\in B_j} a_n x_n\Big\|^p \leq A^p \Delta^p \sum_{j=1}^{\infty} \sup_{n\in B_j} |a_n|^p.$$

Therefore,

$$\Big\|\sum_{n=1}^{\infty} a_n x_n\Big\|^p = \Big\|\sum_{j=1}^{\infty}\sum_{n\in B_j} a_n x_n\Big\|^p \leq \sum_{j=1}^{\infty}\Big\|\sum_{n\in B_j} a_n x_n\Big\|^p \leq A^p\Delta^p \sup_{n\in B_j} |a_n|^p$$

for any sequence of scalars $(a_n)_{n\in\mathbb{N}}$ finitely non zero.

In order to show inequality (2), we use Lemma 3.2 to find "large coefficients" in the coordinates of the elements x_n with respect to the canonical basis $(e_{lk})_{l,k=1}^{\infty}$. Furthermore, we find a $1-1$ map

$$\pi : \mathbb{N} \longrightarrow \mathbb{N}\times\mathbb{N}, \quad n \mapsto \pi(n) = (l,k)$$

so that $|e^*_{\pi(n)}(x_n)| > \beta$ for all $n \in \mathbb{N}$, for some positive constant β not depending on n.

This way, by Lemma 3.1, there is a constant $\rho > 0$ such that

$$\Big\|\sum_{n\in\mathbb{N}} a_n x_n\Big\| \geq \rho\Big\|\sum_{n\in\mathbb{N}} a_n e_{\pi(n)}\Big\| = \rho\Big(\sum_{j=1}^{\infty} \sup_{n\in B_j} |a_n|^p\Big)^{\frac{1}{p}}$$

for any finitely non-zero sequence of scalars $(a_n)_{n\in\mathbb{N}}$. □

Let us introduce now the other space we are dealing with in this section. For $0 < p \leq 1$ fixed, $\ell_p(\ell_2)$ is the space of infinite matrices $(x_{lk})_{l,k=1}^{\infty}$ satisfying

$$\|(x_{lk})_{l,k}\|_p = \Big(\sum_{l=1}^{\infty}\Big(\sum_{k=1}^{\infty}|x_{lk}|^2\Big)^{p/2}\Big)^{1/p} < \infty.$$

For each $0 < p < 1$, $(\ell_p(\ell_2), \|\cdot\|_p)$ is a quasi-Banach space whose Banach envelope is $(\ell_1(\ell_2), \|\cdot\|_1)$ and whose dual space can be identified with $\ell_\infty(\ell_1)$, that is, the Banach space of infinite matrices $a = (a_{lk})_{l,k=1}^\infty$ satisfying

$$\|a\| = \sup_l \Big(\sum_{k=1}^\infty |a_{lk}|^2 \Big)^{1/2} < \infty.$$

The lattice structure induced by the canonical basis on $\ell_p(\ell_2)$ is clearly p-convex.

Let us state the corresponding results of uniqueness of unconditional basis up to permutation for $\ell_p(\ell_2)$. The proof of this result is completely analogous to the proof of the $\ell_p(c_0)$ case:

THEOREM 5.4. (Theorem 2.10 of [**1**], Theorem 3.2 of [**2**]) *Let $(x_n)_{n=1}^\infty$ be a normalized unconditional basis for $\ell_p(\ell_2)$ $(0 < p < 1)$. Then, there exist constants Γ_1, Γ_2 and a partition of $\mathbb{N}$ into mutually disjoint subsets $(L_j)_{j=1}^\infty$ so that*

$$\Gamma_1 \Big(\sum_{j=1}^\infty \big(\sum_{n \in L_j} |a_n|^2 \big)^{p/2} \Big)^{1/p} \le \Big\| \sum_{n=1}^\infty a_n x_n \Big\| \le \Gamma_2 \Big(\sum_{j=1}^\infty \big(\sum_{n \in L_j} |a_n|^2 \big)^{p/2} \Big)^{1/p},$$

for any finitely non-zero sequence of scalars $(a_n)_{n \in \mathbb{N}}$.

6. Uniqueness of the unconditional basis of $\ell_1(\ell_p)$ and $\ell_p(\ell_1)$, $0 < p < 1$.

The main references for this section are [**1**] and [**3**].

For $0 < p < 1$ fixed, $\ell_1(\ell_p)$ is the quasi-Banach space of infinite matrices $(x_{lk})_{l,k=1}^\infty$ satisfying

$$\|(x_{lk})_{l,k}\|_{\ell_1(\ell_p)} = \sum_{l=1}^\infty \Big(\sum_{k=1}^\infty |x_{lk}|^p \Big)^{1/p} < \infty,$$

whereas $\ell_p(\ell_1)$ is the quasi-Banach space of infinite matrices $(x_{lk})_{l,k=1}^\infty$ such that

$$\|(x_{lk})_{l,k}\|_{\ell_p(\ell_1)} = \Big(\sum_{l=1}^\infty \big(\sum_{k=1}^\infty |x_{lk}| \big)^p \Big)^{1/p} < \infty.$$

The Banach envelope of both $\ell_1(\ell_p)$ and $\ell_p(\ell_1)$ $(0 < p < 1)$ is ℓ_1 and their dual spaces can be identified with ℓ_∞.

The lattice structure induced by the canonical basis, $(e_{lk})_{l,k=1}^\infty$, in $\ell_1(\ell_p)$ and $\ell_p(\ell_1)$ $(0 < p < 1)$ is p-convex.

Our goal is to show:

MAIN THEOREM. *Suppose $0 < p < 1$. Let $(x_n)_{n \in \mathbb{N}}$ be a normalized unconditional basis for $\ell_1(\ell_p)$ (respectively $\ell_p(\ell_1)$). Then $(x_n)_{n \in \mathbb{N}}$ is equivalent to a permutation of the canonical basis of $\ell_1(\ell_p)$ (respectively $\ell_p(\ell_1)$).*

The canonical basis of both $\ell_1(\ell_p)$ and $\ell_p(\ell_1)$ $(0 < p < 1)$ is also an unconditional basis of its Banach envelope, ℓ_1, where all normalized unconditional bases are equivalent and, therefore, symmetric:

THEOREM 6.1 (Uniqueness of unconditional basis of ℓ_1). ([**18**]) *Suppose $(x_n)_{n=1}^\infty$ is a normalized K-unconditional basis of ℓ_1. Then, there exists a constant D (depending only on K) so that*

$$D\sum_{n=1}^N |a_n| \le \Big\|\sum_{n=1}^N a_n x_n\Big\| \le \sum_{n=1}^N |a_n|$$

for any $(a_n)_{n=1}^N$ scalars and $N \in \mathbb{N}$.

Everything seemed to indicate that there might be a proof of uniqueness of the unconditional basis up to permutation for $\ell_p(\ell_1)$ $(0<p<1)$ similar to the ones for $\ell_p(c_0)$ and $\ell_p(\ell_2)$ and a completely different one for $\ell_1(\ell_p)$ $(0<p<1)$. The common point was that we were considering infinite direct sums of the only classical Banach spaces with unique normalized symmetric basis in the sense of ℓ_p $(0<p<1)$, and for the first two cases what really mattered was the ruling (strongly absolute) aspect of (the canonical basis of) ℓ_p. All our attempts in that direction failed.

The reason was that Theorem 6.1 does not give any information about the different behaviour of the subsets of an unconditional basis of $\ell_p(\ell_1)$ or $\ell_1(\ell_p)$ $(0<p<1)$ seen as unconditional basic sequences of their Banach envelope, in contrast with what happened in $c_0(\ell_p)$, $\ell_p(c_0)$ and $\ell_p(\ell_2)$, $0<p<1$. In these cases, the corresponding theorems of uniqueness of unconditional basis up to permutation in their Banach envelopes ([**4**]), where the canonical basis is not symmetric, were the starting point of the proofs (see [**17**], [**2**]). Since the canonical basis of ℓ_1 is symmetric we could not approach the proofs in the same way.

Eventually, we took notice of the fact that the lattice structure induced by any unconditional basis in ℓ_1, the Banach envelope of both $\ell_1(\ell_p)$ and $\ell_p(\ell_1)$ $(0<p<1)$, is *anti-Euclidean*, that is, ℓ_1 does not contain ℓ_2^n's as uniformly complemented sublattices. The authors had given in [**6**] a much simpler alternative proof to the uniqueness of unconditional basis up to permutation in the Banach space $c_0(\ell_1)$ using this fact (Corollary 2.5 of [**5**]). Their simplification partly depended on a useful characterization by the same authors of complemented unconditional basic sequences in Banach sequence spaces which are anti-Euclidean (see Theorem 3.5 of [**5**]). We generalized this theorem to natural quasi-Banach spaces (Theorem 3.2 of [**1**], Theorem 2.2 of [**3**]) and were highly rewarded by its importance in obtaining the following simplifications:

First simplification (Corollary 3.4 of [**1**], Corollary 2.4 of [**3**]) *Whenever we have a complemented unconditional basic sequence $(x_n)_{n=1}^\eta$ in either $\ell_1(\ell_p)$ or $\ell_p(\ell_1)$ $(0<p<1)$, we can suppose that $(x_n)_{n=1}^\eta$ is disjointly supported in the canonical basis (η can be either a natural number or infinity).*

Second simplification (Lemma 3.6 of [**1**], Lemma 2.6 of [**3**]) *If $(x_n)_{n=1}^\eta$ is a complemented unconditional basic sequence in either $\ell_1(\ell_p)$ or $\ell_p(\ell_1)$ $(0<p<1)$, we may further assume that supp $x_n^* \subset$ supp x_n and $x_n \ge 0$, $x_n^* \ge 0$.*

Third simplification (Lemma 3.8 of [**1**], Lemma 2.9 of [**3**]) *If $(x_n)_{n=1}^\eta$ is a complemented unconditional basic sequence in either $\ell_1(\ell_p)$ or $\ell_p(\ell_1)$ $(0<p<1)$ we may assume that all of the coordinates of x_n^* are "uniformly large".*

Basically, these results allow us to unravel the form in which any complemented, normalized unconditional basic sequence $(x_n)_{n=1}^\eta$ in either $\ell_1(\ell_p)$ or $\ell_p(\ell_1)$ $(0<p<1)$ can be written in terms of the canonical basis. Now we can establish:

THEOREM 6.2 (Main Theorem for $\ell_1(\ell_p)$). ((Theorem 3.9 of [**1**], Theorem 2.11 of [**3**]) *Let* $(x_n)_{n=1}^{\eta}$ *be a normalized, complemented, unconditional basic sequence in* $\ell_1(\ell_p)$ $(0 < p < 1)$. *Then,* $(x_n)_{n=1}^{\eta}$ *is equivalent to a subbasis of* $(e_{lk})_{l,k=1}^{\infty}$.

PROOF. First, using Lemmas 3.1 and 3.2 we prove that there is a complemented basic sequence $(y_n)_{n=1}^{\eta}$ equivalent to $(x_n)_{n=1}^{\eta}$, whose disjoint supports take an extremely simple form:

$$\begin{aligned} y_n &= \sum_{l\in F_n} e^*_{l\sigma_n(l)}(x_n) e_{l\sigma_n(l)} \\ y^*_n &= \sum_{l\in F_n} c^n_{l\sigma_n(l)} e^*_{l\sigma_n(l)} \quad (\omega^* \text{ convergence}) \end{aligned}$$

where the sets $(F_n)_{n=1}^{\eta}$ need not be mutually disjoint but they are finite. Furthermore, for each n we can assume that

$$c^n_{l\sigma_n(l)} > \frac{1}{2}, \quad \text{for all } l \in F_n.$$

By Lemma 3.1, it would be sufficient to show that $\|y_n\|_\infty > \delta$ for all n, for some positive constant δ independent of n. But, unfortunately, this does not necessarily have to be true for all n. So we proceed as follows:

Using Graph Theory we prove that there is a constant $\delta > 0$ (independent of n) such that, if we consider the set

$$\mathcal{A} = \{n \in \{1, \ldots, \eta\}\,;\, \|y_n\|_\infty < \delta\},$$

then $(y_n)_{n\in\mathcal{A}}$ is equivalent (in $\ell_1(\ell_p)$) to $(e_n)_{n\in\mathcal{A}}$, where $(e_n)_{n=1}^{\infty}$ denotes the canonical basis of ℓ_1. In particular, $(y_n)_{n\in\mathcal{A}}$ is equivalent to a subbasis of the canonical basis of $\ell_1(\ell_p)$.

To finish the proof we just have to observe that the basic sequence

$$\{y_n\,;\, n \notin \mathcal{A}\} = \{y_n\,;\, \|y_n\|_\infty > \delta\}$$

is equivalent to a subbasis of the canonical basis of $\ell_1(\ell_p)$. □

Next, we will see the corresponding results for $\ell_p(\ell_1)$, $0 < p < 1$.

THEOREM 6.3 (Main Theorem for $\ell_p(\ell_1)$). (Theorem 3.12 of [**1**], Theorem 2.14 of [**3**]) *Let* $(x_n)_{n=1}^{\eta}$ *be a normalized, complemented, unconditional basic sequence in* $\ell_p(\ell_1)$ $(0 < p < 1)$. *Then,* $(x_n)_{n=1}^{\eta}$ *is equivalent to a subbasis of* $(e_{lk})_{l,k=1}^{\infty}$.

PROOF. Using "large coefficients" techniques and the fact that the canonical basis of ℓ_p for $0 < p < 1$ is strongly absolute, we prove that $(x_n)_{n=1}^{\eta}$ is equivalent to a complemented, unconditional basic sequence $(y_n)_{n=1}^{\eta}$ with disjoint supports:

$$y_n = \sum_{k\in F_{l_n}^n} e^*_{l_n k}(x_n) e_{l_n k},$$

where, for each $n \in \{1, \cdots, \eta\}$ and $l \in \mathbb{N}$,

$$F_l^n = \{k\,;\, e^*_{lk}(x_n) \neq 0\}.$$

That is, F_l^n is the set consisting of the non-zero entries of the matrix $(e^*_{lk}(x_n))_{l,k=1}^{\infty}$ in the lth row. Then, the support of y_n is contained in just a single row, namely l_n.

To finish the proof, we only have to observe that $(y_n)_{n=1}^{\eta}$ is equivalent to a subbasis of a normalized block basic sequence with respect to the canonical basis given by

$$u_{nl} = \frac{1}{\sum_{k\in F_l^n} e_{lk}^*(x_n)} \sum_{k\in F_l^n} e_{lk}^*(x_n)e_{lk},$$

for each $n \in \{1, \cdots, \eta\}$ and integer l for which $F_l^n \neq \emptyset$. □

References

[1] F. Albiac, *Unconditional basis of quasi-Banach spaces*, Ph. D. thesis, Universidad Pública de Navarra, 2000.

[2] F. Albiac, C. Leránoz, *Uniqueness of unconditional basis of $\ell_p(c_0)$ and $\ell_p(\ell_2)$, $0 < p < 1$*, Studia Math. **150** (2002), 35-52.

[3] F. Albiac, C. Leránoz and N. Kalton *Uniqueness of the unconditional basis of $\ell_1(\ell_p)$ and $\ell_p(\ell_1)$, $0 < p < 1$*. Submitted.

[4] J. Bourgain, P. G. Casazza, J. Lindenstrauss and L. Tzafriri, *Banach spaces with a unique unconditional basis, up to permutation*, Mem. Amer. Math. Soc. No. 322, Providence 1985.

[5] P. G. Casazza and N. J. Kalton, *Uniqueness of unconditional bases in Banach spaces*, Israel J. Math. **103** (1998), 141-176.

[6] P. G. Casazza and N. J. Kalton, *Uniqueness of unconditional bases in c_0-products*, Studia Math. **133** (1999), 275-294.

[7] P. G. Casazza and T. J. Schura, *Tsirelson space*, Springer Lecture Notes 1363, 1989.

[8] I.S. Edelstein and P. Wojtaszczyk, *On projections and unconditional bases in direct sums of Banach spaces*, Studia Math. **56** (1976), 263-276.

[9] W. T. Gowers, *A solution to Banach's hyperplane problem*, Bull. London Math. Soc. **26** (1994), 523-530.

[10] N. J. Kalton, *Orlicz sequence spaces without local convexity*, Math. Proc. Camb. Phil. Soc. **81** (1977), 253-278.

[11] N. J. Kalton, *Convexity conditions on non-locally convex lattices*, Glasgow Math. J. **25** (1984), 141-152.

[12] N. J. Kalton, N. T. Peck and J. W. Roberts, *An F-space Sampler*, London Math. Soc. Lecture Note Ser. 89, Cambridge University Press, 1985.

[13] N. J. Kalton, *Banach envelopes of non-locally convex spaces*, Canad. J. Math. **38** (1986), 65-86.

[14] N. J. Kalton, C. Leránoz and P. Wojtaszczyk, *Uniqueness of unconditional bases in quasi-Banach spaces with applications to Hardy spaces*, Israel J. Math. **72** (1990) 299-311.

[15] G. Köthe and O. Toeplitz, *Lineare Raume mit unendlich vielen Koordinaten und Ringen unendlicher Matrizen*, J. Reine Angew Math. **171** (1934), 193-226.

[16] C. Leránoz, *Uniqueness of unconditional bases in quasi-Banach spaces*, Ph. D. thesis, University of Missouri-Columbia, 1990.

[17] C. Leránoz, *Uniqueness of unconditional basis of $c_0(\ell_p)$, $0 < p < 1$*, Studia Math. **102** (1992), 193-207.

[18] J. Lindenstrauss and A. Pelczynski, *Absolutely summing operators in $\mathcal{L}_p$-spaces and their applications*, Studia Math. **29** (1968), 275-326.

[19] J. Lindenstrauss and L. Tzafriri, *Classical Banach Spaces II, Function spaces*, Springer Verlag, Berlin-Heidelberg-New York 1979.

[20] J. Lindenstrauss and M. Zippin, *Banach spaces with a unique unconditional basis*, J. Functional Analysis **3** (1969), 115-125.

[21] M. Nawrocki and A. Ortynski, *The Mackey topology and complemented subspaces of Lorentz sequence spaces $d(w,p)$ for $0 < p < 1$*, Trans. Amer. Math. Soc. **287** (1985), 713-722.

[22] P. Wojtaszczyk, Uniqueness of unconditional bases in quasi-Banach spaces with applications to Hardy spaces, II, Israel J. Math. **97** (1997), 253–280.

Department of Mathematics, University of Missouri, Columbia MO, 65211 USA
E-mail address: albiac@math.missouri.edu

Department of Mathematics, University of Missouri, Columbia MO, 65211 USA
E-mail address: nigel@math.missouri.edu

Departamento de Matemática e Informática, Universidad Pública de Navarra, 31006 Pamplona, Spain
E-mail address: camino@si.unavarra.es

Contemporary Mathematics
Volume **321**, 2003

A note on the method of minimal vectors

George Androulakis

Abstract: The methods of "minimal vectors" were introduced by Ansari and Enflo and strengthened by Pearcy, in order to prove the existence of hyperinvariant subspaces for certain operators on Hilbert space. In this note we present the method of minimal vectors for operators on super-reflexive Banach spaces and we give a new sufficient condition for the existence of hyperinvariant subspaces of certain operators on these spaces..

1. Introduction

The *Invariant Subspace Problem (I.S.P.)* asks whether there exists a separable infinite dimensional Banach space on which every operator has a non-trivial invariant subspace. By "operator" we always mean "continuous linear map", by "subspace" we mean "closed linear manifold", and by "non-trivial" we mean "different than zero and the whole space". Several negative solutions to the I.S.P. are known [**4**] [**5**] [**13**] [**14**], [**15**], [**16**]. It remains unknown whether the separable Hilbert space is a positive solution to the I.S.P.. There is an extensive literature of results towards a positive solution of the I.S.P. especially in the case of the infinite dimensional separable complex Hilbert space ℓ_2. We only mention Lomonosov's result: every operator which is not a multiple of the identity and commutes with a non-zero compact operator on a complex Banach space has a non-trivial hyperinvariant subspace [**8**]. For surveys on the topic see [**12**] and [**9**]. Recently Ansari and Enflo [**1**] introduced the methods of minimal vectors and gave a new proof of the existence of non-trivial hyperinvariant subspaces of non-zero compact operators on ℓ_2. The method of minimal vectors which was introduced by Enflo, was strengthened by Pearcy [**10**] in order to give a new proof to the following special case of Lomonosov's theorem: every non-zero quasi-nilpotent operator on ℓ_2 which commutes with a non zero compact operator has a non-trivial hyperinvariant subspace. In this note we present the method of minimal vectors of an operator and we carry out two generalizations compared to the existing versions of Ansari-Enflo and Pearcy: Firstly, the operators are defined on a general super-reflexive Banach space rather than the space ℓ_2. This may be proved important if we try to find some Banach space which is a solution to the I.S.P. rather than examining whether ℓ_2 is a solution to the I.S.P.. Secondly, we introduce a property $(\star)$ that an operator may

1991 *Mathematics Subject Classification.* Primary: 47A15, Secondary: 46B03.

This research was partially supported by NSF.

satisfy. If an operator Q commutes with a non-zero compact operator then Q satisfies property ($\star$). Our main result (Theorem 2.2) refers to operators that satisfy property ($\star$) rather than those that commute with a non-zero compact operator. More precisely, we prove that every non-zero quasi-nilpotent operator which satisfies property ($\star$) on a super-reflexive Banach space has a non-trivial hyperinvariant subspace. We ask whether there exist operators which satisfy property ($\star$) but do not have any non-zero compact operator in their commutant. Also we ask whether every operator with no non-trivial invariant subspace must satisfy property ($\star$). If the answer is positive then Theorem 2.2 will imply that every quasi-nilpotent operator on a super-reflexive Banach space has a non-trivial invariant subspace. Then, every strictly singular operator on a super-reflexive Hereditarily Indecomposable complex Banach space has a non-trivial invariant subspace (see [**7**]), and hence the space constructed in [**6**] would provide a positive solution to the I.S.P..

We now recall some standard definitions and results that we shall use in this paper. A Banach space $(X, \|\cdot\|)$ is called *strictly convex* if for every $x, y \in X$ with $\|x\| = \|y\| = \|(x+y)/2\| = 1$ we have that $x = y$. A Banach space $(X, \|\cdot\|)$ is called *uniformly convex* if for every $\varepsilon > 0$ there exists a $\delta = \delta(\varepsilon) > 0$ such that for $x, y \in X$ with $\|x\| = \|y\| = 1$ and $\|\frac{x+y}{2}\| > 1 - \delta$ we have that $\|x - y\| < \varepsilon$. The function $\delta(\varepsilon)$ is called the modulus of uniform convexity of X. The norm of X is *Gâteaux differentiable* if for every $x \in X \backslash \{0\}$ and for every $y \in X$ the limit

$$\lim_{t \to 0} \frac{\|x + ty\| - \|x\|}{t} \tag{1}$$

exists. The Banach space X is called *smooth* if for every $x \in X \backslash \{0\}$ there exists a unique $f \in X^*$ such that $f(x) = \|x\|^2 = \|f\|^2$. We denote the functional f by $(x)^*$. It can be proved that the norm of X is Gâteaux differentiable if and only if X is smooth, in which case the limit in (1) is equal to $\mathrm{Re}\,(x)^*(y)/\|(x)^*\|$. The norm of X is called *Fréchet differentiable* if the limit in (1) exists uniformly for all $y \in X$ with $\|y\| = 1$. The norm of X is called *uniformly smooth* if the limit in (1) exists uniformly for all $x, y \in X$ with $\|x\| = \|y\| = 1$. A Banach space X is called *super-reflexive* if every infinite dimensional space Y which is finitely represented in X must be reflexive. It is proved in [**3**] (see also [**11**]) that every super-reflexive Banach space X can be equivalently renormed to be uniformly convex. It follows from a renorming technique of Asplund [**2**] that a Banach space is super-reflexive if and only if it can be equivalently renormed to be uniformly convex or uniformly smooth or both.

2. Minimal vectors and invariant subspaces

We start by introducing some notations and terminology. If X is a Banach space, $x \in X$ and $\varepsilon > 0$ we denote by $\mathrm{S}(x, \varepsilon)$ (respectively $\mathrm{Ba}(x, \varepsilon)$) the *sphere* (respectively the *closed ball*) of X with center x and radius ε, namely the set $\{y \in X : \|x - y\| = \varepsilon\}$ (respectively the set $\{y \in X : \|x - y\| \leq \varepsilon\}$).

Definition 2.1. *Let X be a Banach space and Q be an operator on X. We say that an operator Q satisfies property ($\star$) if for every $\varepsilon \in (0,1)$ there exists $x_0 \in S(0,1)$ such that for every weakly convergent sequence $(x_n) \subset S(x_0, \varepsilon)$ there exists a subsequence $(x_{n_k})_k$ of (x_n) and a sequence $(K_k) \subset \{Q\}'$ such that*

(a) *$\|K_k\| \leq 1$ and $\|K_k(x_0)\| \geq \frac{1+\varepsilon}{2}$ for all $k \in \mathbb{N}$.*
(b) *$(K_k(x_{n_k}))_k$ converges in norm.*

The purpose of (a) is to ensure that the limit of (b) is non-zero. Indeed, notice that if $Q, \varepsilon, x_0, (K_k)_k, (x_{n_k})_k$ are as in the previous definition then $\|K_k x_0 - K_k x_{n_k}\| \leq \|K_k\| \|x_0 - x_{n_k}\| \leq \varepsilon$, thus $\|K_k x_{n_k}\| \geq \|K_k x_0\| - \|K_k x_0 - K_k x_{n_k}\| \geq (1+\varepsilon)/2 - \varepsilon = (1-\varepsilon)/2 > 0$.

Also notice that for every operator Q if there exists a non-zero compact operator which commutes with Q then Q satisfies property $(\star)$.

Our main result is the following:

THEOREM 2.2. *Let X be a super-reflexive Banach space and Q be a non-zero quasi-nilpotent operator on X which satisfies property $(\star)$. Then Q has a non-trivial hyperinvariant subspace.*

For the proof of this result we use the method of minimal vectors of an operator. Let $(X, \|\cdot\|)$ be a reflexive Banach space, Q be an operator on X with dense range, $x_0 \in X$ with $\|x_0\| = 1$ and $\varepsilon \in (0,1)$. We define a sequence of *minimal vectors of Q with respect to x_0 and ε* to be a sequence $(y_n)_{n\in\mathbb{N}}$ as follows. For every $n \in \mathbb{N}$ the set $Q^{-n}\mathrm{Ba}(x_0, \varepsilon)$ is non-empty (since Q^n has a dense range), closed and convex. Thus there exists $y_n \in Q^{-n}\mathrm{Ba}(x_0,\varepsilon)$ such that

$$\|y_n\| = \inf\{\|y\| : y \in Q^{-n}\mathrm{Ba}(x_0, \varepsilon)\}. \tag{2}$$

Indeed, if $(y_{n,m})_m$ is a sequence in $Q^{-n}\mathrm{Ba}(x_0,\varepsilon)$ with

$$\|y_{n,m}\| \searrow \inf\{\|y\| : y \in Q^{-n}\mathrm{Ba}(x_0, \varepsilon)\}, \tag{3}$$

then $(y_{n,m})_m$ is a subset of $Q^{-n}\mathrm{Ba}(x_0,\varepsilon) \cap \mathrm{Ba}(0, \|y_{n,1}\|)$ which is weakly compact (since it is a closed, convex and bounded subset of a reflexive space). Thus by passing to a subsequence and relabeling we can assume that $(y_{n,m})_m$ converges weakly to some vector y_n. Since the norm is weakly lower semicontinuous, (3) implies (2).

In order to prove Theorem 2.2 we need the following three results whose proofs are postponed. For the first result, notice that if X is a reflexive Banach space, Q is an operator on X with dense range, $x_0 \in X$ with $\|x_0\| = 1$, $\varepsilon \in (0,1)$ and (y_n) is a sequence of minimal vectors of Q with respect to x_0 and ε, then the sequence $((Q^n y_n - x_0)^*)_n$ is bounded, (namely, $\|(Q^n y_n - x_0)^*\| = \|Q^n y_n - x_0\| = \varepsilon$) by the minimality of $\|y_n\|$), thus it has weak* limit points. We want to know that 0 is not a weak* limit point of the sequence $((Q^n y_n - x_0)^*)_n$. The next result yields that this is true provided that the choice of ε is appropriate.

LEMMA 2.3. *Let $(X, \|\cdot\|)$ be a smooth and uniformly convex Banach space and Q be an operator on X with dense range. Then there exists $\varepsilon \in \left[\frac{1}{2}, 1\right)$ such that the following is satisfied: if $x_0 \in X$ with $\|x_0\| = 1$, $(y_n)_n$ is a sequence of minimal vectors of Q with respect to x_0 and ε, and f is a weak* limit point of $((Q^n y_n - x_0)^*)_n$, then $f \neq 0$.*

LEMMA 2.4. *Let X be a reflexive Banach space, Q be a quasi-nilpotent operator on X with dense range, $x_0 \in X$ with $\|x_0\| = 1$, $\varepsilon > 0$, and $(y_n)_{n\in\mathbb{N}}$ be a sequence of minimal vectors of Q with respect to x_0 and ε. Then there exists an increasing sequence $(n_k)_k$ of $\mathbb{N}$ such that*

$$\lim_k \frac{\|y_{n_k - 1}\|}{\|y_{n_k}\|} = 0. \tag{4}$$

For the next Lemma, if X is a Banach space and $f \in X^*$ then $\ker(f)$ denotes the *kernel* of f.

LEMMA 2.5. *Let X be a reflexive smooth Banach space, Q be an operator on X with dense range, $x_0 \in X$ with $\|x_0\| = 1$, $\varepsilon \in (0,1)$, and $(y_n)_{n\in\mathbb{N}}$ be a sequence of minimal vectors of Q with respect to x_0 and ε. Then for all $n \in \mathbb{N}$,*

$$\ker((y_n)^*) \subseteq \ker((Q^n)^*(Q^n y_n - x_0)^*). \tag{5}$$

Now we are ready for the

PROOF OF THEOREM 2.2. Since X is super-reflexive we can assume by our discussion in the previous section, that $(X, \|\cdot\|)$ is smooth and locally uniformly convex. Without loss of generality we assume that Q has a dense range and it is 1-1 (because the range and the kernel of Q are hyperinvariant subspaces of Q). By Lemma 2.3 there exists $\varepsilon \in \left[\frac{1}{2}, 1\right)$ such that the conclusion of the lemma is satisfied. For that ε, since Q satisfies property $(\star)$, let $x_0 \in X$, $\|x_0\| = 1$ such that the statement of the definition of property $(\star)$ is valid for the operator Q. Let $(y_n)_n$ be a sequence of minimal vectors of Q with respect to x_0 and ε. By Lemma 2.4 let $(n_k)_k$ be an increasing subsequence of $\mathbb{N}$ such that (4) is valid. Since X is reflexive, by considering a further subsequence of (n_k) and relabeling we can assume that $(Q^{n_k-1}y_{n_k-1})_k$ converges weakly. By the property $(\star)$ of Q, there exists a subsequence of (n_k) (which, by relabeling, is still called (n_k)) and a sequence $(K_k)_k \subset \{Q\}'$ such that $(K_k Q^{n_k-1} y_{n_k-1})_k$ converges in norm to some vector $w \in X$. By our discussion following the definition of property $(\star)$ we have that $w \neq 0$. Since Q is 1-1 we have that $Qw \neq 0$. We claim that $Y := \{Q\}'(Qw)$ is a non-trivial hyperinvariant subspace for Q. We only need to show that $Y \neq X$. For that reason we let f to be a weak* limit point of $((Q^{n_k} y_{n_k} - x_0)^*)_k$, which is non-zero by Lemma 2.3, and we will show that $Y \subset \ker(f)$. We need to show that if $T \in \{Q\}'$ then $f(TQw) = 0$. Let $T \in \{Q\}'$ and $k \in \mathbb{N}$. Since $\ker((y_{n_k})^*)$ is a 1-codimensional subspace of X and $y_{n_k} \notin \ker((y_{n_k})^*)$ (notice that $(y_{n_k})^*(y_{n_k}) = \|y_{n_k}\|^2 \neq 0$), we have that $X = \operatorname{span}\{y_{n_k}\} \oplus \ker((y_{n_k})^*)$, thus there exists a scalar a_k and $r_k \in \ker((y_{n_k})^*)$ such that

$$TK_k(y_{n_k-1}) = a_k y_{n_k} + r_k. \tag{6}$$

We claim that $a_k \to 0$. Indeed,

$$\begin{aligned}
|a_k|\|y_{n_k}\|^2 &= |(y_{n_k})^*(a_k y_{n_k} + r_k)| \\
&= |(y_{n_k})^* T K_k(y_{n_k-1})| \quad \text{(by (6))} \\
&\leq \|(y_{n_k})^*\|\|T\|\|K_k\|\|y_{n_k-1}\| \\
&\leq \|y_{n_k}\|\|T\|\|y_{n_k-1}\|,
\end{aligned}$$

thus $|a_k| \leq \|T\|\|y_{n_k-1}\|/\|y_{n_k}\| \underset{k\to\infty}{\longrightarrow} 0$.

First apply Q^{n_k} and then $(Q^{n_k}y_{n_k} - x_0)^*$, on (6), to obtain

$$Q^{n_k} T K_k y_{n_k-1} = a_k Q^{n_k} y_{n_k} + Q^{n_k} r_k \quad \text{and}$$

$$(Q^{n_k}y_{n_k} - x_0)^* Q^{n_k} T K_k y_{n_k-1} = a_k (Q^{n_k}y_{n_k} - x_0)^* Q^{n_k} y_{n_k} + (Q^{n_k}y_{n_k} - x_0)^* Q^{n_k} r_k. \tag{7}$$

Since $r_k \in \ker((y_{n_k})^*)$, by Lemma 2.5 we have that $(Q^{n_k}y_{n_k} - x_0)^* Q^{n_k}(r_k) = 0$. Furthermore, since $a_k \to 0$ and $\|Q^{n_k}y_{n_k} - x_0\| = \varepsilon$, we have that the right hand side of (7) tends to zero. Thus by taking limits and noticing that $T, K_k \in \{Q\}'$ for all k, (7) becomes

$$\lim_k (Q^{n_k}y_{n_k} - x_0)^* T Q K_k Q^{n_k-1} y_{n_k-1} = 0. \tag{8}$$

Since $(K_k Q^{n_k-1} y_{n_k-1})_k$ converges in norm to w and f is a weak* limit point of $((Q^{n_k} y_{n_k} - x_0)^*)_k$, (8) yields that $f(TQw) = 0$ which finishes the proof. $\square$

We now turn our attention to the proof of Lemma 2.3. Before presenting the proof of Lemma 2.3 we need the following two results.

SUBLEMMA 2.6. *Let $(X, \|\cdot\|)$ be a smooth and strictly convex Banach space, $x_0 \in X$, $\|x_0\| = 1$, $0 < \varepsilon < 1$ and $w \in S(x_0, \varepsilon)$. The following conditions are equivalent:*

(a) $\|x_0 - \lambda w\| > \varepsilon$ *for all* $\lambda \in [0, 1)$.

(b) *Re* $\frac{(x_0-w)^*}{\|x_0-w\|^2}(x_0) \geq 1$.

PROOF. (a) $\Rightarrow$ (b): Since $\|x_0 - \lambda w\| > \varepsilon = \|x_0 - 1 \cdot w\|$, for all $\lambda \in [0, 1)$, we have that the derivative of the function $f(\lambda) = \|x_0 - \lambda w\|$ at 1 is non-positive. Set $g(\mu) = f(1 - \mu)$. Then $g'(0) = -f'(1)$, thus $g'(0) \geq 0$. Note that $g(\mu) = \|x_0 - (1 - \mu)w\| = \|x_0 - w + \mu w\|$. Thus

$$\begin{aligned} g'(0) &= \operatorname{Re} \frac{(x_0 - w)^*}{\|(x_0 - w)^*\|}(w) = \operatorname{Re} \frac{(x_0 - w)^*}{\|x_0 - w\|}(w) \\ &= \operatorname{Re} \frac{(x_0 - w)^*}{\|x_0 - w\|}(w - x_0 + x_0) = -\frac{\|x_0 - w\|^2}{\|x_0 - w\|} + \operatorname{Re} \frac{(x_0 - w)^*}{\|x_0 - w\|}(x_0) \\ &= -\|x_0 - w\| + \operatorname{Re} \frac{(x_0 - w)^*}{\|x_0 - w\|}(x_0). \end{aligned}$$

Thus $g'(0) \geq 0$ if and only if

$$\operatorname{Re} \frac{(x_0 - w)^*}{\|x_0 - w\|^2}(x_0) \geq 1.$$

(b) $\Rightarrow$ (a): By the proof of (a) $\Rightarrow$ (b), notice that if (b) is valid then $f'(1) \leq 0$. Notice also that f is a strictly convex function (since $(X, \|\cdot\|)$ is strictly convex) with $f(1) = \varepsilon$ (since $w \in \mathrm{S}(x_0, \varepsilon)$). Thus if (b) is valid then $f(\lambda) > \varepsilon$ for $\lambda \in [0, 1)$. $\square$

LEMMA 2.7. *If $(X, \|\cdot\|)$ is a smooth and uniformly convex Banach space, then for every $\eta > 0$ there exists $\varepsilon \in \left[\frac{1}{2}, 1\right)$ such that for all $x_0 \in X$ with $\|x_0\| = 1$ and for all $w \in S(x_0, \varepsilon)$ satisfying $\|x_0 - \lambda w\| > \varepsilon$ for all $\lambda \in [0, 1)$, we have that $\|w\| \leq \eta$.*

PROOF. We start with the

Claim: Let $(X, \|\cdot\|)$ be a uniformly convex Banach space, and $\eta' > 0$. Let $\delta(\cdot)$ denote the modulus of uniform convexity of X. Then for every $x_0 \in X$ with $\|x_0\| = 1$,

$$\sup\{\|x\| \colon x \in \mathrm{S}(x_0, 1),\ \operatorname{Re}(x_0 - x)^*(x_0) > 1 - 2\delta(\eta')\} \leq \eta'. \tag{9}$$

Indeed, if $x \in \mathrm{S}(x_0, 1)$ with $\|x\| \geq \eta'$, then we have that $\|x_0 - (x_0 - x)\| \geq \eta'$, $\|x_0\| = 1$, $\|x_0 - x\| = 1$, hence

$$\begin{aligned} 1 - \delta(\eta') &\geq \left\| \frac{x_0 + (x_0 - x)}{2} \right\| \geq \left| (x_0 - x)^* \left(\frac{x_0 + (x_0 - x)}{2} \right) \right| \\ &\geq \operatorname{Re}(x_0 - x)^* \left(\frac{x_0 + (x_0 - x)}{2} \right) \\ &= \frac{\operatorname{Re}(x_0 - x)^*(x_0) + 1}{2}. \end{aligned}$$

Thus $\operatorname{Re}(x_0 - x)^*(x_0) \leq 1 - 2\delta(\eta')$ which finishes the proof of the Claim.

Let X and η as in the statement of Lemma 2.7. Let ε satisfying

$$\frac{1}{2} \leq \varepsilon, \quad 0 < \frac{1}{\varepsilon} - 1 \leq 2\delta\left(\frac{\eta}{2}\right) \quad \text{and} \quad 1 - \varepsilon \leq \eta/2. \tag{10}$$

Let $w \in \mathrm{S}(x_0, \varepsilon)$ satisfying $\|x_0 - \lambda w\| > \varepsilon$ for all $\lambda \in [0,1)$. By Sublemma 2.6 we have that

$$\operatorname{Re} \frac{(x_0 - w)^*}{\|x_0 - w\|^2}(x_0) \geq 1. \tag{11}$$

Let $x \in \mathrm{S}(x_0, 1)$ with

$$x_0 - w = \varepsilon(x_0 - x). \tag{12}$$

Then

$$\begin{aligned}
\left| \operatorname{Re} \frac{(x_0 - w)^*}{\|x_0 - w\|^2}(x_0) - \operatorname{Re}(x_0 - x)^*(x_0) \right| &\leq \left| \frac{(x_0 - w)^*}{\|x_0 - w\|^2}(x_0) - (x_0 - x)^*(x_0) \right| \\
&\leq \|x_0\| \left\| \frac{(x_0 - w)^*}{\|x_0 - w\|^2} - (x_0 - x)^* \right\| \\
&= \left\| \frac{\varepsilon(x_0 - x)^*}{\|\varepsilon(x_0 - x)\|^2} - (x_0 - x)^* \right\| \quad \text{(by (12))} \\
&\leq \left| \frac{1}{\varepsilon} - 1 \right| \leq 2\delta\left(\frac{\eta}{2}\right) \quad \text{(by (10))}.
\end{aligned}$$

Therefore by (11) we have that $\operatorname{Re}(x_0 - x)^*(x_0) \geq 1 - 2\delta\left(\frac{\eta}{2}\right)$. By (9) we have that $\|x\| \leq \eta/2$. By (12) we have that $w = (1-\varepsilon)x_0 + \varepsilon x$ and thus by the triangle inequality, $\|x\| \leq \eta/2$, and (10) we obtain $\|w\| \leq (1-\varepsilon) + \varepsilon\|x\| \leq (1-\varepsilon) + \|x\| \leq \eta$. □

Proof of Lemma 2.3. Let X be a smooth and uniformly convex Banach space and Q be an operator on X with dense range. For $\eta = \frac{1}{3}$ we choose $\varepsilon \in \left[\frac{1}{2}, 1\right)$ to satisfy the statement of Lemma 2.7.

Let $(y_n)_{n\in\mathbb{N}}$ be a sequence of minimal vectors of Q. For $n \in \mathbb{N}$, by the minimality of $\|y_n\|$ we have that $Q^n y_n \in \mathrm{S}(x_0, \varepsilon)$ and $\|x_0 - \lambda Q^n y_n\| > \varepsilon$ for all $\lambda \in [0,1)$. Thus by Lemma 2.7 we obtain $\|Q^n y_n\| \leq \frac{1}{3}$. Let f be any weak* limit point of the sequence $((Q^n y_n - x_0)^*)_{n\in\mathbb{N}}$. Since $\|x_0\| = 1$, $\|Q^n y_n\| \leq \frac{1}{3}$ and $\|Q^n y_n - x_0\| = \varepsilon \geq 1/2$ we have that $f \neq 0$. Indeed for $n \in \mathbb{N}$,

$$\begin{aligned}
(Q^n y_n - x_0)^*(-x_0) &= (Q^n y_n - x_0)^*(Q^n y_n - x_0 - Q^n y_n) \\
&= (Q^n y_n - x_0)^*(Q^n y_n - x_0) - (Q^n y_n - x_0)^*(Q^n y_n) \\
&= \|Q^n y_n - x_0\|^2 - (Q^n y_n - x_0)^*(Q^n y_n) \\
&\geq \varepsilon^2 - |(Q^n y_n - x_0)^*(Q^n y_n)| \\
&\geq \varepsilon^2 - \|(Q^n y_n - x_0)^*\| \|Q^n y_n\| \\
&\geq \varepsilon^2 - \varepsilon \cdot \frac{1}{3} \\
&\geq \frac{1}{4} - \frac{1}{6} = \frac{1}{12} > 0 \quad \left(\text{since } \frac{1}{2} \leq \varepsilon\right).
\end{aligned}$$

Since f is a weak* limit point of $((Q^n y_n - x_0)^*_n$, we have that $f(-x_0) \geq \frac{1}{12}$, thus $f \neq 0$. □

PROOF OF LEMMA 2.4. If the statement is not true, then there exists a positive number δ such that

$$\frac{\|y_{n-1}\|}{\|y_n\|} \geq \delta \text{ for all } n \in \mathbb{N}.$$

Thus for every integer $n \in \mathbb{N}$, we have

$$\|y_1\| \geq \delta\|y_2\| \geq \delta^2\|y_3\| \geq \cdots \geq \delta^n\|y_{n+1}\|. \tag{13}$$

We have that $\|Qy_1 - x_0\| \leq \varepsilon$. Also we have that $\|Q(Q^n y_{n+1}) - x_0\| = \|Q^{n+1}y_{n+1} - x_0\| \leq \varepsilon$. By the minimality of $\|y_1\|$ we have:

$$\|y_1\| \leq \|Q^n y_{n+1}\| \leq \|Q^n\|\|y_{n+1}\|. \tag{14}$$

By combining (13) and (14) we have $\delta \leq \|Q^n\|^{1/n}$ which is a contradiction since Q is quasi-nilpotent. □

In order to prove Lemma 2.5 we need the following

REMARK 2.8. *Let X be a Banach space, $f \in X^*\backslash\{0\}$ and $g \in X^*$ such that*

$$\text{for all } x \in X, \text{ if } \mathrm{Re}(f(x)) < 0 \text{ then } \mathrm{Re}(g(x)) \geq 0. \tag{15}$$

Then $\ker(f) \subseteq \ker(g)$ and moreover there exists a non-positive real number a such that $g = af$.

PROOF. Let $x_0 \in X$ with $f(x_0) = -1$. Thus $\mathrm{Re}\, g(x_0) \geq 0$. We first claim that $\ker(f) \subseteq \ker(g)$. Indeed, otherwise there exists $x \in \ker(f)\backslash \ker(g)$. Without loss of generality assume that $\mathrm{Re}\, g(x) < -2\mathrm{Re}\, g(x_0)$. Let $x' = x_0 + x$. Then $f(x') = f(x_0) + f(x) = f(x_0)$, hence $\mathrm{Re}\, f(x') = -1 < 0$. We also have $\mathrm{Re}\, g(x') = \mathrm{Re}\, g(x_0) + \mathrm{Re}\, g(x) < -\mathrm{Re}\, g(x_0) \leq 0$ which is a contradiction, proving that $\ker(f) \subseteq \ker(g)$.

Since both $\ker(f), \ker(g)$ are at most 1-codimensional subspaces of X we have that there exists a scalar a such that $g = af$. Since $f(x_0) = -1$ and $\mathrm{RE}\, g(x_0) \geq 0$ we have that $\mathrm{Re}\, a \leq 0$. If $a \notin \mathbb{R}$ then let $a = a_1 + ia_2$ with $a_1, a_2 \in \mathbb{R}$, $a_1 \leq 0$ and $a_2 \neq 0$. Let $x_1 \in X$ with $f(x_1) = -1 + ia_2^{-1}(1 - a_1)$. Then $\mathrm{Re}\, f(x_1) = -1 < 0$ and $\mathrm{Re}\, g(x_1) = \mathrm{Re}\,((a_1 + ia_2)(-1 + ia_2^{-1}(1 - a_1))) = -1 < 0$ which is a contradiction, proving that $a \in \mathbb{R}$. □

Now we are ready for the

PROOF OF LEMMA 2.5. For a fixed $n \in \mathbb{N}$ we prove that the assumption of Remark 2.8 is satisfied for $f = (y_n)^*$ and $g = (Q^n)^*((Q^n y_n - x_0)^*)$. Let $x \in X$ with $\mathrm{Re}\,(y_n)^*(x) < 0$. We claim that $\mathrm{Re}\,(Q^n)^*(Q^n y_n - x_0)^*(x) \geq 0$ i.e. $\mathrm{Re}\,(Q^n y_n - x_0)^*(Q^n x) \geq 0$. Indeed, otherwise, since $\mathrm{Re}\,(Q^n y_n - x_0)^*(Q^n x)/\|(Q^n y_n - x_0)^*\|$ is the derivative of the function

$$t \mapsto \|Q^n y_n - x_0 + tQ^n x\|$$

at 0, we obtain that this function is decreasing for t in a neighborhood of 0. Thus for small $t > 0$ we have

$$\varepsilon = \|Q^n y_n - x_0\| \geq \|Q^n y_n - x_0 + tQ^n x\|$$

i.e.

$$\|Q^n(y_n + tx) - x_0\| \leq \varepsilon.$$

We have by the minimality of $\|y_n\|$ that

$$\|y_n\| \leq \|y_n + tx\| \quad \text{for small} \quad t > 0.$$

Thus the derivative of the function

$$t \mapsto \|y_n + tx\|$$

must be non-negative at 0, i.e. $\mathrm{Re}\,(y_n)^*(x) \geq 0$ which is a contradiction. $\square$

References

[1] Ansari, S.; Enflo, P. *Extremal vectors and invariant subspaces.* Trans. Amer. Math. Soc. **350** (1998), no. 2, 539–558.

[2] Asplund, E. *Averaged norms.* Israel J. Math. **5** 1967 227–233.

[3] Enflo, P. *Banach spaces which can be given an equivalent uniformly convex norm.* Proceedings of the International Symposium on Partial Differential Equations and the Geometry of Normed Linear Spaces (Jerusalem, 1972). Israel J. Math. **13** (1972), 281–288 (1973).

[4] Enflo, P. *On the invariant subspace problem in Banach spaces.* Seminaire Maurey–Schwartz (1975–1976) Espaces L_p, applications radonifiantes et geometrie des espaces de Banach, Exp. Nos. 14-15, 7 pp. Centre Math., cole Polytech., Palaiseau, 1976.

[5] Enflo, P. *On the invariant subspace problem for Banach spaces.* Acta Math. **158** (1987), no. 3-4, 213–313.

[6] Ferenczi, V. *A uniformly convex hereditarily indecomposable Banach space.* Israel J. Math. **102** (1997), 199–225.

[7] Gowers, W.T.; Maurey, B. *The unconditional basic sequence problem.* J. Amer. Math. Soc. **6** (1993), no. 4, 851–874.

[8] Lomonosov, V.I. *Invariant subspaces of the family of operators that commute with a completely continuous operator.* (Russian) Funkcional. Anal. i Priložen. **7** (1973), no. 3, 55–56.

[9] Pearcy, C.; Shields, A. *A survey of the Lomonosov technique in the theory of invariant subspaces*, Topics in operator theory. Edited by C. Pearcy. Mathematical Surveys, No. 13. American Mathematical Society, Providence, R.I., 1974., 219–230.

[10] Pearcy, C. *On the technique of Enflo*, preprint.

[11] Pisier, G. *Martingales with values in uniformly convex spaces.* Israel J. Math. **20** (1975), no. 3-4, 326–350.

[12] Radjavi, H.; Rosenthal, P.*Invariant subspaces.* Ergebnisse der Mathematik und ihrer Grenzgebiete, Band 77. Springer-Verlag, New York-Heidelberg, 1973.

[13] Read, C.J. *A solution to the invariant subspace problem.* Bull. London Math. Soc. **16** (1984), no. 4, 337–401.

[14] Read, C.J. *A solution to the invariant subspace problem on the space* $\ell_1 1$. Bull. London Math. Soc. **17** (1985), no. 4, 305–317.

[15] Read, C.J. *A short proof concerning the invariant subspace problem.* J. London Math. Soc. (2) **34** (1986), no. 2, 335–348.

[16] Read, C.J. *Strictly singular operators and the invariant subspace problem.* Studia Math. **132** (1999), no. 3, 203–226.

Department of Mathematics, University of South Carolina, Columbia, SC 29208. giorgis@math.sc.edu

Contemporary Mathematics
Volume **321**, 2003

The Projective Tensor Product I

Joe Diestel, Jan Fourie, and Johan Swart

ABSTRACT. Permanence properties of the projective tensor product are discussed. We discuss the situation where the projective tensor product of subspaces is a subspace of the projective tensor product of the Banach spaces involved. The role of the Radon-Nikodým property and the approximation property in the duality between the projective and the injective tensor products is outlined. The representation of the projective tensor product in the case where one coordinate is a reflexive Lebesgue space is discussed. Throughout the paper we look at Banach space invariants like the Radon-Nikodým property, the approximation property, weakly sequentially completeness, the Dunford-Pettis property and Pełczyński's property V.

Introduction

The study of tensor products of Banach spaces is more than a half-century into its development. It enjoys a notoriety (derived from its aridity, in part) that is unmatched in functional analysis. This report will no doubt help solidify that notoriety.

As viewed by their primary mover, Alexander Grothendieck, tensor products of Banach spaces are a means of isolating 'natural' classes of operators, the understanding of which are key to understanding the structure of Banach spaces. By themselves tensor products possess a certain elegance often lost in seemingly endless technicalities. We believe their study is still fertile ground worth consideration.

In this report we discuss the projective tensor product of *two* Banach spaces. To be sure we barely broach this topic and have been parochial in the extreme in our choice of what we treat.

A few words about the content of each section are in order.

The first section discusses the situation when X being a closed linear subspace of Y implies that the projective tensor product of X with Z is a subspace of Y's projective product with Z. Famous results of Grothendieck, Jean Bourgain and Bernard Maurey dominate the scene. We also make a few remarks about the permanence of having bases and approximation properties in projective tensor products.

1991 *Mathematics Subject Classification.* Primary 46B28; Secondary 47B10, 46M05.

Much of the work of the first author was done while visiting the Department of Mathematics and Applied Mathematics at the University of Pretoria; sincere thanks must be extended to the members of the faculty and staff at UP for providing a comfortable setting in which to work.

The second section talks briefly about the duality of the projective tensor product with the injective tensor product. The main point here is that if either X^* or Y^* has both the Radon-Nikodym property and the approximation property then the dual of the injective tensor product of X and Y is the projective tensor product of X^* and Y^*.

The main results of the second section come in handy in the third section, wherein the case when one coordinate is a reflexive Lebesgue space is dealt with. Here Qingying Bu's representation theorems for which sequences in X constitute $L^p(0,1)\hat{\otimes}X$ hold center stage. Results of Bu, Paddy Dowling and Dan Lewis affirming positive stability features of $L^p(0,1)\hat{\otimes}X$ are mentioned. In addition we discuss a stunning result of Bu pertaining to mapping properties on absolutely p-summing operators with a Hilbertian domain which use his representation theorem.

In the fourth section we speak briefly to what happens when we take the projective tensor product of two $C(K)$'s; it seems plain that the usual Banach space invariants enjoyed by $C(K)$-spaces do *not* survive taking projective tensor products.

Unquestionably, the Banach-algebraic structure is more amenable to the study of such spaces. In addition to our discussion of $C(K)$'s we speak of a couple of Gilles Pisier's most remarkable results: His K-convexity theorem and his counterexample to Grothendieck's conjecture. The first of these results has lots to say about the structure of the projective tensor product of two *good* Banach spaces; while the second sets serious limitations to the hopes and ambitions of the uninitiated.

Open questions are mentioned throughout.

This report discusses but a small part of what's known about the projective tensor product, in particular, and a miniscule part of the theory of tensor norms. In an effort to further isolate themselves from polite mathematical conversation, the same authors have embarked on a program of trying to expose Grothendieck's Resume using the technology of his time. A skeletal version of what we plan to say can be found in the papers [**11**], [**12**], [**13**], [**14**].

Our notation is by-and-large standard. We usually use X, Y and Z for Banach spaces, E, F for finite dimensional Banach spaces. The dual of a Banach space X will be X^* and the unit ball of X is B_X. Almost any terms we use herein without definition can be found in 'Absoulutely Summing Operators' [**16**].

1. The basics about the projective tensor product

Recall that if X and Y are linear spaces (over the same scalar field, be it $\mathbb{R}$ or $\mathbb{C}$), then their tensor product $X \otimes Y$ is the linear space for which the Universal Mapping Property holds. That is, there is a (natural) linear isomorphism between the space $(X \otimes Y)'$ of all linear functionals on $X \otimes Y$ and the space $B(X, Y)$ of all bilinear functionals on $X \times Y$; the relationship is summarized by $t' \in (X \otimes Y)'$ corresponds to $\phi \in B(X, Y)$ via the formula

$$t'(x \otimes y) = \phi(x, y)$$

for $x \in X, y \in Y$. So, in a sense, $X \otimes Y$ provides a mechanism for linearizing bilinear objects.

What happens when X and Y are Banach spaces? The first hope may well be to keep the Universal Mapping Property, or some facsimile thereof, in place. To do this we norm $X \otimes Y$ via duality with the Banach space $\mathcal{B}(X, Y)$ of all continuous linear functionals on $X \times Y$, where for $\varphi \in \mathcal{B}(X,Y)$

$$\|\varphi\| = \sup\{|\varphi(x,y)| : x \in B_X, y \in B_Y\};$$

if $u \in X \otimes Y$, then

$$\|u\|_\wedge = \sup\{|\sum \varphi(x_i, y_i)|\},$$

where the supremum is taken over all $\varphi \in B_{\mathcal{B}(X,Y)}$ and all representations

$$u = \sum x_i \otimes y_i$$

of u as a (finite) linear combination of elementary tensors $x_i \otimes y_i$.

The result is a norm on $X \otimes Y$ which enjoys the properties listed in our first theorem.

THEOREM 1.1 (Grothendieck [**23**]). *$(X \otimes Y, \|\cdot\|_\wedge)$ is a normed linear space enjoying the following continuous version of the Universal Mapping Property : there is a natural linear isometric isomorphism identifying the dual $(X \otimes Y, \|\cdot\|_\wedge)^*$ of $(X \otimes Y, \|\cdot\|_\wedge)$ with the space $\mathcal{B}(X,Y)$ of bilinear continuous functionals on $X \times Y$; under this isomorphism $t^* \in (X \otimes Y, \|\cdot\|_\wedge)^*$ and $\varphi \in \mathcal{B}(X,Y)$ correspond via the formula*

$$t^*(x \otimes y) = \varphi(x,y).$$

It's important to realize that this norm is *not* the only norm on $X \otimes Y$ that's useful or important; it *is* the only one for which the dual of $X \otimes Y$ is $\mathcal{B}(X,Y)$. Since one role of tensor products is to linearize bilinear objects, it's natural to ask of a norm on $X \otimes Y$ that it satisfy certain "reasonable" conditions. To wit, the norm α should be so that $\alpha(x \otimes y) = \|x\|\|y\|$ and if $x^* \in X^*$ and $y^* \in Y^*$ then $x^* \otimes y^* \in (X \otimes Y, \alpha)^*$ with

$$\|x^* \otimes y^*\|_{(X \otimes Y,\alpha)^*} = \|x^*\|\|y^*\|.$$

In fact, $\|\cdot\|_\wedge$ does all this and more.

THEOREM 1.2 (Grothendieck [**23**]). *The projective tensor norm $\|\cdot\|_\wedge$ is a reasonable norm on $X \otimes Y$; in fact, it's the greatest reasonable norm on $X \otimes Y$ and can, in fact, be written in the form*

$$\|u\|_\wedge = \inf\{\sum \|x_i\|\|y_i\|\},$$

where the infimum is taken over all representations

$$u = \sum x_i \otimes y_i$$

of u as a member of the (algebraic) tensor product $X \otimes Y$ of X and Y.

Our emphasis on 'algebraic' is intentional. Unless either X or Y is finite dimensional, $(X \otimes Y, \|\cdot\|_\wedge)$ is *not* complete. So: complete it. The result is $X \overset{\wedge}{\otimes} Y$ and the norm $\| \;\|_\wedge$ - we don't change our notation a bit. Our elements are a bit less easily described but we've lost little else: the dual of $X \overset{\wedge}{\otimes} Y$ is still $\mathcal{B}(X;Y)$, since the dual of a normed linear space and that of its completion are the same. Just how much we've lost in the description of members of $X \overset{\wedge}{\otimes} Y$ vis-a-vis those of $X \otimes Y$ is encapsulated in the following classical result generally ascribed to Grothendieck.

THEOREM 1.3 (Grothendieck [**25**], p51). *Let X and Y be Banach spaces and $u \in X \overset{\wedge}{\otimes} Y$. Then u is representable in the form*

$$u = \sum_n x_n \otimes y_n$$

where (x_n) is a sequence in $X, (y_n)$ is a sequence in Y and the series converges absolutely. In fact,

$$\|u\|_\wedge = \inf\{\sum_n \|x_n\|\|y_n\|\},$$

where the infimum is taken over all such (finite or infinite) representations of u.

This representation is remarkably flexible and, in the right hands, can be made to play a mathematical tune with amazing clarity.

The idea behind the above representation is easy enough - telescope! Some care need be taken though. It's plain that if $\sum_n x_n \otimes y_n$ is an absolutely convergent series with sum u, then

$$\|u\|_\wedge \leq \sum_n \|x_n\|\|y_n\|.$$

It's rigging things in the form we want that requires care. Let $\varepsilon > 0$ be our margin of error and select $(u_n) \subset X \otimes Y$ so that

$$\|u - u_n\|_\wedge < \varepsilon/2^{n+3}.$$

Write

$$u_1 = \sum_{i=1}^{i(1)} x_i \otimes y_i$$

with

$$\|u_1\|_\wedge \leq \sum_{i=1}^{i(1)} \|x_i\|\|y_i\| \leq \|u_1\|_\wedge + \frac{\varepsilon}{2^4} \leq \|u\|_\wedge + \frac{\varepsilon}{2^3}$$

in mind. Next, notice that

$$\|u_2 - u_1\|_\wedge \leq \|u - u_2\|_\wedge + \|u - u_1\|_\wedge \leq \frac{\varepsilon}{2^5} + \frac{\varepsilon}{2^4} < \frac{\varepsilon}{2^3}$$

so we can represent $u_2 - u_1 \in X \otimes Y$ in the form

$$u_2 - u_1 = \sum_{i=i(1)+1}^{i(2)} x_i \otimes y_i$$

where

$$\|u_2 - u_1\|_\wedge \leq \sum_{i=i(1)+1}^{i(2)} \|x_i\|\|y_i\| \leq \frac{\varepsilon}{2^3}.$$

Now estimate $\|u_3 - u_2\|_\wedge, \|u_4 - u_3\|_\wedge$ and continue the natural and righteous path to finish the proof.

Another basic property enjoyed by the projective tensor norm $\|\cdot\|_\wedge$ is its 'uniform' behavior. More precisely, we have the

THEOREM 1.4. *If X_1, X_2, Y_1 and Y_2 are Banach spaces (same scalars, please) and $u : X_1 \to X_2$ and $v : Y_1 \to Y_2$ are bounded linear operators, then $u \otimes v$ is a bounded linear operator from $X_1 \overset{\wedge}{\otimes} Y_1$ into $X_2 \overset{\wedge}{\otimes} Y_2$ with bound $\|u \otimes v\| \leq \|u\|\|v\|$.*

Regarding the subspace structure of $X \overset{\wedge}{\otimes} Y$, the best we can say in general is that *if Z is a complemented subspace of Y, then $X \overset{\wedge}{\otimes} Z$ is a complemented subspace of $X \overset{\wedge}{\otimes} Y$.* This, too, is quite easy; all we need do is realize that if $P : Y \to Y$ is a bounded linear projection of Y onto Z, then $\mathrm{id}_X \otimes P$ is a bounded linear projection of $X \overset{\wedge}{\otimes} Y$ onto $X \overset{\wedge}{\otimes} Z$.

If we know more then *sometimes* we can say more. For instance we have the following fundamental result of Grothendieck, which again reflects the tensorial nature of the projective norm.

THEOREM 1.5 (Grothendieck [**23**]). *For any Banach spaces X and Y,*

$$\begin{array}{ccccc} & & X \overset{\wedge}{\otimes} Y^{**} & & \\ & \subseteq & & \subseteq & \\ X \overset{\wedge}{\otimes} Y & & & & X^{**} \overset{\wedge}{\otimes} Y^{**} \\ & \subseteq & & \subseteq & \\ & & X^{**} \overset{\wedge}{\otimes} Y & & \end{array}$$

When we refer to the tensorial nature of $\|\cdot\|_\wedge$, we have in mind Grothendieck's theory of *tensor norms.* If one has a method of ascribing to each pair (E, F) of finite dimensional Banach spaces E and F a reasonable cross norm α on $E \otimes F$ in such a way that for finite dimensional spaces E_1, E_2, F_1 and F_2 if $u : E_1 \to E_2$ and $v : F_1 \to F_2$ are linear then $u \otimes v$ has bound $\leq \|u\|\|v\|$ we call the method "α" a tensor norm. Once a tensor norm is at hand one can define α for any tensor product $X \otimes Y$ of Banach spaces: rely on the fact that if $E_1 \subseteq E_2$ then the natural inclusion $i_{E_1,E_2} : E_1 \hookrightarrow E_2$ is of norm ≤ 1; hence, if $u \in X \otimes Y$, then $u \in E \otimes F$ for many finite dimensional spaces $E \subseteq X$ and $F \subseteq Y$ where $\alpha(u)$ gets smaller as E and F get bigger – so $\lim_{(E,F)} \alpha_{E \otimes F}(u)$ makes sense.

The point here is just this: $\|\cdot\|_\wedge$ is a tensor norm and so shares with all tensor norms the property just formulated in Theorem 5. It's also worth emphasizing that it's the tensorial character of the projective norm that highlights the 'local' nature of the process of taking projective products.
One more general result.

THEOREM 1.6 (Randrianantoanina [**40**]). *If X is a closed linear subspace of Y and if there is a linear operator $T : X^* \to Y^*$ of norm one such that $Tx^*|_X = x^*$ for each $x^* \in X^*$, then $Z \overset{\wedge}{\otimes} X$ is a subspace of $Z \overset{\wedge}{\otimes} Y$, regardless of the Banach space Z.*

PROOF. Given such a T we define the bounded linear projection $P : Y^* \to Y^*$ by $Py^* = T \circ (y^*|_X)$; PY^* is just X^* and P has norm one. Moreover, if $u = \sum_{i \leq n} z_i \otimes x_i \in Z \otimes X$ then the Universal Mapping Property tells us that

$$\begin{aligned} \|u\|_{Z \overset{\wedge}{\otimes} X} &= \sup \left\{ \left| T(\sum_{i \leq n} z_i \otimes x_i) \right| : T \in B_{\mathcal{L}(Z;X^*)} \right\} \\ &= \sup \left\{ \left| \sum_{i \leq n} T(z_i)(x_i) \right| : T \in B_{\mathcal{L}(Z;X^*)} \right\} \\ &= \sup \left\{ \left| \sum_{i \leq n} P \circ T(z_i)(x_i) \right| : T \in B_{\mathcal{L}(Z;X^*)} \right\} \\ &= \sup \left\{ \left| \sum_{i \leq n} S(z_i)(x_i) \right| : S \in B_{\mathcal{L}(Z;Y^*)} \right\} \end{aligned}$$

$$= \sup \left\{ \left| S(\sum_{i \le n} z_i \otimes x_i) \right| : S \in B_{\mathcal{L}(Z;Y^*)} \right\}$$

$$= \|u\|_{Z \hat{\otimes} Y}$$

□

REMARK. It's easy to see that the fact that T is a linear Hahn-Banach extension operator is 'overkill', if T is a *bounded* linear operator so that Tx^* extends x^* for each x^*, then the projection P defined at the outset of the above proof is a bounded linear surjection and $Z \overset{\wedge}{\otimes} X$ is still *isomorphic* to a subspace of $Z \overset{\wedge}{\otimes} Y$.

Though we've found no place where the next result can be put to use in the theory of tensor products – at least not in the study of the projective tensor product – we believe it to be likely that the result will, in the right hands, be of considerable use in the study of the projective tensor product. Here's a beautiful result due to S. Heinrich and P. Mankiewicz, [**27**].

THEOREM 1.7. *If S is a* separable *closed linear subspace of the Banach space X, then there is a* separable *closed linear subspace X_0 of X that contains S and a linear operator $T : X_0^* \to X^*$ of norm one so $Tx_0^*|_{X_0} = x_0^*$ for each $x_0^* \in X_0^*$.*

Consequently, regardless of $Z, Z \overset{\wedge}{\otimes} X_0$ is a subspace of $Z \overset{\wedge}{\otimes} X$.

$C(K)$-spaces (called by Grothendieck 'C-spaces') play a special role in the study of any tensor norm. For the projective norm this takes the following form.

THEOREM 1.8. *Suppose Y is a subspace of Z and assume Y is a C space. Then for any Banach space $X, Y \overset{\wedge}{\otimes} X$ is a subspace of $Z \overset{\wedge}{\otimes} X$.*

In fact, Y^{**} is a 1-injective Banach space so that any bounded linear operator into Y^{**} extends to any superspace of its domain in a bounded linear fashion with values *still* in Y^{**} without increase in the bound. In particular, there is an operator $I : Z \to Y^{**}$ such that $\|I\| = 1$ and $I|_Y = id_Y$. If we denote by i the natural inclusion, $i : Y \hookrightarrow Z$, of Y into Z, then here's what we get:

$$Y \overset{\wedge}{\otimes} X \xrightarrow{i \otimes id_X} Z \overset{\wedge}{\otimes} X \xrightarrow{I \otimes id_X} Y^{**} \overset{\wedge}{\otimes} X$$

$$j_Y \otimes id_X$$

where $j_Y : Y \hookrightarrow Y^{**}$ is the natural embedding. Because the projective norm is tensorial, $j_Y \otimes id_Y$ is an isometry. Since each of the operators $i \otimes id_X$ and $I \otimes id_X$ have norm $\le 1, j_Y \otimes id_Y$ passes its isometric behavior on to $i \otimes id_X$.

To say a bit more we recall a sharp, if elementary, observation of Sean Dineen and Richard Timoney: *if Y is a closed linear subspace of the Banach space Z and $|||\cdot|||$ is an equivalent norm on Y (equivalent to that inherited from Z), then there is an equivalent norm $||||\cdot||||$ on Z such that for $y \in Y$,*

$$||||y|||| = |||y|||.$$

Why is this so? Well, if we let $m, M > 0$ be chosen so that

$$mB_Y \subseteq B_{(Y,|||\cdot|||)} \subseteq MB_Y$$

and define K to be

$$K = \overline{co}(B_{(Y,|||\cdot|||)} \cup mB_Z),$$

then K generates a norm $|||\cdot|||$ on Z that's equivalent to Z's original norm. Moreover, since $mB_Y \subseteq B_{(Y,|||\cdot|||)}$, $K \cap Y = B_{(Y,|||\cdot|||)}$ and that's all that need be said.

Of course for tensor products, in particular the projective tensor product, there's a dividend to be collected.

COROLLARY 1.9. *If Y is a subspace of Z and Y is isomorphic to a C-space, then $Y \overset{\wedge}{\otimes} X$ is isomorphic to a subspace of $Z \overset{\wedge}{\otimes} X$ regardless of X.*

THEOREM 1.10. *For any measure μ, the projective tensor product $L^1(\mu) \overset{\wedge}{\otimes} X$ is isometrically isomorphic to the space $L^1_X(\mu)$ (of equivalence classes) of strongly μ-measurable X-valued Bochner integrable functions.*

Consequently, if Y is a closed linear subspace of X, then $L^1(\mu) \overset{\wedge}{\otimes} Y$ is a closed linear subspace of $L^1(\mu) \overset{\wedge}{\otimes} X$.

This result, whose standard and easy proof is found in virtually every book brave enough to discuss tensor products, has a companion, a somewhat anonymous companion, also due to Grothendieck. More precisely, the following is so.

THEOREM 1.11 (Grothendieck [**24**]). *Suppose Z is a Banach space and whenever Y is a closed linear subspace of X, then $Z \overset{\wedge}{\otimes} Y$ is a closed linear subspace of $Z \overset{\wedge}{\otimes} X$. Then there is a measure μ so that Z is isometrically isomorphic to $L^1(\mu)$.*

To be sure, suppose Z is a space with the hypothesized property: whenever Y is a closed linear subspace of X, then $Z \overset{\wedge}{\otimes} Y$ is a closed linear subspace of $Z \overset{\wedge}{\otimes} X$; and let's first notice what constraints this places on Z. Of course, this says that whenever Y is a subspace of X then every member of $(Z \overset{\wedge}{\otimes} Y)^*$ extends, without loss of norm, to a member of $(Z \overset{\wedge}{\otimes} X)^*$. Equivalently, every bilinear continuous functional on $Z\times Y$ extends without loss in norm, to a bilinear continuous functional on $Z \times X$. Interpreting this in terms of operators (and taking obvious advantage of the symmetry of these affairs), every bounded linear operator from Y to Z^* extends to a bounded linear operator from X to Z^* with the same norm. *Z^* is a 1-injective Banach space.*

Now we're in business. The combined work of L. Nachbin [**34**], D. Goodner [**21**], J. Kelley [**28**] and, in case of complex scalars, M. Hasumi [**26**] tells us that Z^* *is isometrically isomorphic to a space $C = C(K)$, where K is compact Hausdorff and extremally disconnected.*

At this juncture a closer look at the lattice structure (be it as a real Banach lattice) of C is needed.

First, we'll look closely at the positive cone C^+

$$C^+ = \{f \in C : f \geq 0\}$$

of $C = Z^*$.

Notice that each of the sets

$$\{f \in C : \|f - g\| \leq 1\}$$

is a closed ball in Z^* and so is weak* compact. Further,

$$C^+ \cap B_C = \bigcap_{g\in C, 0\leq g\leq 1} \{f \in C : \|f - g\| \leq 1\}$$

and so $C^+ \cap B_C$ is weak* compact. The Krein-Šmulian theorem now assures us that *C^+ is weak* closed.* It's a quick hop, step and jump to conclude that the sets of the form $\{g \in C : g \leq f\}$ and $\{h \in C : f \leq h\}$ are weak* closed, too, and from this that order intervals $[g, h]$ in C

$$[g, h] = \{f \in C : g \leq f \leq h\}$$

are weak* compact.

Next we claim that if $f \in C$ then the operator $M_f : C \to C$ given by $M_f(g) = f \cdot g$ is weak*-weak* continuous, that is, is the adjoint of some operator from Z to Z. Here the fact that $C = C(K)$ for an extremally disconnected K comes in handy. In fact, we first test what happens when $f = \chi_F$ for a *clopen* subset F of K. In this case, M_f is a bounded linear projection whose range $M_f(C)$ is weak*-closed– after all,

$$M_f(C) \cap B_C = [-f, f],$$

is a weak* compact set. If we let

$$H_F = \{z \in Z : h(z) = 0 \text{ if } h \in C \text{ has support } \subseteq F\},$$

then H_F is a closed linear subspace of Z. Let q_F be the natural quotient map

$$q_F : Z \to Z/H_F.$$

For $z \in Z$ define $L(z)$ to be the member $(q_F(z), q_{F^c}(z))$ of $(Z/H_F) \oplus (Z/H_{F^c})$; $z \to L(z)$ is a bounded linear operator. Of course, $F \cap F^c = \phi$ ensures us that $H_F \cap H_{F^c} = \{0\}$. Now if $h^* \in (Z/H_F)^*$ and $\tilde{h}^* \in (Z/H_{F^c})^*$, then $h^* \circ q_F \in Z^* = C$ and is supported by F and $\hat{h}^* \circ q_{F^c} \in Z^* = C$ is supported by F^c.

Naturally,

$$L^*(h^*, \tilde{h}^*) = h^* \circ q_F + \tilde{h}^* \circ q_{F^c}.$$

But each member of $Z^* = C$ with support contained in F is an $h^* \circ q_F$ and each member of $Z^* = C$ with support contained in F^c is an $\tilde{h}^* \circ q_{F^c}$; hence, L^* is a surjective linear map. L must be injective. But $L|_{H_F \oplus H_{F^c}}$ is a surjection. $Z = H_F \oplus H_{F^c}$!

If $\pi_F : Z \to H_F$ is the natural bounded linear projection then $\pi_F^* = M_f$ and M_f *is* weak*-weak* continuous.

For general $f \in C$, it's enough to realize that, thanks to the Stone-Weierstrass theorem and K's totally disconnected character, each $f \in C(K)$ is in the norm closed linear span of $\{\chi_F : F \text{ is clopen in } K\}$; from this it follows that M_f is the operator norm limit of linear combinations of M_h's where $h = \chi_H, H$ a clopen subset of K. And that's all that need be said about the weak*-weak* continuity of operators M_f of multiplication from C to C by a member f of C.

Now we take a big step. If we take $z \in Z$ then $z \in Z^{**} = C^*$ and in C^* we can consider $|z|$. In fact, $|z| \in Z$! After all, $z \in Z$ is a weak* continuous linear functional on Z^* and so if $f \in C = Z^*, z \circ M_f$ is a weak*-continuous linear functional on C, too. But for any $g \in C$

$$z \circ M_f(g) = z(fg) = (f \cdot z)(g),$$

where we *must* remember that C^* is a space of measures and so $f \cdot z$ makes sense as a measure. So $f \cdot z \in Z$ if $f \in C$ and $z \in Z$. But thanks to Lusin, $|z|$ is the norm limit in C^* of a sequence of measures of the form $f \cdot z$ for $f \in C$– indeed, the Radon-Nikodym theorem provides us with a member ${}^{d|z|}/_{dz}$ of C^* which is the limit in variation of functions of the form $f \cdot z$ with $f \in C$, a la Lusin! But as a *norm* limit of a sequence in $Z, |z|$ is itself in Z.

The predual Z of C is a closed sublattice of C^* and so, fits within the beautiful mosaic designed by Kakutani and *is* an L-space, that is, an $L^1(\mu)$ for some μ.

With regards to the question of when $X \overset{\wedge}{\otimes} Z$ is a closed linear subspace of $Y \overset{\wedge}{\otimes} Z$ given that X is a closed linear subspace of Y, we wish to point out that if there's a $K > 0$ so that whenever $\varphi \in \mathcal{B}(X, Z)$ there's a $\Phi \in \mathcal{B}(Y, Z)$ such that $\Phi|_{X \times Z} = \varphi$ and $\|\Phi\| \leq K\|\varphi\|$, then $X \overset{\wedge}{\otimes} Z$ is (K-isomorphic to) a closed linear subspace of $Y \overset{\wedge}{\otimes} Z$.

Naturally, extending bilinear functionals is tantamount to extending linear operators, a rare occurrence to be sure. Without some sort of injectivity there appears to be little hope of success. There is however one set of circumstances under which something substantial can be said.

The situation involves the absolutely 2-summing operators of Pietsch [**36**]. The pertinent feature of 2-summing operators?

Extension property: *If X is a closed linear subspace of Y and $u : X \to Z$ is absolutely 2-summing, then there is an absolutely 2-summing operator $U : Y \to Z$ such that $U|_X = u$ and $\pi_2(U) = \pi_2(u)$.*

This remarkable theorem of Pietsch follows from his equally remarkable "Domination Theorem."

For our purpose, the Extension Property has the following consequence: *if X is a closed linear subspace of Y and if every bounded linear operator $u : X \to Z^*$ is absolutely 2-summing, then $X \overset{\wedge}{\otimes} Z$ is a closed linear subspace of $Y \overset{\wedge}{\otimes} Z$.*

Notice the submissive role of Y.

The phenomena encoded in "every bounded linear operator is absolutely 2-summing" is a fundamental one. Since the appearance of Grothendieck's Résumé [**23**] and the Lindenstrauss-Pełczyński deciphering and enlargement [**30**] of the fundamental inequality of that document, many occurrences of this phenomenon have been uncovered and an extensive literature has grown around it. It is most interesting when encountered in natural circumstances. Here are a few such:

- every operator from a $C(K)$-space to an L^p-space is absolutely 2-summing if $1 \leq p \leq 2$.
- every operator from an L^1-space to a Hilbert space is absolutely 2-summing.
- every operator from a $C(K)$-space to a space having cotype 2 is absolutely 2-summing.
- every operator from the disk algebra $A(D)$ to an L^p-space is absolutely 2-summing, if $1 \leq p \leq 2$.

The first two of the above are due in some sense to Grothendieck [**23**] but Lindenstrauss and Pełczyński [**30**] must share in the glory by reformulating Grothendieck's results in the language of absolutely summing operators. The role of spaces of cotype was made clear by Maurey and Pisier [**33**]; the above cited result about operators on $C(K)$ seems to be due to Maurey. The final result is a famous result of Bourgain [**3**].

Let's see how to parlay the theory of cotype and that of summing operators to say something about old friends. Be forewarned: we're doing all the soft work. *None of the above results is even remotely easy.*

$A(D) \overset{\wedge}{\otimes} C(\Pi)$ *is a closed linear subspace of* $C(\Pi) \overset{\wedge}{\otimes} C(\Pi)$. Schematically,

$$\begin{aligned}[A(D) \overset{\wedge}{\otimes} C(\Pi)]^* &= \mathcal{B}(A(D), C(\Pi)) \\ &= \mathcal{L}(A(D); C(\Pi)^*) \\ &= \mathcal{L}(A(D); M(\Pi)),\end{aligned}$$

where Π denotes the unit circle $\{z \in \mathcal{C} : |z| = 1\}$ and $M(\Pi)$ is the space of Borel measures on Π with variation norm; of course, $M(\Pi)$ is an L^1-space so the above string extends to

$$= \pi_2(A(D); M(\Pi))$$

So the discussion above provides for each $\varphi \in [A(D) \overset{\wedge}{\otimes} C(\Pi)]^*$ an extension $\Phi \in [C(\Pi) \overset{\wedge}{\otimes} C(\Pi)]^*$ whose operator version is in fact a member of $\pi_2(C(\Pi); M(\Pi))$. The extension $\Phi \in [C(\Pi) \overset{\wedge}{\otimes} C(\Pi)]^*$ can be chosen so that

$$\|\Phi\| \le K\|\varphi\|$$

where K is the constant such that

$$\|u\| \le \pi_2(u) \le K\|u\|$$

for any $u \in \mathcal{L}(A(D); M(\Pi))$, a constant whose existence is assured us by Bourgain.

The conclusion is safe and secure.

To take the next step we need a bit more fire power; again it's Bourgain who provides us with the key analytic ingredients. Our goal: $A(D) \overset{\wedge}{\otimes} A(D)$ *is a closed linear subspace of* $C(\Pi) \overset{\wedge}{\otimes} C(\Pi)$. Again, we string together some identifications.

$$\begin{aligned}[A(D) \overset{\wedge}{\otimes} A(D)]^* &= \mathcal{B}(A(D), A(D)) \\ &= \mathcal{L}(A(D); A(D)^*)\end{aligned}$$

The F. and M. Riesz theorem describes $A(D)^*$ as $(L^1(\Pi)/H^1_0) \oplus_1 S$, where S is the space of Borel measures on Π that're singular with respect to normalized Lebesgue measure on Π and H^1_0 is the codimension - one subspace of the Hardy space H^1 consisting of those functions that have 0 constant term and "$\oplus_1$" means the sum is taken in the "ℓ^1-sense". If we denote by P_R and P_S the norm-one linear projections of $A(D)^*$ onto $L^1(\Pi)/H^1_0$ and S, respectively, then any $u \in \mathcal{L}(A(D); A(D)^*)$ is the sum of $P_R u$ and $P_S u$. $P_S u : A(D) \to S$ and S is also an L^1-space so $P_S u$ is always absolutely 2-summing regardless of $u \in \mathcal{L}(A(D); A(D)^*)$, What about $P_R u$?

Here it seems we need to dig a bit deeper. If v is any operator from $A(D)$ to $L^1(\Pi)/H^1_0$, then v^* takes $(L^1(\Pi)/H^1_0)^*$ to $A(D)^*$. Now $(L^1(\Pi)/H^1_0)^*$ can be identified with the Hardy space $H^\infty(D)$ of bounded analytic functions on D so v^* takes $H^\infty(D)$ to $A(D)^*$, a space that Bourgain showed has cotype 2. Such an operator is known (thanks to Bourgain!) to factor through a Hilbert space; but if v^* factors through a Hilbert space, then so does v. But $v : A(D) \to L^1(\Pi)/H^1_0$ factoring through a Hilbert space implies v is absolutely 2-summing.The upshot of the previous paragraph is that $P_R u$ is always absolutely 2-summing.

Since $P_R u$ and $P_S u$ are absolutely 2-summing, so is their sum u.

So, every $u : A(D) \to A(D)^*$ is absolutely 2-summing. Our scheme is in effect: there's a $C > 0$ so for each $u \in \mathcal{L}(A(D); A(D)^*)$,

$$\|u\| \leq \pi_2(u) \leq C\|u\|.$$

We can extend any such u to a $U \in \pi_2(C(\Pi); A(D)^*)$, where $\|U\| \leq C\|u\|$. Translating to the bilinear functional setting, each $\psi \in \mathcal{B}(A(D), A(D))$ admits and extension to a $\Psi \in \mathcal{B}(C(\Pi), A(D))$ where $\|\Psi\| \leq C\|\psi\|$. The result is that $A(D) \overset{\wedge}{\otimes} A(D)$ is a closed linear subspace of $C(\Pi) \overset{\wedge}{\otimes} A(D)$, which is – after considerations regarding symmetry – the same as saying that $A(D) \overset{\wedge}{\otimes} A(D)$ is (isomorphic to) a closed linear subspace of $C(\Pi) \overset{\wedge}{\otimes} A(D)$ which is itself (isomorphic to) a subspace of $C(\Pi) \overset{\wedge}{\otimes} C(\Pi)$.

It should be mentioned that using similar ideas and results, Bourgain [**3**] also derived that projective tensor product $H^\infty(D) \overset{\wedge}{\otimes} H^\infty(D)$ *is (isomorphic to) a closed linear subspace of* $L^\infty(\Pi) \overset{\wedge}{\otimes} L^\infty(\Pi)$.

One final result about subspace-stability in the projective tensor product; the result due to Bernard Maurey and is known appropriately enough, as *Maurey's Extension Theorem* [**32**].

THEOREM 1.12. *Suppose X_0 and Y_0 are Banach spaces having type 2. Then there is a constant $M > 0$ so that regardless of the Banach spaces X and Y containing X_0 and Y_0, respectively, any continuous bilinear functional ϕ on $X_0 \times Y_0$ extends to a continuous bilinear functional Φ on $X \times Y$ with $\|\Phi\| \leq M\|\phi\|$*

Duality quickly provides the threatened consequence.

COROLLARY 1.13. *If X_0 and Y_0 are Banach spaces having type 2 and X and Y are Banach spaces containing (isomorphisms of) X_0 and Y_0 respectively, the $X_0 \overset{\wedge}{\otimes} Y_0$ is (isomorphic to) a closed linear subspace of $X \overset{\wedge}{\otimes} Y$.*

THEOREM 1.14. *The projective tensor norm is projective, that is, if Y is a closed linear subspace of Z and $q_Y : Z \to Z/Y$ is the natural quotient map, then regardless of $X, q_Y \otimes id_X$ is a quotient map of $Z \overset{\wedge}{\otimes} X$ onto $(Z/Y) \overset{\wedge}{\otimes} X$.*

PROOF. Let $\varepsilon > 0$ and suppose $\hat{u} \in (Z/Y) \overset{\wedge}{\otimes} X$. Then there are sequences (q_n) and (x_n) in Z/Y and X respectively so that $\|x_n\| \leq 1, \hat{u} = \sum_n q_n \otimes x_n$ and

$$\|\hat{u}\|_{(Z/Y)\hat{\otimes}X} \leq \sum_n \|q_n\|\|x_n\| \leq \|\hat{u}\|_{(Z/Y)\hat{\otimes}X} + \frac{\varepsilon}{2}.$$

For each n there is a $z_n \in Z$ so that $q_Y z_n = q_n$ and

$$\|q_n\|_{Z/Y} \leq \|z_n\| \leq \|q_n\|_{Z/Y} + \varepsilon 2^{-n-1}.$$

Look at $\sum_n z_n \otimes x_n = u \in Z \overset{\wedge}{\otimes} X$. It's plain that $\hat{u}$ is just $(q_Y \otimes id_X)(u)$ and that

$$\begin{aligned}\|\hat{u}\|_{(Z/Y)\hat{\otimes}X} &\leq \|u\|_{Z\hat{\otimes}X} \\ &\leq \sum_n \|z_n\|\|x_n\| \\ &\leq \sum_n (\|q_n\| + \frac{\varepsilon}{2^{n+1}})\|x_n\|\end{aligned}$$

$$\leq \sum_n \|q_n\|\|x_n\| + \frac{\varepsilon}{2}$$
$$\leq \|\hat{u}\|_{(Z/Y)\hat{\otimes}X} + \varepsilon.$$

□

REMARK. In Grothendieck's 'Résumé' you'll find many projective tensor norms, nevertheless, $\|\cdot\|_\wedge$ is still referred to as *the* projective tensor norm. Go figure.

The projective tensor product respects *some* fundamental structures. For instance, we have the following well-known result.

THEOREM 1.15 (Gelbaum-Gil de la Madrid [**19**]). *Suppose X has a Schauder basis (x_n) and Y has a Schauder basis (y_n). Then $(x_m \otimes y_n)$ is a Schauder basis for $X \overset{\wedge}{\otimes} Y$ under the Gelbaum-Gil de la Madrid ordering:*

$$x_1 \otimes y_1, x_1 \otimes y_2, x_2 \otimes y_2, x_2 \otimes y_1, x_1 \otimes y_3, x_2 \otimes y_3, x_3 \otimes y_3, x_3 \otimes y_2, x_3 \otimes y_1, \ldots$$

(here it's advisable to layout the rectangular tableau for $(x_m \otimes y_n)$ and 'see' how the above listing is proceeding).

It seems to be *unknown* if *(x_n) is a basic sequence in X and (y_n) is a basic sequence in Y, then need $(x_m \otimes y_n)$ be a basic sequence in $X \overset{\wedge}{\otimes} Y$ under* any *ordering.*

For the projective tensor product we have the following convenient consequence of Grothendieck's characterization of members of $X \overset{\wedge}{\otimes} Y$ (Theorem 1.3 above):

THEOREM 1.16 (Grothendieck). *Let K be a compact subset of $X \overset{\wedge}{\otimes} Y$. Then there is a compact subset $K_X \subseteq X$ and a compact subset K_Y of Y such that*

$$K \subseteq \overline{aco}(K_X \otimes K_Y),$$

the closed absolutely convex hull of $K_X \otimes K_Y$.

Such a result is a perfect set-up to understand approximation properties in $X \overset{\wedge}{\otimes} Y$.

Suppose K is a norm compact subset of $X \overset{\wedge}{\otimes} Y$. Then, as we just saw, there are norm compact subsets K_X of X and K_Y of Y so that K is contained in the absolutely closed convex hull of $K_X \otimes K_Y$. It's easy to see that if u is a finite rank bounded linear operator on X which approximates the identity uniformly on K_X (say, within ε) and if v is a finite rank bounded linear operator on Y which approximates the identity on K_Y (by ε, too), then $u \otimes v$ is a finite rank bounded linear operator on $X \overset{\wedge}{\otimes} Y$ which approximates the identity on $K_X \otimes K_Y$ with

$$\|(u \otimes v)(x \otimes y) - (x \otimes y)\| \leq (\|u\| sup_{x \in K_X} \|x\| + sup_{y \in K_Y} \|y\|)\varepsilon = \eta$$

It follows that $u \otimes v$ is a finite rank bounded linear operator on $X \overset{\wedge}{\otimes} Y$ that approximates the identity on K within η.

COROLLARY 1.17. *If X and Y have the approximation property, then so does $X \overset{\wedge}{\otimes} Y$.*

COROLLARY 1.18. *If X and Y have the bounded approximation property, then so does $X \overset{\wedge}{\otimes} Y$.*

COROLLARY 1.19. *If X and Y have the metric approximation property, then so does $X \overset{\wedge}{\otimes} Y$.*

2. Duality : $(X \overset{\vee}{\otimes} Y)^* = X^* \overset{\wedge}{\otimes} Y^*$

A good bit of the importance of the projective tensor product derives from its duality with the injective tensor product. Recall that if X and Y are Banach spaces, then the injective tensor norm $|\ |_\vee$ is defined as follows: here $u \in X \otimes Y$ and

$$|u|_\vee = \sup\{|x^* \otimes y^*(u)| : x^* \in B_{X^*}, y^* \in B_{Y^*}\};$$

naturally, if $u = \sum_{i \leq n} x_i \otimes y_i \in X \otimes Y$ then $x^* \otimes y^*(u)$ is just $\Sigma x^*(x_i) y^*(y_i)$. Plainly, the injective tensor norm is so calibrated to 'make do' with a minimal number of continuous functionals and is indeed the least of the crossnorms on $X \otimes Y$. $X \overset{\vee}{\otimes} Y$ denotes the completion of $(X \otimes Y, |\ |_\vee)$.

In case both coordinates are finite dimensional the duality between $X \otimes Y$ and $X^* \otimes Y^*$ is complete.

THEOREM 2.1. *If E is a finite dimensional Banach space then*

$$(E \overset{\wedge}{\otimes} Y)^* = (E^* \overset{\vee}{\otimes} Y^*).$$

Indeed, if $u^* \in E^* \overset{\vee}{\otimes} Y^*$ then $u = \sum_{i \leq n} e_i^* \otimes y_i^*$, and

$$\begin{aligned} |u^*|_\vee &= \sup\{|(e^{**} \otimes y^{**})(u^*)| : e^{**} \in B_{E^{**}}\ , y^{**} \in B_{Y^{**}}\} \\ &= \sup\{|(e \otimes y^{**})(u^*)| : e \in B_E\ , y^{**} \in B_{Y^{**}}\} \\ &= \sup\{|y^{**}(\Sigma_{i \leq n} e_i^*(e) y_i^*)| : e \in B_E, y^{**} \in B_{Y^{**}}\} \\ &= \sup\{|(\Sigma_{i \leq n} e_i^*(e) y_i^*)(y)| : e \in B_E\ , y \in B_Y\} \end{aligned}$$

by Goldstine's theorem. Naturally this last quantity is just

$$= \sup\{|u^*(e, y)| : e \in B_E\ , y \in B_Y\}\ ,$$

which is precisely $\|u^*\|_{\mathcal{B}(E,Y)}$, the norm of u^* as a bilinear continuous functional on $E \times Y$, that is, as a member of $(E \hat{\otimes} Y)^*$.

It follows that the map

$$E^* \overset{\vee}{\otimes} Y^* \hookrightarrow \mathcal{B}(E, Y) = (E \overset{\wedge}{\otimes} Y)^*$$

is an isometry.

It is also a surjection. Here is where the finite-dimensionality of E enters the foray. Let $(e_i, e_i^*)_{i \leq n}$ be a bi-orthogonal system such that E is the span of $\{e_1, \ldots, e_n\}$. Take $\varphi \in \mathcal{B}(E, Y)$.

For $e \in E$ and $y \in Y$ we have

$$\begin{aligned} \varphi(e, y) &= \varphi(\Sigma_{i \leq n} e_i^*(e) e_i, y) \\ &= \Sigma_{i \leq n} e_i^*(e) \varphi(e_i, y) \\ &= (\Sigma_{i \leq n} e_i^* \otimes \varphi_{e_i})(e \otimes y), \end{aligned}$$

where $\varphi_{e_i} \in Y^*$ is just the functional $\varphi_{e_i}(y) = \varphi(e_i, y)$. It follows that if $u_\varphi^* = \Sigma_{i \leq n} e_i^* \otimes \varphi_{e_i}$, then $u_\varphi^* \in E^* \otimes Y^*$ and under the map of the previous paragraph the image of u_φ^* is just φ.

Of course, if E and F are both finite dimensional Banach spaces, then $(E \overset{\wedge}{\otimes} F)^* = (E^* \overset{\vee}{\otimes} F^*)$. But in this case both $E \otimes F$ and $E^* \otimes F^*$ are also finite-dimensional and any norm on either is already complete. So

$$\begin{aligned}(E \overset{\vee}{\otimes} F)^* &= (E^{**} \overset{\vee}{\otimes} F^{**})^* \\ &= (E^* \overset{\wedge}{\otimes} F^*)^{**} \\ &= E^* \overset{\wedge}{\otimes} F^*\end{aligned}$$

by finite-dimensionality.

In other words we have the simple

COROLLARY 2.2. *For finite dimensional Banach spaces E and F $(E \overset{\wedge}{\otimes} F)^* = (E^* \overset{\vee}{\otimes} F^*)$ and $(E \overset{\vee}{\otimes} F)^* = E^* \overset{\wedge}{\otimes} F^*$.*

In case the Banach spaces involved are infinite dimensional more needs to be present to get either of the above statements.

Key to the understanding of any duality results involving the injective tensor norm is the following still-wonderful result of Grothendieck.

THEOREM 2.3 (Grothendieck [**23**]). *A continuous bilinear functional φ on $X \times Y$ defines a member of $(X \overset{\vee}{\otimes} Y)^*$ precisely when there exists a regular Borel measure μ on the product of $(B_{X^*}$, weak$^*)$ and $(B_{Y^*}$, weak$^*)$ such that for any $x \in X$ and $y \in Y$*

$$\varphi(x, y) = \int_{B_{X^*} \times B_{Y^*}} x^*(x) y^*(y) d\mu(x^*, y^*)$$

In this case, the norm of φ as a member of $(X \overset{\vee}{\otimes} Y)^*$ is the variation $|\mu|(B_{X^*} \times B_{Y^*})$ of the measure μ.

This surprisingly easy and direct application of the knowledge of $C(K)$'s duality is profound for its consequences.

Naturally, bilinear functionals φ that define members of $(X \overset{\vee}{\otimes} Y)^*$ are called *integral.* If $u : X \to Y$ is a bounded linear operator we say that u is integral if the functional $\varphi_u(x, y^*) = y^*(u(x))$ is an integral bilinear form on $X \times Y^*$. The integral norm $\|u\|_\wedge$ of u is the norm of $\varphi_u \in (X \overset{\vee}{\otimes} Y^*)$.

In keeping with Grothendieck's Résumé [**23**], we denote by $\mathcal{B}^\wedge(X, Y)$ the space of integral bilinear forms on $X \times Y$ and by $\mathcal{L}^\wedge(X; Y)$ the space of integral linear operators from X to Y. So, for example, Grothendieck's theorem identifies $(X \overset{\vee}{\otimes} Y)^*$ with $\mathcal{B}^\wedge(X, Y)$.

Now the class of integral linear operators is filled with wonders and we collect a few of them here.

THEOREM 2.4 (Grothendieck [**23**]). *Let $u : X \to Y$ be a bounded linear operator. Then the following statements regarding u are equivalent:*

(1) *u is integral.*
(2) *u^* is integral.*
(3) *u^{**} is integral.*
(4) *$j_Y u$ is integral, where $j_Y : Y \hookrightarrow Y^{**}$ is the canonical embedding.*

(5) *u admits a factorization in the form*

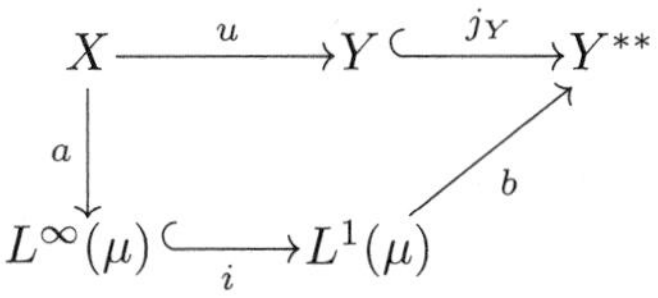

$$\begin{array}{ccccc} X & \xrightarrow{u} & Y & \overset{j_Y}{\hookrightarrow} & Y^{**} \\ \downarrow a & & & \nearrow b & \\ L^\infty(\mu) & \underset{i}{\hookrightarrow} & L^1(\mu) & & \end{array}$$

where μ is a regular Borel measure on some compact Hausdorff space Ω, $i : L^\infty(\mu) \hookrightarrow L^1(\mu)$ is the natural inclusion and $a : X \to L^\infty(\mu)$, $b : L^1(\mu) \to Y^{**}$ are operators of norm ≤ 1.

In case (1) - (5) *hold then*

$$\|u\|_\wedge = \|u^*\|_\wedge = \|u^{**}\|_\wedge = \|j_Y u\|_\wedge = \inf\{\|a\| \ |\mu|(\Omega)\|b\|\}$$

where the infimum is taken over all factorizations a la (5).

Viewed properly, the factorization (5) holds the key to understanding integral operators and to extending the relationship $(X \overset{\vee}{\otimes} Y)^* = X^* \overset{\wedge}{\otimes} Y^*$ beyond its finite-dimensional status.

Indeed, on taking a typical $\varphi \in (X \overset{\vee}{\otimes} Y)^*$ and looking at the natural operator $u_\varphi : X \to Y^*$ associated with φ, $u_\varphi(x)(y) = \varphi(x, y)$, φ's integrability soon translates into that of u_φ with $\|\varphi\|_{\mathcal{B}^\wedge(X,Y)} = \|u_\varphi\|_\wedge$; with this (and noticing that dual spaces are always norm-one complemented in their own second duals) we can factor u_φ as follows

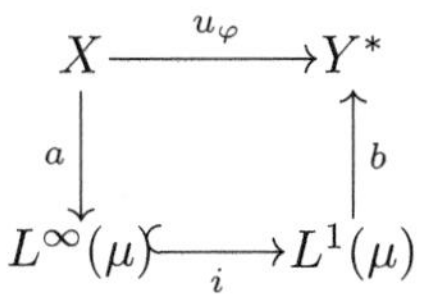

$$\begin{array}{ccc} X & \xrightarrow{u_\varphi} & Y^* \\ \downarrow a & & \uparrow b \\ L^\infty(\mu) & \underset{i}{\hookrightarrow} & L^1(\mu) \end{array}$$

where μ is a regular Borel measure on some compact Hausdorff space Ω, $i : L^\infty(\mu) \hookrightarrow L^1(\mu)$ is the natural inclusion and a and b are bounded linear operators of norm ≤ 1.

It's '$b : L^1(\mu) \to Y^*$' that holds the key to extending the relationship

$$(X \overset{\vee}{\otimes} Y)^* = X^* \overset{\wedge}{\otimes} Y^*$$

beyond its finite dimensional boundaries. If b is 'representable' (so is of the form $b(f) = \text{Bochner-}\int f(w)g(w)d\mu(w)$ for some function $g : \Omega \to Y^*$, which is strongly μ-measurable and essentially μ-bounded) then, in fact, u_φ can be represented in the form

$$u_\phi(x) = \sum_n x_n^*(x) y_n^*$$

where (x_n^*) is a sequence in X^*, (y_n^*) is a sequence in Y^* and

$$\Sigma \|x_n^*\| \ \|y_n^*\| < \infty.$$

This last condition ought to look familiar and indeed it is indicative of what's going on: it tells us that the integral operator $u_\varphi : X \to Y^*$ is actually 'nuclear'. This is close to what the doctor ordered. Close but not quite all. The last, and finishing, touch is to be sure that when we have a nuclear operator from X to Y^* that it be the only operator that gives the natural member $\Sigma x_n \otimes y_n^*$ of $X^* \overset{\wedge}{\otimes} Y^*$. To ensure this we can assume that Y^* has the approximation property and be done with it.

Of course there's a great deal of symmetry present and so we can state the following duality result with some satisfaction.

THEOREM 2.5. *Assume that either X or Y enjoys both the approximation and the Radon-Nikodym properties. Then*

$$(X \overset{\vee}{\otimes} Y)^* = X^* \overset{\wedge}{\otimes} Y^*.$$

There are, by the way, many pieces of the puzzle that is the above statement, pieces that hint that the hypotheses are close to the whole story. The survey 'Vector Measures' [**15**] contains a telling of the story, a story that hasn't changed much in the past quarter of a century.

Of course it's important to realize that among those spaces that have the Radon-Nikodym property are the reflexive spaces and separable duals. Each class deserves its own special corollary.

COROLLARY 2.6. *If X is a reflexive Banach space with the approximation property, then for any Banach space Y*

$$(X \overset{\vee}{\otimes} Y)^* = X^* \overset{\wedge}{\otimes} Y^*$$

COROLLARY 2.7. *If X^* is separable and has the approximation property, then for any Banach space Y*

$$(X \overset{\vee}{\otimes} Y)^* = X^* \overset{\wedge}{\otimes} Y^*.$$

3. $L^p \overset{\wedge}{\otimes} X$, $1 < p < \infty$

As we noted in §1, regardless of the measure $\mu, L^1(\mu) \overset{\wedge}{\otimes} X$ is identifiable with the Lebesgue-Bochner space $L^1_X(\mu)$; what's more, this identification reveals an aspect of L^1-spaces characteristic of them and only them: they're "flat." The fact that any Banach space Z for which $Z \overset{\wedge}{\otimes} X$ is a subspace of $Z \overset{\wedge}{\otimes} Y$ whenever X is a subspace of Y is an $L^1(\mu)$-space says, in particular, that if $1 < p < \infty$ then $L^p(\mu) \overset{\wedge}{\otimes} X$ is likely to be much smaller than $L^p_X(\mu)$, its Lebesgue-Bochner cousin.

In addition to the natural curiosity piqued by their pseudo-classical stature, spaces like $L^p(\mu) \overset{\wedge}{\otimes} X$ do arise in the affairs of analysis.

To be sure, it is a still-wonder-full result of *W. Orlicz* which states that if (f_n) is an unconditionally summable sequence in an L^1-space Y, then $(f_n) \in \ell^2_X$. A. Grothendieck shows more: if X is an L^1-space and (f_n) is an unconditionally summable sequence in X, then (f_n) is actually a sequence in $\ell^2 \overset{\wedge}{\otimes} X$, a much smaller space than ℓ^2_X. This is derived by Grothendieck from his fundamental inequality (and credited by him to J.E. Littlewood).

In his sequel to the Résumé, Grothendieck makes a careful study of the spaces $\ell^p \overset{\wedge}{\otimes} X, \ell^p_X$ and $\ell^p \check{\otimes} X$ with the ideas of A. Dvoretzky and C. Rogers in hand. Among other things he shows that *if $1 < p < \infty$ and dim $X = \infty$, then*

$$\ell^p \overset{\wedge}{\otimes} X \subsetneqq \ell^p_X \subsetneqq \ell^p \check{\otimes} X.$$

It is not difficult to mimic his proofs to see that if $1 < p < \infty$ and X is an infinite dimensional Banach space, then

$$L^p(0,1) \overset{\wedge}{\otimes} X \subsetneqq L^p_X(0,1) \subsetneqq L^p(0,1)\check{\otimes}X.$$

holds as well.

Now quite a bit is known about the sequence spaces ℓ^p_X and their Lebesgue-Bochner function space relatives $L^p_X(0,1)$; similarly, $L^p(0,1)\check{\otimes}X$ is naturally the space of compact linear operators from $L^{p'}(0,1)$ to X (p' is the index conjugate to p) and a lot is known about this space. On the other hand, little seems to be known about $\ell^p \overset{\wedge}{\otimes} X$ or $L^p(0,1) \overset{\wedge}{\otimes} X$ (or for that matter, $L^p(\mu) \overset{\wedge}{\otimes} X$). In this section we discuss some of what is known and indicate how it might be relevant.

We'll start with $\ell^p \overset{\wedge}{\otimes} X$; here $1 < p < \infty$ and X is an infinite dimensional Banach space.

A principal obstruction to studying $\ell^p \overset{\wedge}{\otimes} X$ has been a lack of understanding just which sequences (x_n) in X belong to $\ell^p \overset{\wedge}{\otimes} X$. *Q. Bu* [**7**] removed this obstruction when he proved the following

THEOREM 3.1. *A sequence (x_n) in X belongs to $\ell^p \overset{\wedge}{\otimes} X$ precisely when given any sequence (x^*_n) in X^* such that $\Sigma_n|x^*_n(x)|^{p'} < \infty$ for each $x \in X$ we have*

$$\Sigma_n|x^*_n(x_n)| < \infty.$$

In such a case,

$$\|(x_n)\|_{\ell^p \overset{\wedge}{\otimes} X} = \sup\{\Sigma_n|x^*_n(x_n)|\},$$

where the supremum is taken over all (x^*_n)'s in X^* so that $\Sigma_n|x^*_n(x)|^{p'} \leq 1$ for each $x \in B_X$.

Bu's proof relies on the Riesz Theorem characterizing $C(K)^*$ and the fact that the inclusion $\ell^p \overset{\wedge}{\otimes} X \hookrightarrow \ell^p \overset{\wedge}{\otimes} X^{**}$ is always an isometry, something that lies close to the heart of the projective norm's tensorial character. It was the recognition of just which sequences lie in $\ell^p \overset{\wedge}{\otimes} X$ that soon paid dividends when Bu uncovered some remarkable mapping properties of absolutely p-summing operators with Hilbertian domains. The following result is included with Bu's proof because it is too elegant to exclude.

THEOREM 3.2. *Let $1 < p, q < \infty$ and $u : H \to X$ be an absolutely p-summing operator (with Hilbert space H as domain). Then u takes absolutely q-summable sequences in H into members of $\ell^q \overset{\wedge}{\otimes} X$.*

Bu makes apt use of two old friends: Khinchin's and Kahane's inequalities. Khinchin says that if $1 \leq p < \infty$ then there are constants A_p, B_p so that

$$A_p(\Sigma_i|a_i|^2)^{1/2} \leq \left(\int_0^1 |\Sigma_i r_i(t)a_i|^p dt\right)^{1/p} \leq B_p(\Sigma_i|a_i|^2)^{1/2};$$

Kahane asserts that whenever $1 \leq p, q' < \infty$ there is a constant $K_{p,q'}$ such that regardless of the Banach space X and regardless of $x_1, \ldots, x_n \in X$ we have

$$\left(\int_0^1 \|\Sigma_{i=1}^n r_i(t)x_i\|^{q'} dt\right)^{1/q'} \leq K_{p,q'} \left(\int_0^1 \|\Sigma_{i=1}^n r_i(t)x_i\|^p dt\right)^{1/p}.$$

Both are dependable old friends, good company to keep in a proof.

PROOF. The key is what happens in case $H = \ell_n^2$.

Let $u \in \Pi_p(\ell_n^2; X)$. By Pietsch's Domination theorem [**16**], p. 44, there is a regular Borel probability measure μ on $B_{\ell_n^2}$ such that for any $y \in \ell_n^2$ we have

$$\|uy\| \leq \pi(u) \left(\int_{B_{\ell_n^2}} |\langle y, z\rangle|^p d\mu(z) \right)^{1/p}.$$

Now if $y_1, \ldots, y_m \in \ell_n^2$ and $x_1^*, \ldots, x_m^* \in X^*$ we have

$$y_k = \Sigma_{i=1}^n y_{ki} e_i; \; k = 1, 2, \ldots, m$$

and so

$$\begin{aligned}
\textstyle\sum_{k=1}^m |\langle u y_k, x_k^*\rangle| & \\
&= \sum_{k=1}^m |\langle \sum_{i=1}^n y_{ki} u e_i, x_k^*\rangle| \\
&\leq \sum_{k=1}^m \left(\sum_{i=1}^n |y_{ki}|^2 \right)^{1/2} \cdot \left(\sum_{i=1}^n |\langle u e_i, x_k^*\rangle|^2 \right)^{1/2}
\end{aligned}$$

which by Khinchin is

$$\leq \sum_{k=1}^m \|y_k\| \cdot \frac{1}{A_{q'}} \left(\int_0^1 |\sum_{i=1}^n r_i(t) \langle u e_i, x_k^*\rangle|^{q'} dt \right)$$

which by Hölder is

$$\begin{aligned}
&\leq \frac{1}{A_{q'}} (\sum_k \|y_k\|^q)^{1/q} \left(\sum_k \int_0^1 |\sum_{i=1}^n r_i(t) \langle u e_i, x_k^*\rangle|^{q'} dt \right)^{1/q'} \\
&= \frac{1}{A_{q'}} \|(\|y_k\|)\|_{\ell_m^q} \left(\int_0^1 \sum_k |\langle u(\sum_{i=1}^n r_i(t) e_i), x_k^*\rangle|^{q'} dt \right)^{1/q'} \\
&\leq \frac{1}{A_{q'}} \|(\|y_k\|)\|_{\ell_m^q} \left(\int_0^1 \|u(\sum_{i=1}^n r_i(t) e_i\|^{q'} \sup_{x \in B_X} \|(x_k^*(x))\|_{\ell_m^{q'}}^{q'} dt \right)^{1/q'} \\
&= \frac{1}{A_{q'}} \|(\|y_k\|)\|_{\ell_m^q} \sup_{x \in B_X} \|(x_k^*(x))\|_{\ell_m^{q'}} \cdot \left(\int_0^1 \|u(\sum_{i=1}^n r_i(t) e_i)\|^{q'} dt \right)^{1/q'}
\end{aligned}$$

But Kahane has his constants too and so

$$\begin{aligned}
\left(\textstyle\int_0^1 \|u(\sum_{i=1}^n r_i(t) e_i)\|^{q'} dt \right)^{1/q'} & \\
&\leq K_{p,q'} \left(\int_0^1 \|u(\sum_{i=1}^n r_i(t) e_i)\|^p dt \right)^{1/p} \\
&\leq K_{p,q'} \cdot \Pi_p(u) \left(\int_0^1 \left(\int_{B_{\ell_n^2}} |\langle \sum_{i=1}^n r_i(t) e_i, z\rangle|^p d\mu(z) \right) dt \right)^{1/p}
\end{aligned}$$

$$= K_{p,q'}\Pi_p(u)\left(\int_{B_{\ell_n^2}}\left(\int_0^1 |\langle\sum_{i=1}^n r_i(t)e_i, z\rangle|^p dt\right) d\mu(z)\right)^{1/p}$$

which by the continued good graces of Khinchin and his constants is

$$\leq K_{p,q'}\Pi_p(u)\left(\int_{B_{\ell_n^2}} (B_p(\sum|\langle e_i, z\rangle|^2)^{1/2})^p d\mu(z)\right)^{1/p}$$

$$\leq K_{p,q'}\Pi_p(u)B_p.$$

In sum we have

$$\sum_k |\langle uy_k, x_k^*\rangle| \leq \frac{B_p\cdot K_{p,q'}\cdot\Pi_p(u)}{A_q}\|(\|y_n\|)\|_{\ell_m^q}\sup_{x\in B_X}\|(x_n^*(\cdot))\|_{\ell_m^{q'}}.$$

Reflecting on this soon establishes the theorem. □

A key to unlocking many of the geometric mysteries of the spaces $\ell^p \overset{\wedge}{\otimes} X$ is the notion of a semi-embedding : if Y and Z are Banach spaces and $\sigma : Y \to Z$ is a 1-1, bounded linear operator for which σB_Y is closed in Z then σ is called a *semi-embedding.* It's Bu's representation that allows one to prove the following

THEOREM 3.3. *The natural inclusion $\ell^p \overset{\wedge}{\otimes} X \hookrightarrow \ell^p_X$ is a semi-embedding.*

Now much of the power of semi-embeddings lies in the fact that if Y is semi-embedded in Z and both are separable then the semi-embedding is a Borel equivalence between the Polish spaces B_Y and σB_Y so measure theoretically B_Y is much like B_Z. There's no better evidence of this than F. Delbaen's proof that if Y and Z are separable Banach spaces, $\sigma : Y \to Z$ a semi-embedding and Z has the Radon-Nikodym property, then Y has the Radon-Nikodym property, too. This proves the separable case of the following

THEOREM 3.4. *If X has the Radon-Nikodym property, then so does $\ell^p \overset{\wedge}{\otimes} X$.*

The non-separable case hinges on the following easy fact: if S is a separable closed linear subspace of $\ell^p \overset{\wedge}{\otimes} X$ then there is a separable closed linear subspace X_0 of X such that S is a closed linear subspace of $\ell^p \overset{\wedge}{\otimes} X_0$.

The proper representation of $L^p(0,1) \overset{\wedge}{\otimes} X$ is once again the key to what can be said in an affirmative direction but in this case the "natural" variation of Bu's representation of $\ell^p \overset{\wedge}{\otimes} X$ in the context of X-valued strongly Lebesgue measurable functions doesn't seem to work. Here, what measure theory can't overcome is suitable for Banach space ideas. More precisely, we'll use the fact that $L^p(0,1)$ has an unconditional basis for $1 < p < \infty$.

That $L^p(0,1)$ having an unconditional basis might have such an insightful effect on the structure of $L^p(0,1) \overset{\wedge}{\otimes} X$ ought not be too much of a surprise. One need only look to Dan Lewis's often overlooked study of tensor norms with the Radon-Nikodym property to uncover the following jewel; crucial use is made of the existence of an unconditional basis for $L^p(0,1)$.

THEOREM 3.5 (D.R. Lewis [**29**]). *If X is weakly sequentially complete, then so is $L^p(0,1) \overset{\wedge}{\otimes} X$.*

Let's see how Bu used the existence of an unconditional basis for $L^p(0,1)$ to study the structure of $L^p(0,1) \overset{\wedge}{\otimes} X$.

First we'll be specific about our goal. We want to study the isomorphic structure of $L^p(0,1) \overset{\wedge}{\otimes} X$. This in mind we'll renorm $L^p(0,1)$ so that in the new norm the unconditional basis (u_n) is monotone, normalized and unconditional. This change of norm gives an isomorphic copy of $L^p(0,1)$ and with it an isomorph of $L^p(0,1) \overset{\wedge}{\otimes} X$ as well; since we're amongst friends of tensor products, we'll concentrate on the isomorphic copy. We'll use L^p_{new} to denote $L^p(0,1)$ in this new norm.

Next, we represent $L^p_{\text{new}}(0,1) \overset{\wedge}{\otimes} X$: to start we isolate the sequence space $L^p_{\text{weak}}(Y)$ for a given Banach space Y as the collection of all sequences (y_n) in Y such that $\sum_n y^*(y_n)u_n$ converges in $L^p_{\text{new}}(0,1)$ for each $y^* \in Y^*$; next, we look at $L^p\langle X\rangle$ to be the collection of all sequences (x_n) in X such that

$$\sum_n |x_n^*(x_n)| < \infty$$

for each member (x_n^*) of $L^{p'}_{\text{weak}}(X^*)$. For $(x_n) \in L^p\langle X\rangle$ we define

$$\|(x_n)\|_{L^p\langle X\rangle} = \sup\{\sum_n |x_n^*(x_n)|\}$$

with the supremum being taken over all $(x_n^*) \in L^{p'}_{\text{weak}}(X^*)$ such that

$$\|\sum_n x^{**}(x_n^*)u_n^*\|_{L^{p'}_{\text{weak}}(X^*)} \leq 1$$

where we must keep in mind that (u_n^*) is the unconditional basis of $L^{p'}_{\text{new}}(0,1)$ in its new norm.

Key to our program is the following theorem of Bu.

THEOREM 3.6 (Bu [**6**]). *Let X be a Banach space. Then the spaces $L^p\langle X\rangle$ and $L^p_{new}(0,1) \overset{\wedge}{\otimes} X$ are isometrically isomorphic.*

We outline a proof of this theorem, due to the second author and Ilse Roentgen. It relies, naturally, enough, on our study of the isometric coincidence of $\mathcal{L}^\wedge(L^{p'}_{\text{new}}(0,1);X^*)$ and $L^p_{\text{new}}(0,1) \overset{\wedge}{\otimes} X$, a consequence of the fact that $L^p_{\text{new}}(0,1)$ has both the Radon-Nikodym property and the approximation property.

We list some easily derived ingredients.

- $(x_n) \in L^p\langle X\rangle$ if and only if $\sum_n |x_n^*(x_n)| < \infty$ for each $(x_n^*) \in L^{p'}_{\text{weak}}(X^*)$ which, in turn, occurs if and only if $\sum_n x_n^*(x_n)$ converges for each $(x_n^*) \in L^{p'}_{\text{weak}}(X^*)$. When this happens
$$\|(x_n)\|_{L^p\langle X\rangle} = \sup\left\{|\sum_n x_n^*(x_n)| : (x_n^*) \in B_{L^{p'}_{\text{weak}}(X^*)}\right\}.$$
- Let $u : X \to Y$ be a bounded linear operator and $(x_n) \in L^p_{\text{weak}}(X)$. Then $(ux_n) \in L^p_{\text{weak}}(Y)$ and $\|(ux_n)\|_{L^p_{\text{weak}}(Y)} \leq \|u\| \, \|(x_n)\|_{L^p_{\text{weak}}(X)}$.

Consequently,

- If $(x_n^{***}) \in L^p_{\text{weak}}(X^{***})$ and $j_X : X \hookrightarrow X^{**}$ is the canonical embedding, then $(x_n^{***} \circ j_X) \in L^p_{\text{weak}}(X^*)$ and $\|(x_n^{***} \circ j_X)\|_{L^p_{\text{weak}}(X^*)} \leq \|(x_n^{***})\|_{L^p_{\text{weak}}(X^{***})}$.
- If $(x_n^*) \in L^p_{\text{weak}}(X^*)$, then $(j_{X^*} \circ x_n^*) \in L^p_{\text{weak}}(X^{***})$ and $\|(j_{X^*} \circ x_n^*)\|_{L^p_{\text{weak}}(X^{***})} \leq \|(x_n^*)\|_{L^p_{\text{weak}}(X^*)}$.
- $(x_n) \in L^p\langle X\rangle$ if and only if $(j_X x_n) \in L^p\langle X^{**}\rangle$ and, in this case, $\|(j_X x_n)\|_{L^p\langle X^{**}\rangle} = \|(x_n)\|_{L^p\langle X\rangle}$.

Once again we need to know about the injective tensor product to uncover features of the projective norm.

- If $x_1, \dots, x_n \in X$, then
$$|\sum_{i\leq n} u_i \otimes x_i|_{L^p_{\text{new}}(0,1)\check{\otimes} X} = \|(x_1, x_2, \dots, x_n, 0, 0 \dots)\|_{L^p_{\text{weak}}(X)}.$$

Now we're ready to tackle the proof of Bu's representation of $L^p_{\text{new}}(0,1) \overset{\wedge}{\otimes} X$.

We know that $(L^{p'}_{\text{new}}(0,1)\check{\otimes} X)^*$ is isometrically isomorphic to the space
$$\mathcal{L}^\wedge(L^{p'}_{\text{new}}(0,1); X^*)$$
of all integral linear operators from $L^{p'}_{\text{new}}(0,1)$ to X^*. Under this correspondence, $u \in \mathcal{L}^\wedge(L^{p'}_{\text{new}}(0,1); X^*)$ corresponds to a $Q_u \in (L^{p'}_{\text{new}}(0,1)\check{\otimes} X)^*$ precisely when
$$Q_u(\sum_{k\leq n} u_k^* \otimes x_k) = \sum_{k\leq n} u(u_k^*)(x_k)$$

Now look at the map
$$\Phi : L^p\langle X^*\rangle \to \mathcal{L}^\wedge(L^{p'}_{\text{new}}(0,1); X^*)$$
defined by
$$\Phi\left((x_n^*)\right)(u_k^* \otimes x) = x_k^*(x).$$
It's plain that
$$\begin{aligned}|\Phi\left((x_n^*)\right)(\sum_{k\leq n} u_k^* \otimes x_k)| &= |\sum_{k\leq m} x_k^*(x_k)| \\ &\leq \|(x_n^*)\|_{L^p\langle X^*\rangle}\|(x_1, \dots, x_n, 0, 0, \dots)\|_{L^p_{\text{weak}}(X)} \\ &= \|(x_n^*)\|_{L^p\langle X^*\rangle}\|\sum_{k\leq n} u_k^* \otimes x_k\|_{L^{p'}_{\text{new}}(0,1)\check{\otimes} X}.\end{aligned}$$
Hence $\Phi\left((x_n^*)\right) \in (L^{p'}_{\text{new}}(0,1)\check{\otimes} X)^*$ and
$$\|\Phi\left((x_n^*)\right)\|_{(L^{p'}_{\text{new}}(0,1)\check{\otimes} X)^*} \leq \|(x_n^*)\|_{L^p\langle X^*\rangle}.$$
The map Φ has an inverse $\Psi : (L^{p'}_{\text{new}}(0,1)\check{\otimes} X)^* \to L^p\langle X^*\rangle$ which is defined by
$$\Psi(\xi) = (x_n^*)$$
for $\xi \in (L^{p'}_{\text{new}}(0,1)\check{\otimes} X)^* = \mathcal{B}^\wedge(L^{p'}_{\text{new}}(0,1), X)$, the space of integral bilinear forms on $L^{p'}_{\text{new}}(0,1) \times X$, where
$$\xi(u_j^* \otimes x) = x_j^*(x).$$

For any $x_1, \ldots, x_n \in X$ we have

$$\sum_{k \leq n} |x_k^*(x_k)| \leq \|\xi\| \; \|(x_1, \ldots, x_n, 0, 0, \ldots)\|_{L^p_{\text{weak}}(X)}$$

and so $(x_n^*) \in L^p\langle X^* \rangle$ with $\|(x_n^*)\|_{L^p\langle X^* \rangle} \leq \|\xi\|$.

Ah ha!

$$\mathcal{L}^\wedge(L^{p'}_{\text{new}}(0,1); X^*)$$

and

$$L^p\langle X^* \rangle$$

are isometrically isomorphic. *BUT* we saw earlier that

$$\mathcal{L}^\wedge(L^{p'}_{\text{new}}(0,1); X^*)$$

is just $\mathcal{L}_\wedge(L^{p'}_{\text{new}}(0,1); X^*)$, which is just $L^p_{\text{new}}(0,1) \overset{\wedge}{\otimes} X^*$.

Here we've used the facts that $L^p_{\text{new}}(0,1)$ has the Radon-Nikodym property and the (metric) approximation property.

Alas, we remove the "*" 's from our eyes. Start with noticing that we've shown that

$$\mathcal{L}^\wedge(L^{p'}_{\text{new}}(0,1); X^{**}) = L^p\langle X^{**} \rangle$$

with $u \in \mathcal{L}^\wedge(L^{p'}_{\text{new}}(0,1); X^{**})$ corresponding to $(u(u_n^*)) \in L^p\langle X^{**} \rangle$.

But

$$\begin{aligned} u \in \mathcal{L}^\wedge(L^{p'}_{\text{new}}(0,1); X) &\iff j_X u \in \mathcal{L}^\wedge(L^{p'}_{\text{new}}(0,1); X^{**}) \\ &\iff ((j_X \circ u)(u_n^*)) \in L^p\langle X^{**} \rangle \\ &\iff (u(u_n^*)) \in L^p\langle X \rangle \end{aligned}$$

with each quantity normed precisely like his neighbor.

$L^p\langle X \rangle$ and $L^p_{\text{new}}(0,1) \overset{\wedge}{\otimes} X$ are isometrically isomorphic and so we have a candidate for representing $L^p(0,1) \overset{\wedge}{\otimes} X$. This candidate is an apt one. But to see this one needs to find a substitute for the Lebesgue-Bochner space $L^p_X(0,1)$! One is handy : $L^p_{\text{`strong'}}(X)$ is the collection of all sequences $(x_n)$ in $X$ such that $\sum \|x_n\| u_n$ converges in $L^p(0,1)$ and for $(x_n) \in L^p_{\text{`strong'}}(X)$ we define

$$\|(x_n)\|_{L^p_{\text{`strong'}}(X)} = \|\sum_n \|x_n\| u_n\|_{L^p_{\text{new}}(0,1)}.$$

The cast is on stage and what's missing is the script:

THEOREM 3.7 (Bu [**6**]). *The natural inclusion*

$$L^p_{new}(0,1) \overset{\wedge}{\otimes} X = L^p\langle X \rangle \hookrightarrow L^p_{\text{`strong'}}(X)$$

is a semi-embedding.

The space $L^p_{\text{`strong'}}(X)$ was meant to be a substitute for $L^p_X(0,1)$ and in at least some ways it is. For instance,

THEOREM 3.8 (Bu [**6**]). *If X has the Radon-Nikodym property, then so does $L^p_{\text{`strong'}}(X)$.*

COROLLARY 3.9. *If X has the Radon-Nikodym property, then so does $L^p(0,1) \overset{\wedge}{\otimes} X$.*

This representation of $L^p(0,1) \overset{\wedge}{\otimes} X$ has quite a bit of life to it. Changing pace Bu and Paddy Dowling showed that the space $L^p_{\text{strong}}(X)$ also shares the analytic Radon-Nikodym property with a complex Banach space X and so, since the analytic Radon-Nikodym property is passed from separable spaces to separable spaces that're semi-embedded within we get the

COROLLARY 3.10 (Bu, Dowling [**8**]). *If X is a complex Banach space with the analytic Radon-Nikodym property, then $L^p(0,1) \overset{\wedge}{\otimes} X$ has the analytic Radon-Nikodym property, too.*

Viewed from the vantage point offered us by harmonic analysts, the analytic Radon-Nikodym property is but one of a list of non-trivial Radon-Nikodym properties introduced and studied by Gerry Edgar and Paddy Dowling in the late 80's. We discuss but one such variant: the Sidon-Radon-Nikodym property. To set the table we assume G is a compact metrizable Abelian group with dual group Γ, X is a Banach space and Λ is a subset of Γ, an X-valued Borel measure μ on G that has finite variation is a Λ-measure if

$$0 = \hat{\mu}(\gamma) = \int_G \overline{\gamma(t)} d\mu(t)$$

holds for all $\gamma \notin \Lambda$. X has the Sidon-Radon-Nikodym property if for any G, any Sidon set $\Lambda \subset \Gamma$ that's infinite and any Λ-measure μ with values in X there is an $f \in L^1_X(G)$ so that $\mu(E) = \int_E f d\lambda$, where λ is the normalized Haar measure on G and $L^1(G)$ denotes $L'(\lambda)$.

It's a stunning result of Dowling [**17**] that says *a Banach space X has the Sidon-Radon-Nikodym property precisely when X contains no copy of c_0.*

The Sidon-Radon-Nikodym property dutifully respects semi-embeddings and, you guessed it, Bu and Dowling also showed that if X has the Sidon-Radon-Nikodym property then so does $L^p_{\text{strong}}(X)$.

COROLLARY 3.11 (Bu-Dowling [**8**]). *$L^p(0,1) \overset{\wedge}{\otimes} X$ contains an isomorphic copy of c_0 if and only if X does.*

If X has non-trivial cotype, does $L^p(0,1) \overset{\wedge}{\otimes} X$?

Recall that a Banach space has *non-trivial cotype* if there is a $q : 2 \leq q < \infty$ and a constant $K(X)$ so that regardless of $x_1, \ldots, x_n \in X$ we have

$$(\sum_{k \leq n} \|x_k\|^q)^{1/q} \leq K(X) \left(\int_0^1 \| \sum_{k \leq n} r_k(t) x_k \|^q dt \right)^{1/q}.$$

It is part of the "Great Theorem" of Maurey and Pisier that a Banach space X has non trivial cotype precisely when it does *not* contain ℓ_n^∞'s uniformly.

It is a known and highly non-trivial result of Nicole Tomczak-Jaegermann [**41**] that $L^2(0,1) \overset{\wedge}{\otimes} H$ has cotype 2. Little else is known about cotype in the projective tensor product with an $L^p(0,1)$ coordinate.

4. Some Things That Must Be Said

To put what's been said (especially about $L^p(0,1) \overset{\wedge}{\otimes} X$) in context it's useful to recall certain aspects of the work of Gilles Pisier.

We start with Pisier's discovery of a Banach space P whose injective tensor product $P\check{\otimes}P$ with itself and projective tensor product $P \overset{\wedge}{\otimes} P$ with itself are isomorphic. This space (actually class of spaces) is discussed in his Missouri lecture notes [**39**]as well as in his original paper [**38**]; we had intended to give a discussion of it along with some of its relatives but space (and time) prohibited a reasonable presentation. In any case, Pisier's space(s) show that the projective tensor product can be mysterious indeed.

For the purposes of this report it might be more relevant to discuss the examples of Jean Bourgain and Pisier [**5**]. These spaces are a kind of hybrid to the Pisier examples (using as they do, the magical construction of Sergei Kisliakov of 'spaces of universal disposition') and the Bourgain-Delbaen special $\mathcal{L}_\infty$-spaces [**4**]. In particular, the Bourgain-Pisier paper shows:

- *there exist Banach spaces X and Y so that each is weakly sequentially complete yet $X \overset{\wedge}{\otimes} Y$ contains a copy of* c_0; in fact, X can be taken to be L^1/H_0^1. One might compare this fact with the theorems of Dan Lewis and of Qingying Bu and Paddy Dowling.
- *there exist Banach spaces X and Y each having the Radon-Nikodym property so that $X \overset{\wedge}{\otimes} Y$ contains a copy of* c_0. Again, a comparison with Bu's theorem in case $X = L^p(0,1)$ is natural.

In affairs to do with tensor products (as in many delicate analytic situations) the deft hand of Gilles Pisier is always around to lead us if we will but follow. The remarkable K-convexity theorem of Pisier comes to mind.

Recall that a Banach space X has *non-trivial type* if there exists a p, $1 \le \rho \le 2$ and a $\tau > 0$ so that for any finite number of vectors $x_1, \ldots, x_n \in X$ we have

$$\left(\sum_{k=1}^{n} \|x_k\|^p\right)^{1/p} \le \tau \left(\int_0^1 \|\sum_{k=1}^{n} r_k(t)x_k\|^\rho dt\right)^{1/p.}$$

It is a marvelous theorem of Pisier that says that X has non-trivial type precisely when X does *not* contain uniform copies of ℓ_n^1's, here a space X does contain ℓ_n^1's uniformly if for some constant K there is for each $n \in \mathbb{N}$ an isomorphism $u_n : \ell_n^1 \to X$ such that $\|u_n\|\ \|u_n^{-1}\| \le K$. Pisier's K-convexity theorem [**37**] says that in spaces with non trivial type (and only in such spaces) the following strengthening of Dvoretzky's spherical sections theorem holds true: There is a $K > 0$ so that for each n there is a subspace E in X with dim $E = n$, a linear projection $P_E : X \to X$ with $P_E X = E$ and an isomorphism $u_E : \ell_n^2 \to E$ so that $\|u_E\|\ \|u_E^{-1}\|\ \|P_E\| \le K$. In other words, X has non-trivial type precisely when X contains uniformly complemented copies of ℓ_n^2's.

What's this to do with the projective tensor product?

Well, if X and Y are infinite dimensional Banach spaces, then $X \overset{\wedge}{\otimes} Y$ *never enjoys nontrivial type.* It's plain that we need only consider the case when both X and Y have non-trivial type in which case each of X and Y contain uniformly complemented copies of ℓ_n^2's, suppose (E_n) and (F_n) denote the supposed copies of ℓ_n^2 in X and Y respectively. Then $(E_n \overset{\wedge}{\otimes} F_n)$ is a sequence of uniformly complemented subspace of $X \overset{\wedge}{\otimes} Y$. But E_n and F_n are ℓ_n^2's, Hilbert spaces, and it's a large part of the charm of the projective norm that the projective tensor product of two Hilbert spaces is isomorphic to the space of trace class operators between the spaces with the usual trace-class norm. But along the diagonal of the trace class $\ell_n^2 \overset{\wedge}{\otimes} \ell_n^2$ ($E_n \overset{\wedge}{\otimes} F_n$ in slight disguise) is a norm-one complemented copy of ℓ_n^1 - the closed linear span of $e_1 \otimes e_1, \ldots, e_n \otimes e_n$. In other words, $X \overset{\wedge}{\otimes} Y$ contains uniformly complemented copies of ℓ_n^1's. And so, if we've to believe Pisier (a good habit to be sure), $X \overset{\wedge}{\otimes} Y$ does not have non-trivial type.

Much follows from this. Let us just say that uniformly convexifiable and uniformly smoothable spaces have non-trivial type and so $X \overset{\wedge}{\otimes} Y$ *is never uniformly convexifiable nor is it uniformly smoothable once dim* $X = \infty = dim Y$.

On a positive note : *if* X *and* Y *are Banach algebras, then* $X \overset{\wedge}{\otimes} Y$ *is a Banach algebra as well; it is commutative if* X *and* Y *are and have an identity if both* X *and* Y *do.* This is surprisingly easy. In fact, if $u = \sum_{i \le m} x_i \otimes y_i$ and $v = \sum_{j \le n} x'_j \otimes y'_j$ are members of $X \otimes Y$, then their product is just

$$uv = \Sigma_{i \le m, j \le n} x_i x'_j \otimes y_i y'_2$$

and so

$$\begin{aligned} \|uv\|_\wedge &= \|\Sigma_{i \le m, j \le n} x_i x'_j \otimes y_i y'_j\|_\wedge \\ &\le \Sigma_{i \le m, j \le n} \|x_i\| \; \|x'_j\| \; \|y_i\| \; \|y'_j\| \\ &= (\Sigma_{i \le m} \|x_i\| \; \|y_i\|)(\Sigma_{j \le n} \|x'_j\| \; \|y'_j\|). \end{aligned}$$

It follows that

$$\|uv\|_\wedge \le \|u\|_\wedge \; \|v\|_\wedge.$$

This inequality passes unscathed from $X \otimes Y$ to its completion $X \overset{\wedge}{\otimes} Y$.

It is noteworthy that of the 14 natural tensor norms uncovered by Grothendieck, the projective norm is one of just four that preserve Banach algebra structure. This is a beautiful result of Tim Carne [**10**], worthy of any and all attention paid it.

It is natural to consider other natural structures on Banach spaces and test what happens on taking tensor products. Of course, if X is a Banach lattice, then $L^1(\mu) \overset{\wedge}{\otimes} X$ is too. But this has more to do with $L^1_X(\mu)$ than the projective tensor norm. Generally, one cannot expect Banach lattice structure to survive the projective tensor product; indeed, *if* X *and* Y *are infinite dimensional Banach lattices having non-trivial type, then* $X \overset{\wedge}{\otimes} Y$ *doesn't even have* $LUST$. Again, Pisier's K-convexity theorem tells us why if we'll only listen. Proceeding as above we soon find that if X and Y are infinite dimensional Banach spaces having non-trivial type, then the space $X \overset{\wedge}{\otimes} Y$ contains uniformly complemented copies of $\ell_n^2 \overset{\wedge}{\otimes} \ell_n^2$. Now a close look at the computations and conclusions of the wonder-full paper of Yehoram Gordon and Dan Lewis [**22**] tells us that $X \overset{\wedge}{\otimes} Y$ fails LUST hence, thanks to Maurey's proof that Banach lattices do have lust [**31**], $X \overset{\wedge}{\otimes} Y$ cannot even be isomorphic to a Banach lattice.

Several points to be made here. It's unclear what conditions allow $X \overset{\wedge}{\otimes} Y$ to be isomorphic to a Banach lattice, which conditions ensure $X \overset{\wedge}{\otimes} Y$ has lust or whether *any* of Grothendieck's 14 natural tensor norms preserve lattice structure even isomorphically. It's ironic that in this situation it's $L^1(\mu)$-spaces that have the right geometry to preserve some local structure.

Finally, we end with a very brief discussion of what happens when $C(K)$ is one of the coordinates and, indeed, we concentrate on the case when we're looking at $C(K_1) \overset{\wedge}{\otimes} C(K_2)$.

A fundamental isomorphic invariant enjoyed by all $C(K)$ spaces is the Dunford-Pettis property; X has the *Dunford-Pettis property* if whenever $u : X \to Y$ is a weakly compact linear operator then $\lim_n \|ux_n\| = 0$ whenever weak $\lim_n x_n = 0$. Grothendieck introduced this property, called it as he did in light of the old Dunford-Pettis theorem which asserts that L^1-spaces enjoy the property and proved that $C(K)$-spaces also enjoy the property. What about $C(K_1) \overset{\wedge}{\otimes} C(K_2)$? Does this space have the Dunford-Pettis property?

Well it's not too hard to show that if K_1 and K_2 are scattered (= dispersed, have no perfect subsets) (cf [**35**]), so that $C(K_1)^*$ and $C(K_2)^*$ both have the Schur property, then $[C(K_1)\hat{\otimes}C(K_2)]^*$ enjoys the Schur property (weak and norm sequential converges coincides), too. So *if K_1 and K_2 are scattered, then $C(K_1) \overset{\wedge}{\otimes} C(K_2)$ has the Dunford-Pettis property.*

Otherwise, we have the following:

THEOREM 4.1 (Bombal, Villanueva [**2**]). *If K_1 and K_2 are infinite compact Haussdorff spaces and K_2 is not scattered then $C(K_1) \overset{\wedge}{\otimes} C(K_2)$ fails the Dunford-Pettis property.*

PROOF. The lynch-pin of the proof is the following fact: *if (f_n) is a weakly null sequence in $C(K_1)$ and (g_n) is a bounded sequence in $C(K_2)$, then $(f_n \otimes g_n)$ is weakly null in $C(K_1) \overset{\wedge}{\otimes} C(K_2)$.* (note that in this we assume only that K_1, K_2 are compact Hausdorff spaces.) Look at $\varphi \in [C(K)_1) \overset{\wedge}{\otimes} C(K_2)]^*$, realize φ as a linear operator from $C(K_1)$ to $C(K_2)^*$. Since $C(K_2)^*$ is weakly sequentially complete, every operator from $C(K_1)$ to $C(K_2)^*$ is weakly compact (another wonderful discovery of Grothendieck) and so takes weakly null sequences in $C(K_1)$, like (f_n), into norm null sequences in $C(K_2)^*$. It's easy to see from this that $\lim_{n\to\infty} \varphi(f_n, g_n) = 0$.

Now suppose that both K_1 and K_2 are infinite compact Hausdorff spaces and that K_2 is *not* scattered.

Because K_1 is infinite, $C(K_1)$ contains an isomorphic copy of c_0. It follows that we can find a weakly null sequence (f_n) in $C(K_1)$ and we can find a sequence (μ_n) in $B_{C(K_1)^*}$ so that $\mu_n(f_n) = 1$.

On the other hand, K_2 is *not* scattered and so, as Pełczyński and Semadeni are apt to tell us, we can find a copy of ℓ^1 inside $C(K_2)$. But whenever a Banach space contains a copy of ℓ^1, it also has a quotient map onto ℓ^2; suppose $Q : C(K_2) \twoheadrightarrow \ell^2$ is the said quotient map. Let $(g_n) \subseteq C(K_2)$ be a bounded sequence so that $Q(g_n) = e_n$, where (e_n) is the sequence of unit coordinate vectors in ℓ^2.

Our opening comments alert us to the fact that$(f_n \otimes g_n)$ is a weakly null sequence in $C(K_1) \overset{\wedge}{\otimes} C(K_2)$.

Consider the bilinear map

$$\Phi : C(K_1) \times C(K_2) \to \ell^2$$

defined by

$$\Phi(f, g) = (\mu_n(f)Q(g)_n)_{n\in\mathbb{N}}.$$

It's plain that Φ is a bounded bilinear map and so the Universal Mapping Principle assumes us we can consider Φ to be a bounded linear operator from $C(K_1) \overset{\wedge}{\otimes} C(K_2)$ to ℓ^2; of course, Φ is weakly compact – ℓ^2 is reflexive.

However, $\Phi(f_n \otimes g_n) = e_n$ and (e_n) is *not* norm null. □

Another fundamental invariant enjoyed by all $C(K)$'s is Pełczyński's property V. An operator $u : X \to Y$ is *unconditionally converging* if $\Sigma_n u x_n$ is unconditionally convergent whenever $\Sigma|x^* x_n| < \infty$ for each $x^* \in X^*$; X has *property* V if any unconditionally converging operator $u : X \to Y$ is weakly compact. As noted ahead, all $C(K)$'s have V—this is due to Pełczyński himself. There's a general result due to Emmanuele and Hensgen that goes as follows

THEOREM 4.2 (Emmanuele, Hensgen [**18**]). *If X and Y have property V and if every bounded linear operator $u : X \to Y^*$ is compact, then $X \overset{\wedge}{\otimes} Y$ has V.*

In case the spaces X and Y involved are $C(K)$'s, then this theorem translates to the following:

COROLLARY 4.3 (Bombal, Fernandes, Villaneuva [**1**]). *$C(K_1) \overset{\wedge}{\otimes} C(K_2)$ has property V precisely when at least one of K_1 and K_2 is scattered.*

Many of the topics mentioned in this last miscellaneous section will be discussed in more detail in a sequel.

References

[1] Fernando Bombal, Maite Fernández, and Ignacio Villanueva. Some classes of multilinear operators on $C(K)$ spaces. *Studia Math.*, 148(3):259–273, 2001.

[2] Fernando Bombal and Ignacio Villanueva. On the Dunford-Pettis property of the tensor product of $C(K)$ spaces. *Proc. Amer. Math. Soc.*, 129(5):1359–1363, 2001.

[3] J. Bourgain. New Banach space properties of the disc algebra and H^∞. *Acta Math.*, 152(1-2):1–48, 1984.

[4] J. Bourgain and F. Delbaen. A class of special $\mathcal{L}_\infty$ spaces. *Acta Math.*, 145(3-4):155–176, 1980.

[5] Jean Bourgain and Gilles Pisier. A construction of $\mathcal{L}_\infty$-spaces and related Banach spaces. *Bol. Soc. Brasil. Mat.*, 14(2):109–123, 1983.

[6] Qingying Bu. Observations about the projective tensor product of Banach spaces. II. $L^p(0,1)\widehat{\otimes}X$, $1 < p < \infty$. *Quaest. Math.*, 25(2):209–227, 2002.

[7] Qingying Bu and Joe Diestel. Observations about the projective tensor product of Banach spaces. I. $l^p\widehat{\otimes}X$, $1 < p < \infty$. *Quaest. Math.*, 24(4):519–533, 2001.

[8] Qingying Bu and Patrick N. Dowling. Observations about the projective tensor product of Banach spaces. III. $L^p[0,1]\hat{\otimes}X$, $1 < p < \infty$. *Quaest. Math.*, 25(3):303–310, 2002.

[9] Qingying Bu and Pei-Kee Lin. Radon-Nikodym property for the projective tensor product of Köthe function spaces. To appear.

[10] T. K. Carne. Tensor products and Banach algebras. *J. London Math. Soc. (2)*, 17(3):480–488, 1978.

[11] J. Diestel, J. H. Fourie, and J. Swart. The metric theory of tensor products, part 1: Tensor norms. *Quaestiones Math*, 25:37–72, 2002.

[12] J. Diestel, J. H. Fourie, and J. Swart. The metric theory of tensor products, part 2: Bilinear forms and linear operators of type α. *Quaestiones Math*, 25:73–94, 2002.

[13] J. Diestel, J. H. Fourie, and J. Swart. The metric theory of tensor products, part 3: Vector sequence spaces. *Quaestiones Math*, 25:95–118, 2002.

[14] J. Diestel, J. H. Fourie, and J. Swart. The metric theory of tensor products, part 4: The role of $C(K)$-spaces and L_1-spaces. 2002.

[15] J. Diestel and J. J. Uhl, Jr. *Vector measures*. American Mathematical Society, Providence, R.I., 1977. With a foreword by B. J. Pettis, Mathematical Surveys, No. 15.

[16] Joe Diestel, Hans Jarchow, and Andrew Tonge. *Absolutely summing operators*. Cambridge University Press, Cambridge, 1995.

[17] Patrick N. Dowling. Radon-Nikodým properties associated with subsets of countable discrete abelian groups. *Trans. Amer. Math. Soc.*, 327:879–890, 1991.

[18] G. Emmanuele and W. Hensgen. Property (V) of Pelczyński in projective tensor products. *Proc. Roy. Irish Acad. Sect. A*, 95(2):227–231, 1995.

[19] B. R. Gelbaum and J. Gil de Lamadrid. Bases of tensor products of Banach spaces. *Pacific J. Math.*, 11:1281–1286, 1961.

[20] M. González and J. N. Gutiérrez. The Dunford-Pettis property on tensor products. *Math. Proc. Camb. Phil. Soc.*, 131: 185-192, 2001.

[21] Dwight B. Goodner. Projections in normed linear spaces. *Trans. Amer. Math. Soc.*, 69:89–108, 1950.

[22] Y. Gordon and D. R. Lewis. Absolutely summing operators and local unconditional structures. *Acta Math.*, 133:27–48, 1974.

[23] A. Grothendieck. Résumé de la théorie métrique des produits tensoriels topologiques. *Bol. Soc. Mat. São Paulo*, 8:1–79, 1953/1956.

[24] A. Grothendieck. Une caractérisation vectorielle-métrique des espaces L^1. *Canad. J. Math.*, 7:552–561, 1955.

[25] Alexandre Grothendieck. Produits tensoriels topologiques et espaces nucléaires. *Mem. Amer. Math. Soc.*, 1955(16):140, 1955.

[26] Morisuke Hasumi. The extension property of complex Banach spaces. *Tôhoku Math. J. (2)*, 10:135–142, 1958.

[27] S. Heinrich and P. Mankiewicz. Applications of ultrapowers to the uniform and Lipschitz classification of Banach spaces. *Studia Math.*, 73(3):225–251, 1982.

[28] J. L. Kelley. Banach spaces with the extension property. *Trans. Amer. Math. Soc.*, 72:323–326, 1952.

[29] D. R. Lewis. Duals of tensor products. In *Banach spaces of analytic functions (Proc. Pełczyński Conf., Kent State Univ., Kent, Ohio, 1976)*, pages 57–66. Lecture Notes in Math., Vol. 604. Springer, Berlin, 1977.

[30] J. Lindenstrauss and A. Pełczyński. Absolutely summing operators in L_p-spaces and their applications. *Studia Math.*, 29:275–326, 1968.

[31] B. Maurey. Type et cotype dans les espaces munis de structures locales inconditionnelles. In *Séminaire Maurey-Schwartz 1973–1974: Espaces L^p, applications radonifiantes et géométrie des espaces de Banach, Exp. Nos. 24 et 25*, page 25. Centre de Math., École Polytech., Paris, 1974.

[32] Bernard Maurey. *Théorèmes de factorisation pour les opérateurs linéaires à valeurs dans les espaces L^p*. Société Mathématique de France, Paris, 1974. With an English summary, Astérisque, No. 11.

[33] Bernard Maurey and Gilles Pisier. Séries de variables aléatoires vectorielles indépendantes et propriétés géométriques des espaces de Banach. *Studia Math.*, 58(1):45–90, 1976.

[34] Leopoldo Nachbin. A theorem of the Hahn-Banach type for linear transformations. *Trans. Amer. Math. Soc.*, 68:28–46, 1950.

[35] A. Pełczyński and Z. Semadeni. Spaces of continuous functions. III. Spaces $C(\Omega)$ for Ω without perfect subsets. *Studia Math.*, 18:211–222, 1959.

[36] A. Pietsch. Absolut p-summierende Abbildungen in normierten Räumen. *Studia Math.*, 28:333–353, 1966/1967.

[37] Gilles Pisier. Holomorphic semigroups and the geometry of Banach spaces. *Ann. of Math. (2)*, 115(2):375–392, 1982.

[38] Gilles Pisier. Counterexamples to a conjecture of Grothendieck. *Acta Math.*, 151(3-4):181–208, 1983.

[39] Gilles Pisier. *Factorization of linear operators and geometry of Banach spaces*. Published for the Conference Board of the Mathematical Sciences, Washington, DC, 1986.

[40] Narcisse Randrianantoanina. Complemented copies of l^1 and Pełczyński's property (V*) in Bochner function spaces. *Canad. J. Math.*, 48(3):625–640, 1996.

[41] Nicole Tomczak-Jaegermann. The moduli of smoothness and convexity and the Rademacher averages of trace classes $S_p(1 \leq p < \infty)$. *Studia Math.*, 50:163–182, 1974.

Banach Center, Kent State University, Kent, OH 44242, USA
E-mail address: j_diestel@hotmail.com

School of Computer, Statistical and Math Sciences Potchefstroom Univ, Potchefstroom 2520, South Africa
E-mail address: wskjhf@puknet.puk.ac.za

Dept Maths and Applied Maths, University of Pretoria, Pretoria 0002, South Africa
E-mail address: jswart@math.up.ac.za

Contemporary Mathematics
Volume **321**, 2003

Spectrum of a weakly hypercyclic operator meets the unit circle

S. J. Dilworth and Vladimir G. Troitsky

ABSTRACT. It is shown that every component of the spectrum of a weakly hypercyclic operator meets the unit circle. The proof is based on the lemma that a sequence of vectors in a Banach space whose norms grow at geometrical rate doesn't have zero in its weak closure.

Suppose that T is a bounded operator on a nonzero Banach space X. Given a vector $x \in X$, we say that x is ***hypercyclic*** for T if the orbit $\mathrm{Orb}_T x = \{T^n x\}_n$ is dense in X. Similarly, x is said to be ***weakly hypercyclic*** if $\mathrm{Orb}_T x$ is weakly dense in X. A bounded operator is called ***hypercyclic*** or ***weakly hypercyclic*** if it has a hypercyclic or, respectively, a weakly hypercyclic vector. It is shown in [**CS**] that a weakly hypercyclic vector need not be hypercyclic, and there exist weakly hypercyclic operators which are not hypercyclic. C. Kitai showed in [**K**] that every component of the spectrum of a hypercyclic operator intersects the unit circle. K. Chan and R. Sanders asked in [**CS**] if the spectrum of a weakly hypercyclic operator meets the unit circle. In this note we show that every component of the spectrum of a weakly hypercyclic operator meets the unit circle.

LEMMA 1. *Let X be a Banach space and let $c > 1$. Suppose that $x_n \in X$ satisfies $\|x_n\| \geqslant c^n$ for all $n \geqslant 1$. Then $0 \notin \overline{\{x_n\}_n}^w$.*

PROOF. Let N be the smallest positive integer such that $c^N > 2$. We shall prove that there exist $F_1, \dots, F_N \in X^*$ such that

$$\max_{1 \leqslant k \leqslant N} \bigl|F_k(x_n)\bigr| \geqslant 1 \qquad (n \geqslant 1). \tag{1}$$

Since $\|x_n\| \geqslant c^n$, by replacing x_n by $(c^n/\|x_n\|)x_n$, it suffices to prove (1) for the case in which $\|x_n\| = c^n$ for all $n \geqslant 1$. First suppose that $c > 2$, so that $N = 1$. We have to construct $F_1 \in X^*$ such that $\bigl|F_1(x_n)\bigr| \geqslant 1$ for all $n \geqslant 1$. First choose $f_1 \in X^*$ with $f_1(x_1) = 1$. Then either $\bigl|f_1(x_2)\bigr| < 1$ or $\bigl|f_1(x_2)\bigr| \geqslant 1$. In the former case the Hahn-Banach theorem guarantees the existence of $g_2 \in X^*$ such that $\|g_2\| \leqslant 1/\|x_2\| = c^{-2}$, $\bigl|g_2(x_2)\bigr| = 1 - \bigl|f_1(x_2)\bigr|$, and $\bigl|(f_1 + g_2)(x_2)\bigr| = 1$. In the

2000 *Mathematics Subject Classification.* Primary: 47A16,47A10,47A25.

Key words and phrases. Weakly hypercyclic operator, spectrum, spectral set, weak closure.

The research of the first author was supported by a summer research grant from the University of South Carolina while on sabbatical as a Visiting Scholar at The University of Texas at Austin.

latter case, set $g_2 = 0$. Note that

$$\left|(f_1 + g_2)(x_1)\right| \geqslant 1 - \|g_2\|\|x_1\| \geqslant 1 - c^{-1}.$$

Set $f_2 = f_1 + g_2$. Repeating this argument, we can find $g_3 \in X^*$ such that $\|g_3\| \leqslant 1/\|x_3\| = c^{-3}$ and $\left|(f_2 + g_3)(x_3)\right| \geqslant 1$. Note that

$$\left|(f_2 + g_3)(x_1)\right| \geqslant \left|f_2(x_1)\right| - \|g_3\|\|x_1\| \geqslant 1 - c^{-1} - c^{-2}$$

and also that

$$\left|(f_2 + g_3)(x_2)\right| \geqslant 1 - \|g_3\|\|x_2\| \geqslant 1 - c^{-1}.$$

Set $f_3 = f_2 + g_3$. Continuing in this way we obtain $f_n \in X^*$ such that (setting $g_n = f_n - f_{n-1}$) $\|g_n\| \leqslant c^{-n}$ and

$$\left|f_n(x_k)\right| \geqslant 1 - \sum_{i=1}^{n-k} c^{-i} \qquad (1 \leqslant k \leqslant n). \tag{2}$$

Thus, $\{f_n\}_n$ is norm-convergent in X^* to some $f \in X^*$. From (2), we obtain (since $c > 2$)

$$\left|f(x_k)\right| = \lim_n \left|f_n(x_k)\right| \geqslant 1 - \sum_{i=1}^{\infty} c^{-i} = \frac{c-2}{c-1} > 0.$$

Set $F_1 = (c-1)(c-2)^{-1}f$, to complete the proof in the case $c > 2$.

Now suppose that $1 < c < 2$. Set $\alpha = c^N > 2$. For each $1 \leqslant k \leqslant N$, consider the sequence $y_n = x_{k+(n-1)N}$ $(n \geqslant 1)$. Then $\|y_n\| = (c^k/\alpha)\alpha^n$. Since $\alpha > 2$ there exists $F_k \in X^*$ such that $\left|F_k(y_n)\right| \geqslant 1$ for all $n \geqslant 1$, which proves (1). □

REMARK 2. Recall that a closed subspace Y of X^* is said to be ***norming*** if there exists $C > 0$ such that

$$\|x\| \leqslant C \sup\bigl\{\left|f(x)\right| : f \in Y, \|f\| \leqslant 1\bigr\} \qquad (x \in X).$$

The argument of Lemma 1 easily generalizes to give the following result. Suppose that Y is norming for X and that $\{x_n\}_n$ is a sequence in X satisfying $\|x_n\| \geqslant c^n$, where $c > 1$. Then 0 does not belong to the $\sigma(X, Y)$-closure of $\{x_n\}_n$. In particular, Lemma 1 is valid for the weak-star topology when X is a dual space.

We also make use of the following simple numerical fact. If (t_n) is a sequence in $\mathbb{R}^+ \cup \{\infty\}$, then

$$\limsup_{n\to\infty} \sqrt[n]{t_n} = \inf\{\, \nu > 0 \mid \lim_{n\to\infty} \tfrac{t_n}{\nu^n} = 0 \,\} = \inf\{\, \nu > 0 \mid \limsup_{n\to\infty} \tfrac{t_n}{\nu^n} < \infty \,\}.$$

In particular, if T is a bounded operator with spectral radius r, then the Gelfand formula $\lim_n \sqrt[n]{\|T^n\|} = r$ yields that $\frac{\|T^n\|}{\lambda^n} \to 0$ for every scalar λ with $|\lambda| > r$.

THEOREM 3. *If T is weakly hypercyclic, then every component of $\sigma(T)$ meets $\{z : |z| = 1\}$.*

PROOF. Let x be a weakly hypercyclic vector for T. Let σ be a non-empty component of $\sigma(T)$, denote $\sigma' = \sigma(T)\setminus\sigma$. Denote by X_σ and $X_{\sigma'}$ the corresponding spectral subspaces, then X_σ and $X_{\sigma'}$ are closed, T-invariant, and $X = X_\sigma \oplus X_{\sigma'}$. Also, $\sigma(T_{|X_\sigma}) = \sigma$ and $\sigma(T_{|X_{\sigma'}}) = \sigma'$. Note that σ' might be empty, in which case we have $X_\sigma = X$ and $X_{\sigma'} = \{0\}$.

Denote by P_σ the spectral projection corresponding to σ, then $X_\sigma = \operatorname{Range} P_\sigma$. Denote $y = P_\sigma x$. Without loss of generality, $\|y\| = 1$. Since P_σ is bounded and, therefore, weakly continuous, and $\text{Orb}_T y = P_\sigma(\text{Orb}_T x)$, we conclude that $\text{Orb}_T y$ is weakly dense in X_σ. Thus, y is weakly hypercyclic for $T_{|X_\sigma}$.

Observe that the inclusion $\sigma \subseteq \{z : |z| < 1\}$ is impossible. Indeed, in this case the spectral radius of $T_{|X_\sigma}$ would be less than 1, so that $T^n y \to 0$, which contradicts y being weakly hypercyclic for $T_{|X_\sigma}$.

Finally, we show that the inclusion $\sigma \subseteq \{z : |z| > 1\}$ is equally impossible. In this case $0 \notin \sigma = \sigma(T_{|X_\sigma})$, so that $T_{|X_\sigma}$ is invertible. Denote the inverse by S. Then S is a bounded operator on X_σ and by the Spectral Mapping Theorem

$$\sigma(S) = \{\lambda \mid \lambda^{-1} \in \sigma(T_{|X_\sigma})\} \subset \{z : |z| < 1\}.$$

Therefore, $r(S) < a$ for some $0 < a < 1$. This yields $\lim_n \frac{\|S^n\|}{a^n} = 0$, so that $\|S^n\| \leqslant a^n$ for all sufficiently large n. In particular,

$$1 = \|y\| = \|S^n T^n y\| \leqslant a^n \|T^n y\|,$$

so that $\|T^n y\| \geqslant \frac{1}{a^n}$. Lemma 1 asserts that $0 \notin \overline{\{T^n y\}_n}^w$, which contradicts y being weakly hypercyclic for $T_{|X_\sigma}$. □

PROPOSITION 4. *Suppose that Y is norming for X. If T has a hypercyclic vector for the $\sigma(X, Y)$ topology, then the spectrum of T intersects $\{z : |z| = 1\}$.*

PROOF. Suppose, to derive a contradiction, that $\sigma(T)$ does not intersect the unit circle. We use the notation introduced above with $\sigma = \sigma(T) \cap \{z : |z| < 1\}$ and $\sigma' = \sigma(T) \setminus \sigma$. Let x be a hypercyclic vector for the $\sigma(X, Y)$-topology. Then $x = y + z$, where $y \in X_\sigma$ and $z \in X_{\sigma'}$. Since $\|T^n y\| \to 0$, it follows easily that z is hypercyclic, and hence that $z \neq 0$. But then there exists $c > 1$ such that $\|T^n z\| \geqslant c^n$ for all sufficiently large n, which contradicts Remark 2. □

References

[CS] Kit C. Chan and Rebecca Sanders. A weakly hypercyclic operator that is not norm hypercyclic. To appear in J. Operator Theory.

[K] Carol Kitai. *Invariant Closed Sets for Linear Operators.* PhD thesis, Univ. of Toronto, 1982.

DEPARTMENT OF MATHEMATICS, UNIVERSITY OF SOUTH CAROLINA, COLUMBIA, SC 29208. USA.

Current address: Department of Mathematics, The University of Texas at Austin, Austin, Texas 78712. USA

DEPARTMENT OF MATHEMATICAL AND STATISTICAL SCIENCES, 632 CAB, UNIVERSITY OF ALBERTA, EDMONTON, AB T6G 2G1. CANADA.

E-mail address: dilworth@math.sc.edu, vtroitsky@math.ualberta.ca

Contemporary Mathematics
Volume **321**, 2003

The Dynamics of Cohyponormal Operators

Nathan S. Feldman

ABSTRACT. In this expository article we survey some of the recent work on the dynamics of linear operators, and in particular for cohyponormal operators. We give several examples using the adjoints of Bergman operators and also state several open questions. The overall theme is that the spectral properties of a hyponormal operator S determine the dynamics of its adjoint S^*. The final section of the paper gives some examples of "complicated" orbits.

1. Introduction

In what follows X will denote a separable complex Banach space and $\mathcal{H}$, a separable complex Hilbert space; $\mathcal{B}(X)$ will denote the algebra of all bounded linear operators on X. Studying the dynamics of linear operators entails a study of the behavior of orbits. We shall see that it is natural and important to study the orbits of both points and of subsets. If $T \in \mathcal{B}(X)$ and $x \in X$, then the orbit of x under T is $\text{Orb}(T, x) = \{x, Tx, T^2x, \dots\}$. If $C \subseteq X$, then the orbit of C under T is $\text{Orb}(T, C) = \bigcup\{C, T(C), T^2(C), \dots\} = \bigcup_{x \in C} \text{Orb}(T, x)$.

While the term "chaotic" has a very precise meaning, the type of "chaotic behavior" we shall consider for an operator is that of having a "small" subset with dense orbit. Of course, the smallest (nonempty) subset would be a single point. If an operator has a point with dense orbit, it is called *hypercyclic*. An example where orbits of subsets play an important role is in the well known fact that an operator is hypercyclic if and only if every open set has dense orbit. An operator is called *chaotic* (Devaney's definition) if it is hypercyclic and has a dense set of periodic points (x is a periodic point for T if $T^n x = x$ for some $n \geq 1$).

Other types of "small" subsets that we shall consider include finite sets, countable discrete sets, finite dimensional subspaces, and bounded sets. An operator is called *n-supercyclic* if it has an n-dimensional subspace with dense orbit and a 1-supercyclic operator is generally called *supercyclic*. Also an operator is *countably hypercyclic* if it has a bounded, countable, separated set with dense orbit.

It is surprising that a linear operator can actually be hypercyclic; the first example was constructed by Rolewicz [**27**] in 1969. He showed that if B is the backward shift on $\ell^2(\mathbb{N})$, then λB is hypercyclic if and only if $|\lambda| > 1$. Since that

2000 *Mathematics Subject Classification.* Primary: 47B20, 47A16.
Key words and phrases. Hyponormal, Subnormal, Hypercyclic, Supercyclic, Orbits.
Supported in part by NSF grant DMS-9970376.

time, a "Hypercyclicity Criterion" has been developed independently by Kitai [**24**] and Gethner and Shapiro [**19**]. This criterion has been used to show that hypercyclic operators arise within the classes of composition operators [**6**], weighted shifts [**29**], adjoints of multiplication operators [**20**], and adjoints of subnormal and hyponormal operators [**18**].

An operator $S \in \mathcal{B}(\mathcal{H})$ is *subnormal* if it has a normal extension. An operator $T \in \mathcal{B}(\mathcal{H})$ is *hyponormal* if its self-commutator is positive, i.e. $[T^*, T] = T^*T - TT^* \geq 0$. This is equivalent to requiring that $\|T^*x\| \leq \|Tx\|$ for all $x \in \mathcal{H}$. An operator is called cosubnormal (resp. cohyponormal) if its adjoint is subnormal (resp. hyponormal). It is well known that every subnormal operator is hyponormal ([**9**]). A good example of a subnormal operator is the Bergman operator. For a bounded open set G in the complex plane, the Bergman space on G, denoted by $L^2_a(G)$, is the space of all analytic functions on G that are square integrable with respect to area measure on G. The operator S of multiplication by z on $L^2_a(G)$ is a subnormal operator, and hence also hyponormal. For further information on Bergman operators, subnormal operators and hyponormal operators, see Conway [**9**].

While a subnormal operator, or even a hyponormal operator, has very tame dynamics [**4**], their adjoints can be hypercyclic, and even chaotic. Furthermore, since pure cosubnormal operators are always cyclic (see Feldman [**13**]) it is natural to inquire about stronger forms of cyclicity for cosubnormal operators. The following theorem is a summary—and a special case—of the type of theorems discussed throughout the paper. Notice that the theme of the following theorem is that the spectral properties of S determine the dynamics of S^*.

In what follows a *region* is an open connected set.

THEOREM 1.1. *Suppose that $\{D_n\}$ is a bounded collection of bounded regions in $\mathbb{C}$. For each n, let $S_n = M_z$ on $L^2_a(D_n)$, and let $S = \bigoplus_n S_n$. Then the following hold:*

1. *S^* is hypercyclic if and only if $D_n \cap \{z : |z| = 1\} \neq \emptyset$ for all n.*
2. *S^* is supercyclic if and only if there exists a circle $\Gamma_\rho = \{z : |z| = \rho\}$, $\rho \geq 0$, such that either:*
   (a) For every n, $\operatorname{cl} D_n$ intersects Γ_ρ and $\operatorname{int} \Gamma_\rho$, or
   (b) For every n, $\operatorname{cl} D_n$ intersects Γ_ρ and $\operatorname{ext} \Gamma_\rho$.
3. *S^* is n-supercyclic if and only if S^* can be expressed as the direct sum of n operators $T_1, \ldots, T_n$ where each T_k is supercyclic and has the form $T_k = \bigoplus_{n \in C_k} S_n^*$ where $C_k \subseteq \mathbb{N}$.*
4. *S^* has a bounded set with dense orbit if and only if $D_n \cap \{z : |z| > 1\} \neq \emptyset$ for all n.*
5. *S^* is countably hypercyclic if and only if $D_n \cap \{z : |z| > 1\} \neq \emptyset$ for all n and there exists an n such that $D_n \cap \{z : |z| < 1\} \neq \emptyset$.*

It is natural to ask just how complicated can orbits of linear operators be? In section 6 of this paper, we present some examples of "interesting" orbits; for instance, orbits dense in Cantor sets, Julia sets and other geometrically complicated sets. In fact, we show that orbits for linear operators can be as complicated as the orbits of any continuous function on a compact metric space! We also define what a chaotic point is and show that several operators have a dense set of chaotic points. We also discuss examples of ϵ-dense orbits and weakly dense orbits that are not dense.

2. General Theory of Hypercyclic Operators

Here we outline the general theory of hypercyclic operators, summarize results that are known for all hypercyclic operators and state some open questions. We only give sketches of proofs, but provide a few important examples. We begin with some elementary necessary conditions for hypercyclicity, due to Carol Kitai [**24**].

PROPOSITION 2.1. *If $T \in \mathcal{B}(X)$ is hypercyclic, then the following hold:*

1. $\sup_n \|T^n\| = \infty$.
2. *Every component of $\sigma(T)$ intersects the unit circle.*
3. *If T is invertible, then T^{-1} is also hypercyclic.*
4. *T^* has no eigenvalues.*
5. *$ind(T-\lambda) \geq 0$ for all $\lambda \in \sigma(T) \setminus \sigma_e(T)$.*

The following two results are the main tools used to show that an operator is hypercyclic.

PROPOSITION 2.2. *If $T \in \mathcal{B}(X)$, then T is hypercyclic if and only if every open set in X has dense orbit under T. Equivalently, for any two open sets U, V in X, there exists an $n \geq 0$ such that $T^n(U) \cap V \neq \emptyset$. Furthermore, if T is hypercyclic, then T has a dense G_δ set of hypercyclic vectors.*

PROOF. Simply use the following fact: If $\{V_n\}_{n=1}^\infty$ is a countable basis for the topology of X, then the set of hypercyclic vectors for T is $\bigcap_{n=1}^\infty \bigcup_{k=0}^\infty (T^k)^{-1}(V_n)$, where $(T^k)^{-1}(V_n)$ means the inverse image of the set V_n under T^k. Thus, if every open set has dense orbit, then for each n, $\bigcup_{k=0}^\infty (T^k)^{-1}(V_n)$ is a dense open set, hence by the Baire Category Theorem $\bigcap_{n=1}^\infty \bigcup_{k=0}^\infty (T^k)^{-1}(V_n)$ is a dense G_δ. Hence, there exists a dense G_δ set of hypercyclic vectors. □

The following criterion is due independently to Kitai [**24**] and Gethner and Shapiro [**19**].

THEOREM 2.3 (The Hypercyclicity Criterion). *Suppose that $T \in \mathcal{B}(X)$. If there exists two dense subsets Y and Z in X and a sequence $n_k \to \infty$ such that:*

1. $T^{n_k}x \to 0$ *for every $x \in Y$, and*
2. *There exists functions $B_{n_k} : Z \to X$ such that for every $x \in Z$, $B_{n_k}x \to 0$ and $T^{n_k}B_{n_k}x \to x$,*

then T is hypercyclic.

PROOF. Suppose that U and V are two nonempty open sets in X. Let $x \in Y \cap U$ and $y \in Z \cap V$. Consider the vectors $z_k = x + B_{n_k}y$. Then $z_k \to x$ and $T^{n_k}z_k \to y$ as $k \to \infty$. So, for all large values of k, $z_k \in U$ and $T^{n_k}z_k \in V$. Thus $T^{n_k}(U) \cap V \neq \emptyset$, for all large k; thus by Proposition 2.2, T is hypercyclic □

EXAMPLE 2.4 (Rolewicz, 1969 [**27**]). If $T = \lambda B$ where B denotes the Backward shift on $\ell^2(\mathbb{N})$ and $|\lambda| > 1$, then T is hypercyclic.

PROOF. We will use the Hypercyclicity Criterion. Let Y be the set of all vectors in $\ell^2(\mathbb{N})$ with only finitely many non-zero coordinates. Let $Z = \ell^2(\mathbb{N})$ and define $B_n : Z \to \ell^2(\mathbb{N})$ by $B_n(x_0, x_1, \dots) = \frac{1}{\lambda^n}(0, 0, \dots, 0, x_0, x_1, \dots)$, where there are n zeros in front of x_0. Then the conditions of the Hypercyclicity Criterion are easily verified. So, T is hypercyclic. □

The following result is of fundamental importance in the study of hypercyclicity. Recall that a set is said to be *somewhere dense* if its closure has nonempty interior, otherwise it is called nowhere dense. In 2001, Bourdon & Feldman [**5**] proved that somewhere dense orbits must be dense. This Theorem gives a unified approach to several important results about hypercyclicity.

THEOREM 2.5 (Somewhere Dense Theorem). *Suppose $T \in \mathcal{B}(X)$ and $x \in X$. If $Orb(T,x)$ is somewhere dense in X, then $Orb(T,x)$ is dense in X.*

In 1995 Ansari [**1**] answered a long standing question by giving a truly original proof that powers of hypercyclic operators are still hypercyclic. While her proof can now be simplified using the Somewhere Dense Theorem, she introduced important tools and ideas into hypercyclicity that played a role in proving the Somewhere Dense Theorem.

COROLLARY 2.6 (Ansari). *If T is hypercyclic, then T^n is hypercyclic. Furthermore, T and T^n have the same set of hypercyclic vectors.*

PROOF. We shall prove the $n = 2$ case. Suppose that $\text{Orb}(T,x)$ is dense. Now notice that $\text{Orb}(T,x) = \{T^{2n}x\} \cup \{T^{2n}(Tx)\} = \text{Orb}(T^2,x) \cup \text{Orb}(T^2,Tx)$. Hence taking closures we have that $X = \text{clOrb}(T^2,x) \cup \text{clOrb}(T^2,Tx)$. Thus either $\text{Orb}(T^2,x)$ or $\text{Orb}(T^2,Tx)$ is somewhere dense. Hence, by Theorem 2.5, either $\text{Orb}(T^2,x)$ or $\text{Orb}(T^2,Tx)$ is dense. Thus T^2 is hypercyclic. If $\text{Orb}(T^2,x)$ is dense, then x is a hypercyclic vector for T^2. If $\text{Orb}(T^2,Tx)$ is dense, then since T has dense range, $T\text{Orb}(T^2,Tx)$ is also dense and $T\text{Orb}(T^2,Tx) \subseteq \text{Orb}(T^2,x)$. Thus $\text{Orb}(T^2,x)$ is dense, so in this case x is also a hypercyclic vector for T^2. □

An operator is said to be *finitely hypercyclic* if there exists a finite set with dense orbit. In 1992 Herrero conjectured that finitely hypercyclic operators were actually hypercyclic. His conjecture was proven independently by Costakis [**10**] and Peris [**25**]. It now follows easily from the Somewhere Dense Theorem.

COROLLARY 2.7 (Herrero's Conjecture). *If F is a finite set and $Orb(T,F)$ is dense, then T is hypercyclic. In fact, there exists an $x \in F$ such that $Orb(T,x)$ is dense.*

PROOF. If $F = \{x_1, \dots, x_n\}$. Then $\text{Orb}(T,F) = \bigcup_{k=1}^{n} \text{Orb}(T,x_k)$. Since $\text{Orb}(T,F)$ is dense, it follows that $\text{Orb}(T,x_k)$ is somewhere dense for some k. Hence $\text{Orb}(T,x_k)$ is actually dense for that value of k. In particular, T is hypercyclic. □

We now state some general questions about hypercyclic operators. The first two are motivated by Ansari's Theorem that T^n is hypercyclic whenever T is hypercyclic.

QUESTION 2.8 (Bounded Steps). If T is hypercyclic and $\{n_k\}$ is a sequence of natural numbers such that $0 < n_{k+1} - n_k \leq M$ for all k, then is there a vector x such that $\{T^{n_k}x\}$ is dense?

In [**11**] it is proven that if T satisifes the Hypercyclicity criterion, then the above question has an affirmative answer. However, Alfredo Peris has shown (private communication) that there exists a vector x and a sequence $\{n_k\}$ with bounded steps such that if T is twice the Backward shift, then $\{T^n x\}$ is dense but $\{T^{n_k}x\}$ is not dense.

The next question asks which functions preserve hypercyclicity? It is known that the functions $f(z) = z^n$ and $f(z) = \frac{1}{z}$ preserve hypercyclicity. The fact that $f(z) = \frac{1}{z}$ preserves hypercyclicity, means that "if T is hypercyclic, then T^{-1} is also hypercyclic". This is an elementary result, it follows from Proposition 2.2, and is more generally true for any homeomorphism on a complete separable metric space. However, the fact that $f(z) = z^n$ preserves hypercyclicity - namely "If T is hypercyclic, then T^n is also hypercyclic" is a nontrivial result due to Ansari (see Corollary 2.6), and was the first major result on hypercyclicity, and in general is not true for continuous maps on metric spaces.

QUESTION 2.9 (Functions of Hypercyclic Operators). If T is hypercyclic and f is a function analytic on an open set G satisfying $\sigma(T) \subseteq G$, $f(\mathbb{D} \cap G) \subseteq \mathbb{D}$ and $f((\mathbb{C} \setminus \mathbb{D}) \cap G) \subseteq (\mathbb{C} \setminus \mathbb{D})$, then is $f(T)$ also hypercyclic?

The answer to Question 2.9 is yes when T is cohyponormal, see Corollary 3.3, and when T has an ample supply of eigenvalues. See [**2**] and [**21**] for results along these lines. In [**7**], Bourdon & Shapiro considered functions of the Backward shift on the Bergman space - which is itself hypercyclic, unlike the backward shift on the Hardy space. They show that describing which functions preserve the hypercyclicity of the backward Bergman shift is non-trivial and involves some very nice function theory.

However, for general hypercyclic operators very little is known about Question 2.9. Only recently was it shown that $f(z) = e^{i\theta}z$, $\theta \in \mathbb{R}$, preserves hypercyclicity (see [**12**]). Some non-trivial examples to consider that are not understood in general include, $f(z) = \frac{z-a}{1-\bar{a}z}$, $a \in \mathbb{D}$, and $f(z) = e^{-\frac{1+z}{1-z}}$.

QUESTION 2.10 (Herrero's Question). If T is hypercyclic, then is $T \oplus T$ hypercyclic?

Bes and Peris [**3**] proved that Herrero's question is equivalent to the following question.

QUESTION 2.11 (Necessity of the Hypercyclicity Criterion). If T is hypercyclic, then does T satisfy the hypothesis of the Hypercyclicity Criterion?

It is easy to see that $T \oplus T$ is hypercyclic (and thus T satisfies the Hypercyclicity Criterion) if and only if for any four open sets U_1, V_1 and U_2, V_2, there exists an integer $n \geq 0$ such that $T^n(U_i) \cap V_i \neq \emptyset$ for $i = 1, 2$—this should be contrasted with Proposition 2.2. Every known hypercyclic operator—including those that are weighted shifts, composition operators, or cohyponormal operators—do satisfy the Hypercyclicity Criterion and thus satisfy that their inflation (direct sum with itself) is hypercyclic.

QUESTION 2.12 (Invariant Subset Question). If T is a bounded linear operator on a separable Hilbert space, then does T have a non-trivial invariant closed set? Equivalently, is there a non-zero vector $x \in \mathcal{H}$ such that the orbit of x is not dense?

It was shown in 1988 by Read [**26**] that there exists Banach space operators with no non-trivial invariant closed sets. In particular, such operators exist on ℓ^1.

It is natural to ask if the invariant subspace problem is equivalent to the invariant subset problem.

QUESTION 2.13 (Invariant subspaces -vs- Invariant subsets). If every bounded linear operator on a separable Hilbert space $\mathcal{H}$ has a non-trivial invariant closed set, then does every bounded linear operator on $\mathcal{H}$ also have a non-trivial invariant subspace? That is, are the Invariant subspace and Invariant subset problems equivalent?

If $A \in \mathcal{B}(X)$ and $B \in \mathcal{B}(Y)$ where X and Y are Banach spaces, then A and B are *topologically conjugate* if there exists a homeomorphism $h : X \to Y$ such that $h^{-1} \circ B \circ h = A$. This is the standard equivalence used in dynamical systems theory to say that two continuous functions on metric spaces have the "same" dynamics.

QUESTION 2.14. When are two bounded linear operators topologically conjugate? What operator theory properties are preserved under topologically conjugacy? Is every linear operator on ℓ^1 topologically conjugate to a linear operator on ℓ^2?

Notice that an affirmative answer to the last question would answer the invariant subset problem and thus the invariant subspace problem on Hilbert space.

3. Hypercyclic Cohyponormal Operators

If S is a bounded linear operator, then a part of S is an operator obtained by restricting S to an invariant subspace, say $S|\mathcal{M}$. A part of the spectrum of S is the spectrum of a part of S, that is, $\sigma(S|\mathcal{M})$, where $\mathcal{M}$ is an invariant subspace for S. In 2000, Feldman, Miller and Miller [**18**] characterized the hyponormal operators S with hypercyclic adjoints in terms of the parts of the spectrum of S.

THEOREM 3.1. *If S is a hyponormal operator, then S^* is hypercyclic if and only if for every hyperinvariant subspace $\mathcal{M}$ of S, $\sigma(S|\mathcal{M}) \cap \{z : |z| < 1\} \neq \emptyset$ and $\sigma(S|\mathcal{M}) \cap \{z : |z| > 1\} \neq \emptyset$.*

In [**18**] what was actually proven is that for any subdecomposable operator S, if every hyperinvariant part of the spectrum of S intersects both the interior and exterior of the unit circle, then S^* is hypercyclic. Furthermore, for hyponormal operators (which are subdecomposable), this condition becomes both necessary and sufficient.

EXAMPLE 3.2. Suppose that $\{D_n\}$ is a bounded collection of bounded regions in $\mathbb{C}$. For each n, let $S_n = M_z$ on $L^2_a(D_n)$, and let $S = \bigoplus_n S_n$. Then S is a pure subnormal operator and S^* is hypercyclic if and only if $D_n \cap \partial\mathbb{D} \neq \emptyset$ for all n.

COROLLARY 3.3.

1. *If $\{T_n\}$ is a bounded sequence of cohyponormal hypercyclic operators, then $\bigoplus_n T_n$ is also hypercyclic.*
2. *If T is a cohyponormal hypercyclic operator and f is a function analytic on an open set G satisfying $\sigma(T) \subseteq G$, $f(\mathbb{D} \cap G) \subseteq \mathbb{D}$ and $f((\mathbb{C} \backslash \mathbb{D}) \cap G) \subseteq (\mathbb{C} \backslash \mathbb{D})$, then $f(T)$ is also hypercyclic.*

4. n-Supercyclic Operators

An operator T is said to be n-supercyclic if there exists an n-dimensional subspace whose orbit is dense. A 1-supercyclic operator is simply called supercyclic. A simple example of a supercyclic operator is a multiple of a hypercyclic operator. Supercyclicity is a well-known property introduced in 1974 by Hilden and Wallen [**23**], where they proved that every backward unilateral weighted shift is supercyclic. Their result gives examples of supercyclic operators for which no multiple is hypercyclic. The idea of n-supercyclicity was introduced by Feldman [**14**] in 2001.

In 1999 Salas [**30**] developed a Supercyclicity Criterion, similar to the Hypercyclicity Criterion, that could be used to show an operator is supercyclic.

THEOREM 4.1 (The Supercyclicity Criterion). *Suppose that $T \in \mathcal{B}(X)$. If there is a sequence $n_k \to \infty$ and dense sets Y and Z and functions $B_{n_k} : Z \to X$ such that:*

1. *If $z \in Z$, then $T^{n_k} B_{n_k} z \to z$ as $k \to \infty$, and*
2. *If $y \in Y$ and $z \in Z$, then $\|T^{n_k} y\| \|B_{n_k} z\| \to 0$ as $k \to \infty$;*

then T is supercyclic.

In [**18**], Feldman, Miller, and Miller used a refined (but equivalent) version of the Supercyclicity Criterion to characterize which cohyponormal operators are supercyclic. As with hypercyclicity the characterization involves the parts of the spectrum.

THEOREM 4.2 (Supercyclic Cohyponormal Operators). *If S is a pure hyponormal operator, then S^* is supercyclic if and only if there exists a circle $\Gamma_\rho = \{z : |z| = \rho\}$, $\rho \geq 0$, such that either:*

(a) For every hyperinvariant subspace $\mathcal{M}$ of S, $\sigma(S|\mathcal{M})$ intersects Γ_ρ and $\operatorname{int} \Gamma_\rho$, or

(b) For every hyperinvariant subspace $\mathcal{M}$ of S, $\sigma(S|\mathcal{M})$ intersects Γ_ρ and $\operatorname{ext} \Gamma_\rho$.

If condition (a) holds, then we say that S^* is ρ-inner and if (b) holds then we say that S^* is ρ-outer. The radius ρ, which is not necessarily unique, is called a supercyclicity radius for S^*.

COROLLARY 4.3 (Direct Sums of Supercyclic Cohyponormal Operators).

If $\{S_n\}$ is a bounded sequence of pure hyponormal operators such that S_n^* is supercyclic for every n, then $\bigoplus_n S_n^*$ is supercyclic if and only if the operators S_n^* all have a common supercyclicity radius, ρ, and all have the same type (ρ-inner or ρ-outer).

EXAMPLE 4.4. Suppose that $\{D_n\}$ is a bounded collection of bounded regions in $\mathbb{C}$. For each n, let $S_n = M_z$ on $L^2_a(D_n)$, and let $S = \bigoplus_n S_n$. Then S^* is supercyclic if and only if there exists a circle $\Gamma_\rho = \{z : |z| = \rho\}$, $\rho \geq 0$, such that either:

(a) For every n, $\operatorname{cl} D_n$ intersects Γ_ρ and $\operatorname{int} \Gamma_\rho$, or

(b) For every n, $\operatorname{cl} D_n$ intersects Γ_ρ and $\operatorname{ext} \Gamma_\rho$.

Specific Examples

1. (An inner (resp. outer) example) If each D_n is a disk that is internally (resp. externally) tangent to the unit circle and their radii converge to zero, then S^* is supercyclic, inner (resp. outer) and has supercyclicity radius 1.
2. If $0 \in \operatorname{cl} D_n$ for each n and $diam(D_n) \to 0$, then S^* is supercyclic, outer, and has supercyclicity radius 0.

We will now see how we can use supercyclic operators to construct n-supercyclic operators. The following theorem appeared in Feldman [**14**].

THEOREM 4.5 (Creating n-Supercyclic Operators). *If $T_1, \dots, T_n$ are each operators that satisfy the hypothesis of the Supercyclicity Criterion with respect to the same sequence $\{n_k\}$, then $\bigoplus_{k=1}^n T_k$ is n-supercyclic.*

It was shown in [**18**] that supercyclic cohyponormal operators all satisfy the Supercyclicity Criterion with respect to the sequence $n_k = k$. Thus we have the following corollary from Feldman [**14**].

COROLLARY 4.6. *If $T_1, \dots, T_n$ are each supercyclic cohyponormal operators, then $\bigoplus_{k=1}^n T_k$ is n-supercyclic.*

EXAMPLE 4.7. (a) If $D_1, \dots, D_n$ are bounded regions and $S_k = M_z$ on $L_a^2(D_k)$, then $\bigoplus_{k=1}^n S_k^*$ is n-supercyclic.

(b) If $S = M_z$ on $L_a^2(D)$ where D is a bounded open set with a finite number of components, say n components, then S^* is n-supercyclic. (This is a special case of (a) where the regions are all disjoint.)

Now we need to know some necessary conditions for n-supercyclicity, so that we can say that certain operators in the previous example are not supercyclic. We present a necessary condition, due to Feldman [**14**] that is remarkably similar to the fact that every component of the spectrum of a hypercyclic operator must intersect the unit circle. However, since multiples of supercyclic operators are still supercyclic, one wouldn't expect their spectra to intersect the unit circle, but some circle centered at the origin.

THEOREM 4.8 (The Circle Theorem). *If $T \in \mathcal{B}(\mathcal{H})$ is n-supercyclic, then there are n circles $\Gamma_i = \{z : |z| = r_i\}$, $r_i \geq 0$, $i = 1, \dots, n$ such that for every invariant subspace $\mathcal{M}$ of T^*, we have $\sigma(T^*|\mathcal{M}) \cap (\bigcup_{i=1}^n \Gamma_i) \neq \emptyset$.*

In particular, every component of the spectrum of T intersects $\bigcup_{i=1}^n \Gamma_i$.

The above theorem says that if T is n-supercyclic, then there exists n circles such that every part of the spectrum of T^* intersects at least one of these n circles. With this Circle Theorem we can now construct operators that have n-dimensional subspaces with dense orbit, but no $(n-1)$-dimensional subspaces with dense orbit.

EXAMPLE 4.9. Suppose that $\{D_k\}$ is a bounded collection of bounded regions and $S_k = M_z$ on $L_a^2(D_k)$. Let $S = \bigoplus_k S_k$. If S^* is n-supercyclic, then there exists n circles $\Gamma_1, \dots, \Gamma_n$ centered at the origin such that for each k, $\mathrm{cl}D_k \cap (\bigcup_{i=1}^n \Gamma_i) \neq \emptyset$.

EXAMPLE 4.10. Suppose that D_k, $k = 1, \dots, n$ is the open disk with center at k and radius $1/4$. If $S_k = M_z$ on $L_a^2(D_k)$, then $\bigoplus_{k=1}^n S_k^*$ is n-supercyclic, but not $(n-1)$-supercyclic.

THEOREM 4.11. *Suppose that $\{D_k\}_{k=1}^\infty$ is a bounded collection of bounded regions and $S_k = M_z$ on $L_a^2(D_k)$. Let $S = \bigoplus_{k=1}^\infty S_k$. Then S^* is n-supercyclic if and only if S^* can be expressed as the direct sum of n operators $T_1, \dots, T_n$ where each T_k is supercyclic and has the form $T_k = \bigoplus_{j \in \mathcal{C}_k} S_j^*$ where $\mathcal{C}_k \subseteq \mathbb{N}$ and $\mathbb{N} = \bigcup_{k=1}^n \mathcal{C}_k$.*

The previous theorem says that for $S = \bigoplus_k S_k$, S^* is n-supercyclic if and only if we can group the operators $\{S_k\}$ into n different collections, namely $\{S_j : j \in \mathcal{C}_k\}$ for $k = 1, \dots, n$ in such a way that the direct sum of the operators in each collection is supercyclic. Then we may appeal to Corollary 4.6.

COROLLARY 4.12. *Suppose that $\{D_k\}$ is a bounded collection of bounded regions and $S_k = M_z$ on $L_a^2(D_k)$. Let $S = \bigoplus_k S_k$. If there exists n circles $\Gamma_1, \dots, \Gamma_n$*

centered at the origin such that for each k, $\mathrm{cl}\, D_k \cap \bigcup_{i=1}^n \Gamma_i \neq \emptyset$, then S^* is $2n$-supercyclic.

PROOF. For $1 \leq k \leq n$, let $\mathcal{C}_k = \{j : \mathrm{cl}D_j \cap \Gamma_k \neq \emptyset$ and $\mathrm{cl}D_j \cap \mathrm{int}\Gamma_k \neq \emptyset\}$. Also for $1 \leq k \leq n$ let $\mathcal{C}'_k = \{j : \mathrm{cl}D_j \cap \Gamma_k \neq \emptyset$ and $\mathrm{cl}D_j \cap \mathrm{ext}\Gamma_k \neq \emptyset\}$. Then by hypothesis, $\mathbb{N} = \bigcup_{k=1}^n \mathcal{C}_k \cup \bigcup_{k=1}^n \mathcal{C}'_k$. Let us also suppose that all the sets $\{\mathcal{C}_k\} \cup \{\mathcal{C}'_k\}$ are disjoint (note we may have to shrink some of the sets to achieve this, but we can disjointify them). Then for each k, $T_k := \bigoplus_{j \in \mathcal{C}_k} S_j^*$ and $T'_k := \bigoplus_{j \in \mathcal{C}'_k} S_j^*$ are both supercyclic (see Example 4.4). Thus by Corollary 4.6, $S^* = (\bigoplus_{k=1}^n T_k) \oplus (\bigoplus_{k=1}^n T'_k)$ is $2n$-supercyclic. □

EXAMPLE 4.13. Suppose that $\{D_k\}$ is a bounded collection of bounded regions and $S_k = M_z$ on $L_a^2(D_k)$. Let $S = \bigoplus_k S_k$. If there exists n circles $\Gamma_1, \ldots, \Gamma_n$ centered at the origin such that for each region D_k, there exists a j such that $\mathrm{cl}D_k \cap \Gamma_j \neq \emptyset$ and $\mathrm{cl}D_k \cap \mathrm{int}\Gamma_j \neq \emptyset$, then S^* is n-supercyclic.

QUESTION 4.14. If T is n-supercyclic, and T^* has no eigenvalues, then must T be cyclic?

QUESTION 4.15. If S is a pure cohyponormal operator and there exists n circles $\Gamma_1, \ldots, \Gamma_n$ centered at the origin such that for each hyperinvariant subspace $\mathcal{M}$ of S, $\sigma(S|\mathcal{M}) \cap \bigcup_{i=1}^n \Gamma_i \neq \emptyset$, then is S^* $2n$-supercyclic?

Feldman has answered the above question affirmatively for $n = 1$, see [**14**].

THEOREM 4.16. *Suppose that S is a pure hyponormal operator and there exists a circle $\Gamma = \{z : |z| = r\}$, $r > 0$, such that for every hyperinvariant subspace $\mathcal{M}$ of S, $\sigma(S|\mathcal{M}) \cap \Gamma \neq \emptyset$, then S^* is 2-supercyclic.*

QUESTION 4.17. Are there weighted shifts that are n-supercyclic and not supercyclic? If so, can we characterize the n-supercyclic weighted shifts?

Feldman [**14**] showed that a normal operator cannot be n-supercyclic, also an operator T such that T^* has an open set of eigenvalues cannot be n-supercyclic. It is therefore natural to expect a similar result for subnormal and hyponormal operators.

QUESTION 4.18. Can a subnormal or hyponormal operator be n-supercyclic?

Feldman [**14**] also defined ∞-supercyclic operators and showed that there are ∞-supercyclic operators that are not n-supercyclic for any $n < \infty$. There are subnormal operators that are ∞-supercyclic because every bilateral weighted shift is ∞-supercyclic. Thus, in fact, The bilateral weighted shift, a unitary operator, is ∞-supercyclic.

5. Countably Hypercyclic Operators

We proved in Corollary 2.7 that if $T \in \mathcal{B}(X)$ and there is a finite set with dense orbit, then T must actually be hypercyclic. That is, a "finitely hypercyclic" operator is actually hypercyclic. Several people, including V. Miller and W. Wogen, asked whether there is a meaningful definition of "countably hypercyclic operators", and if so, is a countably hypercyclic operator necessarily hypercyclic? Notice that *every* operator has a countable set with dense orbit, namely choose a countable dense set. Thus any reasonable definition of "countable hypercyclicity" should require an operator to have a countable set with dense orbit, but the countable sets should be small in some sense.

In [**16**] Feldman gave a reasonable definition of countable hypercyclicity and proved that there are countably hypercyclic operators that are not hypercyclic. Thus the class of countably hypercyclic operators is a non-trivial class of operators; and as we shall see contains some very natural cohyponormal operators. Furthermore, Feldman was able to give a spectral characterization of the cohyponormal operators that are countably hypercyclic.

Following [**16**], we say that an operator $T \in \mathcal{B}(X)$ is *countably hypercyclic* if there exists a bounded, countable, separated set C with dense orbit. Recall that a set $C \subseteq X$ is separated if there exists an $\epsilon > 0$ such that $\|x - y\| \geq \epsilon$ for all $x, y \in C$ with $x \neq y$. The fact that C is required to be bounded and separated excludes certain trivial operators from being countably hypercyclic (see [**16**]). It also forces certain orbits of T to be unbounded, and others to cluster at zero, hence T must have some non-trivial dynamics. In particular we have the following theorem (see [**16**]).

THEOREM 5.1. *If T is a hyponormal operator, then T is not countably hypercyclic.*

Feldman [**16**] also proved the following result, which shows that for certain operators countable hypercyclicity does imply hypercyclicity.

THEOREM 5.2. *If T is a backward unilateral weighted shift that is countably hypercyclic, then T is hypercyclic.*

Paul Bourdon has proven (private communication) that a countably hypercyclic operator with some regularity must be hypercyclic.

THEOREM 5.3 (Bourdon). *If T is countably hypercyclic and there is a dense set of vectors whose orbits converge, then T is hypercyclic.*

Some simple necessary conditions for countable hypercyclicity are in the following proposition, see [**16**]. In fact, the proposition considers the more general situation of an operator that has a bounded set with dense orbit. Notice that twice the identity is such an operator!

PROPOSITION 5.4. *If $T \in \mathcal{B}(X)$ and T has a bounded set with dense orbit, then the following hold:*

1. $\sup_n \|T^n\| = \infty$.
2. *Every component of $\sigma(T)$ must intersect $\{z : |z| \geq 1\}$.*
3. *T^* has no eigenvalues in $\{z : |z| \leq 1\}$.*
4. *T^* has no bounded orbits.*
5. *If T is countably hypercyclic, then $\sigma(T) \cap \partial\mathbb{D} \neq \emptyset$.*

We shall see that every component of $\sigma(T)$ need not intersect the unit circle. The amazing discovery made in [**16**] is that there is a "Criterion" very similar to the Hypercyclicity Crition—in fact, almost identical—that can be used to show that an operator is countably hypercyclic.

THEOREM 5.5 (The Countable Hypercyclicity Criterion). *Let $T \in \mathcal{B}(X)$. If there exists two subsets Y and Z in X, with dim span$Y = \infty$ and Z dense, and a sequence of integers $n_k \to \infty$ such that:*

1. *$T^{n_k}x \to 0$ for every $x \in Y$, and*
2. *There exists functions $B_{n_k} : Z \to X$ such that for every $x \in Z$, $B_{n_k}x \to 0$ and $T^{n_k}B_{n_k}x \to x$,*

then T is countably hypercyclic.

With this criterion we are able to characterize the cohyponormal operators that are countably hypercyclic. Notice that the only difference between the Countable Hypercyclicity Criterion and the Hypercyclicity Criterion is that here the set Y is not required to be dense, but only to span an infinite dimensional space. Another difference lies in the fact that while the proof of the Hypercyclicity Criterion relies on the Baire Category Theorem, the proof of the Countable Hypercyclicity Criterion does not.

THEOREM 5.6 (Countably Hypercyclic Cohyponormal Operators). *Let S be a pure hyponormal operator. Then S^* is countably hypercyclic if and only if $\sigma(S)\cap\{z : |z| < 1\} \neq \emptyset$ and for every hyperinvariant subspace $\mathcal{M}$ of S, $\sigma(S|\mathcal{M}) \cap \{z : |z| > 1\} \neq \emptyset$.*

Thus for S^* to be countably hypercyclic, one needs every part of the spectrum of S to intersect $\{z : |z| > 1\}$ and at least one part of the spectrum of S should intersect $\{z : |z| < 1\}$. Since every part of the spectrum of S need not intersect $\{z : |z| < 1\}$, we see that S^* need not be hypercyclic.

EXAMPLE 5.7. Suppose that $\{D_n\}$ is a bounded collection of bounded regions in $\mathbb{C}$. For each n, let $S_n = M_z$ on $L^2_a(D_n)$, and let $S = \bigoplus_n S_n$. Then S^* is countably hypercyclic if and only if $D_n \cap \{z : |z| > 1\} \neq \emptyset$ for all n and there exists an n such that $D_n \cap \{z : |z| < 1\} \neq \emptyset$.

In [**16**] Feldman also characterized the cohyponormal operators that have a bounded set with dense orbit. One can easily check that an operator has a bounded set with dense orbit if and only if the unit ball in the space has dense orbit.

THEOREM 5.8 (Bounded sets with Dense Orbit). *If S is a hyponormal operator, then S^* has a bounded set with dense orbit if and only if for every hyperinvariant subspace $\mathcal{M}$ of S, $\sigma(S|\mathcal{M}) \cap \{z : |z| > 1\} \neq \emptyset$.*

EXAMPLE 5.9. Suppose that $\{D_n\}$ is a bounded collection of bounded regions in $\mathbb{C}$. For each n, let $S_n = M_z$ on $L^2_a(D_n)$, and let $S = \bigoplus_n S_n$. Then S^* has a bounded set with dense orbit if and only if $D_n \cap \{z : |z| > 1\} \neq \emptyset$ for all n.

COROLLARY 5.10. *If T is an invertible cohyponormal operator and T and T^{-1} both have bounded sets with dense orbit, then T is hypercyclic.*

However, it was also shown in [**16**], that there is an invertible bilateral weighted shift such that T and T^{-1} have bounded sets with dense orbit, yet T is not hypercyclic. In fact Alfredo Peris has shown (private communication) that there is an invertible bilateral weighted shift T such that T and T^{-1} are both countably hypercyclic, and yet T is not hypercyclic.

The following corollary shows that if there is a set C that is bounded and bounded away from zero with dense orbit under a cohyponormal operator, then in fact there is a bounded, countable, separated set with dense orbit.

COROLLARY 5.11. *If S is a pure hyponormal operator, then S^* is countably hypercyclic if and only if S^* has a bounded set, which is also bounded away from zero, that has dense orbit.*

QUESTION 5.12. If $T \in \mathcal{B}(X)$ has a bounded set, that is also bounded away from zero, with dense orbit, then must T be countably hypercyclic?

EXAMPLE 5.13. Suppose that $T_1, T_2 \in \mathcal{B}(X)$ and T_1 satisfies the Countably Hypercyclic Criterion and the spectrum of T_2 is contained in $\{z : |z| > 1\}$, then $T_1 \oplus T_2$ also satisfies the Countably Hypercyclic Criterion.

It follows, for example, that if T_1 is twice the Backward shift on $\ell^2(\mathbb{N})$ and T_2 is any operator satisfying $\sigma(T_2) \subseteq \{z : |z| > 1\}$, then $T_1 \oplus T_2$ is countably hypercyclic. In particular, countably hypercyclic operators need not be cyclic, and their spectral properties in the exterior of the unit disk can be arbitrary.

6. Interesting Orbits

In this section we shall give some interesting and complicated examples of orbits of linear operators. Perhaps the most complicated orbits are those that are dense in the whole space. Here we will present some other complicated types of orbits. It follows from Theorem 2.5 that if an orbit is somewhere dense, then it must be everywhere dense. Thus, if we are looking for some interesting behavior of orbits that are not everywhere dense, they must be nowhere dense. For instance, one may ask if an orbit for a linear operator can be dense in a Cantor set (i.e. a compact, totally disconnected, perfect set), or some other geometrically complicated set. Can the closure of an orbit have any prescribed Hausdorff dimension? Surprisingly, the answers to these questions are YES!

In [**17**] Feldman proved that linear operators can have orbits that are just as complicated as the orbits of any continuous function. Recall that two continuous functions $f : X \to X$ and $g : K \to K$ on metric spaces are called topologically conjugate if there exists a homeomorphism $h : X \to K$ such that $h \circ f \circ h^{-1} = g$. Topological conjugacy is the standard equivalence used to say that two functions have the "same" dynamics.

THEOREM 6.1. *If $f : X \to X$ is a continuous function on a compact metric space X and T is twice the backward shift of infinite multiplicity, then there is an invariant compact set K for T such that $T|K$ is topologically conjugate to f.*

It was also proven in [**17**] that under certain conditions, if f is Lipschitz, then the conjugating homoemorphism may also be chosen to be bi-Lipschitz. It follows then, for example, that one may show that linear operators can have orbits whose closures are bi-Lipschitz homeomorphic to Cantor sets or Julia sets and that there are orbits whose closures have any prescribed Hausdorff dimension, because such orbits exist for Lipschitz continuous functions.

6.1. Chaotic Points. While an orbit may be called predictable if it is convergent, periodic, or converging to a periodic orbit, or exhibits some other moderate behavior. What is a chaotic orbit? Recall that a continuous function $f : K \to K$ on a complete separable metric space K is said to be *transitive* if for any two open sets U, V in K, there exists an $n \geq 0$ such that $f^n(U) \cap V \neq \emptyset$, where f^n denotes the nth iterate of f. Also, f is *chaotic* if f is transitive and has a dense set of periodic points.

We shall say that $x \in K$ is a *chaotic point* for f if x is not a periodic point and $f|\text{clOrb}(f, x)$ is chaotic. We shall say that x is a *weakly chaotic point*, if x is not a periodic point and $f|\text{clOrb}(f, x)$ is transitive.

The following proposition is an easy exercise that we leave to the reader.

PROPOSITION 6.2. *If $f : K \to K$ is a continuous function on a complete separable metric space K, then the following are equivalent for a point $x \in K$,*

1. *x is a weakly chaotic point for f.*
2. *$clOrb(f, x)$ is a perfect set, that is, it has no isolated points.*
3. *For each $k \geq 0$, $cl\{f^n(x) : n \geq k\} = clOrb(f, x)$.*

It follows that if $T \in \mathcal{B}(X)$, then an $x \in X$ is a chaotic point for T if $\text{clOrb}(T, x)$ is a perfect set and T has a dense set of periodic points in $\text{clOrb}(T, x)$.

Can an operator have a chaotic point that does not have a dense orbit? Yes!

EXAMPLE 6.3. There exists an operator T on a separable Hilbert space $\mathcal{H}$ such that:

1. T has a dense set of periodic points,
2. T has a dense set of chaotic points,
3. T is cyclic,

but, T is not hypercyclic.

PROOF. The operator is the adjoint of a composition operator, $T = C_\phi^*$ acting on the Sobolev space $W^{1,2}(0,1)$, where ϕ is the "Tent map", $\phi(x) = 1 - |x - (1/2)|$. One easily checks that ϕ induces a bounded composition operator on $W^{1,2}(0,1)$ and using the well-known fact that $\phi : [0,1] \to [0,1]$ is chaotic gives the result. □

A doubly infinite orbit for a linear operator T is a sequence $\{x_k\}_{k=-\infty}^{\infty}$ such that $Tx_k = x_{k+1}$ for all $k \in \mathbb{Z}$.

THEOREM 6.4. *If $T \in \mathcal{B}(X)$ and T has a doubly infinite orbit $\{x_k\}_{k=-\infty}^{\infty}$ such that $\sum_{k=-\infty}^{\infty} \|x_k\| < \infty$, then T has a chaotic point whose orbit has compact closure. Furthermore, if $\{x_k\}_{k=-\infty}^{\infty}$ has dense linear span in X, then T has a dense set of chaotic points whose orbits have compact closure.*

REMARK. It follows easily from the Hypercyclicity Criterion (Theorem 2.3) that if T has a doubly infinite orbit $\{x_k\}_{k=-\infty}^{\infty}$ with dense linear span such that $\|x_k\| \to 0$ as $|k| \to \infty$, then T is hypercyclic. Thus the above condition implies hypercyclicity.

EXAMPLE 6.5. (a) If T is a backward unilateral weighted shift with weight sequence $\{w_0, w_1, \dots\}$ and if $\sum_{n=1}^{\infty} \frac{1}{w_0 w_1 \cdots w_n} < \infty$, then $x = e_0$ has a doubly infinite orbit that is norm summable and has dense linear span. Thus T has a dense set of chaotic points with orbits having compact closure. Recall that T is hypercyclic if and only if $\sup_n (w_0 w_1 \cdots w_n) = \infty$.

(b) If $T = \lambda B$ where B is the Backward shift on $\ell^2(\mathbb{N})$ and $|\lambda| > 1$, then the above condition is satisfied. Thus T has a dense set of chaotic points having orbits with compact closure.

(c) If $S = M_z$ on $L_a^2(D)$ where $\{z : |z| \leq 1\} \subseteq D$, then S^* has a dense set of chaotic points. (consider the doubly infinite orbit generated by the reproducing kernel for the point 0.)

QUESTION 6.6. Can we characterize the cohyponormal operators (or weighted shifts, or other classes of operators) that have a dense set of chaotic points?

QUESTION 6.7. Does a chaotic linear operator have a chaotic point whose orbit is not dense?

6.2. ϵ-Dense Oribts. If $T \in \mathcal{B}(X)$ and $\epsilon > 0$, we shall say that $\text{Orb}(T, x)$ is ϵ-dense, if for every $y \in X$ there exists an n such that $\|y - T^n x\| \leq \epsilon$. In [**15**] Feldman proved the following results.

THEOREM 6.8. *If $T \in \mathcal{B}(X)$ and T has an ϵ-dense orbit for some $\epsilon > 0$, then T is hypercyclic.*

EXAMPLE 6.9. If T is twice the backward shift on $\ell^2(\mathbb{N})$, then for each $\epsilon > 0$, T has an orbit that is ϵ-dense, but is not dense. In fact, for each $\epsilon > 0$, there are vectors $x, y \in \ell^2(\mathbb{N})$ such that $\text{Orb}(T, x)$ is dense, and $\|T^n x - T^n y\| \leq \epsilon$ for all $n \geq 0$, yet $\text{Orb}(T, y)$ is not dense.

6.3. Weakly Dense Orbits. It is natural to ask if $T \in \mathcal{B}(X)$ and $x \in X$ is such that $\text{Orb}(T, x)$ is dense in X in the weak topology, then must $\text{Orb}(T, x)$ be dense in X in the norm topology? Or must T itself be hypercylic, that is have some other vector with dense orbit? Feldman raised this question in [**15**].

EXAMPLE 6.10 (Feldman). If T is twice the backward shift on $\ell^2(\mathbb{N})$, then T has an orbit that is weakly dense but not norm dense.

Very recently Kit Chan and Rebecca Sanders [**8**] gave an example of a bilateral weighted shift T that is weakly hypercyclic - meaning T has a vector whose orbit is dense in the weak topology, but T is not hypercyclic. In fact the shift $Te_n = w_n e_{n-1}$ with weights $w_n = 1$ when $n \leq 0$ and $w_n = 2$ when $n > 0$ is weakly hypercyclic, but not hypercyclic. In fact, notice that T^{-1} has norm one! Hence T^{-1} is not even weakly hypercyclic (the Baire Category Theorem cannot be used with the weak topology so inverses of weakly hypercyclic operators need not be weakly hypercyclic). Notice that T is in fact cohyponormal! Thus there are cohyponormal operators that are weakly hypercyclic, but not hypercyclic.

QUESTION 6.11. Can we characterize the cohyponormal operators that are weakly hypercyclic?

Notice that having a weakly dense orbit is different than having an orbit that is weakly sequentially dense. In fact it follows easily from Theorem 3.1 that if a cohyponormal operator has a weakly sequentially dense orbit, then it must be hypercyclic.

7. Final Remarks

Another nice survey paper on the dynamics of linear operators, emphasizing the dynamics of composition operators, is given by J.H. Shapiro, *Notes on Dynamics of Linear Operators.* These are unpublished lecture notes available at http://www.math.msu.edu/~shapiro. Also, the author's papers are all available at http://www.wlu.edu/~feldmanN.

References

[1] S. I. Ansari, 'Hypercyclic and cyclic vectors', *J. Funct. Anal.* 128 (1995), 374–383.

[2] T. Bermdez and V.G. Miller, *On operators T such that $f(T)$ is hypercyclic*, Integral Equations Operator Theory **37** (2000), no. 3, 332–340.

[3] J. Bes and A. Peris, *Hereditarily hypercyclic operators* J. Funct. Anal. **167** (1999), no. 1, 94–112.

[4] P.S. Bourdon, *Orbits of Hyponormal Operators*, Mich. Math. J. **44** (1997), 345–353.

[5] P.S. Bourdon and N. S. Feldman, *Somewhere dense orbits are everywhere dense*, To appear in Indiana Univ. Math. J.

[6] P. S. Bourdon and J. H. Shapiro, *Cyclic Phenomena for Composition Operators*, Memoirs of the AMS, 125, AMS, Providence, RI, 1997.

[7] P. S. Bourdon and J. H. Shapiro, *Hypercyclic operators that commute with the Bergman backward shift*, Trans. Amer. Math. Soc. **352** (2000), no. 11, 5293–5316.

[8] K. C. Chan and R. Sanders, *A weakly hypercyclic operator that is not norm hypercyclic*, preprint.

[9] J.B. Conway, *The Theory of Subnormal Operators*, Amer. Math. Soc., Providence, RI, 1991.

[10] G. Costakis, *On a conjecture of D. Herrero concerning hypercyclic operators*, C. R. Acad. Sci. Paris **330** (2000), 179–182.

[11] F. León-Saavedra, *Notes about the Hypercyclicity Criterion*, preprint.

[12] F. León-Saavedra and V. Müller, *Rotations of hypercyclic and supercyclic operators*, preprint.

[13] N.S. Feldman, *Pure Subnormal Operators have Cyclic Adjoints*, J. Funct. Anal. **162** (1999), no. 2, 379 - 399.

[14] N.S. Feldman *N-Supercyclic Operators*, Studia Math. **151** (2002), 141–159.

[15] N.S. Feldman *Perturbations of Hypercyclic Vectors*, to appear in J. Math. Anal. Appl.

[16] N.S. Feldman *Countably Hypercyclic Operators*, to appear in J. Operator Theory.

[17] N.S. Feldman *Linear Chaos*, preprint.

[18] N.S. Feldman, T.L. Miller, and V.G. Miller, *Hypercyclic and Supercyclic Cohyponormal Operators*, Acta Sci. Math. (Szeged) **68** (2002), 303–328.

[19] R. M. Gethner and J. H. Shapiro, *Universal vectors for operators on spaces of holomorphic functions*, Proc. AMS, **100** (1987), 281–288.

[20] G. Godefroy and J. H. Shapiro, *Operators with dense invariant cyclic manifolds*, J. Func. Anal., **98** (1991), 229–269.

[21] G. Herzog and C. Schmoeger, *On operators T such that $f(T)$ is hypercyclic*, Studia Math. **108** (1994), no. 3, 209–216.

[22] D. A. Herrero, 'Hypercyclic operators and chaos', *J. Operator Theory* 28 (1992), 93–103.

[23] H. M. Hilden and L. J. Wallen, *Some cyclic and non–cyclic vectors of certain operators*, Ind. Univ. Math. J., **23** (1974), 557–565.

[24] C. Kitai, *Invariant closed sets for linear operators*, Dissertation, Univ. of Toronto, 1982.

[25] A. Peris, *Multihypercyclic operators are hypercyclic*, Math. Z. **236** (2001), 779–786.

[26] C. Read, *The invariant subspace problem for a class of Banach spaces, 2: Hypercyclic operators*, Israel J. Math. **63** (1988), 1–40.

[27] S. Rolewicz, *On orbits of elements*, Studia Math., **32** (1969), 17–22.

[28] J.H. Shapiro, *Notes on Dynamics of Linear Operators*, unpublished lecture notes, available at http://www.math.msu.edu/~shapiro

[29] H. N. Salas, *Hypercyclic weighted shifts*, Trans AMS, **347** (1995), 993–1004.

[30] H. N. Salas, *Supercyclicity and weighted shifts*, Studia Math. **135** (1999), no. 1, 55–74.

DEPT. OF MATHEMATICS, WASHINGTON AND LEE UNIVERSITY, LEXINGTON VA 24450
E-mail address: `feldmanN@wlu.edu`

Contemporary Mathematics
Volume **321**, 2003

Hilbert-Schmidt Composition Operators on Dirichlet Spaces

Eva A. Gallardo-Gutiérrez and María J. González

ABSTRACT. In this note we show that analytic self-maps φ of the unit disk inducing Hilbert-Schmidt composition operators C_φ on the weighted Dirichlet space $\mathcal{D}_\alpha$ satisfy that the set $E_\varphi = \{e^{i\theta} \in \partial\mathbb{D} : \ |\varphi(e^{i\theta})| = 1\}$ has zero α-capacity.

1. Introduction

Let $\mathbb{D}$ denote the open unit disc of the complex plane. If φ is an analytic function on $\mathbb{D}$ with $\varphi(\mathbb{D}) \subset \mathbb{D}$, the composition operator induced by φ is defined by

$$C_\varphi f = f \circ \varphi$$

for all holomorphic function f on $\mathbb{D}$. Composition operators have been extensively studied in last four decades. One of the main points is that they connect two important areas of mathematics: operator theory and analytic function theory. Actually, the aim of the works on composition operators has been to discover geometric properties of φ that allow to determine functional analytic properties of C_φ and viceversa. In particular, if $\partial\mathbb{D}$ denotes the boundary of $\mathbb{D}$, the analysis of the set

$$E_\varphi = \{e^{i\theta} \in \partial\mathbb{D} : \ |\varphi(e^{i\theta})| = 1\}$$

has played an important role in the results on composition operators. Here $\varphi(e^{i\theta}) = \lim_{r\to 1^-} \varphi(re^{i\theta})$, where the limit exists a. e. by Fatou's radial limit theorem (see [**4**], for instance).

The first result related to the set E_φ traces back to 1969 with H. J. Schwartz's work [**8**]. He proved that a necessary condition for C_φ to be compact on the Hardy space $\mathcal{H}^2$ is that E_φ has Lebesgue measure zero. After that, other authors have studied problems relating operators properties of C_φ to the size of E_φ. For a comprehensive treatment of such problems, we refer to C. C. Cowen and B. D. MacCluer's book [**3**].

2000 *Mathematics Subject Classification.* Primary 30C85,47B38.

Key words and phrases. Composition operator, Dirichlet space, Weighted Dirichlet space, Logarithmic capacity, α-capacity .

Partially supported by Plan Nacional I+D grant no. BFM2000-0360, DGICYT grant PB98-0872, CIRIT grant 1998SRG00052, Junta de Andalucía FQM-260 and by the Plan Propio de la Universidad de Cádiz.

The focus of this work is the size of E_φ whenever φ induces a Hilbert-Schmidt composition operator on weighted Dirichlet spaces $\mathcal{D}_\alpha$. These spaces along with the concept of α-capacity of a set are introduced in Section 2. After discussing that there is a close connection between the functions in $\mathcal{D}_\alpha$ and α-capacity, we show the main result

Theorem: *If C_φ is Hilbert-Schmidt on $\mathcal{D}_\alpha$, then the α-capacity of E_φ is zero.*

The above result generalizes a previous one proved by the authors for the Dirichlet space (see [**5**]).

2. Preliminaries

2.1. Weighted Dirichlet spaces. Let A denotes the normalized Lebesgue area measure on the unit disc $\mathbb{D}$, i. e., $dA(z) = \frac{1}{\pi}\, dx\, dy$. If $-1 < \alpha < 1$ the weighted Dirichlet space $\mathcal{D}_\alpha$ consists of analytic functions f on $\mathbb{D}$ such that the norm

$$\|f\|_\alpha^2 = |f(0)|^2 + \int_{\mathbb{D}} |f'(z)|^2 (1-|z|^2)^\alpha dA(z)$$

is finite. Note that $\mathcal{D}_0$ is the Dirichlet space $\mathcal{D}$, and allowing $\alpha = 1$ the norm obtained is equivalent to the usual one in the Hardy space $\mathcal{H}^2$, so $\mathcal{D}_1$ is the Hardy space.

Observe that as α decreases the functions in $\mathcal{D}_\alpha$ are of slower growth when approaching to the boundary of the unit disc. Moreover, if $\alpha_1 < \alpha_2$, then $\mathcal{D}_{\alpha_1}$ is strictly contained in $\mathcal{D}_{\alpha_2}$.

2.2. Capacity of a set. Let μ be a measure of bounded support S_μ. The α-potential of the measure μ is defined by

$$u_\alpha^\mu(x) = \begin{cases} \displaystyle\int \log \frac{1}{|x-y|}\, d\mu(y), & \alpha = 0; \\[2ex] \displaystyle\int \frac{d\mu(y)}{|x-y|^\alpha}, & 0 < \alpha < 1. \end{cases}$$

If $I_\alpha(\mu)$ denotes the energy integral of μ,

$$I_\alpha(\mu) = \int u_\alpha^\mu d\,\mu(x),$$

then the α-capacity of a bounded Borel set E is defined by

$$C_\alpha(E) = \{\inf I(\mu)\}^{-1}$$

where the infimum is taken over all positive measures μ with total mass 1 and support S_μ contained in E. For $\alpha = 0$, the α-capacity is also called logarithmic capacity.

Note that there exist Borel sets E of Lebesgue measure zero and α-capacity positive for any $0 \le \alpha < 1$. In addition, observe that if $C_{\alpha_1}(E) = 0$ then $C_{\alpha_2}(E) = 0$ for any $\alpha_2 > \alpha_1$. For more about capacities see [**2**] and [**6**].

On the other hand, if $0 \le \alpha < 1$ there is a close relation between functions in $\mathcal{D}_\alpha$ and α-capacity. In 1939, Beurling [**1**] proved that if f belongs to the Dirichlet space $\mathcal{D}_0$, $\alpha = 0$, then the radial limits $f(e^{i\theta}) = \lim_{r\to 1^-} f(re^{i\theta})$ exist except on a set of logarithmic capacity zero (see also [**2**]). In 1965 H. Wallin [**9**] generalized this result for $0 < \alpha < 1$, that is, if $f \in \mathcal{D}_\alpha$ then $f(e^{i\theta})$ is defined except on a set of α-capacity zero.

2.3. Hilbert-Schmidt composition operators. Recall that a linear operator T on a Hilbert space $\mathcal{H}$ is said to be Hilbert-Schmidt if there exists an orthonormal basis $\{e_n\}$ in $\mathcal{H}$ such that the series

$$\sum_{n=1}^{\infty} \|Te_n\|^2 \tag{1}$$

is convergent. It is easy to see that, in such a case, condition (1) does not depend on the particular choice of the orthonormal basis $\{e_n\}$. Clearly, every Hilbert-Schmidt operator is bounded.

First, let us give the following characterization of Hilbert-Schmidt composition operators on weighted Dirichlet spaces

PROPOSITION 2.1. *Let φ be an analytic self-map of $\mathbb{D}$. Then C_φ is Hilbert-Schmidt on $\mathcal{D}_\alpha$ if and only if*

$$\int_{\mathbb{D}} \frac{|\varphi'(z)|^2}{(1-|\varphi(z)|^2)^{\alpha+2}} (1-|z|^2)^\alpha dA(z) < \infty.$$

The proof of proposition 2.1 just follows by considering that $\{z^n/(n+1)^{(1-\alpha)/2}\}$ is an orthonormal basis in $\mathcal{D}_\alpha$ and Stirling's formula.

3. Proof of the Main Result

Let us suppose that φ induces a Hilbert-Schmidt composition operator on $\mathcal{D}_\alpha$. If $-1 < \alpha < 0$ then the supremum norm $\|\varphi\|_\infty < 1$ (see [**3**], Theorem 4.5) and therefore, E_φ is the empty set and there is nothing to discuss.

Suppose that $0 \le \alpha < 1$. Let us consider u the positive harmonic function

$$u(z) = \mathrm{Re} \frac{1+\varphi(z)}{1-\varphi(z)}.$$

If $z = x + iy$ and $\nabla u(z)$ denotes the the gradient vector $\nabla u(z) = (\partial u/\partial x, \partial u/\partial y)$, a computation using proposition 2.1 yields that the integral

$$\int_{\mathbb{D}} \frac{|\nabla u(z)|^2}{u(z)^{\alpha+2}} \frac{(1-|z|^2)^\alpha}{|1-\varphi(z)|^{2\alpha}} dA(z)$$

is finite. Therefore, the integral

$$\int_{\mathbb{D}} \frac{|\nabla u(z)|^2}{u(z)^{\alpha+2}} (1-|z|^2)^\alpha dA(z)$$

is also finite, or equivalently,

$$\int_{\mathbb{D}} |\nabla(u^{-\alpha/2})|^2 (1-|z|^2)^\alpha dA < \infty. \tag{2}$$

From here we construct an analytic function f belonging to $\mathcal{D}_\alpha$ to which is applied Wallin's result.

Let $P_r(\theta)$ denote the Poisson kernel and consider v the harmonic function

$$v(z) = v(re^{i\theta}) = \frac{1}{2\pi} \int_{-\pi}^{\pi} u^{-\alpha/2}(t)\, P_r(\theta - t)\, dt.$$

Since harmonic functions minimize the energy integral, we deduce that the integral

$$\int_{\mathbb{D}} |\nabla v(z)|^2 (1-|z|^2)^\alpha dA(z)$$

is finite. Let f be an analytic function on the unit disc such that $f(0)=0$ and $\mathrm{Re}\, f = v$. From what we have just showed, it follows that f belongs to the weighted Dirichlet space $\mathcal{D}_\alpha$. Moreover, the radial limits satisfy

$$\lim_{r\to 1^{-1}} \mathrm{Re}\, f(re^{i\theta}) = u^{-\alpha/2}(e^{i\theta}).$$

Now Wallin's Theorem states that the radial limits $f(e^{i\theta}) = \lim_{r\to 1^-} f(re^{i\theta})$ exist except on a set of α-capacity zero, and so do $\mathrm{Re}\, f(e^{i\theta})$. Therefore, the set

$$\{e^{i\theta} \in \partial\mathbb{D} :\ u^{-\alpha/2}(e^{i\theta}) = \infty\}$$

has α-capacity zero. The desired result follows since this set coincides with E_φ.

REMARK 3.1. *Main theorem is only a necessary condition. In fact, the function* $\varphi(z) = (1+z)/2$ *induces a non Hilbert-Schmidt composition operator on any weighted Dirichlet space* $\mathcal{D}_\alpha$ *and the set* E_φ *only has one element.*

References

[1] A. Beurling, *Ensembles exceptionnels*, Acta Math. **72** (1939) 1-13.

[2] L. Carleson, *Selected problems on exceptional sets*, D. Van Nostrand Company, Inc. 1966.

[3] C. C. Cowen and B. D. MacCluer, *Composition Operators on Spaces of Analytic Functions*, CRC Press, 1995.

[4] P. L. Duren, *Theory of* $\mathcal{H}^p$ *spaces*, Academic Press, New York, 1970.

[5] E. A. Gallardo-Gutiérrez and M. J. González, *Exceptional sets and Hilbert-Schmidt composition operators*, J. Functional Analysis (to appear).

[6] P. Koosis, *The logarithmic integral II*, Cambridge University Press, 1992.

[7] B. D. MacCluer and J. H. Shapiro, *Angular derivatives and compact composition operators on the Hardy and Bergman spaces*, Canadian J. Math. **38** (1986) 878–906.

[8] H. J. Schwartz, *Composition Operators on* $\mathcal{H}^p$, Thesis, University of Toledo, 1969.

[9] H. Wallin, *On the existence of boundary values of a class of Beppo Levi functions*, Trans. Amer. Math. Soc. **120** (1965) 510–525.

DEPARTAMENTO DE MATEMÁTICAS, UNIVERSIDAD DE CÁDIZ, APARTADO 40, 11510 PUERTO REAL (CÁDIZ), SPAIN.

E-mail address: eva.gallardo@uca.es

DEPARTAMENTO DE MATEMÁTICAS, UNIVERSIDAD DE CÁDIZ, APARTADO 40, 11510 PUERTO REAL (CÁDIZ), SPAIN.

E-mail address: majose.gonzalez@uca.es

Contemporary Mathematics
Volume **321**, 2003

A remark on sectorial operators with an H^∞–calculus

N. J. Kalton

Abstract. We construct examples of sectorial operators admitting an H^∞-calculus so that the angle of sectoriality and the angle of the H^∞–calculus are different.

1. Introduction

Let X be a complex Banach space. A *sectorial* operator A on X is a one-one closed operator with dense domain and range such that the resolvent operator $R(\lambda, A) = (\lambda - A)^{-1}$ is defined and bounded outside a sector $|\arg \lambda| \leq \phi$ and further satisfies an estimate

$$\|\lambda R(\lambda, A)\| \leq C \qquad |\arg \lambda| \geq \phi. \tag{1.1}$$

The infimum of all ϕ so that (1.1) holds is denoted by $\omega(A)$. Let us recall that a closed operator is of *type* ω if its resolvent is well-defined outside a sector and satisfies an estimate of type (1.1). Such an operator becomes sectorial if in addition we have that $\lim_{t\to 0-} tR(t, A)x = 0$ and $\lim_{t\to -\infty} tR(t, A)x = x$ for every $x \in X$.

If A is sectorial it is possible to define a functional calculus for certain functions bounded and analytic on a sector $\Sigma_\phi = \{\lambda : |\arg \lambda| < \phi\}$ where $\phi > \omega(A)$. We refer to [**2**] for details. We say that A admits an $H^\infty(\Sigma_\phi)$–calculus if $f(A)$ is a bounded operator for every $f \in H^\infty(\Sigma_\phi)$. If A admits an H^∞–calculus for some $0 < \phi < \pi$ we define $\omega_H(A)$ to be the infimum of all such ϕ.

A basic result due to McIntosh [**4**] is that if X is a Hilbert space and A admits an H^∞–calculus for some angle then $\omega_H(A) = \omega(A)$. In [**2**] the question is asked whether this is true in an arbitrary Banach

2000 *Mathematics Subject Classification.* Primary: 47A60.
The author was supported by NSF grant DMS-9870027.

space. There is an example in [**2**] (Example 5.5) which almost answers this question negatively; it is, however, not a sectorial operator because it fails to have dense range.

The object of this note is to give a natural counterexample to the question in [**2**]. For $0 < \theta < \pi$ we construct a sectorial operator with $\omega(A) = 0$ and $\omega_H(A) = \theta$. By an interpolation argument we show that we can choose X to be uniformly convex.

Unfortunately we do not know an example on an explicit space such as L_p when $1 < p < \infty$ and $p \neq 2$.

2. The examples

We start with the space $L_2(\mathbb{R})$. It will be convenient to norm this space by

$$\|f\|_0^2 = 2\pi \int_{-\infty}^{\infty} |f(x)|^2 dx = \int_{-\infty}^{\infty} |\hat{f}(\xi)|^2 d\xi$$

where $\hat{f}$ is the Fourier transform. The identity follows by Plancherel's theorem. On this space we define a sectorial operator A by

$$Af(x) = e^x f(x)$$

with domain $\mathcal{D}(A) = \{f : \ e^x f(x) \in L_2\}$. It is clear that A is sectorial with $\omega(A) = 0$. In fact A has an H^∞-calculus and $\omega_H(A) = 0$.

For $\theta > 0$ we define a Euclidean norm on L_2 by

$$\|f\|_\theta^2 = \int_{-\infty}^{\infty} e^{-2\theta|\xi|} |\hat{f}(\xi)|^2 d\xi.$$

Let $\mathcal{H}_\theta$ be the completion of L_2 with respect to this (weaker) norm.

If $f \in L_2$ then $A^{is} f(x) = e^{isx} f(x)$ so that if $g = A^{is} f$ then $\hat{g}(\xi) = \hat{f}(\xi - s)$. Hence

$$\|A^{is} f\|_\theta \leq e^{\theta|s|} \|f\|_\theta. \tag{2.1}$$

We now wish to show that A induces a sectorial operator on $\mathcal{H}_\theta$. We do this by simply checking that the appropriate resolvent operators extend boundedly and satisfy the necessary bounds. To be precise if for some $0 < \phi < \pi$ we show that the operators $\lambda R(\lambda, A) = \lambda(\lambda - A)^{-1}$ extend to be bounded on $\mathcal{H}_\theta$ and if further

$$\sup_{|\arg \lambda| \geq \phi} \|\lambda R(\lambda, A)\|_{\mathcal{H}_\theta} < \infty$$

then the operator A defined with domain $(I + A)^{-1}(\mathcal{H}_\theta)$ and range $A(I + A)^{-1}(\mathcal{H}_\theta)$ is necessarily sectorial with $\omega(A) \leq \phi$. The facts that the domain and range are dense and A is one-one follow quickly once one notes

$$\lim_{t\to 0+} tA(I+tA)^{-1}f = \lim_{t\to\infty}(I+tA)^{-1}f = 0 \qquad f\in\mathcal{H}_\theta.$$

This follows easily from the bounds on the resolvent and the fact it is true on the dense subset L_2 of $\mathcal{H}_\theta$. This principle will be used several times for different completions of L_2.

The appropriate bounds on the resolvent follow from (2.1) by a method similar to that of the proof of the Dore-Venni Theorem [**3**]. The argument only requires that a Hilbert space has the (UMD)-property, but in the next Lemma we give a slightly more general result.

LEMMA 2.1. *There exists a constant C so that if $m\in L^1\cap L^\infty(\mathbb{R})$ satisfies*

$$\int_{-\infty}^{\infty} |\hat{m}(\xi)|e^{\theta|\xi|}d\xi < \infty$$

then for $f\in L_2(\mathbb{R})$

$$\|mf\|_\theta \le C\left(\|m\|_\infty + \int_{|\xi|\ge 1}|\hat{m}(\xi)|\,e^{\theta|\xi|}\,d\xi\right)\|f\|_\theta. \tag{2.2}$$

PROOF. Let us split $m=m_0+m_1$ where $\hat{m}_0=\hat{m}\chi_{[-1,1]}$. Note that

$$\|m_1\|_\infty \le C_0\int_{|\xi|\ge 1}|\hat{m}(\xi)|\,e^{\theta|\xi|}\,d\xi$$

where $C_0=C_0(\theta)$. Hence

$$\|m_0\|_\infty \le C_1\left(\|m\|_\infty + \int_{|\xi|\ge 1}|\hat{m}(\xi)|\,e^{\theta|\xi|}\,d\xi\right). \tag{2.3}$$

Now if $f\in L_2$

$$m_0 f = \frac{1}{2\pi}\int_{-\infty}^{\infty}\hat{m}_0(s)A^{is}f\,ds$$

as a Bochner integral in $L_2(\mathbb{R})$. Hence

$$A^{-it}(m_0 f) = \frac{1}{2\pi}\int_{-\infty}^{\infty}\hat{m}_0(s)A^{i(s-t)}f\,ds.$$

Let $F,G\in L_2(\mathbb{R};L_2)$ be defined by $F(t)=A^{-it}(m_0f)\chi_{[-1,1]}$ and $G(t)=A^{-it}f\chi_{[-2,2]}$. Then by the above $F=(2\pi)^{-1}\hat{m}_0 * G$ and so $\|F\|\le$

$\|m_0\|_\infty\|G\|$. Hence

$$\|m_0 f\|_\theta \le e^\theta \left(\int_{-1}^{1} \|A^{it}(m_0 f)\|_\theta^2 dt\right)^{\frac{1}{2}}$$
$$\le e^\theta \|m_0\|_\infty \left(\int_{-2}^{2} \|A^{it} f\|_\theta^2 dt\right)^{\frac{1}{2}}$$
$$\le 2e^{3\theta}\|m_0\|_\infty \|f\|_\theta,$$

where the last estimate follows from the fact that $\|A^{it}f\|_\theta \le e^{2\theta}\|f\|_\theta$ for $|t| \le 2$. In view of (2.3) we have

$$\|m_0 f\|_\theta \le C_2 \left(\|m\|_\infty + \int_{|\xi|\ge 1} |\hat{m}(\xi)| e^{\theta|\xi|} d\xi\right) \|f\|_\theta, \tag{2.4}$$

where $C_2 = C_2(\theta)$. On the other hand

$$m_1 f = \int_{|s|\ge 1} \hat{m}(s) A^{is} f ds$$

so that

$$\|m_1 f\|_\theta \le \left(\int_{|s|\ge 1} |\hat{m}(s)| e^{\theta|s|} ds\right) \|f\|_\theta.$$

Combining with (2.4) gives the Lemma. □

LEMMA 2.2. *A naturally extends to a sectorial operator on $\mathcal{H}_\theta$, which has an H^∞–calculus with $\omega(A) = \omega_H(A) = \theta$.*

PROOF. Let us start from the formula

$$\int_{-\infty}^{\infty} \frac{e^{zx}}{1+e^x} dx = \frac{\pi}{\sin \pi z} \qquad 0 < \Re z < 1.$$

Hence if $t \in \mathbb{R}$

$$\int_{-\infty}^{\infty} \frac{e^{zx}}{e^t+e^x} dx = \frac{\pi e^{t(z-1)}}{\sin \pi z} \qquad 0 < \Re z < 1.$$

By analytic continuation we obtain that for any w in the complex plane with the negative real axis removed,

$$\int_{-\infty}^{\infty} \frac{e^{zx}}{w+e^x} dx = \frac{\pi w^{z-1}}{\sin \pi z} \qquad 0 < \Re z < 1.$$

Now let $m_{a,w}(x) = w^{1-a}e^{ax}(w+e^x)^{-1}$ where $0 < a < 1$. Then

$$\hat{m}_{a,w}(\xi) = \frac{\pi w^{-i\xi}}{\sin \pi(a - i\xi)}.$$

It follows from Lemma 2.1 that we have a uniform estimate

$$\|m_{a,w} f\|_\theta \le C\|f\|_\theta \qquad f \in L_2$$

as long as $|\arg w| + \theta < \pi - \delta$ for some $\delta > 0$. Here C depends on δ but not on a. We can let $a \to 0$ and deduce a similar estimate for $m_{0,w} = w(w + e^x)^{-1}$. Hence if we consider the resolvent operators

$$R(\lambda, A) = (\lambda - A)^{-1}$$

we obtain a uniform bound

$$\|\lambda R(\lambda, A)f\|_\theta \leq C\|f\|_\theta \qquad f \in L_2$$

as long as $|\arg \lambda| \geq \theta + \delta$ for some $\delta > 0$. This implies that we can naturally extend A to be sectorial on $\mathcal{H}_\theta$ and $\omega(A) \leq \theta$. Now, by the result of McIntosh [**4**] since $\mathcal{H}_\theta$ is a Hilbert space (2.1) implies that A admits an H^∞–calculus and $\omega_H(A) = \omega(A)$. □

We now introduce a new space by defining the norm

$$\|f\|_{X_\theta} := \sup_{a \in \mathbb{R}} \|f\chi_{(-\infty,a]}\|_\theta.$$

The space X_θ is defined as the completion of L_2 with respect to this norm. Note for $f \in L_2$ we have

$$\|f\|_\theta \leq \|f\|_{X_\theta} \leq \|f\|_0.$$

For $a \neq 0$ and $m, n \in \mathbb{N}$ we define the operator $E(m, n, a)$ on L_2 by

$$E(m, n, a)f(x) = \frac{1}{\sqrt{n}} \sum_{k=1}^{n} f(x - mka).$$

LEMMA 2.3. *For any $f \in L_2$ we have*

$$\lim_{n\to\infty} \limsup_{m\to\infty} \|E(m, n, a)f\|_{X_\theta} = \|f\|_\theta.$$

PROOF. Suppose $\alpha = 0$ or $\alpha = \theta$. First we note that

$$\|E(m, n, a)f\|_\alpha \leq \sqrt{n}\|f\|_\alpha \qquad f \in L_2(\mathbb{R}). \tag{2.5}$$

Now fix n, a and let $g_m = E(m, n, a)f$. Then

$$\hat{g}_m(\xi) = \frac{1}{\sqrt{n}} \hat{f}(\xi) \sum_{k=1}^{n} e^{-imka}.$$

Hence

$$\|g_m\|_\alpha^2 = \frac{1}{n} \int_{-\infty}^{\infty} (\sum_{j=1}^{n} \sum_{k=1}^{n} e^{i(j-k)ma}) |\hat{f}(\xi)|^2 e^{-2\alpha|\xi|} d\xi.$$

By the Riemann-Lebesgue Lemma we obtain

$$\lim_{m\to\infty} \|E(m, n, a)f\|_\alpha = \|f\|_\alpha. \tag{2.6}$$

Now suppose $f \in L_2$ and $\epsilon > 0$. Fix M so large that

$$\|f - f\chi_{[-M,M]}\|_0 < \epsilon.$$

Let $f_0 = f\chi_{[-M,M]}$ and $f_1 = f - f_0$.

If $m > 2M|a|^{-1}$ then any $t \in \mathbb{R}$ falls in the support of at most one of the functions $f_0(x - mka)$ for $k = 1, 2, \ldots$. Hence for any n we have for some $0 \le k \le n$,

$$\|\chi_{(-\infty,t)}E(m,n,a)f_0\|_\theta \le (k/n)^{\frac{1}{2}}\|E(m,k,a)f_0\|_\theta + n^{-\frac{1}{2}}\|f_0\|_0.$$

(If $k = 0$ we interpret $E(m,0,a)f$ as 0). This shows that

$$\|E(m,n,a)f_0\|_{X_\theta} \le \max_{0\le k\le n}(k/n)^{\frac{1}{2}}\|E(m,k,a)f_0\|_\theta + n^{-\frac{1}{2}}\|f_0\|_0.$$

In view of (2.5) and (2.6) this gives

$$\limsup_{m\to\infty}\|E(m,n,a)f_0\|_{X_\theta} \le \|f_0\|_\theta + n^{-\frac{1}{2}}\|f_0\|_0. \tag{2.7}$$

On the other hand

$$\limsup_{m\to\infty}\|E(m,n,a)f_1\|_0 = \|f_1\|_0 < \epsilon$$

so that combining with (2.7) gives

$$\limsup_{m\to\infty}\|E(m,n,a)f\|_{X_\theta} \le \|f_0\|_\theta + n^{-\frac{1}{2}}\|f_0\|_0 + \epsilon. \tag{2.8}$$

Since $\|f_0\|_\theta \le \|f\|_\theta + \|f_1\|_\theta < \|f\|_\theta + \epsilon$ since obtain

$$\limsup_{n\to\infty}\limsup_{m\to\infty}\|E(m,n,a)f\|_{X_\theta} \le \|f\|_\theta + 2\epsilon.$$

Since the $X_\theta-$norm is larger than the norm $\|\cdot\|_\theta$ this equation and (2.6) imply the conclusion. □

THEOREM 2.4. *The operator A on X_θ is sectorial and admits an $H^\infty-$calculus but $\omega(A) = 0$ and $\omega_H(A) = \theta$.*

PROOF. For $\lambda \in \mathbb{C} \setminus \mathbb{R}_+$ let $m_\lambda(x) = \lambda(\lambda - e^x)^{-1}$. Then for $f \in L_2$

$$m_\lambda f = \int_{-\infty}^{\infty} \frac{\lambda e^x}{(\lambda - e^x)^2} f\chi_{(-\infty,x)}dx$$

as a Bochner integral in L_2. Hence if $\psi = \arg\lambda$,

$$\|m_\lambda f\|_{X_\theta} \le \|f\|_{X_\theta}\int_{-\infty}^{\infty}\frac{|\lambda|e^x}{|\lambda - e^x|^2}dx.$$

Now

$$\int_{-\infty}^{\infty} \frac{|\lambda|e^x}{|\lambda - e^x|^2}dx = \int_0^{\infty} \frac{|\lambda|}{|t-\lambda|^2}dt$$
$$= \int_0^{\infty} |t - e^{i\psi}|^{-2}dt.$$

Now reasoning as before we can deduce that $\lim_{t\to 0+} tA(I+tA)^{-1}f = \lim_{t\to\infty}(I+tA)^{-1}f = 0$ for $f \in X_\theta$ by a density argument since it is true for $f \in L_2$. It follows that A is sectorial on X_θ and $\omega(A) = 0$.

For any $m \in L^\infty(\mathbb{R})$ note that if $f \to mf$ extends to a bounded operator on $\mathcal{H}_\theta$ then that for $f \in L_2$ we have

$$\|mf\|_{X_\theta} = \sup_{-\infty<t<\infty} \|mf\chi_{(-\infty,t)}\|_\theta \leq C\|f\|_{X_\theta}.$$

It follows that on X_θ, A has an $H^\infty-$calculus and $\omega_H(A) \leq \theta$. It remains to show that $\omega_H(A) \geq \theta$. To do this, we show that for any s, $\|A^{is}\|_{X_\theta} = \|A^{is}\|_{\mathcal{H}_\theta} = e^{\theta|s|}$.

Suppose $s > 0$ and let $a = 2\pi/s$. For any $f \in L_2$ and $m, n \in \mathbb{N}$, we note that

$$\|A^{is}E(m,n,a)f\|_{X_\theta} \leq \|A^{is}\|_{X_\theta}\|E(m,n,a)f\|_{X_\theta}.$$

Note by choice of a we have $A^{is}E(m,n,a)f = E(m,n,a)A^{is}f$ and so by Lemma 2.3

$$\|A^{is}f\|_\theta \leq \|A^{is}\|_{X_\theta}\|f\|_\theta$$

and this shows $\|A^{is}\|_{X_\theta} = \|A^{is}\|_{\mathcal{H}_\theta}$ and completes the proof. □

We conclude by showing that we can use this example to produce a similar example modelled on a super-reflexive space. For this we will use complex interpolation. For $0 < \tau < 1$ we consider the complex interpolation space $X_{\theta,\tau} = [L_2, X_\theta]_\tau$. Let us recall the definition of this space. Let $\mathcal{S}$ denote the strip $0 < \Re z < 1$. We consider the vector space $\mathcal{F}$ of all bounded continuous functions $F : \overline{\mathcal{S}} \to X_\theta$ which are analytic on $\mathcal{S}$ and such that $F(it) \in L_2$ for $-\infty < t < \infty$ and $t \to F(it)$ is continuous into L_2. We norm $\mathcal{F}$ by

$$\|F\|_{\mathcal{F}} = \max(\sup_{-\infty<t<\infty} \|F(it)\|_0, \sup_{-\infty<t<\infty} \|F(1+it)\|_{X_\theta}).$$

We then define $X_{\theta,\tau}$ to be the space of all $f \in X_\theta$ such that for some $F \in \mathcal{F}$ we have $F(\tau) = f$ under the norm

$$\|f\|_{X_{\theta,\tau}} = \inf\{\|F\|_{\mathcal{F}} : F(\tau) = f\}.$$

We will need the following fact about complex interpolation. Let $P : \partial\mathcal{S} \times \mathcal{S} \to \mathbb{R}$ be the Poisson kernel for the strip. Given τ let $h_0(t) = P(it, \tau)$ and $h_1(t) = P(1+it, \tau)$. Thus the measure on $\partial\mathcal{S}$ given by

$h_0(t)dt$ on the line $i\mathbb{R}$ and $h_1(t)dt$ on the line $1+i\mathbb{R}$ is harmonic measure for the point τ. Then h_0, h_1 are non-negative continuous functions in $L_1(\mathbb{R})$ with

$$\int_{-\infty}^{\infty} (h_0(t) + h_1(t))\, dt = 1$$

such that if $F \in \mathcal{F}$ then

$$\|F(\tau)\|_{X_{\theta,\tau}} \leq \int_{-\infty}^{\infty} (\|F(it)\|_0 h_0(t) + \|F(1+it)\|_{X_\theta} h_1(t))\, dt. \tag{2.9}$$

This estimate goes back to Calderón [**1**].

It follows immediately by interpolation that A induces a sectorial operator on $X_{\theta,\tau}$ with $\omega(A) = 0$. Indeed (1.1) for any $\phi > 0$ is immediate and we can deduce that

$$\lim_{t\to 0+} tA(I+tA)^{-1}f = \lim_{t\to\infty} (I+tA)^{-1}f = 0$$

for every $f \in X_{\theta,\tau}$ either by a standard density argument or by the remarks above. Indeed if F is admissible then, for example, we have

$$\lim_{t\to 0+} \|tA(I+tA)^{-1}F(is)\|_0 = \lim_{t\to 0+} \|tA(I+tA)^{-1}F(1+is)\|_{X_\theta} = 0$$

if $-\infty < s < \infty$ and so by (2.9) and the Dominated Convergence Theorem

$$\lim_{t\to 0} \|tA(1+tA)^{-1}F(\tau)\|_{X_{\theta,\tau}} = 0.$$

Interpolation also quickly yields that the sectorial operator A on $X_{\theta,\tau}$ has an $H^\infty-$calculus with $\omega_H(A) \leq \theta$. The spaces $X_{\theta,\tau}$ for $0 < \tau < 1$ are uniformly convex (and thus super-reflexive). We now show that on these spaces we also have $\omega_H(A) > \omega(A)$.

PROPOSITION 2.5. *On $X_{\theta,\tau}$ we have $\omega_H(A) = \tau\theta$.*

PROOF. By interpolation we have $\|A^{is}\|_{X_{\theta,\tau}} \leq e^{\tau\theta|s|}$. We shall show that $\|A^{is}\|_{X_{\theta,\tau}} = e^{\tau\theta|s|}$ and by Theorem 5.4 of [**2**] this will imply that $\omega_H(A) = \tau\theta$.

We need the fact that if $f \in L_2$ then $\|f\|_{X_{\theta,\tau}} \geq \|f\|_{\tau\theta}$. This follows immediately from the fact that $\|f\|_{X_\theta} \geq \|f\|_\theta$ and $[L_2, \mathcal{H}_\theta]_\tau = \mathcal{H}_{\tau\theta}$.

Fix $s \neq 0$ and let $a = 2\pi/s$.

Suppose $f \in L_2$ is such that

$$\int |\hat{f}(\xi)|^2 e^{2\theta|\xi|} d\xi < \infty.$$

Define $F : \mathcal{S} \to L_2$ be defined by

$$\widehat{F(z)}(\xi) = e^{\theta(z-\tau)} \hat{f}(\xi).$$

Thus F extends continuously to ∂S and $\|F(it)\|_0 = \|F(1+it)\|_\theta = \|f\|_{\tau\theta}$. Then using (2.9)

$$\|E(m,n,a)f\|_{X_{\theta,\tau}} \le$$
$$\int_{-\infty}^{\infty} \|E(m,n,a)F(it)\|_0 h_0(t) + \|E(m,n,a)F(1+it)\|_{X_\theta} h_1(t)\, dt.$$

If we fix n and let $m \to \infty$ we can use the Dominated Convergence Theorem and (2.7) to deduce that

$$\limsup_{m\to\infty} \|E(m,n,a)f\|_{X_{\theta,\tau}} \le$$
$$\int_{-\infty}^{\infty} (\|F(it)\|_0 h_0(t) + \|F(1+it)\|_\theta h_1(t) + n^{-\frac{1}{2}}\|F(1+it)\|_0 h_1(t))dt.$$

By (2.6) we have

$$\lim_{m\to\infty} \|E(m,n,a)f\|_{\tau\theta} = \|f\|_{\tau\theta}.$$

Hence, letting $n \to \infty$ we obtain

$$\lim_{n\to\infty} \limsup_{m\to\infty} \|E(m,n,a)f\|_{X_{\theta,\tau}} = \|f\|_{\tau\theta}.$$

Thus the analogue of Lemma 2.3 holds at least for f in a dense subset of L_2 (which is itself dense in $X_{\theta,\tau}$.) Hence arguing as before in Theorem 2.4 we obtain that

$$\|A^{is}\|_{X_{\theta,\tau}} \ge \|A^{is}\|_{\mathcal{H}_{\tau\theta}} = e^{\tau\theta|s|}.$$

This completes the proof. □

References

[1] A.P. Calderón, Intermediate spaces and interpolation, the complex method, Studia Math. 24 (1964) 113-190.

[2] M. Cowling, I. Doust, A. McIntosh and A. Yagi, Banach space operators with a bounded H^∞-calculus, J. Austral. Math. Soc. 60 (1996) 51-89.

[3] G. Dore and A. Venni, On the closedness of the sum of two closed operators, Math. Z. 196 (1987), 189-201.

[4] A. McIntosh, Operators which have an H_∞ functional calculus, Miniconference on operator theory and partial differential equations (North Ryde, 1986), 210–231, Proc. Centre Math. Anal. Austral. Nat. Univ., 14, Austral. Nat. Univ., Canberra, 1986.

Department of Mathematics, University of Missouri, Columbia, MO 65211

E-mail address: nigel@math.missouri.edu

Contemporary Mathematics
Volume **321**, 2003

Borel Injective Tensor Product and Convolution of Vector Measures and Their Weak Convergence

Jun Kawabe

ABSTRACT. The purpose of the paper is to show the existence and uniqueness of the Borel injective tensor product of two Banach space-valued vector measures and the validity of a Fubini-type theorem. Thanks to this result, the convolution of vector measures on a topological semigroup is defined as the measure induced by their Borel injective tensor product and the semigroup operation. It is also proved the joint weak continuity of Borel injective tensor products or convolutions of vector measures.

1. Introduction

In the usual set-theoretical theory, the notion of a product measure is of central importance. In the case of Borel measures on topological spaces S and T, we are immediately confronted with the problem that the Borel σ-field $\mathcal{B}(S \times T)$ is in general larger than the product σ-field $\mathcal{B}(S) \times \mathcal{B}(T)$; see [**29**, Corollary to Lemma I.1.1]. Our main purpose of the paper is to show the existence and uniqueness of the Borel injective tensor product of two Banach space-valued vector measures and the validity of a Fubini-type theorem.

Let X and Y be Banach spaces. Denote by $X\hat{\otimes}_\varepsilon Y$ the injective tensor product of X and Y. We say that vector measures $\mu : \mathcal{B}(S) \to X$ and $\nu : \mathcal{B}(T) \to Y$ have their *Borel injective tensor product* if there is a unique vector measure $\lambda : \mathcal{B}(S \times T) \to X\hat{\otimes}_\varepsilon Y$ such that $\lambda(A \times B) = \mu(A) \otimes \nu(B)$ for all $A \in \mathcal{B}(S)$ and $B \in \mathcal{B}(T)$. In Section 2 we shall formulate some notation and results which are needed in the sequel, and it is shown in Section 3 that if two Borel vector measures are τ-smooth, one of them with some separability condition with respect to the other, then they have their τ-smooth Borel injective tensor product; see Theorem 3.5. This type of problems has been already investigated in [**11, 12, 16**] in more general settings. However, it should be remarked that the vector measures in those papers are always supposed to be of bounded variation and be defined on compact or locally compact spaces.

2000 *Mathematics Subject Classification.* Primary 28B05, 28C15; Secondary 46G10.

Key words and phrases. Borel injective tensor product of vector measures, convolution of vector measures, weak convergence, Bartle bilinear integration.

The author is supported by Grant-in-Aid for General Scientific Research No. 13640162, Ministry of Education, Culture, Sports, Science and Technology, Japan.

In Section 4 we shall show the joint weak continuity of Borel injective tensor products of Banach lattice-valued positive vector measures. This result has already been proved in [**21**, Theorem 5.4] with the additional assumption that the spaces S and T, on which the vector measures are defined, satisfy $\mathcal{B}(S \times T) = \mathcal{B}(S) \times \mathcal{B}(T)$.

In Section 5, the convolution of Banach algebra-valued vector measures on a topological semigroup is defined as the measure induced by their Borel injective tensor product and the semigroup operation. The paper concludes with showing the joint weak continuity of convolutions of positive vector measures with values in special Banach algebras.

2. Preliminaries

All the topological spaces and uniform spaces in this paper are Hausdorff and the scalar fields of Banach spaces are taken to be the field $\mathbb{R}$ of all real numbers. Denote by $\mathbb{N}$ the set of all natural numbers.

Let T be a topological space and $\mathcal{B}(T)$ the σ-field of all Borel subsets of T, that is, the σ-field generated by the open subsets of T. Let X be a Banach space and X^* a topological dual of X. Denote by $\boldsymbol{B}_X$ and $\boldsymbol{B}_{X^*}$ the closed unit balls of X and X^*, respectively. A countably additive set function $\nu : \mathcal{B}(T) \to X$ is called a *vector measure*. The *semivariation* of ν is the set function $\|\nu\|(B) := \sup\{|x^*\nu|(B) : x^* \in \boldsymbol{B}_{X^*}\}$, where $B \in \mathcal{B}(T)$ and $|x^*\nu|(\cdot)$ is the total variation of the real measure $x^*\nu$. Then $\|\nu\|(T) < \infty$ [**2**, Lemma 2.2].

We define several notions of regularity for vector measures. A vector measure $\nu : \mathcal{B}(T) \to X$ is said to be *Radon* if for each $\eta > 0$ and $B \in \mathcal{B}(T)$, there exists a compact subset K of T with $K \subset B$ such that $\|\nu\|(B - K) < \eta$, and it is said to be *tight* if the condition is satisfied for $B = T$. We say that ν is *τ-smooth* if for every increasing net $\{G_\alpha\}_{\alpha\in\Gamma}$ of open subsets of T with $G = \bigcup_{\alpha\in\Gamma} G_\alpha$, we have $\lim_{\alpha\in\Gamma} \|\nu\|(G - G_\alpha) = 0$. It follows that ν is Radon (respectively, tight, τ-smooth) if and only if for each $x^* \in X^*$, $x^*\nu$ is Radon (respectively, tight, τ-smooth). In fact, this is a consequence of the Rybakov theorem [**9**, Theorem IX.2.2], which ensures that there exists $x_0^* \in X^*$ for which $x_0^*\nu$ and ν are mutually absolutely continuous. Thus, all the regularity properties which are valid for positive finite measures remain true of vector measures.

In this paper, we shall employ two types of integration which will be briefly explained below. Let $\nu : \mathcal{B}(T) \to X$ be a vector measure. Denote by χ_B the indicator function of a set B. A $\|\nu\|$-null set is a set $B \in \mathcal{B}(T)$ for which $\|\nu\|(B) = 0$; the term $\|\nu\|$-*almost everywhere* (for short, $\|\nu\|$-a.e.) refers to the complement of a $\|\nu\|$-null set. Given a $\mathcal{B}(T)$-simple function of the form $f = \sum_{i=1}^k a_i\chi_{B_i}$ with $a_1, \ldots, a_k \in \mathbb{R}, B_1, \ldots, B_k$ are pairwise disjoint sets in $\mathcal{B}(T)$, and $k \in \mathbb{N}$, define its integral $\int_B f d\nu$ over a set $B \in \mathcal{B}(T)$ by $\int_B f d\nu := \sum_{i=1}^k a_i\nu(B_i \cap B)$. We say that a $\mathcal{B}(T)$-measurable function $f : T \to \mathbb{R}$ is *ν-integrable* if there is a sequence $\{f_n\}$ of $\mathcal{B}(T)$-simple functions converging $\|\nu\|$-almost everywhere to f such that the sequence $\left\{\int_B f_n d\nu\right\}$ converges in the norm of X for each $B \in \mathcal{B}(T)$. This limit $\int_B f d\nu$ does not depend on the choice of such $\mathcal{B}(T)$-simple functions $f_n, n \in \mathbb{N}$. By the Orlicz-Pettis theorem [**9**, Corollary I.4.4], the indefinite integral $B \mapsto \int_B f d\nu$ is countably additive. For further properties of this integral see [**2, 14**].

Next we define the Bartle bilinear integration in our setting. Let X, Y and Z be Banach spaces with a continuous bilinear mapping $b : X \times Y \to Z$. Let $\nu : \mathcal{B}(T) \to Y$ be a vector measure. The *semivariation of ν with respect to b* is the set function $\|\nu\|_b(B) := \sup \| \sum_{i=1}^k b(x_i, \nu(B_i))\|$, where B is a subset of T and the supremum is taken for all finite families $\{B_i\}_{i=1}^k$ of pairwise disjoint sets in $\mathcal{B}(T)$ contained in B and all families $\{x_i\}_{i=1}^k$ of elements in $\boldsymbol{B}_X$. In what follows, we shall assume that ν has the (*)-*property with respect to* b, that is, there is a positive finite measure q on $\mathcal{B}(T)$ such that $q(B) \to 0$ if and only if $\|\nu\|_b(B) \to 0$ [**3**, Definition 2]. Then $\|\nu\|_b(T) < \infty$ [**27**, Lemma 1]. It will be verified below that this assumption is automatically satisfied whenever we use this integration in this paper.

A $\|\nu\|_b$-null set is a subset B of T for which $\|\nu\|_b(B) = 0$; the term $\|\nu\|_b$-*almost everywhere* (for short, $\|\nu\|_b$-a.e.) refers to the complement of a $\|\nu\|_b$-null set. Given an X-valued $\mathcal{B}(T)$-simple function $\varphi = \sum_{i=1}^k x_i \chi_{B_i}$ with $x_1, \dots, x_k \in X$, $B_1, \dots, B_k$ are pairwise disjoint sets in $\mathcal{B}(T)$, and $k \in \mathbb{N}$, define its integral $\int_B b(\varphi, d\nu)$ over a set $B \in \mathcal{B}(T)$ by $\int_B b(\varphi, d\nu) := \sum_{i=1}^k b(x_i, \nu(B_i \cap B))$. We say that a vector function $\varphi : T \to X$ is (ν, b)-*measurable* if there is a sequence $\{\varphi_n\}$ of X-valued $\mathcal{B}(T)$-simple functions converging $\|\nu\|_b$-almost everywhere to φ. The function φ is said to be (ν, b)-*integrable* if there is a sequence $\{\varphi_n\}$ of X-valued $\mathcal{B}(T)$-simple functions converging $\|\nu\|_b$-almost everywhere to φ such that the sequence $\{\int_B b(\varphi_n, d\nu)\}$ converges in the norm of Z for each $B \in \mathcal{B}(T)$. This limit $\int_B b(\varphi, d\nu)$ does not depend on the choice of such X-valued $\mathcal{B}(T)$-simple functions $\varphi_n, n \in \mathbb{N}$, and the indefinite integral $B \mapsto \int_B b(\varphi, d\nu)$ is a Z-valued vector measure on $\mathcal{B}(T)$.

Since we always assume that ν has the (*)-property with respect to b, we make free use of the (*)-theorems of [**3**, Section 4]. Among others we shall frequently use the following theorem and refer to it as the *Bounded convergence theorem*; see [**3**, Theorem 7] and the statement after [**3**, (*) Theorem 10]. For further properties of this integral see [**3, 15**].

PROPOSITION 2.1 (Bounded Convergence Theorem). *Let $\{\varphi_n\}$ be a sequence of (ν, b)-integrable X-valued functions on T which converges $\|\nu\|_b$-almost everywhere to an X-valued function φ on T . If there is a $M > 0$ such that $\|\varphi_n(t)\| \le M$ for $\|\nu\|_b$-almost everywhere $t \in T$, then φ is (ν, b)-integrable over any set $B \in \mathcal{B}(T)$ and $\int_B b(\varphi, d\nu) = \lim_{n\to\infty} \int_B b(\varphi_n, d\nu)$.*

3. Borel injective tensor product of vector measures

Throughout this section, we assume that S and T are topological spaces and X and Y are Banach spaces. Denote by U_{X^*} and U_{Y^*} the closed unit balls $\boldsymbol{B}_{X^*}$ and $\boldsymbol{B}_{Y^*}$ equipped with the relative topologies induced by the weak* topologies on X^* and Y^*, respectively. Then, they are compact spaces by the Alaoglu theorem [**14**, Theorem V.4.2]. Denote by $X\hat{\otimes}_\varepsilon Y$ the injective tensor product of X and Y with injective norm

$$\|z\|_\varepsilon := \sup \left\{ \left| \sum_{i=1}^k x^*(x_i) y^*(y_i) \right| : x^* \in \boldsymbol{B}_{X^*}, y^* \in \boldsymbol{B}_{Y^*} \right\},$$

where $\sum_{i=1}^k x_i \otimes y_i$ is any representation of $z \in X \otimes Y$. If $x^* \in X^*$ and $y^* \in Y^*$, then there is a unique bounded linear functional on $X\hat{\otimes}_\varepsilon Y$, which is called the *tensor product* of x^* and y^* and denoted by $x^* \otimes y^*$, such that $(x^* \otimes y^*)(x \otimes y) = x^*(x) y^*(y)$ for every $x \in X$ and $y \in Y$. Then it is easily verified that the equation $\|x^* \otimes y^*\| = \|x^*\| \|y^*\|$ holds and for each $z \in X\hat{\otimes}_\varepsilon Y$ the function $(x^*, y^*) \in U_{X^*} \times U_{Y^*} \mapsto (x^* \otimes y^*)(z)$ is bounded and continuous. In the proofs of Proposition 3.2 and Theorem 3.5 we shall use the fact that any element of $(X\hat{\otimes}_\varepsilon Y)^*$ is of integral type, that is, for each $z^* \in (X\hat{\otimes}_\varepsilon Y)^*$, there exists a positive finite Radon measure r on $U_{X^*} \times U_{Y^*}$ with $r(U_{X^*} \times U_{Y^*}) = \|z^*\|$ such that for any $z \in X\hat{\otimes}_\varepsilon Y$,

$$z^*(z) = \int_{U_{X^*} \times U_{Y^*}} (x^* \otimes y^*)(z) r(dx^*, dy^*). \tag{3.1}$$

For the above results and further information on the injective tensor product of Banach spaces see [**9**, Chapter VIII].

Let $\varepsilon : X \times Y \to X\hat{\otimes}_\varepsilon Y$ be the bilinear mapping defined by $\varepsilon(x, y) = x \otimes y$ for all $x \in X$ and $y \in Y$. Let $\nu : \mathcal{B}(T) \to Y$ be a vector measure. Then the semivariation $\|\nu\|_\varepsilon$ with respect to ε coincides with the usual semivariation $\|\nu\|$ on $\mathcal{B}(T)$, so that ν has the (*)-property with respect to ε [**30**, Remark 2]. In this case we shall write $\int_B \varphi \otimes d\nu := \int_B \varepsilon(\varphi, d\nu)$ for every (ν, ε)-integrable function $\varphi : T \to X$ and $B \in \mathcal{B}(T)$.

We begin by giving a useful criterion for the τ-smoothness of vector measures with values in $X\hat{\otimes}_\varepsilon Y$.

LEMMA 3.1. *Let $\lambda : \mathcal{B}(S \times T) \to X\hat{\otimes}_\varepsilon Y$ be a vector measure. Let $E \in \mathcal{B}(S \times T)$. Then the real function $(x^*, y^*) \in U_{X^*} \times U_{Y^*} \mapsto |(x^* \otimes y^*)\lambda|(E)$ is bounded and lower semicontinuous.*

PROOF. For each $(x^*, y^*) \in U_{X^*} \times U_{Y^*}$, put $h_E(x^*, y^*) := |(x^* \otimes y^*)\lambda|(E)$. It is readily seen that $|h_E(x^*, y^*)| \leq \|\lambda\|(S \times T) < \infty$ for every $(x^*, y^*) \in U_{X^*} \times U_{Y^*}$, so that h_E is bounded.

The lower semicontinuity of h_E follows from the fact that h_E is a supremum of the family of continuous functions of the form $h_{\mathcal{E}}(x^*, y^*) = \sum_{F \in \mathcal{E}} |(x^* \otimes y^*)\lambda(F)|$, where $\mathcal{E}$ is a finite $\mathcal{B}(S \times T)$-measurable partition of E. □

PROPOSITION 3.2. *Let $\lambda : \mathcal{B}(S \times T) \to X\hat{\otimes}_\varepsilon Y$ be a vector measure. Then λ is τ-smooth if $(x^* \otimes y^*)\lambda$ is τ-smooth for every $x^* \in X^*$ and $y^* \in Y^*$.*

PROOF. Let $E \in \mathcal{B}(S \times T)$ and $\mathcal{E}$ be a finite $\mathcal{B}(S \times T)$-measurable partition of E. Fix $z^* \in (X\hat{\otimes}_\varepsilon Y)^*$ and take a positive finite Radon measure r on $U_{X^*} \times U_{Y^*}$ with $r(U_{X^*} \times U_{Y^*}) = \|z^*\|$ satisfying (3.1). Then

$$\begin{aligned} \sum_{F \in \mathcal{E}} |z^*\lambda(F)| &= \sum_{F \in \mathcal{E}} \left| \int_{U_{X^*} \times U_{Y^*}} (x^* \otimes y^*)\lambda(F) r(dx^*, dy^*) \right| \\ &\leq \int_{U_{X^*} \times U_{Y^*}} \sum_{F \in \mathcal{E}} |(x^* \otimes y^*)\lambda(F)| r(dx^*, dy^*). \end{aligned} \tag{3.2}$$

By Lemma 3.1 the function $(x^*, y^*) \mapsto |(x^* \otimes y^*)\lambda|(E)$ is r-integrable, it follows from (3.2) that

$$|z^*\lambda|(E) \leq \int_{U_{X^*} \times U_{Y^*}} |(x^* \otimes y^*)\lambda|(E) r(dx^*, dy^*), \quad E \in \mathcal{B}(S \times T). \tag{3.3}$$

Let $\{G_\alpha\}_{\alpha \in \Gamma}$ be an increasing net of open subsets of $S \times T$ with $G = \bigcup_{\alpha \in \Gamma} G_\alpha$. Then the τ-smoothness of $(x^* \otimes y^*)\lambda$ implies $|(x^* \otimes y^*)\lambda|(G) = \sup_{\alpha \in \Gamma} |(x^* \otimes y^*)\lambda|(G_\alpha)$, so that by (3.3) and [**1**, Theorem 2.1.5 and Remark 2.1.6], we have

$$|z^*\lambda|(G - G_\alpha) \leq \int_{U_{X^*} \times U_{Y^*}} |(x^* \otimes y^*)\lambda|(G - G_\alpha) r(dx^*, dy^*) \to 0,$$

since the functions $(x^*, y^*) \mapsto |(x^* \otimes y^*)\lambda|(G_\alpha)$ are bounded and lower semicontinuous by Lemma 3.1. Thus, $z^*\lambda$ is τ-smooth for every $z^* \in (X\hat{\otimes}_\varepsilon Y)^*$, and this proves that λ is τ-smooth. □

Our goal in this section will be to show the existence and uniqueness of the Borel injective tensor product of two Banach space-valued vector measures and the validity of a Fubini-type theorem; see Theorems 3.5 and 3.7 and Proposition 3.9. This has been already accomplished in [**6**, Lemma 2] and [**29**, Theorem I.4.1 and its Corollary] for τ-smooth or Radon probability measures, and the result can be extended to τ-smooth or Radon positive finite measures verbatim (see [**1**, Corollary 2.1.11 and Theorem 2.1.12]).

To accomplish our goal we start to extend the above results to real measures. In what follows, denote by $B(S)$ the set of all Borel measurable (that is, $\mathcal{B}(S)$-measurable), bounded real functions on S. Similar definitions are made for $B(T)$ and $B(S \times T)$. For each $E \in \mathcal{B}(S \times T)$ and $t \in T$ put $E^t := \{s \in S : (s,t) \in E\}$. Then $E^t \in \mathcal{B}(S)$. Since every τ-smooth real measure can be represented by the difference of two τ-smooth positive finite measures (the Jordan decomposition), we can prove the following result without any difficulty, and leave its proof to the reader.

PROPOSITION 3.3. *If real measures $p : \mathcal{B}(S) \to \mathbb{R}$ and $q : \mathcal{B}(T) \to \mathbb{R}$ are τ-smooth, then there is a unique τ-smooth real measure $r : \mathcal{B}(S \times T) \to \mathbb{R}$ such that*

$$r(A \times B) = p(A) \cdot q(B) \quad \textit{for all } A \in \mathcal{B}(S) \textit{ and } B \in \mathcal{B}(T). \tag{3.4}$$

Then r is called the Borel product of p and q and denoted somewhat ambiguously by $p \times q$. Further, the following assertions hold:

(i) *Let $h \in B(S \times T)$. Then the two functions*

$$s \in S \mapsto \int_T h(s,t)q(dt) \quad \textit{and} \quad t \in T \mapsto \int_S h(s,t)p(ds) \tag{3.5}$$

are Borel measurable and

$$\int_{S\times T} hd(p \times q) = \int_S \int_T h(s,t)q(dt)p(ds) = \int_T \int_S h(s,t)p(ds)q(dt). \tag{3.6}$$

(ii) *The restriction of $p \times q$ to the product σ-field $\mathcal{B}(S) \times \mathcal{B}(T)$ is a product measure of p and q in the usual sense.*

(iii) $|p \times q| = |p| \times |q|$ *on $\mathcal{B}(S \times T)$.*

(iv) *If p and q are Radon, so is $p \times q$.*

Let $\mu : \mathcal{B}(S) \to X$ and $\nu : \mathcal{B}(T) \to Y$ be vector measures. We say that μ has the *separability condition with respect to* ν if for each $E \in \mathcal{B}(S \times T)$ the vector function $t \in T \mapsto \mu(E^t)$ is $\|\nu\|$-essentially separably valued, that is, there exists a set $N \in \mathcal{B}(T)$ with $\|\nu\|(N) = 0$ such that the set $\{\mu(E^t) : t \in T - N\}$ is a separable subset of X.

REMARK 3.4. Let $\mu : \mathcal{B}(S) \to X$ and $\nu : \mathcal{B}(T) \to Y$ be vector measures. Then μ has the separability condition with respect to ν if any of the following assumptions is valid; see [**16**, page 125].

(i) μ has separable range.

(ii) μ has relatively compact range.

(iii) μ is given by a σ-finite positive measure p on $\mathcal{B}(S)$ and a p-measurable, Pettis integrable vector function $\varphi : S \to X$ as the Pettis integral $\mu(A) = \int_A \varphi dp$ for all $A \in \mathcal{B}(S)$ [**26**, page 363].
(iv) μ is of bounded variation and X has the Radon-Nikodým property [**9**, Definition III.1.3].
(v) There is a countable subset $\mathcal{D}$ of $\mathcal{B}(S)$ with the property that for any $\eta > 0$ and $A \in \mathcal{B}(S)$ there exists $D \in \mathcal{D}$ such that $\|\mu\|(A \triangle D) < \eta$, where $A \triangle D := (A - D) \cup (D - A)$.
(vi) The spaces S and T satisfy $\mathcal{B}(S \times T) = \mathcal{B}(S) \times \mathcal{B}(T)$ [**5**, Theorem 2].
(vii) One of the spaces S or T has a countable base of open sets [**18**, Theorem 8.1].
(viii) The spaces S and T are locally compact and μ and ν are τ-smooth [**28**, P16, page xiii] and [**30**, Theorem 2.2].

The following theorem is our first goal in this paper, and it extends [**30**, Theorem 2.2].

THEOREM 3.5. *Let $\mu : \mathcal{B}(S) \to X$ and $\nu : \mathcal{B}(T) \to Y$ be τ-smooth vector measures. If one of them has the separability condition with respect to the other, then there is a unique τ-smooth vector measure $\lambda : \mathcal{B}(S \times T) \to X\hat{\otimes}_\varepsilon Y$ such that*

$$\lambda(A \times B) = \mu(A) \otimes \nu(B) \quad \textit{for all } A \in \mathcal{B}(S) \textit{ and } B \in \mathcal{B}(T). \tag{3.7}$$

Further we have the following properties.

(i) *$(x^* \otimes y^*)\lambda = x^*\mu \times y^*\nu$ and $|(x^* \otimes y^*)\lambda| = |x^*\mu| \times |y^*\nu|$ for every $x^* \in X^*$ and $y^* \in Y^*$.*
(ii) *$\|\lambda\|(A \times B) = \|\mu\|(A) \cdot \|\nu\|(B)$ for every $A \in \mathcal{B}(S)$ and $B \in \mathcal{B}(T)$.*

PROOF. We first assume that μ has the separability condition with respect to ν. Fix $E \in \mathcal{B}(S \times T)$ for a moment and put $\varphi(t) := \mu(E^t)$ for all $t \in T$. Then φ is bounded. By Proposition 3.3, $x^*\varphi$ is Borel measurable for each $x^* \in X^*$, since $x^*\mu$ is τ-smooth. Then it follows from [**3**, Theorem 4] and the Pettis measurability theorem [**9**, Theorem II.1.2] that φ is (ν, ε)-integrable.

Thanks to this integrability condition, one can define a vector-valued set function $\lambda(E) := \int_T \mu(E^t) \otimes \nu(dt)$ for all $E \in \mathcal{B}(S \times T)$. It is obvious that $\lambda : \mathcal{B}(S \times T) \to X\hat{\otimes}_\varepsilon Y$ satisfies (3.7). The countable additivity of λ follows from the Bounded convergence theorem.

Let $x^* \in X^*$ and $y^* \in Y^*$. Since $x^*\mu$ and $y^*\nu$ are τ-smooth, by Proposition 3.3 there is a unique τ-smooth Borel product $x^*\mu \times y^*\nu$ such that $|x^*\mu \times y^*\nu| = |x^*\mu| \times |y^*\nu|$ on $\mathcal{B}(S \times T)$. Then, by [**15**, p. 327] and Proposition 3.3, for each $E \in \mathcal{B}(S \times T)$ we have $(x^* \otimes y^*)\lambda(E) = (x^* \otimes y^*)\left(\int_T \mu(E^t) \otimes \nu(dt)\right) = \int_T x^*\mu(E^t) y^*\nu(dt) = (x^*\mu \times y^*\nu)(E)$. This proves (i), so that the τ-smoothness of λ follows from Proposition 3.2. The uniqueness of λ follows from the corresponding uniqueness in the scalar case.

The proof of (ii) is as follows. Let $A \in \mathcal{B}(S)$ and $B \in \mathcal{B}(T)$. For any $x^* \in \boldsymbol{B}_{X^*}$ and $y^* \in \boldsymbol{B}_{Y^*}$, $|x^*\mu|(A) \cdot |y^*\nu|(B) = |(x^* \otimes y^*)\lambda|(A \times B) \leq \|\lambda\|(A \times B)$, so that the inequality $\|\mu\|(A) \cdot \|\nu\|(B) \leq \|\lambda\|(A \times B)$ holds.

To prove the reverse inequality, fix $z^* \in (X\hat{\otimes}_\varepsilon Y)^*$ and take a positive finite Radon measure r on $U_{X^*} \times U_{Y^*}$ with $r(U_{X^*} \times U_{Y^*}) = \|z^*\|$ satisfying (3.1). Then by the same argument as that in Proposition 3.2, we have

$$|z^*\lambda|(A\times B)\le\int_{U_{X^*}\times U_{Y^*}}|(x^*\otimes y^*)\lambda|(A\times B)r(dx^*,dy^*). \tag{3.8}$$

Since $|(x^*\otimes y^*)\lambda|(A\times B)\le\|\mu\|(A)\cdot\|\nu\|(B)$ for all $(x^*,y^*)\in U_{X^*}\times U_{Y^*}$, it follows from (3.8) that $|z^*\lambda|(A\times B)\le\|\mu\|(A)\cdot\|\nu\|(B)\cdot\|z^*\|$. Since $z^*\in(X\hat{\otimes}_\varepsilon Y)^*$ is arbitrary, we obtain the reverse inequality $\|\lambda\|(A\times B)\le\|\mu\|(A)\cdot\|\nu\|(B)$.

Next we assume that ν has the separability condition with respect to μ. By the preceding argument, there is a unique τ-smooth vector measure $\rho:\mathcal{B}(S\times T)\to Y\hat{\otimes}_\varepsilon X$ which satisfies $\rho(A\times B)=\nu(B)\otimes\mu(A)$ for all $A\in\mathcal{B}(S)$ and $B\in\mathcal{B}(T)$ and $(y^*\otimes x^*)\rho=x^*\mu\times y^*\nu$ for every $x^*\in X^*$ and $y^*\in Y^*$. Let κ be the canonical isometric isomorphism from $Y\hat{\otimes}_\varepsilon X$ onto $X\hat{\otimes}_\varepsilon Y$ satisfying $\kappa(y\otimes x)=x\otimes y$ for all $x\in X$ and $y\in Y$. It is easy to see that $(x^*\otimes y^*)\kappa=y^*\otimes x^*$ for all $x^*\in X^*$ and $y^*\in Y^*$. Put $\lambda(E):=\kappa(\rho(E))$ for all $E\in\mathcal{B}(S\times T)$. Then we can easily verify that $\lambda:\mathcal{B}(S\times T)\to X\hat{\otimes}_\varepsilon Y$ is a vector measure and satisfies (3.7). For each $x^*\in X^*$ and $y^*\in Y^*$, $(x^*\otimes y^*)\lambda=(x^*\otimes y^*)\kappa(\rho)=(y^*\otimes x^*)\rho=x^*\mu\times y^*\nu$ on $\mathcal{B}(S\times T)$, so that the τ-smoothness and uniqueness of λ, together with properties (i) and (ii), can be proved in the same way. □

The proof of the following result is a slight modification of that of [**29**, Exercise I.4.2, page 71], and so we omit it.

Proposition 3.6. *Let Z be a Banach space. Assume that a τ-smooth vector measure $\lambda:\mathcal{B}(S\times T)\to Z$ satisfies the following two conditions:*

(i) *For each $\eta>0$ and $A\in\mathcal{B}(S)$, there is a compact subset K of S with $K\subset A$ such that $\|\lambda\|((A-K)\times T)<\eta$.*
(ii) *For each $\eta>0$ and $B\in\mathcal{B}(T)$, there is a compact subset L of T with $L\subset B$ such that $\|\lambda\|(S\times(B-L))<\eta$.*

Then λ is Radon.

Theorem 3.7. *Let $\mu:\mathcal{B}(S)\to X$ and $\nu:\mathcal{B}(T)\to Y$ be Radon vector measures. If one of them has the separability condition with respect to the other, then there is a unique Radon vector measure $\lambda:\mathcal{B}(S\times T)\to X\hat{\otimes}_\varepsilon Y$ such that $\lambda(A\times B)=\mu(A)\otimes\nu(B)$ for all $A\in\mathcal{B}(S)$ and $B\in\mathcal{B}(T)$. Further properties (i) and (ii) in Theorem 3.5 are valid.*

Proof. This can be proved by Theorem 3.5 and Proposition 3.6. □

Let $\mu:\mathcal{B}(S)\to X$ and $\nu:\mathcal{B}(T)\to Y$ be vector measures. We say that μ has a *proper Borel injective tensor product with respect to* ν if there is a unique vector measure $\lambda:\mathcal{B}(S\times T)\to X\hat{\otimes}_\varepsilon Y$ such that $\lambda(A\times B)=\mu(A)\otimes\nu(B)$ for all $A\in\mathcal{B}(S)$ and $B\in\mathcal{B}(T)$, and the following two conditions are valid;

(i) for each $E\in\mathcal{B}(S\times T)$, the function $t\in T\mapsto\mu(E^t)$ is (ν,ε)-integrable and
(ii) the primitive Fubini theorem holds:

$$\lambda(E)=\int_T\mu(E^t)\otimes\nu(dt),\quad E\in\mathcal{B}(S\times T). \tag{3.9}$$

Then λ is called a *proper Borel injective tensor product of the ordered pair* μ, ν. It should be remarked that this is not a symmetrical definition for μ and ν, contrary to the case of the Borel injective tensor product. By the proof of Theorem 3.5 and Theorem 3.7 we have

COROLLARY 3.8. *Let $\mu : \mathcal{B}(S) \to X$ and $\nu : \mathcal{B}(T) \to Y$ be τ-smooth (respectively, Radon) vector measures. Assume that μ has the separability condition with respect to ν. Then μ has a τ-smooth (respectively, Radon) proper Borel injective tensor product with respect to ν.*

The following result shows that a Fubini-type theorem for bounded Borel measurable functions is valid for proper Borel injective tensor products. The Fubini theorem for real measures involves two pairs of iterated integrals; however, in our present context, only one of the pairs makes sense.

PROPOSITION 3.9. *Let $\mu : \mathcal{B}(S) \to X$ and $\nu : \mathcal{B}(T) \to Y$ be vector measures. Assume that μ has a proper Borel injective tensor product λ with respect to ν. Let $h \in B(S \times T)$. Then the following assertions are valid:*

(i) *For each $t \in T$, the function $s \in S \mapsto h(s,t)$ is μ-integrable,*
(ii) *the vector function $t \in T \mapsto \int_S h(s,t)\mu(ds)$ is (ν, ε)-integrable, and*
(iii) $\int_{S\times T} h d\lambda = \int_T \left\{\int_S h(s,t)\mu(ds)\right\} \otimes \nu(dt)$.

PROOF. (i) follows from [**2**, Theorem 2.6 (c)].

(ii) and (iii): Let $\{h_n\}$ be a sequence of $\mathcal{B}(S \times T)$-simple functions with $|h_n| \leq |h|$ converging uniformly to h on $S \times T$. Then by [**2**, Theorem 2.6 (c)], h_n and h are λ-integrable and $\int_{S\times T} h_n d\lambda \to \int_{S\times T} hd\lambda$. Further, for each $t \in T$, $h_n(\cdot, t)$ and $h(\cdot, t)$ are μ-integrable, and $\varphi_n(t) := \int_S h_n(s,t)\mu(ds)$ converges uniformly to $\varphi(t) := \int_S h(s,t)\mu(ds)$ on T.

By assumption, each φ_n is (ν, ε)-integrable and $\{\varphi_n\}$ is uniformly bounded on T. Hence it follows from the Bounded convergence theorem that φ is (ν, ε)-integrable and $\int_T \varphi_n \otimes d\nu \to \int_T \varphi \otimes d\nu$. Thus the proof ends if we observe the easily verified fact that $\int_T \varphi_n \otimes d\nu = \int_{S\times T} h_n d\lambda$ for every $n \in \mathbb{N}$. □

The notion that ν has a proper Borel injective tensor product with respect to μ can be introduced in a similar fashion and the corresponding results are valid by obvious modifications.

4. Weak convergence of injective tensor products of vector measures

In this section, we shall consider the weak convergence of proper Borel injective tensor products of Banach lattice-valued positive vector measures. The importance of the facts in section 3 will be apparent here. Otherwise the weak convergence of vector measures in the product space would not make any sense.

Throughout this section, we assume that X and Y are Banach lattices such that their injective tensor product $X\hat{\otimes}_\varepsilon Y$ is also a Banach lattice and satisfies the positivity condition (P_ε): $x \otimes y \geq 0$ for every $x \in X$ with $x \geq 0$ and $y \in Y$ with $y \geq 0$. We refer to [**23**] for the basic theory of Banach lattices. In general, the injective tensor product $X\hat{\otimes}_\varepsilon Y$ may not be a vector lattice for the natural ordering. However, for instance, the following Banach lattices satisfy the above assumption; see examples in [**23**, pages 274–276] and [**17**, page 90].

EXAMPLE 4.1. (1) Let Ω be a compact space and Y be any Banach lattice. Denote by $C(\Omega, Y)$ the Banach lattice with its canonical ordering of all (bounded) continuous functions $\varphi : \Omega \to Y$. We write $C(\Omega) := C(\Omega, \mathbb{R})$. Then $C(\Omega)\hat{\otimes}_\varepsilon Y$ is isometrically lattice isomorphic to the Banach lattice $C(\Omega, Y)$. Especially, when $Y = C(\Omega')$ for some compact space Ω', $C(\Omega)\hat{\otimes}_\varepsilon C(\Omega')$ is isometrically lattice isomorphic to $C(\Omega \times \Omega')$.

(2) Let Λ be a locally compact space and Y be any Banach lattice. Denote by $C_0(\Lambda, Y)$ the Banach lattice with its canonical ordering of all (bounded) continuous functions $\varphi : \Lambda \to Y$ such that for every $\eta > 0$, the set $\{\omega \in \Lambda : \|\varphi(\omega)\| \geq \eta\}$ is compact. We write $C_0(\Lambda) := C_0(\Lambda, \mathbb{R})$. Then $C_0(\Lambda)\hat{\otimes}_\varepsilon Y$ is isometrically lattice isomorphic to $C_0(\Lambda, Y)$. Especially, when $Y = C_0(\Lambda')$ for some locally compact space Λ', $C_0(\Lambda)\hat{\otimes}_\varepsilon C_0(\Lambda')$ is isometrically lattice isomorphic to $C_0(\Lambda \times \Lambda')$.

(3) Let $(\Phi, \mathcal{A}, p)$ be a measure space and Y be any Banach lattice. Denote by $L^\infty(\Phi, Y)$ the Banach lattice of all (equivalent classes of) p-essentially bounded measurable functions $\varphi : \Phi \to Y$ with its canonical ordering. We write $L^\infty(\Phi) := L^\infty(\Phi, \mathbb{R})$. Then $L^\infty(\Phi)\hat{\otimes}_\varepsilon Y$ is a Banach lattice. However, in general, $L^\infty(\Phi)\hat{\otimes}_\varepsilon Y$ is a proper closed subset of $L^\infty(\Phi, Y)$.

Let S be a topological space. We say that a vector measure $\mu : \mathcal{B}(S) \to X$ is *positive* if $\mu(A) \geq 0$ for every $A \in \mathcal{B}(S)$. Then for any positive vector measure μ, the equation $\|\mu\|(A) = \|\mu(A)\|$ holds for all $A \in \mathcal{B}(S)$ [**25**, Lemma 1.1]. We begin by showing the positivity of the proper Borel injective tensor product.

LEMMA 4.2. *Let T be a topological space. Let $\nu : \mathcal{B}(T) \to Y$ be a positive vector measure. Assume that a vector function $\varphi : T \to X$ is bounded and (ν, ε)-measurable, so that it is (ν, ε)-integrable. If $\varphi \geq 0$ then $\int_T \varphi \otimes d\nu \geq 0$.*

PROOF. By [**10**, Theorem II.6.2] there is a sequence $\{\varphi_n\}$ of X-valued $\mathcal{B}(T)$-simple functions on T such that $\varphi_n(t) \to \varphi(t)$ $\|\nu\|_\varepsilon$-a.e. and $\|\varphi_n(t)\| \leq \|\varphi(t)\|$ for all $t \in T$. Denote by $|x|$ the modulus of an element x of X. Then $\{|\varphi_n(t)|\}$ also converges to $|\varphi(t)| = \varphi(t)$ $\|\nu\|_\varepsilon$-a.e. and is uniformly bounded on T, so that it follows from the Bounded convergence theorem that

$$\int_T \varphi \otimes d\nu = \lim_{n\to\infty} \int_T |\varphi_n| \otimes d\nu. \tag{4.1}$$

Fix $n \in \mathbb{N}$. Since φ_n is an X-valued $\mathcal{B}(T)$-simple function, we put $\varphi_n = \sum_{i=1}^k x_i \chi_{B_i}$, where $x_1, \ldots, x_k \in X$, $B_1, \ldots, B_k$ are pairwise disjoint sets in $\mathcal{B}(T)$, and $k \in \mathbb{N}$. Then $|\varphi_n| = \sum_{i=1}^k |x_i| \chi_{B_i}$, and it follows from the positivity condition (P_ε) that

$$\int_T |\varphi_n| \otimes d\nu = \sum_{i=1}^k |x_i| \otimes \nu(B_i) \geq 0. \tag{4.2}$$

Since the set of all positive elements in $X\hat{\otimes}_\varepsilon Y$ is closed, (4.1) and (4.2) imply that $\int_T \varphi \otimes d\nu \geq 0$. □

PROPOSITION 4.3. *Let S and T be topological spaces. Let $\mu : \mathcal{B}(S) \to X$ and $\nu : \mathcal{B}(T) \to Y$ be vector measures. Assume that μ has a proper Borel injective tensor product λ with respect to ν. If μ and ν are positive, then so is λ.*

PROOF. Let $E \in \mathcal{B}(S \times T)$. By assumption, the vector function $t \in T \mapsto \varphi(t) := \mu(E^t)$ is (ν, ε)-integrable and $\lambda(E) = \int_T \varphi \otimes d\nu$. Then φ is bounded and $\varphi(t) \geq 0$ for all $t \in T$ by the positivity of μ. Thus it follows from Lemma 4.2 and the positivity of ν that $\lambda(E) = \int_T \varphi \otimes d\nu \geq 0$. This proves the positivity of λ. □

Let S and T be topological spaces. Let $C(S)$ be the Banach space of all bounded continuous real functions on S with norm $\|f\|_\infty := \sup_{s\in S} |f(s)|$. Denote by $\mathcal{M}^+(S, X)$ the set of all positive vector measures $\mu : \mathcal{B}(S) \to X$, and denote by $\mathcal{M}_t^+(S,X)$ the set of all $\mu \in \mathcal{M}^+(S,X)$ which are tight. Similar definitions are made for $\mathcal{M}^+(T,Y)$ and $\mathcal{M}_t^+(T,Y)$.

We recall the definition of weak convergence of vector measures. Let $\{\mu_\alpha\}_{\alpha\in\Gamma}$ be a net in $\mathcal{M}^+(S,X)$ and $\mu \in \mathcal{M}^+(S,X)$. We say that $\{\mu_\alpha\}_{\alpha\in\Gamma}$ *converges weakly* to μ, and write $\mu_\alpha \xrightarrow{w} \mu$, if for each $f \in C(S)$ we have $\int_S f d\mu = \lim_{\alpha\in\Gamma} \int_S f d\mu_\alpha$ in the norm of X; see [**8, 19, 20, 21, 25**]. The following result shows the joint weak continuity of proper Borel injective tensor products of positive vector measures. Except for assuming the tightness of vector measures it extends previous results for probability measures; see [**4**, Theorem 3.2] and [**29**, Proposition I.4.1].

THEOREM 4.4. *Let S and T be completely regular spaces. Let $\{\mu_\alpha\}_{\alpha\in\Gamma}$ be a net in $\mathcal{M}^+(S,X)$ and $\mu \in \mathcal{M}_t^+(S,X)$. Let $\{\nu_\alpha\}_{\alpha\in\Gamma}$ be a net in $\mathcal{M}_t^+(T,Y)$ and $\nu \in \mathcal{M}_t^+(T,Y)$. Assume that μ_α and μ have τ-smooth proper Borel injective tensor products λ_α and λ with respect to ν_α and ν, respectively. Further assume that ν is τ-smooth. If $\mu_\alpha \xrightarrow{w} \mu$ and $\nu_\alpha \xrightarrow{w} \nu$ then $\lambda_\alpha \xrightarrow{w} \lambda$.*

REMARK 4.5. Theorem 4.4 has already been proved in [**21**] with the additional assumption $\mathcal{B}(S\times T) = \mathcal{B}(S)\times\mathcal{B}(T)$. However, the result obtained here applies to many cases which are not covered by Theorem 5.4 of [**21**]; see Remark 3.4 and Corollary 3.8. Similar results are given in [**19**] for not necessarily positive vector measures with values in certain nuclear spaces.

To prove the theorem, we need the following two results which seem to be of some interest. Let S and T be uniform spaces with the uniformities $\mathcal{U}_S$ and $\mathcal{U}_T$, respectively. Let $S\times T$ be a uniform space with the product uniformity. Let X be a Banach space. Denote by $U(T,X)$ the Banach space of all bounded uniformly continuous vector functions $\varphi : T \to X$ with norm $\|\varphi\|_\infty := \sup_{t\in T}\|\varphi(t)\|$. We write $U(T) = U(T,\mathbb{R})$. Similar definitions are made for $U(S)$, $U(S\times T)$, and so on. For the proofs of the following two results see [**21**, Theorem 4.1 and Proposition 5.1].

THEOREM 4.6 (Diagonal Convergence Theorem). *Let T be a uniform space with the uniformity $\mathcal{U}_T$. Let X be a Banach space and Y a Banach lattice. Consider a net $\{\varphi_\alpha\}_{\alpha\in\Gamma} \subset U(T,X)$ and $\varphi \in U(T,X)$ satisfying the following conditions:*

(i) *$\varphi_\alpha(t) \to \varphi(t)$ for every $t \in T$,*
(ii) *$\{\varphi_\alpha\}_{\alpha\in\Gamma}$ is uniformly bounded, that is, $\sup_{\alpha\in\Gamma}\|\varphi_\alpha\|_\infty < \infty$, and*
(iii) *$\{\varphi_\alpha\}_{\alpha\in\Gamma}$ is uniformly equicontinuous on T, that is, for any $\eta > 0$, there exists $V \in \mathcal{U}_T$ such that $\sup_{\alpha\in\Gamma}\|\varphi_\alpha(t)-\varphi_\alpha(t')\| < \eta$ whenever $(t,t') \in V$.*

Given a net $\{\nu_\alpha\}_{\alpha\in\Gamma} \subset \mathcal{M}_t^+(T,Y)$ and a τ-smooth measure $\nu \in \mathcal{M}_t^+(T,Y)$, if $\int_T g d\nu = \lim_{\alpha\in\Gamma}\int_T g d\nu_\alpha$ for every $g \in U(T)$, then $\int_T \varphi \otimes d\nu = \lim_{\alpha\in\Gamma}\int_T \varphi_\alpha \otimes d\nu_\alpha$.

PROPOSITION 4.7. *Let S be a uniform space. Let X be a Banach lattice. Let $\{\mu_\alpha\}_{\alpha\in\Gamma}$ be a net in $\mathcal{M}^+(S,X)$ and $\mu \in \mathcal{M}^+(S,X)$. Assume that μ is tight. Then the following two conditions are equivalent:*

(i) *$\int_S f d\mu_\alpha \to \int_S f d\mu$ for every $f \in U(S)$.*
(ii) *$\int_S f d\mu_\alpha \to \int_S f d\mu$ for every $f \in C(S)$.*

We start to prove Theorem 4.4. Since every completely regular topology is the uniform topology of some uniform structure (see [**22**, Corollary 6.17]), one may consider that S and T are uniform spaces, and $S\times T$ is given their product uniformity. We may also assume without loss of generality that $\{\mu_\alpha\}_{\alpha\in\Gamma}$ and $\{\nu_\alpha\}_{\alpha\in\Gamma}$ are uniformly bounded, that is, $\sup_{\alpha\in\Gamma}\|\mu_\alpha\|(S) < \infty$ and $\sup_{\alpha\in\Gamma}\|\nu_\alpha\|(T) < \infty$.

Let $h \in U(S \times T)$. By assumption and Proposition 3.9 the vector functions $\varphi_\alpha(t) := \int_S h(s,t)\mu_\alpha(ds)$ and $\varphi(t) := \int_S h(s,t)\mu(ds)$ on T are (ν, ε)-integrable, and satisfy the equations $\int_{S\times T} hd\lambda_\alpha = \int_T \varphi_\alpha \otimes d\nu_\alpha$ and $\int_{S\times T} hd\lambda = \int_T \varphi \otimes d\nu$. Then it is easy to verify that the net $\{\varphi_\alpha\}_{\alpha\in\Gamma}$ and φ satisfy the conditions (i), (ii) and (iii) of Theorem 4.6, so that

$$\int_{S\times T} hd\lambda_\alpha = \int_T \varphi_\alpha \otimes d\nu_\alpha \to \int_T \varphi \otimes d\nu = \int_{S\times T} hd\lambda. \tag{4.3}$$

Since μ_α, ν_α, μ and ν are positive, so are λ_α and λ by Proposition 4.3. Since μ and ν are tight by assumption, it is easy to prove that λ is also tight. Thus it follows from Proposition 4.7 and (4.3) that $\int_{S\times T} hd\lambda_\alpha \to \int_{S\times T} hd\lambda$ for every $h \in C(S\times T)$, and this proves that $\lambda_\alpha \xrightarrow{w} \lambda$.

5. Weak convergence of convolutions of vector measures

In this section, we shall define the convolution of Banach algebra-valued vector measures on a topological semigroup and consider their weak convergence.

Let X be a Banach algebra with the multiplication $(x,y) \in X \times X \mapsto xy$. We write $m(x,y) = xy$. Then $m : X \times X \to X$ is a continuous bilinear mapping. Denote by $\tilde{m}$ the *linearization* of m, that is, $\tilde{m}$ is a mapping from the algebraic tensor product $X \otimes X$ into X defined by $\tilde{m}(u) = \sum_{i=1}^k m(x_i, y_i)$ for $u = \sum_{i=1}^k x_i \otimes y_i$ in $X \otimes X$. This is well-defined since the value of $\tilde{m}(u)$ does not depend on the representation of u. We suppose throughout this section that the mapping $\tilde{m}$ satisfies the condition (M_ε): $\|\tilde{m}(u)\| \leq \|u\|_\varepsilon$ for all $u \in X \otimes X$, which was also supposed in [**30**, Section 3]. In this case, $\tilde{m}$ has a unique continuous extension, which we also denote by $\tilde{m}$; thus $\tilde{m} : X \hat{\otimes}_\varepsilon X \to X$ is a continuous linear mapping with $\|\tilde{m}\| \leq 1$. Although the condition (M_ε) is a fairly strong assumption, it is satisfied, for instance, X is the Banach algebra $C(\Omega)$ with a compact space Ω or the Banach algebra $C_0(\Lambda)$ with a locally compact space Λ; see Example 4.1.

Let T be a topological space. Let $\nu : \mathcal{B}(T) \to X$ be a vector measure. Then the inequality $\|\nu\|_m(B) \leq \|\nu\|_\varepsilon(B)$ holds for all subsets of T [**30**, Theorem 3.1]. Since every vector measure $\nu : \mathcal{B}(T) \to X$ has the (*)-property with respect to the tensor mapping $\varepsilon(x,y) = x \otimes y$ [**30**, Remark 2], it follows from the above inequality and [**13**, Lemma 2] that ν also has the (*)-property with respect to m. In what follows, we shall write $\int_B \varphi d\nu := \int_B m(\varphi, d\nu)$ for every (ν, m)-integrable function $\varphi : T \to X$ and $B \in \mathcal{B}(T)$. The proof of the following result is the same as that of [**30**, Theorem 3.1] and is omitted.

PROPOSITION 5.1. *Let S and T be topological spaces. Let $\mu : \mathcal{B}(S) \to X$ and $\nu : \mathcal{B}(T) \to X$ be vector measures. Assume that μ has a proper Borel injective tensor product λ with respect to ν. Let $h \in B(S\times T)$. Then the following assertions are valid:*

(i) *The vector function $t \in T \mapsto \int_S h(s,t)\mu(ds)$ is (ν, m)-integrable, and*
(ii) $\tilde{m}\left(\int_{S\times T} hd\lambda\right) = \int_T \int_S h(s,t)\mu(ds)\nu(dt)$.

From now throughout this paper, assume that S is a topological semigroup with a completely regular topology, that is, S is a completely regular space with a jointly continuous, associative, semigroup operation $(s,t) \in S \times S \mapsto st \in S$. We write $\theta(s,t) := st$. Then θ is measurable with respect to $\mathcal{B}(S \times S)$ and $\mathcal{B}(S)$. The following theorem extends [**30**, Theorem 3.2] and shows that the convolution of X-valued vector measures can be defined as the measure induced by their proper Borel injective tensor product and the semigroup operation.

THEOREM 5.2. *Let $\mu : \mathcal{B}(S) \to X$ and $\nu : \mathcal{B}(S) \to X$ be vector measures. Assume that μ has a τ-smooth proper Borel injective tensor product with respect to ν. Then there is a unique τ-smooth vector measure $\gamma : \mathcal{B}(S) \to X$ such that*

$$\int_S f d\gamma = \int_S \int_S f(st)\mu(ds)\nu(dt) \quad \textit{for all } f \in C(S). \tag{5.1}$$

Then γ is called the convolution of the ordered pair μ, ν and denoted by $\mu * \nu$. Further, (5.1) *continues to hold for every $f \in B(S)$.*

PROOF. Let λ be a τ-smooth proper Borel injective tensor product of the ordered pair μ, ν. Put $\gamma(A) := \widetilde{m}\left(\lambda(\theta^{-1}(A))\right)$ for all $A \in \mathcal{B}(S)$. Then it is readily seen that $\gamma : \mathcal{B}(S) \to X$ is a vector measure and satisfies the inequality

$$\|\gamma\|(A) \leq \|\lambda\|(\theta^{-1}(A)) \quad \text{for all } A \in \mathcal{B}(S). \tag{5.2}$$

We first prove that γ is τ-smooth. Let $\{G_\alpha\}_{\alpha\in\Gamma}$ be an increasing net of open subsets of S with $G = \bigcup_{\alpha\in\Gamma} G_\alpha$. Then $\{\theta^{-1}(G_\alpha)\}_{\alpha\in\Gamma}$ is also an increasing net of open subsets of $S \times S$ with $\theta^{-1}(G) = \bigcup_{\alpha\in\Gamma} \theta^{-1}(G_\alpha)$. Thus it follows from (5.2) and the τ-smoothness of λ that $\|\gamma\|(G - G_\alpha) \leq \|\lambda\|\left(\theta^{-1}(G) - \theta^{-1}(G_\alpha)\right) \to 0$, and this proves that γ is τ-smooth.

The uniqueness of γ follows from [**29**, Corollary 2 to Proposition I.3.8].

Let $f \in B(S)$. It follows from Proposition 5.1 that $\widetilde{m}\left(\int_{S\times S} f(st)\lambda(ds, dt)\right) = \int_S \int_S f(st)\mu(ds)\nu(dt)$. Thus, to prove that (5.1) continues to hold for the function f, it suffices to show that $\int_S f d\gamma = \widetilde{m}\left(\int_{S\times S} f(st)\lambda(ds, dt)\right)$.

Let $\{f_n\}$ be a sequence of $\mathcal{B}(S)$-simple functions converging uniformly to f on S. Put $h_n(s,t) := f_n(st)$ and $h(s,t) := f(st)$. Then $\{h_n\}$ is a sequence of $\mathcal{B}(S \times S)$-simple functions converging uniformly to h on $S \times S$. Thus it follows from [**2**, Theorem 2.6 (c)] that $\int_S f d\gamma = \lim_{n\to\infty} \int_S f_n d\gamma$ and $\int_{S\times S} h d\lambda = \lim_{n\to\infty} \int_S h_n d\lambda$. Since $\int_S f_n d\gamma = \widetilde{m}\left(\int_{S\times S} h_n d\lambda\right)$ for every $n \in \mathbb{N}$, it follows from the continuity of $\widetilde{m}$ that

$$\int_S f d\gamma = \lim_{n\to\infty} \int_S f_n d\gamma = \lim_{n\to\infty} \widetilde{m}\left(\int_{S\times S} h_n d\lambda\right) = \widetilde{m}\left(\int_{S\times S} h d\lambda\right),$$

and the proof is complete. □

In the following, we shall assume that the Banach algebra X is one of the spaces $C(\Omega)$ with a compact space Ω or $C_0(\Lambda)$ with a locally compact space Λ. Then X satisfies the condition (M_ε). Further, X is a Banach lattice such that $X\hat{\otimes}_\varepsilon X$ is also a Banach lattice and satisfies the positivity condition (P_ε) in Section 4. We recall that $C(\Omega)$ and $C_0(\Lambda)$ are separable if Ω is metrizable and Λ is separable and metrizable; see [**14**, Exercise IV.13.16] and [**24**, Theorem II.6 in Part I, page 111]. The proof of the following proposition is so close to that of Proposition 4.3 that we omit it.

PROPOSITION 5.3. *Let X be one of the spaces $C(\Omega)$ with a compact space Ω or $C_0(\Lambda)$ with a locally compact space Λ. Let $\mu : \mathcal{B}(S) \to X$ and $\nu : \mathcal{B}(S) \to X$ be vector measures. Assume that μ has a τ-smooth proper Borel injective tensor* *product with respect to ν (this condition is satisfied if, for instance, μ and ν are τ-smooth, Ω is metrizable, and Λ is separable and metrizable by Corollary 3.8). If μ and ν are positive then so is the convolution $\mu * \nu$.*

The following theorem shows the joint weak convergence of convolutions of vector measures with values in special Banach algebras $C(\Omega)$ or $C_0(\Lambda)$. Except for assuming the tightness of vector measures, it extends previous results for probability measures; see [**7**, Corollary] and [**29**, Proposition I.4.5].

THEOREM 5.4. *Let X be one of the spaces $C(\Omega)$ with a compact space Ω or $C_0(\Lambda)$ with a locally compact space Λ. Let $\{\mu_\alpha\}_{\alpha\in\Gamma}$ be a net in $\mathcal{M}^+(S,X)$ and $\mu \in \mathcal{M}_t^+(S,X)$. Let $\{\nu_\alpha\}_{\alpha\in\Gamma}$ be a net in $\mathcal{M}_t^+(S,X)$ and $\nu \in \mathcal{M}_t^+(S,X)$. Assume that μ_α and μ have τ-smooth proper Borel injective tensor products with respect to ν_α and ν, respectively. Further assume that ν is τ-smooth (these conditions are satisfied if, for instance, μ_α, ν_α, μ and ν are all Radon, Ω is metrizable, and Λ is separable and metrizable by Corollary 3.8). If $\mu_\alpha \xrightarrow{w} \mu$ and $\nu_\alpha \xrightarrow{w} \nu$ then $\mu_\alpha * \nu_\alpha \xrightarrow{w} \mu * \nu$.*

PROOF. Let λ_α and λ be τ-smooth proper Borel injective tensor products of the ordered pairs μ_α, ν_α and μ, ν, respectively. Fix $f \in C(S)$ and put $h(s,t) := f(st)$ for all $(s,t) \in S \times S$. Then $h \in C(S \times S)$, so that it follows from Theorem 4.4 that $\int_{S\times S} hd\lambda_\alpha \to \int_{S\times S} hd\lambda$. On the other hand, it follows from Proposition 5.1 and Theorem 5.2 that $\widetilde{m}\left(\int_{S\times S} hd\lambda_\alpha\right) = \int_S fd(\mu_\alpha * \nu_\alpha)$ and $\widetilde{m}\left(\int_{S\times S} hd\lambda\right) = \int_S fd(\mu * \nu)$. Thus it follows from the continuity of $\widetilde{m}$ that $\int_S fd(\mu_\alpha * \nu_\alpha) \to \int_S fd(\mu * \nu)$, and this proves $\mu_\alpha * \nu_\alpha \xrightarrow{w} \mu * \nu$. □

References

1. C. Berg, J. P. R. Christensen and P. Ressel, *Harmonic analysis on semigroups*, Springer-Verlag, New York, 1984.
2. R. G. Bartle, N. Dunford and J. T. Schwartz, *Weak compactness and vector measures*, Canad. J. Math. **7** (1955), 289–305.
3. R. G. Bartle, *A general bilinear vector integral*, Studia Math. **15** (1956), 337–352.
4. P. Billingsley, *Convergence of probability measures*, John Wiley & Sons, New York, 1968.
5. R. Rao Chivukula and A. S. Sastry, *Product vector measures via Bartle integrals*, J. Math. Anal. Appl. **96** (1983), 180–195.
6. I. Csiszár, *Some problems concerning measures on topological spaces and convolutions of measures on topological groups*, Les Probabilités sur les Structures Algébriques, Clermont-Ferrand, 1969, 75–96. Colloques Internationaux du CNRS, Paris, 1970.
7. ———, *On the weak* continuity of convolution in a convolution algebra over an arbitrary topological group*, Studia Sci. Math. Hungar. **6** (1971), 27–40.
8. M. Dekiert, *Kompaktheit, Fortsetzbarkeit und Konvergenz von Vektormassen*, Dissertation, University of Essen, 1991.
9. J. Diestel and J. J. Uhl, Jr., *Vector measures*, Amer. Math. Soc., Math. Surveys No. 15, Providence, R. I., 1977.
10. N. Dinculeanu, *Vector measures*, Pergamon Press, Oxford, 1967.
11. M. Duchoň, *A convolution algebra of vector-valued measures on a compact Abelian semigroup*, Rev. Roum. Math. Pures et Appl. **16** (1971), 1467–1476.
12. ———, *The Fubini theorem and convolution of vector-valued measures*, Mat. časop. **23** (1973), 170–178.
13. ———, *Product of dominated vector measures*, Math. Slovaca **27** (1977), 293–301.
14. N. Dunford and J. T. Schwartz, *Linear operators*, part I: general theory, John Wiley & Sons, 1957.
15. F. J. Freniche and J. C. García-Vázquez, *The Bartle bilinear integration and Carleman operators*, J. Math. Anal. Appl. **240** (1999), 324–339.

16. J. E. Huneycutt, Jr., *Products and convolution of vector valued set functions*, Studia Math. **41** (1972), 119–129.
17. A. Grothendieck, *Produits tensoriels topologiques et espaces nucléaires*, Mem. Amer. Math. Soc. No. 16, Providence, R. I., 1955.
18. R. A. Johnson, *On product measures and Fubini's theorem in locally compact spaces*, Trans. Amer. Math. Soc. **123** (1966), 112–129.
19. J. Kawabe, *Weak convergence of tensor products of vector measures with values in nuclear spaces*, Bull. Austral. Math. Soc. **59** (1999), 449–458.
20. ———, *A type of Strassen's theorem for positive vector measures with values in dual spaces*, Proc. Amer. Math. Soc. **128** (2000), 3291–3300.
21. ———, *Joint continuity of injective tensor products of positive vector measures in Banach lattices*, to appear in J. Austral. Math. Soc. Ser. A.
22. J. L. Kelley, *General topology*, Van Nostrand, New York, 1955.
23. H. H. Schaefer, *Banach lattices and positive operators*, Springer-Verlag, Berlin, 1974.
24. L. Schwartz, *Radon measures on arbitrary topological spaces and cylindrical measures*, Tata Institute of Fundamental Research, Oxford University Press, 1973.
25. R. M. Shortt, *Strassen's theorem for vector measures*, Proc. Amer. Math. Soc. **122** (1994), 811–820.
26. C. Swartz, *The product of vector-valued measures*, Bull. Austral. Math. Soc. **8** (1973), 359–366.
27. ———, *A generalization of a theorem of Duchon on products of vector measures*, J. Math. Anal. Appl. **51** (1975), 621–628.
28. F. Topsøe, *Topology and measure*, Lecture Notes in Math. **133**, Springer-Verlag, New York, 1970.
29. N. N. Vakhania, V. I. Tarieladze and S. A. Chobanyan, *Probability distributions on Banach spaces*, D. Reidel, 1987.
30. A. J. White, *Convolution of vector measures*, Proc. Roy. Soc. Edinburgh Sect. A **73** (1974/75), 117–135.

Department of Mathematics, Shinshu University, Wakasato, Nagano 380-8553, Japan
E-mail address: `jkawabe@gipwc.shinshu-u.ac.jp`

Contemporary Mathematics
Volume **321**, 2003

On linear operator pencils and inclusions of images of balls

V.A. Khatskevich and V.S. Shulman

1. Introduction.

Working with the generalized operator fractional-linear transformations and Abel-Schroeder equations (see [**3**], [**4**], [**5**], [**6**]), the authors recognized the necessity to consider spectral properties of linear operator pencils $K \to A+BK$, where A, B are arbitrary operators and K is a contraction or a uniform contraction. Doing this they were constantly forced to consider some conditions of subordination for pairs of operators on Banach spaces, which were expressed either in the form of norm inequalities (of the type $\|Ax\| \leq \|Bx\|$) or in the form of the inclusions of the images of unit balls (open balls, closed balls or closures of the images). So it was naturally to look at these subordination conditions more attentively. It has appeared that the subject, being very general and quite elementary, is not completely trivial and deserves a more or less systematic study. The first section of the present paper is devoted to such a study, the second one contains some applications to operator pencils. Applications to the operator fractional-linear transformations will be given elsewhere.

The authors are very indebted to Vladimir Fonf for many valuable suggestions and information.

2. Inclusions for the images of balls.

For an absolutely convex subset E of a linear space Z we denote by E' its "linear closure": $E' = \cap_{\varepsilon>0}(1+\varepsilon)E = \{z \in Z : tz \in E$, for all $t \in (0;1)\}$

2000 Mathematics Subject Classification. 47L07, 39B42

Clearly $E \subset E'$ and

$$AE' \subset (AE)' \tag{1}$$

for any linear operator $A : Z \to X$ where X is another linear space. We will consider only the case that Z, X are Banach spaces. Clearly $E' \subset \overline{E}$ (the closure of E); moreover $E' = \overline{E}$ if $interior(E) \neq \emptyset$

In particular $(\mathcal{S}_Z)' = \overline{\mathcal{S}_Z}$ where $\mathcal{S}_Z$ is the open unit ball of Z. Hence

$$A\overline{\mathcal{S}_Z} \subset (A\mathcal{S}_Z)'. \tag{2}$$

Another easy consequence is the equality $E' = \overline{E}$, for all absolutely convex subsets in finite-dimensional spaces.

LEMMA 1. *(i) Let $A : Z \to X$ be a bounded linear operator. If* $\ker A$ *is reflexive then*

$$A\overline{\mathcal{S}_Z} = (A\mathcal{S}_Z)'. \tag{3}$$

(ii) If Z is non-reflexive then, for any X, there is $A \in \mathcal{B}(Z, X)$ such that (3) fails.

PROOF. (i) Let $x \in (A\mathcal{S}_Z)'$. Clearly $x = Az$, for some $z \in Z$. Since the subspace $\ker A$ is reflexive it is proximinal in Z. It follows that z can be chosen with $\|z\| = dist(z, \ker A)$. Hence $\|tz\| = \operatorname{dist}(tz, \ker A)$, for any $t \in (0; 1)$. Since $tx \in A\mathcal{S}_Z$ there is $z_1 \in \mathcal{S}_Z$ with $tx = Az_1$. It follows that $tz - z_1 \in \ker A$ whence $\|tz\| \leq \|z_1\| < 1$. This shows that $\|z\| \leq 1$, $x \in A\overline{\mathcal{S}_Z}$. We proved the inclusion $(A\mathcal{S}_Z)' \subset A\overline{\mathcal{S}_Z}$; the converse was proved above.

(ii) It suffices to prove the assertion for $X = \mathcal{C}$. James Theorem [**1**] there is $f \in Z^*(= \mathcal{B}(Z, X))$ with $\|f\| \notin f(\overline{\mathcal{S}_Z})$. On the other hand $t\|f\| \in f(\mathcal{S}_Z)$, for any $t \in (0; 1)$, and $\|f\| \in f(\mathcal{S}_Z)'$. □

We saw in the proof of Lemma 1 (ii) the examples of operators A for which $A\overline{\mathcal{S}} \neq (A\mathcal{S})'$ (this is possible only if $\ker A$ is non-reflexive). The following example exhibits an injective operator for which $(A\mathcal{S})' \neq \overline{A\mathcal{S}}$.

EXAMPLE 2. Let $Z = c_0$ and X - any of the spaces $c_0, l^p(1 < p < \infty)$. Elements of c_0 we write as rows: $\overrightarrow{z} = (z_1, z_2, \ldots)$. Define $A \in \mathcal{B}(Z, X)$ by $(A\overrightarrow{z})_k = (z_k - z_{k+1})/k$. Clearly AZ consists of all $\overrightarrow{x} \in X$ for which $\sum_{k=1}^{\infty} kx_k$ converges (then $\overrightarrow{x} = A\overrightarrow{z}$ with $z_n = \sum_{k=n}^{\infty} kx_k$) and

$$A\mathcal{S}_Z = \{\overrightarrow{x} \in X : |\sum_{k=n}^{\infty} kx_k| < 1, n \in \mathcal{N}\}$$

Define elements $\overrightarrow{x}, \overrightarrow{x}^{(m)} \in X$ by $x_k = (-1)^k/(k+1)$, $x_k^{(m)} = x_k$, for $k \leq m$, and 0 otherwise. Then $\overrightarrow{x}^{(m)} \in A\mathcal{S}_Z$, $\overrightarrow{x}^{(m)} \to \overrightarrow{x}$. Thus $\overrightarrow{x} \in \overline{A\mathcal{S}_Z}$ but $\overrightarrow{x} \notin (A\mathcal{S}_Z)'$ because $\overrightarrow{x} \notin AZ$. ◇

It is easy to see that

$$(A\mathcal{S}_Z)' \subset AZ \cap \overline{A\mathcal{S}_Z}.$$

We will find the conditions under which the converse inclusion holds.

LEMMA 3. *The equality*

$$(A\mathcal{S}_Z)' = AZ \cap \overline{A\mathcal{S}_Z} \tag{4}$$

holds iff

$$\operatorname{dist}(z, \ker A) = \operatorname{dist}(z, \ker A^{**}) \tag{5}$$

for any $z \in Z$.

PROOF. Let (5) hold and let $x \in AZ \cap \overline{A\mathcal{S}_Z}$. Then $x = Az, z \in Z$ and $x = \lim Az_n, \|z_n\| < 1$. Choosing a *-weakly convergent subnet we have that z_n *-weakly tends to $z^{**} \in Z^{**}, \|z^{**}\| \le 1, A^{**}z^{**} = x = A^{**}z, z^{**} - z \in \ker(A^{**}), dist(z, \ker(A^{**}) \le 1$. Hence $dist(z, \ker(A)) \le 1$. Thus, for any $t \in (0;1)$, $\operatorname{dist}(tz, \ker A) \le 1$, $tz \in \mathcal{S}_Z + \ker A$, $tx \in A\mathcal{S}_Z, x \in (A\mathcal{S}_Z)'$.

Conversely if (5) fails we take $z \in Z$ with

$$dist(z, \ker A) > 1 > dist(z, \ker A^{**}).$$

Hence $z - z^{**} \in \ker(A^{**})$, for some $z^{**}, \|z^{**}\| < 1$. Since $\mathcal{S}_Z$ is *-weakly dense in $\mathcal{S}_{Z^{**}}$, there are $z_n \in \mathcal{S}_Z$ such that $z_n \to z^{**}$ *-weakly. Hence $Az_n \to A^{**}z^{**} = A^{**}z = Az$. It follows that Az belongs to the weak closure of $A\mathcal{S}_Z$ which is the same as the norm closure. On the other hand $Az \notin (A\mathcal{S}_Z)'$ (otherwise $tAz = Az_t, \|z_t\| < 1, tz - z_t \in \ker A, \operatorname{dist}(tz, \ker A) < 1, \operatorname{dist}(z, \ker A) \le 1$, a contradiction). Therefore (4) fails. □

Note that if $\ker A = 0$ then (5) means that

$$\ker(A^{**}) \perp_{\|\cdot\|} Z \tag{6}$$

where the norm-orthogonality of a subspace L_1 to a subspace L_2 means that $\|v\| \le \|u+v\|$, for any $u \in L_1, v \in L_2$. Taking into account Lemma 1 we obtain the following

COROLLARY 4. . Let $\ker A = 0$. Then the equality

$$A\overline{\mathcal{S}_Z} = AZ \cap \overline{A\mathcal{S}_Z} \tag{7}$$

is equivalent to the condition (6).

For $A \in \mathcal{B}(Z, X)$ let us denote by $\widetilde{A}$ the operator from $Z/\ker(A)$ to X, acting by the rule: $\widetilde{A}(\overline{z}) = Az$, $(z \in \overline{z} \in Z/\ker(A))$. The proof of Lemma 1(i) shows actually that

$$(AS_Z)' = \widetilde{A}(\overline{\mathcal{S}}_{Z/\ker(A)}). \tag{8}$$

It will be convenient also, for any quotient space Z/Z_1, to denote by $\mathcal{S}^0_{Z/Z_1}$ the image of $\overline{\mathcal{S}_Z}$ in Z/Z_1 under the canonical surjection. Clearly $\mathcal{S}^0_{Z/Z_1} = \overline{\mathcal{S}_{Z/Z_1}}$ if Z_1 is reflexive. This shows that (8) immediately implies Lemma 1(i).

Note also that $A\overline{\mathcal{S}_Z} = \widetilde{A}\mathcal{S}^0_{Z/\ker(A)}$ and $A\mathcal{S}_Z = \widetilde{A}\mathcal{S}_{Z/\ker(A)}$.

Let now $B : Y \to X$ be another bounded operator. We consider the conditions

$$B\overline{\mathcal{S}_Y} \subset A\overline{\mathcal{S}_Z} \tag{a}$$

$$B\mathcal{S}_Y \subset A\mathcal{S}_Z \tag{b}$$

$$(B\overline{\mathcal{S}_Y})' \subset (A\overline{\mathcal{S}_Z})' \tag{b'}$$

$$\overline{B\mathcal{S}_Y} \subset \overline{A\mathcal{S}_Z} \tag{c}$$

THEOREM 5. *(i) (a) $\Rightarrow$ (b) $\Leftrightarrow$ (b′) $\Rightarrow$ (c)*
(ii) If $\ker A$ *is reflexive then (a) $\Leftrightarrow$ (b)*
(iii) If Z is reflexive then all four conditions are equivalent.

PROOF. The implications $(b) \Rightarrow (b')$, $(b) \Rightarrow (c)$ are evident.

$(a) \Rightarrow (b')$: If (a) holds then by (3) $B\mathcal{S}_Y \subset (A\overline{\mathcal{S}_Z})'$. Hence $(B\overline{\mathcal{S}_Y})' \subset B\mathcal{S}_Y \subset (A\overline{\mathcal{S}_Z})'$.

$(b') \Rightarrow (b)$: If (b') holds then, by (8),

$$B\mathcal{S}_Y \subset \widetilde{A}(\overline{\mathcal{S}}_{Z/\ker(A)}). \tag{9}$$

It follows that $BY \subset \widetilde{A}(Z/\ker(A))$. Since $\widetilde{A}$ is injective, for any $y \in Y$ there is a unique $\overline{z} \in Z/\ker(A)$ with $By = \widetilde{A}\overline{z}$. Setting $\overline{z} = Vy$ we obtain a linear map $V : Y \to Z/\ker(A)$ such that

$$B = \widetilde{A}V. \tag{10}$$

It follows now from (9) that

$$V\mathcal{S}_Y \subset \overline{\mathcal{S}_{Z/\ker(A)}}.$$

In other words V is bounded and $\|V\| \leq 1$. We have now $V\mathcal{S}_Y \subset \mathcal{S}_{Z/\ker(A)}$, $B\mathcal{S}_Y = \widetilde{A}V\mathcal{S}_Y \subset \widetilde{A}\mathcal{S}_{Z/\ker(A)} = A\mathcal{S}_Z$. This proves (i). Clearly (ii) follows from (i) and Lemma 1(ii). Indeed if (b) holds and $\ker(A)$ is reflexive then by (i) $B\overline{\mathcal{S}}_Y \subset (B\mathcal{S}_Y)' \subset (A\mathcal{S}_Z)' = A\overline{\mathcal{S}}_Z$.

To prove (iii) note that if Z is reflexive then $A\overline{\mathcal{S}}_Z$ is closed and (c) means that $\overline{B\mathcal{S}_Y} \subset A\overline{\mathcal{S}}_Z$ which automatically implies (a). □

REMARK 6. (i) It is natural to ask what happens if we impose the reflexivity condition on X or Y instead of Z.

Repeating the arguments of the proof of Lemma 1(ii) we see that even for one-dimensional X and Y (b) does not imply (a). Furthermore (c) does not imply (b) for one-dimensional Y and reflexive X. Indeed let $Z = c_0$, $X = l^p(p > 1)$, $A \in \mathcal{B}(Z, X)$ as in the example 2, $Y = \mathcal{C}, B(\lambda) = \lambda x_0$, for $x_0 \in \overline{A\mathcal{S}_Z} \backslash (A\mathcal{S}_Z)'$. Then $\overline{B\mathcal{S}_Y} \subset \overline{A\mathcal{S}_Z}$ but $(B\mathcal{S}_Y)' = (B\overline{\mathcal{S}}_Y)' \not\subseteq (A\mathcal{S}_Z)'$.

On the other hand $(c) \Rightarrow (b)$ if X is finite-dimensional, because in this case $(A\mathcal{S}_Z)' = \overline{A\mathcal{S}_Z}$ (see (1.2)).

(ii) The example in (i) shows that (c) does not imply (b) even if we assume that B is bounded from below ($\|By\| \geq C\|y\|, C > 0$). It should be noted that if B^* is bounded from below then $(c) \Rightarrow (b)$. Indeed the condition $\|B^*x^*\| \geq C\|x^*\|, x^* \in X^*$ gives, by Hahn-Banach Theorem (see the proof of Theorem 8 below) that $\overline{B\mathcal{S}_Y} \supset C\overline{\mathcal{S}_X}$. If (c) holds then $\overline{A\mathcal{S}_Z} \supset C\overline{\mathcal{S}_X}$. It follows in a usual way that for any $x \in \overline{A\mathcal{S}_Z}$ and any $\varepsilon > 0$ there is $z \in (1+\varepsilon)\mathcal{S}_Z$ with $x = Az$. This means that $\overline{A\mathcal{S}_Z} = (A\mathcal{S}_Z)'$ and our claim follows. For $B = 1$ see **[2]** .

For operators in Hilbert spaces the condition (a) is usually reformulated in terms of factorization (see, for example, [**7**]). Let us do this in the general case. We preserve the notation of Theorem 5.

THEOREM 7. *(i) The condition (b) is equivalent to the existence of an operator $V : Y \to Z/\ker(A)$ satisfying (10) with $\|V\| \leq 1$.*

(ii) The condition (a) is equivalent to the existence of an operator V satisfying (10) and

$$V\overline{\mathcal{S}}_Y \subset \mathcal{S}^o_{Z/\ker(A)} \tag{11}$$

PROOF. (i) was established in the proof of Theorem 5(i). Let us prove (ii). If (a) holds then, by Theorem 5 and (i), (10) is true with $\|V\| \leq 1$. Thus (a) can be rewritten in the form

$$\widetilde{A}V\overline{\mathcal{S}_Y} \subset A\overline{\mathcal{S}_Z} = \widetilde{A}\mathcal{S}^o_{Z/\ker(A)}$$

Since $\widetilde{A}$ is injective this implies (9).

Conversely if V satisfies (10) and (11) then $B\overline{\mathcal{S}_Y} = \widetilde{A}V\overline{\mathcal{S}_Y} \subset \widetilde{A}\mathcal{S}^o_{Z/\ker(A)} = A\overline{\mathcal{S}_Z}$. □

Let us now turn to the dual conditions.

THEOREM 8. *(i) Inclusion (c) is equivalent to the condition*

$$\|B^*x^*\| \leq \|A^*x^*\| \tag{12}$$

for any $x^* \in X^*$.

(ii) If operators A, B *are adjoint to some operators* $A_1 : X_1 \to Z_1$ *and* $B_1 : X_1 \to Y_1$ *(where* $X_1^* = X, Y_1^* = Y, Z_1^* = Z$*) then each of the inclusions (a), (b), (c) is equivalent to the condition*

$$\|A_1x_1\| \leq \|B_1x_1\| \tag{13}$$

for all $x_1 \in X_1$.

PROOF. (i) If (c) holds then $\|B^*\ x^*\| = \sup_{y\in\mathcal{S}_Y} |\langle B^*x^*, y\rangle| = \sup\{|\langle x^*, x\rangle| : x \in B\mathcal{S}_Y\} = \sup\ \{|\langle x^*,x\rangle| : x \in \overline{B\mathcal{S}_Y}\} \leq \sup\{|\langle x^*,x\rangle| : x \in \overline{A\mathcal{S}_Z}\} = \|A^*x^*\|$.

Conversely, if (c) fails then there is $x_0 \in \overline{B\mathcal{S}_Y} \setminus \overline{A\mathcal{S}_Z}$. By Hahn-Banach Theorem there is $x^* \in X^*$ such that $|\langle x_0^*,x_0\rangle| > 1$, and $|\langle x_0^*,x\rangle| \leq 1$, for all $x \in \overline{A\mathcal{S}_Z}$. Hence $\sup\{|\langle x^*,x\rangle| : x \in \overline{B\mathcal{S}_Y}\} > \sup\{|\langle x^*,x\rangle| : x \in \overline{A\mathcal{S}_Z}\}$, $\|B^*x^*\| > \|A^*x^*\|$.

(ii) The equivalence of (13) and (c) can be proved as above. Now $(c) \Rightarrow (a)$ because $A\overline{\mathcal{S}_Z}$ is $\sigma(X, X_1)$-compact (being a continuous image of a $\sigma(Z, Z_1)$-compact set $\overline{\mathcal{S}_Z}$) whence $A\overline{\mathcal{S}_Z} = \overline{A\mathcal{S}_Z}$. □

Now we can obtain a factorization criterion for (c):

COROLLARY 9. The condition (c) of Theorem 5 is equivalent to the existence of an operator $V : Y \to Z^{**}/\ker(A^{**})$ such that $\|V\| \leq 1$ and

$$i_XB = \widetilde{A^{**}}V \tag{14}$$

where $i_X : X \to X^{**}$ is the canonical injection.

PROOF. If (14) holds then $B\mathcal{S}_Y = \widetilde{A^{**}}V\mathcal{S}_Y \subset \widetilde{A^{**}}\mathcal{S}_{Z^{**}/\ker(A^{**})} = A^{**}\mathcal{S}_{Z^{**}}$ whence $\overline{B\mathcal{S}_Y} \subset \overline{A^{**}\mathcal{S}_{Z^{**}}}$ and, by *-weak continuity, $\overline{B^{**}\mathcal{S}_{Y^{**}}} \subset \overline{A^{**}\mathcal{S}_{Z^{**}}}$. By Theorem 8(ii) this is equivalent to (12) which is in its turn equivalent to (c) by Theorem 8(i).

Conversely the condition (c), that is (12), is equivalent , by Theorem 8(ii) to the inclusion $B^{**}\mathcal{S}_{Y^{**}} \subset A^{**}\mathcal{S}_{Z^{**}}$. Applying Theorem 7(i) we get a contraction $V : Y^{**} \to Z^{**}/\ker(A^{**})$ satisfying the equality $B^{**} = A^{**}V_1$. It remains only to set $V = V_1|Y$. □

REMARK 10. One can be interested in the conditions under which the "mixed" inclusions hold - for example, $(B\mathcal{S})' \subset \overline{A\mathcal{S}}$ and so on. We gathered the known to us conditions in the following table:

$\subset$	$A\mathcal{S}_Z$	$A\overline{\mathcal{S}_Z}$	$(A\mathcal{S}_Z)'$	$\overline{A\mathcal{S}_Z}$
$B\mathcal{S}_Y$	(b)	(b)	(b)	(c)
$B\overline{\mathcal{S}_Y}$	(b_1)	(a)	(b)	(c)
$(B\mathcal{S}_Y)'$	(b_2)	(a_1)	(b)	(c)
$\overline{B\mathcal{S}_Y}$	(b_3)	(a_2)	(b_4)	(c)

Besides the implications $(a) \Rightarrow (b) \Rightarrow (c)$, all implications $\Rightarrow$ (left implies right) and $\Uparrow$ (low implies upper) are also true. In particular, (b_3) is the strongest condition, (c) is the weakest one.

The condition (a_1) is equivalent to the following: $\overline{\widetilde{V}\mathcal{S}_{Y/\ker(B)}} \subset \mathcal{S}^0_{Z/\ker(A)}$.

The condition (a_2) is equivalent to the inclusion $\overline{\widetilde{A}V\mathcal{S}_Y} \subset \widetilde{A}\mathcal{S}^0_{Z/\ker(A)}$.

The condition (b_4) means that $\overline{\widetilde{A}V\mathcal{S}_Y} \subset \overline{\widetilde{A}\mathcal{S}_{Z/\ker(A)}}$.

The condition (b_1) is equivalent to the condition that V is a strict contraction ($\|Vy\| < 1$ for all $y \in \overline{\mathcal{S}_Y}$).

The condition (b_2) means that $\widetilde{V}$ is a strict contraction.

The arguments in the proof of Theorem 5 do not use the completeness of the spaces X, Y. This allows one to change X by AZ and, considering the operator $B_1 = B|_{B^{-1}(AZ)}$, to obtain the following result.

COROLLARY 11. For the conditions

$$B\overline{\mathcal{S}_Y} \cap AZ \subset A\overline{\mathcal{S}_Z} \qquad \text{(a}*\text{)}$$

$$cs_Y \cap AZ \subset A\mathcal{S}_Z \qquad \text{(b}*\text{)}$$

$$\mathcal{S}_Y)' \cap AZ \subset (A\mathcal{S}_Z)' \qquad \text{(b}'*\text{)}$$

one has: $(a*) \Rightarrow (b*) \Leftrightarrow (b'*)$. If $\ker(A)$ is reflexive then $(b*) \Rightarrow (a*)$. The conditions $(b*), (b'*)$ are equivalent to the following one:

$$xi = B\eta \Rightarrow \|\xi\|_{Z/\ker(A)} \le \|\eta\|_Y \qquad (*).$$

3. Operator pencils

For Banach spaces X, Y let us write $\mathcal{K}$ or $\mathcal{K}(X,Y)$ instead of $\mathcal{S}_{\mathcal{B}(X,Y)}$ and denote by $\overline{\mathcal{K}}$ its closure. Thus $\overline{\mathcal{K}}$ is the set of all contractions and $\mathcal{K}$ the set of all uniform contractions from X to Y. Let also $\mathcal{K}_s$ denote the set of all strict contractions: $K \in \mathcal{K}_s$ if $\|Kx\| < \|x\|$, for $x \neq 0$. We say that an operator K is a spectrally uniform contraction if $\rho(TK) < 1$, for any $T \in \overline{\mathcal{K}(Y,X)}$ (here ρ means the spectral radius). The set of all spectrally uniform contractions will be denoted by $\mathcal{K}_{su}$.

LEMMA 12. *(i)* $\mathcal{K} \subset \mathcal{K}_{su} \subset \mathcal{K}_s$.
(ii) If X, Y *are Hilbert spaces then* $\mathcal{K}_{su} = \mathcal{K}$.
(iii) If X *is non-reflexive then* $\mathcal{K} \neq \mathcal{K}_{su}$.
(iv) $V \in \mathcal{K}_{su}(X,Y) \Leftrightarrow V^* \in \mathcal{K}_{su}(Y^*,X^*)$.
(v) $V^* \in \mathcal{K}_s(Y^*,X^*) \Rightarrow V \in \mathcal{K}_s(X,Y)$ *and the converse implication holds only for reflexive* X, Y.

PROOF. The first inclusion in (i) is evident. If $K \notin \mathcal{K}_s$ then $\|Kx\| = \|x\|$ for some $x \in X, \|x\| = 1$. Take $f \in Y^*$ with $\|f\| = f(Kx) = 1$ and set $T = x \otimes f$. then $TKx = x, \rho(TK) = 1, K \notin \mathcal{K}_{su}$.

To prove (ii) take $T = K/\|K\|$ then $\rho(KT) = \|K\|$.

If X is non-reflexive then $x \otimes f \in \mathcal{K}_{su}$ if $\|x\| = \|f\| = 1$ and f does not attain its norm. This proves (iii).

(iv) is evident.

If $V \notin \mathcal{K}_s(X,Y)$ then $1 = \|x\| = \|Vx\|$, for some $x \in X$. Let $f \in Y^*$ be such that $f(Vx) = \|f\| = 1$, then $\|V^*f\| = \|f\|, V^* \notin \mathcal{K}_s(Y^*, X^*)$. □

Question. It would be very interesting to know if the equality $\mathcal{K} = \mathcal{K}_{su}$ holds for all reflexive X, Y.

For $A \in \mathcal{B}(Z,X), B \in \mathcal{B}(Y,X)$ we consider the map $K \mapsto A + BK$ from $\mathcal{B}(Z,Y)$ to $\mathcal{B}(Z,X)$. It is an analogue (and extension) of the usual pencil $A + \lambda B, \lambda \in \mathcal{C}$. The restriction of this map to $\mathcal{K}(Z,Y)$ (respectively to $\overline{\mathcal{K}(Z,Y)}$) will be denoted by $\mathcal{P}_{A,B}$ (by $\overline{\mathcal{P}}_{A,B}$). Let us say that a pencil has some property (for example is invertible or injective) if all its values have this property.

THEOREM 13. *(i)* $\mathcal{P}_{A,B}$ *is invertible iff* A *is invertible and* $\|A^{-1}B\| \leq 1$.

(ii) $\overline{\mathcal{P}}_{A,B}$ *is invertible iff* A *is invertible and* $A^{-1}B \in \mathcal{K}_{su}(Y,Z)$.

PROOF. Set $V = A^{-1}B$. Since $A + BK = A(1 + VK)$ we may consider the pencils $\mathcal{P}_{1,V}$ and $\overline{\mathcal{P}}_{1,V}$.

If $\|V\| \leq 1$ then clearly $1 + VK$ is invertible for $K \in \mathcal{K}$. Conversely if $\|V\| > 1$ then there is $y \in Y$ such that $\|y\| = 1, \|Vy\| > 1$. Hence there is $f \in \mathcal{S}_{Z^*}$ with $f(Vy) = -1$. Setting $K = y \otimes f$ we obtain that $\|K\| < 1$, $VK = Vy \otimes f$, $(1 + VK)(Vy) = 0$. Part (i) is proved.

If $\overline{\mathcal{P}}_{1,V}$ is invertible then, for any $K \in \overline{\mathcal{K}}$, $-1 \notin \sigma(VK)$. Hence $\lambda \notin \sigma(VK)$, for any $\lambda \in \mathcal{C}, |\lambda| \geq 1$ (otherwise change K by $-\lambda^{-1}K$). So $\rho(VK) < 1, V \in \mathcal{K}_{su}(X,Y)$. The converse statement is evident. □

THEOREM 14. *(i)* $\mathcal{P}_{A,B}$ *is injective iff* A *is injective and some (= any) of the conditions* $(a*), (b*), (*)$ *holds.*

(ii)$\overline{\mathcal{P}}_{A,B}$ *is injective iff* A *is injective and*

$$B\overline{\mathcal{S}_Y} \cap AZ \subset A\mathcal{S}_Z. \tag{15}$$

PROOF. (i) Suppose that $\mathcal{P}_{A,B}$ is injective. If $(*)$ fails then $Az = By$, for some $z \in Z, y \in Y$ with $\|y\| < \|z\|$. Choose $K \in \mathcal{K}$ with $y = -Kz$. Then $(A + BK)z = 0$, a contradiction. Conversely if $(*)$ holds and $z \in \ker(A + BK)$ then $Az = By$, for $y = -Kz, \|y\| < \|z\|$, a contradiction.

As follows from Corollary 11 the conditions $(a*)$ and $(b*)$ are equivalent to $(*)$ if $\ker A = 0$.

(ii) A similar argument shows that the injectivity of $\overline{\mathcal{P}}_{A,B}$ is equivalent to the condition: $Az = By \Rightarrow \|z\| < \|y\|$ which in its turn is equivalent to (15) (see the proof of Corollary 11). □

COROLLARY 15. If A is invertible and $\mathcal{P}_{A,B}$ is injective then $\mathcal{P}_{A,B}$ is invertible.

PROOF. If A is invertible then $(*)$ means that $\|A^{-1}B\| \leq 1$. □

It is clear that the conditions (a),(b) imply respectively $(a*)$ and $(b*)$ and, as a consequence, imply injectivity of $\mathcal{P}_{A,B}$, if A is injective. Let us show that (c) also implies the injectivity of $\mathcal{P}_{A,B}$ if A is strictly injective (this means that $\ker(A^{**}) \perp Z$ in Z^{**}).

THEOREM 16. *(i) If A is strictly injective then (c) implies injectivity of $\mathcal{P}_{A,B}$. (ii) If A is not strictly injective then there is B such that (c) holds but $\mathcal{P}_{A,B}$ is not injective.*

PROOF. (i) If A is strictly injective and (c) holds then, by Lemma 3

$$(B\mathcal{S}_Y)' \cap AZ \subset \overline{B\mathcal{S}_Y} \subset \overline{A\mathcal{S}_Z} \cap AZ = (A\mathcal{S}_Z)'$$

which implies injectivity of $\mathcal{P}_{A,B}$ by Proposition 14.

(ii) If A is not strictly injective then there is $z \in Z$ such that $\|z\| = a > 1, \operatorname{dist}(z, \ker(A^{**})) < 1$. Choose z^{**} such that $z - z^{**} \in \ker(A^{**}), \|z^{**}\| = 1$.

Let $Y = \mathcal{C}z^{**}$, $V : Y \to Z^{**}$ the identity inclusion, $B = A^{**}V$. Then

$$\|B^*x^*\| = \|V^*A^{***}x^*\| \leq \|A^*x^*\|,$$

for any $x^* \in X^*$, so (c) holds. Furthermore since $A^{**}z^{**} = Az$ we have $A^{**}z^{**} \in \overline{B\mathcal{S}_Y} \cap AZ$. On the other hand $A^{**}z^{**} \notin A\overline{\mathcal{S}_Z}$ because otherwise $A^{**}z^{**} = Az_1, z_1 \in \mathcal{S}_Z$ whence $Az = Az_1, z = z_1$, a contradiction (we used the injectivity of A; if A is not injective there is nothing to prove). We see that $(a*)$ does not hold and $\mathcal{P}_{A,B}$ is not injective. □

Let us say that an operator is *coinjective* if its adjoint is injective (equivalently, if its image is dense). An operator is *biinjective* if it is injective and coinjective.

THEOREM 17. *(i) $\mathcal{P}_{A,B}$ is coinjective if A is coinjective and (c) holds.*

(ii) If

$$\|B^*x^*\| < \|A^*x^*\|, \tag{16}$$

for $x^* \neq 0$ then $\overline{\mathcal{P}}_{A,B}$ is coinjective.

PROOF. Clearly if there is $K \in \mathcal{K}$ with $(A^* + K^*B^*)x^* = 0$ then $\|A^*x\| < \|B^*x^*\|$ (or both parts are zero). Thus (c) implies coinjectivity. Conversely if $\|A^*x^*\| < \|B^*x^*\|$ then one easily finds a rank one operator $K \in \mathcal{K}$ with $K^*B^*x^* = A^*x^*, x^* \in \ker(A^* - K^*B^*)$. This proves (i). The proof of (ii) is similar. □

COROLLARY 18. If Y is reflexive then (16) is also necessary for the coinjectivity of $\overline{\mathcal{P}}_{A,B}$.

PROOF. If (2.2) fails then $\|B^*x^*\| \geq \|A^*x^*\| = 1$, for some $x^* \in X^*$. Since Y is reflexive there is $y \in Y$ such that $\|y\| \leq 1, B^*x^*(y) = 1$. For $K = y \otimes (A^*x^*)$, we get: $K^*B^*x^* = A^*x^*$ and the operator $A - BK$ is not coinjective. □

COROLLARY 19. (i) $\mathcal{P}_{A,B}$ is biinjective iff A is biinjective and the conditions (c) and $(a*)$ hold.

(ii) For reflexive Z (more generally, for strictly injective A) the pencil $\mathcal{P}_{A,B}$ is biinjective iff (c) holds.

PROOF. (i) follows from Theorems 14 and 17. In (ii) we should only prove that if A is strictly injective and (c) holds then $\mathcal{P}_{A,B}$ is injective. But this was proved in Theorem 16. □

References

[1] R.E. Edwards, *Functional Analysis*, 1965.

[2] W. Rudin, *Functional Analysis*, N.Y., 1973.

[3] V.A. Khatskevich, V.A. Senderov and V.S. Shulman, *On generalized fractional-linear operator transforms*, International Conference in Operator Theory and its Applications, dedicated to the Memory of A.V.Strauss, (Ul'yanovsk, June 23-29, 2001).

[4] V. Khatskevich, S. Reich and D. Shojkhet, *Schroeder's functional equation and the Koenigs embedding property*, Nonlinear analysis, **47** (2001), 3977–3988.

[5] V.A. Khatskevich, V.A. Senderov, *The Abel-Schroeder equations for linear fractional mappings of operator balls*, Far East J. Math. Sci(FJMS), **3**(1), (2001), 1–13.

[6] V.A. Khatskevich, V.A. Senderov, *Powers of plus-operators*, Integr. Equations Oper.Th., **15** (1992), 784–795.

[7] Yu.L. Shmulian, *On divisibility in the class of plus-operators*, Mat. Zametki, **74** (1967), 516–525.

DEPARTMENT OF APPLIED MATHEMATICS, ORT BRAUDE COLLEGE, COLLEGE CAMPUS, P.O.BOX, 78, KARMIEL 21982, ISRAEL

E-mail address: victor_kh@hotmail.com

DEPARTMENT OF MATHEMATICS, VOLOGDA STATE TECHNICAL UNIVERSITY, 15 LENINA STR., VOLOGDA 160000, RUSSIA

E-mail address: shulman_v@yahoo.com

Contemporary Mathematics
Volume **321**, 2003

Ordinal indices and ℓ^1-spreading models

Denny H. Leung and Wee-Kee Tang

ABSTRACT. Some new ordinal indices on sequences of continuous functions are defined. It is shown that they are related to the existence of higher order ℓ^1-spreading models.

1. Introduction

The celebrated Rosenthal's ℓ^1-theorem [**10**] says that a uniformly bounded sequence of functions with a large degree of oscillation has a subsequence equivalent to the ℓ^1-basis. This result was subsequently extended to finite dimensional ℓ^1-sequences (ℓ^1-spreading models) with the help of ordinal oscillation indices for sequence of continuous functions. (See [**1**], [**4**], and [**7**].) We refer the reader to §3 of [**8**] for a survey of results regarding ordinal indices of functions and ℓ^1-spreading models. In this paper, we present several more oscillation type indices and show that they govern the existence of ℓ^1_α-spreading models. The results presented here overlap with those in [**7**]. However, our approach is different. The use of Rademacher systems goes back more directly to Rosenthal's approach to the ℓ^1-theorem. Our main result is a more precise version of Theorem 3.1 in [**1**] (see Theorem 3.9). It is also a close relative of Theorem 3.1 in [**7**] and Theorem 3.8 in [**8**]. More on ordinal indices associated with sequences of continuous functions can be found in [**5**], [**6**] and [**3**].

Our notation is standard. If L is an infinite subset of $\mathbb{N}$, we denote the set of all infinite subsets of L by $[L]$. Given subsets I, J of $\mathbb{N}$, we write $I < J$ to mean that $\max I < \min J$ if $I, J \neq \emptyset$. We also set $\emptyset < J$ and $I < \emptyset$ for any I, J. If $I = \{n\}$, we write $n < J$ for short. Similar definitions hold for the symbol "$\leq$". A crucial role is played by the Schreier families $(\mathcal{S}_\alpha)_{\alpha<\omega_1}$. Let $\mathcal{S}_0 = \{\{n\} : n \in \mathbb{N}\} \cup \{\emptyset\}$. If α is a countable ordinal and $\mathcal{S}_\alpha$ is defined, let

$$\mathcal{S}_{\alpha+1} = \left\{ \bigcup_{j=1}^{k} I_j : I_1 < I_2 < \ldots < I_k,\ k \leq \min I_1,\ I_j \in \mathcal{S}_\alpha \right\}.$$

If α is a countable limit ordinal, fix a sequence of ordinals (α_n) strictly increasing to α and define

$$\mathcal{S}_\alpha = \{I : I \in \mathcal{S}_{\alpha_n} \text{ for some } n \leq \min I\}.$$

2000 *Mathematics Subject Classification.* Primary 46E15, 03E15; Secondary 26A21, 54C30.

Key words and phrases. Baire-1 functions, Oscillation indices, Rademacher systems, ℓ^1-spreading models.

If α is a countable ordinal and $L = (\ell_n) \in [\mathbb{N}]$, where (ℓ_n) is a strictly increasing sequence, we let

$$\mathcal{S}_\alpha [L] = \{\{\ell_n : n \in J\} : J \in \mathcal{S}_\alpha\}.$$

We will have occasion to use an extended form of the Schreier families. Suppose that $\alpha = \omega^{\alpha_1} \cdot m_1 + ... + \omega^{\alpha_k} \cdot m_k$ in Cantor normal form and $L \in [\mathbb{N}]$, then we say that $I = \left(I_j^i\right)_{i=1}^{k}{}_{j=1}^{m_i} \in \mathcal{T}_\alpha [L]$ if

$$I_1^k < I_2^k < ... < I_{m_k}^k < ... < I_1^1 < ... < I_{m_1}^1$$

and $I_j^i \in \mathcal{S}_{\alpha_i} [L]$, $1 \leq j \leq m_i, 1 \leq i \leq k$. For simplicity, we denote $\mathcal{T}_\alpha [\mathbb{N}]$ by $\mathcal{T}_\alpha$. If f is a real-valued function defined on a set K and $r \in \mathbb{R}$, we write

$$[f < r] = \{x \in K : f(x) < r\}.$$

The set $[f > r]$ is defined similarly. The greatest integer function is denoted by $\lfloor \cdot \rfloor$. Finally, the cardinality of a set A is denoted by $|A|$.

2. Oscillation Indices

Let K be a Polish space, i.e., a separable completely metrizable topological space. Denote by $\mathcal{C}$ the collection of all closed subsets of K. A *derivation* on $\mathcal{C}$ is a map $\mathcal{D} : \mathcal{C} \to \mathcal{C}$ such that $\mathcal{D}(F) \subseteq F$ for every $F \in \mathcal{C}$. A derivation map may be iterated inductively by defining $\mathcal{D}^{\alpha+1}(F) = \mathcal{D}(\mathcal{D}^\alpha (F))$ for any ordinal $\alpha < \omega_1$ and $\mathcal{D}^\alpha (F) = \bigcap_{\beta<\alpha} \mathcal{D}^\beta (F)$ for any limit ordinal $\alpha < \omega_1$. By the Cantor-Baire Stationary Principle, there exists a countable ordinal α such that $\mathcal{D}^\beta (F)$ is constant for all $\beta \geq \alpha$. The least such ordinal α is called the *index* associated with the derivation $\mathcal{D}$.

Henceforth, fix a uniformly bounded sequence of functions (f_n) in $C_b(K)$, the space of bounded continuous functions on K, endowed with the supremum norm. We do not assume that (f_n) is pointwise convergent unless otherwise stated.

Recall a well-known family of derivations introduced by Alspach and Argyros [**1**]. Suppose that (f_n) is a pointwise convergent sequence. Let

$$A_{n,m}^+ = \{x \in K : f_n(x) - f_m(x) > \epsilon\}$$

and

$$A_{n,m}^- = \{x \in K : f_n(x) - f_m(x) < -\epsilon\}.$$

The family of derivations $\mathcal{O}$, parametrized by $\varepsilon > 0$, is defined as follows:

$$\begin{aligned}\mathcal{O}(\epsilon, (f_n), H) = \{x \in H :\ & \text{for every neighborhood } U \text{ of } x \text{ there is an} \\ & N \in \mathbb{N} \text{ such that for all } n \geq N \text{ there exists } M \in \mathbb{N} \text{ such that} \\ & \cap_{m \geq M} A_{n,m}^+ \cap H \cap U \neq \emptyset \text{ or} \\ & \cap_{m \geq M} A_{n,m}^- \cap H \cap U \neq \emptyset\}.\end{aligned}$$

Alspach and Argyros showed that the index associated with $\mathcal{O}(\epsilon, (f_n), \cdot)$ has implications for the finite dimensional ℓ^1-structure of the sequence (f_n).

More recently, Kiriakouli [**6**] proved similar results with regard to the "convergence index" (cf. [**5**]). This index is associated with a family of derivations $\mathcal{G}$, whose definition we now recall. For any $\varepsilon > 0$ and $H \in \mathcal{C}$, let

$$\mathcal{D}_1(\varepsilon, (f_n), H) = \{x \in H : \text{ for every neighborhood } U \text{ of } x, \text{ there exist } m, n \in \mathbb{N} \text{ and } x' \in U \cap H \text{ such that } |f_n(x') - f_m(x')| > \varepsilon\}.$$

Then define $\mathcal{G}(\varepsilon, (f_n), H) = \bigcap_{k \geq 1} \mathcal{D}_1\left(\varepsilon, (f_n)_{n \geq k}, H\right)$. In this paper, we study several families of derivations closely related to $\mathcal{G}$. For any $L \in [\mathbb{N}]$, let $(f_n)_L$ be the subsequence $(f_n)_{n \in L}$. Also define $N = N(\varepsilon) = \left\lfloor 1 + \frac{1}{\varepsilon} \sup \|f_n\| \right\rfloor$ and $F_N = \{j : |j| \leq N\}$ for any $\varepsilon > 0$. We consider the following families of derivations. For any $\varepsilon > 0$ and any $H \in \mathcal{C}$, let

$$\mathcal{D}_2(\varepsilon, (f_n), H) = \{x \in H : \text{ for every neighborhood } U \text{ of } x, \text{ there exist } n \in \mathbb{N} \text{ and } j \in F_N \text{ such that } [f_n < j\varepsilon] \cap U \cap H \neq \emptyset \text{ and } [f_n > (j+1)\varepsilon] \cap U \cap H \neq \emptyset\},$$

$$\mathcal{P}(\varepsilon, (f_n), H) = \bigcap_{L \in [\mathbb{N}]} \mathcal{D}_1(\varepsilon, (f_n)_L, H) = \bigcap_{L \in [\mathbb{N}]} \mathcal{G}(\varepsilon, (f_n)_L, H),$$

$$\mathcal{Q}(\varepsilon, (f_n), H) = \bigcap_{L \in [\mathbb{N}]} \mathcal{D}_2(\varepsilon, (f_n)_L, H).$$

We first observe the relationships among the families of derivations considered.

PROPOSITION 2.1. *The families of derivations $\mathcal{P}$ and $\mathcal{Q}$ are equivalent in the sense that for any $\varepsilon > 0$, any $H \in \mathcal{C}$, and any $\alpha < \omega_1$,*

$$\mathcal{Q}^\alpha(4\varepsilon, (f_n), H) \subseteq \mathcal{P}^\alpha(\varepsilon, (f_n), H) \subseteq \mathcal{Q}^\alpha(\frac{\varepsilon}{8}, (f_n), H).$$

Moreover, if (f_n) is pointwise convergent, then these are equivalent to the family $\mathcal{O}$:

$$\mathcal{P}^\alpha(2\varepsilon, (f_n), H) \subseteq \mathcal{O}^\alpha(\varepsilon, (f_n), H) \subseteq \mathcal{P}^\alpha(\varepsilon, (f_n), H).$$

PROOF. It suffices to show that both statements hold for the case $\alpha = 1$.

Let $x \in \mathcal{Q}(4\varepsilon, (f_n), H)$ and $L \in [\mathbb{N}]$. By the boundedness of $(f_n(x))_{n \in L}$, there exists $M \in [L]$ such that $(f_n(x))_{n \in M}$ converges. Say $a = \lim_{n \in M} f_n(x)$. Pick $m \in M$ so that $|f_m(x) - a| < \varepsilon/2$. Suppose U is a neighborhood of x. By continuity of f_m at x, there exists a neighborhood $V \subseteq U$ of x such that $|f_m(y) - a| < \varepsilon$ for all $y \in V \cap H$. Since $x \in \mathcal{D}_2(4\varepsilon, (f_n)_L, H)$, there exist $n \in L$ and $j \in F_{N(4\varepsilon)}$ such that

$$[f_n < 4j\varepsilon] \cap V \cap H \neq \emptyset \text{ and } [f_n > 4(j+1)\varepsilon] \cap V \cap H \neq \emptyset.$$

Therefore, there exists $x' \in V \cap H \subseteq U \cap H$ such that $|f_n(x') - a| > 2\varepsilon$. Thus

$$|f_m(x') - f_n(x')| \geq |f_n(x') - a| - |f_m(x') - a| > \varepsilon.$$

Hence, $x \in \mathcal{D}_1(\varepsilon, (f_n)_L, H)$. Since $L \in [\mathbb{N}]$ is arbitrary, $x \in \mathcal{P}(\varepsilon, (f_n), H)$.

Now suppose $x \in \mathcal{P}(\varepsilon, (f_n), H)$ and let $L \in [\mathbb{N}]$. As above, there exists $M \in [L]$ such that $a = \lim_{n \in M} f_n(x)$ exists. Suppose U is a neighborhood of x. Since $x \in \mathcal{D}_1(\varepsilon, (f_n)_P, H)$ for all $P \in [M]$, there exist $m_1 < n_1 < m_2 < n_2 < \ldots$ in M and $x'_k \in U \cap H$, $k \in \mathbb{N}$, such that

$$|f_{n_k}(x'_k) - f_{m_k}(x'_k)| > \varepsilon \text{ for all } k \in \mathbb{N}.$$

Therefore for each k, either

$$\left|f_{n_k}(x'_k) - a\right| > \tfrac{\varepsilon}{2} \text{ or } \left|f_{m_k}(x'_k) - a\right| > \tfrac{\varepsilon}{2}.$$

Thus for all k large enough, either

$$\left|f_{n_k}(x'_k) - f_{n_k}(x)\right| > \tfrac{\varepsilon}{4} \text{ or } \left|f_{m_k}(x'_k) - f_{m_k}(x)\right| > \tfrac{\varepsilon}{4}.$$

In either case, there exist $n \in L$ and $|j| \leq N\left(\frac{\varepsilon}{8}\right)$ such that

$$\left[f_n < \tfrac{j\varepsilon}{8}\right] \cap U \cap H \neq \emptyset \text{ and } \left[f_n > \tfrac{(j+1)\varepsilon}{8}\right] \cap U \cap H \neq \emptyset.$$

Hence $x \in \mathcal{D}_2\left(\frac{\varepsilon}{8}, (f_n)_L, H\right)$. Since $L \in [\mathbb{N}]$ is arbitrary,

$$x \in \bigcap_{L \in [\mathbb{N}]} \mathcal{D}_2\left(\tfrac{\varepsilon}{8}, (f_n)_L, H\right) = \mathcal{Q}\left(\tfrac{\varepsilon}{8}, (f_n), H\right).$$

Suppose (f_n) is pointwise convergent. We shall now prove the following:

$$\mathcal{P}(2\varepsilon, (f_n), H) \subseteq \mathcal{O}(\varepsilon, (f_n), H) \subseteq \mathcal{P}(\varepsilon, (f_n), H).$$

Suppose $x \notin \mathcal{O}(\epsilon, (f_n), H)$. There exists a neighborhood U of x such that for all $N \in \mathbb{N}$, there exists $n \geq N$ such that for all $M \in \mathbb{N}$,

$$\cap_{m \geq M} A^+_{n,m} \cap U \cap H = \emptyset \text{ and } \cap_{m \geq M} A^-_{n,m} \cap U \cap H = \emptyset.$$

Hence there exists $L \in [\mathbb{N}]$ such that for all $n \in L$ and all $M \in \mathbb{N}$,

$$\cap_{m \geq M} A^+_{n,m} \cap U \cap H = \emptyset \text{ and } \cap_{m \geq M} A^-_{n,m} \cap U \cap H = \emptyset.$$

Fix $n \in L$ and $x' \in U \cap H$. For all $M \in \mathbb{N}$,

$$x' \notin \cap_{m \geq M} A^+_{n,m} \text{ and } x' \notin \cap_{m \geq M} A^-_{n,m}$$

imply that there are m_1, $m_2 \geq M$ such that

$$f_{m_2}(x') - \varepsilon \leq f_n(x') \leq f_{m_1}(x') + \varepsilon.$$

Since (f_k) converges pointwise to a function f,

$$f(x') - \varepsilon \leq f_n(x') \leq f(x') + \varepsilon.$$

Therefore for all m, $n \in L$, and all $x' \in U \cap H$,

$$|f_m(x') - f_n(x')| \leq |f_m(x') - f(x')| + |f(x') - f_n(x')| \leq 2\varepsilon.$$

This shows that $x \notin \mathcal{D}_1(2\epsilon, (f_n)_L, H)$. Consequently, $x \notin \mathcal{P}(2\varepsilon, (f_n), H)$. This proves the first inclusion.

Now suppose $x \in \mathcal{O}(\epsilon, (f_n), H)$ and $L \in [\mathbb{N}]$. Let U be a neighborhood of x. Choose $N \in \mathbb{N}$ such that for all $n \geq N$, there exists $M = M(n) \in \mathbb{N}$ such that

$$\cap_{m \geq M} A^+_{n,m} \cap U \cap H \neq \emptyset \text{ or } \cap_{m \geq M} A^-_{n,m} \cap U \cap H \neq \emptyset.$$

It follows that if $m, n \in L$, $n \geq N$, $m \geq M(n)$, then there exists $x \in U \cap H$ such that $|f_n(x) - f_m(x)| > \varepsilon$. Hence $x \in \mathcal{D}_1(\varepsilon, (f_n)_L, H)$. Since $L \in [\mathbb{N}]$ is arbitrary, $x \in \bigcap_{L \in [\mathbb{N}]} \mathcal{D}_1(\varepsilon, (f_n)_L, H) = \mathcal{P}(\varepsilon, (f_n), H)$. □

3. Rademacher Systems

The aim of this section is to show that if $\mathcal{Q}^{\alpha}(\varepsilon,(f_n),K)\neq\emptyset$, then (f_n) contains a subsequence with a complex finite dimensional ℓ^1-structure (determined by the ordinal α). This is carried out by proving that certain subsequences of (f_n) exhibit the characteristic Rademacher functions-like behavior. The direct application of the assumption that $\mathcal{Q}^{\alpha}(\varepsilon,(f_n),K)\neq\emptyset$ only yields Rademacher-like functions with their "bumps" shifted vertically in a non-coherent way (see e.g. Proposition 3.2). The key is to observe that the shifts for each function are collapsible to the same level whenever α is a limit ordinal (Proposition 3.4 and Theorems 3.6 and 3.8). The proof of the next lemma is left to the reader. For a subset H of K, denote its interior by $\operatorname{int}H$.

LEMMA 3.1. *Let α be a countable ordinal, $\varepsilon>0$, and let H be a closed subset of K. Then*

$$\operatorname{int}H\cap\mathcal{Q}^{\alpha}(\varepsilon,(f_n),K)\subseteq\mathcal{Q}^{\alpha}(\varepsilon,(f_n),H).$$

In the following, let T and T_k denote respectively the infinite and finite dyadic trees

$$T=\bigcup_{j=0}^{\infty}\{0,1\}^j \text{ and } T_k=\bigcup_{j=0}^{k}\{0,1\}^j.$$

Recall that $F_N=\{j:|j|\le N\}$, where $N=N(\varepsilon)$.

PROPOSITION 3.2. *Suppose that $\mathcal{Q}^{k}(\varepsilon,(f_n),H)\neq\emptyset$ for some $k\in\mathbb{N}\cup\{0\}$. For all $M\in[\mathbb{N}]$, there exist $L\in[M]$ and $\Phi:T\to F_N$ such that if $I=\{n_1<n_2<...<n_k\}\subseteq L$, $(\varepsilon_j)_{j=1}^k\in\{0,1\}^k$, then there exists $x\in H$ such that for $1\le j\le k$,*

$$f_{n_j}(x)\begin{array}{ll}<\varepsilon\Phi((\varepsilon_1,\varepsilon_2,...,\varepsilon_{j-1})) & \text{if } \varepsilon_j=0,\\ >\varepsilon(\Phi((\varepsilon_1,\varepsilon_2,...,\varepsilon_{j-1}))+1) & \text{if } \varepsilon_j=1.\end{array}$$

($\Phi((\varepsilon_1,\varepsilon_2,...,\varepsilon_{j-1}))=\Phi(\emptyset)$ if $j=1$.)

PROOF. The statement holds vacuously if $k=0$. Assume that it holds for some $k\in\mathbb{N}$. Let $M\in[\mathbb{N}]$ be given and suppose $x\in\mathcal{Q}^{k+1}(\varepsilon,(f_n),H)$. Then $x\in\mathcal{D}_2(\varepsilon,(f_n)_M,Q^k)$, where $Q^k=\mathcal{Q}^k(\varepsilon,(f_n),H)$. Hence there exist $m_1\in M$, $j_1\in F_N$ such that

$$[f_{m_1}<j_1\varepsilon]\cap Q^k\neq\emptyset \text{ and } [f_{m_1}>(j_1+1)\varepsilon]\cap Q^k\neq\emptyset.$$

Let $x_0\in[f_{m_1}<j_1\varepsilon]\cap Q^k$ and $x_1\in[f_{m_1}>(j_1+1)\varepsilon]\cap Q^k$. Choose closed neighborhoods U_0 and U_1 of x_0 and x_1 respectively such that

$$U_0\subseteq[f_{m_1}<j_1\varepsilon] \text{ and } U_1\subseteq[f_{m_1}>(j_1+1)\varepsilon].$$

By Lemma 3.1, $x_0\in\mathcal{Q}^k(\varepsilon,(f_n),U_0)$ and $x_1\in\mathcal{Q}^k(\varepsilon,(f_n),U_1)$. From the inductive hypothesis, we obtain $L_1\in[M]$, $L_1>m_1$ and $\Psi_0',\Psi_1':T\to F_N$ such that if $I'=\{n_1<...<n_k\}\subseteq L_1$, $(\varepsilon_j)_{j=1}^k\in\{0,1\}^k$, then there exist $x_0'\in U_0$ and $x_1'\in U_1$ such that for $1\le j\le k$, $i=0,1$,

$$f_{n_j}(x_i') \begin{cases} < \varepsilon \Psi_i'((\varepsilon_1, \varepsilon_2, ..., \varepsilon_{j-1})) & \text{if } \varepsilon_j = 0, \\ > \varepsilon(\Psi_i'((\varepsilon_1, \varepsilon_2, ..., \varepsilon_{j-1})) + 1) & \text{if } \varepsilon_j = 1. \end{cases}$$

Define $\Phi_1 : T \to F_N$ by

$$\Phi_1(\emptyset) = j_1$$
$$\Phi_1((\varepsilon_1, \varepsilon_2, ..., \varepsilon_j)) = \Psi_{\varepsilon_1}'((\varepsilon_2, ..., \varepsilon_j)), \ j \geq 1.$$

Then if $I = \{n_1 < n_2 < ... < n_{k+1}\}$ with $n_1 = m_1$, $I \setminus \{n_1\} \subseteq L_1$ and $(\varepsilon_j)_{j=1}^{k+1} \in \{0,1\}^{k+1}$, there exists $x \in H$ such that

$$f_{n_j}(x) \begin{cases} < \varepsilon \Phi_1((\varepsilon_1, \varepsilon_2, ..., \varepsilon_{j-1})) & \text{if } \varepsilon_j = 0, \\ > \varepsilon(\Phi_1((\varepsilon_1, \varepsilon_2, ..., \varepsilon_{j-1})) + 1) & \text{if } \varepsilon_j = 1, \end{cases}$$

$1 \leq j \leq k+1$. Repeating the process, we obtain an infinite sequence of triples (L_i, m_i, Φ_i):

$$M = L_0 \supseteq L_1 \supseteq L_2 \supseteq ...,$$
$$m_1 < m_2 < m_3 < ..., \text{ and}$$
$$\Phi_1, \Phi_2, \Phi_3, ... \ : T \to F_N$$

such that

- $m_i < L_i$, $m_i \in L_{i-1}$ for all $i \in \mathbb{N}$,
- if $I = \{n_1 < n_2 < ... < n_{k+1}\}$, where $n_1 = m_i$ and $I \setminus \{n_1\} \subseteq L_i$, and $(\varepsilon_j)_{j=1}^{k+1} \in \{0,1\}^{k+1}$,

then there exists $x \in H$ such that

$$f_{n_j}(x) \begin{cases} < \varepsilon \Phi_i((\varepsilon_1, \varepsilon_2, ..., \varepsilon_{j-1})) & \text{if } \varepsilon_j = 0, \\ > \varepsilon(\Phi_i((\varepsilon_1, \varepsilon_2, ..., \varepsilon_{j-1})) + 1) & \text{if } \varepsilon_j = 1, \end{cases}$$

$1 \leq j \leq k+1$. Choose $p_1 < p_2 < p_3 < ...$ such that $\Phi_{p_i|T_k} = \Phi_{p_{i+1}|T_k}$ for all $i \in \mathbb{N}$. Let $L = \{m_{p_1}, m_{p_2}, ...\} \in [M]$ and $\Phi = \Phi_{p_1}$. Given $I = \{n_1 < ... < n_{k+1}\} \subseteq L$, say $n_1 = m_{p_\ell}$. Then $I \setminus \{n_1\} \subseteq L_{m_{p_\ell}}$. Therefore, for all $(\varepsilon_j)_{j=1}^{k+1} \in \{0,1\}^{k+1}$, there exists $x \in H$ such that

$$f_{n_j}(x) \begin{cases} < \varepsilon \Phi_{p_\ell}((\varepsilon_1, \varepsilon_2, ..., \varepsilon_{j-1})) = \varepsilon \Phi((\varepsilon_1, \varepsilon_2, ..., \varepsilon_{j-1})) & \text{if } \varepsilon_j = 0, \\ > \varepsilon(\Phi_{p_\ell}((\varepsilon_1, \varepsilon_2, ..., \varepsilon_{j-1})) + 1) = \varepsilon(\Phi((\varepsilon_1, \varepsilon_2, ..., \varepsilon_{j-1})) + 1) & \text{if } \varepsilon_j = 1, \end{cases}$$

$1 \leq j \leq k+1$. □

PROPOSITION 3.3. *Suppose $Q^\omega(\varepsilon, (f_n), H) \neq \emptyset$. For all $M \in [\mathbb{N}]$, there exist $L \in [M]$ and $\Phi : T \to F_N$ such that if $I = \{n_1 < n_2 < ... < n_k\} \in \mathcal{S}_1[L]$, $(\varepsilon_j)_{j=1}^{k} \in \{0,1\}^k$, then there exists $x \in H$ such that for $1 \leq j \leq k$,*

$$f_{n_j}(x) \begin{cases} < \varepsilon \Phi((\varepsilon_1, \varepsilon_2, ..., \varepsilon_{j-1})) & \textit{if } \varepsilon_j = 0, \\ > \varepsilon(\Phi((\varepsilon_1, \varepsilon_2, ..., \varepsilon_{j-1})) + 1) & \textit{if } \varepsilon_j = 1. \end{cases}$$

PROOF. Applying the last proposition, we obtain infinite sets $M \equiv L_0 \supseteq L_1 ...$ and $\Phi_1, \Phi_2, ... : T \to F_N$ such that if $I = \{n_1 < n_2 < ... < n_k\} \subseteq L_k$, and $(\varepsilon_j)_{j=1}^k \in \{0,1\}^k$, then there exists $x \in H$ such that for $1 \leq j \leq k$,

$$f_{n_j}(x) \begin{cases} < \varepsilon \Phi_k((\varepsilon_1, \varepsilon_2, ..., \varepsilon_{j-1})) & \text{if } \varepsilon_j = 0, \\ > \varepsilon(\Phi_k((\varepsilon_1, \varepsilon_2, ..., \varepsilon_{j-1})) + 1) & \text{if } \varepsilon_j = 1. \end{cases}$$

Choose $\ell_1 < \ell_2 < \ell_3 < ...$ and $p_1 < p_2 < p_3 < ...$ such that $\ell_i \in L_{p_i}$ for all $i \in \mathbb{N}$ and

$$\Phi_{p_i|T_{i-1}} = \Phi_{p_{i+1}|T_{i-1}} \text{ for all } i \in \mathbb{N}.$$

Define $L = \{\ell_1, \ell_2, \ell_3, ...\}$ and $\Phi : T \to F_N$ by

$$\Phi_{|T_{i-1}} = \Phi_{p_i|T_{i-1}} \text{ for all } i \in \mathbb{N}.$$

Then $L \in [M]$. Assume that $I = \{n_1 < n_2 < ... < n_k\} \in \mathcal{S}_1[L]$ and $(\varepsilon_j)_{j=1}^k \in \{0,1\}^k$ are given. If $n_1 = \ell_{i_0}$, then $k \leq i_0 \leq p_{i_0}$ and $I \subseteq L_{p_{i_0}}$. Choose arbitrary elements $n_{k+1} < n_{k+2} < ... < n_{p_{i_0}}$ in $L_{p_{i_0}}$ so that $n_{k+1} > n_k$. Also choose $\varepsilon_{k+1}, \varepsilon_{k+2}, ..., \varepsilon_{p_{i_0}} \in \{0,1\}$ arbitrarily. By the choice of $L_{p_{i_0}}$ and $\Phi_{p_{i_0}}$, there exists $x \in H$ such that for $1 \leq j \leq p_{i_0}$,

$$f_{n_j}(x) \begin{cases} < \varepsilon \Phi_{p_{i_0}}((\varepsilon_1, \varepsilon_2, ..., \varepsilon_{j-1})) & \text{if } \varepsilon_j = 0, \\ > \varepsilon\left(\Phi_{p_{i_0}}((\varepsilon_1, \varepsilon_2, ..., \varepsilon_{j-1})) + 1\right) & \text{if } \varepsilon_j = 1. \end{cases}$$

As $\Phi_{p_{i_0}}((\varepsilon_1, \varepsilon_2, ..., \varepsilon_{j-1})) = \Phi((\varepsilon_1, \varepsilon_2, ..., \varepsilon_{j-1}))$ for $1 \leq j \leq k$, we have the desired result. □

The next proposition is a simple consequence of the pigeonhole principle. It says that a finite-valued function on an infinite dyadic tree is constant on some dyadic subtree. This observation permits us to uniformize the Rademacher shifts obtained in Proposition 3.3. Given a node $\varphi = (\varepsilon_1, ..., \varepsilon_j) \in T$, we let the length of φ be $|\varphi| = j$. The empty node $\emptyset$ has length 0. The node $(\varepsilon_1, ..., \varepsilon_j, \varepsilon)$ is denoted by (φ, ε), $\varepsilon = 0, 1$. If $\psi = (\delta_1, ..., \delta_k) \in T$, $k \geq j$, and $\varepsilon_i = \delta_i$, $1 \leq i \leq j$, then we say that $\psi \geq \varphi$.

PROPOSITION 3.4. *Let X be a finite set. For any $\Phi : T \to X$, there exist $\Psi : T \to T$ and $p \in X$ such that*

(a) $\Phi(\Psi(\varphi)) = p$ *for* $\varphi \in T$,

(b) $|\varphi_1| \leq |\varphi_2|$ *implies* $|\Psi(\varphi_1)| \leq |\Psi(\varphi_2)|$,

(c) $\Psi((\varphi, 0)) \geq (\Psi(\varphi), 0)$ *and* $\Psi((\varphi, 1)) \geq (\Psi(\varphi), 1)$ *for all* $\varphi \in T$.

PROOF. It suffices to prove the proposition when X is a 2-element set, say $X = \{a, b\}$, $a \neq b$. We consider two cases.

Case 1. There exist $\varphi \in T$, $|\varphi| < \ell_0 < \ell_1 < \ell_2 < ...$ such that for all $\psi \geq \varphi$, $|\psi| \in \{\ell_i : i \in \mathbb{N} \cup \{0\}\}$, $\Phi(\psi) = a$.

Define Ψ as follows: Let $\Psi(\emptyset)$ be an arbitrary node such that $\Psi(\emptyset) \geq \varphi$ and $|\Psi(\emptyset)| = \ell_0$. If $\Psi(\psi)$ has been defined, let $\Psi((\psi, 0))$ and $\Psi((\psi, 1))$ be any two distinct nodes such that $\Psi((\psi, \varepsilon)) \geq (\Psi(\psi), \varepsilon)$ and $|\Psi((\psi, \varepsilon))| = \ell_{|\psi|+1}, \varepsilon = 0, 1$. It is easy to verify that Ψ satisfies the proposition with $p = a$.

Case 2. For all $\varphi \in T$, there exists ℓ_φ such that for all $\ell \geq \ell_\varphi$, there exists $\psi \geq \varphi$, $|\psi| = \ell$ such that $\Phi(\psi) = b$.

In particular there exists $\varphi_0 \in T$ such that $\Phi(\varphi_0) = b$. Define $\Psi(\emptyset) = \varphi_0$. Assume that $\Psi(\varphi)$ has been defined for all $\varphi \in T$ with $|\varphi| = m$. Choose ℓ such that $\ell \geq \ell_{(\Psi(\varphi),\varepsilon)}$ for all $\varphi \in T$ with $|\varphi| = m$ and $\varepsilon = 0, 1$. By the case assumption, for any φ such that $|\varphi| = m$, we can choose $\Psi((\varphi,\varepsilon))$, $\varepsilon = 0, 1$ so that $\Psi((\varphi,\varepsilon)) \geq (\Psi(\varphi),\varepsilon)$, $|\Psi((\varphi,\varepsilon))| = \ell$, and $\Phi(\Psi((\varphi,\varepsilon))) = b$. It is easy to verify that Ψ satisfies the proposition with $p = b$. □

DEFINITION 3.5. Let $(f_\alpha)_{\alpha\in A}$ be a finite set of real-valued functions and $\varepsilon > 0$. We say that $(f_\alpha)_{\alpha\in A}$ is an ε-Rademacher system on a (not necessarily closed) subset H of K if for any $(\varepsilon_\alpha)_{\alpha\in A} \subseteq \{0,1\}$,

$$H \cap \bigcap_{\{\varepsilon_\alpha=0\}} [f_\alpha < 0] \cap \bigcap_{\{\varepsilon_\alpha=1\}} [f_\alpha > \varepsilon] \neq \emptyset.$$

THEOREM 3.6. *Suppose $\mathcal{Q}^\omega(\varepsilon,(f_n),H) \neq \emptyset$. For all $M \in [\mathbb{N}]$, there exist $L \in [M]$ and $p \in F_N$ such that for all $I \in \mathcal{S}_1[L]$, $(f_n - p\varepsilon)_{n\in I}$ is an ε-Rademacher system on H.*

PROOF. Let $M \in [\mathbb{N}]$ be given. Choose $L_0 \in [M]$ and $\Phi : T \to F_N$ according to Proposition 3.3. Then apply Theorem 3.4 with Φ and $X = F_N$ to obtain Ψ and $p \in F_N$. For all $k \in \mathbb{N}\cup\{0\}$, let $i_k = |\Psi(\varphi)|$ for any $\varphi \in T$ with $|\varphi| = k$. It is possible to choose $L \in [L_0]$ such that for all $I \in \mathcal{S}_1[L]$ with $|I| = k$, there exists $J = \{\ell_1 < \ell_2 < ... < \ell_{i_{k-1}+1}\} \in \mathcal{S}_1[L_0]$ such that $I = \{\ell_{i_0+1}, \ell_{i_1+1}, ..., \ell_{i_{k-1}+1}\}$. Now suppose $I = \{n_1 < ... < n_k\} \in \mathcal{S}_1[L]$. Choose J as above. Given $(\varepsilon_j)_{j=1}^k \in \{0,1\}^k$, choose $(\delta_j)_{j=1}^{i_{k-1}+1} \in \{0,1\}^{i_{k-1}+1}$ so that $(\delta_1,\delta_2,...,\delta_{i_0}) = \Psi(\emptyset)$, $(\delta_1,\delta_2,...,\delta_{i_j}) = \Psi((\varepsilon_1,...,\varepsilon_j))$, $1 \leq j \leq k-1$, and $\delta_{i_{k-1}+1} = \varepsilon_k$. There exists $x \in H$ such that for $1 \leq j \leq i_{k-1}+1$,

$$f_{\ell_j}(x) \begin{cases} < \varepsilon\Phi((\delta_1,\delta_2,...,\delta_{j-1})) & \text{if } \delta_j = 0, \\ > \varepsilon(\Phi((\delta_1,\delta_2,...,\delta_{j-1}))+1) & \text{if } \delta_j = 1. \end{cases}$$

Hence, for $1 \leq j < k$,

$$f_{\ell_{i_{j-1}+1}}(x) \begin{cases} < \varepsilon\Phi\left((\delta_1,\delta_2,...,\delta_{i_{j-1}})\right) & \text{if } \delta_{i_{j-1}+1} = 0, \\ > \varepsilon\left(\Phi\left((\delta_1,\delta_2,...,\delta_{i_{j-1}})\right)+1\right) & \text{if } \delta_{i_{j-1}+1} = 1. \end{cases}$$

Note that for $1 \leq j < k$,

$$(\delta_1,\delta_2,...,\delta_{i_j}) = \Psi((\varepsilon_1,...,\varepsilon_j)) \geq (\Psi((\varepsilon_1,...,\varepsilon_{j-1})),\varepsilon_j) = (\delta_1,...,\delta_{i_{j-1}},\varepsilon_j).$$

Thus $\delta_{i_{j-1}+1} = \varepsilon_j$. Hence for $1 \leq j < k$

$$f_{n_j}(x) = f_{\ell_{i_{j-1}+1}}(x) \begin{cases} < \varepsilon\Phi(\Psi(\varepsilon_1,\varepsilon_2,...,\varepsilon_j)) = p\varepsilon & \text{if } \varepsilon_j = 0, \\ > \varepsilon(\Phi(\Psi(\varepsilon_1,\varepsilon_2,...,\varepsilon_j))+1) = (p+1)\varepsilon & \text{if } \varepsilon_j = 1. \end{cases}$$

Therefore,

$$x \in H \cap \bigcap_{\{\varepsilon_j=0\}} \left[f_{n_j} - p\varepsilon < 0\right] \cap \bigcap_{\{\varepsilon_j=1\}} \left[f_{n_j} - p\varepsilon > \varepsilon\right].$$

Hence $(f_n - p\varepsilon)_{n\in I}$ is an ε-Rademacher system on H. □

In the proof of the next theorem, we will use the following simple lemma whose proof we omit.

LEMMA 3.7. *Let L and M be infinite subsets of $\mathbb{N}$. Suppose there exists $n \in \mathbb{N}$ such that $L \cap [n, \infty) \subseteq M \cap [n, \infty)$ and $|L \cap [1, n)| \leq |M \cap [1, n)|$. If α is a countable ordinal and $n \leq I \in \mathcal{S}_\alpha [L]$, then $I \in \mathcal{S}_\alpha [M]$.*

THEOREM 3.8. *Suppose $\mathcal{Q}^\alpha (\varepsilon, (f_n), H) \neq \emptyset$, where α is a countable limit ordinal whose Cantor normal form is $\omega^{\alpha_1} \cdot m_1 + ... + \omega^{\alpha_k} \cdot m_k$. Then for all $M \in [\mathbb{N}]$, there exist $L \in [M]$ and $\left(p_j^i\right)_{j=1}^{m_i}{}_{i=1}^{k} \subseteq F_N$ such that for all $\left(I_j^i\right)_{j=1}^{m_i}{}_{i=1}^{k} \in \mathcal{T}_\alpha [L]$, $\bigcup_{i=1}^k \bigcup_{j=1}^{m_i} \left\{f_n - p_j^i \varepsilon : n \in I_j^i\right\}$ is an ε-Rademacher system on H.*

PROOF. The proof is by induction on all countable limit ordinals α. The result holds for $\alpha = \omega$ by Theorem 3.6. Now suppose α is a countable limit ordinal and the result holds for all limit ordinals $< \alpha$. Write $\alpha = \omega^{\alpha_1} \cdot m_1 + ... + \omega^{\alpha_k} \cdot m_k$ in Cantor normal form. Let $M \in [\mathbb{N}]$ be given.

Case 1. α_k is a successor and $\alpha_k > 1$.

For each $r \in \mathbb{N}$, the inductive hypothesis applies to the limit ordinal

$$\eta_r = \omega^{\alpha_1} \cdot m_1 + ... + \omega^{\alpha_k} \cdot (m_k - 1) + \omega^{\alpha_k - 1} \cdot r < \alpha.$$

Let

$$m_i (r) = \begin{cases} m_i & 1 \leq i < k, \\ m_k - 1 \quad \text{if} & i = k, \\ r & i = k + 1. \end{cases}$$

Then there exist infinite sets $M \supseteq L_1 \supseteq L_2 \supseteq ...$ and $\left(p_j^i (r)\right)_{j=1}^{m_i(r)}{}_{i=1}^{k+1} \subseteq F_N$ such that for all $\left(I_j^i\right)_{j=1}^{m_i(r)}{}_{i=1}^{k+1} \in \mathcal{T}_{\eta_r} [L_r]$, $\bigcup_{i=1}^{k+1} \bigcup_{j=1}^{m_i(r)} \left\{f_n - p_j^i (r) \varepsilon : n \in I_j^i\right\}$ is an ε-Rademacher system on H. Choose strictly increasing sequences (r_t) and (t_s) such that

$$\begin{aligned} &p_j^i (r_t) = p_j^i (r_{t+1}), \ 1 \leq j \leq m_i (r_t), \ 1 \leq i \leq k, \ t \in \mathbb{N}, \\ &p_j^{k+1} (r_t) = p_j^{k+1} (r_{t+1}), \ 1 \leq j \leq t, \text{ and} \\ &p_{t_s}^{k+1} (r_{t_s}) = p_{t_{s+1}}^{k+1} \left(r_{t_{s+1}}\right), \ s \in \mathbb{N}. \end{aligned}$$

Define $p_j^i = p_j^i (r_1)$, $1 \leq i \leq k$ and $p = p_{t_1}^{k+1} (r_{t_1})$. Choose $\ell_1 < \ell_2 < \ell_3 < ...$ such that $\ell_s \in L_{r_{t_s}}$ for all $s \in \mathbb{N}$ and $\left|\left\{\ell \in L_{r_{t_s}} : \ell \leq \ell_s\right\}\right| \geq s$ for all $s \in \mathbb{N}$. Let $L = \{\ell_1, \ell_2, \ell_3, ...\}$. Suppose $I \in \mathcal{T}_\alpha [L]$, then we may write

$$I = (I_0^k, (I_j^k)_{j=1}^{m_k - 1}, (I_j^{k-1})_{j=1}^{m_{k-1}}, ..., (I_j^1)_{j=1}^{m_1}),$$

where $I_j^i \in \mathcal{S}_{\alpha_i} [L]$. Say $\min I_0^k = \ell_{s_0}$. Then $I_j^i \subseteq L_{r_{t_{s_0}}}$ and $I_0^k \in (\mathcal{S}_{\alpha_k - 1} [L])^q$ for some $q \leq s_0$. By Lemma 3.7, we see that

$$(I_0^k, (I_j^k)_{j=1}^{m_k - 1}, (I_j^{k-1})_{j=1}^{m_{k-1}}, ..., (I_j^1)_{j=1}^{m_1}) \in \mathcal{T}_{\eta_q} [L_{r_{t_{s_0}}}].$$

We may write I_0^k as $J_1 \cup J_2 \cup ... \cup J_{t_{s_0}}$, where $J_1 < J_2 < ... < J_{t_{s_0}}$, $J_1, ..., J_{t_{s_0}} \in \mathcal{S}_{\alpha_k - 1} \left[L_{r_{t_{s_0}}}\right]$ and $J_u = \emptyset$ if $u \notin \{t_1, t_2, ..., t_{s_0}\}$. By the choice of $L_{r_{t_{s_0}}}$

$$\bigcup_{j=1}^{r_{t_{s_0}}} \{f_n - p_j^{k+1}(r_{t_{s_0}})\varepsilon : n \in J_j\} \cup \bigcup_{i=1}^{k} \bigcup_{j=1}^{m_i(r_{t_{s_0}})} \{f_n - p_j^i(r_{t_{s_0}})\varepsilon : n \in I_j^i\}$$

is an ε-Rademacher system on H. But for $1 \leq j \leq s_0$, $p_{t_j}^{k+1}(r_{t_{s_0}}) = p_{t_j}^{k+1}(r_{t_j}) = p_{t_1}^{k+1}(r_{t_1}) = p$ and $J_u = \emptyset$ if $u \notin \{t_1, t_2, ..., t_{s_0}\}$. Similarly, $p_j^i(r_{t_{s_0}}) = p_j^i(r_1) = p_j^i$ for $1 \leq i \leq k$, $1 \leq j \leq m_i(r_{t_{s_0}})$. Hence

$$\bigcup_{n \in I_0^k} \{f_n - p\varepsilon\} \cup \bigcup_{i=1}^{k} \bigcup_{j=1}^{m_i(r_{t_{s_0}})} \{f_n - p_j^i \varepsilon : n \in I_j^i\}$$

is an ε-Rademacher system on H.

Case 2. α_k is a limit ordinal.

Let (ρ_r) be the sequence of ordinals used to define $\mathcal{S}_{\alpha_k}$ and let

$$\eta_r = \omega^{\alpha_1} \cdot m_1 + ... + \omega^{\alpha_k} \cdot (m_k - 1) + \omega^{\rho_r}, \ r \in \mathbb{N}.$$

Define

$$m_i' = \begin{cases} m_i & 1 \leq i < k, \\ m_k - 1 \quad \text{if} & i = k, \\ 1 & i = k+1. \end{cases}$$

Applying the inductive hypothesis to η_r yields infinite sets $M \supseteq L_1 \supseteq L_2 \supseteq ...$ and $\left(\left(p_j^i(r)\right)_{j=1}^{m_i'}\right)_{i=1}^{k+1} \subseteq F_N$ such that for all $\left(\left(I_j^i\right)_{j=1}^{m_i'}\right)_{i=1}^{k+1} \in \mathcal{T}_{\eta_r}[L_r]$,

$$\bigcup_{i=1}^{k+1} \bigcup_{j=1}^{m_i'} \{f_n - p_j^i(r)\varepsilon : n \in I_j^i\}$$

is an ε-Rademacher system on H. Choose $r_1 < r_2 < r_3 < ...$ such that

$$p_j^i(r_t) = p_j^i(r_{t+1}), \ 1 \leq j \leq m_i', \ 1 \leq i \leq k+1, \ t \in \mathbb{N}.$$

Define $p_j^i = p_j^i(r_1)$, $1 \leq j \leq m_i'$, $1 \leq i \leq k+1$. Choose $\ell_1 < \ell_2 < \ell_3 < ...$ such that

(a) $\ell_s \in L_{r_s}$ for all $s \in \mathbb{N}$,

(b) $|\{\ell \in L_{r_s} : \ell \leq \ell_s\}| \geq s$ for all $s \in \mathbb{N}$,

(c) $\ell_s \leq I \in \bigcup_{r \leq s} \mathcal{S}_{\rho_r}[L_{r_s}] \Longrightarrow I \in \mathcal{S}_{\rho_{r_s}}[L_{r_s}]$. ([**9**, Proposition 3.2]).

Let $L = \{\ell_1, \ell_2, \ell_3, ...\}$. Suppose $\left(\left(I_j^i\right)_{j=1}^{m_i}\right)_{i=1}^{k} \in \mathcal{T}_\alpha[L]$. Say $\min I_1^k = \ell_{s_0}$. Then $I_j^i \subseteq L_{r_{s_0}}$ for all i, j. By (b) and Lemma 3.7, $\left(\left(I_j^i\right)_{j=1}^{m_i}\right)_{i=1}^{k} \in \mathcal{T}_\alpha\left[L_{r_{s_0}}\right]$. Also $\ell_{s_0} \leq I_1^k \in \mathcal{S}_{\alpha_k}\left[L_{r_{s_0}}\right] \Rightarrow \ell_{s_0} \leq I_1^k \in \bigcup_{r \leq s_0} \mathcal{S}_{\rho_r}\left[L_{r_{s_0}}\right]$. Hence $I_1^k \in \mathcal{S}_{\rho_{r_{s_0}}}\left[L_{r_{s_0}}\right]$ by (c). Thus $\left(\left(I_j^i\right)_{i=1}^{k}\right)_{j=1}^{m_i} \in \mathcal{T}_{\eta_{r_{s_0}}}\left[L_{r_{s_0}}\right]$. By the choice of $L_{r_{s_0}}$,

$$\{f_n - p_1^{k+1}\varepsilon : n \in I_1^k\} \cup \bigcup_{j=2}^{m_k} \{f_n - p_{j-1}^k \varepsilon : n \in I_j^k\} \cup \bigcup_{i=1}^{k-1} \bigcup_{j=1}^{m_i} \{f_n - p_j^i \varepsilon : n \in I_j^i\}$$

is an ε-Rademacher system on H.

Case 3. $\alpha_k = 1$.

Let $\alpha' = \omega^{\alpha_1} \cdot m_1 + ... + \omega^{\alpha_k} \cdot (m_k - 1)$, so that $\alpha = \alpha' + \omega$. Then $\mathcal{Q}^\omega(\varepsilon, (f_n), Q^{\alpha'})$ is nonempty, where $Q^{\alpha'} = \mathcal{Q}^{\alpha'}(\varepsilon, (f_n), H)$. Let $M \in [\mathbb{N}]$ be given. By Theorem 3.6, there exists $L_0 \in [M]$ and $p \in F_N$ such that for all $I \in \mathcal{S}_1[L_0]$, $(f_n - p\varepsilon)_{n \in I}$ is an ε-Rademacher system on $Q^{\alpha'}$. Set $m_i' = m_i$ if $1 \leq i < k$ and $m_k' = m_k - 1$.

Claim. *For all $q \in \mathbb{N}$ and all $P \in [L_0]$, there exist $N' \in [P]$ and $p_j^i : \{0,1\}^q \to F_N$, $1 \leq i \leq k$, $1 \leq j \leq m_i'$, such that for all $I = \{n_1 < ... < n_q\} \subseteq$*

N', *all* $\left(\left(I_j^i\right)_{j=1}^{m_i'}\right)_{i=1}^{k} \in \mathcal{T}_{\alpha'}[N']$ *with* $I < I_1^k$, *and all* $\varphi = (\varepsilon_1, \varepsilon_2, ..., \varepsilon_q) \in \{0,1\}^q$, $\bigcup_{i=1}^k \bigcup_{j=1}^{m_i'} \left\{f_n - p_j^i(\varphi)\varepsilon : n \in I_j^i\right\}$ *is an* ε*-Rademacher system on*

$$\bigcap_{\{\varepsilon_\ell=0\}} [f_{n_\ell} < p\varepsilon] \cap \bigcap_{\{\varepsilon_\ell=1\}} [f_{n_\ell} > (p+1)\varepsilon].$$

Proof of Claim:

Let $P \in [L_0]$ and $q \in \mathbb{N}$. Pick $n_1 < n_2 < ... < n_q$ in P such that $\{n_1, n_2, ..., n_q\} \in \mathcal{S}_1[P]$. Then $\{n_1, n_2, ..., n_q\} \in \mathcal{S}_1[L_0]$. Hence $(f_{n_\ell} - p\varepsilon)_{\ell=1}^q$ is an ε-Rademacher system on $Q^{\alpha'}$. Given $\varphi = (\varepsilon_1, \varepsilon_2, ..., \varepsilon_q) \in \{0,1\}^q$,

$$\bigcap_{\{\varepsilon_\ell=0\}} [f_{n_\ell} < p\varepsilon] \cap \bigcap_{\{\varepsilon_\ell=1\}} [f_{n_\ell} > (p+1)\varepsilon] \cap Q^{\alpha'} \neq \emptyset.$$

Thus for each $\varphi \in \{0,1\}^q$, there is an open set U_φ such that $U_\varphi \cap Q^{\alpha'} \neq \emptyset$ and $\overline{U_\varphi} \subseteq \bigcap_{\{\varepsilon_\ell=0\}} [f_{n_\ell} < p\varepsilon] \cap \bigcap_{\{\varepsilon_\ell=1\}} [f_{n_\ell} > (p+1)\varepsilon]$. By Lemma 3.1, $\mathcal{Q}^{\alpha'}\left(\varepsilon, (f_n), \overline{U_\varphi}\right) \neq \emptyset$. Applying the inductive hypothesis, there exist $N_1 \in [P]$, $N_1 > n_q$ and $p_j^i : \{0,1\} \to F_N$, $1 \leq i \leq k$, $1 \leq j \leq m_i'$ such that for all $\left(\left(I_j^i\right)_{j=1}^{m_i'}\right)_{i=1}^{k} \in \mathcal{T}_{\alpha'}[N_1]$, and all $\varphi \in \{0,1\}^q$, $\bigcup_{i=1}^k \bigcup_{j=1}^{m_i'} \left\{f_n - p_j^i(\varphi)\varepsilon : n \in I_j^i\right\}$ is an ε-Rademacher system on $\overline{U_\varphi}$. Pick $n_{q+1} \in N_1$ such that $|[1, n_{q+1}] \cap N_1| \geq q+1$. Continue in the same manner to choose infinite sets $P \supseteq N_1 \supseteq N_2 \supseteq ...$ and $n_1 < ... < n_q < n_{q+1} < ...$ such that

(a) $n_1, n_2, ..., n_q \in P$ and $n_{q+r} \in N_r$, $r \in \mathbb{N}$,
(b) $n_{q+r-1} < N_r$, $r \in \mathbb{N}$,
(c) $|[1, n_{q+r}] \cap N_r| \geq q + r$, $r \in \mathbb{N}$, and,
(d) for all $I = \left\{n_{s_1} < ... < n_{s_q}\right\} \subseteq \{n_1, ..., n_{q+r-1}\}$, $r \in \mathbb{N}$, there exist functions $p_j^i(r, I) : \{0,1\}^q \to F_N$ such that for all $\left(\left(I_j^i\right)_{j=1}^{m_i'}\right)_{i=1}^{k} \in \mathcal{T}_{\alpha'}[N_r]$ and all $\varphi = (\varepsilon_1, \varepsilon_2, ..., \varepsilon_q) \in \{0,1\}^q$, $\bigcup_{i=1}^k \bigcup_{j=1}^{m_i'} \left\{f_n - p_j^i(r, I)(\varphi)\varepsilon : n \in I_j^i\right\}$ is an ε-Rademacher system on $\bigcap_{\{\varepsilon_\ell=0\}} [f_{n_{s_\ell}} < p\varepsilon] \cap \bigcap_{\{\varepsilon_\ell=1\}} [f_{n_{s_\ell}} > (p+1)\varepsilon]$.

Now let $N'' = \{n_1, n_2, n_3, ...\}$. If $I = \left\{n_{s_1} < n_{s_2} < ... < n_{s_q}\right\}$ and $\left(\left(I_j^i\right)_{j=1}^{m_i'}\right)_{i=1}^{k} \in \mathcal{T}_{\alpha'}[N'']$ with $I < I_1^k$, then $I_j^i \subseteq N_{s_q-q+1}$ by (a). It follows from (c) and Lemma 3.7 that $\left(\left(I_j^i\right)_{j=1}^{m_i'}\right)_{i=1}^{k} \in \mathcal{T}_{\alpha'}\left[N_{s_q-q+1}\right]$. Thus, for all $\varphi = (\varepsilon_1, ..., \varepsilon_q) \in \{0,1\}^q$,

$$\bigcup_{i=1}^{k} \bigcup_{j=1}^{m_i'} \left\{f_n - p_j^i(s_q - q + 1, I)(\varphi)\varepsilon : n \in I_j^i\right\}$$

is an ε-Rademacher system on $\bigcap_{\{\varepsilon_\ell=0\}} [f_{n_{s_\ell}} < p\varepsilon] \cap \bigcap_{\{\varepsilon_\ell=1\}} [f_{n_{s_\ell}} > (p+1)\varepsilon]$. By Ramsey's Theorem, there exists $N' \in [N'']$ such that

$$p_j^i(s_q - q + 1, I) = p_j^i\left(s_q' - q + 1, J\right)$$

for all $1 \leq j \leq m_i', 1 \leq i \leq k$, and all subsets $I = \left\{n_{s_1} < ... < n_{s_q}\right\}$ and $J = \left\{n_{s_1'} < ... < n_{s_q'}\right\}$ of N'. Denote by p_j^i the common value. It is now clear that if $I = \left\{n_{s_1} < ... < n_{s_q}\right\} \subseteq N'$, $\left(\left(I_j^i\right)_{j=1}^{m_i'}\right)_{i=1}^{k} \in \mathcal{T}_{\alpha'}[N']$, $I < I_1^k$ and $\varphi = (\varepsilon_1, \varepsilon_2, ..., \varepsilon_q) \in$

$\{0,1\}^q$, then

$$\bigcup_{i=1}^{k}\bigcup_{j=1}^{m_i'}\left\{f_n - p_j^i(\varphi)\,\varepsilon : n \in I_j^i\right\}$$

is an ε-Rademacher system on $\bigcap_{\{\varepsilon_\ell=0\}}[f_{n_{s_\ell}} < p\varepsilon] \cap \bigcap_{\{\varepsilon_\ell=1\}}[f_{n_{s_\ell}} > (p+1)\varepsilon]$. This proves the claim.

Applying the claim, we can find infinite sets $L_0 \supseteq L_1 \supseteq \dots$ and $p_j^i(q) : \{0,1\}^q \to F_N$, $q \in \mathbb{N}$, such that for all $I \subseteq L_q$ and $\left(I_j^i\right)_{j=1}^{m_i'}{}_{i=1}^{k} \in \mathcal{T}_{\alpha'}[L_q]$, with $I = \{n_1 < \dots < n_q\} < I_1^k$, $\bigcup_{i=1}^k \bigcup_{j=1}^{m_i'}\left\{f_n - p_j^i(q)(\varphi)\,\varepsilon : n \in I_j^i\right\}$ is an ε-Rademacher system on $\bigcap_{\{\varepsilon_\ell=0\}}[f_{n_\ell} < p\varepsilon] \cap \bigcap_{\{\varepsilon_\ell=1\}}[f_{n_\ell} > (p+1)\varepsilon]$ for all $\varphi = (\varepsilon_1, \varepsilon_2, \dots, \varepsilon_q) \in \{0,1\}^q$. Define $\Phi : T \to F_N^{m_1+m_2+\dots+m_{k-1}+m_k-1}$ by $\Phi(\varphi) = \left(p_j^i(r)(\varphi)\right)_{j=1}^{m_i'}{}_{i=1}^{k}$ for all $|\varphi| = r \in \mathbb{N}$, with $\Phi(\emptyset)$ defined arbitrarily. By Proposition 3.4, there exists $\Psi : T \to T$ and $\left(p_j^i\right)_{j=1}^{m_i'}{}_{i=1}^{k}$ such that

(a) $\Phi(\Psi(\varphi)) = \left(p_j^i\right)$ for all $\varphi \in T$,
(b) $|\Psi(\varphi_1)| \le |\Psi(\varphi_2)|$ whenever $|\varphi_1| \le |\varphi_2|$,
(c) for all $\varphi \in T$, $\Psi((\varphi, \varepsilon)) \ge (\Psi(\varphi), \varepsilon)$, $\varepsilon = 0, 1$.

Without loss of generality, assume that $|\Psi(\emptyset)| > 0$. Let $t_q = |\Psi(\varphi)|$ for any φ such that $|\varphi| = q$. Choose $\ell_1 < \ell_2 < \ell_3 < \dots$ such that $\ell_q \in L_{t_q}$ for all $q \in \mathbb{N}$, $|[1, \ell_1) \cap L_{t_1}| \ge t_0$, and $\left|(\ell_q, \ell_{q+1}) \cap L_{t_{q+1}}\right| \ge t_q$ for all $q \in \mathbb{N}$. Let $L = \{\ell_1, \ell_2, \ell_3, \dots\}$. Then $L \in [M]$. Now suppose $I \in \mathcal{S}_1[L]$, $|I| = q$, $\left(I_j^i\right)_{j=1}^{m_i'}{}_{i=1}^{k} \in \mathcal{T}_{\alpha'}[L]$ and $I < I_1^k$. With the given choice of L, one can choose $J = \left\{u_1 < u_2 < \dots < u_{t_q}\right\} \subseteq L_{t_q}$, such that $J < I_1^k$ and $I = \left\{u_{t_0+1}, u_{t_1+1}, \dots, u_{t_{q-1}+1}\right\}$. Also note that $|[1, \ell_q] \cap L| = q \le t_{q-1} + 1 \le \left|[1, \ell_q] \cap L_{t_q}\right|$. Hence $\left(I_j^i\right)_{j=1}^{m_i'}{}_{i=1}^{k} \in \mathcal{T}_{\alpha'}\left[L_{t_q}\right]$ by Lemma 3.7. By choice of L_{t_q},

$$\bigcup_{i=1}^{k}\bigcup_{j=1}^{m_i'}\left\{f_n - p_j^i(t_q)(\varphi)\,\varepsilon : n \in I_j^i\right\}$$

is an ε-Rademacher system on $\bigcap_{\{\varepsilon_k=0\}}[f_{u_k} < p\varepsilon] \cap \bigcap_{\{\varepsilon_k=1\}}[f_{u_k} > (p+1)\varepsilon]$ for all $\varphi = \left(\varepsilon_1, \varepsilon_2, \dots, \varepsilon_{t_q}\right) \in \{0,1\}^{t_q}$. If $\psi = (\delta_1, \delta_2, \dots, \delta_q) \in \{0,1\}^q$, write $\varphi = \Psi(\psi)$ as $\varphi = \left(\varepsilon_1, \varepsilon_2, \dots, \varepsilon_{t_q}\right) \in \{0,1\}^{t_q}$. Note that in particular, $\varepsilon_{t_{r-1}+1} = \delta_r$, $1 \le r \le q$. Thus

$$\bigcup_{i=1}^{k}\bigcup_{j=1}^{m_i'}\left\{f_n - p_j^i(t_q)(\Psi(\psi))\,\varepsilon : n \in I_j^i\right\} = \bigcup_{i=1}^{k}\bigcup_{j=1}^{m_i'}\left\{f_n - p_j^i\varepsilon : n \in I_j^i\right\}$$

is an ε-Rademacher system on

$$\begin{aligned}
&\bigcap_{\{\varepsilon_k=0\}} [f_{u_k} < p\varepsilon] \cap \bigcap_{\{\varepsilon_k=1\}} [f_{u_k} > (p+1)\varepsilon] \\
&\subseteq \bigcap_{\{\varepsilon_{t_{r-1}+1}=0\}} \left[f_{u_{t_{r-1}+1}} < p\varepsilon\right] \cap \bigcap_{\{\varepsilon_{t_{r-1}+1}=1\}} \left[f_{u_{t_{r-1}+1}} > (p+1)\varepsilon\right] \\
&= \bigcap_{\{\delta_r=0\}} \left[f_{u_{t_{r-1}+1}} < p\varepsilon\right] \cap \bigcap_{\{\delta_r=1\}} \left[f_{u_{t_{r-1}+1}} > (p+1)\varepsilon\right].
\end{aligned}$$

Therefore, $\{f_n - p\varepsilon : n \in I\} \cup \bigcup_{i=1}^{k} \bigcup_{j=1}^{m_i'} \{f_n - p_j^i\varepsilon : n \in I_j^i\}$ is an ε-Rademacher system on H. □

In view of the equivalence of the families of derivations $\mathcal{Q}$ and $\mathcal{O}$ when (f_n) is pointwise convergent, the next theorem is a strengthening of Theorem 3.1 in [**1**]. If α is a countable ordinal and $\varepsilon > 0$, a bounded sequence (f_n) in $C_b(K)$ is said to be an ℓ^1-$\mathcal{T}_\alpha$-spreading model with constant ε if for all $(a_n) \in c_{00}$ and all $I \in \mathcal{T}_\alpha$,

$$\left\| \sum_{n\in I} a_n f_n \right\| \geq \varepsilon \sum_{n\in I} |a_n|.$$

An $\ell^1 - \mathcal{T}_{\omega^\alpha}-$ spreading model is usually called an $\ell^1_\alpha-$spreading model in the literature [**2**].

THEOREM 3.9. *Suppose $\mathcal{Q}^\alpha(\varepsilon, (f_n), K) \neq \emptyset$, where α is a limit ordinal. Then there exists $L \in [\mathbb{N}]$ such that $(f_n)_L$ is an ℓ^1-$\mathcal{T}_\alpha$-spreading model with constant $\frac{\varepsilon}{2}$.*

PROOF. Suppose $\mathcal{Q}^\alpha(\varepsilon, (f_n), K) \neq \emptyset$, for some limit ordinal α. By Theorem 3.8, there exists $L \in [\mathbb{N}]$ and $\left(p_j^i\right)_{j=1}^{m_i}{}_{i=1}^{k} \subseteq F_N$ such that for all $\left(I_j^i\right)_{j=1}^{m_i}{}_{i=1}^{k} \in \mathcal{T}_\alpha[L]$, $\bigcup_{i=1}^{k} \bigcup_{j=1}^{m_i} \{f_n - p_j^i\varepsilon : n \in I_j^i\}$ is an ε-Rademacher system on H. Let $(a_n) \in c_{00}$ and $I = \left(I_j^i\right)_{j=1}^{m_i}{}_{i=1}^{k} \in \mathcal{T}_\alpha[L]$ be given. Define, for $n \in I$,

$$\varepsilon_n = \begin{cases} 0 & \text{if } a_n < 0 \\ 1 & \text{if } a_n \geq 0. \end{cases}$$

Since $\bigcup_{i=1}^{k} \bigcup_{j=1}^{m_i} \{f_n - p_j^i\varepsilon : n \in I_j^i\}$ is an ε-Rademacher system on H, there exist (note that i, j are completely determined by n)

$$x \in H \cap \bigcap_{\{\varepsilon_n=0\}} \{f_n - p_j^i\varepsilon < 0\} \cap \bigcap_{\{\varepsilon_n=1\}} \{f_n - p_j^i\varepsilon \geq \varepsilon\}$$

and

$$x' \in H \cap \bigcap_{\{\varepsilon_n=1\}} \{f_n - p_j^i\varepsilon < 0\} \cap \bigcap_{\{\varepsilon_n=0\}} \{f_n - p_j^i\varepsilon \geq \varepsilon\}.$$

Then

$$\begin{aligned}\left|\sum_{n\in I} a_n f_n(x) - \sum_{n\in I} a_n f_n(x')\right| &\geq \sum_{n\in I} a_n \left[f_n(x) - f_n(x')\right] \\ &= \sum_{n\in I} a_n \left[\left(f_n(x) - p_j^i\varepsilon\right) - \left(f_n(x') - p_j^i\varepsilon\right)\right] \\ &\geq \varepsilon \sum_{n\in I} |a_n|.\end{aligned}$$

Hence $\left\|\sum_{n\in I} a_n f_n\right\| \geq \dfrac{\varepsilon}{2}\sum_{n\in I} |a_n|$, as required. □

If K is compact metric, then the following partial converse to Theorem 3.9 holds. The proof is similar to that of Theorem 3.2 in [**7**] and will be omitted.

THEOREM 3.10. *Let (f_n) be a uniformly bounded sequence of continuous functions on a compact metric space K. Suppose there exists $\varepsilon > 0$ such that (f_n) is an ℓ^1_α-spreading model with constant ε. Then there exists $M \in [\mathbb{N}]$ such that*

$$\mathcal{P}^{\omega^\alpha}\left(\tfrac{\varepsilon}{3}, (f_n)_M, K\right) \neq \emptyset.$$

References

[1] D. E. ALSPACH AND S. A. ARGYROS, Complexity of weakly null sequences, Diss. Math. **321** (1992), 1-44.

[2] S. A. ARGYROS, S. MERCOURAKIS AND A. TSARPALIAS, Convex unconditionality and summability of weakly null sequences, Israel J. Math. **107** (1998), 157-193.

[3] DENNY H. LEUNG AND WEE-KEE TANG, On Functions of Baire Class One, preprint.

[4] R. HAYDON, E. ODELL AND H. P. ROSENTHAL, On certain classes of Baire-1 functions with applications to Banach space theory, Functional Analysis Proceedings, The University of Texas at Austin 1987-89, Lecture Notes in Math., vol. **1470**, Springer-Verlag, New York, 1991, pp. 1-35.

[5] A. S. KECHRIS AND A. LOUVEAU, A classification of Baire-1 functions, Trans. Amer. Math. Soc. **318**(1990), 209-236.

[6] P. KIRIAKOULI, A classification of Baire-1 functions, Trans. Amer. Math. Soc. **351**(1999), 4599-4609.

[7] P. KIRIAKOULI, Characterizations of spreading models of ℓ^1, Comment. Math. Univ. Carolinae **41, 1**(2000), 79-95.

[8] S. MERCOURAKIS AND S. NEGREPONTIS, Banach spaces and topology II, Recent Progress in General Topology, M. Hušek and J. van Mill, eds., Elsevier Science Publisher B. V., 1992, pp. 493–536.

[9] EDWARD ODELL, NICOLE TOMCZAK-JAEGERMANN, AND ROY WAGNER, Proximity to ℓ_1 and distortion in asymptotic ℓ_1 spaces, J. Funct. Anal. **150**(1997), 101-145.

[10] H. P. ROSENTHAL, A characterization of Banach spaces containing ℓ_1. Proc. Nat. Acad. Sci. (U.S.A.) **71**, 2411-2413(1974).

DEPARTMENT OF MATHEMATICS, NATIONAL UNIVERSITY OF SINGAPORE, SINGAPORE 117543
E-mail address: matlhh@nus.edu.sg

MATHEMATICS AND MATHEMATICS EDUCATION, NATIONAL INSTITUTE OF EDUCATION, NANYANG TECHNOLOGICAL UNIVERSITY, 1 NANYANG WALK, SINGAPORE 637616
E-mail address: wktang@nie.edu.sg

Contemporary Mathematics
Volume **321**, 2003

Characterizing the Existence of Local Smith Forms for $\mathcal{C}^\infty$ Families of Matrix Operators

Julián López-Gómez and Carlos Mora-Corral

ABSTRACT. This paper characterizes whether or not the lengths of all Jordan chains of an arbitrary $\mathcal{C}^\infty$ family of square matrices $\mathfrak{L}(\lambda)$ are uniformly bounded from above. Consequently, the problem of ascertaining whether or not a local Smith form at a given singular value —eigenvalue— λ_0 of $\mathfrak{L}(\lambda)$ exists is solved. Actually, a local Smith form exists if and only if λ_0 is an algebraic eigenvalue of $\mathfrak{L}(\lambda)$. The main technical tools to obtain this characterization are the construction of the local Smith form of P. J. Rabier [**8**], and the transversalization theory of J. Esquinas and J. López-Gómez [**3, 2, 1, 6**].

1. Introduction

Generalized Jordan chains were introduced to develop a spectral theory in the context of matrix polynomial pencils (e.g., I. Gohberg et al. [**4**] and the references therein), and have shown to be a powerful technical tool in a number of areas; among them, for studying the structure of the solution set of broad classes of linear systems of differential equations (cf. J. T. Wloka et al. [**9**]) and for analyzing nonlinear eigenvalue problems in bifurcation theory (cf. P. J. Rabier [**8**]).

Closely related to generalized Jordan chains is the so-called *Smith form.* In dealing with polynomial families $\mathfrak{L}(\lambda)$ the existence of a *local Smith form* can be easily deduced from the *global Smith form* of I. Gohberg et al. [**4**, Chapter S1]. Actually, these techniques allowed P. J. Rabier [**8**, Section 4] to build up a *local Smith form* for families of class $\mathcal{C}^\infty$ regularity at any eigenvalue λ_0 where no generalized Jordan chain can be continued indefinitely; this is what occurs, for example, when $\mathfrak{L}(\lambda)$ is of polynomial type and it is invertible at some point.

In this note it will be shown that $\mathfrak{L}(\lambda)$ possesses a local Smith form at λ_0 if and only if λ_0 is an algebraic eigenvalue of $\mathfrak{L}(\lambda)$. By an algebraic eigenvalue it is meant any isolated eigenvalue λ_0 —singular value of $\mathfrak{L}(\lambda)$— for which $\mathfrak{L}^{-1}(\lambda)$, $\lambda \simeq \lambda_0$,

---

2000 *Mathematics Subject Classification.* Primary 47A56; Secondary 47A10, 47A11.

Key words and phrases. Jordan chains. Local Smith form. Algebraic eigenvalue. Algebraic multiplicities.

Both authors have been supported by the Spanish Ministry of Science and Technology under Grant BFM2000-0797.

The second author has been supported by the Spanish MECD under Grant AP2000-3316.

$\lambda \neq \lambda_0$, satisfies the following estimate

$$\|\mathfrak{L}^{-1}(\lambda)\| \leq \frac{C}{|\lambda - \lambda_0|^\nu} \tag{1.1}$$

for some constants $C > 0$ and $\nu \geq 1$. The least integer $\nu \geq 1$ for which (1.1) holds true is referred to as the order of the algebraic eigenvalue λ_0. The following fundamental characterization holds true.

THEOREM 1.1. *Suppose* $\mathbb{K} \in \{\mathbb{R}, \mathbb{C}\}$, $\Omega \subset \mathbb{K}$ *is an open subset,* $\lambda_0 \in \Omega$, $m \geq 1$ *and* $\mathfrak{L} : \Omega \to \mathcal{L}(\mathbb{K}^m)$ *is a map of class* $\mathcal{C}^\infty$ *such that* $n := \dim N[\mathfrak{L}(\lambda_0)] \geq 1$. *Then, the following conditions are equivalent:*

- $\mathfrak{L}(\lambda)$ *possesses a local Smith form at* λ_0.
- $\det \mathfrak{L}(\lambda)$ *has a zero of finite order at* $\lambda = \lambda_0$.
- λ_0 *is an algebraic eigenvalue of* $\mathfrak{L}(\lambda)$.
- $L > 0$ *exists such that every Jordan chain of* $\mathfrak{L}(\lambda)$ *at* λ_0 *has length less than* L.
- *No Jordan chain of* $\mathfrak{L}(\lambda)$ *at* λ_0 *can be continued indefinitely.*

Moreover, the local Smith form is unique. Equivalently, the partial multiplicities introduced by P. J. Rabier [**8**] *as well as the algebraic multiplicity of* [**6**] *are fixed. Actually, the algebraic multiplicity equals the order of the determinant of* $\mathfrak{L}(\lambda)$ *at* λ_0 *and, therefore, an easily computable algorithm —e.g., the determinant— to ascertain whether or not all Jordan chains have bounded length is at our disposal.*

It should be noted that even in the case when $\det \mathfrak{L}(\lambda)$ has a zero of finite order at $\lambda = \lambda_0$, which is not always true as the example in Remark 2.4 shows, the fact that λ_0 is an algebraic eigenvalue of $\mathfrak{L}(\lambda)$ cannot be obtained straight away from the identity

$$\det \mathfrak{L}^{-1}(\lambda) \det \mathfrak{L}(\lambda) = 1 .$$

Rather, the estimate for the norm follows from the determinant estimate through a not very well known estimate attributable to T. Kato [**5**].

Our proof of Theorem 1.1 is based upon the construction of the local Smith form carried out by P. J. Rabier [**8**] and the *transversalization theorem* of J. López-Gómez [**6**], where it was shown that $\mathfrak{L}(\lambda)$ can be transversalized at λ_0 if and only if λ_0 is an algebraic eigenvalue of $\mathfrak{L}(\lambda)$. Actually, transversalizing the family $\mathfrak{L}(\lambda)$ at λ_0 enables us to prove Theorem 1.1. Given an integer $k \geq 1$, it is said that λ_0 is a k-transversal eigenvalue of the family $\mathfrak{L}(\lambda)$ if

$$\bigoplus_{j=1}^{k} \mathfrak{L}_j(N[\mathfrak{L}_0] \cap \cdots \cap N[\mathfrak{L}_{j-1}]) \oplus R[\mathfrak{L}_0] = \mathbb{K}^m$$

and

$$\mathfrak{L}_k(N[\mathfrak{L}_0] \cap \cdots \cap N[\mathfrak{L}_{k-1}]) \neq [0] ,$$

where we have denoted

$$\mathfrak{L}_j := \frac{1}{j!} \frac{d^j \mathfrak{L}}{d\lambda^j}(\lambda_0) , \qquad j \geq 0 .$$

The integer $k = k(\lambda_0)$ is called the order of transversality of $\mathfrak{L}(\lambda)$ at λ_0. Subsequently, given any linear operator T, $N[T]$ and $R[T]$ will stand for the null space and the image of T, respectively. Also, given any Banach space Z, I_Z will stand for the identity map of Z.

It is said that $\mathfrak{L}(\lambda)$ can be transversalized at λ_0 if a family of isomorphisms $\Phi(\lambda) : \mathbb{K}^m \to \mathbb{K}^m$, $\lambda \simeq \lambda_0$, exists for which λ_0 is a k-transversal eigenvalue of the new family

$$\mathfrak{L}^{\Phi}(\lambda) := \mathfrak{L}(\lambda)\Phi(\lambda)\,, \qquad \lambda \simeq \lambda_0\,.$$

In such case, according to the abstract theory of [**6**, Chapters 4, 5], $k = k(\lambda_0, \Phi)$ equals the order of λ_0 as an algebraic eigenvalue of $\mathfrak{L}(\lambda)$ and, actually, Φ can be taken of polynomial type.

The key point of our transversalization approach to the proof of Theorem 1.1 relies in the existence of a very efficient finite algorithm for constructing complete sets of generalized Jordan chains of $\mathfrak{L}(\lambda)$ at any k-transversal eigenvalue λ_0 (cf. Section 5 for further details). Such construction should clarify some crucial connections between the several concepts of algebraic multiplicities considered in this work.

Quite comfortably, our algorithm shows that λ_0 is an algebraic eigenvalue of λ_0 if and only if $\mathfrak{L}(\lambda)$ possesses a local Smith form at λ_0.

This paper is distributed as follows. In Sections 2, 3 we collect some basic concepts and results from P. J. Rabier [**8**]. In Section 4 we collect the main results of the transversality theory developed by J. Esquinas and J. López-Gómez [**3, 2, 1, 6**]. In Section 5 we give the algorithm for constructing a complete set of generalized Jordan chains, simultaneously obtaining some fundamental spectral properties of great interest in their own right. Finally, in Section 6 we will bring together all this information to prove Theorem 1.1.

It must be emphasized that we are dealing with some invariants of algebraic and/or analytic nature that provide us with any change of the Brouwer topological degree of $\mathfrak{L}(\lambda)$ as the parameter λ crosses a given eigenvalue λ_0 (cf. [**6**, Chapter 5] and R. J. Magnus [**7**] for further details). Consequently, this paper deals with very sharp properties of one of the most important topological invariants in nonlinear analysis, though, in this occasion, our attention has not been focused into this face of the problem.

2. Generalized Jordan chains

Throughout this paper, $\mathbb{K} \in \{\mathbb{R}\,, \mathbb{C}\}$, $\Omega \subset \mathbb{K}$ is an open subset such that $\lambda_0 \in \Omega$, and, for some integer $m \geq 1$,

$$\mathfrak{L} \,:\, \Omega \to \mathcal{L}(\mathbb{K}^m)$$

is a map of class $\mathcal{C}^\infty$ such that

$$n := \dim N[\mathfrak{L}(\lambda_0)] \geq 1\,.$$

Subsequently, $\mathcal{L}(\mathbb{K}^m)$ stands for the space of linear mappings from $\mathbb{K}^m$ to $\mathbb{K}^m$, which will be identified with the space of $m \times m$ matrices. We denote

$$\mathfrak{L}_j := \frac{1}{j!}\frac{d^j\mathfrak{L}}{d\lambda^j}(\lambda_0)\,, \qquad j \geq 0\,.$$

The following concept is crucial for the theory that we are going to develop in this paper. Also, it has been used to study the solution set of some general classes of linear systems of differential equations with constant coefficients (e.g., [**9**]).

DEFINITION 2.1 (Jordan chain). Let $X := \mathbb{K}^m$. Any ordered finite set of $s+1$ vectors $(x_0, \dots, x_s) \in X^{s+1}$ with $s \geq 0$, $x_0 \neq 0$, and

$$\sum_{i=0}^{j} \mathfrak{L}_i x_{j-i} = 0\,, \qquad 0 \leq j \leq s\,, \tag{2.1}$$

is said to be a **Jordan chain of length** $s+1$ of the family $\mathfrak{L}(\lambda)$ at λ_0 —originating at x_0—.

REMARK 2.2. If $(x_0, \dots, x_s)$ is a Jordan chain, then so is $(x_0, \dots, x_p)$ for any $0 \leq p \leq s$.

DEFINITION 2.3 (Maximal chain and rank). A Jordan chain $(x_0, \dots, x_s)$ is said to be **maximal** if $x \in \mathbb{K}^m$ does not exist for which $(x_0, \dots, x_s, x)$ is a Jordan chain of length $s+2$. The **rank** of an eigenvector $x_0 \in N[\mathfrak{L}_0] \setminus \{0\}$, denoted in the sequel by $\operatorname{rank}(x_0)$, is defined as the maximum length of all Jordan chains originating at x_0 if such lengths are bounded, and $\operatorname{rank}(x_0) := \infty$ when the set of lengths of all Jordan chains originating at x_0 is unbounded.

REMARK 2.4. Very simple examples show that, in general, the rank of an eigenvector is not necessarily finite. For example, if $m = 1$, $\lambda_0 = 0$, and

$$\mathfrak{L}(\lambda) := \begin{cases} e^{\frac{-1}{\lambda^2}} & \text{if } \lambda \in \mathbb{R} \setminus \{0\}\,, \\ 0 & \text{if } \lambda = 0\,, \end{cases} \tag{2.2}$$

then $\mathfrak{L}_j = 0$ for each $j \geq 0$, and, hence, $\operatorname{rank}(x_0) = \infty$ for all $x_0 \in \mathbb{R} \setminus \{0\}$, since for each $n \geq 1$,

$$\underbrace{(x_0, \dots, x_0)}_{n}$$

provides us with a Jordan chain of length n of $\mathfrak{L}(\lambda)$ at $\lambda_0 = 0$.

3. The local Smith form

Assume that the length of all Jordan chains of $\mathfrak{L}(\lambda)$ at λ_0 is uniformly bounded from above by some natural number $k_1 \in \mathbb{N}$. Without loss of generality, k_1 can be assumed to be optimal, i.e., $x_{0,1} \in N[\mathfrak{L}_0] \setminus \{0\}$ exists for which

$$\operatorname{rank}(x_{0,1}) = k_1\,.$$

Now, we will follow the construction of P. J. Rabier [**8**, pp. 906]. Proceeding inductively, having chosen $i-1$ linearly independent vectors,

$$x_{0,1}\,, \dots, x_{0,i-1}\,, \qquad 2 \leq i \leq n = \dim N[\mathfrak{L}_0]\,,$$

pick up

$$x_{0,i} \in C_i := N[\mathfrak{L}_0] \setminus \operatorname{span}[x_{0,1}, \dots, x_{0,i-1}]$$

such that

$$\operatorname{rank}(x_{0,i}) = k_i := \max\{\operatorname{rank}(x) \ : \ x \in C_i\} \leq k_{i-1}\,.$$

This process can be repeated until n elements $x_{0,1}, \dots, x_{0,n}$ have been selected. Now, for each $i \in \{1, \dots, n\}$, let

$$(x_{0,i},\, x_{1,i}, \dots, x_{k_i-1,i})$$

be any Jordan chain of length k_i originating at $x_{0,i}$ of $\mathfrak{L}(\lambda)$ at λ_0. The ordered family

$$[x_{0,1}, \dots, x_{k_1-1,1}; \cdots ;\ x_{0,i}, \dots, x_{k_i-1,i};\ \cdots ; x_{0,n}, \dots, x_{k_n-1,n}]$$

is said to be a **canonical set of Jordan chains** of $\mathfrak{L}(\lambda)$ at λ_0. By construction,

$$k_1 \geq \cdots \geq k_n \geq 1 .$$

Moreover, the following result is satisfied (cf. [**8**, pp. 907]).

PROPOSITION 3.1. *The integers $k_1, \ldots, k_n$ are independent of the canonical set of Jordan chains of $\mathfrak{L}(\lambda)$ at λ_0.*

Therefore, the following concept of multiplicity attributable to P. J. Rabier [**8**] is consistent.

DEFINITION 3.2 (Algebraic multiplicity of P. J. Rabier). The integers $k_1, \ldots, k_n$ are called the partial multiplicities of $\mathfrak{L}(\lambda)$ at λ_0. The number

$$m[\mathfrak{L}(\lambda); \lambda_0] := \sum_{j=1}^{n} k_j$$

will be referred to as Rabier's multiplicity of $\mathfrak{L}(\lambda)$ at λ_0.

The next result follows from Corollary 2.1 in [**8**, pp. 908].

PROPOSITION 3.3. *Let $\mathfrak{E} \in \mathcal{C}^\infty(\Omega; \mathcal{L}(\mathbb{K}^m))$, $\mathfrak{F} \in \mathcal{C}^\infty(\Omega; \mathcal{L}(\mathbb{K}^m))$ two operator families such that $\mathfrak{E}(\lambda_0)$ and $\mathfrak{F}(\lambda_0)$ are isomorphisms. Then, Rabier's partial multiplicities of $\mathfrak{L}(\lambda)$ at λ_0 equal those of the new family*

$$\mathfrak{L}^{\mathfrak{E},\mathfrak{F}}(\lambda) := \mathfrak{E}(\lambda)\mathfrak{L}(\lambda)\mathfrak{F}(\lambda) , \qquad \lambda \in \Omega .$$

In particular,

$$m[\mathfrak{L}(\lambda); \lambda_0] = m[\mathfrak{L}^{\mathfrak{E},\mathfrak{F}}(\lambda); \lambda_0] .$$

To conclude this section we collect a result coming from [**8**, Section 4] that shows the existence of a *local Smith form*; the existence of a *global Smith form* for polynomial families follows from the theory of [**4**, Chapter S1].

THEOREM 3.4 (Local Smith form). *Assume that no Jordan chain of $\mathfrak{L}(\lambda)$ at λ_0 can be continued indefinitely. Then, $\mathfrak{L}(\lambda)$ has a local Smith form at λ_0, i.e., $\mathfrak{L}(\lambda)$ admits a decomposition of the following type*

$$\mathfrak{L}(\lambda) = \mathfrak{E}(\lambda)\mathfrak{D}(\lambda)\mathfrak{F}(\lambda) , \qquad \lambda \simeq \lambda_0 ,$$

where $\mathfrak{E}(\lambda)$ are $\mathfrak{F}(\lambda)$ are $\mathcal{C}^\infty$ matrices of order m such that $\mathfrak{E}(\lambda_0)$ and $\mathfrak{F}(\lambda_0)$ are isomorphisms, and $\mathfrak{D}(\lambda)$ is a diagonal matrix of the form

$$\mathfrak{D}(\lambda) = \operatorname{diag}\{(\lambda - \lambda_0)^{k_1}, \ldots, (\lambda - \lambda_0)^{k_n}, 1, \ldots, 1\} ,$$

where

$$k_1 \geq \cdots \geq k_n \geq 1$$

are Rabier's partial multiplicities of $\mathfrak{L}(\lambda)$ at λ_0.

4. Transversal eigenvalues. Transversalization theorem

This section collects the main results concerning the algebraic multiplicity introduced in [**3, 2, 1**] and further extended in [**6**]. It should be remarked that all results of [**6**, Chapters 4, 5] are valid working in $\mathbb{C}$, instead of in $\mathbb{R}$.

DEFINITION 4.1 (Concept of transversal eigenvalue). Given an integer $k \geq 1$, it is said that λ_0 is a k-transversal eigenvalue of the family $\mathfrak{L}(\lambda)$ if

$$\bigoplus_{j=1}^{k} \mathfrak{L}_j(N[\mathfrak{L}_0] \cap \cdots \cap N[\mathfrak{L}_{j-1}]) \oplus R[\mathfrak{L}_0] = \mathbb{K}^m$$

and

$$\mathfrak{L}_k(N[\mathfrak{L}_0] \cap \cdots \cap N[\mathfrak{L}_{k-1}]) \neq [0]\,. \tag{4.1}$$

The integer $k = k(\lambda_0)$ is called the order of transversality of $\mathfrak{L}(\lambda)$ at λ_0. Thanks to (4.1), it is well defined.

PROPOSITION 4.2. *Suppose λ_0 is a k-transversal eigenvalue of $\mathfrak{L}(\lambda)$ and set, for each $1 \leq i \leq k$,*

$$n_i := \dim N[\mathfrak{L}_0] \cap \cdots \cap N[\mathfrak{L}_i]\,, \quad \ell_i := \dim \mathfrak{L}_i(N[\mathfrak{L}_0] \cap \cdots \cap N[\mathfrak{L}_{i-1}])\,.$$

Then,

$$n = \ell_1 + \cdots + \ell_k \qquad \text{and} \quad n_k = 0\,.$$

In particular,

$$N[\mathfrak{L}_0] \cap \cdots \cap N[\mathfrak{L}_k] = [0]\,. \tag{4.2}$$

PROOF. Note that $n = \dim N[\mathfrak{L}_0]$ and that, for each $1 \leq i \leq k$,

$$\bigcap_{j=0}^{i} N[\mathfrak{L}_j] \subset \bigcap_{j=0}^{i-1} N[\mathfrak{L}_j] \subset N[\mathfrak{L}_0]\,.$$

Let X_i be a complement of $N[\mathfrak{L}_0] \cap \cdots \cap N[\mathfrak{L}_i]$ in $N[\mathfrak{L}_0] \cap \cdots \cap N[\mathfrak{L}_{i-1}]$. Then,

$$N[\mathfrak{L}_0] \cap \cdots \cap N[\mathfrak{L}_{i-1}] = X_i \oplus (N[\mathfrak{L}_0] \cap \cdots \cap N[\mathfrak{L}_i])\,, \qquad 1 \leq i \leq k\,,$$

and, hence,

$$N[\mathfrak{L}_0] = X_1 \oplus \cdots \oplus X_k \oplus (N[\mathfrak{L}_0] \cap \cdots \cap N[\mathfrak{L}_k])\,.$$

Moreover, by construction,

$$\mathfrak{L}_i(N[\mathfrak{L}_0] \cap \cdots \cap N[\mathfrak{L}_{i-1}]) = \mathfrak{L}_i(X_i)\,, \qquad 1 \leq i \leq k\,,$$

and

$$\dim X_i = \dim \mathfrak{L}_i(X_i)\,, \qquad 1 \leq i \leq k\,,$$

since each of the operators

$$\mathfrak{L}_i|_{X_i} \;:\; X_i \to \mathfrak{L}_i(X_i)\,, \qquad 1 \leq i \leq k\,,$$

is an isomorphism.

On the other hand, since λ_0 is a k-transversal eigenvalue of $\mathfrak{L}(\lambda)$, we have

$$\mathfrak{L}_1(X_1) \oplus \cdots \oplus \mathfrak{L}_k(X_k) \oplus R[\mathfrak{L}_0] = \mathbb{K}^m$$

and, hence,

$$\dim N[\mathfrak{L}_0] = \operatorname{codim} R[\mathfrak{L}_0] = \sum_{i=1}^{k} \dim \mathfrak{L}_i(X_i) = \sum_{i=1}^{k} \dim X_i\,.$$

Therefore,

$$N[\mathfrak{L}_0] = X_1 \oplus \cdots \oplus X_k$$

and, in particular,

$$N[\mathfrak{L}_0] \cap \cdots \cap N[\mathfrak{L}_k] = [0]$$

and, hence, $n_k = 0$. Finally, taking into account that the sum of the dimensions of the kernel and the image equals the one of the domain for each of the operators

$$\mathfrak{L}_i : \bigcap_{j=0}^{i-1} N[\mathfrak{L}_j] \to \mathbb{K}^m , \qquad 1 \le i \le k ,$$

gives

$$n = n_1 + \ell_1 , \qquad n_{i-1} = n_i + \ell_i , \quad 2 \le i \le k ,$$

and, therefore,

$$n = \ell_1 + \cdots + \ell_k ,$$

since $n_k = 0$. This concludes the proof. □

DEFINITION 4.3 (Multiplicity of J. Esquinas et al. [**3**]). Suppose λ_0 is a k-transversal eigenvalue of $\mathfrak{L}(\lambda)$. Then, the algebraic multiplicity of $\mathfrak{L}(\lambda)$ at λ_0 is defined through

$$\chi[\mathfrak{L}(\lambda); \lambda_0] := \sum_{j=1}^{k} j\, \ell_j .$$

Although transversal eigenvalues do not arise very often in applications — except in problems where strong symmetries occur—, a wide class of families $\mathfrak{L}(\lambda)$ can be *transversalized* by means of a change of variable. To state the precise result we need to introduce the following concept.

DEFINITION 4.4 (Algebraic eigenvalue). The eigenvalue λ_0 is said to be an algebraic eigenvalue of $\mathfrak{L}(\lambda)$ if there exist $\delta > 0$, $C > 0$ and an integer $\nu \ge 1$ such that for each λ satisfying $0 < |\lambda - \lambda_0| < \delta$ the operator $\mathfrak{L}(\lambda)$ is an isomorphism and

$$\left\|\mathfrak{L}^{-1}(\lambda)\right\| \le \frac{C}{|\lambda - \lambda_0|^{\nu}} .$$

The least integer $\nu \ge 1$ for which this estimate holds true is called the **order** of λ_0.

It should be noted that $\lambda_0 = 0$ is not an algebraic eigenvalue of the family $\mathfrak{L}(\lambda)$ defined by (2.2). The following result allows us to extend the concept of multiplicity introduced by Definition 4.3 to cover the case of general families of operators (cf. [**6**, Chapters 4, 5]).

THEOREM 4.5 (Transversalizing families). *Let $\mathfrak{L} \in \mathcal{C}^\infty(\Omega; \mathcal{L}(\mathbb{K}^m))$ such that $\mathfrak{L}(\lambda_0)$ is not an isomorphism. Then, for each integer $k \ge 1$ there is a polynomial family of operators*

$$\Phi^k : \mathbb{K} \to \mathcal{L}(\mathbb{K}^m) ,$$

depending upon k, $\mathfrak{L}$ and λ_0, such that $\Phi^k(\lambda_0) = I_{\mathbb{K}^m}$ and for which the new family of operators

$$\mathfrak{L}^{\Phi^k}(\lambda) := \mathfrak{L}(\lambda)\Phi^k(\lambda) , \qquad \lambda \in \Omega ,$$

satisfies

$$Y_k := \bigoplus_{j=1}^{k} \mathfrak{L}_j^{\Phi^k}(N[\mathfrak{L}_0^{\Phi^k}] \cap \cdots \cap N[\mathfrak{L}_{j-1}^{\Phi^k}]) \oplus R[\mathfrak{L}_0^{\Phi^k}] \subset \mathbb{K}^m .$$

Moreover, the following assertions are true:

1. *λ_0 is a ν-transversal eigenvalue of $\mathfrak{L}^{\Phi^\nu}(\lambda)$ if λ_0 is an algebraic eigenvalue of $\mathfrak{L}(\lambda)$ of order $\nu \ge 1$.*

2. *Y_k is a proper subspace of $\mathbb{K}^m$ for all $k \geq 1$ if λ_0 is not an algebraic eigenvalue of $\mathfrak{L}(\lambda)$.*
3. *Suppose Φ, $\Psi : \Omega \to \mathcal{L}(\mathbb{K}^m)$ are two $\mathcal{C}^\infty$ operator families satisfying the following:*
   (a) *$\Phi(\lambda_0)$ and $\Psi(\lambda_0)$ are isomorphisms,*
   (b) *λ_0 is a k_1-transversal eigenvalue of $\mathfrak{L}^\Phi(\lambda) := \mathfrak{L}(\lambda)\Phi(\lambda)$ for some $k_1 \geq 1$,*
   (c) *λ_0 is a k_2-transversal eigenvalue of $\mathfrak{L}^\Psi(\lambda) := \mathfrak{L}(\lambda)\Psi(\lambda)$ for some $k_2 \geq 1$.*

   Then, $k_1 = k_2$ and, for each $1 \leq j \leq k_1 = k_2$,
$$\dim \mathfrak{L}_j^\Phi(\bigcap_{i=0}^{j-1} N[\mathfrak{L}_i^\Phi]) = \dim \mathfrak{L}_j^\Psi(\bigcap_{i=0}^{j-1} N[\mathfrak{L}_i^\Psi]) .$$
   In particular, $\chi[\mathfrak{L}^\Phi(\lambda); \lambda_0] = \chi[\mathfrak{L}^\Psi(\lambda); \lambda_0]$.

Therefore, Theorem 4.5 shows the consistency of the following concept of multiplicity which extends the one introduced by Definition 4.3.

DEFINITION 4.6. Suppose λ_0 is an algebraic eigenvalue of $\mathfrak{L}(\lambda)$ of order $\nu \geq 1$. Then, the algebraic multiplicity of $\mathfrak{L}(\lambda)$ at λ_0 is defined through
$$\chi[\mathfrak{L}(\lambda); \lambda_0] := \chi[\mathfrak{L}^{\Phi^\nu}(\lambda); \lambda_0] ,$$
where $\mathfrak{L}^{\Phi^\nu}(\lambda)$ is any of the families whose existence is guaranteed by Theorem 4.5.

5. Building up canonical sets of Jordan chains

The following results provide us with a general recipe for constructing Jordan chains and canonical sets of Jordan chains of $\mathfrak{L}(\lambda)$ at transversal eigenvalues.

PROPOSITION 5.1. *Let $X := \mathbb{K}^m$. Suppose $k \geq 1$ and λ_0 is a k-transversal eigenvalue of $\mathfrak{L}(\lambda)$. Then, for any integer $s \geq 0$ and $(x_0, \ldots, x_s) \in X^{s+1}$, the ordered set $(x_0, \ldots, x_s)$ is a Jordan chain of length $s+1$ of $\mathfrak{L}(\lambda)$ originating at x_0 if, and only if, $x_0 \neq 0$ and*
$$x_j \in \bigcap_{i=0}^{s-j} N[\mathfrak{L}_i] , \qquad 0 \leq j \leq s . \tag{5.1}$$
Henceforth, $s < k$ and, therefore, no Jordan chain of length greater than k is available.

PROOF. Suppose $s \geq 0$, $x_0 \neq 0$ and (5.1). Then,
$$x_i \in N[\mathfrak{L}_{j-i}] , \qquad 0 \leq i \leq j \leq s ,$$
and, hence,
$$\sum_{i=0}^{j} \mathfrak{L}_{j-i} x_i = 0 , \qquad 0 \leq j \leq s . \tag{5.2}$$
Therefore, $(x_0, \ldots, x_s)$ is a Jordan chain of $\mathfrak{L}(\lambda)$ of length $s+1$.

Now, suppose $s \geq 0$ and $(x_0, \ldots, x_s)$ is a Jordan chain of length $s+1$. Then, $x_0 \neq 0$ and (5.2) is satisfied. To prove (5.1) we will proceed by induction on s. It follows from (5.2) that $\mathfrak{L}_0 x_0 = 0$ and, hence, (5.1) holds true for $s = 0$.

Assume that

$$y_j \in \bigcap_{i=0}^{s-1-j} N[\mathfrak{L}_i]\,, \qquad 0 \le j \le s-1\,,$$

for any Jordan chain $(y_0, \dots, y_{s-1})$ of length s, and let $(x_0, \dots, x_s)$ be a Jordan chain of length $s+1$. Then, thanks to Remark 2.2, $(x_0, \dots, x_{s-1})$ provides us with a Jordan chain of length s and, hence, thanks to the induction hypothesis,

$$x_j \in \bigcap_{i=0}^{s-1-j} N[\mathfrak{L}_i]\,, \qquad 0 \le j \le s-1\,. \tag{5.3}$$

On the other hand, particularizing (5.2) at $j = s$ gives

$$\mathfrak{L}_0 x_s + \sum_{i=1}^{s} \mathfrak{L}_i x_{s-i} = 0\,. \tag{5.4}$$

Thanks to (5.3), we obtain

$$\mathfrak{L}_0 x_s \in R[\mathfrak{L}_0]\,, \qquad \mathfrak{L}_i x_{s-i} \in \mathfrak{L}_i \left(N[\mathfrak{L}_0] \cap \cdots \cap N[\mathfrak{L}_{i-1}]\right)\,, \qquad 1 \le i \le s\,.$$

Thus, since λ_0 is a k-transversal eigenvalue of $\mathfrak{L}(\lambda)$, (5.4) implies

$$\mathfrak{L}_0 x_s = 0\,, \qquad \mathfrak{L}_i x_{s-i} = 0\,, \qquad 1 \le i \le s\,.$$

In other words,

$$\mathfrak{L}_{s-j}\, x_j = 0\,, \qquad 0 \le j \le s\,.$$

It should be noted that the previous argument also works out when $s > k$, due to (4.2). Combining these relations with (5.3) gives

$$x_j \in \bigcap_{i=0}^{s-j} N[\mathfrak{L}_i]\,, \qquad 0 \le j \le s\,,$$

and, therefore, (5.1) is true.

Finally, we will show that $\mathfrak{L}(\lambda)$ cannot have a Jordan chain of length greater than k at λ_0. Indeed, if it has some, then, according to Remark 2.2, it has a Jordan chain of length $k+1$, say $(x_0, \dots, x_k)$. Then, $x_0 \neq 0$ and, thanks to (5.1), $x_0 \in N[\mathfrak{L}_0] \cap \cdots \cap N[\mathfrak{L}_k]$. On the other hand, since λ_0 is a k-transversal eigenvalue of $\mathfrak{L}(\lambda)$, we have $N[\mathfrak{L}_0] \cap \cdots \cap N[\mathfrak{L}_k] = [0]$ and, therefore, $x_0 = 0$. This contradiction concludes the proof. □

The next result characterizes the canonical sets of Jordan chains of $\mathfrak{L}(\lambda)$ at any transversal eigenvalue.

THEOREM 5.2. *Suppose $k \ge 1$ and λ_0 is a k-transversal eigenvalue of $\mathfrak{L}(\lambda)$. Set*

$$\ell_j := \dim \mathfrak{L}_j(N[\mathfrak{L}_0] \cap \cdots \cap N[\mathfrak{L}_{j-1}])\,, \qquad 1 \le j \le k\,.$$

Then, every canonical set of Jordan chains of $\mathfrak{L}(\lambda)$ at λ_0 is formed by ℓ_j Jordan chains of length j, for each $j \in \{1, \dots, k\}$. In other words, the partial multiplicities of P. J. Rabier are given through

$$\underbrace{\underbrace{k, \dots, k}_{\ell_k}, \underbrace{k-1, \dots, k-1}_{\ell_{k-1}}, \dots, \underbrace{2, \dots, 2}_{\ell_2}, \underbrace{1, \dots, 1}_{\ell_1}}_{n = \ell_k + \ell_{k-1} + \cdots + \ell_2 + \ell_1}.$$

PROOF. Thanks to Proposition 5.1, we already know that $\mathfrak{L}(\lambda)$ does not admit a Jordan chain of length greater than k. As in the proof of Proposition 4.2, for each $1 \leq i \leq k$ we consider a subspace $X_i \subset \mathbb{K}^m$ such that

$$N[\mathfrak{L}_0] \cap \cdots \cap N[\mathfrak{L}_{i-1}] = X_i \oplus (N[\mathfrak{L}_0] \cap \cdots \cap N[\mathfrak{L}_i]) \,, \qquad 1 \leq i \leq k \,.$$

Then,

$$\ell_i = \dim X_i = \dim \mathfrak{L}_i(X_i) \,, \qquad 1 \leq i \leq k \,,$$
$$N[\mathfrak{L}_0] = X_1 \oplus \cdots \oplus X_k \,, \qquad \mathfrak{L}_1(X_1) \oplus \cdots \oplus \mathfrak{L}_k(X_k) \oplus R[\mathfrak{L}_0] = \mathbb{K}^m \,,$$

and

$$n = \ell_1 + \cdots + \ell_k \,.$$

Let $\{u_1, \ldots, u_{\ell_k}\}$ be any basis of X_k. Then,

$$X_k = \operatorname{span}[u_1, \ldots, u_{\ell_k}] \,.$$

Let $\{u_{\ell_k+1}, \ldots, u_{\ell_k+\ell_{k-1}}\}$ denote any basis of X_{k-1}. Then,

$$X_k \oplus X_{k-1} = \operatorname{span}[u_1, \ldots, u_{\ell_k+\ell_{k-1}}] \,.$$

Arguing recursively, it is rather clear that this procedure provides us with a basis of $N[\mathfrak{L}_0]$, say $\{u_1, \ldots, u_n\}$. Thanks to Proposition 5.1, it is easy to see that the following ordered set provides us with a canonical set of Jordan chains of $\mathfrak{L}(\lambda)$ at λ_0:

$$\begin{aligned}
&[\,\underbrace{u_1, 0, \ldots, 0}_{k}; \cdots ; \underbrace{u_{\ell_k}, 0, \ldots, 0}_{k}; \\
&\underbrace{u_{\ell_k+1}, 0, \ldots, 0}_{k-1}; \cdots ; \underbrace{u_{\ell_k+\ell_{k-1}}, 0, \ldots, 0}_{k-1}; \cdots ; \\
&\underbrace{u_{\ell_k+\cdots+\ell_3+1}, 0}_{2}; \cdots ; \underbrace{u_{\ell_k+\cdots+\ell_3+\ell_2}, 0}_{2}; \underbrace{u_{\ell_k+\cdots+\ell_3+\ell_2+1}}_{1}; \cdots ; \underbrace{u_{\ell_k+\cdots+\ell_2+\ell_1}}_{1}\,] \,.
\end{aligned}$$

Therefore, it consists of ℓ_k Jordan chains of length k, ℓ_{k-1} Jordan chains of length $k-1$,..., and ℓ_1 Jordan chains of length 1. Finally, Proposition 3.1 concludes the proof. □

As an easy consequence from Theorem 3.4 the following result is obtained.

COROLLARY 5.3. *Suppose $k \geq 1$ and λ_0 is a k-transversal eigenvalue of $\mathfrak{L}(\lambda)$. Then, $\mathfrak{L}(\lambda)$ has a local Smith form*

$$\mathfrak{L}(\lambda) = \mathfrak{E}(\lambda)\mathfrak{D}(\lambda)\mathfrak{F}(\lambda)$$

with $\mathfrak{E}(\lambda_0)$ and $\mathfrak{F}(\lambda_0)$ isomorphisms, and

$$\mathfrak{D}(\lambda) = \operatorname{diag}\{\underbrace{(\lambda-\lambda_0)^k, \ldots, (\lambda-\lambda_0)^k}_{\ell_k}, \ldots, \underbrace{\lambda-\lambda_0, \ldots, \lambda-\lambda_0}_{\ell_1}, \underbrace{1, \ldots, 1}_{m-n}\}$$

In other words, the partial multiplicities of $\mathfrak{L}(\lambda)$ at λ_0 are

$$\underbrace{k, \ldots, k}_{\ell_k}, \underbrace{k-1, \ldots, k-1}_{\ell_{k-1}}, \ldots, \underbrace{1, \ldots, 1}_{\ell_1} .$$

In particular, $\det \mathfrak{L}(\lambda)$ *has a zero of order*

$$m[\mathfrak{L}(\lambda); \lambda_0] = \sum_{j=1}^{k} j\,\ell_j = \chi[\mathfrak{L}(\lambda); \lambda_0]$$

at the eigenvalue λ_0.

When dealing with general families, for which λ_0 is not necessarily a transversal eigenvalue, the following result provides us with the most crucial relationships between the algebraic invariants introduced in Sections 3, 4.

THEOREM 5.4. *Suppose* λ_0 *is an algebraic eigenvalue of* $\mathfrak{L}(\lambda)$ *of order* $k \geq 1$ *and let*

$$\Phi : \mathbb{K} \to \mathcal{L}(\mathbb{K}^m)$$

be any polynomial with $\Phi(\lambda_0) = I_{\mathbb{K}^m}$ *for which* λ_0 *is a* k*-transversal eigenvalue of the new family*

$$\mathfrak{L}^\Phi(\lambda) := \mathfrak{L}(\lambda)\Phi(\lambda), \qquad \lambda \in \Omega.$$

Then, the lengths of the Jordan chains of $\mathfrak{L}^\Phi(\lambda)$ *at* λ_0 *in any canonical set equal those of* $\mathfrak{L}(\lambda)$ *at* λ_0. *Therefore, setting*

$$\ell_j := \dim \mathfrak{L}_j^\Phi(N[\mathfrak{L}_0^\Phi] \cap \cdots \cap N[\mathfrak{L}_{j-1}^\Phi]), \qquad 1 \leq j \leq k,$$

any canonical set of Jordan chains of $\mathfrak{L}(\lambda)$ *possesses* ℓ_k *chains of length* k, ℓ_{k-1} *chains of length* $k-1$, ..., *and* ℓ_1 *chains of length* 1.

In particular, $\mathfrak{L}(\lambda)$ *has a local Smith form*

$$\mathfrak{L}(\lambda) = \mathfrak{E}(\lambda)\mathfrak{D}(\lambda)\mathfrak{F}(\lambda),$$

where $\mathfrak{E}(\lambda_0)$ *and* $\mathfrak{F}(\lambda_0)$ *are isomorphisms, and*

$$\mathfrak{D}(\lambda) = \operatorname{diag}\{\underbrace{(\lambda-\lambda_0)^k, \ldots, (\lambda-\lambda_0)^k}_{\ell_k}, \ldots, \underbrace{\lambda-\lambda_0, \ldots, \lambda-\lambda_0}_{\ell_1}, \underbrace{1, \ldots, 1}_{m-n}\}. \tag{5.5}$$

PROOF. The existence of $\Phi(\lambda)$ is guaranteed by Theorem 4.5, as well as the independence of ℓ_j on the transversalizing family $\Phi(\lambda)$. Thanks to Theorem 5.2, any canonical set of Jordan chains of $\mathfrak{L}^\Phi(\lambda)$ consists of ℓ_j Jordan chains of length j for each $1 \leq j \leq k$. Thanks to Propositions 3.1, 3.3, any canonical set of Jordan chains of $\mathfrak{L}(\lambda)$ consists of ℓ_j Jordan chains of length j for each $1 \leq j \leq k$. This completes the proof of the first part of the statement.

Thanks to Corollary 5.3, $\mathfrak{L}^\Phi(\lambda)$ has a local Smith form

$$\mathfrak{L}^\Phi(\lambda) = \mathfrak{E}(\lambda)\mathfrak{D}(\lambda)\mathfrak{F}(\lambda)$$

where $\mathfrak{D}(\lambda)$ is given by (5.5) and $\mathfrak{E}(\lambda_0)$, $\mathfrak{F}(\lambda_0)$ are isomorphisms. Therefore, for $\lambda \simeq \lambda_0$, we have that

$$\mathfrak{L}(\lambda) = \mathfrak{L}^\Phi(\lambda)\Phi^{-1}(\lambda) = \mathfrak{E}(\lambda)\mathfrak{D}(\lambda)\mathfrak{F}(\lambda)\Phi^{-1}(\lambda),$$

which concludes the proof, since $\mathfrak{F}(\lambda_0)\Phi^{-1}(\lambda_0) = \mathfrak{F}(\lambda_0)$ is an isomorphism. □

6. Characterizing the existence of the local Smith form

In complete agreement with the counterexample given in Remark 2.4, the main result of this paper reads as follows.

THEOREM 6.1. *The following conditions are equivalent:*

C1. $\mathfrak{L}(\lambda)$ *possesses a local Smith form at* λ_0.

C2. $\det \mathfrak{L}(\lambda)$ *has a zero of finite order at* $\lambda = \lambda_0$.

C3. λ_0 *is an algebraic eigenvalue of* $\mathfrak{L}(\lambda)$.

C4. $L > 0$ *exists such that every Jordan chain of* $\mathfrak{L}(\lambda)$ *at* λ_0 *has length bounded above by* L.

C5. *No Jordan chain of $\mathfrak{L}(\lambda)$ at λ_0 can be continued indefinitely.*
C6. *The algebraic multiplicities introduced in Sections 3, 4 are well defined and finite.*

PROOF. Assume $\mathfrak{L}(\lambda)$ is of class $\mathcal{C}^\infty$. To prove the equivalence of the five conditions we will proceed by steps.

C1 implies C2: If $\mathfrak{L}(\lambda)$ possesses a local Smith form,

$$\mathfrak{L}(\lambda) = \mathfrak{E}(\lambda)\mathfrak{D}(\lambda)\mathfrak{F}(\lambda)$$

where

$$\mathfrak{D}(\lambda) = \operatorname{diag}\{(\lambda-\lambda_0)^{k_1}, \dots, (\lambda-\lambda_0)^{k_n}, 1, \dots, 1\}, \qquad k_1 \geq \cdots \geq k_n \geq 1,$$

and $\mathfrak{E}(\lambda)$, $\mathfrak{F}(\lambda)$ are of class $\mathcal{C}^\infty$ and invertible at λ_0, then

$$\det \mathfrak{L}(\lambda) = (\lambda-\lambda_0)^{k_1+\cdots+k_n} \det \mathfrak{E}(\lambda) \det \mathfrak{F}(\lambda) = \eta(\lambda-\lambda_0)^{k_1+\cdots+k_n}[1+o(1)]$$

as $\lambda \to \lambda_0$, where

$$\eta = \det \mathfrak{E}(\lambda_0) \det \mathfrak{F}(\lambda_0) \neq 0 .$$

C2 implies C3: Suppose $\det \mathfrak{L}(\lambda)$ has a zero of finite order, say χ, at $\lambda = \lambda_0$. Then, there exists a $\mathcal{C}^\infty$ function $a(\lambda)$, $\lambda \simeq \lambda_0$, such that

$$a(\lambda_0) \neq 0, \qquad \det \mathfrak{L}(\lambda) = (\lambda-\lambda_0)^\chi a(\lambda), \qquad \lambda \simeq \lambda_0 . \tag{6.1}$$

On the other hand, thanks to Kato's inequality (e.g., [**5**, pp. 28], or [**6**, Lemma 4.4.2]), for any fixed norm in $\mathbb{K}^m$, $\|\cdot\|$, there exists a universal constant $\gamma > 0$ such that for each invertible $T \in \mathcal{L}(\mathbb{K}^m)$ the following estimate is satisfied

$$\|T^{-1}\| \leq \gamma \frac{\|T\|^{m-1}}{|\det T|} . \tag{6.2}$$

Thus, combining (6.2) with (6.1), it is easy to see that a constant $C > 0$ exists for which

$$\|\mathfrak{L}^{-1}(\lambda)\| \leq \frac{C}{|\lambda-\lambda_0|^\chi}$$

for any λ in some perforated neighborhood of λ_0. Therefore, λ_0 is an algebraic eigenvalue of $\mathfrak{L}(\lambda)$ of order $k \leq \chi$.

C3 implies C4: Suppose λ_0 is an algebraic eigenvalue of $\mathfrak{L}(\lambda)$ of order k. Then, thanks to Theorem 5.4, the length of any Jordan chain of $\mathfrak{L}(\lambda)$ at λ_0 is bounded above by k.

C4 implies C5 is obvious, and **C5 implies C1** is Theorem 3.4.

Finally, by the construction already done in the proof of Theorem 5.4, **C3 implies C6**. Conversely, the fact that **C6 implies C2** follows from P. J. Rabier [**8**], Corollary 5.3, Theorem 5.4 and the definition of χ itself. This completes the proof of the theorem. □

Acknowledgements. We sincerely thank to the referee for his/her useful comments and suggestions.

References

[1] J. Esquinas, Optimal multiplicity in local bifurcation theory, II: General case, *J. Diff. Eqns.* **75** (1988), 206-215.
[2] J. Esquinas and J, López-Gómez, Optimal multiplicity in local bifurcation theory, I: Generalized generic eigenvalues, *J. Diff. Eqns.* **71** (1988), 72-92.
[3] J. Esquinas and J. López-Gómez, Resultados óptimos en teoría de bifurcación y aplicaciones, *Actas del IX Congreso de Ecuaciones Diferenciales y Aplicaciones*, Universidad de Valladolid, 1986, 159–162.
[4] I. Gohberg, P. Lancaster and L. Rodman, *Matrix polynomials*, Computer Science and Applied Mathematics, Academic Press, New York 1982.
[5] T. Kato, *Perturbation Theory for Linear Operators*, Classics in Mathematics, Springer, Berlin 1995.
[6] J. López-Gómez, *Spectral Theory and Nonlinear Functional Analysis,* CRC Press, Chapman and Hall RNM 426, Boca Raton 2001.
[7] R. J. Magnus, A Generalization of Multiplicity and the Problem of Bifurcation. *Proc. London Math. Soc.* (3) **32** (1976) 251-278.
[8] P. J. Rabier, Generalized Jordan Chains and Two Bifurcation Theorems of Krasnoselskii. *Nonl. Anal. T. M. A.* **13** (1989), 903-934.
[9] J. T. Wloka, B. Rowley and B. Lawruk, *Boundary Value Problems for Elliptic Systems,* Cambridge University Press, New York 1995.

Departamento de Matemática Aplicada. Universidad Complutense de Madrid. 28040-Madrid. Spain
E-mail address: Lopez_Gomez@mat.ucm.es

Departamento de Matemática Aplicada. Universidad Complutense de Madrid. 28040-Madrid. Spain
E-mail address: Carlos_Mora@mat.ucm.es

Contemporary Mathematics
Volume **321**, 2003

Convex Geometry of Coxeter-invariant Polyhedra

Nicholas McCarthy, David Ogilvie, Nahum Zobin, and Veronica Zobin

ABSTRACT. We study the facial structure of convex polyhedra invariant under the natural action of a Coxeter group. The results are applied to the study of faces of maximal dimension of orbihedra related to some non-Coxeter groups.

1. Introduction

Let G be a finite Coxeter group naturally acting on a finite dimensional real space V. We study the geometry of convex G-invariant polyhedra.

The simplest convex G-invariant polyhedron is a G-**orbihedron** $\mathrm{Co}_G\, x$ – the convex hull of the G-orbit of x, $x \in V$. Geometric properties of G-orbihedra play important roles in many problems, ranging from Topology and Algebra to Operator Theory and Statistics– see, e.g., [**1, 8, 9, 10, 12, 14, 17, 18, 19**]. G-orbihedra may be viewed as building blocks of general G-invariant convex polyhedra – every such polyhedron may be represented as the convex hull of a finite number of G-orbihedra.

We study the facial structure of a convex G-invariant polyhedron. It is natural to start with faces of maximal dimension. One can always introduce a G-invariant bilinear symmetric positive definite form on V, so we may assume that V is Euclidean and that G is a subgroup of the orthogonal group. The most simple and fundamental geometric characteristic of such face is its normal vector which can be identified with an extreme vector of the polar polyhedron.

Normals to faces of maximal dimension for the simplest G-invariant polyhedron — a G-orbihedron — can be completely described in convenient geometric terms, see [**19**], or Section 3 below. Many deep problems require very precise understanding of both the generic structure of such normals and the sorts of degenerations that may occur when the vector x approaches special subsets of V. In the present paper we use this description to study the geometric structure of faces of all dimensions for G-orbihedra.

As it has been already mentioned, every G-invariant convex polyhedron can be represented as a convex hull of finitely many G-orbihedra. The minimal number of the required G-polyhedra is a natural measure of complexity of the G-invariant

2000 *Mathematics Subject Classification.* Primary 52B15; Secondary 20F55, 52B05.

A part of this research was conducted during the summer of 1999, N.M. and D.O. were student participants in the College of William and Mary's Research Experiences for Undergraduates program, N.Z. was a mentor participant in this program, all partially supported by NSF REU grant DMS-96-19577. N. Zobin was also partially supported by a Summer Research Grant at the College of William and Mary.

convex polyhedron. If this number is small (compared to the dimension of the space) then only vectors of very specific structure can serve as normals to faces of the maximal dimension. If this number is large (greater than or equal to the dimension of the space), then any nonzero vector can be a normal to a face of a G-invariant convex polyhedron.

As soon as we depart from the natural representation of a Coxeter group, the problem of description of the convex structure of the related orbihedra becomes much more difficult. For example, consider a Coxeter group G naturally acting on V, and let $G^2 = G \times G$ act on $V \otimes V$ in the usual tensor way:

$$(g_1, g_2)(v_1 \otimes v_2) = (g_1 v_1) \otimes (g_2 v_2).$$

Note that this action on $V \otimes V$ is not generated by reflections across hyperplanes. Nonetheless, G^2 is a Coxeter group, but its natural representation is on $V \bigoplus V$:

$$(g_1, g_2)(v_1 \oplus v_2) = (g_1 v_1) \oplus (g_2 v_2).$$

Preliminary computer experiments (C.K. Li, I. Spitkovsky and N. Zobin) show that the structure of normals to faces of the orbihedra related to the tensor action of G^2 may be quite wild even if $\dim V = 3$. This is not too surprising — see [**2**] for a study of closely related topics from the Complexity Theory viewpoint.

Nevertheless, for groups of operators close to Coxeter ones it is still possible to obtain rather detailed results concerning the geometric structure of the related orbihedra. Consider a finite group K of operators, acting on V. It may contain reflections across hyperplanes, so consider the subgroup G generated by all such reflections in K. Assume that G acts effectively (i.e., without nontrivial fixed vectors) on V, so G is a Coxeter subgroup. The description of K-orbihedra can be reduced to a description of G-invariant convex polyhedra. If the index of G in K is small compared to the dimension of V (in this case K should be called a **quasi-Coxeter** group) then we can use the Coxeter machinery, which makes it possible to describe the geometric structure of K-orbihedra. In particular, we describe the normals to faces of maximal dimension for $S_2(G)$-orbihedra, where S_2 is the group of permutations of $\{1, 2\}$, G is a Coxeter group acting on V, and the group $S_2(G) = S_2 \times G^2$ acts on the space $V^2 = V \bigoplus V$ as follows:

$$(\sigma, g_1, g_2)(v_1 \oplus v_2) = (g_1 v_{\sigma(1)}) \oplus (g_2 v_{\sigma(2)}).$$

Actually, it was this problem that stimulated the whole project. Group $S_2(G)$ has an index 2 Coxeter subgroup G^2, so K is quasi-Coxeter. In the case when $G = B_2$ this problem was studied and solved by the last two authors (see [**21**]), using vastly different methods which seemingly cannot be extended even to B_m with greater m. Though the present paper is completely independent of [**21**] the results and ideas from [**21**] were very helpful to us. In particular, the idea of consideration of a Coxeter subgroup already appeared in [**21**] though played there a rather technical role.

The paper is organized as follows: Section 2 contains a brief introduction to Coxeter groups adjusted to our needs, in Section 3 we present old and new results concerning the structure of normals to the faces of G-orbihedra of maximal dimension. We describe the faces of maximal dimension adjacent to a given vertex of a G-orbihedron, and as a corollary obtain some known results about simplicial orbihedra. In Section 4 we complement results of the previous section by a description of faces of lower dimensions. Section 5 is devoted to description of faces of maximal dimension for general G-invariant convex polyhedra. We also briefly discuss applications of these results to some problems of linear algebra. In Section 6 we apply these results to investigate the geometric structure of K-orbihedra for quasi-Coxeter groups K, and in particular, for the group $S_2(G)$. Section 7 contains a brief introduction to the duality approach to Operator Interpolation, its goal is to explain why the geometric results of the preceding sections are important in this field.

Acknowledgments. We are thankful to Igor Dolgachev who pointed to one of us that the questions we consider may have applications to toroidal geometry, and to Ilya Spitkovsky for valuable discussions.

2. A Brief Review of Coxeter Groups

Let us address several facts concerning the theory of Coxeter groups. For greater detail, consult [**3**], [**5**], or [**11**]. Let G be a group of linear operators on a finite dimensional real space V. Then G is called a **Coxeter group** if it is finite, generated by reflections across hyperplanes, and acts **effectively** (i.e., if $gx = x$ for all $g \in G$ then $x = 0$). Again, one can always introduce a G-invariant bilinear symmetric positive definite form on V, turning V into a Euclidean space, and making all operators from G orthogonal. So we always assume that this has already been done. By definition, a Coxeter group is a group of linear operators, so it has a preferred representation which is called the **natural representation** or the **natural action**. One can describe Coxeter groups in pure group-theoretic terms, namely, in terms of generators and relations, see, e.g., [**5**].

2.1. Roots and weights. Consider the set $\mathcal{M}_G$ of all **mirrors** — hyperplanes H such that the orthogonal reflection across H belongs to G. These mirrors divide V into connected components, each one a simplicial cone. The closures of these cones are called **Weyl chambers** of G. Weyl chambers are **fundamental domains** of G, i.e., every G-**orbit** $\operatorname{Orb}_G x = \{gx : g \in G\}$ intersects every Weyl chamber at exactly one point, let this point be denoted by $x^* = x^*(C, G)$.

Fix a Weyl chamber C. A **wall** of C is a $(\dim V - 1)$-dimensional face of C, contained in a mirror. Reflections across the walls of C (i.e., across the related mirrors) generate the whole group G. The finiteness of G implies that the angle between any two walls of C must equal π/k for some natural $k \geq 2$.

For every wall W_i of C, let n_i be the **root** – a specially scaled normal vector to W_i pointing inwards with respect to C. It is convenient for us to choose all roots to be unit vectors (the standard normalization of roots is different, see, e.g., [**5**]). Since C is a simplicial cone, for each wall W_j there exists a unique extreme ray of C not lying on W_j. Let ω_j be a vector pointing in the direction of this extreme ray, so that $\langle n_i, \omega_j \rangle = c_j \delta_{ij}$, $c_j > 0$. Each ω_j is called a **fundamental weight** of G. Note that we prefer not to normalize fundamental weights, for the standard normalization see [**5**]. Let $\mathcal{R}_G$ and $\mathcal{W}_G$ denote the sets of all roots and, respectively, the set of all weights of G (i.e., associated with all Weyl chambers). Since group G acts **(simply) transitively** on the set of its Weyl chambers (i.e., for any two Weyl chambers C_1, C_2 there exists (exactly one) $g \in G$ such that $gC_1 = C_2$), then

$$\mathcal{R}_G = \bigcup_i \operatorname{Orb}_G n_i, \quad \mathcal{W}_G = \bigcup_i \operatorname{Orb}_G \omega_i.$$

2.2. Coxeter graphs. There is a graph $\Gamma(G)$ (called the **Coxeter graph**) assigned to each Coxeter group. Fix a Weyl chamber C. The set ver (G) of vertices of the graph is in a one-to-one correspondence with the set of walls of C. Two vertices of this graph are connected with an edge if and only if the angle between the related walls is $\pi/k, k \geq 3$. The number $k-2$ is the multiplicity of this edge. Obviously, the Coxeter graph does not depend upon the choice of the Weyl chamber. In particular, every wall (but not the mirror containing this wall !) of any Weyl chamber is associated with a vertex of $\Gamma(G)$, and walls transformed one into another by the action of G are associated with the same vertex. Similarly, each weight ω is associated with a vertex $\pi(\omega)$ of the Coxeter graph $\Gamma(G)$. Obviously, $\pi(g\omega) = \pi(\omega)$ for every $g \in G$. So, $\pi(\omega)$ actually depends only upon the G-orbit of ω. Every vertex from ver (G) determines the G-orbit of exactly one weight (up to a positive factor).

A Coxeter group is irreducible if and only if its Coxeter graph is connected.

Notably, a Coxeter graph completely determines its Coxeter group, so if Γ is a Coxeter graph, let $G(\Gamma)$ denote the related Coxeter group.

There exists a full classification of connected Coxeter graphs, which implies a full classification of irreducible Coxeter groups. It worth noting that a reducible Coxeter group G is naturally isomorphic to the direct product of irreducible Coxeter groups $G(j)$ whose Coxeter graphs are the components j of $\Gamma(G)$, independently acting on mutually orthogonal subspaces $V(j)$. Let $J(G)$ denote the set of components of $\Gamma(G)$. Then

$$G = \prod_{j \in J(G)} G(j), \quad V = \bigoplus_{j \in J(G)} V(j),$$

and if

$$g = (g(j))_{j \in J(G)} \in \prod_{j \in J(G)} G(j), \quad v = \bigoplus_{j \in J(G)} v(j) \in \bigoplus_{j \in J(G)} V(j),$$

then

$$gv = \bigoplus_{j \in J(G)} g(j)v(j) \in \bigoplus_{j \in J(G)} V(j).$$

2.3. Supports and stabilizers. Fix a Weyl chamber C, let $x^* = x^*(C, G)$ be the unique vector in $C \cap \operatorname{Orb}_G x$. Since $x^* \in C$ and C is a simplicial cone, then there exists a unique decomposition of x^* into a positive linear combination of the related fundamental weights:

$$x^* = \sum_i \lambda_i \omega_i, \quad \lambda_i \geq 0.$$

Let us introduce the **support** of x as follows:

$$\operatorname{supp}_G x = \{\pi_i \in \operatorname{ver}(G) : \lambda_i > 0\}.$$

In other words, a vertex π_i of the Coxeter graph $\Gamma(G)$ belongs to $\operatorname{supp}_G x$ if x^* does not belong to the related wall W_i. One can easily show that $\operatorname{supp}_G x$ does not depend upon the choice of the Weyl chamber C. In fact, $\operatorname{supp}_G x$ depends only upon the G-orbit of x, therefore the notation $\operatorname{supp}_G i$ is meaningful for a G-orbit i. Note that

$$\operatorname{supp}_G x = \emptyset \quad \text{if and only if} \quad x = 0.$$

Let

$$J(G, x) = \{j \in J(G) : \operatorname{supp}_G x \text{ intersects } j\}.$$

Combining the definition of $\operatorname{supp}_G x$ with the description of the action of a reducible Coxeter group we see that

$$\operatorname{span}(\operatorname{Orb}_G x) = \bigoplus_{j \in J(G,x)} V(j).$$

Now let B be an arbitrary subset of V. Define

$$\operatorname{supp}_G B = \bigcup_{x \in B} \operatorname{supp}_G x.$$

In particular, we shall need the **carrier set** of a G-invariant convex polyhedron U which we define as

$$\operatorname{Carr}_G U = \operatorname{supp}_G \operatorname{Extr} U,$$

where $\operatorname{Extr} U$ denotes the set of extreme vectors (= vertices) of the polyhedron U.

For a subset $A \subset V$ consider the **stabilizer subgroup**

$$\operatorname{Stab}_G A = \{g \in G : \forall x \in A \quad gx = x\}.$$

This subgroup is generated by reflections across the mirrors containing A. It has only obvious fixed vectors, namely those in V^A — the intersection of all mirrors containing A. If A is not contained in any mirror then we put $V^A = V$. The orthogonal complement of V^A is obviously $(\operatorname{Stab}_G A)$-invariant. If we restrict the action of the subgroup $\operatorname{Stab}_G A$ to the subspace $V_A = (V^A)^\perp$, it will act there effectively, and therefore it will become a Coxeter group on V_A. Let G_A denote this Coxeter group:

$$G_A = \operatorname{Stab}_G A|_{V_A}.$$

Let proj_A, proj^A denote the orthogonal projectors onto V_A, V^A, respectively. Obviously, $\mathbf{I} = \operatorname{proj}_A + \operatorname{proj}^A$, $\operatorname{proj}_A \operatorname{proj}^A = \operatorname{proj}^A \operatorname{proj}_A = 0$.

There exists a useful connection between the orthogonal projector proj^A and the stabilizer subgroup $\operatorname{Stab}_G A$. For any finite group K of linear operators acting on V consider the K-**averaging** operator

$$\operatorname{av}_K = (1/\operatorname{card} K) \sum_{g \in K} g.$$

One can easily show that the range of the K-averaging operator is exactly the set of fixed vectors of K.

LEMMA 2.1. *Let G be a Coxeter group. Then*

$$\operatorname{proj}^A = \operatorname{av}_{\operatorname{Stab}_G A}.$$

PROOF. For $x \in V^A$ it is obvious that $\operatorname{av}_{\operatorname{Stab}_G A} x = x$. Since every vector in the range of $\operatorname{av}_{\operatorname{Stab}_G A}|_{V_A} = \operatorname{av}_{G_A}$ is obviously fixed by the action of $\operatorname{Stab}_G A$, and since $\operatorname{Stab}_G A$ acts effectively on V_A, then $\operatorname{av}_{\operatorname{Stab}_G A}|_{V_A} = 0$, which proves the Lemma. □

COROLLARY 2.2. Let U be a convex G-invariant set. Then

$$\operatorname{proj}^A U = U \cap V^A.$$

LEMMA 2.3. *Let G be a Coxeter group. The origin is a relatively interior point of $\operatorname{Co}_G x$ for every nonzero $x \in V$.*

PROOF. Assuming that 0 is not in the relative interior of $\operatorname{Co}_G x$, we find a nonzero vector $b \in \operatorname{span} \operatorname{Orb}_G x$ such that $\langle b, gx \rangle \geq 0$ for all $g \in G$. Since G acts effectively then, by Lemma 2.1, $\operatorname{av}_G = 0$, so $0 = \operatorname{av}_G x = (1/\operatorname{card} G) \sum_{g \in G} gx$, and then $\langle b, gx \rangle = 0$ for all $g \in G$. This means that $b \perp \operatorname{Orb}_G x$, and since $b \in \operatorname{span} \operatorname{Orb}_G x$, we conclude that $b = 0$, contrary to the assumption. □

COROLLARY 2.4. Let G be an irreducible Coxeter group. The origin is an interior point of $\operatorname{Co}_G x$ for every nonzero $x \in V$.

LEMMA 2.5. *Let G be a Coxeter group. Then*

$$\operatorname{Co}_{\operatorname{Stab}_G A} x = \operatorname{proj}^A x + \operatorname{Co}_{G_A} \operatorname{proj}_A x.$$

In particular, $\operatorname{Co}_{\operatorname{Stab}_G A} x$ *is a polyhedron in an affine plane of dimension* $\leq \dim V_A$.

Let κ be a subset of the set $\operatorname{ver}(G)$ of vertices of the graph $\Gamma(G)$. Let $\Gamma(G) \setminus \kappa$ denote the graph obtained from $\Gamma(G)$ by erasing the vertices from κ together with the edges adjacent to these vertices.

The following three useful results follow almost immediately from the definitions and the above mentioned facts.

LEMMA 2.6.

$$\Gamma(G_A) = \Gamma(G) \setminus \operatorname{supp}_G A.$$

LEMMA 2.7.

$$\operatorname{supp}_{G_A} \operatorname{proj}_A x = \operatorname{supp}_G x \setminus \operatorname{supp}_G A.$$

COROLLARY 2.8.

$$\mathcal{W}_{G_A} = (\operatorname{proj}_A \mathcal{W}_G) \setminus \{0\}.$$

Let

$$m_G(x, y) = \sup\{\langle gx, y \rangle : g \in G\}.$$

Obviously, $m_G(x, y)$ depends only upon the G-orbits of x and y, so the notation $m_G(x, i)$ is meaningful for $x \in V$ and i a G-orbit, and for x and i both G-orbits.

LEMMA 2.9 ([**19**]). *Vectors x and y belong to the same Weyl chamber if and only if*

$$m_G(x, y) = \langle x, y \rangle.$$

Moreover, in this case

$$\{z \in \operatorname{Orb}_G x : \langle z, y \rangle = m_G(x, y)\} = \operatorname{Orb}_{\operatorname{Stab}_G y} x.$$

One can easily deduce from Lemma 2.9 that $m_G(x, y) \geq 0$, and $m_G(x, y) = 0$ if and only if there are no components of $\Gamma(G)$ intersecting both $\operatorname{supp}_G x$ and $\operatorname{supp}_G y$. In particular, $m_G(x, y) > 0$ for an irreducible group G and nonzero x and y.

3. Convex Structure of Coxeter Orbihedra

3.1. Polyhedra of full dimension and their polars. Let G be a finite Coxeter group, maybe reducible. An orbihedron $\operatorname{Co}_G x$ is **of full dimension** (i.e., is not a subset in a proper subspace) if and only if x is not in a proper G-invariant subspace. This happens if and only if $J(G, x) = J(G)$, i.e., $\operatorname{supp}_G x$ intersects every component of $\Gamma(G)$. Let us agree that if $\operatorname{Co}_G x$ is not of full dimension then we regard it as a polyhedron in the subspace $\operatorname{span} \operatorname{Orb}_G x = \bigoplus_{j \in J(G,x)} V(j)$, and we consider its faces of maximal dimension (= of codimension 1) in this subspace. This means that we actually consider the orbihedron with respect to the group

$$G[x] = G|_{\mathrm{span}\,(\mathrm{Orb}_G\, x)}.$$

Let $V^{[x]} = V^{\mathrm{Orb}_G\, x} = \mathrm{span}\,(\mathrm{Orb}_G\, x)$. Obviously, $\mathrm{Co}_G\, x = \mathrm{Co}_{(G[x])}\, x$ is of full dimension in $V^{[x]}$. Also, one can easily see that $\Gamma(G[x])$ is the disjoint union of the components of $\Gamma(G)$, intersecting with $\mathrm{supp}_G\, x$:

$$\Gamma(G[x]) = \bigsqcup \{j : j \in J(G,x)\}.$$

The reason for our desire to consider only polyhedra of full dimension is explained in the next paragraph.

For a subset $U \subset V$ let

$$U^\circ = \{y \in V : \forall\ x \in U \ \ \langle x, y\rangle \leq 1\}.$$

U° is called the **polar set** of U, it is convex, closed and contains the origin. Obviously, $U^\circ = (\mathrm{conv}\ U)^\circ$, where $\mathrm{conv}\ U$ denotes the convex hull of U. If U contains the origin, then, by the Bipolar Theorem, $(U^\circ)^\circ$ is the closed convex hull of U. If U is a G-invariant convex polyhedron for a Coxeter group G then U contains the origin and therefore

$$U = (U^\circ)^\circ = \{y \in V : \forall\ z \in U^\circ \ \ \langle y, z\rangle \leq 1\},$$

so we get a description of the polyhedron U in terms of linear inequalities. It is possible to switch to the smallest possible set of inequalities in this description. If the G-invariant convex polyhedron U is of full dimension, then, by Lemma 2.3 it contains the origin as an interior point and therefore its polar set U° is a compact polyhedron. So, by the Krein–Milman Theorem, it is the convex hull of the set $\mathrm{Extr}\,(U^\circ)$ of its extreme vectors,

$$U = (U^\circ)^\circ = (\mathrm{Extr}\,(U^\circ))^\circ = \{y \in V : \forall\ z \in \mathrm{Extr}\,(U^\circ) \ \ \langle y, z\rangle \leq 1\},$$

and this is obviously the smallest possible set of linear inequalities describing the polyhedron U. Affine hyperplanes $\{y \in V : \langle y, z\rangle = 1\}$, $z \in \mathrm{Extr}\,(U^\circ)$, carry codimension 1 faces of U, so the set $\mathrm{Extr}\,(U^\circ)$ is the set of normals to faces of U of codimension 1.

Let us note that if $\Gamma(G[x])$ is not connected (= if group $G[x]$ is reducible) then every G-orbihedron has a natural product structure:

$$\mathrm{Co}_G\, x = \prod_{j \in J(G,x)} \mathrm{Co}_{G(j)}\, \mathrm{proj}\,(j) x,$$

where $G(j)$ denotes the irreducible Coxeter group whose graph is $j \in J(G,x)$, $\mathrm{proj}\,(j)$ denotes the orthogonal projection onto the subspace $V(j)$ where $G(j)$ naturally acts.

3.2. Orbihedra — faces of codimension 1. The following result is an immediate corollary of Lemma 2.9.

LEMMA 3.1. *If* $y \in \mathrm{Extr}\,(\mathrm{Co}_G\, x)^\circ$ *then the codimension* 1 *face*

$$\Phi(y) = \{z \in \mathrm{Co}_G\, x : \langle z, y\rangle = 1\}$$

of $\mathrm{Co}_G\, x$ *coincides with* $\mathrm{Co}_{\mathrm{Stab}_G\, y}\, g_0 x$, *where* $g_0 x$ *is any vector from* $\mathrm{Orb}_G\, x$ *belonging to this face.*

Now we can describe the set $\operatorname{Extr}(\operatorname{Co}_G x)^\circ$.

THEOREM 3.2 ([**19**]). *Let G be a Coxeter group naturally acting on V. Then*

$$\operatorname{Extr}(\operatorname{Co}_G x)^\circ = \{y/m_G(x,y) \in V : \operatorname{supp}_G y \text{ consists of one vertex,}$$
$$\text{belonging to } \Gamma(G[x]), \text{ and } \operatorname{supp}_G x \text{ intersects every component of}$$
$$\Gamma(G[x]) \setminus \operatorname{supp}_G y\}$$

PROOF. It is easy to see that $z \in \operatorname{Extr}(\operatorname{Co}_G x)^\circ$ if and only if the set $\{gx : \langle gx, z\rangle = m_G(x,z) = 1\}$ spans the whole space $V^{[x]} = \operatorname{span} \operatorname{Orb}_G x$. Recall that by Lemma 3.1, $\{gx : \langle gx, z\rangle = m_G(x,z) = 1\} = \operatorname{Orb}_{\operatorname{Stab}_G z}\, g_0 x$, $\langle g_0 x, z\rangle = m_G(x,z) = 1$. By Lemma 2.5,

$$\operatorname{Co}_{\operatorname{Stab}_G z}\, g_0 x = \operatorname{proj}^z g_0 x + \operatorname{Co}_{G_z} \operatorname{proj}_z g_0 x,$$

so $z \in \operatorname{Extr}(\operatorname{Co}_G x)^\circ$ if and only if $\operatorname{Co}_{G_z} \operatorname{proj}_z g_0 x$ is of dimension $\dim V^{[x]} - 1$. This happens if and only if $\Gamma((G[x])_z)$ has exactly $\dim(V^{[x]}) - 1$ vertices. Since, by Lemma 2.7, $\Gamma((G[x])_z) = \Gamma(G[x]) \setminus \operatorname{supp}_G z$, this can happen if and only if $\operatorname{supp}_G z$ consists of one vertex, belonging to $\Gamma(G[x])$, and $\operatorname{supp}_{(G[x])_z} \operatorname{proj}_z g_0 x$ intersects every component of the graph $\Gamma(\operatorname{Stab}_{G[x]} z)$. By Lemma 2.7, $\operatorname{supp}_{(G[x])_z} \operatorname{proj}_z g_0 x = \operatorname{supp}_G x \setminus \operatorname{supp}_G z$, and by Lemma 2.6, $\Gamma(\operatorname{Stab}_{G[x]} z) = \Gamma(G[x]) \setminus \operatorname{supp}_G z)$. The Theorem is proven. □

Since the only vectors having one-vertex supports are weights, and since $m_G(x,\omega) > 0$ for $\omega \in \mathcal{W}_{G[x]} = \mathcal{W}_G \cap V^{[x]}$, we see that $z \in \operatorname{Extr}(\operatorname{Co}_G x)^\circ$ if and only if

$$z = \omega/m_G(x,\omega),\ \omega \in \mathcal{W}_G,\ \pi(\omega) \in \Gamma(G[x]),$$

and

$$\operatorname{supp}_G x \text{ intersects every component of } \Gamma(G[x]) \setminus \{\pi(\omega)\}.$$

According to our agreement, we disregard all weights ω such that $\pi(\omega) \notin \Gamma(G[x])$, i.e., such that $m_G(x,\omega) = 0$.

Combining the previous results, we arrive to the following description.

THEOREM 3.3. *Let G be a Coxeter group naturally acting in V. For every codimension 1 face Φ of $\operatorname{Co}_G x$ there exists a unique vector $\omega = \omega(\Phi) \in \mathcal{W}_{G[x]}$, such that:*

(i) $\operatorname{supp}_G x$ intersects every component of $\Gamma(G[x]) \setminus \operatorname{supp}_G \omega$,

(ii) $\Phi = \operatorname{Co}_{\operatorname{Stab}_G \omega}\, g_0 x$, where $g_0 \in G$ is such that $g_0 x$ and ω belong to one Weyl chamber.

Moreover, for every $\omega \in \mathcal{W}_{G[x]}$, satisfying (i), the set Φ defined in (ii) is a codimension 1 face of $\operatorname{Co}_G x$.

COROLLARY 3.4. Let G be an irreducible Coxeter group. Let ω be a weight such that $\pi(\omega)$ is an end vertex of $\Gamma(G)$. Then $\omega \notin \operatorname{Extr}(\operatorname{Co}_G x)^\circ$ if and only if $\operatorname{supp}_G x = \operatorname{supp}_G \omega$.

Using the remark preceding Lemma 3.1, we conclude that a face Φ of $\operatorname{Co}_G x$ has a natural product structure if the graph $\Gamma(G) \setminus \{\pi(\omega(\Phi))\}$ is not connected. So, if group G is irreducible, then the only faces Φ of $\operatorname{Co}_G x$ not having the natural product structure are those for which $\pi(\omega(\Phi))$ is an end vertex of $\Gamma(G)$.

3.3. Counting vertices of orbihedra. It is not difficult to find $\operatorname{card}_G x$ — the number of vertices in $\operatorname{Co}_G x$ (= the number of distinct vectors in $\operatorname{Orb}_G x$). It follows from the definition of the stabilizer subgroup that

$$\operatorname{card}_G x = \frac{\operatorname{card} G}{\operatorname{card} \operatorname{Stab}_G x} = \frac{\operatorname{card} G}{\operatorname{card} G_x}.$$

For every irreducible Coxeter group G the number card G is well known and may be found, e.g., in [**5**]. For a reducible group G we know that $G = \prod_{j\in J(G)} G(j)$ where $J(G)$ is the set of components of $\Gamma(G)$ and $G(j)$ denotes the irreducible Coxeter group whose graph is the component j. Therefore,

$$\operatorname{card} G = \prod_{j\in J(G)} \operatorname{card} G(j).$$

Since $\Gamma(G_x) = \Gamma(G) \setminus \operatorname{supp}_G x$, we can compute the number $\operatorname{card}_G x$ in convenient geometric terms.

Since $\operatorname{Co}_G x$ is of full dimension in $V^{[x]}$, then

$$\operatorname{card}_G x \geq 1 + \dim V^{[x]} = 1 + \operatorname{card} \operatorname{ver}(G_x).$$

So, for an irreducible group G we have $\operatorname{card}_G x \geq 1 + \dim V$.

LEMMA 3.5. *Let G be an irreducible Coxeter group. Then* $\operatorname{card}_G x = 1 + \dim V$ *if and only if* $G = A_n$ *and* $\operatorname{supp}_G x$ *is an end vertex of* $\Gamma(G)$.

PROOF. The "if" part can be verified directly: the orbit of the vector $(1, 1, \cdots, 1, -n)$ in the n-dimensional subspace $\{(x_1, x_2, \cdots, x_{n+1}) \in \mathbf{R}^{n+1} : \sum_i x_i = 0\}$ under the action of permutations consists of exactly $(n + 1)$ vectors. Let us concentrate on the "only if" part. If $\operatorname{card}_G x = 1 + \dim V$, then $\operatorname{Co}_G x$ is a simplex in V, so $(\operatorname{Co}_G x)^\circ$ is also a simplex, so the set $\operatorname{Extr}(\operatorname{Co}_G x)^\circ$ consists of $(1+\dim V)$ vectors, therefore it contains exactly one G-orbit. But $\operatorname{Extr}(\operatorname{Co}_G x)^\circ$ always contains the orbit of a weight associated with an end vertex of the Coxeter graph, therefore, $(\operatorname{Co}_G x)^\circ = \operatorname{Co}_G \omega$, $\operatorname{supp}_G \omega$ is an end vertex . Since $\operatorname{Co}_G x = (\operatorname{Co}_G \omega)^\circ$ and the latter is a simplex, we conclude that $\operatorname{supp}_G x$ is also an end vertex, and, besides, $\Gamma(G)$ has exactly two end vertices. Next, the group G cannot contain $-\mathbf{I}$ because a simplex cannot be central symmetric. It is known from the classification (see [**5**]) that A_n is the only irreducible Coxeter group possessing all these properties. □

3.4. Counting faces adjacent to a vertex. Let us obtain a more explicit description of the faces of $\operatorname{Co}_G x$ adjacent to the vertex x. We may assume that $\operatorname{supp}_G x$ intersects all components of $\Gamma(G)$, so the G-orbihedron $\operatorname{Co}_G x$ is of full dimension.

Fix a Weyl chamber C, assume that $x \in C$. Let ω_j, $j = 1, 2, \cdots, \dim V$, be the fundamental weights belonging to C. Then, by Theorem 3.2 the extreme vectors of $(\operatorname{Co}_G x)^\circ$, associated with the faces of maximal dimension adjacent to x are precisely those from the $\operatorname{Stab}_G x$-orbits of the vectors $\omega_j/m_G(x, \omega_j)$ such that $\operatorname{supp}_G x$ intersects every component of $\Gamma(G) \setminus \{\pi(\omega_j)\}$.

Let us introduce some terminology. We fix vector x, all forthcoming notions depend upon x. We say that a vertex π of $\Gamma(G)$ is **admissible** (more precisely, it should be called x-admissible, but we shall skip the label x if it does not create ambiguity) if $\operatorname{supp}_G x$ intersects every component of $\Gamma(G) \setminus \{\pi\}$. So, a vertex π of $\Gamma(G)$ is admissible if and only if the related vector $\omega_j/m_G(x, \omega_j)$, $\pi = \pi(\omega_j)$, defines a codimension 1 face of $\operatorname{Co}_G x$ adjacent to x. The number of vectors in the $(\operatorname{Stab}_G x)$-orbit of ω_j is called the **multiplicity** of the admissible vertex $\pi = \pi(\omega_j)$. So, the sum of multiplicities of all admissible vertices gives the number of codimension 1 faces of $\operatorname{Co}_G x$ adjacent to x. We shall list all admissible vertices, together with their multiplicities, in convenient geometric terms.

LEMMA 3.6. *The multiplicity of an admissible vertex from* $\operatorname{supp}_G x$ *is equal to* 1. *The multiplicity of an admissible vertex* π *belonging to a component* γ *of* $\Gamma(G) \setminus \operatorname{supp}_G x$ *is equal to* $1 + (\operatorname{card} G(\gamma)/\operatorname{card} G(\gamma \setminus \{\pi\}))$.

A vertex $\pi \in \operatorname{supp}_G x$ is called **interior** (or, better, x-interior) if it is adjacent only to vertices from $\operatorname{supp}_G x$. All non-interior vertices of $\operatorname{supp}_G x$ are called **boundary**.

LEMMA 3.7. *Every interior vertex of* $\operatorname{supp}_G x$ *is admissible.*

Let γ be a component of $\Gamma(G) \setminus \operatorname{supp}_G x$. All end vertices of $\Gamma(G)$, belonging to γ, are called the **principal** vertices of γ.

LEMMA 3.8. *Every principal vertex* π *of a component* γ *of* $\Gamma(G) \setminus \operatorname{supp}_G x$ *is admissible.*

Since $\operatorname{supp}_G x$ intersects every component of $\Gamma(G)$ then for every component of $\Gamma(G) \setminus \operatorname{supp}_G x$ there exists a vertex from $\operatorname{supp}_G x$ adjacent to this component. We say that γ is **acceptable** (or, better, x-acceptable) if there is exactly one vertex from $\operatorname{supp}_G x$ adjacent to γ.

LEMMA 3.9. *Every acceptable component must contain at least one principal vertex.*

PROOF. Indeed, if to assume the opposite, then each end vertex π of γ is not an end vertex of $\Gamma(G)$. The graph γ must have end vertices, since a Coxeter graph cannot have cycles — see [**5**]. Therefore it must be adjacent to at least two other vertices of $\Gamma(G)$, but since π is an end vertex of γ, then at most one of these neighboring vertices is in γ, so at least one of them is in $\operatorname{supp}_G x$. Since γ is acceptable, then exactly one of the neighboring vertices is in $\operatorname{supp}_G x$. So, there are exactly two neighboring vertices, therefore π is adjacent to another vertex of γ. Therefore γ must have another end vertex (again, no cycles!). Repeating the same argument, we find another vertex from $\operatorname{supp}_G x$ adjacent to γ, which contradicts the acceptability. □

So, the set of end vertices of an acceptable component consists of principal vertices plus, maybe, one non-principal vertex adjacent to $\operatorname{supp}_G x$. All end vertices of an acceptable component are principal if and only if the component consists of one principal vertex.

COROLLARY 3.10. The only admissible vertices of an acceptable component are the principal vertices.

COROLLARY 3.11. The only non-admissible boundary vertices are those adjacent to at least one acceptable component.

LEMMA 3.12. *A non-acceptable component* γ *either contains no principal vertices or contains exactly one principal vertex and one branching vertex of* $\Gamma(G)$.

PROOF. By the definition of a non-acceptable component, there exist at least two vertices from $\operatorname{supp}_G x$, adjacent to γ. This means that there are two options:

(1) there are at least two end vertices π_1, π_2 of γ adjacent to $\operatorname{supp}_G x$,

(2) there is an end vertex π of γ adjacent to at least two vertices of $\operatorname{supp}_G x$.

Consider the first option. Connected graph γ has at least two end vertices, therefore for every end vertex of γ there exists another vertex of γ adjacent to it. Therefore π_1, π_2 are not principal. If γ contains a principal vertex then γ has at least three end vertices. But a connected Coxeter graph cannot have more than three end vertices, so γ has exactly three end vertices, including exactly one principal vertex. Since γ has three end vertices, then, according to the classification of connected Coxeter graphs, it must have a branching vertex. Also, according to the classification, a connected Coxeter graph cannot have more than one branching vertex, so the statement is true in this situation.

Consider the second option. The vertex π is not principal. Therefore, if γ contains a principal vertex, then this principal vertex is an end vertex of γ, different from π. Since γ is connected, then π must be adjacent to at least one other vertex of γ. Therefore π is a branching vertex of the component of $\Gamma(G)$, containing γ. This component of $\Gamma(G)$ is a connected Coxeter graph, so it cannot have other branching vertices. If γ contains more than one principal vertex then one of the vertices of γ must be branching in γ. But π is not branching in γ, and it is the only branching vertex of the component of $\Gamma(G)$, containing γ. Therefore the statement is true in this situation as well. □

If a non-acceptable component γ does not contain principal vertices then we call all vertices of γ **regular**.

Let a non-acceptable component γ contain a principal vertex π and a branching vertex ρ of $\Gamma(G)$. Consider the vertices along the simple path in γ connecting π to ρ (including both). Let us call these vertices irregular, all other vertices of γ are called **regular**. Note that if ρ is an end vertex of γ (this is exactly the case (2) in the proof of Lemma 3.12) then all vertices of γ are irregular.

COROLLARY 3.13. The regular vertices of non-acceptable components are the only admissible non-principal vertices in these components.

Combining the statements of this subsection, we arrive to the following result:

THEOREM 3.14. *Let G be a Coxeter group, naturally acting on V. Fix $x \in V$ such that* $\operatorname{supp}_G x$ *intersects every component of* $\Gamma(G)$. *The following is a complete list of admissible vertices and their multiplicities:*

(i) every interior vertex from $\operatorname{supp}_G x$; *its multiplicity equals* 1,

(ii) every boundary vertex from $\operatorname{supp}_G x$, *except of those adjacent to at least one acceptable component of* $\Gamma(G) \setminus \operatorname{supp}_G x$; *its multiplicity equals* 1,

(iii) every principal vertex π of a component γ of $\Gamma(G) \setminus \operatorname{supp}_G x$; *its multiplicity equals* $1 + (\operatorname{card} G(\gamma)/\operatorname{card} G(\gamma \setminus \{\pi\}))$;

(iv) every regular vertex π of a non-acceptable component γ; its multiplicity equals $1 + (\operatorname{card} G(\gamma)/\operatorname{card} G(\gamma \setminus \{\pi\}))$.

3.5. Simplicial orbihedra. A version of the next result is known (and is important in construction of special toroidal varieties) — it is due to Klyachko and Voskresenskii (Theorem 4 in [**17**]). They formulate it as a criterion of simpliciality of a cone obtained from a Weyl chamber by the action of a stabilizer group. We obtain this result as a direct corollary of Theorem 3.14.

A full dimensional G-orbihedron is called **simplicial** if there are exactly $\dim V$ faces of maximal dimension adjacent to every vertex of this polyhedron. One can easily see that if G is irreducible and $\operatorname{supp}_G x = \operatorname{ver}(G)$ then $\operatorname{Co}_G x$ is simplicial. It is not hard to present examples of non-simplicial G-orbihedra. A natural question is:

for which $x \in V$ the related G-orbihedron is simplicial?

A Coxeter graph is said to be of A_n **type** if it has no branching vertices and has no multiple edges.

COROLLARY 3.15. Let G be an irreducible Coxeter group. Then $\operatorname{Co}_G x$ is simplicial if and only if the following is true:

(i) the graph $\Gamma(G) \setminus \operatorname{supp}_G x$ is of A_n type,

(ii) every component of $\Gamma(G) \setminus \operatorname{supp}_G x$ contains an end vertex of $\Gamma(G)$ (a principal vertex),

(iii) there are no vertices of $\operatorname{supp}_G x$ adjacent to more than one component of $\Gamma(G) \setminus \operatorname{supp}_G x$.

PROOF. Due to explanations preceding Lemma 3.6 we only need to find out when the sum of multiplicities of all admissible vertices is exactly $\dim V = \operatorname{card} \operatorname{ver}(G)$.

First, all vertices from $\operatorname{supp}_G x$ except of those adjacent to acceptable components, are admissible and have multiplicities 1. To each non-admissible vertex of $\operatorname{supp}_G x$ we assign an acceptable component γ adjacent to this vertex, and distinct non-admissible vertices of $\operatorname{supp}_G x$ get distinct acceptable components, due to the definitions. It may happen that there remains an acceptable component not assigned to any non-admissible vertex of $\operatorname{supp}_G x$. Every principal vertex of an acceptable component γ is admissible, with multiplicity at least $1 + \operatorname{card} \operatorname{ver}(\gamma)$. The multiplicity is exactly $1 + \operatorname{card} \operatorname{ver}(\gamma)$ if and only if γ is of A_n type, due to Lemma 3.5. So, the sum of multiplicities of admissible vertices in an acceptable component is greater or equal to the number of vertices in this component plus one = the number of vertices adjacent to this component (we agree that vertices belonging to the component are also adjacent to it). Therefore the sum of multiplicities of admissible vertices in all acceptable components is greater or equal to the number of vertices in these components plus the number of non-admissible vertices in $\operatorname{supp}_G x$, with equality if and only if all acceptable components are of A_n type, and none of non-admissible vertices from $\operatorname{supp}_G x$ is adjacent to more than one acceptable component.

Every non-acceptable component γ of $\Gamma(G) \setminus \operatorname{supp}_G x$ has admissible vertices each of multiplicity at least $1 + \operatorname{card} \operatorname{ver}(\gamma) > \operatorname{card} \operatorname{ver}(\gamma)$. So, the sum of multiplicities of admissible vertices in all non-acceptable components is strictly greater than the number of vertices in these components.

Therefore the sum of multiplicities of all admissible vertices equals to the overall number of vertices if and only if all components of $\Gamma(G) \setminus \operatorname{supp}_G x$ are acceptable, all are of A_n type, and none of vertices from $\operatorname{supp}_G x$ is adjacent to more than one component of $\Gamma(G) \setminus \operatorname{supp}_G x$. □

4. Coxeter Orbihedra: faces of lower dimensions

Now we generalize Theorem 3.3 to obtain a complete description of all faces (not necessarily of codimension 1) of a G-orbihedron $\operatorname{Co}_G x$.

Each face ϕ of U of codimension 2 is a codimension 1 face in a face Φ of U of codimension 1, and, further, each face of U of codimension k is a codimension 1 face in a face of U of codimension $k-1$.

Theorem 3.3 provides a description of faces of $\mathrm{Co}_G\, x$ of codimension 1 in $V^{[x]}$. A face Φ of $\mathrm{Co}_G\, x$ of codimension 1 in $V^{[x]}$, containing x, is nothing else but $\mathrm{Co}_{\mathrm{Stab}_G\, \omega}\, x$, where $\omega \in \mathcal{W}_{G[x]}$ is such that $\mathrm{supp}_G\, x$ intersects every component of $\Gamma(G[x]) \setminus \{\pi(\omega)\}$. The face Φ is a full dimension convex subset of an affine hyperplane in $V^{[x]}$ orthogonal to ω. The faces of Φ are codimension 2 (in $V^{[x]}$) faces of $\mathrm{Co}_G\, x$. Let us project Φ onto the codimension 1 subspace $(\omega)^\perp$ in $V^{[x]}$. Then $\mathrm{proj}_\omega \Phi = \mathrm{Co}_{G_\omega} \mathrm{proj}_\omega x$. So, we are in the situation of a Coxeter group G_ω, and we can describe a face ψ of $\mathrm{proj}_\omega \Phi$, containing $\mathrm{proj}_\omega x$, with the help of Theorem 3.3:

$$\psi = \mathrm{Co}_{\mathrm{Stab}_{G_\omega} \kappa}\, \mathrm{proj}_\omega x,$$

where $\mathrm{supp}_{G_\omega} \kappa$ consists of one vertex of $\Gamma(G[x]_\omega)$ and $\mathrm{supp}_{G_\omega} \mathrm{proj}_\omega x$ intersects every component of $\Gamma(G[x]_\omega) \setminus \mathrm{supp}_{G_\omega} \kappa$, and $\mathrm{proj}_\omega x$ and κ are in one Weyl chamber of $G[x]_\omega$.

Keep in mind that $\Gamma(G[x]_z) = \Gamma(G[x]) \setminus \mathrm{supp}_G\, z$, $\mathrm{supp}_{G[x]_z} \mathrm{proj}_z x = \mathrm{supp}_{G[x]} x \setminus \mathrm{supp}_G\, z$, and $\kappa = \mathrm{proj}_\omega \tau$, $\tau \in \mathcal{W}_{G[x]}$ (Lemmas 2.6, 2.7, Corollary 2.8). Then $\mathrm{Stab}_{G_\omega} \kappa = \mathrm{Stab}_G \{\omega, \tau\}$. Repeating the same argument, we arrive to the following result:

THEOREM 4.1. *Let G be a Coxeter group naturally acting on V. For every codimension k face ϕ of the G-orbihedron $\mathrm{Co}_G\, x$ there exists a unique set $\Omega = \Omega(\phi) \subset \mathcal{W}_G$, $\mathrm{card}\, \Omega = k$, of fundamental weights, belonging to the same Weyl chamber C, such that*

(i) $\mathrm{supp}_G\, x$ intersects every component of $\Gamma(G[x]) \setminus \mathrm{supp}_G\, \Omega$,

(ii) $\phi = \mathrm{Co}_{\mathrm{Stab}_G\, \Omega}\, x^*(C, G)$.

Moreover, for every set $\Omega \subset \mathcal{W}_G \cap C$, satisfying (i), the set ϕ defined in (ii) is a codimension $\mathrm{card}\, \Omega$ face of $\mathrm{Co}_G\, x$.

The set of all faces of a convex polyhedron is naturally partially ordered by the inclusion relation. We say that two convex polyhedra in V are **facially isomorphic** if their sets of faces are isomorphic as partially ordered sets. Obviously, such an isomorphism must preserve the dimensions of the faces and the number of vertices on the corresponding faces.

COROLLARY 4.2. Let G be a Coxeter group, naturally acting in V. Two G-orbihedra $\mathrm{Co}_G\, x$ and $\mathrm{Co}_G\, y$ are facially isomorphic if and only if there exists an automorphism of the graph $\Gamma(G)$ transforming $\mathrm{supp}_G\, x$ into $\mathrm{supp}_G\, y$.

5. Coxeter-invariant Convex Polyhedra

Let G be a Coxeter group naturally acting on V. Consider a general convex G-invariant polyhedron U. Then the set $\mathrm{Extr}\, U$ of extreme points of U is also G-invariant, and therefore it is fibered into G-orbits. Let $I_G(U)$ denote the set of G-orbits in $\mathrm{Extr}\, U$. Then

$$U = \mathrm{conv} \bigcup_{i \in I_G(U)} i = \mathrm{conv} \bigcup_{\mathrm{Orb}_G\, x \in I_G(U)} \mathrm{Co}_G\, x.$$

Let

$$N_G(U) = \mathrm{card}\, I_G(U).$$

This number is a measure of complexity of the polyhedron U, in particular, if $N_G(U) = 1$, then U is a G-orbihedron. Let us refer to a G-invariant convex polyhedron U such that $N_G(U) = n$, as to a (G, n)-**polytope**. So, a G-orbihedron will also be called a $(G, 1)$-polytope.

Recall that

$$\operatorname{Carr}_G U = \operatorname{supp}_G \operatorname{Extr} U = \bigcup_{i \in I_G(U)} \operatorname{supp}_G i.$$

It is not hard to see that a polyhedron U is of full dimension if and only if the set $\operatorname{Carr}_G U$ intersects every component of $\Gamma(G)$. As before we may assume that the polyhedron U is of full dimension. If not, we switch to the subspace span U and to the group $G[U] = G|_{\text{span } U}$. Again, $\Gamma(G[U])$ consists of those components of $\Gamma(G)$ which intersect $\operatorname{Carr}_G U$.

Obviously,

$$U^\circ = \bigcap_{i \in I_G(U)} i^\circ = \bigcap_{i \in I_G(U)} (\operatorname{conv} i)^\circ,$$

and if we are looking for the extreme points of U° we have to determine the extreme points of this intersection. The extreme points of each of the sets $(\operatorname{conv} i)^\circ$ are already described, they all are weights, i.e., their supports consist of one vertex. Fix a Weyl chamber C. Since the set U° is G-invariant, it is sufficient to find only extreme points of U° that are in C. Obviously, $C \cap \operatorname{Extr}(U^\circ) \subset \operatorname{Extr}(C \cap U^\circ)$, but these sets may be different. Note that for every i the set $C \cap (\operatorname{conv} i)^\circ$ is in fact the simplex

$$S_C(i) = \{\sum_j \lambda_j \omega_j : \lambda_j \geq 0, \sum_j \lambda_j m_G(i, \omega_j) \leq 1\}.$$

Here $\omega_j, j = 1, 2, \cdots, \dim V$, are the fundamental weights belonging to C. Obviously, the origin is a vertex of $S_C(i)$, let us call this vertex a trivial vertex. The non-trivial vertices of $S_C(i)$ are the vectors $\omega_j / m_G(i, \omega_j), j = 1, 2, \cdots, \dim V$. This simplex $S_C(i)$ is cut off the Weyl chamber C by the affine hyperplane

$$\Pi_C(i) = \{\sum_j \lambda_j \omega_j : \sum_j \lambda_j m_G(i, \omega_j) = 1\}.$$

THEOREM 5.1. *Let G be a Coxeter group naturally acting in V, and let U be a (G, n)-polytope.*

(i) If $y \in \operatorname{Extr}(U^\circ)$ then $\operatorname{supp}_G y$ consists of no more than n vertices;

(ii) There cannot exist two distinct vectors in $C \cap \operatorname{Extr}(U^\circ)$ with coinciding supports consisting of exactly n vertices;

(iii) If $y \in \operatorname{Extr}(U^\circ)$ then $\operatorname{Carr}_G U$ intersects every component of $\Gamma(G) \setminus \operatorname{supp}_G y$.

PROOF. Fix a Weyl chamber C and assume that $y \in C$. The point $y \in C$ is an extreme point of U°, therefore it has to be an extreme point of $C \cap (U^\circ)$, so it is the intersection of $\dim V$ linearly independent boundary hyperplanes of $C \cap (U^\circ)$. All boundary hyperplanes of this set are either the walls of C or the hyperplanes $\Pi_C(i)$, $i = 1, 2, \ldots, n$. Therefore y belongs to no more than n affine hyperplanes $\Pi_C(i)$, hence it belongs to no less than $\dim V - n$ walls of C. So, y does not belong to at most n walls, which proves (i).

To prove (ii), it suffices to note that if y has a support of n vertices then it belongs to exactly $\dim V - n$ walls of C. Therefore, it must belong to all of n affine hyperplanes $\Pi_C(i)$. Since y is an extreme vector, these $\dim V$ hyperplanes have only one common point. Therefore, any other vector from $\mathrm{Extr}\,(C \cap U^\circ)$, having the same support, must coincide with this one.

To prove (iii), note that if $x_i, i = 1, 2, \ldots, n$, are representatives of n pairwise distinct G-orbits constituting the set of extreme vectors of our (G, n)-polytope, then the set $\{gx_i : g \in G, i = 1, 2, \ldots, n : \langle y, gx_i \rangle = m_G(y, x_i) = 1\}$ must span the whole space V. Therefore, by Lemma 2.9 for each i all vectors gx_i on the face must belong to the same $\mathrm{Stab}_G\, y$-orbit. So the orthogonal projections of these orbits to V_y must span the whole space V_y. Therefore $\bigcup_i \mathrm{supp}_{G_y} \mathrm{proj}_y\, x_i$ must intersect every component of $\Gamma(G_y) = \Gamma(G) \setminus \mathrm{supp}_G\, y$. Recalling Lemma 2.7, we see that

$$\bigcup_i \mathrm{supp}_{G_y} \mathrm{proj}_y\, x_i = \bigcup_i (\mathrm{supp}_G\, x_i \setminus \mathrm{supp}_G\, y) = \mathrm{Carr}_G\, U \setminus \mathrm{supp}_G\, y,$$

so the proof is completed. □

The above result can be strengthened and complemented with a description of the codimension 1 face of U, associated with a given extreme vector of U°. For a convex G-invariant polyhedron U and for $y \in V$ let

$$m_G(U, y) = \max_{j \in I_G(U)} m_G(j, y),$$
$$I_G(U, y) = \{i \in I_G(U) : m_G(i, y) = m_G(U, y)\},$$
$$\mathrm{Carr}_G(U, y) = \bigcup_{i \in I_G(U,y)} \mathrm{supp}_G\, i.$$

As before, V^y denotes the intersection of all mirrors containing y. Recall that $U \cap V^y = \mathrm{proj}^y\, U$.

THEOREM 5.2. *Let G be a Coxeter group naturally acting in V. For every codimension 1 face Φ of a (G, n)-polytope U there exists a unique subset $\gamma \subset \mathrm{ver}\,(G[U])$ and a unique vector y, $\mathrm{supp}_G\, y = \gamma$, such that*

(i) $\mathrm{card}\, \gamma \leq n$,

(ii) *y is a normal to a codimension 1 face of the polyhedron $U \cap V^y$ in the subspace V^y,*

(iii) *$\mathrm{Carr}_G(U, y)$ intersects every component of $\Gamma(G[U]) \setminus \gamma$,*

(iv) *let $x_i \in i$ be such that $\langle x_i, y \rangle = m_G(i, y)$. Then*

$$\Phi = \mathrm{conv} \bigcup_{i \in I_G(U,y)} \mathrm{Co}_{\mathrm{Stab}_G\, y}\, x_i.$$

So, Φ is a $(\mathrm{Stab}_G\, y, k)$-polytope, where $k \leq \mathrm{card}\, I_G(U, y)$.

Moreover, for every $\gamma \subset \mathrm{ver}\,(G[U])$, $y \in V$, $\mathrm{supp}_G\, y = \gamma$, satisfying (i) — (iii), the set Φ defined in (iv), is a codimension 1 face of U.

Note that if $n = 1$ then $\mathrm{Carr}_G\, U = \mathrm{Carr}_G(U, y)$ for any y, so (iii) is formulated in terms of $\mathrm{Carr}_G\, U$ and γ only, and there always exists a unique y satisfying (ii). So, in this case Theorem 5.2 reduces to Theorem 3.3. This means that (i) and (iii) deliver a full description of $\mathrm{Extr}\,(U^\circ)$ in terms of $\mathrm{Carr}_G\, U$ and γ for $n = N_G(U) = 1$. Such a description is not possible for $n \geq 2$. We present some counterexamples in Theorem 5.10 below. Note that if $N_G(U) \geq \dim V$, then condition (i) in Theorem 5.1 is satisfied by any vector in V. We show that in this case any vector from V can serve as a vector from $\mathrm{Extr}\,(U^\circ)$ for a $(G, \dim V)$-polytope U, see Theorem 5.10 below.

Let us now describe the elements of Extr (U°) having the minimal and maximal supports.

First, we describe all one-vertex supported elements of Extr (U°) for a G-invariant convex polytope U.

COROLLARY 5.3. Let U be a G-invariant convex polytope. Let π be a vertex of $\Gamma(G[U])$. The vector $\omega/m_G(U,\omega)$, $\pi(\omega)=\pi$, belongs to Extr (U°) if and only if the set $\operatorname{Carr}_G(U,\omega)$ intersects every component of $\Gamma(G[U])\setminus\{\pi\}$.

Now let us describe the vectors from Extr (U°) whose supports consist of the maximal possible number of vertices, namely, of $N_G(U)$ vertices.

COROLLARY 5.4. Let U be a (G,n)-polytope, $n\le \dim V$. Choose $\gamma\subset \operatorname{ver}(G[U])$ such that card $\gamma=n$. There exists a unique G-orbit $i\in I_G(U)$, $\operatorname{supp}_G i=\gamma$ if and only if

(i) the linear system

$$\sum_{j\in\gamma}\lambda_j m_G(i,\omega_j)=1,\quad i=1,2,\dots,n,\quad i\in I_G(U),$$

has a unique solution, and all entries of this solution are positive,

(ii) $\operatorname{supp}_G U$ intersects every component of $\Gamma(G[U])\setminus\gamma$.

5.1. $(G,2)$**-polytopes.** Let U be a $(G,2)$-polytope. Then all vectors from Extr (U°) have supports consisting of one or two vertices of $\Gamma(G)$. It is easy to describe these supports in rather explicit geometric terms. Let $i_\pm$ denote the two G-orbits constituting $I_G(U)$. For $\pi\in\operatorname{ver}(G[U])$ choose $\omega\in\mathcal{W}_G$ such that $\pi(\omega)=\pi$, and let

$$\mu(\pi)=\frac{m_G(i_+,\omega)}{m_G(i_-,\omega)}.$$

Since $\pi\in\operatorname{ver}(G[U])$, it cannot happen that the numerator and the denominator of the fraction defining $\mu(\pi)$ are both zero, so $0\le\mu(\pi)\le\infty$. Note that $\mu(\pi)$ does not depend upon the choice of ω, $\pi(\omega)=\pi$.

Consider the **canonical partition** of the set ver $(G[U])$ into the following three subsets:

$$I_+=\{\pi\in\operatorname{ver}(G[U]):\mu(\pi)>1\},$$

$$I_-=\{\pi\in\operatorname{ver}(G[U]):\mu(\pi)<1\},$$

$$I_0=\{\pi\in\operatorname{ver}(G[U]):\mu(\pi)=1\}.$$

Considering the geometry of the related lines, one can easily verify that a linear system

$$\lambda_0a_{00}+\lambda_1a_{01}=1$$

$$\lambda_0a_{10}+\lambda_1a_{11}=1$$

with positive coefficients $a_{ij}, 0\le i,j,\le 1$, has a unique solution with positive entries if and only if

$$\left(\frac{a_{01}}{a_{11}}-1\right)\left(\frac{a_{00}}{a_{10}}-1\right)<0.$$

The above considerations lead to the following result:

COROLLARY 5.5. Let U be a $(G,2)$-polytope. Consider the canonical partition I_+, I_-, I_0 of the set $\mathrm{ver}\,(G[U])$, associated with the two G-orbits $i_\pm$ constituting $I_G(U)$.

Choose a vertex $\pi \in \Gamma(G[U])$. There exists a vector $z \in \mathrm{Extr}\,(U^\circ)$ such that $\mathrm{supp}_G\, z = \{\pi\}$ if and only if **one of the following** conditions is satisfied:

(i) if $\pi \in I_\pm$, then $\mathrm{supp}_G\, i_\pm$ intersects every component of $\Gamma(G[U]) \setminus \{\pi\}$,

(ii) if $\pi \in I_0$, then $\mathrm{Carr}_G\, U = \mathrm{supp}_G\, i_+ \cup \mathrm{supp}_G\, i_-$ intersects every component of $\Gamma(G[U]) \setminus \{\pi\}$,

Choose two distinct vertices π, κ in $\Gamma(G[U])$. There exists a vector $z \in \mathrm{Extr}\,(U^\circ)$ such that $\mathrm{supp}_G\, z = \{\pi, \kappa\}$ if and only if **both of the following** conditions are satisfied:

(iii) $\mathrm{Carr}_G\, U$ intersects every component of $\Gamma(G[U]) \setminus \{\pi, \kappa\}$,

(iv) one of the vertices π, κ belongs to the set I_+, the other belongs to the set I_-.

If (iii - iv) hold then $z = g(\lambda_0\omega + \lambda_1\rho)$, where $g \in G$, and ω, ρ are fundamental weights (belonging to the same Weyl chamber) such that $\mathrm{supp}_G\, \omega = \{\pi\}, \mathrm{supp}_G\, \rho = \{\kappa\}$. Here (λ_0, λ_1) is the unique (positive) solution of the linear system

$$\lambda_0 m_G(i_+, \omega) + \lambda_1 m_G(i_+, \rho) = 1$$

$$\lambda_0 m_G(i_-, \omega) + \lambda_1 m_G(i_-, \rho) = 1.$$

The above vectors z form an exhaustive list of elements of $\mathrm{Extr}\,(U^\circ)$.

5.2. A_n-invariant polytopes and spectra of Hermitian matrices. It has been known for quite a long time that the geometry of A_n-orbihedra is very important for many natural problems related to the spectral theory of Hermitian operators. Recently there was a breakthrough, due mostly to A.A. Klyachko, in an old problem of description of the possible spectra of sums of Hermitian matrices with given spectra (see [**9, 12**]). Here we present some simple remarks related to such problems.

Let $\alpha = (\alpha_1, \cdots, \alpha_n\} \in \mathbf{R}^n$, let diag α denote the diagonal matrix having α as its diagonal. So, diag is a real linear mapping from $\mathbf{R}^n$ to the real linear space of Hermitian $n \times n$ matrices. Let

$$\mathrm{Orb}_U\, \alpha = \{u(\mathrm{diag}\ \alpha)u^* : u \in U(n)\}$$

be the set of Hermitian matrices, whose spectrum is $\{\alpha_1, \cdots, \alpha_n\}$. Here $U(n)$ denotes the unitary group acting on $\mathbf{C}^n$. For any Hermitian $n \times n$ matrix A let Diag A denote the diagonal of A, viewed as a vector from $\mathbf{R}^n$. So, Diag is a real linear mapping from the real linear space of Hermitian $n \times n$ matrices to $\mathbf{R}^n$. Certainly, Diag (diag α) $= \alpha$. To simplify the formulations we restrict ourselves to Hermitian matrices with zero trace, the space of such $n \times n$ matrices is denoted by $H_0(n)$. Note that $iH_0(n)$ is the Lie algebra of the Lie group $U(n)$, and the action of $U(n)$ on $H_0(n)$ is adjoint action of a Lie group on its Lie algebra. Also, let V_{n-1} denote the $(n-1)$-dimensional subspace of $\mathbf{R}^n$ consisting of vectors with the zero sum of coordinates. Obviously, if $\alpha \in V_{n-1}$ then $\mathrm{Orb}_U\, \mathrm{diag}\,(\alpha) \subset H_0(n)$. For a matrix $A \in H_0(n)$ let Spec $A = \{\lambda_1, \cdots, \lambda_n\} \subset \mathbf{R}$ denote its spectrum, let spec A denote the set of vectors $(\lambda_{\sigma(1)}, \cdots, \lambda_{\sigma(n)}) \in V_{n-1}$, where σ runs over all permutations of $\{1, 2, \cdots, n\}$. So, spec A is an A_{n-1}-orbit. We treat spec as a mapping from $H_0(n)$ to the space of A_{n-1}-orbits. The following result, due to I. Schur and A. Horn (see [**10**]), establishes a beautiful connection between A_{n-1}-orbihedra and spectra of Hermitian matrices with zero trace:

THEOREM 5.6. *Let $\alpha \in V_{n-1}$. Then*

$$\mathrm{Diag}\,(\mathrm{Orb}_{\,U}\,\mathrm{diag}\,(\alpha)) = \mathrm{Co}_{\,A_{n-1}}\,\alpha.$$

In other words, for any $A \in H_0(n)$ $\mathrm{Diag\ Orb}_{\,U}\,A$ is a convex polyhedron and spec A *is the set of its extreme vectors:*

$$\mathrm{Extr}\,(\mathrm{Diag\ Orb}_{\,U}\,A) = \mathrm{spec}\,A.$$

Let $\alpha, \beta \in V_{n-1}$. An important problem going back to H. Weyl is to compute the set

$$\mathrm{spec}\,(\mathrm{Orb}_{\,U}\,\mathrm{diag}\,\alpha + \mathrm{Orb}_{\,U}\,\mathrm{diag}\,\beta).$$

After important contributions by H. Weyl, Ky Fan, V.B. Lidskii, H. Wielandt, A. Horn, and others, this problem was recently solved by A. Klyachko [**12**]. The ideas of this solution came from Representation Theory and Algebraic Geometry. It should be noted that the connections of this problem with representation theory of Lie groups were known for at least 50 years.

We present two simple results, using the ideas of the preceding sections.

THEOREM 5.7. *Let $\alpha, \beta \in V_{n-1}$. Then*

$$\mathrm{spec}\,(\mathrm{conv\ Orb}_{\,U}\,\mathrm{diag}\,\alpha + \mathrm{conv\ Orb}_{\,U}\,\mathrm{diag}\,\beta) = \mathrm{Co}_{\,A_{n-1}}(\alpha^* + \beta^*).$$

PROOF. The set $\mathrm{spec}\,(\mathrm{conv\ Orb}_{\,U}\,\mathrm{diag}\,\alpha + \mathrm{conv\ Orb}_{\,U}\,\mathrm{diag}\,\beta)$ is convex – one can verify this by a straightforward computation. Using the definitions and the Schur-Horn Theorem, we obtain

$$\begin{aligned}
&\mathrm{spec}\,(\mathrm{conv\ Orb}_{\,U}\,\mathrm{diag}\,\alpha + \mathrm{conv\ Orb}_{\,U}\,\mathrm{diag}\,\beta)\\
&\subset \mathrm{Diag}\,(\mathrm{conv\ Orb}_{\,U}\,\mathrm{diag}\,\alpha + \mathrm{conv\ Orb}_{\,U}\,\mathrm{diag}\,\beta)\\
&= \mathrm{conv}\,(\mathrm{Diag}\,(\mathrm{Orb}_{\,U}\,\mathrm{diag}\,\alpha) + \mathrm{Diag}\,(\mathrm{Orb}_{\,U}\,\mathrm{diag}\,\beta))\\
&= \mathrm{conv}\,(\mathrm{Co}_{\,A_{n-1}}\,\alpha + \mathrm{Co}_{\,A_{n-1}}\,\beta)\\
&= \mathrm{Co}_{\,A_{n-1}}\,\alpha + \mathrm{Co}_{\,A_{n-1}}\,\beta = \mathrm{Co}_{\,A_{n-1}}(\alpha^* + \beta^*).
\end{aligned}$$

To prove the last equality, we compute the polar sets of both convex sets and verify that they coincide:

$$\begin{aligned}
(\mathrm{Co}_{\,A_{n-1}}\,\alpha + \mathrm{Co}_{\,A_{n-1}}\,\beta)^\circ &= \{z \in V_{n-1} : \forall g, h \in A_{n-1}\ \langle z, g\alpha + h\beta\rangle \le 1\}\\
&= \{z \in V_{n-1} : \langle z^*, \alpha^*\rangle + \langle z^*, \beta^*\rangle \le 1\}\\
&= \{z \in V_{n-1} : \forall g \in A_{n-1}\ \langle z, g(\alpha^* + \beta^*)\rangle \le 1\}\\
&= (\mathrm{Co}_{\,A_{n-1}}(\alpha^* + \beta^*))^\circ.
\end{aligned}$$

On the other hand, vectors $g(\alpha^* + \beta^*)$, $g \in A_{n-1}$, (which are the extreme vectors of $\mathrm{Co}_{\,A_{n-1}}(\alpha^* + \beta^*)$) obviously belong to $\mathrm{spec}\,(\mathrm{Orb}_{\,U}\,\mathrm{diag}\,\alpha + \mathrm{Orb}_{\,U}\,\mathrm{diag}\,\beta)$. So, since we know that

$$\mathrm{spec}\,(\mathrm{conv\ Orb}_{\,U}\,\mathrm{diag}\,\alpha + \mathrm{conv\ Orb}_{\,U}\,\mathrm{diag}\,\beta)$$

is convex then the assertion follows. □

Let us reformulate the result choosing the following Weyl chamber $C = \{x_1 \ge x_2 \ge \cdots \ge x_n\} \subset V_{n-1}$ and letting $x^* = x^*(A_{n-1}, C)$. Note that the related fundamental weights are the orthogonal projections of the vectors $(1, 1, \cdots, 1, 0, \cdots, 0)$ (k ones, $1 \le k \le n-1$) onto V_{n-1}.

COROLLARY 5.8. Let $\alpha, \beta \in V_{n-1}$. Then

$$\gamma \in \operatorname{spec}(\operatorname{conv} \operatorname{Orb}_U \operatorname{diag} \alpha + \operatorname{conv} \operatorname{Orb}_U \operatorname{diag} \beta)$$

if and only if

$$\forall k,\ k \leq n-1,\ \sum_{i=1}^{k} (\gamma^*)_i \leq \sum_{i=1}^{k} (\alpha^*)_i + \sum_{i=1}^{k} (\beta^*)_i.$$

Essentially repeating the considerations in the proof of the previous Theorem, we arrive to the following result:

THEOREM 5.9. *Let $\alpha, \beta \in V_{n-1}$. Then*

$$\operatorname{spec}(\operatorname{conv}(\operatorname{Orb}_U \operatorname{diag} \alpha \bigcup \operatorname{Orb}_U \operatorname{diag} \beta))$$

is the $(A_{n-1}, 2)$-polytope $\operatorname{conv}(\operatorname{Orb}_{A_{n-1}} \alpha \bigcup \operatorname{Orb}_{A_{n-1}} \beta)$.

Using the description of the extreme vectors of polars of $(G, 2)$-polytopes we can describe the vectors from

$$\operatorname{spec}(\operatorname{conv}(\operatorname{Orb}_U \operatorname{diag} \alpha \bigcup \operatorname{Orb}_U \operatorname{diag} \beta)$$

in terms of a system of linear inequalities.

5.3. $(G, 2)$-polytopes – some counterexamples.

THEOREM 5.10. *Let G be an irreducible Coxeter group.*

(a) There exist $(G, 2)$-polytopes such that all elements from $\operatorname{Extr}(U^\circ)$ *have the same one-vertex support, provided* $\dim V \geq 3$.

(b) If $\Gamma(G)$ is not branching and $\dim V \geq 4$, *then there exist two distinct $(G, 2)$-polytopes U_1, U_2 such that* $\operatorname{Carr}_G U_1 = \operatorname{Carr}_G U_2$, *all elements of* $\operatorname{Extr}(U_1^\circ)$ *have the same one-vertex support π, all elements of* $\operatorname{Extr}(U_2^\circ)$ *have the same one-vertex support κ, but $\pi \neq \kappa$.*

(c) For every $z \in V$ such that $\operatorname{supp}_G z = \operatorname{ver}(G)$ *there exists a $(G, \dim V)$-polytope U such that $z \in$* $\operatorname{Extr}(U^\circ)$.

PROOF. (a) If $\Gamma(G)$ is not branching, then let ω be a non-extremal fundamental weight (i.e., the related vertex $\pi = \pi(\omega)$ is not one of the end vertices π_1, π_2 of the Coxeter graph $\Gamma(G)$). Then

$$\operatorname{Extr}(Co_G \omega)^\circ = \operatorname{Orb}_G \frac{\omega_1}{m_G(\omega, \omega_1)} \bigcup \operatorname{Orb}_G \frac{\omega_2}{m_G(\omega, \omega_2)}$$

where $\pi(\omega_i) = \pi_i$, $i = 1, 2$, – this immediately follows from Theorem 3.2. Then $U = (\operatorname{Co}_G \omega)^\circ$ is a $(G, 2)$-polytope, $\operatorname{Carr}_G U = \{\pi_1, \pi_2\}$, and

$$\operatorname{Extr}(U^\circ) = \operatorname{Extr}(\operatorname{Co}_G \omega) = \operatorname{Orb}_G \omega_1.$$

So, all elements of $\operatorname{Extr}(U^\circ)$ have the same one-vertex support $\pi(\omega)$. If $\Gamma(G)$ is branching, choose π to be one of the three end vertices of $\Gamma(G)$.

(b) In the previous construction choose two distinct fundamental weights ω, τ, associated to non-end vertices of the Coxeter graph. It is possible since $\operatorname{card} \operatorname{ver}(G) = \dim V \geq 4$.

(c) Put $U = (\operatorname{Co}_G z)^\circ$. Obviously, $\operatorname{Extr}(U^\circ) = \operatorname{Extr}(\operatorname{Co}_G z) = \operatorname{Orb}_G z$. On the other hand, according to Theorem 3.2,

$$\operatorname{Extr} U = \{\omega / m_G(\omega, z) : \operatorname{supp}_G z \text{ intersects every component of}$$

$$\Gamma(G) \setminus \{\pi(\omega)\}\} = \bigcup_{\pi(\omega)\in \mathrm{ver}\,(G)} \mathrm{Orb}_G\, \omega / m_G(\omega, z)$$

so $N_G(U) = \mathrm{card\ ver}\,(G) = \dim V$. □

6. Orbihedra for quasi-Coxeter groups

6.1. Quasi-Coxeter groups. Let us start with an important example:

DEFINITION 6.1. Let $V = M_{n,m}(\mathbf{R})$, the set of matrices having n columns of m real entries each. Define $B_{n,m}$ as the group of operators acting on V by permuting the columns, permuting elements in each individual column, and performing sign changes on any number of entries.

Groups $B_{n,m}$ naturally arise as symmetry groups for the so called **mixed norms**: choose a B_n-invariant norm $l : \mathbf{R}^n \to \mathbf{R}_+$, and a B_m-invariant norm $L : \mathbf{R}^m \to \mathbf{R}_+$, treat an element $A \in M_{n,m}(\mathbf{R})$ as a string $(a_1, a_2, \cdots, a_n)$ of n vectors from $\mathbf{R}^m$, then define the mixed norm (lL) as follows:

$$(lL)(A) = l(L(a_1), L(a_2), \cdots, L(a_n)).$$

The unit balls of such norms are important examples of $B_{n,m}$-invariant convex bodies.

Notably, $B_{n,m}$ is not generated by reflections across hyperplanes, but it does have a close relationship to the Coxeter group B_m. A reformulation of the above definition makes this relationship more evident. Specifically, $B_{n,m} = S_n(B_m)$ where $S_n(G)$ has the following definition. Let S_n denote the group of permutations of the set $\{1, 2, \cdots, n\}$.

DEFINITION 6.2. Let G be a Coxeter group naturally acting on V. Define

$$S_n(G) = S_n \times G^n = \{\sigma \times (\prod_{i=1}^{n} g_i) : \sigma \in S_n, g_i \in G,\ 1 \le i \le n\}.$$

Group $S_n(G)$ acts on V^n as follows:

$$(\sigma \times (\prod_{i=1}^{n} g_i))(v_1, v_2, \ldots, v_n) = (g_1 v_{\sigma(1)}, g_2 v_{\sigma(2)}, \ldots, g_n v_{\sigma(n)}).$$

In other terms, the action of $S_n(G)$ on V^n is induced by the natural action of G on V.

Again, group $S_n(G)$ is not generated by reflections across hyperplanes, but it has a Coxeter subgroup G^n.

Now consider a more general situation: let K be a finite group of operators acting on a real finite dimensional space V. Consider all reflections across hyperplanes contained in K. Let $G = G(K)$ denote the subgroup generated by all these reflections. Assume that G acts effectively on V. Then G is called a **Coxeter subgroup** of K.

Let $G\backslash K = \{Gk = \{gk : g \in G\} : k \in K\}$ be the left homogeneous space. For every $x \in V$ we have

$$\mathrm{Orb}_K\, x = \bigcup_{Gk \in G\backslash K} \mathrm{Orb}_G\, kx.$$

Therefore

$$\mathrm{Co}_K\, x = \mathrm{conv} \bigcup_{Gk \in G\backslash K} \mathrm{Co}_G\, kx.$$

So, every K-orbihedron is a convex G-invariant polyhedron, and $I_G(\mathrm{Co}_K\, x) \leq \mathrm{card}\,(G\backslash K)$. The number $\mathrm{card}\,(G\backslash K) = (K:G)$ is called the index of the subgroup G in the group K.

Consider a finite group K of operators acting on a real finite dimensional space V containing a Coxeter subgroup (= an effectively acting subgroup generated by reflections across hyperplanes) of index smaller that $\dim V$. We call such group K a **quasi-Coxeter** group.

The techniques presented in the previous sections allows to study the convex structure of K-orbihedra for quasi-Coxeter groups K.

6.2. $S_2(G)$-orbihedra. Consider the case $K = S_2(G)$, where G is an irreducible Coxeter group naturally acting on V (irreducibility is actually not very important, but this assumption simplifies some formulations). This group contains an index 2 Coxeter subgroup G^2, naturally acting on V^2 :

$$(g_0, g_1)(v_+, v_-) = (g_0 v_+, g_1 v_-).$$

There is an obvious action of S_2 on V^2 by permutations of the components. Let

$$\sigma(v_+, v_-) = (v_-, v_+).$$

We call this operator σ a **flip**. Let $v = (v_+, v_-) \in V^2$. Consider a $S_2(G)$-orbihedron (which is also a $(G^2, 2)$-polytope)

$$U = \mathrm{Co}_{S_2(G)}\, v = \mathrm{conv}\,(\mathrm{Co}_{G^2}\, v \bigcup \mathrm{Co}_{G^2}\, \sigma(v)).$$

Corollary 5.5 provides a complete description of the set $\mathrm{Extr}\,(U^\circ)$, but we can simplify this description by incorporating flip symmetries of $\Gamma(G^2)$ (to be defined in the next paragraph) and U.

The Coxeter graph $\Gamma(G^2)$ is the disjoint union of two copies Γ_+, Γ_- of the connected graph $\Gamma(G)$:

$$\Gamma(G^2) = \Gamma_+ \bigsqcup \Gamma_-.$$

There is a natural automorphism of $\Gamma(G^2)$, interchanging Γ_+ and Γ_-. Slightly abusing notation, we call this automorphism a **flip**. So, the graph $\Gamma(G^2)$ is also flip-invariant. The carrier set $\mathrm{Carr}_{G^2}\, U$ is also flip-invariant. Therefore, it intersects both components of $\Gamma(G^2)$, so $G^2[U] = G^2$.

Let C be a Weyl chamber for G, then $C^2 = C \times C$ is a Weyl chamber for G^2. Let $\omega_j \in C \cap \mathcal{W}_G$, $1 \leq j \leq \dim V$, be the fundamental weights. Let $\pi_j, 1 \leq j \leq \dim V$, denote the related vertices of $\Gamma(G)$. Then $\omega_j^+ = (\omega_j, 0)$, $\omega_j^- = (0, \omega_j)$, $1 \leq j \leq \dim V$, are the fundamental weights of G^2, belonging to C^2. Let π_j^i, $1 \leq j \leq \dim V$, denote the related vertices of Γ_i, $i = \pm$, so we have $\pi(\omega_j^i) = \pi_j^i$. We wish to describe the extreme vectors of U° in terms of group G rather than in terms of group G^2.

Let

$$\nu_j = \frac{m_G(v_+, \omega_j)}{m_G(v_-, \omega_j)},\ 1 \leq j \leq \dim V.$$

Since $\Gamma(G)$ is connected then the numerator and the denominator of this fraction can both vanish if and only if $v_+ = v_- = 0$, which is obviously excluded. Consider the canonical partition of $\mathrm{ver}\,(G)$ associated with vectors v_+ and v_- :

$$J_+ = \{\pi_j \in \mathrm{ver}\,(G) : \nu_j > 1\},$$

$$J_- = \{\pi_j \in \mathrm{ver}\,(G) : \nu_j < 1\},$$

$$J_0 = \{\pi_j \in \operatorname{ver}(G) : \nu_j = 1\}.$$

Let $(J_k)^i = \{\pi_j^i \in \operatorname{ver}(G^2) : \pi_j \in J_k\}$, $i = \pm$, $k = 0, \pm$.

As before,

$$\mu(\pi_j^i) = \frac{m_{G^2}(v, \omega_j^i)}{m_{G^2}(\sigma(v), \omega_j^i)}, \quad i = \pm,\ 1 \le j \le \dim V.$$

Obviously,

$$m_{G^2}(v, \omega_j^i) = m_G(v_i, \omega_j), \ m_{G^2}(\sigma(v), \omega_j^i) = m_G(v_{-i}, \omega_j), \ i = \pm,\ 1 \le j \le \dim V.$$

Therefore the following is true for the canonical partition of $\operatorname{ver}(G^2)$, associated with vectors v and $\sigma(v)$:

$$\begin{aligned} I_+ &= \{\pi_j^+ \in \operatorname{ver}(G^2) : \nu_j > 1\} \bigcup \{\pi_j^- \in \operatorname{ver}(G^2) : \nu_j < 1\} \\ &= (J_+)^+ \bigcup (J_-)^-, \\ I_- &= \{\pi_j^+ \in \operatorname{ver}(G^2) : \nu_j < 1\} \bigcup \{\pi_j^- \in \operatorname{ver}(G^2) : \nu_j > 1\} \\ &= (J_+)^- \bigcup (J_-)^+, \\ I_0 &= \{\pi_j^i \in \operatorname{ver}(G^2) : \nu_j = 1,\ i = \pm\} = (J_0)^+ \bigcup (J_0)^-. \end{aligned}$$

The flip maps the sets I_+ and I_- one onto another and leaves the set I_0 invariant.

Applying Corollary 5.5 we arrive to the following result:

THEOREM 6.3. *Let $U = \operatorname{Co}_{S_2(G)} v$, where G is an irreducible Coxeter group, naturally acting on space V, and $v = (v_+, v_-) \in V^2$. The set $\operatorname{Extr}(U^\circ)$ can be described as follows:*

1. *Take $\pi_j \in J_s$, $s = \pm$. There exists $z \in \operatorname{Extr}(U^\circ)$, $\operatorname{supp}_{G^2} z = \pi_j^i$, $i = +$ or $-$, if and only if the following is true:*

(i) $\operatorname{supp}_G v_s$ intersects every component of $\Gamma(G) \setminus \{\pi_j\}$, and $v_{-s} \neq 0$.

In this case $z = g\omega_j^i / m_G(v_s, \omega_j)$, $g \in G$.

2. *Take $\pi_j \in J_0$. There exists $z \in \operatorname{Extr}(U^\circ)$, $\operatorname{supp}_{G^2} z = \pi_j^i$, $i = +$ or $-$, if and only if the following is true:*

(ii) $\operatorname{supp}_G v_+ \cup \operatorname{supp}_G v_-$ intersects every component of $\Gamma(G) \setminus \{\pi_j\}$.

In this case $z = g\omega_j^i / m_G(v_0, \omega_j)$, $g \in G$.

3. *Take $\pi_j, \pi_k \in \operatorname{ver}(G)$, $j \neq k$. There exists $z \in \operatorname{Extr}(U^\circ)$, $\operatorname{supp}_{G^2} z = \{\pi_j^i, \pi_k^i\}$, $i = +$ or $-$, if and only if the following is true:*

(iii) $\operatorname{supp}_G v_+ \cup \operatorname{supp}_G v_-$ intersects every component of $\Gamma(G) \setminus \{\pi_j, \pi_k\}$.

(iv) one of the vertices π_j, π_k belongs to J_+, the other - to J_-.

In this case $z = g(\lambda_0 \omega_j^i + \lambda_1 \omega_k^i)$, $g \in G$, $i = \pm$, (λ_0, λ_1) is the unique (positive) solution of the linear system

$$\lambda_0 m_G(v_+, \omega_j) + \lambda_1 m_G(v_+, \omega_k) = 1$$

$$\lambda_0 m_G(v_-, \omega_j) + \lambda_1 m_G(v_-, \omega_k) = 1.$$

4. *Take $\pi_j, \pi_k \in \operatorname{ver}(G)$ (the case $j = k$ is not excluded here). There exists $z \in \operatorname{Extr}(U^\circ)$, $\operatorname{supp}_{G^2} z = \{\pi_j^i, \pi_k^{-i}\}$, $i = +$ or $-$, if and only if the following is true:*

(v) $\operatorname{supp}_G v_+ \cup \operatorname{supp}_G v_-$ intersects every component of of the graphs $\Gamma(G) \setminus \{\pi_j\}$ and $\Gamma(G) \setminus \{\pi_k\}$,

(vi) both of the vertices π_j, π_k belong to J_i, $i = +$ or $-$.

In this case $z = g(\lambda_0\omega_j^i + \lambda_1\omega_k^{-i})$, $g \in G$, $i = \pm$, (λ_0, λ_1) *is the unique (positive) solution of the linear system*

$$\lambda_0 m_G(v_+, \omega_j) + \lambda_1 m_G(v_-, \omega_k) = 1$$
$$\lambda_0 m_G(v_-, \omega_j) + \lambda_1 m_G(v_+, \omega_k) = 1.$$

The above is an exhaustive list of elements of $\operatorname{Extr}(\operatorname{Co}_{S_2(G)} v)^\circ$.

A less explicit form of this result for $G = B_2$ was obtained in [**21**] by a hard (non-computer) computation, based on an algorithm calculating the extreme rays of a polyhedral cone defined by a system of linear inequalities (this algorithm is known as the Chernikova's algorithm, or the Double Description Method).

These results have an application to the Operator Interpolation Theory in the spirit of [**19, 18, 20**], which we discuss in the next sections.

7. Operator Interpolation

Our initial interest in the convex geometry of orbihedra was motivated by an approach to operator interpolation developed by the last two authors, for a complete exposition see [**18, 19**]. Let us briefly describe the main features of this approach.

Let $G \subset O(V)$ be a subgroup of orthogonal operators on a real finite dimensional Euclidean space V. We wish to describe the **envelope** of G (denoted $\operatorname{env}(G)$) — the set of linear operators on V transforming every G-invariant convex closed set into itself:

$$\operatorname{env}(G) = \{T \in \operatorname{End} V : TU \subset U \text{ for every closed convex } G\text{-invariant } U \subset V\}.$$

Obviously, $\operatorname{env}(G)$ is a convex closed semigroup of linear operators, containing the convex hull of the group G.

A collection $\{U_\alpha, \alpha \in A\}$ of G-invariant convex closed sets is called G-**sufficient** if

$$T \in \operatorname{env}(G) \iff \forall \alpha \in A \quad TU_\alpha \subset U_\alpha.$$

We would like to describe some natural G-sufficient collections. A collection consisting of G-orbihedra is called a **simple collection**. A collection consisting of polar sets of G-orbihedra is called a **dual simple collection**.

Example (Calderon–Mityagin Theorem). Let G be the Coxeter group B_n. It acts on $\mathbf{R}^n$ as follows:

$$(x_1, x_2, \cdots, x_n) \mapsto (s_1 x_{\sigma(1)}, s_2 x_{\sigma(2)}, \cdots, s_n x_{\sigma(n)}),$$

where σ is a permutation of $\{1, 2, \cdots, n\}$, and $s_k = \pm 1$. Let

$$U_1 = \{(x_k) \in \mathbf{R}^n : \sum_k |x_k| \le 1\},$$
$$U_\infty = \{(x_k) \in \mathbf{R}^n : \max_k |x_k| \le 1\}.$$

A finite dimensional version of the celebrated Calderon-Mityagin interpolation theorem (see [**7, 16**]) asserts that if a linear operator $T : V \to V$ is such that $TU_1 \subset U_1$ and $TU_\infty \subset U_\infty$ then $TU \subset U$ for every closed convex B_n-invariant $U \subset V$ (in other words, the contraction property of T with respect to U_1 and U_∞ can be interpolated to all closed convex B_n-invariant bodies). In our terms this means that the collection U_1, U_∞ is a (both simple and dual simple) B_n-sufficient collection. It was shown in [**19**] that this collection is actually the smallest B_n-sufficient collection, more precisely, it is a subset of the Hausdorff closure of any B_n-sufficient collection (up to scaling).

There is a natural duality between the spaces End V of linear operators in V and the tensor product space $V \otimes V$:

$$(T, \sum x_i \otimes y_i) = \sum \langle T y_i, x_i \rangle.$$

Therefore there is a natural notion of the polar set, in particular,

$$G^\circ = \{S \in V \otimes V : \forall g \in G (g, S) \leq 1\}.$$

Sufficient collections can be described in terms of the following sets in $V \otimes V$:

$$\mathcal{A}_G = \{x \otimes y \in V \otimes V : m_G(x, y) \leq 1\} = G^\circ \bigcap \{\text{rank 1 tensors}\},$$

$$K_G = \text{conv } \mathcal{A}_G.$$

Since $\mathcal{A}_G$ is obviously closed and $V \otimes V$ is finite dimensional, then K_G is closed. One can show (see [**19**]) that K_G is bounded if and only if G acts irreducibly, which we assume henceforth. Let Extr K_G denote the set of extreme elements of K_G. One can show that

$$\text{Extr } K_G \subset \mathcal{A}_G.$$

Since K_G is a compact convex set in a finite dimensional space then, by the Krein–Milman Theorem and the Caratheodory Theorem,

$$K_G = \text{conv Extr } K_G.$$

Note that K_G is invariant with respect to the following **tensor flip**: $x \otimes y \to y \otimes x$. Therefore the set Extr K_G is also flip-invariant.

For every set $U \subset V$ let

$$S(U) = \{x \otimes y \in V \otimes V : x \in U, y \in U^\circ\}.$$

Obviously, if U is convex and G-invariant then $S(U) \subset \mathcal{A}_G$. For any $x \otimes y \in \mathcal{A}_G$ (i.e., such that $m_G(x, y) \leq 1$) we have $x \otimes y \in S(\text{Co}_G\, x)$ and $x \otimes y \in S((\text{Co}_G\, y)^\circ)$.

It is not hard to see that a collection $(U_\alpha : \alpha \in A)$ of G-invariant convex closed sets is G-sufficient if and only if

$$\text{Extr } K_G \subset \overline{\bigcup_{\alpha \in A} S(U_\alpha)}.$$

This observation leads to the following constructions: let

$$\mathcal{N}_G = \{x \in V : \exists y \in V, x \otimes y \in \text{Extr } K_G\},$$

and let us consider the following **simple canonical collection**:

$$\mathcal{C}_G = \{\text{Co}_G\, x : x \in \mathcal{N}_G\},$$

and the following **dual simple canonical collection**:

$$\mathcal{C}_G^\circ = \{(\text{Co}_G\, x)^\circ : x \in \mathcal{N}_G\}.$$

It is easy to see that each of the canonical collections is G-sufficient, and each of them is minimal in some natural sense — see [**19, 18**].

7.1. Non-canonical sufficient collections. It is often very difficult to compute the sets Extr K_G and $\mathcal{N}_G$. Therefore it is interesting to find larger sets and construct larger G-sufficient collections.

The set

$$\mathcal{K}_G = \{x \otimes y \in V \otimes V : x \in \text{Extr}\,(\text{Co}_G\, y)^\circ, y \in \text{Extr}\,(\text{Co}_G\, x)^\circ\}$$

is a very natural set of this type, it is contained in $\mathcal{A}_G$ and contains Extr K_G (in fact, we have no examples of groups with Extr $K_G \neq \mathcal{K}_G$, though believe that such examples do exist). Let

$$\widehat{\mathcal{N}}_G = \{x \in V : \exists y \in V,\ x \otimes y \in \mathcal{K}_G\},$$

and we arrive to the **quasi-canonical** G-sufficient collections

$$\widehat{\mathcal{C}}_G = \{\text{Co}_G\, x : x \in \widehat{\mathcal{N}}_G\}$$

and

$$\widehat{\mathcal{C}}_G^\circ = \{(\text{Co}_G\, x)^\circ : x \in \widehat{\mathcal{N}}_G\}.$$

Let us construct several other G-sufficient collections. Let $\mathcal{N}_0(G) = V$, and let

$$\mathcal{N}_{s+1}(G) = \bigcup\{\text{Extr}\,(\text{Co}_G\, z)^\circ :\ z \in \bigcup_{w \in \mathcal{N}_s(G)} \text{Extr}\,(\text{Co}_G\, w)^\circ\}.$$

It is not hard to show that

$$V = \mathcal{N}_0(G) \supset \mathcal{N}_1(G) \supset \mathcal{N}_2(G) \supset \cdots \supset \widehat{\mathcal{N}}_G \supset \mathcal{N}_G.$$

Let

$$\mathcal{C}_s(G) = \{\text{Co}_G\, x : x \in \mathcal{N}_s(G)\}.$$

Obviously,

$$\mathcal{C}_1(G) \supset \mathcal{C}_2(G) \supset \cdots \supset \widehat{\mathcal{C}}_G \supset \mathcal{C}_G.$$

Therefore, all these collections are G-sufficient (but not minimal, if different from $\mathcal{C}_G$). The actual construction of these collection heavily depends upon the knowledge of the convex structure of G-orbihedra. This was our initial motivation for the study of these problems.

As it was shown in [**19**], the equality $\mathcal{C}_1(G) = \mathcal{C}_G$ is equivalent to the fact that "interpolation in the canonical collection is described by the real method", see [**19**] for details. Such assertions are important in Operator Interpolation. It actually means that there exist very simple decompositions of elements of $\mathcal{A}_G$ into convex combinations of elements of Extr K_G, and all convex G-invariant bodies can be obtained from the bodies of the canonical collection by rather simple constructions (by the so called real method). In particular, in this case one may interpolate not only linear operators but also many non-linear ones.

7.2. Canonical collections for Coxeter groups. If G is an irreducible Coxeter group then the set Extr K_G was explicitly computed in terms of weights (see [**19**]):

$$\text{Extr}\,K_G$$
$$= \{\frac{\omega \otimes \tau}{m_G(\omega,\tau)} : \omega, \tau \in \mathcal{W}_G, \pi(\omega), \pi(\tau) \text{ are distinct end vertices of } \Gamma(G)\}.$$

Elements of Extr K_G are called **Birkhoff's tensors** — see [**6, 15**] for an explanation how the Birkhoff's tensors are related to the Birkhoff's description ([**4**]) of the extreme points of the set of doubly stochastic matrices. So, for every irreducible

Coxeter group the set Extr (env $G)^\circ$ is explicitly computed. In fact conv G = env G if the Coxeter graph $\Gamma(G)$ is not branching (this is proven in [**6, 15**] for all irreducible Coxeter groups with non-branching graphs with the only exception of the group H_4 for which this assertion is still a conjecture). As for the case when $\Gamma(G)$ is branching, it was shown in [**15**] that conv $G \neq$ env G. These results were recently applied in [**13**] to a description of linear isomorphisms of the convex hulls of Coxeter groups.

Thus in the case of an irreducible Coxeter group we have

$$\mathcal{N}_G = \{\omega \in \mathcal{W}_G : \pi(\omega) \text{ is an end vertex of } \Gamma(G)\}.$$

Also, one can show that $\mathcal{C}_1 = \mathcal{C}_G$ for all irreducible Coxeter groups, so the interpolation here is "described by the real method" — see [**7**] for $G = B_n$ (even in the infinite dimensional setting), and [**19**] for all other irreducible Coxeter groups. Moreover, in this case the canonical collections have some additional nice extremal properties — see [**18, 19**].

In the case of $G = B_n$ both canonical collections coincide with the collection U_1, U_∞ described above.

7.3. Sufficient collections for non-Coxeter groups. All above notions and constructions may be generalized to the case when G is a bounded semigroup of operators on V — see [**20**]. In particular, we may consider the semigroup of operators contracting every mixed norm on $V = \mathbf{R}^{n\times m}$. It is possible to compute the canonical collections for this semigroup, see [**18**]. As it was mentioned before, the mixed norms are $B_{n,m}$-invariant. Not all $B_{n,m}$-invariant norms are mixed norms — one can present counterexamples. The group $B_{n,m}$ is not a Coxeter group, and our initial goal was to construct canonical (or, at least, quasi-canonical) collections for this group. A calculation of the quasi-canonical collections for $G = B_{2,2}$ was carried out in [**21**]. It was based on the calculation of the extreme vectors of the polar sets of $B_{2,2}$-orbihedra. Since we now have a rather detailed description of these vectors for groups $S_2(G)$, we can calculate quasi-canonical and even canonical collections for these groups. This calculation is rather lengthy and we plan to discuss it in a separate publication.

References

[1] Atiyah, M.F., Angular momentum, convex polyhedra and algebraic geometry, *Proc. Edinburgh Math. Soc.* **26** (1983), 121–138.

[2] Barvinok, A.I., Combinatorial complexity of orbits in representations of the symmetric group, *Advances in Soviet Math.* **9** (1992), 161–182.

[3] Benson, C., and Grove, L., *Finite Reflection Groups*, 2nd ed., Springer-Verlag, New York, 1985.

[4] Birkhoff, G., Tres observaciones sobre el algebra lineal, *University Nac. Tucuman Rev. Ser.* A5 (1946),147–150.

[5] Bourbaki, N., *Groupes et Algebres de Lie*, Ch. IV–VI, Hermann, Paris, 1968.

[6] Brandman, J., Fowler, J., Lins, B., Spitkovsky, I., and Zobin, N., Convex hulls of Coxeter groups, in *"Function Spaces, Interpolation Theory and Related Topics"*, Walter de Gruyter, Berlin – New York, 2002, 213–240.

[7] Calderon, A.P., Spaces between L^1 and L^∞ and the theorem of Marcinkiewicz, *Studia Math.* **26** (1966), 229–273.

[8] Eaton, M.L., and Perlman, M.D., Reflection groups, generalized Schur functions and the geometry of majorization, *Annals of Probability* **56** (1977), 829–860.

[9] Fulton, W., Eigenvalues, invariant factors, highest weights, and Schubert calculus, *Bull. Amer. Math. Soc.* **37** (2000), 209–249.

[10] Horn, A., Doubly stochastic matrices and the diagonal of a rotation matrix, *Amer. J. Math.* **76** (1954), 620–630.

[11] Humphreys, J., *Reflection Groups and Coxeter Groups*, Cambridge University Press, Cambridge, 1990.

[12] Klyachko, A.A., Stable bundles, representation theory and Hermitian operators, *Selecta Math.* **4** (1998), 419–445.

[13] Li, C.K., Spitkovsky, I., and Zobin, N., Finite reflection groups and linear preserver problems, 22 pp., to appear in *Rocky Mount. Math. J.*, 2002.

[14] Marshall, A.W., and Olkin, I., *Inequalities: Theory of Majorization and its Applications*, Academic Press, New York - London, 1979.

[15] McCarthy, N., Ogilvie, D., Spitkovsky, I., and Zobin, N., Birkhoff's Theorem and convex hulls of Coxeter groups, *Linear Algebra and Appl.* **347** (2002), 219–231.

[16] Mityagin, B.S., An interpolation theorem for modular spaces, *Matem. Sbornik* **66:4** (1965), 473–482, (in Russian). English translation: in *Lecture Notes in Math.* **1070** Springer, New York, 1983.

[17] Voskresenskii, V.E., and Klyachko, A.A., Toroidal Fano varieties and systems of roots, *Math. USSR Izvestiya* **24:2** (1985), 221–244.

[18] Zobin, N., and Zobina, V., Duality in operator spaces and problems of interpolation of operators, *Pitman Research Notes in Math.* **257** (1992), 123–144.

[19] Zobin, N., and Zobina, V., Coxeter groups and interpolation of operators, *Integral Equations and Operator Theory* **18** (1994), 335–367.

[20] Zobin, N., and Zobina, V., A general theory of sufficient collections of norms with a prescribed semigroup of contractions, *Operator Theory: Advances and Applications* **73** (1994), 397–416.

[21] Zobin, N., and Zobin, V., Geometric structure of $B_{2,2}$-orbihedra and interpolation of operators, *Linear and Multilinear Algebra*, **48** (2000), 67–91.

Massachusetts Institute of Technology
E-mail address: namccart@mit.edu

University of Chicago
E-mail address: david.ogilvie@midway.uchicago.edu

College of William and Mary
E-mail address: zobin@math.wm.edu

College of William and Mary
E-mail address: veronica@math.wm.edu

Contemporary Mathematics
Volume **321**, 2003

Commutators on bounded symmetric domains in $\mathbb{C}^n$

Jie Miao

ABSTRACT. Let D be a bounded symmetric domain in $\mathbb{C}^n$. Let M_f denote the multiplication operator given by $M_f[g](z) = (fg)(z)$ and let $P(K)$ denote the integral operator with kernel K given by $P(K)[g](z) = \int_D K(z,w)\,g(w)\,dV(w)$. Let K_a denote the Bergman reproducing kernel of holomorphic functions on D. If K_a satisfies an integral condition on D and if K satisfies certain upper and lower bound conditions, then we can characterize the boundedness and compactness of $M_fP(K) - P(K)M_f$ on $L^p(D,dV)$ for $p \in [2,\infty)$.

1. Introduction

Let D denote a bounded symmetric domain in $\mathbb{C}^n$ and let dV denote the usual Lebesgue measure on D, normalized so that $V(D) = 1$. For $p \in [1,\infty]$, let L^p denote $L^p(D,dV)$ and $\|g\|_p$ denote the usual L^p norm of $g \in L^p$. Let L^2_a denote the holomorphic Bergman space which consists of all holomorphic functions on D that are also in L^2. Let L^2_h denote the pluriharmonic Bergman space which consists of all pluriharmonic functions on D that are also in L^2. Both L^2_a and L^2_h are closed in L^2. The orthogonal projections from L^2 onto L^2_a and from L^2 onto L^2_h are denoted by P_a and P_h, respectively. There exist functions K_a and K_h called the Bergman reproducing kernels of L^2_a and L^2_h such that

$$P_a[g](z) = \int_D K_a(z,w)g(w)\,dV(w), \quad P_h[g](z) = \int_D K_h(z,w)g(w)\,dV(w)$$

for all $g \in L^2$ and $z \in D$. The reproducing kernel K_a has been well studied. The following are some well-known properties for K_a that we will need (see [**4**]):

(1) $K_a(z,0) = K_a(0,z) = 1$;
(2) $K_a(z,z) > 0$ and $K_a(z,z) \to \infty$ as $z \to \partial D$;
(3) $K_a(z,w) = \overline{K_a(w,z)}$;
(4) $K_a(z,w)$ is a continuous function on $D \times \overline{D}$;
(5) $K_a(z,w)^{-1}$ is a continuous function on $\mathbb{C}^n \times \mathbb{C}^n$;
(6) There is a positive number t_D such that

$$\int_D K_a(w,w)^\epsilon\,dV(w) < \infty \iff \epsilon < t_D. \tag{1.1}$$

Here $z \to \partial D$ means that $\inf_{w\in\partial D}|z-w| \to 0$.

1991 *Mathematics Subject Classification.* Primary 47G10, 47B32; Secondary 46E30.

Key words and phrases. commutator, reproducing kernel, bounded operator, compact operator.

If D is the open unit ball, then it is well known that

$$K_a(z,w)=\frac{1}{(1-\langle z,w\rangle)^{n+1}}.$$

In this case, $t_D=1/(n+1)$. If D is the open unit polydisk, then the reproducing kernel K_a is the product of the reproducing kernels of the unit disk. In this case, $t_D=1/2$.

Throughout the paper we shall assume that the reproducing kernel K_a satisfies the following condition: for every $\epsilon<t_D$ and $s>0$, there is a constant $c>0$ such that

$$\int_D K_a(w,w)^{\epsilon}|K_a(z,w)|^{1-\epsilon+s}\,dV(w)\le c\,K_a(z,z)^s \tag{1.2}$$

for all $z\in D$. The domains for which (1.2) holds include the open unit ball and the open unit polydisk. See Proposition 1.4.10 of [**12**] for the unit ball and also Theorem 4.1 of [**7**] for some generalization.

For a measurable function K defined on $D\times D$, let $P(K)$ denote the integral operator with kernel K given by

$$P(K)[g](z)=\int_D K(z,w)g(w)\,dV(w).$$

We will need the following two conditions on K:

(a) there is a constant $c_1>0$ such that

$$|K(z,w)|\le c_1|K_a(z,w)|$$

for all z, $w\in D$;

(b) there are a constant $c_2>0$ and a number $r>0$ such that

$$|K(z,z)|\ge c_2\,K_a(z,z),\quad \left|\frac{K(w,u)}{K(z,z)}-1\right|\le\frac{1}{4}$$

for all $z\in D$ and w, $u\in E(z,r)$, where $E(z,r)$ is the Bergman ball centered at z of radius r as defined in [**4**].

In the case of the unit disk, conditions similar to (a) and (b) were introduced in [**11**] in which they were called conditions (4) and (5). It is easy to see that (b) (in the context of the unit disk) is slightly weaker than (5) of [**11**]. The way the condition (b) is stated makes it much easier to verify than (5) of [**11**].

Given $f\in L^1$ and an integral operator P, we study when the commutator M_fP-PM_f is bounded or compact on L^p, where $M_f[g]=fg$. A complete characterization for the boundedness or compactness of $M_fP_a-P_aM_f$ on L^2 was obtained in [**5**]. The corresponding question concerning $M_fP_h-P_hM_f$, which is related to Hankel operators on harmonic Bergman spaces, was studied in [**11**] and [**13**] when the domain D is the unit disk. See [**3**] for more information on harmonic Bergman spaces.

The main results of this paper can be stated as follows. The definitions of the spaces BMO^p and VMO^p will be given in the next section.

THEOREM 1.1. *Let $p\in[2,\infty)$ and $f\in L^p$. Suppose K satisfies the conditions (a) and (b) and suppose $P=P(K)$. Then*

(i) M_fP-PM_f is bounded on L^p if and only if $f\in\mathrm{BMO}^p$.

(ii) M_fP-PM_f is compact on L^p if and only if $f\in\mathrm{VMO}^p$.

THEOREM 1.2. *Let $p \in [2, \infty)$ and $f \in L^p$. Then*
(i) $M_f P_h - P_h M_f$ is bounded on L^p if and only if $f \in \mathrm{BMO}^p$.
(ii) $M_f P_h - P_h M_f$ is compact on L^p if and only if $f \in \mathrm{VMO}^p$.

THEOREM 1.3. *Let $p \in [2, \infty)$ and $f \in L^p$ and λ, $\mu \in \mathbb{C}$. Suppose $|\lambda| + |\mu| > 0$ and $P = P(\lambda \operatorname{Re} K_a + \mu \operatorname{Im} K_a)$. Then*
(i) $M_f P - P M_f$ is bounded on L^p if and only if $f \in \mathrm{BMO}^p$.
(ii) $M_f P - P M_f$ is compact on L^p if and only if $f \in \mathrm{VMO}^p$.

Theorem 1.1 extends Theorems 1 and 2 of [**11**] from the unit disk to more general domains in $\mathbb{C}^n$. The proof of Theorem 1.1 will be given in Section 3. The ideas to prove that if $f \in \mathrm{BMO}^p$ (or VMO^p), then $M_f P - P M_f$ is bounded (or compact) on L^p are similar to those used in [**5**]. However the proof for the other direction calls for new techniques. The proof of Theorem 3 of [**5**] relies on some techniques (e.g. $g \in L^2_a \Longrightarrow K_a(z, w)^{-1} g(z) \in L^2_a$ for every $w \in D$) that even can not be applied to the pluriharmonic reproducing kernel $K_h(z, w)$ because the set of pluriharmonic functions is not closed under multiplication. Furthermore K_h can vanish on $D \times D$. See [**10**] for some discussion about this.

Theorems 1.2 and 1.3 are consequences of Theorem 1.1. Their proofs will be given in Section 4. Theorem 1.2 extends Corollary 3 of [**11**] from the unit disk to more general domains. To prove Theorems 1.2, we will show that K_h satisfies both (a) and (b). The method that we use to show this fact in this paper (which is Lemma 4.3) is significantly easier than the one used in [**11**] (which is Proposition of [**11**]). To prove Theorem 1.3, we need to show that Theorem 1.1 remains valid for $P = P(\operatorname{Im} K_a)$, although $\operatorname{Im} K_a$ does not satisfies the first inequality of (b).

2. Preliminary results

Let $p \in [1, \infty)$. For z, $w \in D$, let $\beta(z, w)$ denote the Bergman distance between z and w. It is well known that for every fixed $z \in D$, $\beta(z, w) \to \infty$ as $w \to \partial D$. For $z \in D$ and $r \in (0, \infty)$, the Bergman ball centered at z of radius r is denoted by

$$E(z, r) = \{w \in D : \beta(w, z) \leq r\}.$$

For any measurable set $S \subset \mathbb{C}^n$, let $|S|$ denote $V(S)$. Let

$$\widehat{f}(z, r) = \frac{1}{|E(z, r)|} \int_{E(z,r)} f(w)\, dV(w)$$

$$\mathrm{MO}_p(f, z, r) = \frac{1}{|E(z, r)|} \int_{E(z,r)} |f(w) - \widehat{f}(z, r)|^p\, dV(w).$$

We define BMO^p_r and its subspace VMO^p_r as follows:

$$\mathrm{BMO}^p_r = \{f \in L^p : \sup_{z \in D} \mathrm{MO}_p(f, z, r) < \infty\}$$

$$\mathrm{VMO}^p_r = \{f \in L^p : \mathrm{MO}_p(f, z, r) \to 0 \text{ as } z \to \partial D\}.$$

Hölder's inequality shows that if $p \leq q$, then $\mathrm{BMO}^q_r \subseteq \mathrm{BMO}^p_r$ and $\mathrm{VMO}^q_r \subseteq \mathrm{VMO}^p_r$. When $p = 2$, BMO^2_r and VMO^2_r have been well studied in [**5**].

Let

$$\mathrm{OS}(f, z) = \sup_{w \in E(z,1)} |f(z) - f(w)|.$$

The space BO and its subspace VO are defined in [**5**] as follows:

$$\mathrm{BO} = \{f \in C(D) : \sup_{z \in D} \mathrm{OS}(f, z) < \infty\}$$

$$\mathrm{VO}_\partial = \{f \in C(D) : \mathrm{OS}(f,z) \to 0 \text{ as } z \to \partial D\}.$$

By Corollary to Theorem 13 in [**5**], if we replace $E(z,1)$ by $E(z,r)$ in the definition of $\mathrm{OS}(f,z)$, then the corresponding BO and VO will be the same.

Let

$$\mathrm{MV}_p(f,z) = \frac{1}{|E(z,1)|}\int_{E(z,1)} |f(w)|^p\, dV(w).$$

We define F^p and its subspace F^p_∂ as follows:

$$\mathrm{F}^p = \{f \in L^p : \sup_{z\in D} \mathrm{MV}_p(f,z) < \infty\}$$

$$\mathrm{F}^p_\partial = \{f \in L^p : \mathrm{MV}_p(f,z) \to 0 \text{ as } z \to \partial D\}.$$

Hölder's inequality implies that $\mathrm{F}^q \subseteq \mathrm{F}^p$ and $\mathrm{F}^q_\partial \subseteq \mathrm{F}^p_\partial$ for $p \le q$. If we replace $E(z,1)$ by $E(z,r)$ in the definition of $\mathrm{MV}_p(f,z)$, then the corresponding F^p and F^p_∂ will be the same (see Theorem 7 of [**14**]).

Different choices of r also yield the same BMO^p_r and VMO^p_r. If D is the unit disk, this follows immediately from Lemma 1 of [**11**]. Lemma 1 of [**11**] can be extended to any bounded symmetric domains with an identical proof. Henceforth we shall use BMO^p and VMO^p to denote BMO^p_r and VMO^p_r, respectively.

Norms on these spaces are defined by

$$\|f\|_{\mathrm{BMO}^p} = \sup_{z\in D} \mathrm{MO}_p(f,z,1)^{1/p}$$

$$\|f\|_{\mathrm{BO}} = \sup_{z\in D} \mathrm{OS}(f,z)$$

$$\|f\|_{\mathrm{F}^p} = \sup_{z\in D} \mathrm{MV}_p(f,z)^{1/p}.$$

Throughout the paper, we shall follow the practice of allowing c to denote a positive constant whose value may change from line to line, but does not depend on variables under consideration.

Lemma 2.1. *Let $p \in [1,\infty)$ and $\widehat{f}(z) = \widehat{f}(z,1)$.*

(i) If $f \in \mathrm{BMO}^p$, then $\widehat{f} \in \mathrm{BO}$ and $f - \widehat{f} \in \mathrm{F}^p$. Moreover there is a constant c such that

$$|\widehat{f}(z) - \widehat{f}(w)| \le c\,\|f\|_{\mathrm{BMO}^p}(1 + \beta(z,w))$$

for all z, $w \in D$ and

$$\|f - \widehat{f}\|_{\mathrm{F}^p} \le c\,\|f\|_{\mathrm{BMO}^p}.$$

(ii) If $f \in \mathrm{VMO}^p$, then $\widehat{f} \in \mathrm{VO}$ and $f - \widehat{f} \in \mathrm{F}^p_\partial$. Moreover given $\epsilon > 0$, there is $R > 0$ (depending on ϵ and $\|f\|_{\mathrm{BMO}^p}$) such that

$$|\widehat{f}(z) - \widehat{f}(w)| \le \epsilon(1 + \beta(z,w))$$

for all $z \in D$ and $w \in E(0,R)^c$.

Proof. The proof of Lemma 1 of [**11**] and Theorem 13 of [**5**] give (i).

To prove (ii), let $f \in \mathrm{VMO}^p$. It also follows from the proof of Lemma 1 of [**11**] that $\widehat{f} \in \mathrm{VO}$ and $f - \widehat{f} \in \mathrm{F}^p_\partial$. Given $\epsilon > 0$, there is $R_1 > 0$ such that if $\beta(0,w) \ge R_1$, then

$$\mathrm{OS}(\widehat{f}, w) \le \epsilon/2.$$

Let γ be the geodesic arc between z and w of length $\beta(z,w)$. Let $m = [[\beta(z,w)]]$, where $[[\cdot]]$ denotes the greatest integer function. Let us divide γ into $m+1$ subdivisions of equal length and let us denote the endpoints of the j-th subdivision by ζ_j and ζ_{j+1} for $j = 1, \cdots, m+1$ ($\zeta_1 = z$ and $\zeta_{m+1} = w$). It is clear that $|\beta(\zeta_j) - \beta(\zeta_{j+1})| \leq 1$ for all j. Thus

$$|\widehat{f}(z) - \widehat{f}(w)| \leq \sum_{j=1}^{m+1} |\widehat{f}(\zeta_j) - \widehat{f}(\zeta_{j+1})| \leq \sup_{\zeta \in \gamma} \mathrm{OS}(\widehat{f}, \zeta)(m+1).$$

Let $\beta(0, w) > R_1$. If $\beta(z, w) < \beta(0, w) - R_1$, then entire γ lies outside $E(0, R_1)$ and thus

$$|\widehat{f}(z) - \widehat{f}(w)| \leq (\epsilon/2)(m+1) \leq (\epsilon/2)(1 + \beta(z, w)).$$

By (i), if λ, $\mu \in E(0, R_1)$, then

$$|\widehat{f}(\lambda) - \widehat{f}(\mu)| \leq c \, \|f\|_{\mathrm{BMO}^p}(1 + \beta(0, \lambda) + \beta(0, \mu)) \leq c \, \|f\|_{\mathrm{BMO}^p}(1 + 2R_1).$$

If $\beta(z, w) \geq \beta(0, w) - R_1$, then a portion of γ may lie inside $E(0, R_1)$ and thus

$$|\widehat{f}(z) - \widehat{f}(w)| \leq c \, \|f\|_{\mathrm{BMO}^p}(1 + 2R_1) + (\epsilon/2)(1 + \beta(z, w)).$$

Therefore if $\beta(0, w) > R$ with R large enough, then

$$|\widehat{f}(z) - \widehat{f}(w)| \leq \left[\frac{c \, \|f\|_{\mathrm{BMO}^p}(1 + 2R_1)}{\beta(0, w) - R_1} + \frac{\epsilon}{2} \right] (1 + \beta(z, w)) \leq \epsilon(1 + \beta(z, w))$$

for all $z \in D$. This finishes the proof of the lemma. □

See the proof of Theorem 21 of [**5**] for the following lemma.

LEMMA 2.2. *For any $\delta > 0$, there is a constant c depending on δ such that*

$$\beta(z, w) \leq c \, K_a(z, z)^\delta K_a(w, w)^\delta |K_a(z, w)|^{-2\delta}$$

for all z, $w \in D$.

The following lemma follows from (1.2).

LEMMA 2.3. *Let $p \in (1, \infty)$ and let $(1/p') + (/p) = 1$. If ϵ is a positive number such that $\epsilon < \min\{t_D/p, t_D/p'\}$, then there is a constant c such that*

$$\int_D K_a(w, w)^{\epsilon p'} |K_a(z, w)| \, dV(w) \leq c \, K_a(z, z)^{\epsilon p'},$$

$$\int_D K_a(w, w)^{\epsilon p} |K_a(z, w)| \, dV(w) \leq c \, K_a(z, z)^{\epsilon p}$$

for all $z \in D$.

LEMMA 2.4. *Let $p \in (1, \infty)$ and let $(1/p') + (/p) = 1$. If ϵ is a positive number such that $\epsilon < \min\{t_D/p, t_D/p'\}$, then there is a constant c such that*

$$\int_D K_a(w, w)^{\epsilon p'} \beta(z, w) |K_a(z, w)| \, dV(w) \leq c \, K_a(z, z)^{\epsilon p'},$$

$$\int_D K_a(w, w)^{\epsilon p} \beta(z, w) |K_a(z, w)| \, dV(w) \leq c \, K_a(z, z)^{\epsilon p}$$

for all $z \in D$.

PROOF. Choose $\delta > 0$ small enough such that $\delta + \epsilon p' < t_D$, $\delta + \epsilon p < t_D$, $\delta < \epsilon p'$, and $\delta < \epsilon p$. By Lemma 2.2, we have

$$K_a(w, w)^{\epsilon p'} \beta(z, w) |K_a(z, w)| \leq c \, K_a(z, z)^\delta K_a(w, w)^{\delta + \epsilon p'} |K_a(z, w)|^{1 - 2\delta}$$

$$K_a(w, w)^{\epsilon p} \beta(z, w) |K_a(z, w)| \leq c \, K_a(z, z)^\delta K_a(w, w)^{\delta + \epsilon p} |K_a(z, w)|^{1 - 2\delta}.$$

The desired inequalities now follow from (1.2) easily. □

If $f \in \mathrm{F}^p$ and $\delta \in \mathbb{R}$, then by Lemma 8 of [**4**], there is a constant c such that

$$\int_{E(z,1)} K_a(w,w)^\delta |f(w)|^p \, dV(w) \le c \, \|f\|_{F^p}^p |E(z,1)|^{1-\delta}$$

for all $z \in D$.

The following lemma follows from Theorem 7 of [**14**].

LEMMA 2.5. *Let $p \in [1,\infty)$ and $\delta < t_D$. If $f \in \mathrm{F}^p$, then there is a constant c such that*

$$\int_D |g(w)|^p K_a(w,w)^\delta |f(w)|^p \, dV(w) \le c \, \|f\|_{F^p}^p \int_D |g(w)|^p K_a(w,w)^\delta \, dV(w)$$

for all holomorphic functions g on D.

Lemma 2.5 implies that if $f \in F^p$, then $\|f\|_p \le c \, \|f\|_{\mathrm{F}^p}$.

3. Proof of Theorem 1.1

For $p \in (1,\infty)$, we shall use p' to denote the number such that $(1/p')+(1/p) = 1$.
Sufficiency: We prove that if K satisfies the condition (a), then

(1) $f \in \mathrm{BMO}^p \Longrightarrow M_f P - P M_f$ is bounded on L^p;

(2) $f \in \mathrm{VMO}^p \Longrightarrow M_f P - P M_f$ is compact on L^p.

Let $f_1(z) = \widehat{f}(z,1)$ and $f_2(z) = f(z) - \widehat{f}(z,1)$. Then

$$M_f P - P M_f = (M_{f_1} P - P M_{f_1}) + (M_{f_2} P - P M_{f_2}).$$

To prove (1), suppose that $f \in \mathrm{BMO}^p$ for some $p \in [2,\infty)$. It follows from Lemma 2.1 that $f_1 \in \mathrm{BO}$ and $f_2 \in \mathrm{F}^p$. We will show that both $M_{f_1} P - P M_{f_1}$ and $M_{f_2} P - P M_{f_2}$ are bounded on L^p.

By Lemma 2.1,

$$|(f_1(z) - f_1(w)) K(z,w)| \le c \, \|f\|_{\mathrm{BMO}^p} (1 + \beta(z,w)) |K_a(z,w)|$$

for all z, $w \in D$. It follows from

$$(M_{f_1} P - P M_{f_1})[g](z) = \int_D (f_1(z) - f_1(w)) K(z,w) g(w) \, dV(w),$$

Lemmas 2.3 and 2.4, and Schur's Lemma (Lemma 3.1 of [**8**]) that $M_{f_1} P - P M_{f_1}$ is bounded on L^p.

In order to show that $M_{f_2} P - P M_{f_2}$ is bounded on L^p, we first show that if $q \in (1,\infty)$ and $h \in \mathrm{F}^q$, then $M_h P$ is bounded on L^q. For $g \in L^q$, we have

$$|M_h P[g](z)|^q \le c \, |h(z)|^q \left[\int_D |K_a(z,w)| |g(w)| \, dV(w) \right]^q .$$

Choose $\epsilon > 0$ such that $\epsilon < \min\{t_D/q, t_D/q'\}$. Hölder's inequality shows that

$$\left[\int_D |K_a(z,w)| |g(w)| \, dV(w) \right]^q$$

is less than or equal to

$$\int_D |g(w)|^q K_a(w,w)^{-\epsilon q} |K_a(z,w)| \, dV(w) \left[\int_D K_a(w,w)^{\epsilon q'} |K_a(z,w)| \, dV(w) \right]^{q-1} .$$

It follows from Lemma 2.3 that

$$|M_h P[g](z)|^q \le c \, |h(z)|^q K_a(z,z)^{\epsilon q} \int_D |g(w)|^q K_a(w,w)^{-\epsilon q} |K_a(z,w)| \, dV(w).$$

Fubini's Theorem gives

$$\|M_h P[g]\|_q^q \leq c \int_D |g(w)|^q K_a(w,w)^{-\epsilon q} \int_D |h(z)|^q K_a(z,z)^{\epsilon q} |K_a(z,w)|\, dV(z)\, dV(w).$$

Since $K_a(z,w)$ is holomorphic in z, by Lemma 2.5 we get

$$\|M_h P[g]\|_q^q \leq c\,\|h\|_{\mathrm{F}^q}^q \int_D |g(w)|^q K_a(w,w)^{-\epsilon q} \int_D K_a(z,z)^{\epsilon q} |K_a(z,w)|\, dV(z)\, dV(w).$$

By Lemma 2.3, it is easy to see that

$$\|M_h P\|_q \leq c\,\|h\|_{\mathrm{F}^q}. \tag{3.1}$$

Here $\|M_h P\|_q$ denotes the operator norm of $M_h P$ on L^q. Since $f_2 \in \mathrm{F}^p$, $M_{f_2}P$ is bounded on L^p. Write $PM_{f_2} = (M_{\bar{f}_2}P^*)^*$, where P^* is the adjoint of P given by

$$P^*[g](z) = \int_D \overline{K(w,z)} g(w)\, dV(w).$$

Since $\bar{f}_2 \in \mathrm{F}^p \subseteq \mathrm{F}^{p'}$ and the kernel of P^* also satisfies (a),$M_{\bar{f}_2}P^*$ is bounded on $L^{p'}$. Thus $PM_{f_2} = (M_{\bar{h}}P^*)^*$ is bounded on L^p, showing that $M_{f_2}P - PM_{f_2}$ is bounded on L^p. This finishes the proof of (1).

To prove (2), suppose that $f \in \mathrm{VMO}^p$ for some $p \in [2,\infty)$. It follows from Lemma 2.1 that $f_1 \in \mathrm{VO}$ and $f_2 \in \mathrm{F}^p_\partial$. We will show that both $M_{f_1}P - PM_{f_1}$ and $M_{f_2}P - PM_{f_2}$ are compact on L^p. For $R > 0$, let I_R denote the integral operator defined by

$$I_R[g](z) = \int_D (f_1(z) - f_1(w))\chi_{\{\beta(0,w)\leq R\}}(z,w) K(z,w) g(w)\, dV(w).$$

We know that $K_a(z,w)$ is continuous on $\overline{D} \times E(0,R)$. Then

$$|K(z,w)| \leq c\,|K_a(z,w)| \leq c$$

for all $z \in D$ and $w \in E(0,R)$ (here the constant c depends on R). Choose $\epsilon > 0$ such that $\epsilon p < t_D$. By Lemma 2.2, we have for all $z \in D$ and $w \in E(0,R)$,

$$\beta(z,w) \leq c\,K_a(z,z)^\epsilon K_a(w,w)^\epsilon |K_a(z,w)|^{-2\epsilon} \leq c\,K_a(z,z)^\epsilon$$

(since K_a^{-1} is continuous on $\mathbb{C}^n \times \mathbb{C}^n$) and thus

$$|(f_1(z) - f_1(w))K(z,w)| \leq c\,\|f\|_{\mathrm{BMO}^p}(1 + \beta(z,w)) \leq c\,\|f\|_{\mathrm{BMO}^p}(1 + K_a(z,z)^\epsilon).$$

It follows from (1.1) that

$$\int_D \left[\int_D |(f_1(z) - f_1(w))\chi_{\{\beta(0,w)\leq R\}}(z,w) K(z,w)|^p\, dV(z)\right]^{p'/p} dV(w) < \infty.$$

Thus for each $R > 0$, I_R is compact on L^p (see exercise 7 on page 181 of [**6**]). Given $\epsilon > 0$, by Lemma 2.1, there is $R > 0$ such that

$$|(f_1(z) - f_1(w))\chi_{\{\beta(0,w)>R\}}(z,w) K(z,w)| \leq c\,\epsilon(1 + \beta(z,w))|K_a(z,w)|$$

for all z, $w \in D$. It follows from

$$(M_{f_1}P - PM_{f_1} - I_R)[g](z) = \int_D (f_1(z) - f_1(w))\chi_{\{\beta(0,w)>R\}}(z,w) K(z,w) g(w)\, dV(w),$$

Lemmas 2.3 and 2.4, and Schur's Lemma that

$$\|M_{f_1}P - PM_{f_1} - I_R\|_p \leq c\,\epsilon.$$

Therefore $\|M_{f_1}P - PM_{f_1} - I_R\|_p \to 0$ as $R \to \infty$. This shows that $M_{f_1}P - PM_{f_1}$ is compact on L^p.

To show that $M_{f_2}P - PM_{f_2}$ is compact on L^p, let

$$f_{2,R}(z) = f_2(z)\chi_{\{\beta(0,z)\leq R\}}(z)$$

for $R > 0$. Then

$$M_{f_{2,R}}P[g](z) = \int_D f_{2,R}(z)K(z,w)g(w)\,dV(w)$$

and

$$|f_{2,R}(z)K(z,w)| \leq c\,|f_{2,R}(z)K_a(z,w)| \leq c\,|f_2(z)\chi_{\{\beta(0,z)\leq R\}}(z)| \leq c\,|f_2(z)|$$

for all $z, w \in D$ (here the constant c depends on R). It follows that

$$\int_D \left[\int_D |f_{2,R}(z)K(z,w)|^p\,dV(z)\right]^{p'/p} dV(w) \leq c\left[\int_D |f_2(z)|^p\,dV(z)\right]^{p'/p} \leq c\,\|f_2\|_{\mathrm{F}^p}^{p'} < \infty.$$

Thus for each $R > 0$, $M_{f_{2,R}}P$ is compact on L^p. By (3.1), we have

$$\|M_{f_2}P - M_{f_{2,R}}P\|_p \leq c\,\|f_2 - f_{2,R}\|_{\mathrm{F}^p}.$$

Since $f_2 \in \mathrm{F}^p_\partial$, it is easy to see that $\|f_2 - f_{2,R}\|_{\mathrm{F}^p} \to 0$ as $R \to \infty$. Therefore $M_{f_2}P$ is compact on L^p. Since $\bar{f}_2 \in \mathrm{F}^p_\partial \subseteq \mathrm{F}^{p'}_\partial$, the argument above also shows that $M_{\bar{f}_2}P^*$ is compact on $L^{p'}$. Thus $M_{f_2}P - PM_{f_2} = M_{f_2}P - (M_{\bar{f}_2}P^*)^*$ is compact on L^p. This finishes the proof of (2).

Necessity: We prove that if $f \in L^p$ and if K satisfies the condition (b), then

(3) $M_fP - PM_f$ is bounded on $L^p \Longrightarrow f \in \mathrm{BMO}^p$;

(4) $M_fP - PM_f$ is compact on $L^p \Longrightarrow f \in \mathrm{VMO}^p$.

To prove (3), suppose that $M_fP - PM_f$ is bounded on L^p. For z, w, $u \in D$, let

$$R(z,w,u) = \frac{K(w,u)}{K(z,z)} - 1.$$

By (b), there is $r > 0$ such that

$$|R(z,w,u)| \leq 1/4 \tag{3.2}$$

for all $z \in D$ and w, $u \in E(z,r)$. We shall use this r for the rest of the proof.

Replacing 1 by $K(w,u)K(z,z)^{-1} - R(z,w,u)$ in the inner integral of

$$\mathrm{MO}_p(f,z,r) = \frac{1}{|E(z,r)|^{p+1}}\int_{E(z,r)}\left|\int_{E(z,r)} (f(w) - f(u))\cdot 1\,dV(u)\right|^p dV(w)$$

and using the inequality $|x-y|^p \leq 2^{p-1}(|x|^p + |y|^p)$, we get

$$\mathrm{MO}_p(f,z,r) \leq I_1 + I_2,$$

where

$$I_1 = \frac{2^{p-1}}{|E(z,r)|^{p+1}}\int_{E(z,r)}\left|\int_{E(z,r)} (f(y) - f(z))\frac{K(w,u)}{K(z,z)}\,dV(u)\right|^p dV(w)$$

$$I_2 = \frac{2^{p-1}}{|E(z,r)|^{p+1}}\int_{E(z,r)}\left|\int_{E(z,r)} (f(w) - f(u))R(z,w,u)\,dV(u)\right|^p dV(w).$$

By (b) and Lemma 8 of [**4**], we have

$$|E(z,r)||K(z,z)| \geq c\,|E(z,r)|K_a(z,z) \geq c$$

for all $z \in D$. Thus

$$\begin{aligned} I_1 &= \frac{2^{p-1}}{|E(z,r)|^p|K(z,z)|^p} \int_{E(z,r)} \left| \int_{E(z,r)} (f(w)-f(u)) \frac{K(w,u)}{|E(z,r)|^{1/p}}\, dV(u) \right|^p dV(w) \\ &\leq c \int_D \left| \int_D (f(w)-f(u))K(w,u) \frac{\chi_{E(z,r)}(u)}{|E(z,r)|^{1/p}}\, dV(u) \right|^p dV(w) \\ &= c\,\|(M_fP - PM_f)[\chi_{E(z,r)}/|E(z,r)|^{1/p}]\|_p^p. \end{aligned}$$

By (3.2) and Hölder's inequality, we have

$$\begin{aligned} I_2 &\leq \frac{2^{p-1}}{4^p|E(z,r)|^{p+1}} \int_{E(z,r)} \left[\int_{E(z,r)} |f(w)-f(u)|\, dV(u) \right]^p dV(w) \\ &\leq \frac{1}{2^{p+1}|E(z,r)|^2} \int_{E(z,r)} \int_{E(z,r)} |f(w)-f(u)|^p\, dV(u)\, dV(w). \end{aligned}$$

For almost every w, $u \in D$,

$$|f(w)-f(u)|^p \leq 2^{p-1}|f(w)-\hat{f}(z,r)|^p + 2^{p-1}|f(u)-\hat{f}(z,r)|^p.$$

This gives

$$\begin{aligned} I_2 &\leq \frac{1}{2|E(z,r)|} \int_{E(z,r)} |f(w)-\hat{f}(z,r)|^p\, dV(w) \\ &= \frac{\mathrm{MO}_p(f,z,r)}{2}. \end{aligned}$$

It now follows from $\mathrm{MO}_p(f,z,r) \leq I_1 + I_2$ that

$$\mathrm{MO}_p(f,z,r) \leq c\,\|(M_fP - PM_f)[\chi_{E(z,r)}/|E(z,r)|^{1/p}]\|_p^p. \tag{3.3}$$

Since

$$\|\chi_{E(z,r)}/|E(z,r)|^{1/p}\|_p = 1,$$

we have

$$\sup_{z\in D} \mathrm{MO}_p(f,z,r) \leq c\,\|M_fP - PM_f\|_p^p < \infty.$$

This shows that $f \in \mathrm{BMO}^p$ and finishes the proof of (3).

To prove (4), suppose that $M_fP - PM_f$ is compact on L^p. For every $g \in L^{p'}(B)$, by Hölder's inequality, we have

$$\frac{1}{|E(z,r)|} \left| \int_{E(z,r)} \chi_{E(z,r)}(w)\overline{g(w)}\, dV(w) \right|^p \leq \left[\int_{E(z,r)} |g(w)|^{p'}\, dV(w) \right]^{p-1} \to 0$$

as $z \to \partial D$ (since $|E(z,r)| \to 0$ as $z \to \partial D$). This implies that

$$\chi_{E(z,r)}/|E(z,r)|^{1/p} \to 0$$

weakly in L^p as $z \to \partial D$. Thus by (3.3), we have

$$\mathrm{MO}_p(f,z,r) \to 0$$

as $z \to \partial D$. This shows that $f \in \mathrm{VMO}^p$ and finishes the proof of (4).

4. Proofs of Theorems 1.2 and 1.3

For each $z \in D$, let φ_z denote the Möbius transformation of D with the following properties:

(1) $\varphi_z(z) = 0$;

(2) $\varphi_z \circ \varphi_z$ is the identity map of D.

The following is the well-known transformation law for the reproducing kernel K_a:

$$K_a(\varphi_z(w), \varphi_z(u))(J_c\varphi_z)(w)\overline{(J_c\varphi_z)(u)} = K_a(w, u), \tag{4.1}$$

where $(J_c\varphi_z)(\cdot)$ denotes the determinant of the complex Jacobian of φ_z. There is $\eta_z \in \mathbb{C}$ depending only on z with $|\eta_z| = 1$ such that

$$(J_c\varphi_z)(w) = \eta_z(K_a(z, z))^{-1/2}K_a(w, z). \tag{4.2}$$

See pages 926-927 of [**4**] for more details about (4.1) and (4.2).

LEMMA 4.1. *The reproducing kernel K_a satisfies the conditions (a) and (b).*

PROOF. We only need to show that K_a satisfies the second inequality of (b), i.e. there is $r > 0$ such that

$$\left|\frac{K_a(w, u)}{K_a(z, z)} - 1\right| \leq 1/4$$

for all $z \in D$ and w, $u \in E(z, r)$.

Let $w = \varphi_z(\lambda)$ and $u = \varphi_z(\mu)$. Since the Bergman distance is invariant under any Möbius transformation, we have

$$\beta(\varphi_z(\lambda), z) = \beta(\lambda, 0), \ \beta(\varphi_z(\mu), z) = \beta(\mu, 0).$$

Then $w = \varphi_z(\lambda)$, $u = \varphi_z(\mu) \in E(z, r)$ if and only if λ, $\mu \in E(0, r)$. If $w = \varphi_z(\lambda)$ and $u = \varphi_z(\mu)$, then by (4.1) and (4.2)

$$\begin{aligned}\left|\frac{K_a(w, u)}{K_a(z, z)} - 1\right| &= \left|\frac{K_a(\lambda, \mu)}{(J_c\varphi_z(\lambda))\overline{(J_c\varphi_z(\mu))}K_a(z, z)} - 1\right| \\ &= \left|\frac{K_a(\lambda, \mu)}{K_a(\lambda, z)\overline{K_a(\mu, z)}} - 1\right|.\end{aligned}$$

It is clear that the function $K_a(\lambda, \mu)(K_a(\lambda, z)\overline{K_a(\mu, z)})^{-1} - 1$ (with variable (λ, μ, z)) is continuous on $D \times D \times \overline{D}$ and hence uniformly continuous on $E(0, r) \times E(0, r) \times \overline{D}$ for any $r > 0$. The value of this function is 0 at $(0, 0, z)$ for any $z \in D$. Since the topology induced by the Bergman metric β is the usual topology, we conclude that K_a satisfies the desired inequality if r is small enough. □

The following lemma was proved in [**13**] (page 357) when D is the unit disk. The method used there is based on an explicit orthonormal basis for L^2_h and a direct computation. We provide a different proof for more general domain D.

LEMMA 4.2. *For z, $w \in D$,*

$$K_h(z, w) = 2\operatorname{Re} K_a(z, w) - 1.$$

PROOF. We only need to show that

$$P_h[g](z) = \int_D (2\operatorname{Re} K_a(z, w) - 1)g(w)\, dV(w)$$

for all $g \in L^2$. We first show that

$$g(z) = \int_D (2\operatorname{Re} K_a(z,w) - 1) g(w)\, dV(w).$$

for all $g \in L^2_h$. If $g \in L^2_h$ is real-valued, then by Theorem 4.4.9 in [**12**] (see also page 64 in [**12**]), there is a holomorphic function h such that $g = \operatorname{Re} h$. Let us assume for a moment that $h \in L^2$. Thus

$$h(z) = \int_D K_a(z,w) h(w)\, dV(w).$$

Since $\overline{K_a(z,w)} h(w)$ is holomorphic in w, it is harmonic in w. It follows from the mean value property that

$$h(0) = \overline{K_a(z,0)} h(0) = \int_D \overline{K_a(z,w)} h(w)\, dV(w).$$

Adding the real parts of the last two equations gives

$$g(z) + g(0) = 2\int_D \operatorname{Re} K_a(z,w) g(w)\, dV(w).$$

Since $g(0) = \int_D g(w)\, dV(w)$, this gives

$$g(z) = \int_D (2\operatorname{Re} K_a(z,w) - 1) g(w)\, dV(w)$$

for all real-valued (hence all) $g \in L^2_h$.

For any $g \in L^2$, there are unique $g_1 \in L^2_h$ and $g_2 \in (L^2_h)^\perp$ such that $g = g_1 + g_2$. Thus

$$P_h[g](z) = g_1(z) = \int_D (2\operatorname{Re} K_a(z,w) - 1) g_1(w)\, dV(w).$$

It follows from

$$\int_D (2\operatorname{Re} K_a(z,w) - 1) g_2(w)\, dV(w) = 0$$

that

$$P_h[g] = \int_D (2\operatorname{Re} K_a(z,w) - 1) g(w)\, dV(w).$$

To finish the proof, we need to show that if $g \in L^2_h$ is real valued, then the holomorphic function h such that $\operatorname{Re} h = g$ is also in L^2. We use an argument from [**2**]. For a function f on D and $0 < t < 1$, let $f_t(z) = f(tz)$. Then h_t is a bounded holomorphic function on D and $\operatorname{Re} h_t = g_t$ for every $t \in (0,1)$. It follows from the mean value property that $\overline{h_t(0)} = P_a(\overline{h_t})(z)$ for all $z \in D$. Thus

$$\|h_t + \overline{h_t(0)}\|_2 = \|P_a(h_t + \overline{h_t})\|_2 = 2\|P_a(\operatorname{Re} h_t)\|_2 \le 2\|\operatorname{Re} h_t\|_2 = 2\|g_t\|_2.$$

Taking limits as $t \to 1$ gives $\|h + \overline{h(0)}\|_2 \le 2\|g\|_2$, showing that $h \in L^2$. □

LEMMA 4.3. *The reproducing kernel K_h satisfies the conditions (a) and (b).*

PROOF. Since $K_a(z,w)^{-1}$ is continuous on $\mathbb{C} \times \mathbb{C}$, there is a constant c such that $|K_a(z,w)| \ge c$ for all z, $w \in D$. Thus

$$|K_h(z,w)| = |2\operatorname{Re} K_a(z,w) - 1| \le 2|K_a(z,w)| + 1 \le (2 + 1/c)|K_a(z,w)|$$

for all z, $w \in D$. This shows that K_h satisfies (a). It follows from

$$K_a(z,z) = \|K_a(\cdot,z)\|_2 \ge \left|\int_D K_a(w,z)\, dV(w)\right| = |K_a(0,z)| = 1$$

that

$$K_h(z,z) = 2K_a(z,z) - 1 \geq K_a(z,z)$$

for all $z \in D$. This shows that K_h satisfies the first inequality of (b). According to the proof of Lemma 4.1, there is $r > 0$ such that

$$\left|\frac{K_a(w,u)}{K_a(z,z)} - 1\right| \leq \frac{1}{8}$$

for all $z \in D$ and $w,\ u \in E(z,r)$. Thus

$$\begin{aligned} |K_h(w,u) - K_h(z,z)| &= 2|\operatorname{Re} K_a(w,u) - \operatorname{Re} K_a(z,z)| \\ &\leq 2|K_a(w,u) - K_a(z,z)| \\ &\leq (1/4)K_a(z,z) \\ &\leq (1/4)K_h(z,z) \end{aligned}$$

for all $z \in D$ and $w,\ u \in E(z,r)$, showing that K_h satisfies the second inequality of (b). □

Proof of Theorem 1.2. It follows immediately from Theorem 1.1 and Lemma 4.3. □

Let P_r denote $P(\operatorname{Re} K_a)$ and let P_i denote $P(\operatorname{Im} K_a)$. The proof of Lemma 4.3 shows that $\operatorname{Re} K_a$ satisfies both (a) and (b). However, $\operatorname{Im} K_a$ does not satisfy (b) because $\operatorname{Im} K_a(z,z) = 0$ for all $z \in D$.

In order to prove Theorem 1.3, we need three more lemmas.

LEMMA 4.4. *If $\lambda,\ \mu \in \mathbb{C}$ and $\lambda \neq 0$, then $\lambda \operatorname{Re} K_a + \mu \operatorname{Im} K_a$ satisfies the conditions (a) and (b).*

PROOF. Let $K(z,w) = \lambda \operatorname{Re} K_a(z,w) + \mu \operatorname{Im} K_a(z,w)$. It is clear that K satisfies (a). It is also clear that K satisfies the first inequality of (b). Let $c(\lambda,\mu) = |\lambda|/4(|\lambda| + |\mu|)$. According to the proof of Lemma 4.1, there is $r > 0$ such that

$$|K_a(w,u) - K_a(z,z)| \leq c(\lambda,\mu)K_a(z,z)$$

for all $z \in D$ and $w,\ u \in E(z,r)$. It follows that

$$|\operatorname{Re} K_a(w,u) - K_a(z,z)| \leq |K_a(w,u) - K_a(z,z)| \leq c(\lambda,\mu)K_a(z,z)$$

and

$$|\operatorname{Im} K_a(w,u)| \leq |K_a(w,u) - K_a(z,z)| \leq c(\lambda,\mu)K_a(z,z)$$

for all $z \in D$ and $w,\ u \in E(z,r)$. Thus

$$\begin{aligned} |K(w,u) - K(z,z)| &\leq |\lambda||\operatorname{Re} K_a(w,u) - K_a(z,z)| + |\mu||\operatorname{Im} K_a(w,u)| \\ &\leq (|\lambda| + |\mu|)c(\lambda,\mu)K_a(z,z) \\ &= (1/4)|K(z,z)| \end{aligned}$$

for all $z \in D$ and $w,\ u \in E(z,r)$. This shows that K satisfies the second inequality of (b). □

LEMMA 4.5. *$2P_i^2[g](z) + P_r[g](z) - P_r[g](0) = 0$ for all $g \in L^2$ and $z \in D$.*

PROOF. Without loss of generality, we may assume that $g \in L^2$ is real valued. Thus both $P_r[g]$ and $P_i[g]$ are real valued. Computing the real parts of

$$P_a[g](z) = \int_D K_a(z,w)P_a[g](w)\,dV(w)$$

gives

$$P_r[g](z) = \int_D (\operatorname{Re} K_a(z,w)P_r[g](w) - \operatorname{Im} K_a(z,w)P_i[g](w))\,dV(w).$$

Thus

$$P_i^2[g](z) = -P_r[g](z) + \int_D \operatorname{Re} K_a(z,w)P_r[g](w)\,dV(w).$$

Lemma 4.2 gives

$$\int_D \operatorname{Re} K_a(z,w)P_r[g](w)\,dV(w) = \frac{P_r[g](z) + P_r[g](0)}{2}.$$

The desired identity now follows from the last two equations. □

LEMMA 4.6. *Let $p \in [2,\infty)$ and $f \in L^p$. Then the integral operator defined by*

$$I[g](z) = \int_D (f(z) - f(w))g(w)\,dV(w)$$

is compact on L^p.

PROOF. We only need to verify that

$$c(f) = \int_D \left[\int_D |f(z) - f(w)|^p\,dV(z)\right]^{p'/p} dV(w) < \infty.$$

It follows from the inequality $|x-y|^p \le 2^{p-1}(|x|^p + |y|^p)$ that

$$c(f) \le 2\int_D \left[\int_D (|f(z)|^p + |f(w)|^p)\,dV(z)\right]^{p'/p} dV(w).$$

It follows from the inequality $|x+y|^{p'/p} \le |x|^{p'/p} + |y|^{p'/p}$ (since $p'/p \le 1$) that

$$c(f) \le 2\left[\int_D |f(z)|^p\,dV(z)\right]^{p'/p} + 2\int_D |f(w)|^{p'}\,dV(w) < \infty.$$

□

Proof of Theorem 1.3. If $\lambda \neq 0$, then the conclusion of the theorem follows from Theorem 1.1 and Lemma 4.4.

If $\lambda = 0$, then without loss of generality we may assume that $\mu = 1$ and thus $P = P_i$. The proof of Theorem 1.1 indicates that we only need to show that if $M_fP_i - P_iM_f$ is bounded (or compact) on L^p, then $f \in \mathrm{BMO}^p$ (or VMO^p).

Let P_j denote $(1/2)P(\operatorname{Re} K_a - 1)$. Lemma 4.5 implies that $P_i^2 + P_j = 0$ on L^2. Thus

$$-(M_fP_i - P_iM_f)P_i - P_i(M_fP_i - P_iM_f) = -M_fP_i^2 + P_i^2M_f = M_fP_j - P_jM_f.$$

Since P_i is bounded on L^p (which follows from Lemma 2.3), we conclude that if $M_fP_i - P_iM_f$ is bounded (or compact) on L^p, then $M_fP_j - P_jM_f$ is bounded (or compact) on L^p.

Lemma 4.6 shows that $(M_fP_r - P_rM_f) - 2(M_fP_j - P_jM_f)$ is compact on L^p. Thus if $M_fP_i - P_iM_f$ is bounded (or compact) on L^p, then $M_fP_r - P_rM_f$ is bounded (or compact) on L^p. It now follows from Theorem 1.1 that $f \in \mathrm{BMO}^p$ (or VMO^p) if $M_f P_i - P_i M_f$ is bounded (or compact) on L^p. $\square$

5. More examples

All kernel functions that are discussed in Section 4 are pluriharmonic functions (in each variable) and they are closely related to the reproducing kernel K_a. Let us give two more examples based on K_a.

A Positive Kernel. Let $K_p(z,w) = |K_a(z,w)|$ for z, $w \in D$. It is obvious that K_p satisfies (a) and the first inequality of (b). It follows from

$$\left|\frac{K_p(w,u)}{K_p(z,z)} - 1\right| = \left|\left|\frac{K_a(w,u)}{K_a(z,z)}\right| - 1\right| \leq \left|\frac{K_a(w,u)}{K_a(z,z)} - 1\right|$$

and Lemma 4.1 that K_p satisfies the second inequality of (b).

A Discontinuous Kernel. Let $K_d(z,w) = K_a(z,w)\chi_{\{\beta(z,w)\leq 2r\}}(z,w)$ for z, $w \in D$, where r is a positive number such that

$$\left|\frac{K_a(w,u)}{K_a(z,z)} - 1\right| \leq \frac{1}{4}$$

for all $z \in D$ and w, $u \in E(z,r)$. It is clear that K_d satisfies (a) and the first inequality of (b). If w, $u \in E(z,r)$, then $\beta(w,u) \leq 2r$. Thus

$$\left|\frac{K_d(w,u)}{K_d(z,z)} - 1\right| = \left|\frac{K_a(w,u)}{K_a(z,z)} - 1\right| \leq 1/4$$

for all $z \in D$ and w, $u \in E(z,r)$, showing that K_d satisfies the second inequality of (b).

References

[1] S. Axler, *The Bergman space, the Bloch space, and commutators of multiplication operators,* Duke Math. J. 53 (1986), 315-332.

[2] S. Axler, *Bergman spaces and their operators,* Surveys of Some Results in Operator Theory, vol. 1, (J. B. Conway and B. B. Morrel, editors), Pitman Res. Notes Math. Ser. vol. 171, 1988, 1-50.

[3] S. Axler, P. Bourdon, and W. Ramey, *Harmonic function theory,* Second Edition, Springer-Verlag, New York, 2001.

[4] C. A. Berger, L. A. Coburn, and K. H. Zhu, *Function theory on Cartan domains and the Berezin-Toeplitz symbol calculus,* Amer. J. Math. 110 (1988), 921-953.

[5] D. Békollé, C. A. Berger, L. A. Coburn, and K. H. Zhu, BMO *and the Bergman metric on bounded symmetric domains,* J. Funct. Anal. 93 (1990), 310-350.

[6] J. B. Conway, *A Course in Functional analysis*, Springer-Verlag, New York, 1985.

[7] J. Faraut, A. Koranyi, *Function spaces and reproducing kernels on bounded symmetric domains*, J. Funct. Anal. 88 (1990), 64-89.

[8] F. Forelli and W. Rudin, *Projections on spaces of holomorphis functions in balls*, Indiana Univ. Math. J. 24 (1974), 593-602.

[9] J. Miao, *Toeplitz operators on harmonic Bergman spaces,* Integral Equations and Operator Theory 27 (1997), 426-438.

[10] J. Miao, *Reproducing kernels for harmonic Bergman spaces of the unit ball,* Monatsh. Math. 125 (1998), 25-35.

[11] J. Miao, *Hankel type operators on the unit disk,* Studia Mathematica 146 (2001), 55-67.

[12] W. Rudin, *Function theory in the unit ball of $\mathbb{C}^n$*, Springer-Verlag, New York/Berlin, 1980.

[13] Z. Wu, *Operators on harmonic Bergman spaces,* Integral Equations and Operator Theory 24 (1996), 352-371.

[14] K. Zhu, *Positive Toeplitz operators on weighted Bergman spaces of bounded symmetric domains*, J. Operator Theory 20 (1998), 329-357.

Department of Computer Science and Mathematics, P.O. Box 70, Arkansas State University, AR 72467

E-mail address: miao@csm.astate.edu

Contemporary Mathematics
Volume **321**, 2003

Growth Conditions and Decomposable Extensions

T. L. Miller, V. G. Miller and M. M. Neumann

Abstract. Classical results of Colojoară and Foiaş give sufficient conditions for an invertible Banach space operator T to be decomposable or generalized scalar in terms of rates of growth of $\|T^n\|$, $n \in \mathbb{Z}$. In this note, we show that a large class of operators T admits invertible extensions whose powers grow comparably to those of T. Therefore, growth conditions for these operators analogous to those of Colojoară and Foiaş imply the existence of decomposable or generalized scalar extensions. Our results are exemplified in the case of weighted shifts and certain operators on Hilbert spaces.

Introduction and motivation. In the general setting of bounded linear operators on Banach spaces, this note addresses the classical problem of finding extensions with good spectral decomposition properties. In the case of Hilbert spaces, an important example is, of course, provided by the theory of subnormal operators. Here we are seeking extensions within the class of decomposable operators.

Recall from [**3**] and [**8**] that a bounded linear operator $T \in \mathcal{L}(X)$ on a complex Banach space X is *decomposable* if, for every open cover $\{U, V\}$ of $\mathbb{C}$, there exist T-invariant closed linear subspaces Y and Z of X such that $X = Y + Z$, $\sigma(T|_Y) \subseteq U$, and $\sigma(T|_Z) \subseteq V$. Note that, by [**8,** 1.2.23], this simple definition of decomposability is equivalent to the classical notion introduced by Foiaş.

An important subclass of the decomposable operators is formed by the *generalized scalar operators*, i.e., those operators which admit a continuous functional calculus on the Fréchet algebra $\mathcal{E}(\mathbb{C})$ of all infinitely differentiable complex-valued functions on $\mathbb{C}$. Of particular interest are the generalized scalar operators with spectrum contained in the unit circle $\mathbb{T}$. By a classical result due to Colojoară and Foiaş, these are exactly the operators with a continuous functional calculus on $\mathcal{E}(\mathbb{T})$, the Fréchet algebra of all C^∞–functions on $\mathbb{T}$. Moreover, the $\mathcal{E}(\mathbb{T})$–scalar operators are also characterized as those invertible operators $T \in \mathcal{L}(X)$ which satisfy, with some constants $K, s > 0$, the polynomial growth condition $\|T^n\| \leq K\,|n|^s$ for all non-zero integers $n \in \mathbb{Z}$; see [**3,** 5.1] or [**8,** 1.5.12]. In fact, by [**3,** 5.1.4], this growth condition implies that T has a continuous functional calculus on $C^m(\mathbb{T})$ for every integer $m > s + 1$.

An illuminating case is that of isometries. Evidently, all invertible isometries on a Banach space satisfy the preceding growth condition, and hence are $\mathcal{E}(\mathbb{T})$–scalar.

2000 *Mathematics Subject Classification.* Primary 47A11, 47B40; Secondary 47B37.

More precisely, as noted in [**8,** 1.6.7], an isometry is invertible if and only if it is $\mathcal{E}(\mathbb{T})$–scalar, or, equivalently, generalized scalar, decomposable, or the quotient of a decomposable operator by a closed invariant subspace. On the other hand, by a result due to Douglas [**6**], recorded in [**8,** 1.6.6], every isometry may be extended to an invertible isometry, and hence is, in an obvious sense, $\mathcal{E}(\mathbb{T})$–subscalar.

One of the main issues of the present note is the characterization of $\mathcal{E}(\mathbb{T})$–subscalar operators in terms of a suitable growth condition. For an arbitrary operator $T \in \mathcal{L}(X)$ on a non-zero Banach space X, the *lower bound of* T is defined to be $\kappa(T) := \inf\{\|Tx\| : \|x\| = 1\}$. Clearly, $\kappa(T) = \|T^{-1}\|^{-1}$ provided that T is invertible. Thus, by the result of Colojoară and Foiaş mentioned above, every $\mathcal{E}(\mathbb{T})$-subscalar operator $T \in \mathcal{L}(X)$ satisfies, for some $s > 0$, the polynomial growth condition

$$P(s) \qquad \frac{1}{K\,n^s} \le \kappa(T^n) \le \|T^n\| \le K\,n^s$$

for some constant $K > 0$ and all $n \in \mathbb{N}$. In the opposite direction, we note that, by [**8,** 1.6.2], condition $P(s)$ for an arbitrary operator $T \in \mathcal{L}(X)$ ensures that the approximate point spectrum $\sigma_{ap}(T)$ is contained in the unit circle, and that $\sigma_{ap}(T) = \mathbb{T}$ and $\sigma(T) = \overline{\mathbb{D}}$, the closed unit disc, provided that T is not invertible. Since, by [**8,** 1.3.2 and 2.5.5], the identity $\sigma_{ap}(T) = \sigma(T)$ holds whenever T is the quotient of a decomposable operator, we are led to the following result.

PROPOSITION 1. *Suppose that the operator $T \in \mathcal{L}(X)$ satisfies condition $P(s)$ for some $s > 0$. Then the following assertions are equivalent:*
(i) *T is invertible;*
(ii) $\sigma(T) \subseteq \mathbb{T}$;
(iii) *T is the quotient of a decomposable operator;*
(iv) *T is decomposable;*
(v) *T is $\mathcal{E}(\mathbb{T})$-scalar.*
Moreover, if T is not invertible, then $\sigma_{ap}(T) = \mathbb{T}$ and $\sigma(T) = \overline{\mathbb{D}}$.

Consequently, the problem of characterizing $\mathcal{E}(\mathbb{T})$–subscalar operators by the polynomial growth condition turns out to be equivalent to the problem of extending an arbitrary operator that satisfies $P(s)$ for some $s > 0$ to an invertible operator that satisfies $P(t)$ for some $t > 0$.

A similar problem arises for operators $T \in \mathcal{L}(X)$ which satisfy, for some $s > 0$ with $0 < s < 1$, the exponential growth condition

$$E(s) \qquad \frac{1}{K\,e^{n^s}} \le \kappa(T^n) \le \|T^n\| \le K\,e^{n^s}$$

for some constant $K > 0$ and all $n \in \mathbb{N}$, or, more generally, the Beurling condition

$$(B) \qquad \sum_{n=1}^{\infty} \frac{1}{n^2}\left(\left|\log \kappa(T^n)\right| + \left|\log \|T^n\|\right|\right) < \infty.$$

In fact, by another result of Colojoară and Foiaş, [**3,** 5.3.2] or [**8,** 4.4.7], condition (B) for an invertible operator $T \in \mathcal{L}(X)$ implies that T is decomposable, and even super–decomposable in the sense of [**8,** 1.4.1]; i.e., for every open cover $\{U, V\}$ of $\mathbb{C}$, there exists an operator $R \in \mathcal{L}(X)$ commuting with T so that $\sigma(T|\overline{R(X)}) \subseteq U$

and $\sigma(T|\overline{(I-R)(X)}) \subseteq V$. It follows that Proposition 1 remains valid for operators with property (B), once assertion (v) is replaced by the condition of super-decomposability. Thus, to prove that condition (B) for an arbitrary operator T implies that T has a decomposable extension, it suffices to show that every operator with property (B) has an invertible extension with property (B).

For problems of this type, the extension procedures from the theory of analytic functional models are not directly applicable. Indeed, the extension of hyponormal operators to generalized scalar operators due to Putinar [**9,** 3.4.3 and 3.4.5] preserves the spectrum, while the extension of operators with Bishop's property (β) to decomposable operators in the Albrecht–Eschmeier functional model [**8,** 2.4.3 and 2.4.4] increases the spectrum. For a discussion of growth conditions and decomposable extensions based on the theory of functional models, we refer to [**8,** 1.7].

In this note, we present an elementary construction that provides, for certain large classes of operators, extensions to invertible operators while preserving the polynomial and exponential growth conditions mentioned above. Moreover, for the class of unilateral weighted right shifts on the sequence space $\ell^p(\mathbb{N}_0)$ for arbitrary p with $1 \leq p < \infty$, our method leads to very simple decomposable extensions within the class of bilateral weighted right shifts on $\ell^p(\mathbb{Z})$, and to a complete characterization of $\mathcal{E}(\mathbb{T})$–subscalarity in terms of the underlying weight sequence.

In the Hilbert space case, $\ell^2(\mathbb{N}_0)$, this characterization was recently obtained by Didas, [**5,** 4.1.3], based on the theory of functional models and topological tensor products from [**7**]. Our more elementary approach is inspired by a construction of Bercovici and Petrović, who showed that an operator $T \in \mathcal{L}(X)$ is the compression of an $\mathcal{E}(\mathbb{T})$–scalar operator if and only if T satisfies, for some $s > 0$, the one-sided growth condition $\|T^n\| = \mathcal{O}(n^s)$ as $n \to \infty$, [**2**].

The Construction. If $T \in \mathcal{L}(X)$ satisfies condition (B) above, then it follows from [**8,** 1.6.2] that $\sigma_{ap}(T) \subseteq \mathbb{T}$; in particular, T is injective and $T^n X$ is closed for every natural number n. We make the additional assumption that TX is complemented in X.

PROPOSITION 2. *A Banach space operator $T \in \mathcal{L}(X)$ that is bounded below has complemented range if and only if T has a left inverse $L \in \mathcal{L}(X)$. In this case, $X = T^n X \oplus \ker L^n$, and $T^n L^n$ is the projection onto $T^n X$ with kernel* $\ker L^n$*; moreover,*

$$\ker L^{n+1} = \ker L + T \ker L^n = \ker L + T \ker L + \cdots + T^n \ker L$$

for every $n \geq 1$.

PROOF. For the sake of completeness, we sketch a proof. Since T is bounded below, $\operatorname{ran} T$ is closed, and if $LT = I$, then TL is idempotent with $\operatorname{ran} TL = \operatorname{ran} T$ and $\ker TL = \ker L$. Thus there is a one-to-one correspondence between complementary subspaces of $\operatorname{ran} T$ and left inverses of T. Fix a left inverse L of T, let $Q = TL$, and, for every $n \geq 1$, let $Y_n = \ker L + T \ker L^{n-1}$. Clearly, Q is the projection of X onto TX along Y_1, and, more generally, $T^n L^n$ is the projection onto $T^n X$ with kernel $\ker L^n$. Moreover, since T is injective, if $Y_n = \ker L^n$ for some n, then

$$L^{n+1}x = 0 \Leftrightarrow Lx \in Y_n \Leftrightarrow TLx \in TY_n \Leftrightarrow x = (I-Q)x + TLx \in Y_1 + TY_n = Y_{n+1},$$

and therefore $Y_{n+1} = \ker L^{n+1}$. This establishes the first of the displayed identities, and the second follows from the first by an obvious induction argument. $\square$

Now suppose that the operator $T \in \mathcal{L}(X)$ has a left inverse L, and let $Q = TL$ and $Y = \ker L$. For $1 \le p < \infty$, let Z^p be the space of sequences $(z_k)_{k\ge 0}$ such that $z_0 \in X$, $z_k \in Y$ for every $k \ge 1$, and $\sum_{k\ge 0} \|z_k\|^p < \infty$. Define $S(z_k)_{k\ge 0} = (Tz_0 + z_1, z_2, \dots)$. So

$$S \sim \begin{pmatrix} T & I & 0 & 0 & \cdots \\ 0 & 0 & I & 0 & \cdots \\ & \ddots & \ddots & \ddots & \ddots \end{pmatrix}.$$

Clearly, S is continuous, X is S-invariant, and $S|_X = T$. Also, $(0, 0, \dots) = (Tz_0 + z_1, z_2, \dots)$ if and only if $Tz_0 = -z_1 \in \operatorname{ran}(T) \cap Y$ and $z_k = 0$ for all $k \ge 2$, and so S is injective. Since $TL = Q$,

$$S(Lz_0, (I-Q)z_0, z_1, z_2, \dots) = (TLz_0 + (I-Q)z_0, z_1, z_2, \dots) = (z_0, z_1, z_2, \dots);$$

therefore S is surjective, and

$$S^{-1} \sim \begin{pmatrix} L & 0 & 0 & 0 & \cdots \\ I-Q & 0 & 0 & 0 & \cdots \\ 0 & I & 0 & 0 & \cdots \\ 0 & \ddots & \ddots & \ddots & \ddots \end{pmatrix}.$$

THEOREM 3. *Suppose that the operator $T \in \mathcal{L}(X)$ has a left inverse L for which the sequence of projections $Q_n := T^n L^n$ is uniformly bounded. Let $Y = \ker L$ and $S \in \mathcal{L}(Z^p)$ be defined as above. Then S satisfies the growth condition $P(s+1)$ if $P(s)$ holds for T, and S satisfies $E(t)$ for all t, $s < t < 1$, provided that T satisfies $E(s)$. Consequently, S is $\mathcal{E}(\mathbb{T})$–scalar in the first case and super–decomposable in the second.*

PROOF. Since the result is trivial when T is surjective, we may assume that $Q \ne I$. Let us also assume that $p > 1$; a similar argument applies to the case $p = 1$. Let q be the conjugate exponent for p, so $\frac{1}{p} + \frac{1}{q} = 1$. Observe that

$$S^n \sim \begin{pmatrix} T^n & T^{n-1} & \cdots & T & I & 0 & \cdots \\ 0 & & \cdots & 0 & I & 0 & \cdots \\ 0 & & \cdots & & 0 & I & \ddots \\ \vdots & & & & & \ddots & \ddots \end{pmatrix} \text{ and } S^{-n} \sim \begin{pmatrix} L^n & 0 & 0 & \cdots \\ (I-Q)L^{n-1} & 0 & 0 & \ddots \\ \vdots & \vdots & \vdots & \ddots \\ I-Q & 0 & 0 & \ddots \\ 0 & I & \ddots & \ddots \\ \vdots & \ddots & \ddots & \ddots \end{pmatrix}.$$

Therefore, for $(z_k)_{k\ge 0} \in Z^p$, we see that

$$\begin{aligned} \|S^n(z_k)_{k\ge 0}\|^p &= \left\| \sum_{k=0}^n T^{n-k} z_k \right\|^p + \sum_{k=n+1}^\infty \|z_k\|^p \\ &\le \left(\sum_{k=0}^n \|T^{n-k}\| \, \|z_k\| \right)^p + \sum_{k=n+1}^\infty \|z_k\|^p \end{aligned}$$

$$\leq \left(\sum_{k=0}^{n} \|T^{n-k}\|^q\right)^{p/q} \left(\sum_{k=0}^{n} \|z_k\|^p\right) + \sum_{k=n+1}^{\infty} \|z_k\|^p$$
$$\leq \left(\sum_{k=0}^{n} \|T^k\|^q\right)^{p/q} \sum_{k=0}^{\infty} \|z_k\|^p.$$

Thus $\|S^n\| \leq \left(\sum_{k=0}^{n} \|T^k\|^q\right)^{1/q} \leq \sum_{k=0}^{n} \|T^k\|$. If T satisfies condition $P(s)$ or $E(s)$, then it follows that there is a constant K so that for every $n \geq 1$, $\|S^n\|$ is bounded by, respectively, $K\,n^{s+1}$ or $K\,e^{n^t}$ for all t, $s < t < 1$.

The assumption of uniform boundedness of the projections $Q_n = T^n L^n$ enables us control the growth of $\|L^n\|$. Specifically, if $\|Q_n\| \leq M$ for all n, then $\|L^n\| \leq \kappa(T^n)^{-1}\|Q_n\| \leq M\kappa(T^n)^{-1}$ for every $n \geq 1$. Now we can obtain the desired estimate for $\|S^{-n}\|$. Let $(z_k)_{k\geq 0} \in Z^p$ and compute:

$$\|S^{-n}(z_k)_{k\geq 0}\|^p = \|L^n z_0\|^p + \sum_{k=1}^{n} \|(I-Q)L^{n-k} z_0\|^p + \sum_{k=1}^{\infty} \|z_k\|^p$$
$$\leq \|I-Q\|^p \sum_{k=0}^{n} \|L^{n-k} z_0\|^p + \sum_{k=1}^{\infty} \|z_k\|^p$$
$$\leq \|I-Q\|^p \left(\sum_{k=0}^{n} \|L^k\|^p\right) \|z_0\|^p + \sum_{k=1}^{\infty} \|z_k\|^p$$
$$\leq \|I-Q\|^p \left(\sum_{k=0}^{n} \|L^k\|^p\right) \sum_{k=0}^{\infty} \|z_k\|^p$$
$$\leq \|I-Q\|^p \left(\sum_{k=0}^{n} M^p \kappa(T^k)^{-p}\right) \sum_{k=0}^{\infty} \|z_k\|^p.$$

Therefore, $\|S^{-n}\| \leq M\|I-Q\| \left(\sum_{k=0}^{n} \kappa(T^k)^{-p}\right)^{1/p} \leq M\|I-Q\| \sum_{k=0}^{n} \kappa(T^k)^{-1}$, and so lower bounds for $\kappa(T^n)$ give upper bounds for $\|S^{-n}\|$ of the same kind. The theorem now follows from the classical results of Colojoară and Foiaş, [**3,** 5.3.2 and 5.3.4]. □

Of course, the assumption in Theorem 3 that the sequence of projections $Q_n = T^n L^n$ be uniformly bounded can be relaxed. Specifically, if T satisfies $P(s)$, and if $\|Q_n\| = \mathcal{O}(n^s)$, then the extension S is $\mathcal{E}(\mathbb{T})$–scalar. Also, if T satisfies $E(s)$ for some $0 < s < 1$ and if $\|Q_n\| = \mathcal{O}(e^{n^s})$, then S is super–decomposable. The stronger assumption of the theorem in fact holds in many cases however. Two examples are given below.

An important special case of the theorem is given in the following result, which includes a partial solution of [**8,** 6.1.15]. Condition (D) below was introduced by Didas, who showed, with completely different methods, that for operators with (D), the growth condition $P(s)$ for some $s > 0$ characterizes $\mathcal{E}(\mathbb{T})$–subscalarity, [**5,** 2.2.11]. If X is a Hilbert space and if $T \in \mathcal{L}(X)$ is bounded below, then it is natural to choose Y to be the orthogonal complement of $\operatorname{ran} T$, $Y = \ker L = \ker T^*$, and to consider the Hilbert space extension of T, namely $S \in \mathcal{L}(Z^2)$.

COROLLARY 4. *Suppose that the Hilbert space operator $T \in \mathcal{L}(X)$ is bounded below and satisfies*

$$(D) \qquad T^*T^{n+1}X \subseteq T^nX$$

for every $n \in \mathbb{N}$. Then:

(i) *The extension $S \in \mathcal{L}(Z^2)$ is $\mathcal{E}(\mathbb{T})$–scalar provided that T satisfies $P(s)$ for some $s > 0$; in particular, T is $\mathcal{E}(\mathbb{T})$–subscalar if and only if T satisfies $P(s)$ for some $s > 0$.*

(ii) *$S \in \mathcal{L}(Z^2)$ is super–decomposable if T satisfies $E(s)$ for some s, $0 < s < 1$.*

PROOF. We wish to show that if $Y = \ker L = \ker T^*$, then the assumption (D) implies that $\ker L^n = \ker T^{*n}$ for all n. In this case, Q_n is the orthogonal projection onto T^nX, and Theorem 3 applies. Suppose that $x \in X$ and $y \in \ker T^*$. Then for every k, $0 \le k < n$, $\langle T^k y, T^n x\rangle = \langle y, T^{*k}T^n x\rangle = 0$ because $T^{*k}T^n x \in TX$ by (D). Since x is arbitrary, it follows that $T^k \ker L = T^k \ker T^* \subseteq \ker T^{*n}$ and therefore, by Proposition 2, $\ker L^n = \ker L + T\ker L + \cdots + T^{n-1}\ker L \subseteq \ker T^{*n}$. Since $T^nX \oplus \ker T^{*n} = X = T^nX \oplus \ker L^n$, and $\ker L^n \subseteq \ker T^{*n}$, equality must obtain. □

Notice that the Didas condition (D) says simply that the spaces $\operatorname{ran} T^n$ are all T^*T–invariant. Examples of Hilbert space operators T satisfying (D) include normal operators, since $T^*T = TT^*$ in this case, and isometries, since $T^*T = I$. A common generalization is given by the class of quasinormal operators: T is quasinormal provided that T^*T commutes with T, and so quasinormality evidently implies (D). It is well known that quasinormal operators are subnormal; in fact, a Hilbert space operator T is subnormal if and only if T has a quasinormal extension; see [**4**, II.1.7 and II.1.9]. Condition (D) generally fails for subnormal operators though. An easy example is the operator on H^2 defined by $f(z) \mapsto (z-\lambda)f(z)$ for any λ satisfying $0 < |\lambda| < 1$.

Moreover, every weighted right shift T on $\ell^2(\mathbb{N}_0)$ such that $\kappa(T) > 0$ satisfies condition (D), since $\operatorname{ran} T^n = \{(x_k)_{k\ge 0} : x_k = 0 \text{ for all } k < n\}$ and T^* is a left shift. A similar idea also applies to weighted right shifts on $\ell^p(\mathbb{N}_0)$, $1 \le p < \infty$. Proposition 5 below is based on the fact that a weighted right shift has as a left inverse a weighted left shift. For the convenience of the reader, we recall some of the basic facts regarding weighted shifts below.

Let $\mathbb{K}$ denote either the set of integers $\mathbb{Z}$ or the set $\mathbb{N}_0 = \{0, 1, 2, \dots\}$, and let $(e_k)_{k\in\mathbb{K}}$ denote the canonical basis for $\ell^p(\mathbb{K})$, $1 \le p < \infty$. Given a bounded sequence $(\omega_n)_{n\in\mathbb{K}}$ of strictly positive weights, we define the corresponding weighted right shift on $\ell^p(\mathbb{K})$, $1 \le p < \infty$, by $Te_n = \omega_n e_{n+1}$, for every $n \in \mathbb{K}$. Weighted shifts acting on $\ell^p(\mathbb{N}_0)$ and $\ell^p(\mathbb{Z})$ are referred to as unilateral and bilateral, respectively. Upper and lower bounds of T^n are given in terms of products of the weights: if $\alpha_n := \omega_0 \cdots \omega_{n-1}$ for $n > 0$, $\alpha_0 := 1$ and, in the bilateral case, $\alpha_n := (\omega_n \cdots \omega_{-1})^{-1}$ for $n < 0$, then $T^n e_k = \alpha_{n+k}/\alpha_k\, e_{n+k}$ for all $n \ge 0$ and $k \in \mathbb{K}$, and therefore $\|T^n\| = \sup_{k\in\mathbb{K}} \frac{\alpha_{n+k}}{\alpha_k}$ and $\kappa(T^n) = \inf_{k\in\mathbb{K}} \frac{\alpha_{n+k}}{\alpha_k}$ for all $n \ge 0$. We refer the reader to Shields's article [**11**] for a survey of the theory of weighted shifts on Hilbert spaces; for unilateral weighted shifts on $\ell^p(\mathbb{N})$, $p \ne 2$, see [**8**, 1.6] and [**10**] as well.

Let T be an unilateral right shift on $\ell^p(\mathbb{N}_0)$ with weight sequence $(\omega_n)_{n\ge 0}$ such that $\kappa(T) = \inf_k \omega_k > 0$. If $Y = \ker L$ is chosen to be $Y = \operatorname{span}(e_0)$, then the invertible extension $S \in \mathcal{L}(Z^p)$ of T has a particularly simple representation. The space $Z^p = \ell^p(\mathbb{N}_0) \oplus Y \oplus Y \oplus \cdots$ is isometrically isomorphic to $\ell^p(\mathbb{Z})$ via

$$\big((x_k)_{k\geq 0}, z_1\, e_0, z_2\, e_0, \dots\big) \leftrightarrow (\dots, z_2, z_1, \boxed{x_0}, x_1, x_2, \dots),$$

where the boxed entry indicates the 0th component of the doubly infinite sequence in $\ell^p(\mathbb{Z})$. In this representation, the invertible extension of T is just the bilateral shift $S : \ell^p(\mathbb{Z}) \to \ell^p(\mathbb{Z})$ given by

$$\begin{aligned} S(\dots, x_{-n}, x_{-n+1}, \dots, x_{-1}, \boxed{x_0}, x_1, \dots, x_n, \dots) \\ = (\dots, x_{-n-1}, x_{-n}, \dots, x_{-2}, \boxed{x_{-1}}, \omega_0\, x_0, \dots, \omega_{n-1}\, x_{n-1}, \dots). \end{aligned}$$

PROPOSITION 5. *Suppose that T is a unilateral weighted right shift on $\ell^p(\mathbb{N}_0)$ with weight sequence $(\omega_n)_{n\geq 0}$, and define a bilateral shift S on $\ell^p(\mathbb{Z})$ by*

$$Se_n = \begin{cases} \omega_n e_{n+1} & \text{if } n \geq 0, \text{ and} \\ e_{n+1} & \text{if } n < 0. \end{cases}$$

Then S is $\mathcal{E}(\mathbb{T})$–scalar if and only if there exist constants $K, s > 0$ such that

$$P(s) \qquad \frac{1}{K\, n^s} \leq \inf_{k\geq 0} \omega_k \cdots \omega_{n+k-1} \leq \sup_{k\geq 0} \omega_k \cdots \omega_{n+k-1} \leq K\, n^s$$

for every $n \geq 1$, and in this case, S satisfies the same growth condition $P(s)$. Similarly, S is super–decomposable provided that for some K and s, $0 < s < 1$,

$$E(s) \qquad \frac{1}{K\, e^{n^s}} \leq \inf_{k\geq 0} \omega_k \cdots \omega_{n+k-1} \leq \sup_{k\geq 0} \omega_k \cdots \omega_{n+k-1} \leq K\, e^{n^s}$$

for all $n \geq 1$, and S satisfies the same growth condition $E(s)$ as well.

In particular, T is $\mathcal{E}(\mathbb{T})$–subscalar precisely when T satisfies condition $P(s)$ for some $s > 0$, while condition $E(s)$ for some s with $0 < s < 1$ ensures that T has Bishop's property (β).

PROOF. Rather than appeal to Theorem 3, we estimate $\|S^n\|$ and $\kappa(S^n)$ directly. As before, let $\alpha_0 = 1$ and $\alpha_n = \omega_0 \cdots \omega_{n-1}$ if $n \geq 1$. Let $(\omega'_k)_{k\in\mathbb{Z}}$ denote the weight sequence for S and let $(\alpha'_k)_{k\in\mathbb{Z}}$ be the corresponding sequence of weight products. Then, for every $n \geq 0$,

$$\alpha'_{n+k}/\alpha'_k = \begin{cases} 1 & \text{if } k < -n, \\ \alpha_{n+k} & \text{if } -n \leq k < 0, \\ \alpha_{n+k}/\alpha_k & \text{if } k \geq 0. \end{cases}$$

Therefore,

$$\|S^n\| = \sup_{k\in\mathbb{Z}} \alpha'_{n+k}/\alpha'_k \leq \max\{\alpha_j\}_{j=0}^{n-1} \cup \{\|T^n\|\} \leq \max_{0\leq j\leq n} \|T^n\|,$$

and similarly,

$$\kappa(S^n) = \inf_{k\in\mathbb{Z}} \alpha'_{n+k}/\alpha'_k \geq \min\{\alpha_j\}_{j=0}^{n-1} \cup \{\kappa(T^n)\} \geq \min_{0\leq j\leq n} \kappa(T^n).$$

Thus, if T satisfies $P(s)$ for some $s > 0$ (resp. $E(s)$, $0 < s < 1$), then S satisfies the same condition $P(s)$, (resp. $E(s)$, $0 < s < 1$). The remaining statements now follow from Proposition 1. ☐

Of course, bilateral shifts satisfying a growth condition $P(s)$ or (B) are invertible, and so the classical results of Colojoară and Foiaş apply directly. By [**8,** 1.6.14], a unilateral weighted shift T is decomposable only in the trivial case that $\sigma(T) = \{0\}$, and is never generalized scalar by [**8,** 1.5.10]. While the $\mathcal{E}(\mathbb{T})$–subscalar weighted shifts are characterized by polynomial growth, $P(s)$, no growth condition can possibly characterize the subscalar unilateral weighted shifts. Indeed, every weighted shift T with non-decreasing weight sequence $(\omega_n)_{n\in\mathbb{K}}$ on $\ell^2(\mathbb{K})$, is hyponormal, [**11**], and therefore subscalar, [**8,** 3.4.3 and 3.4.5].

Two classical examples to which the last proposition applies are the Bergman and Dirichlet shifts. The multiplication operator $M_z f(z) := z\,f(z)$ on the space of analytic, square-integrable functions on the unit disc is subnormal and unitarily equivalent to the unilateral shift $Te_n := \sqrt{\frac{n+1}{n+2}}\,e_{n+1}$, on $\ell^2(\mathbb{N}_0)$. The Dirichlet shift corresponds to the unilateral shift $Te_n := \sqrt{\frac{n+2}{n+1}}\,e_{n+1}$ for all $n \geq 0$. In both cases it is clear that the growth condition $P(s)$ holds for $s = 1/2$. Thus, by Proposition 5, the Bergman and Dirichlet shifts each admits a bilateral shift extension with a functional calculus on $C^2(\mathbb{T})$. In contrast, the spectrum of the minimal normal extension of the Bergman shift, $Nf(z) := z\,f(z)$ on $L^2(\mathbb{D}, \pi^{-1}dA)$, is the closed unit disc $\overline{\mathbb{D}}$.

We conclude with three open questions.

1. Does the polynomial growth condition $P(s)$ characterize $\mathcal{E}(\mathbb{T})$–subscalar operators?
2. Does the Beurling condition (B) imply Bishop's property (β)? In other words, is every operator which satisfies (B) similar to the restriction of a decomposable operator to a closed invariant subspace?
3. Does every unilateral weighted right shift with convergent weight sequence $(\omega_n)_{n\geq 0}$ have Bishop's property (β)?

It seems that the first question is open even in the case of Hilbert space operators, and the second is unresolved even for unilateral weighted right shifts. For partial results regarding the third question, we refer to [**10**].

References

1. E. Albrecht and J. Eschmeier, *Analytic functional models and local spectral theory*, Proc. London Math. Soc. (3) **75** (1997), 323–348.
2. H. Bercovici and S. Petrović, *Generalized scalar operators as dilations*, Proc. Amer. Math. Soc. **123** (1995), 2173–2180.
3. I. Colojoară and C. Foiaş, *Theory of Generalized Spectral Operators*, Gordon and Breach, New York, 1968.
4. J. B. Conway, *The Theory of Subnormal Operators*, Mathematical Surveys and Monographs **36**, Amer. Math. Soc., Providence, RI, 1991.
5. M. Didas, *$\mathcal{E}(\mathbb{T}^n)$–subscalar n–tuples and the Cesàro operator on H^p*, Annales Universitatis Saraviensis, Series Mathematicae **10** (2000), 285–335.
6. R. G. Douglas, *On extending commutative semigroups of isometries*, Bull. London Math. Soc. **1** (1969), 157–159.
7. J. Eschmeier and M. Putinar, *Spectral Decompositions and Analytic Sheaves*, Clarendon Press, Oxford, 1996.
8. K. B. Laursen and M. M. Neumann, *An Introduction to Local Spectral Theory*, Clarendon Press, Oxford, 2000.
9. M. Martin and M. Putinar, *Lectures on Hyponormal Operators*, Birkhäuser, Basel, 1989.
10. T. L. Miller, V. G. Miller and M. M. Neumann, *Local spectral properties of weighted shifts*, preprint.

11. A. L. Shields, *Weighted shift operators and analytic function theory*, Topics in Operator Theory (C. Pearcy, ed.), Mathematical Surveys **13**, Amer. Math. Soc., Providence, RI, 1974, pp. 49–128.

Department of Mathematics and Statistics, Mississippi State University, Drawer MA, Mississippi State, MS 39762, USA

E-mail address: miller@math.msstate.edu, vivien@math.msstate.edu, and neumann@math.msstate.edu

Contemporary Mathematics
Volume **321**, 2003

Differences of Composition Operators

Jennifer Moorhouse and Carl Toews

ABSTRACT. We consider composition operators acting on the Hardy space and all the standard weighted Bergman spaces, and present a sufficient condition for the difference of two such operators to be compact. The proof technique lends itself to estimating norms of composition operator differences, and this extension is used to draw some conclusions about the topological structure of the set of composition operators. In particular, a conjecture of Joel Shapiro and Carl Sundberg is answered negatively.

Let D be the open unit disk in the complex plane, B a Banach space of analytic functions on D, and $\mathcal{C}(B)$ the set of bounded composition operators with analytic symbol acting on B. The component structure of $\mathcal{C}(H^p)$ in the topology induced by the operator norm was first studied by Earl Berkson in [B]; subsequent investigation extended to the component structure of $\mathcal{C}(B)$ for a variety of different Banach spaces B in a variety of different topologies (see [SS],[M],[MOZ],and [HJM].) In the setting of the norm topology on $\mathcal{C}(H^2)$, Joel Shapiro and Carl Sundberg conjecture that two operators C_ϕ and C_ψ lie in the same component if and only if they have compact difference (see [SS]). In [S], Jonathan Shapiro investigates compact composition operator differences in the setting of the Hardy space using Aleksandrov measures, and in [G], Tom Goebeler characterizes the same for composition operators acting on mixed Hardy spaces. Here we use different techniques to explore compact difference for composition operators acting on a range of spaces, including the Hardy space and all the standard weighted Bergman spaces. We provide a sufficient condition for compactness in terms of weighted composition operators acting between these spaces, and use this condition to provide a counterexample to the Shapiro-Sundberg conjecture.

Recall that the Hardy space H^2 consists of those functions f, analytic on the unit disk, which satisfy $\|f\|_{H^2}^2 := \lim_{r\to 1}\int_{\partial D}|f(r\zeta)|^2 d\sigma(\zeta) < \infty$, where σ is normalized Lebesgue measure on the boundary of the disk. For $\alpha > -1$, the standard weighted Bergman space A_α^2 is the set of functions analytic on the disk with $\|f\|_{A_\alpha^2}^2 := \int_D |f(z)|^2 d\lambda_\alpha(z) < \infty$, where $d\lambda_\alpha(z)$ is the weighted area measure $\frac{(\alpha+1)}{\pi}(1-|z|^2)^\alpha dA(z)$. Using power series (see [CM]), it is not hard to show that

2000 *Mathematics Subject Classification.* 47B33.

Key words and phrases. composition operator, Hardy space, Bergman space.

This paper is part of the authors' doctoral work, written at the University of Virginia under the direction of Thomas Kriete and Barbara MacCluer.

if $f \in A^2_\alpha$, then $f' \in A^2_{\alpha+2}$ with $\|f'\|_{A^2_{\alpha+2}} \leq M_\alpha \|f\|_{A^2_\alpha}$ for some constant M_α that does not depend on f. Similarly, if $f \in H^2$, then $f' \in A^2_1$ and $\|f'\|_{A^2_1} \leq M_H \|f\|_{H^2}$ for a constant M_H independent of f.

We will make use of the following, whose proof is a slight modification of the proof of Proposition 3.11 in [CM].

PROPOSITION 1. *Let $\mathcal{X}$ and $\mathcal{Y}$ each represent either the Hardy space or a standard weighted Bergman space. Then a finite sum of weighted composition operators $\sum w_i C_{\phi_i}$ is compact from $\mathcal{X}$ to $\mathcal{Y}$ if and only if whenever $\{f_n\}$ is bounded in $\mathcal{X}$ and $f_n \to 0$ uniformly on compact subsets of D, then $(\sum w_i C_{\phi_i}) f_n \to 0$ in $\mathcal{Y}$.*

THEOREM 1. *Let ϕ and ψ be analytic self maps of the disk, and define $\phi_s(z) := s\phi(z) + (1-s)\psi(z)$ for $0 \leq s \leq 1$. Let $w(z)$ denote the bounded analytic function $\phi(z) - \psi(z)$. If the weighted composition operators $wC_{\phi_s} : A^2_{\alpha+2} \to A^2_\alpha$ are uniformly norm bounded in s and, moreover, compact for each s, then $C_\phi - C_\psi$ is compact from A^2_α to A^2_α. Further, the result holds on the Hardy space, provided the operators $wC_{\phi_s} : A^2_1 \to H^2$ satisfy the given conditions.*

PROOF. First consider the weighted Bergman spaces. Fix an $f \in A^2_\alpha$ and write

$$(C_\phi - C_\psi)f(z) = f(\phi(z)) - f(\psi(z)) = w(z) \int_0^1 f'(\phi_s(z))ds,$$

whence

$$\begin{aligned}
\|(C_\phi - C_\psi)f\|^2_{A^2_\alpha} &= \int_D |(C_\phi - C_\psi)f(z)|^2 \, d\lambda_\alpha(z) \\
&= \int_D \left| w(z) \int_0^1 f'(\phi_s(z))ds \right|^2 d\lambda_\alpha(z) \\
&\leq \int_D \int_0^1 |w(z)f'(\phi_s(z))|^2 \, ds d\lambda_\alpha(z) \\
&= \int_0^1 \|wC_{\phi_s} f'\|^2_{A^2_\alpha} ds.
\end{aligned}$$

If f_n is any bounded sequence in A^2_α converging almost uniformly to zero, then f'_n is a bounded sequence in $A^2_{\alpha+2}$ which converges almost uniformly to zero; the uniform boundedness of the norms of wC_{ϕ_s} acting from $A^2_{\alpha+2}$ to A^2_α now permits us to apply Lebesgue's Dominated Convergence Theorem to conclude:

$$\begin{aligned}
\limsup_{n\to\infty} \|(C_\phi - C_\psi)f_n\|^2_{A^2_\alpha} &\leq \limsup_{n\to\infty} \int_0^1 \|wC_{\phi_s} f'_n\|^2_{A^2_\alpha} ds \\
&= \int_0^1 \limsup_{n\to\infty} \|wC_{\phi_s} f'_n\|^2_{A^2_\alpha} ds \\
&= 0
\end{aligned}$$

where the last line follows from the compactness of $wC_{\phi_s} : A^2_{\alpha+2} \to A^2_\alpha$ and the "only if" part of Proposition 1. The compactness of $C_\phi - C_\psi : A^2_\alpha \to A^2_\alpha$ now follows from the "if" part of the same.

For the case of the Hardy space, replace integration over D by the integral that defines the Hardy space norm; the same calculations then apply. □

Remark: Since the hypothesis of uniform norm boundedness was only used to apply Dominated Covergence in the second line of the last display, it is clear that this assumption can be weakened. Also note that the same techniques can be used to show that $C_{\phi_s} - C_{\phi_t}$ is compact for any $0 \leq s \leq t \leq 1$.

The utility of this condition lies in the fact that the boundedness and compactness of these operators lend themselves to a simple Carleson-measure characterization. Recall that if μ is a positive, finite Borel measure on the open disk D, then $A^2_\alpha \subset L^2(\mu)$ if and only if μ is an $\alpha-$*Carleson measure*, i.e if and only if

$$\|\mu\|_\alpha := \sup_{S(\zeta,\delta)} \frac{\mu(S(\zeta,\delta))}{\delta^{\alpha+2}} < \infty, \tag{1}$$

where the supremum is over all Carleson sets $S(\zeta,\delta) := \{z \in D : |z-\zeta| < \delta, \zeta \in \partial D\}$. In this case, the inclusion map $I_\alpha : A^2_\alpha \to L^2(\mu)$ is bounded with norm comparable to $\|\mu\|_\alpha$. Analogously, we say that μ is a *compact* $\alpha-$*Carleson measure* if the ratio in (1) goes to 0 uniformly in ζ as $\delta \to 0$, and it can be shown that this is equivalent to the map I_α being compact (see [MS], Theorem 4.3).

To apply these results to the weighted composition operator $wC_\phi : A^2_\alpha \to A^2_\beta$, where ϕ is an analytic self map of D and w is a bounded analytic function on D, employ a formal change of variable (see [H], pg. 163, Theorem C) to write

$$\begin{aligned}\|wC_\phi f\|^2_\beta &= \int_D |w(z)|^2 |f(\phi(z))|^2 d\lambda_\beta(z) \\ &= \int_D |f(z)|^2 d(|w|^2\lambda_\beta\phi^{-1})(z) \\ &= \|f\|^2_{L^2(|w|^2\lambda_\beta\phi^{-1})},\end{aligned}$$

where $|w|^2\lambda_\beta\phi^{-1}$ is the measure on the open disk D that assigns to each Borel set F the measure

$$|w|^2\lambda_\beta\phi^{-1}(F) := \int_{\phi^{-1}(F)} |w(z)|^2 d\lambda_\beta(z) \tag{2}$$

It follows that $wC_\phi : A^2_\alpha \to A^2_\beta$ will be bounded (respectively compact) if and only if $|w|^2\lambda_\beta\phi^{-1}$ is an $\alpha-$Carleson (compact α-Carleson) measure (cf. [MS], Corollary 4.4).

In the case where wC_ϕ acts from a weighted Bergman space to the Hardy space, we define a measure $|w|^2\sigma\phi^{-1}$ on the closed disk $\overline{D}$ that assigns to each Borel set F the measure

$$|w|^2\sigma\phi^{-1}(F) := \int_{\phi^{*-1}(F)} |w^*(\zeta)|^2 d\sigma(\zeta) \tag{3}$$

where ϕ^* and w^* denote the radial limit functions of ϕ and w. If ϕ has radial limits of modulus strictly less than 1 almost everywhere, this measure is actually supported on the open unit disk. Supposing this to be the case, the same comments as above show that the map $I_\alpha : A^2_\alpha \to L^2(|w|^2\sigma\phi^{-1})$ is bounded (compact) if and only if $|w|^2\sigma\phi^{-1}$ is an $\alpha-$Carleson (compact $\alpha-$Carleson) measure. Assuming for the moment that p is a polynomial in A^2_α, a formal change of variable yields

$$\begin{aligned}\|wC_\phi p\|^2_{H^2} &= \lim_{r\to 1}\int_{\partial D}|w(r\zeta)|^2\cdot|p(\phi(r\zeta))|^2 d\sigma(\zeta)\\ &= \int_{\partial D}|w^*(\zeta)|^2\cdot|p(\phi^*(\zeta))|^2 d\sigma(\zeta)\\ &= \int_D |p(z)|^2 d(|w|^2\sigma\phi^{-1})(z)\\ &= \|p\|^2_{L^2(|w|^2\sigma\phi^{-1})}\end{aligned}$$

where the second equality is justified by the fact that the integrand belongs to H^2 (see [D].) Since the polynomials are dense in A^2_α, and for each polynomial $\|wC_\phi p\|_{H^2} = \|p\|_{L^2(|w|^2\sigma\phi^{-1})}$, it follows that for ϕ with radial limits of modulus strictly less than one almost everywhere, $wC_\phi : A^2_\alpha \to H^2$ is bounded (compact) if and only if $|w|^2\sigma\phi^{-1}$ is an $\alpha-$Carleson (compact $\alpha-$Carleson) measure.

Note that if ϕ has radial limits of modulus one on a set of positive measure, then $wC_\phi : A^2_\alpha \to H^2$ can be bounded only if w is uniformly zero. This can be shown using the fact that each weighted Bergman space contains functions f such that for each curve Γ in D tending to ∂D, f assumes arbitrarily large values at points of Γ (see [BES]). It is easy to see that if f is such a function and for some $\zeta \in \partial D$ both $|\phi^*(\zeta)| = 1$ and $w^*(\zeta) \neq 0$, then $wf\circ\phi$ cannot have a radial limit at ζ. In particular, if ϕ has radial limits of modulus 1 on a set E of positive measure, then either $w \equiv 0$ or $w^* \neq 0$ almost everywhere on E; in the latter case, $wf\circ\phi$ has no radial limits on a set of positive measure, i.e. $wf\circ\phi \notin H^2$.

We summarize this discussion in the following proposition. The case where wC_ϕ acts from A^2_α to itself is included as a special case of (a); we include (c) to cover the case $wC_\phi : H^2 \to H^2$. Although here we have not explicitly touched on (c), it follows directly from the proof of Theorem 3.12 in [CM].

PROPOSITION 2. *Let $\phi : D \to D$ be analytic, and suppose w is a bounded analytic function, not identically zero on D. Let the measures $|w|^2\lambda_\beta\phi^{-1}$ and $|w|^2\sigma\phi^{-1}$ be as in* (2) *and* (3)*, respectively. Then:*
(a) $wC_\phi : A^2_\alpha \to A^2_\beta$ is bounded if and only if

$$\left\||w|^2\lambda_\beta\phi^{-1}\right\|_\alpha := \sup_{S(\zeta,\delta)}\frac{|w|^2\lambda_\beta\phi^{-1}(S(\zeta,\delta))}{\delta^{\alpha+2}} < \infty$$

(b) $wC_\phi : A^2_\alpha \to H^2$ is bounded if and only ϕ has radial limits of modulus strictly less than 1 almost everywhere and

$$\left\||w|^2\sigma\phi^{-1}\right\|_\alpha := \sup_{S(\zeta,\delta)}\frac{|w|^2\sigma\phi^{-1}(S(\zeta,\delta))}{\delta^{\alpha+2}} < \infty$$

(c) $wC_\phi : H^2 \to H^2$ is bounded if and only if

$$\left\||w|^2\sigma\phi^{-1}\right\|_{H^2} := \sup_{S(\zeta,\delta)}\frac{|w|^2\sigma\phi^{-1}(S(\zeta,\delta))}{\delta} < \infty.$$

In all cases, the supremum is comparable to the norm of wC_ϕ acting on the appropriate spaces, and if the displayed quotient goes to 0 uniformly in ζ as $\delta \to 0$, then wC_ϕ is compact on these spaces.

Example: Consider $\phi(z) = (1+z)/2$, and $\psi(z) = \phi(z) + t(z-1)^b$, where $b \geq 2$ and t is small, as in [CM] pg. 337. (In particular, we may take $t < 1/128$.) It is known that on both the Hardy space and weighted Bergman spaces, $C_\phi - C_\psi$ is compact for $b \geq 2.5$ and not compact for $b = 2$. (See [CM], Ex. 9.3.3 for the Hardy space case, from which the Bergman case follows readily.) We use Theorem 1 and the Carleson conditions of Proposition 2 to resolve the intermediate cases.

We begin with the Bergman cases. By Theorem 1, $C_\phi - C_\psi : A^2_\alpha \to A^2_\alpha$ will be compact if the weighted composition operators $wC_{\phi_s} : A^2_{\alpha+2} \to A^2_\alpha$ are compact and uniformly norm bounded in s for $s \in [0,1]$, where

$$w(z) = t(z-1)^b \tag{4}$$

and

$$\phi_s(z) = \frac{(1+z)}{2} + t(1-s)(z-1)^b \tag{5}$$

We will verify compactness and uniform boundedness by explictly computing the Carleson quotient of Proposition 2(a).

In what follows, M represents a constant whose value may change from line to line, but which is always independent of ζ, δ and s. To estimate the integral defining $|w|^2 \lambda_\alpha \phi_s^{-1}(S(\zeta,\delta))$, note that if $z \in \phi_s^{-1}(S(\zeta,\delta))$, then $1 - |\phi_s(z)| \leq \delta$, and simple calculations show that

$$1 - |\phi_s(z)| \geq \frac{1}{16}|z-1|^2.$$

A little algebra now reveals that on $\phi_s^{-1}(S(\zeta,\delta))$,

$$|w(z)|^2 = t^2|z-1|^{2b} \leq t^2(16(1-|\phi_s(z)|))^b \leq M\delta^b. \tag{6}$$

Moreover, setting $w = 1$ and $\alpha = \beta$ in Proposition 2(a) shows that

$$\int_{\phi_s^{-1}(S(\zeta,\delta))} d\lambda_\alpha(z) \leq M\delta^{\alpha+2}\|C_{\phi_s}\|_\alpha, \tag{7}$$

where $\|C_{\phi_s}\|_\alpha$ denotes the norm of C_{ϕ_s} acting from A^2_α to itself and M is a constant depending only on α. Now it is known that for any analytic self map ϕ of the disk, the norm $\|C_\phi\|_\alpha$ is less than some multiple of a power of $\frac{1}{1-|\phi(0)|}$, where the multiple and the power depend only on α. Since for all s, $|\phi_s(0)| \leq 1/2 + t$ which is bounded uniformly away from 1, the norms $\|C_{\phi_s}\|_\alpha$ are uniformly bounded in s.

Putting everything together we have, for fixed δ,

$$\begin{aligned}
\sup_\zeta \frac{|w|^2\lambda_\alpha \phi_s^{-1}(S(\zeta,\delta))}{\delta^{\alpha+4}} &= \sup_\zeta \frac{\int_{\phi_s^{-1}(S(\zeta,\delta))} |w(z)|^2 d\lambda_\alpha(z)}{\delta^{\alpha+4}} \\
&\leq \frac{M\delta^b}{\delta^{\alpha+4}} \int_{\phi_s^{-1}(S(\zeta,\delta))} d\lambda_\alpha(z) && \text{(by 6)} \\
&\leq M\delta^{b-2} && \text{(by 7)} \qquad (8)
\end{aligned}$$

where the constant M in the last line depends only on α. We conclude that the operators $wC_{\phi_s} : A^2_{\alpha+2} \to A^2_\alpha$ are uniformly bounded in s for $b \geq 2$, and compact for $b > 2$. It follows that $C_\phi - C_\psi : A^2_\alpha \to A^2_\alpha$ is compact for $b > 2$.

For the Hardy space, note that ϕ_s has radial limits of modulus strictly less than one almost everywhere, whence Proposition 2(b) shows that $wC_{\phi_s} : A^2_1 \to H^2$ will be bounded or compact as $|w|^2\sigma\phi_s^{-1}$ is a bounded or compact 1−Carleson measure.

Now estimate $|w^*(z)|$ for $z \in \phi_s^{*-1}(S(\zeta,\delta))$ by exactly the same technique that led to (6), and use Proposition 2(c) to show, as in (7), that

$$\int_{\phi_s^{*-1}(S(\zeta,\delta))} d\sigma(\zeta) \leq M\delta \|C_{\phi_s}\|_{H^2},$$

where $\|C_{\phi_s}\|_{H^2}$ denotes the norm of C_ϕ acting from H^2 to itself. Finally, invoke the uniform boundedness of these norms just as above to extract the same conclusion as in (8):

$$\sup_{\zeta} \frac{|w|^2 \sigma \phi_s^{-1}(S(\zeta,\delta))}{\delta^3} \leq M\delta^{b-2} \tag{9}$$

It follows that the $wC_{\phi_s} : A_1^2 \to H^2$ are uniformly bounded in s for $b \geq 2$ and compact for $b > 2$. We conclude that $C_\phi - C_\psi : H^2 \to H^2$ is compact for $b > 2$.

We now use the proof of Theorem 1 to obtain a sufficient condition for two composition operators to be path connected.

COROLLARY 1. *Let ϕ and ψ be analytic self maps of the disk, and define $\phi_s(z) := s\phi(z) + (1-s)\psi(z)$ for $0 \leq s \leq 1$. Let $w(z)$ denote the bounded analytic function $\phi(z) - \psi(z)$. If the operators $wC_{\phi_s} : A_{\alpha+2}^2 \to A_\alpha^2$ are uniformly norm bounded in s, then C_{ϕ_s} is an arc of composition operators in $\mathcal{C}(A_\alpha^2)$. The same conclusion holds in $\mathcal{C}(H^2)$, provided the operators $wC_{\phi_s} : A_1^2 \to H^2$ are uniformly norm bounded.*

PROOF. This follows from the calculations of Theorem 1 and the observation that

$$(C_{\phi_s} - C_{\phi_t})f(z) = (\phi(z) - \psi(z)) \int_s^t f'(\phi_r(z))dr.$$

In particular, if f is a unit vector in A_α^2, then

$$\begin{aligned}
\|(C_{\phi_s} - C_{\phi_t})f\|_{A_\alpha^2}^2 &= \int_D |(C_{\phi_s} - C_{\phi_t})f(z)|^2 d\lambda_\alpha(z) \\
&= \int_D \left|(\phi(z) - \psi(z)) \int_s^t f'(\phi_r(z))dr\right|^2 d\lambda_\alpha(z) \\
&\leq \int_D \int_s^t |w(z)f'(\phi_r(z))|^2 dr d\lambda_\alpha(z) \\
&= \int_s^t \|wC_{\phi_r} f'\|_{A_\alpha^2}^2 dr.
\end{aligned}$$

By hypothesis, the integrand in the last line is bounded independently of f, by which the conclusion follows for the weighted Bergman spaces. For the Hardy space, start with a unit vector $f \in H^2$, replace the integral over D by the integral that defines the Hardy space norm, and proceed identically. □

COROLLARY 2. *On both the Hardy space and any of the standard weighted Bergman spaces, there exist composition operators whose difference is not compact, yet which are arc-connected.*

PROOF. Consider the ϕ and ψ of the Example. As noted, $C_\phi - C_\psi$ is not compact for $b = 2$. However, the Carleson measure calculations in this example (in particular, lines (8) and (9)) show that for $b = 2$, the hypotheses of Corollary 1 are satisfied. The result follows. $\square$

Final Remarks: It is shown in [MOZ] that the maps of Corollary 2 induce an arc in $C(H^\infty)$ under the supremum norm. Using this fact, it is possible to obtain an alternative proof to the Hardy space case of Corollary 2 via an interpolation argument, since $(H^1, H^\infty)_\theta = H^2$ for $\theta = 1/2$ in the Calderon method of complex interpolation.

After this paper was accepted for publication, the authors were made aware of a preprint by Paul Bourdon [Bo] giving many such examples on the Hardy space. Namely, Bourdon shows that if ϕ and ψ are any two linear fractional maps with finite angular derivative at a point ζ on the boundary of the disk, having the same first order data but different second derivatives at ζ, then $C_\phi - C_\psi$ is not compact, but C_ϕ and C_ψ are arc connected in the operator norm topology.

Acknowledgement: The authors would like to thank the referee for helpful suggestions.

References

[B] E. Berkson, *Composition operators isolated in the uniform operator topology*, Proc. Amer. Math. Soc, **81**, (1981), 230-232.

[BES] F. Bagemihl, P. Erdös, W. Seidel, *Sur quelques propriétés frontières des fonctions holomorphes définies par certains produits dans le cercle-unité*, Ann. Sci. Ecole Norm. Sup. (3) **70** (1953), 135-147.

[Bo] P. Bourdon, *Components of Linear Fractional Composition Operators*, preprint.

[CM] C. Cowen and B.D. MacCluer, *Composition Operators on Spaces of Analytic Functions*, CRC Press, Boca Raton, 1995.

[D] P. L. Duren, *Theory of H^p-Spaces*, Dover Publications, New York, 2000.

[G] T. Goebeler, *Composition operators acting between Hardy spaces*, Integral Equations and Operator Theory, **41** (2001), 389-395.

[H] P. Halmos, *Measure Theory*, Springer-Verlag, New York, 1974.

[HJM] H. Hunziker, H. Jarchow, and V. Mascioni, *Some topologies on the space of analytic self-maps of the unit disk,* Geometry of Banach Spaces (Strobl, 1989), Cambridge University Press, Cambridge, 1990, 133-148.

[M] B.D. MacCluer, *Components in the space of composition operators*, Integral Equations Operator Theory **12**, (1989), 725-738.

[MOZ] B.D. MacCluer, S. Ohno, and R. Zhao, *Topological structure of the space of composition operators on H^∞*, Integral Equations and Operator Theory, **40** (2001), no. 4, 481-494.

[MS] B.D. MacCluer and J.H. Shapiro, *Angular derivatives and compact composition operators on the Hardy and Bergman spaces.* Canad. J. Math. **38** (1986), 878-906.

[SS] J.H. Shapiro and C. Sundberg, *Isolation amongst the composition operators*, Pacific J. Math. **145** (1990), 117-152.

[S] J. Shapiro, *Aleksandrov measures used in essential norm inequalities for composition operators*, J. Operator Theory, **40** (1998), no.1, 133-146.

University of Virginia, Dept. of Mathematics, Charlottesville, VA 22904
E-mail address: jlh5t@virginia.edu

E-mail address: cat4n@virginia.edu

Contemporary Mathematics
Volume **321**, 2003

Complex vs real variables for real 3-homogeneous polynomials on ℓ_1^2: A Counterexample.

Gustavo A. Muñoz

ABSTRACT. It was proved in [**8**] that homogeneous polynomials of degree > 3 cannot be generally complexified with preservation of their norms, even if we use the Bochnak complexification. However 2-homogeneous polynomials can be complexified with preservation of their norms when using the Bochnak complexification (see [**8**]). In this paper we show that 3-homogeneous polynomials cannot be generally complexified with preservation of their norms. In particular, for the special 3-homogeneous polynomial $P(x,y) = \alpha x^3 + \beta xy^2$, $\alpha, \beta \in \mathbb{R}$, we find the best constant $M > 1$ so that $\|\widetilde{P}\|_{\ell_1^2} \leq M\|P\|_{\ell_1^2}$, where $\widetilde{P}(z,w) = \alpha z^3 + \beta zw^2$ is the complex version of P.

1. Introduction.

If P is a 3-homogeneous polynomial on the real ℓ_1^2 and $\widetilde{P}$ is the same polynomial on the complex ℓ_1^2, how large the ratio $\|\widetilde{P}\|_{\ell_1^2}/\|P\|_{\ell_1^2}$ can be? Note that here

$$\|P\|_{\ell_1^2} = \sup\{|P(x,y)| : (x,y) \in \mathbb{R}^2, |x| + |y| = 1\}$$

and

$$\|\widetilde{P}\|_{\ell_1^2} = \sup\{|\widetilde{P}(z,w)| : (z,w) \in \mathbb{C}^2, |z| + |w| = 1\}.$$

In other words $\|P\|_{\ell_1^2}$ and $\|\widetilde{P}\|_{\ell_1^2}$ denote the norms of P and $\widetilde{P}$ as polynomials on the real and the complex versions of ℓ_1^2 respectively. One should expect that the ratio is in general greater than 1. Surprisingly enough, the above ratio for the case of 2-homogeneous polynomials is always 1, but for 3-homogeneous polynomials the ratio can be strictly greater than 1. The interest of this specific problem rests on an open question related to the possibility of complexifying 3-homogeneous polynomials with preservation of their norms (see [**6**, Remark 3.36]). We motivate more specifically this problem in Section 1.4 of the introduction. For convenience we present first an introduction to complexifications of real Banach spaces and polynomials.

2000 *Mathematics Subject Classification.* Primary: 47H60; Secondary: 46B99.
The author was supported by DGICYT PB96-0607.

1.1. Complexification of real Banach spaces: If $\{e_1, e_2\}$ is the canonical basis of $\mathbb{R}^2$, the tensor product $\widetilde{E} = E \otimes \mathbb{R}^2$ turns into a complex vector space, namely the complexification of E, if we define the addition operation by

$$(x \otimes e_1 + y \otimes e_2) + (u \otimes e_1 + v \otimes e_2) := (x+u) \otimes e_1 + (y+v) \otimes e_2,$$

where $x, y, u, v \in E$, and scalar multiplication by

$$(\alpha + i\beta)(x \otimes e_1 + y \otimes e_2) := (\alpha x - \beta y) \otimes e_1 + (\beta x + \alpha y) \otimes e_2,$$

for $x, y \in E$ and $\alpha, \beta \in \mathbb{R}$. Notice that $x \mapsto x \otimes e_1$ is an injective real linear map of E into $\widetilde{E}$ and therefore we can view the space E as a real subspace of $\widetilde{E}$. One can also prove easily that $\widetilde{E} = E \oplus iE$. This identification justifies the use of the familiar notation $x + iy$ in place of $x \otimes e_1 + y \otimes e_2$.

If we have a real Banach space E, its norm can be extended in infinitely many ways to a complex complete norm on $\widetilde{E}$. If E is a real Banach space with norm $\|\cdot\|_E$ (we just write $\|\cdot\|$ when there is no possibility of confusion), we say that a complex norm $\|\cdot\|_{\widetilde{E}}$ on $\widetilde{E}$ is a *natural complexification norm* for $\|\cdot\|_E$ if (see for instance [**8**]):

(i) $\|x\|_{\widetilde{E}} = \|x\|_E \quad \forall x \in E$, and
(ii) $\|x + iy\|_{\widetilde{E}} = \|x - iy\|_{\widetilde{E}} \quad \forall x, y \in E$.

The problem we are dealing with in this paper can be stated in terms of a remarkable natural complexification norm named after J. Bochnak [**1**] (see also [**2**]). If E is a real Banach space and π denotes the projective tensor norm on $E \otimes \ell_2^2$, then the mapping

$$\|x + iy\|_B := \|x \otimes e_1 + y \otimes e_2\|_\pi = \inf\{\sum \|x_k\|_E \cdot \|\lambda_k\|_{\ell_2^2}\},$$

for every $x, y \in E$, where the infimum is taken over all the representations $\sum x_k \otimes \lambda_k$ of $x \otimes e_1 + y \otimes e_2$, with $x_k \in E$ and $\lambda_k \in \ell_2^2$, is indeed a natural complexification norm in the sense of the previous definition (see for instance [**8**, Proposition 9]). Among the properties of the Bochnak norm we underline the fact that it coincides with the lattice complexification of $L_1(\mu)$-spaces. In particular, if (x_1, x_2), (y_1, y_2) are elements in ℓ_1^2 and $z_j = x_j + iy_j$, $j = 1, 2$, then $\|(x_1, x_2) + i(y_1, y_2)\|_B = \|(z_1, z_2)\|_{\ell_1^2} = |z_1| + |z_2|$.

We refer to [**4**], [**6**], [**7**], [**8**] and the references therein for a more complete exposition on complexifications and other related topics.

1.2. Complexification of polynomials on a real Banach space: If E and F are real or complex Banach spaces, a map $P : E \to F$ is a (continuous) n-*homogeneous polynomial* if there is a (continuous) n-linear mapping $L : E^n \to F$ for which $P(x) = L(x, \ldots, x)$ for all $x \in E$. From now on we consider only continuous operators. If P and L are as above then we define their norms as

$$\|P\| = \sup_{\|x\|=1} |P(x)| \quad \text{and} \quad \|L\| = \sup_{\substack{\|x_k\|=1 \\ 1 \le k \le n}} |L(x_1, \ldots, x_n)|.$$

It is well-known that if P is an n-homogeneous polynomial between E and F, then there is a unique symmetric n-linear mapping $\check{P}$ such that $P(x) = \check{P}(x, \ldots, x)$, for every $x \in E$. The mapping $\check{P}$ is often called the blossom or polar of P. Consult [**3**] for a more complete exposition on polynomials between real Banach spaces.

On the other hand, it is well-known that if P is an n-homogeneous polynomial from E into F and L is an n-linear mapping from E^n into F, then there is a unique

n-homogeneous polynomial $\widetilde{P}$ from $\widetilde{E}$ into $\widetilde{F}$ and a unique n-linear mapping $\widetilde{L}$ from $\widetilde{E}^n$ into $\widetilde{F}$ that extend P and L respectively. The operators $\widetilde{P}$ and $\widetilde{L}$ are called the complexifications of P and L respectively (see for instance [**8**] for more details). If $P(x_1, \ldots, x_m)$ is a real polynomial on $\mathbb{R}^m$, then its complexification is the polynomial $P(z_1, \ldots, z_m)$, i.e., the polynomial P regarded as a polynomial on $\mathbb{C}^m$.

Given a real Banach space E, one important feature of the Bochnak norm is that it is the largest natural complexification norm that can be defined on $\widetilde{E}$ (see for instance the comment just after Proposition 10 in [**8**]). Because of this, the Bochnak norm seems to be the best candidate in minimizing the ratios $\|\widetilde{P}\|/\|P\|$ and $\|\widetilde{L}\|/\|L\|$, where P is a homogeneous polynomial and L is its associated symmetric multilinear mapping between two real Banach spaces.

From now on, $\|\widetilde{P}\|_B$ and $\|\widetilde{L}\|_B$ represent respectively the norms of $\widetilde{P}$ and $\widetilde{L}$ under the Bochnak complexification. It has been proved by J. Bochnak [**1**] that $\|\widetilde{L}\|_B = \|L\|$ for every multilinear mappings between real Banach spaces. However the same nice result does not hold in general for homogeneous polynomials as we shall see later.

1.3. Restatement of our problem using the terminology of complexifications: If P is a 3-homogeneous polynomial on $\mathbb{R}^2$, then we want to find the best constant in the inequality $\|\widetilde{P}\|_B \leq K\|P\|_{\ell_1^2}$, where $\|P\|_{\ell_1^2}$ is the norm of P as a polynomial on ℓ_1^2. Although we do not give the best value for K, we solve a particular case of our problem, namely if P has the form $P(x, y) = \alpha x^3 + \beta xy^2$, where α, β are real numbers, then we find the best constant M in $\|\widetilde{P}\|_B \leq M\|P\|_{\ell_1^2}$. The constat M turns out to be strictly greater than 1, in particular $M \approx 1.075481$. This shows that 3-homogeneous polynomials cannot be complexified with preservation of their norms in general. It was shown in [**8**] that polynomials of degree ≥ 4 cannot be generally extended with preservation of their norms, but the 2-homogeneous polynomials preserve their norm provided we use a 2-dominating natural complexification norm, i.e., a norm $\|\cdot\|_{\widetilde{E}}$ satisfying $\|x + iy\|_{\widetilde{E}} \geq \sqrt{\|x\|^2 + \|y\|^2}$, for every $x, y \in E$. The Bochnak norm is obviously 2-dominating. The problem for 3-homogeneous polynomials remained open. The importance of the question for 3-homogeneous polynomials rests on a nice characterization for the real extremal polynomials using complexifications.

1.4. Extremal polynomials. If P is an n-homogeneous polynomial between two real or complex Banach spaces E and F, then it is well-known (see for instance [**3**, pg. 10]) that

$$\|\check{P}\| \leq \frac{n^n}{n!}\|P\|. \tag{1.1}$$

The constant $n^n/n!$ in (1.1) is optimal. We say that a scalar-valued n-homogeneous polynomial P on a Banach space E (real or complex) is extremal if

(a) there are vectors $x_1, \ldots, x_n$ of norm 1 such that $\check{P}(x_1, \ldots, x_n) = \|\check{P}\|$, and
(b) $\|\check{P}\| = (n^n/n!)\|P\|$.

In other words, P is extremal if $\check{P}$ achieves its norm and equality is attained in (1.1).

It has been proved by Y. Sarantopoulos [**9**] that if there exists an extremal polynomial P of degree n on an n-dimensional complex Banach space E, then

$E = \ell_1^n$ and P is a complex multiple of the Nachbin polynomial of degree n, Φ_n, defined by $\Phi_n(z_1, \ldots, z_n) = z_1 \cdots z_n$, for every $(z_1, \ldots, z_n) \in \mathbb{C}^n$. P. Kirwan, Y. Sarantopoulos and A. Tonge [**5**] have found a similar characterization of the extremal polynomials of degree n on a real Banach space of dimension n, for $n = 2, 3$. However, for real Banach spaces of dimension $n > 3$ there is a large supply of extremal polynomials other than the real multiples of Φ_n on $\mathbb{R}^n$. As it has been observed in [**5**], the characterization of the real extremal polynomials for $n = 2$ can be obtained using a complexification argument and the corresponding result for complex polynomials. Indeed, if E is a 2-dimensional real Banach space, P is an extremal 2-homogeneous polynomial on E and we set $L = \check{P}$, then $\|P\| = 2\|L\|$. Since the norms of 2-homogeneous polynomials and multilinear mappings are preserved when complexified using the Bochnak norm, we have that $\|\widetilde{P}\|_B = \|P\| = 2\|L\| = 2\|\widetilde{L}\|_B$, and since $\widetilde{L}$ is the polar of $\widetilde{P}$, then $\widetilde{P}$ is extremal on $(\widetilde{E})_B$. Hence, by the result of Y. Sarantopoulos in [**9**], $(\widetilde{E})_B$ is the complex space ℓ_1^2 and $\widetilde{P}(z_1, z_2) = cz_1z_2$, for every $(z_1, z_2) \in \mathbb{C}^2$, where $c \in \mathbb{C}$. Now it is immediate that E must be the real space ℓ_1^2, $P(x_1, x_2) = cx_1x_2$, for every $(x_1, x_2) \in \mathbb{R}^2$ and $c \in \mathbb{R}$. If we could prove that all 3-homogeneous polynomials on a real Banach space can be complexified with preservation of their norms using the Bochnak complexification, then we could use a similar argument to characterize the extremal 3-homogeneous polynomials on a 3-dimensional real Banach space (note that this characterization has been already done using completely different methods in [**5**]). However this is not possible as we show below.

2. Main results.

If $P(x, y) = \alpha x^3 + \beta xy^2$, with $\alpha, \beta \in \mathbb{R}$, we define $|P|$ to be the polynomial

$$|P|(x, y) = |\alpha|x^3 + |\beta|xy^2.$$

LEMMA 2.1. *If $P(x, y) = \alpha x^3 + \beta xy^2$, then $\|\widetilde{P}\|_B = \||P|\|_{\ell_1^2}$.*

PROOF. Let $x = (x_1, x_2)$, $y = (y_1, y_2)$ and $z_j = x_j + iy_j$, $j = 1, 2$. If $\|x + iy\|_B = 1$ then $|z_1| + |z_2| = 1$. Therefore

$$|\widetilde{P}(x + iy)| = |P(z_1, z_2)| = |\alpha z_1^3 + \beta z_1 z_2^2| \leq |\alpha||z_1|^3 + |\beta||z_1||z_2|^2 \leq \||P|\|_{\ell_1^2}.$$

This shows that $\|\widetilde{P}\|_B \leq \||P|\|_{\ell_1^2}$.

On the other hand, let $(x_0, y_0) \in \ell_1^2$ such that $|x_0| + |y_0| = 1$ and $|P|(x_0, y_0) = \||P|\|_{\ell_1^2}$. Let $x = (x_0, y_0 \cos\theta)$, $y = (0, y_0 \sin\theta)$, $z_1 = x_0$ and $z_2 = y_0 e^{i\theta}$ so that $x + iy = (z_1, z_2)$ and $\|x + iy\|_B = |z_1| + |z_2| = |x_0| + |y_0| = 1$. Then

$$|\widetilde{P}(x + iy)| = |P(z_1, z_2)| = |\alpha z_1^3 + \beta z_1 z_2^2| = |\alpha x_0^3 + \beta x_0 y_0^2 e^{i2\theta}|.$$

If $\alpha x_0^3 = 0$ or $\beta x_0 y_0^2 = 0$ then it is obvious that $|\widetilde{P}(x + iy)| = \||P|\|_{\ell_1^2}$. If $\alpha x_0^3 \neq 0$ and $\beta x_0 y_0^2 \neq 0$ take θ such that $e^{i2\theta} = \operatorname{sign} \frac{\alpha x_0^3}{\beta x_0 y_0^2}$. Then

$$|\widetilde{P}(x + iy)| = |\alpha||x_0|^3 + |\beta||x_0||y_0|^2 = \big||\alpha|x_0^3 + |\beta|x_0y_0^2\big| = \||P|\|_{\ell_1^2}.$$

We conclude that $\|\widetilde{P}\|_B = \||P|\|_{\ell_1^2}$. □

PROPOSITION 2.2. *Let $P(x,y) = \alpha x^3 + \beta xy^2$. Let a_0 and a_1 be the numbers*

$$a_0 = -\frac{3\sqrt[3]{(207+48\sqrt{3})^2} + 21\sqrt[3]{207+48\sqrt{3}} + 99}{8\sqrt[3]{207+48\sqrt{3}}} \approx -6.97685. \tag{2.1}$$

and

$$a_1 = 3 + 2\sqrt{3} \approx 6.46410.$$

If $\alpha \neq 0$ we define a to be β/α. Then

$$\|P\|_{\ell_1^2} = \begin{cases} \frac{4}{27}|\beta| & \text{if } \alpha = 0 \\ |\alpha| & \text{if } \alpha \neq 0 \text{ and } a_0 \leq a \leq a_1 \\ \frac{2a^2|\alpha|}{a^2+9a-(a-3)\sqrt{a^2-3a}} & \text{if } \alpha \neq 0 \text{ and } a > a_1 \\ \frac{2a^2|\alpha|}{|a^2+9a+(a-3)\sqrt{a^2-3a}|} & \text{if } \alpha \neq 0 \text{ and } a < a_0. \end{cases}$$

PROOF. We consider first the case where $\alpha = 0$. Then $P(x,y) = \beta xy^2$ and using elementary calculus we get

$$\|P\|_{\ell_1^2} = |\beta| \sup_{|x|+|y|=1} |x||y|^2 = |\beta| \sup_{|x|\leq 1} |x|(1-|x|)^2 = |\beta| \sup_{0\leq t\leq 1} t(1-t)^2 = \frac{4}{27}|\beta|.$$

Now suppose that $\alpha \neq 0$. Then $P(x,y) = \alpha(x^3 + axy^2)$, where $a = \beta/\alpha$. Therefore

$$\begin{aligned} \|P\|_{\ell_1^2} &= \sup_{|x|+|y|=1} |\alpha||x^3 + axy^2| = \sup_{|x|+|y|=1} |\alpha||x|\big||x|^2 + a|y|^2\big| \\ &= \sup_{|x|\leq 1} |\alpha||x|\big||x|^2 + a(1-|x|)^2\big| = \sup_{0\leq t\leq 1} |\alpha||t^3 + at(1-t)^2| \\ &= |\alpha| \sup_{0\leq t\leq 1} \big|(1+a)t^3 - 2at^2 + at\big|. \end{aligned}$$

The extremes of the polynomial $p(t) = (1+a)t^3 - 2at^2 + at$ on $[0,1]$ are achieved at the end points 0 or 1, or at the critical points of $p(t)$ lying in $[0,1]$. These critical points are obtained from the solutions to the equation $p'(t) = 0$, i.e.:

$$3(1+a)t^2 - 4at + a = 0.$$

The two algebraic solutions are

$$t_1 = \frac{2a + \sqrt{a^2-3a}}{3(1+a)} \quad \text{and} \quad t_2 = \frac{2a - \sqrt{a^2-3a}}{3(1+a)}.$$

It is straightforward that t_1 and t_2 are real only if $a \leq 0$ or $a \geq 3$. In other words, if $0 < a < 3$ then

$$\|P\|_{\ell_1^2} = |\alpha| \sup_{0\leq t\leq 1} |p(t)| = |\alpha| \max\{|p(0)|, |p(1)|\} = |\alpha|.$$

On the other hand it is easily proved that

$$0 \leq t_1 = \frac{2a + \sqrt{a^2-3a}}{3(1+a)} \leq 1 \quad \text{if} \quad a \leq 0 \quad \text{or} \quad a \geq 3 \quad \text{and}$$

$$0 \leq t_2 = \frac{2a - \sqrt{a^2-3a}}{3(1+a)} \leq 1 \quad \text{if} \quad a \geq 3.$$

Therefore

$$\|P\|_{\ell_1^2} = |\alpha| \max\{|p(0)|, |p(1)|, |p(t_1)|\} \quad \text{if} \quad a \leq 0 \quad \text{and}$$
$$\|P\|_{\ell_1^2} = |\alpha| \max\{|p(0)|, |p(1)|, |p(t_1)|, |p(t_2)|\} \quad \text{if} \quad a \geq 3.$$

But $p(0) = 0$, $p(1) = 1$ and doing some elementary calculations we get

$$p(t_1) = \frac{2a^2}{a^2 + 9a + (a-3)\sqrt{a^2 - 3a}} \quad \text{and} \quad p(t_2) = \frac{2a^2}{a^2 + 9a - (a-3)\sqrt{a^2 - 3a}}.$$

Looking at the expressions for $p(t_1)$ and $p(t_2)$ it is immediate that for $a \geq 3$ we have that $p(t_2) \geq p(t_1) > 0$. It is also easy to check that $p(t_1) < 1$ for $a \geq 3$. Therefore, if $\alpha \neq 0$ and $a \notin (0, 3)$ we have

$$\|P\|_{\ell_1^2} = |\alpha| \max\{1, |p(t_1)|\} = |\alpha| \max\left\{1, \frac{2a^2}{|a^2 + 9a + (a-3)\sqrt{a^2 - 3a}|}\right\} \quad \text{if} \quad a \leq 0$$

and

$$\|P\|_{\ell_1^2} = |\alpha| \max\{1, |p(t_2)|\} = |\alpha| \max\left\{1, \frac{2a^2}{a^2 + 9a - (a-3)\sqrt{a^2 - 3a}}\right\} \quad \text{if} \quad a \geq 3.$$

Let us consider first the case where $a \leq 0$. If $a \leq 0$ then $\|P\|_{\ell_1^2} = |\alpha|$ if and only if

$$\frac{2a^2}{|a^2 + 9a + (a-3)\sqrt{a^2 - 3a}|} \leq 1. \tag{2.2}$$

Doing some complicated though basic calculations one can show that for $a \leq 0$, equality in (2.2) is achieved only at

$$a_0 = -\frac{3\sqrt[3]{(207 + 48\sqrt{3})^2} + 21\sqrt[3]{207 + 48\sqrt{3}} + 99}{8\sqrt[3]{207 + 48\sqrt{3}}} \approx -6.97685.$$

For convenience one can use a Math package like Maple to obtain a_0. With this in mind it is straightforward to check that for $a \leq 0$, (2.2) holds if $a_0 \leq a \leq 0$ and it does not hold if $a < a_0$. Therefore if $a \leq 0$

$$\|P\|_{\ell_1^2} = \begin{cases} |\alpha| & \text{if} \quad \alpha \neq 0 \text{ and } a_0 \leq a \leq 0 \\ \frac{2a^2|\alpha|}{|a^2+9a+(a-3)\sqrt{a^2-3a}|} & \text{if} \quad \alpha \neq 0 \text{ and } a < a_0. \end{cases}$$

We proceed in a similar way in order to obtain a more explicit formula for $\|P\|_{\ell_1^2}$ when $a \geq 3$. We know that if $a \geq 3$, $\|P\|_{\ell_1^2} = |\alpha|$ if and only if

$$\frac{2a^2}{a^2 + 9a - (a-3)\sqrt{a^2 - 3a}} \leq 1. \tag{2.3}$$

If $a \geq 3$, it is elementary to show that equality is achieved in (2.3) only at $a_1 = 3+2\sqrt{3}$. Having this in mind it is easy to check that (2.3) holds for $3 \leq a \leq 3+2\sqrt{3}$ and it does not hold for $a > 3 + 2\sqrt{3}$. Therefore, if $a \geq 3$

$$\|P\|_{\ell_1^2} = \begin{cases} |\alpha| & \text{if} \quad \alpha \neq 0 \text{ and } 3 \leq a \leq a_1 \\ \frac{2a^2|\alpha|}{a^2+9a-(a-3)\sqrt{a^2-3a}} & \text{if} \quad \alpha \neq 0 \text{ and } a > a_1. \end{cases}$$

This concludes the proof. □

PROPOSITION 2.3. *If* $f(a) = \frac{2a^2}{a^2+9a-(a-3)\sqrt{a^2-3a}}$ *and* a_0 *is defined by (2.1), then for every polynomial* $P(x, y) = \alpha x^3 + \beta xy^2$ *we have*

$$\|\widetilde{P}\|_B \leq f(|a_0|)\|P\|_{\ell_1^2}. \tag{2.4}$$

Moreover, the constant $f(|a_0|) \approx 1.075481$ *is sharp since equality in (2.4) is achieved for* $P(x, y) = x^3 + a_0 xy^2$.

PROOF. By Lemma 2.1, $\|\widetilde{P}\|_B = \||P|\|_{\ell_1^2}$. Hence if $\alpha = 0$ or $a \geq 0$, $\|\widetilde{P}\|_B = \|P\|_{\ell_1^2}$. Suppose now that $\alpha \neq 0$ and define $h(a) = \frac{\|\widetilde{P}\|_B}{\|P\|_{\ell_1^2}}$ for $a \leq 0$. Then according to Lemma 2.1 and Proposition 2.2 we have

$$h(a) = \begin{cases} 1 & \text{if} \quad \alpha \neq 0 \text{ and } -a_1 \leq a \leq 0 \\ \frac{2a^2}{a^2+9|a|-(|a|-3)\sqrt{a^2-3|a|}} & \text{if} \quad \alpha \neq 0 \text{ and } a_0 \leq a < -a_1 \\ \frac{\left|a^2+9a+(a-3)\sqrt{a^2-3a}\right|}{a^2+9|a|-(|a|-3)\sqrt{a^2-3|a|}} & \text{if} \quad \alpha \neq 0 \text{ and } a < a_0. \end{cases} \tag{2.5}$$

Recall that $a_0 \approx -6.97685$ and $a_1 \approx 6.46410$. It is clear that the best constant M in $\|\widetilde{P}\|_B \leq M\|P\|_{\ell_1^2}$ for P of the form $P(x, y) = \alpha x^3 + \beta xy^2$ is given by

$$M = \sup_{a \leq 0} h(a).$$

But the mapping $h(a)$ attains its maximum at a_0 and hence

$$M = \frac{2a_0^2}{a_0^2 + 9|a_0| - (|a_0| - 3)\sqrt{a_0^2 - 3|a_0|}} = f(|a_0|) \approx 1.075481.$$

The following picture obtained with a graphing package shows the graph of $h(a)$:

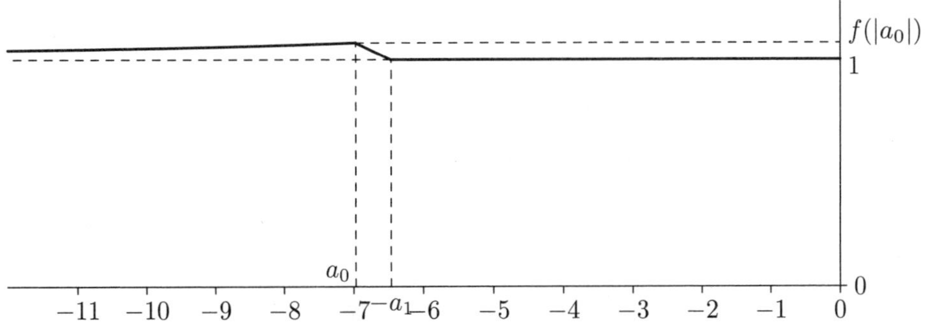

□

REMARK 2.4. If K is the best constant in the inequality $\|\widetilde{P}\|_B \leq K\|P\|$, where P is any real 3-homogeneous polynomial on ℓ_1^2, it would be interesting to know whether $K = f(|a_0|) \approx 1.075481$.

The following example shows that complexifications cannot be used in order to characterize the extremal 3-homogeneous polynomials on a real Banach space of dimension 3.

EXAMPLE 2.5. If $Q : \ell_1^3 \to \mathbb{R}$ is defined by $Q(x, y, z) = x^3 - 7xy^2$ and h is as in (2.5), then

$$\|\widetilde{Q}\|_B = \frac{7 + 5\sqrt{70}}{56 - 4\sqrt{7}}\|Q\|_{\ell_1^3}, \tag{2.6}$$

where

$$h(-7) = \frac{7 + 5\sqrt{70}}{56 - 4\sqrt{7}} \approx 1.075214 > 1.$$

PROOF. If $P(x,y) = Q(x,y,0)$ for every $(x,y) \in \ell_1^2$, then it is easy to prove that $\|Q\|_{\ell_1^3} = \|P\|_{\ell_1^2}$ and $\|\widetilde{Q}\|_B = \|\widetilde{P}\|_B$. On the other hand, by (2.5) we have that

$$\|\widetilde{Q}\|_B = \|\widetilde{P}\|_B = h(-7)\|P\|_{\ell_1^2} = h(-7)\|Q\|_{\ell_1^3},$$

from which (2.6) follows. □

References

[1] J. BOCHNAK. Analytic functions in Banach spaces. *Studia Math.*, **35**(1970), 273-292.

[2] J. BOCHNAK AND J. SICIAK. Polynomials and multilinear mappings in topological vector spaces. *Studia Math.*, **39**(1971), 59-76.

[3] S. DINEEN. Complex analysis on infinite dimensional spaces. *Springer Monographs in Mathematics, Springer-Verlag, Berlin,* 1999.

[4] P. KIRWAN. Complexifications of multilinear and polynomials, *Math. Nachr.*, **231**(2001), 39-68.

[5] P. KIRWAN, Y. SARANTOPOULOS AND A. TONGE. Extremal homogeneous polynomials on real normed spaces. *J. Approx. Theory,* **97**(1999), 201-213.

[6] G. A. MUÑOZ. Two Problems on Real Banach Spaces: Complexifications and Bernstein-Markov Type Inequalities, *Ph.D. thesis, Universidad Complutense de Madrid, Spain,* 1999.

[7] G. A. MUÑOZ. Complexifications of polynomials and multilinear maps on real Banach spaces, in: Function Spaces, The Fifth Conference (Poznan, 1998), H. Hudzik & L. Skrzypczak (ed.), *Lecture Notes in Pure and Applied Mathematics, Vol. 213, Marcel Dekker,* 2000, 389-406.

[8] G. A. MUÑOZ, Y. SARANTOPOULOS AND A. TONGE. Complexifications of real Banach spaces, polynomials and multilinear maps. *Studia Math.* **134**(1999), 1-33.

[9] Y. SARANTOPOULOS. Extremal multilinear forms on Banach spaces. *Proc. Amer. Math. Soc.* **99**(1987), 340-346.

(Gustavo A. Muñoz) FACULTAD DE MATEMÁTICAS, DEPARTAMENTO DE ANÁLISIS, UNIVERSIDAD COMPLUTENSE DE MADRID, MADRID 28040, SPAIN.

E-mail address: gustavo_fernandez@mat.ucm.es

Contemporary Mathematics
Volume **321**, 2003

The Fourier-Stieltjes and Fourier algebras for locally compact groupoids

Alan L. T. Paterson

ABSTRACT. The Fourier-Stieltjes and Fourier algebras $B(G)$, $A(G)$ for a general locally compact group G, first studied by P. Eymard, have played an important role in harmonic analysis and in the study of the operator algebras generated by G. Recently, there has been interest in developing versions of these algebras for locally compact *groupoids*, justification for this being that, just as in the group case, the algebras should play a useful role in the study of groupoid operator algebras. Versions of these algebras for the locally compact groupoid case appear in three related theories: (1) a measured groupoid theory (J. Renault), (2) a Borel theory (A. Ramsay and M. Walter), and (3) a continuous theory (A. Paterson). The present paper is expository in character. For motivational reasons, it starts with a description of the theory of $B(G)$, $A(G)$ in the locally compact group case, before discussing these three related theories. Some open questions are also raised.

1. $A(G)$ and $B(G)$ for locally compact groups

I am grateful to the organizers of the *2001 Conference on Banach spaces* at the University of Memphis for the opportunity to participate in that exciting conference.

We first specify some notation. If X is a locally compact Hausdorff space, then $C(X)$ is the algebra of bounded, continuous, complex-valued functions on X. The space of functions in $C(X)$ that vanish at ∞ is denoted by $C_0(X)$, while $C_c(X)$ is the space of functions in $C(X)$ with compact support. The space of complex, bounded, regular Borel measures on X is denoted by $M(X)$.

The Banach spaces $B(G)$, $A(G)$ (where G is a locally compact groupoid) that we will consider in this paper arise naturally in the group case in non-commutative harmonic analysis and duality theory. (See later in this section.) When G is a locally compact group, $B(G)$ and $A(G)$ are just the (much studied) Fourier-Stieltjes and Fourier algebras. The need to have available versions of these Banach algebras for the case of a locally compact groupoid is a consequence of the fact that many of the operator algebras of present day interest, especially in non-commutative geometry, come from groupoid, and not (in any obvious way) from group, representations, so that having available versions of $B(G)$, $A(G)$ in the groupoid case would provide the resources for extending to that case the properties of group operator algebras that depend on $B(G)$, $A(G)$.

1991 *Mathematics Subject Classification*. Primary: 22A22, 22D25, 43A25, 43A35, 46L87; Secondary: 20L05, 42A38, 46L08.

We start by discussing these algebras in the locally compact group case. Let us, for the present, specialize even further by letting G be a locally compact *abelian* group with character space $\widehat{G}$. An element of $\widehat{G}$ is a continuous homomorphism $t : G \to \mathbf{T}$, and $\widehat{G}$ is a locally compact abelian group with pointwise product and the topology of uniform convergence on compacta. The Fourier transform $f \to \hat{f}$ takes $f \in L^1(G)$ into $C_0(\widehat{G})$, where

$$\hat{f}(t) = \int f(x)\overline{t(x)}\,dx$$

where dx is a left Haar measure on G. The inverse Fourier transform $\mu \to \check{\mu}$ takes $M(\widehat{G})$ back into $C(G)$ where

$$\check{\mu}(x) = \int \hat{x}(t)\,d\mu(t).$$

For example, when $G = \mathbb{R}$, we have $\widehat{G} = \mathbb{R}$ where $t \in \widehat{G}$ is associated with the character $x \to e^{ixt}$. In that case

$$\hat{f}(t) = \int f(x)e^{-ixt}\,dx, \qquad \check{\mu}(x) = \int e^{itx}\,d\mu(t).$$

This is just the usual Fourier transform.

Now $M(\widehat{G})$ is a convolution Banach algebra, and contains $L^1(\widehat{G})$ as a closed ideal. The *Fourier-Stieltjes algebra* $B(G)$ is defined to be $M(\widehat{G})^{\vee}$ while the *Fourier algebra*, $A(G)$, is defined to be $L^1(\widehat{G})^{\vee}$. These are algebras of continuous bounded functions on G under pointwise product, with $A(G) \subset C_0(G)$ an ideal in $B(G)$. Then $B(G)$ is a Banach algebra under the $M(\widehat{G})$ norm, and $A(G) \cong L^1(\widehat{G})$ is a closed ideal in $B(G)$. There is a substantial literature on $A(G)$ in the abelian case - see for example [**71**]. We note some important properties of $A(G)$. Firstly, the set of functions of the form $f * g$, where $f, g \in C_c(G)$, is dense in $A(G)$. Next, $A(G)$ is a dense subalgebra of $C_0(G)$ under the sup-norm. Further, $\|.\|_\infty \leq \|.\|$ on $A(G)$. Then the Gelfand space (character space) of $A(G) = L^1(\widehat{G})$ is G, the elements of G acting by point evaluation. Lastly, $A(G)$ has a bounded approximate identity.

To extend the notions of $A(G), B(G)$ to the non-abelian case, where duality theory is more complicated, we need to be able to interpret the norms on these algebras without reference to a dual space $\widehat{G}$. This was done by Eymard ([**11**]) and will now be described. It provides an excellent example of *quantization*, a process in which one procedes from the commutative situation to the non-commutative by replacing functions by Hilbert space operators. The following four spaces in the commutative case will be replaced by their natural "non-commutative" versions:

$$A(G) = L^1(\widehat{G}), B(G) = M(\widehat{G}), C_0(\widehat{G}), L^\infty(\widehat{G}).$$

These versions will involve G only, so that no duality comes in explicitly.

For general G, we define the left regular representation π_2 of G on $L^2(G)$ by: $\pi_2(x)f(t) = f(x^{-1}t)$. There is also the universal representation π_{univ} of G on a Hilbert space H_{univ}. Every unitary representation of G determines by integration a non-degenerate *-representation of $C_c(G)$. The norm closure of $\pi_2(C_c(G))$ is called the *reduced* C^*-algebra $C^*_{red}(G)$ of G, while that of $\pi_{univ}(C_c(G))$ is called the *universal* C^*-algebra $C^*(G)$ of G. The von Neumann algebra generated by $C^*_{red}(G) \subset B(L^2(G))$ is denoted by $VN(G)$.

Suppose now that G is abelian. The key to the desired quantization is *Plancherel's theorem*: $f \to \hat{f}$ **is an isometry from** $L^2(G)$ **onto** $L^2(\widehat{G})$**.** This induces an isomorphism Φ from $B(L^2(\widehat{G}))$ onto $B(L^2(G))$. Regard $C_0(\widehat{G}), L^\infty(\widehat{G})$ as C^*-subalgebras of $B(L^2(\widehat{G}))$ by having them act as multiplication operators on $L^2(\widehat{G})$. Then Φ identifies $C_0(\widehat{G})$ with $C^*_{red}(G)$ and $L^\infty(\widehat{G})$ with $VN(G)$. The predual $A(G) = L^1(\widehat{G})$ of $L^\infty(\widehat{G})$ goes over to the predual $VN(G)_*$ of $VN(G)$.

For a general locally compact group G, this inspire the definition of $A(G)$: we define $A(G)$ to be $VN(G)_*$. We now turn to $B(G)$. In the abelian case, since the irreducible representations are the characters, it is obvious that $C_0(\widehat{G}) = C^*(G)$, and so, using the Riesz representation theorem, $B(G) = M(\widehat{G}) = C_0(\widehat{G})^*$ is just the Banach space dual $C^*(G)^*$. This inspires the definition of $B(G)$ for general G: we define $B(G)$ to be $C^*(G)^*$.

In a natural way, we can (and will) interpret $A(G)$ as a subspace of $B(G)$. Indeed, since $C^*_{red}(G)$ is a homomorphic image of $C^*(G)$, dualizing ennables us to regard $C^*_{red}(G)$ as a closed subspace of $C^*(G)^*$. Next, since $C^*_{red}(G)$ is a weakly dense subspace of $VN(G)$, we can identify $A(G)$ with a closed subspace of $C^*_{red}(G)^*$ and hence with a closed subspace of $C^*(G)^* = B(G)$.

The space $B(G)$ can be regarded as a subspace of $C(G)$ as follows. Every $\phi \in B(G)$ is a *coefficient* of π_{univ}, i.e. there exist vectors $\xi, \eta \in H_{univ}$ such that for all $f \in C_c(G)$, we have $\phi(f) = \langle \pi_{univ}(f)\xi, \eta\rangle$ ([**8**, 12.1.3]). (Alternatively, $B(G)$ is the set of coefficients arising from all Hilbert space representations of G.) Regard $\phi \in C(G)$ by defining:

$$\phi(x) = (\pi_{univ}(x)\xi, \eta).$$

Then by integration, the two versions of ϕ determine each other, and so we can identify $B(G)$ with a subspace of $C(G)$.

Following J. Renault ([**75**])), we will write $\phi \in C(G)$ as: $\phi = (\xi, \eta)$. The function ϕ is called *positive definite* ($\phi \in P(G)$) if we can take $\xi = \eta$. This property can be abstractly characterized for *phi* $\in C(G)$ as follows:

$$\iint \phi(y^{-1}x) f(y)\overline{f(x)}\, d\lambda(x)\, d\lambda(y) \geq 0 \tag{1.1}$$

for all $f \in C_c(G)$, where λ is a left Haar measure on G. Every element of $B(G)$ is (by polarization) a linear combination of positive definite functions. In the abelian case, $P(G)$ is identified with the set of positive measures in $M(\widehat{G})$ (Bochner's theorem ([**36**, (33.3)])).

The $C^*(G)^*$ norm on $B(G)$ can ([**11**]) be usefully expressed in terms of coefficients: $\|\phi\|$ is just $\inf \|\xi\|\|\eta\|$, the inf being taken over all pairs ξ, η for which $\phi = (\xi, \eta)$. (There is another possible way of defining the norm of $B(G)$ that will be considered later in the groupoid context.) Using direct sums and tensor products of representations of G, one obtains that $B(G)$ is a commutative Banach algebra.

Turning to $A(G)$, it can be defined as the closure of the subspace of $B(G)$ spanned by the coefficients of π_2. In fact, Eymard shows that $A(G)$ is exactly the set of functions $(f, g) = \overline{g} * f^\dagger$ where $f, g \in L^2(G)$. (Here, for $f \in L^2(G)$, $f^\dagger(x) = f(x^{-1}), f^*(x) = \overline{f(x^{-1})}$.) An important result of Godement ([**23**],[**8**, Theorem 13.8.6]) says that if $\phi \in L^2(G) \cap P(G)$, then $\phi = f * f^*$ for some $f \in L^2(G)$. It follows that $B(G) \cap C_c(G)$ is a dense subspace of $A(G)$, and $A(G)$ is a closed ideal in $B(G)$. In particular, $A(G)$ is a commutative Banach algebra. Eymard proves ([**11**]) his remarkable duality theorem for a general locally compact group, viz. *the character space of* $A(G)$ *is identified with* G, *the elements of* G *acting as characters by point evaluation.* Walter ([**80**]) showed that both $A(G)$ and $B(G)$ as Banach algebras determine the group G. The important properties of $A(G)$, given in the paragraph earlier, defining $B(G)$ in the abelian case, all hold except that $A(G)$ need not have a bounded approximate identity. (We will return to this below in our discussion of *amenability.*)

Since the work of Eymard, substantial progress has been made in developing the theory of $B(G)$, $A(G)$, and before leaving the group situation and turning to locally compact groupoids, we will very briefly describe some of the main themes of this progress and cite some of the papers involved. I am grateful to Brian Forrest for helpful information about the literature on $A(G)$. The account given below is not comprehensive, and the present writer apologizes to all authors whose worthy contributions have inadvertently been omitted. For useful surveys of the field, the reader is referred to [**68**] (for work up to 1984), and [**51**].

A striking feature of the theory is the relation that the amenability of G has to properties of $A(G)$. Recall that a locally compact group G is called *amenable* if there exists an invariant mean on $L^\infty(G)$. There are many other characterizations of amenable groups, and the rich phenomenon of amenability is discussed in detail in [**68, 63**]. Leptin ([**56**]) showed that the Banach algebra $A(G)$ has a bounded approximate identity if and only if G is amenable. Further ([**6, 34, 59, 72, 57**]) the multiplier algebra of $A(G)$ is canonically isomorphic to $B(G)$ if and only if G is amenable. There are also L_p- versions of $A(G)$, $B(G)$ due to Herz ([**33**]) which have been further investigated (e.g. in [**68, 24, 25**]).

Banach algebra amenability of Fourier algebras has proved intriguing. B. E. Johnson ([**43**]) showed that the amenability of G does not entail the amenability of the Banach algebra $A(G)$. (In fact, he showed that it fails for the compact group $SU(2, \mathbb{C})$.) Now $A(G)$ is more than a Banach algebra - it is a completely contractive Banach algebra, and a theory of operator amenability can be developed in a natural way. Indeed, while amenability for a Banach algebra A means ([**42**]) that every bounded derivation from A into a dual Banach A-module is inner, so operator amenability for a completely contractive Banach algebra A means that every completely bounded derivation from A into the dual of an operator A-bimodule is inner. Remarkably, Ruan ([**76**]) showed that G is operator amenable if and only if its Fourier algebra $A(G)$ is operator amenable. See the book by Effros and Ruan for a comprehensive treatment of this and related theorems ([**9**, Ch. 16]). The theory has been extended to Kac algebras ([**77, 47**]). We note that in the operator space context, $A(G)$ is regarded as the "convolution algebra of the dual quantum group" ([**10**]).

On the Banach-Hochschild cohomology of $A(G)$, the reader is referred to the papers [**52, 54, 53, 12, 15**], and for information on topological centers, to [**1, 53**]. (The topological center of a Banach algebra A is the set of weak*-bicontinuous elements of A^{**}.) The dual and second dual spaces of $A(G)$ are studied in [**25, 27, 28, 29, 30, 14, 18, 21**]. On the ideal structure of Fourier algebras, see [**13, 16, 83**]. For other studies of $A(G)$, see [**19, 22, 26, 37, 38, 39, 40, 41, 44, 45, 55**].

2. Locally compact groupoids

Accounts of the theory of locally compact groupoids are given in the books of Jean Renault ([**73**]) and the present writer ([**63**]). (See also the CBMS conference lectures by Paul Muhly ([**60**]).) We summarize here the basic theory that will be needed in our discussion of $B(G)$, $A(G)$ in the groupoid case. (There is also an important theory of Lie groupoids - this is discussed in [**50, 58, 63**].)

A *groupoid* is most simply defined as a small category with inverses. Spelled out axiomatically, a groupoid is a set G together with a subset $G^2 \subset G \times G$, a "product map" $m : G^2 \to G$, where we write $m(a,b) = ab$, and an inverse map $i : G \to G$, where we write $i(a) = a^{-1}$ and where $(a^{-1})^{-1} = a$ for all $a \in G$, such that:

1. if $(a,b), (b,c) \in G^2$, then $(ab,c), (a,bc) \in G^2$ and
$$(ab)c = a(bc);$$
2. $(b, b^{-1}) \in G^2$ for all $b \in G$, and if (a,b) belongs to G^2, then
$$a^{-1}(ab) = b \qquad (ab)b^{-1} = a.$$

We define the *range* and *source* maps $r : G \to G^0$, $s : G \to G^0$ by setting $r(x) = xx^{-1}, s(x) = x^{-1}x$. The *unit space* G^0 is defined to be $r(G)$ $(= s(G))$, or equivalently, the set of idempotents u in G. In the category interpretation of a groupoid, G^0 is the class of *objects* of the category alnd each x is a morphism from $s(x)$ to $r(x)$. Each of the maps r, s fibers the groupoid G over G^0 with fibers $\{G^u\}, \{G_u\}$, so that $G^u = r^{-1}(\{u\})$ and $G_u = s^{-1}(\{u\})$. Note that $(x,y) \in G^2$ if and only if $s(x) = r(y)$.

Intuitively, then, a groupoid is a set with a partially defined product and with inverses so that the usual group axioms hold whenever they make sense. Groupoids give the algebra of local symmetry while groups give that of global symmetry. (See, for example, [**81**]). Groupoids are becoming more and more prominent in analysis. Connes's non-commutative geometry ([**4**]) makes extensive use of them and this has been an important motivation for their study. Examples of groupoids are:

(a) locally compact groups
(b) equivalence relations
(c) tangent bundles
(d) the tangent groupoid (e.g. [**4**])
(e) holonomy groupoids for foliations (e.g. [**4**])
(f) Poisson groupoids (e.g. [**82**])
(g) transformation groups (e.g. [**60**])
(h) graph groupoids (e.g. [**48, 65**])

As a simple, helpful example of a groupoid, consider (b) above. Let R be an equivalence relation on a set X. Then R is a groupoid under the following operations: $(x,y)(y,z) = (x,z), (x,y)^{-1} = (y,x)$. Here, $G^0 = X$ (=diagonal of $X \times X$) and $r((x,y)) = x, s((x,y)) = y$. So $R^2 = \{((x,y),(y,z)) : (x,y),(y,z) \in R\}$. When $R = X \times X$, then R is called a *trivial* groupoid.

A special case of a trivial groupoid is $R = R_n = \{1,2,\ldots,n\} \times \{1,2,\ldots,n\}$. (So every i is equivalent to every j.) Identify $(i,j) \in R_n$ with the matrix unit e_{ij}. Then the groupoid R_n is just matrix multiplication except that we only multiply

e_{ij}, e_{kl} when $k = j$, and $(e_{ij})^{-1} = e_{ji}$. We do not really lose anything by restricting the multiplication in this way, since the pairs e_{ij}, e_{kl} excluded from groupoid multiplication just give the 0 product in normal algebra anyway.

For a groupoid G to be a *locally compact* groupoid means what one would expect. The groupoid G is required[1] to be a (second countable) locally compact Hausdorff space, and the product and inversion maps are required to be continuous. Each G^u, as well as the unit space G^0, is closed in G. What replaces left Haar measure on G is a system of measures λ^u $(u \in G^0)$, where each λ^u is a positive regular Borel measure on G^u with support equal to G^u. In addition, the λ^u's are required to vary continuously (when integrated against any $f \in C_c(G)$) and to form an *invariant* family in the sense that for each x, the map $y \to xy$ is a measure preserving homeomorphism from $G^{s(x)}$ onto $G^{r(x)}$. Such a system $\{\lambda^u\}$ is called a *left Haar system* for G. Not all locally compact groupoids possess a left Haar system, and even if there exists a left Haar system, it need not be unique. However, most of the locally compact groupoids that arise in practice have natural left Haar systems. All locally compact groupoids in the present survey are assumed to have a left Haar system.The presence of a left Haar system on G has topological implications: it implies that the range map $r : G \to G^0$ is open.

For such a G, the vector space $C_c(G)$ is a convolution *-algebra, where for $f, g \in C_c(G)$:

$$f * g(x) = \int f(t)g(t^{-1}x)\, d\lambda^{r(x)}(t), \qquad f^*(x) = \overline{f(x^{-1})}.$$

We take $C^*(G)$ to be the enveloping C^*-algebra of $C_c(G)$ (representations are required to be continuous in the inductive limit topology). Equivalently, it is the completion of $\pi_{univ}(C_c(G))$ where π_{univ} is the universal representation of $C_c(G)$. For example, if $G = R_n$, then $C^*(G)$ is easy to guess! - it is just the finite dimensional matrix algebra $C_c(G) = M_n$, the span of the e_{ij}'s.

The class of locally compact groupoids which corresponds to discrete groups in the group category is that of the *r-discrete* groupoids. We can think of such a groupoid G as one for which the unit space G^0 is a locally compact Hausdorff space and the fibers G^u are discrete with local sections. The latter condition is more precisely expressed as saying that every $x \in G$ has an open neighborhood U for which $r(U)$ is open in G and the map $r_{|U}$ is a homeomorphism from U onto $r(U)$. (The unit space G^0 is therefore *open* in G.) The canonical left Haar system for G is that for which each λ^u is counting measure on G^u. Of course, every discrete groupoid is r-discrete. Other examples are transformation groupoids for which the acting group is discrete, and certain kinds of holonomy groupoids.

We will require in §5 the notion of a *bisection* a of a locally compact groupoid G. Here a is a pair of homeomorphisms $u \to a^u \in G^u$, $u \to a_u \in G_u$ from G^0 onto a subset A of G. So A is simultaneously an r-section and an s-section and determines and is determined by the bisection a. The set of bisections of G forms, by multiplying and inverting the A's, a group, and is denoted by Γ. If G is a locally compact group, then trivially $\Gamma = G$.

---

[1]In a number of contexts, e.g. that involving the holonomy groupoids for foliations, the Hausdorff condition is too strong, and locally Hausdorff groupoids need to be considered. For a discussion of the theory of such groupoids, the reader is referred to [**64, 46**]. Only locally compact Hausdorff groupoids will be considered in the present paper.

As in the case of a locally compact group, there is a reduced C^*-algebra $C^*_{red}(G)$ for G which is defined as follows. For each $u \in G^0$, we first define a representation π_u of $C_c(G)$ on the Hilbert space $L^2(G, \lambda_u)$. To this end, regard $C_c(G)$ as a dense subspace of $L^2(G, \lambda_u)$ and define for $f \in C_c(G), \xi \in C_c(G)$,

$$\pi_u(f)(\xi) = f * \xi \in C_c(G). \tag{2.1}$$

Then $\pi_u(f)$ extends to a bounded linear operator on $L^2(G, \lambda_u)$. The reduced C^*-algebra-norm on $C_c(G)$ is then (e.g. [**64**, p.108]) defined by:

$$\|f\|_{red} = \sup_{u \in G^0} \|\pi_u(f)\|$$

and $C^*_{red}(G)$ is defined to be the completion of $C_c(G)$ under this norm.

We next recall some details concerning the disintegration of representations of $C_c(G)$. The theorem is due to J. Renault ([**74**]). A detailed account of the theorem is given in the book of Paul Muhly ([**60**]). Let Φ be a representation of $C_c(G)$, continuous in the inductive limit topology, on a Hilbert space H. Then Φ disintegrates as follows.

There is a probability measure μ on G^0 which is *quasi-invariant* (in the sense defined below). Associated with μ is a positive regular Borel measure ν on G defined by: $\nu = \int \lambda^u \, d\mu(u)$. (As in [**70**], we sometimes write λ^μ in place of ν.) The measure ν^{-1} is the image under ν by inversion: precisely, $\nu^{-1}(E) = \nu(E^{-1})$. There is also a measure ν^2 on G^2 given by: $\nu^2 = \iint \lambda_u \times \lambda^u \, d\mu(u)$. The quasi-invariance of μ just means that ν is equivalent to ν^{-1}. The *modular function* D is defined as the Radon-Nikodym derivative $d\nu/d\nu^{-1}$. The function D can be taken to be Borel ([**32, 69, 60**]), and satisfies the properties: $D(x^{-1}) = D(x)^{-1}$ ν-almost everywhere, and $D(xy) = D(x)D(y)$ ν^2-almost everywhere. Let ν_0 be the measure on G given by: $d\nu_0 = D^{-1/2} d\nu$.

Next, there exists (e.g. [**64**, p.91]) a measurable Hilbert bundle $(G^0, \mathfrak{K}, \mu)$ ($\mathfrak{K} = \{\mathfrak{K}^u\}_{u \in G^0}$) and a *G-representation* L on $\mathfrak{K}$. This means that each $L(x)$ ($x \in G$) is a linear isometry from $\mathfrak{K}^{s(x)}$ onto $\mathfrak{K}^{r(x)}$ and $L(x)$ is the identity map if $x \in G^0$. Further, the map $x \to L(x)$ is multiplicative ν^2-almost everywhere and inverse preserving ν-almost everywhere. Lastly, for every pair ξ, η of square integrable sections of $\mathfrak{K}$, it is required that the function $x \to (L(x)\xi(s(x)), \eta(r(x)))$ be ν-measurable. The representation Φ of $C_c(G)$ is then given by:

$$\langle \Phi(f)\xi, \eta \rangle = \int f(x)(L(x)\xi(s(x)), \eta(r(x))) \, d\nu_0(x). \tag{2.2}$$

We will refer to the triple $(\mu, \mathfrak{K}, L)$ as a *measurable G-Hilbert bundle*. The notion of a *continuous* G-Hilbert bundle, that will be needed in §5, is defined in the obvious way. (In particular, no quasi-invariant measure μ on G^0 is required.)

Significant progress has been made in recent years in developing theories of the Fourier-Stieltjes and Fourier algebras for locally compact groupoids, and these theories will be described in the rest of the paper. The main papers involved are [**75, 70, 66**], and they will be considered in turn. As we will see, the theories relate to one another.

3. The Fourier-Stieltjes and Fourier algebras for a measured groupoid

This section discusses the paper [**75**] of Jean Renault that deals with the Fourier-Stieltjes and Fourier algebras of a *measured groupoid*. We can take the latter to mean ([**69**]) that G is a locally compact groupoid with left Haar system

$\{\lambda^u\}$ and fixed quasi-invariant measure μ with modular function D. The theory is a far reaching extension of the discussion of these algebras in §1 for locally compact groups, and as in the group case, has an operator algebraic character. (In the group case, $G = G^e$ where e is the identity of G, λ^e is a left Haar measure and μ is the point mass at e.)

As we saw in §1, the Fourier-Stieltjes algebra $B(G)$ of a locally compact group is the space of coefficients (ξ, η) arising from the Hilbert space representations of G. In the groupoid case, the Fourier-Stieltjes algebra $B_\mu(G)$ is defined to be the space of coefficients $\phi = (\xi, \eta)$ where ξ, η are L^∞-sections for some measurable G-Hilbert bundle $(\mu, \mathfrak{K}, L)$. So for $x \in G$,

$$\phi(x) = (L(x)\xi(s(x)), \eta(r(x))).$$

Clearly, ϕ belongs to $L^\infty(G) = L^\infty(G, \nu)$.

As in the group case, the set $P_\mu(G)$ of positive definite functions in $L^\infty(G)$ plays an important role. A function $\phi \in L^\infty(G)$ is called *positive definite* if and only if for all $u \in G^0$,

$$\iint \phi(y^{-1}x) f(y) \overline{f(x)}\, d\lambda^u(x)\, d\lambda^u(y) \geq 0.$$

An argument of Ramsay and Walter ([**70**]) gives that ϕ is positive definite if and only if ϕ is of the form (ξ, ξ) for some ξ. By polarization, $B_\mu(G)$ is the span of $P_\mu(G)$.

Also, as in the group case, the norm $\|\phi\|$ of $\phi \in B_\mu(G)$ is defined to be $\inf \|\xi\|\|\eta\|$ over all representations $\phi = (\xi, \eta)$. Using a groupoid version of Paulsen's "off-diagonalization technique" ([**67**, Th. 7.3], [**9**, Th. 5.3.2]), Renault shows that $B_\mu(G)$ is a commutative Banach algebra.

The Fourier algebra $A_\mu(G)$ is defined to be the closed linear span in $B_\mu(G)$ of the coefficients of the *regular* representation of G on the G-Hilbert bundle $\{L^2(G^u)\}_{u \in G^0}$, the G-action being given by left translation. (We will meet the continuous version of this Hilbert bundle again in §5.) Renault shows that $A_\mu(G)$ is a closed ideal in $B_\mu(G)$. The *Hilbertian* functions of Grothendieck ([**31**]) give an example of a $B_\mu(G)$ (with G a trivial groupoid).

Renault then goes on to examine duality theory using operator algebra and operator space techniques. Corresponding to the universal C^*-algebra $C^*(G)$ in the group case is the universal $C^*_\mu(G)$ in the measured groupoid case. The latter is the completion of $C_c(G)$ under the largest C^*-norm coming from some measurable G-Hilbert bundle $(\mu, \mathfrak{K}, L)$. The reduced C^*-algebra $C^*_{red}(G)$ and the von Neumann algebra $VN(G)$ (both depending on μ) are defined using the regular representation in the natural way.

In the group case, we saw that $B(G) = C^*(G)^*$. In the measured groupoid case, the situation is more subtle. Some operator space notions are required. Recall that any Hilbert space H can be regarded as a operator space by identifying it in the obvious way with a subspace of $\mathfrak{B}(\mathbb{C}, H)$: each $\xi \in H$ is identified with the map $a \to a\xi$ ($a \in \mathbb{C}$). Further, trivially, H^* is an operator space as a subspace of $\mathfrak{B}(H, \mathbb{C})$. In the measured groupoid situation, Renault shows that the operator spaces $L^2(G^0)$ and $C^*_\mu(G)$ are completely contractive left $L^\infty(G^0)$ modules, and $L^2(G^0)^*$ is a completely contractive right $L^\infty(G^0)$ module. If E is a right and F is a left A-operator module (A a C^*-algebra) then the Haagerup tensor norm is determined on the algebraic tensor product $E \odot_A F$ by setting $\|u\| = \sum_{i=1}^n \|e_i\|\|f_i\|$

over all representations $u = \sum_{i=1}^{n} e_i \otimes_A f_i$. The completion ([**2**, Ch. 2]) $E \otimes_A F$ of $E \odot_A F$ is called the *module Haagerup tensor product* of E and F over A.

One then forms the module Haagerup tensor product $X(G) = L^2(G^0)^* \otimes C^*_\mu(G) \otimes L^2(G^0)$ over $L^\infty(G^0)$. Renault proves that $X(G)^* = B_\mu(G)$. Under this identification, each $\phi = (\xi, \eta)$ goes over to the linear functional $a^* \otimes f \otimes b \to \int \overline{a \circ r}(\phi f) b \circ s \, d\nu \ (f \in C_c(G))$.

Renault also gives a characterization of $A_\mu(G)$ which is the groupoid version of the group result (§1) that $A(G) = VN(G)_*$. In fact the predual $VN(G)_*$ of $VN(G)$ is completely isometric to the module Haagerup tensor product $L^2(G^0)^* \otimes A_\mu(G) \otimes L^2(G^0)$ over $L^\infty(G^0)$. He also shows that the analogue of the group multiplier result, described in §1, holds, viz. *if the measure groupoid G is amenable (which can be defined as saying that the trivial representation is weakly contained in the regular representation) then $B_\mu(G)$ is the multiplier algebra of $A_\mu(G)$.* He then uses these results on $A_\mu(G), B_\mu(G)$ to investigate absolute Fourier multipliers for r-discrete groupoids, generalizing results of Pisier and Varopoulos.

Leptin's result (§1) relating the amenability of a locally compact group to the existence of a bounded approximate identity in the Fourier algebra has been generalized to the measured groupoid context by Jean-Michel Vallin ([**78, 79**]) using Hopf-von Neumann bimodule structures.

4. The Borel Fourier-Stieltjes algebra

In this section we describe some of the results of the paper [**70**] by Ramsay and Walter. The setting for the paper is that of a locally compact groupoid G. The groupoid is not considered to be a measured groupoid, so that a quasi-invariant measure μ on G^0 is not specified in advance. We will also consider briefly a paper by K. Oty at the end of the section.

Since in the group case, $B(G)$ is the span of the set $P(G)$ of continuous positive definite functions, it would be natural to use this as the definition of $B(G)$ in the groupoid case. We take this to be the definition of $B(G)$ in §5. However, there are examples ([**70**, §7]) in which the span of the continuous positive definite functions on G is not complete in the norm (defined below) on $B(G)$. Instead, Ramsay and Walter consider the span $\mathcal{B}(G)$ of the set $\mathcal{P}(G)$ of bounded *Borel* positive definite functions on G. There are examples of *continuous* functions in $\mathcal{B}(G)$ which cannot be expressed as linear combination of elements of $P(G)$. In $\mathcal{B}(G)$, we identify two functions that are λ^μ-equal almost everywhere for every quasi-invariant measure μ on G^0.

The norm on $\mathcal{B}(G)$ is defined in terms of a remarkable *completely bounded multiplier norm* on the C^*-algebra $M^*(G)$. (This norm does not seem to have been considered in theory of the Fourier-Stieltjes algebra in the group case.) In more detail, the universal representation π_{univ} extends canonically to a representation of the convolution algebra $B_c(G)$ of compactly supported, bounded Borel functions on G. Let $M^*(G)$ be the completion of $\pi_{univ}(B_c(G))$. Then $M^*(G)$ is a C^*-algebra. Each $\phi \in \mathcal{B}(G)$ acts as a multiplier T_ϕ on $M^*(G)$ by extending its action from $\pi_{univ}(B_c(G))$ by continuity. Here, for $f \in B_c(G)$, $T_\phi \pi_{univ}(f) = \pi_{univ}(\phi f)$ (ϕf pointwise multiplication on G). Ramsay and Walter show that T_ϕ is a completely bounded operator on $M^*(G)$. They define $\|\phi\|_{cb} = \|T\|_{cb}$, and show that $\mathcal{B}(G)$ is a Banach algebra under $\|.\|_{cb}$. (Below, we will identify the version of $\|.\|_{cb}$ for a locally compact abelian group in the continuous context.)

Ramsay and Walter give a partial analogue of the locally compact groupoid result (§1) that $B(G) = C^*(G)^*$. Let X be the one point compactification of G^0. They construct two C^*-algebras $C^*(G, X)$, $M^*(R, X)$ which are $C(X)$-submodules. Here R is the orbit equivalence relation on G^0: so $u \sim v$ if and only if there exists $x \in G$ such that $r(x) = u, s(x) = v$. They show using a technical argument that each $\phi \in \mathcal{B}(G)$ gives a completely bounded $C(X)$-bimodule map $S_\phi : C^*(G, X) \to M^*(R, X)$. (When G is a group, then S_ϕ is just the canonical linear functional $\phi : C^*(G) \to \mathbb{C}$.) It would be interesting to know how this result relates to the corresponding duality result involving $X(G)$ for measured groupoids in §3.

Karla Oty ([**62**]) investigates the space $\mathcal{B}(G) \cap C(G)$ and the set $P(G) \subset \mathcal{P}(G)$ of continuous positive definite functions. She shows among other results that $P(G)$ separates the points of G, and that if G is r-discrete then we do not have to work with equivalence classes in $\mathcal{B}(G)$. Further, if the set of cardinalities of the G^u's ($u \in G^0$) is bounded by a positive integer, then $\mathcal{B}(G) \cap C(G) = C(G)$.

5. The continuous Fourier-Stieltjes and Fourier algebras

In this section, we consider some of the results in the paper [**66**] by the present writer. In contrast to the measurable theory of Renault, considered in §3, and the Borel theory of Ramsay and Walter, considered in §4, the theory of this section is a continuous one. For motivation, in §3, in the context of a measured groupoid, $A_\mu(G)$ was defined as the closed linear span in $B_\mu(G)$ of the coefficients of the regular G-Hilbert bundle $\{L^2(G^u)\}$. Now this is a *continuous* Hilbert bundle, the continuous sections being determined in the obvious way by $C_c(G)$. So it is natural to consider the coefficients of *continuous* G-Hilbert bundles as our space $B(G)$. The elements of $B(G)$ are then all continuous as in the group case. As we will see, the ideas and techniques from [**75, 70**] play a fundamental role in the theory. We will have occasion to consider Hilbert modules, and so in accordance with the usual practice (e.g. [**49**]), for the rest of the paper, Hilbert spaces and modules will be taken to be conjugate linear in the first variable. (Because of this, some of the expressions (e.g. that for (f, g)) in the theory have to be modified.)

We now define the Fourier-Stieltjes algebra $B(G)$ in more detail. Given a continuous G-Hilbert bundle $\mathfrak{H}$, we consider the Banach space Δ_b of continuous, bounded sections of $\mathfrak{H}$. For $\xi, \eta \in \Delta_b$, define (as in §3) the *coefficient* $(\xi, \eta) \in C(G)$ by: $(\xi, \eta)(x) = (L_x\xi(s(x)), \eta(r(x)))$ where $x \to L_x$ is the G-action on $\mathfrak{H}$. Then $B(G)$ is defined to be the set of all such coefficients, coming from all possible continuous G-Hilbert bundles. As in [**75**], $B(G)$ is an algebra over $\mathbb{C}$ and the norm of $\phi \in B(G)$ is defined to be $\inf \|\xi\|\|\eta\|$, the inf being taken over all representations $\phi = (\xi, \eta)$. Then $B(G) \subset C(G)$, and $\|.\|_\infty \leq \|.\|$. An argument similar to that of [**75**] shows that $B(G)$ is a commutative Banach algebra.

From an argument of [**70**], $P(G) \subset B(G)$. Clearly, using polarization for sections of continuous Hilbert bundles, every $\phi \in B(G)$ is a linear combination of continuous positive definite elements of $B(G)$ and conversely. From the discussion in §4, we see that $\mathcal{B}(G) \cap C(G) \neq B(G)$ in general. Intuitively, $\mathcal{B}(G)$ captures the representation theory of G but we need to consider Borel functions. On the other hand, $B(G)$ may not capture all of the representation theory of G (though it does capture at least the regular representation) but has the advantage that the functions involved, are, as in the group case, continuous. (Our $B(G)$ is the same as Oty's $B_1(G)$ ([**62**]).)

We defined the norm on $B(G)$ using Hilbert bundles in a way similar to that in which Renault defined the norm on $B_{\mu}(G)$, and it is natural to enquire if the method of Ramsay and Walter for norming $\mathcal{B}(G)$ can also be applied in the present context. This is in fact the case. The difference is that we regard $\phi \in B(G)$ as a multiplier on $C_c(G)$, this action extending to a completely bounded map T_{ϕ} on $C^*(G)$ instead of on $M^*(G)$. We can then define a norm $\|.\|_{cb}$ on $B(G)$ by taking $\|\phi\|_{cb} = \|T_{\phi}\|_{cb}$. Following along the same lines as the corresponding argument in [**70**], one can show that $B(G)$ is a Banach algebra under $\|.\|_{cb}$. Since it is not difficult to show that $\|.\|_{cb} \leq \|.\|$, it follows by Banach's isomorphism theorem that the two norms are equivalent on $B(G)$. It is an open question if the two norms actually coincide in general. Using a result of Paulsen on Schur multipliers ([**67**]), they do coincide when $G = R_n$, the trivial groupoid on n elements (§2).

Let us sketch here how one can show that the two norms are the same when G is a locally compact abelian group. As in §1, we can regard ϕ as a measure $\mu \in M(\widehat{G})$. Recalling that $C^*(G) = C_0(\widehat{G})$, the map T_{ϕ} goes over to the map $T_{\mu} : C_0(\widehat{G}) \to C_0(\widehat{G})$ where

$$T_{\mu}(g)(x) = \mu * g(x) = \int g(t^{-1}x)\, d\mu(t). \tag{5.1}$$

Trivially, $\|T_{\mu}\| \leq \|\mu\|$. With e the identity of $\widehat{G}$ and using (5.1),

$$\|T_{\mu}\| \geq \sup_{\|g\|_{\infty}=1} | T_{\mu}g(e) | = \|\mu\|.$$

So $\|T_{\phi}\| = \|\mu\| = \|\phi\|$. Since $C^*(G)$ is abelian, $\|\phi\|_{cb} = \|T_{\phi}\|$ ([**67**, Th. 3.8]).

For the rest of the discussion, we can use either norm on $B(G)$. We have chosen to use $\|.\|$ since in the setting of $A(G)$ (below), the continuous G-Hilbert bundle $L^2(G) = \{L^2(G^u)\}$ is conveniently available for norm estimates.

Let E^2 be the space of continuous sections of $L^2(G)$ that vanish at ∞. Let $D = C_0(G^0)$. Then E^2 is a Hilbert D-module, with inner product given by: $\langle \xi, \eta \rangle(u) = (\xi(u), \eta(u))$ and left module action by: $(\xi a)(u) = a(u)\xi(u)$ $(a \in D, \xi \in E^2)$. (A good source of information about Hilbert A-modules is the book [**49**] by Lance.)

It is easy to check that if $\xi, \eta \in E^2$, then $\langle \xi, \eta \rangle = \eta * \xi^*$. For $F \in C_c(G)$, define $R_F : C_c(G) \to C_c(G)$ by right convolution: $R_F f = f * F$. Then the map $F \to R_F$ is a *-antirepresentation from $C_c(G)$ into $\mathfrak{L}(E^2)$, the C^*-algebra of adjointable linear maps on E^2, and the closure of its image in $\mathfrak{L}(E^2)$ is canonically isomorphic to $C^*_{red}(G)$. As in the group case, we can define $VN(G)$ to be the commutant of $C^*_{red}(G)$ in $\mathfrak{B}(E^2)$. It is to be stressed that $VN(G)$ is not a von Neumann algebra in general - this is not even the case for $G = R_n$. However, $VN(G)$ is always a Banach algebra, and is strongly closed in $\mathfrak{B}(E^2)$.

As in the group case, the algebra $VN(G)$ plays a useful role in the theory. In particular, if G is r-discrete then the fact that $C^*_{red}(G) \subset \mathfrak{L}(E^2)$ can be used to show that the natural version of Godement's theorem (§1) holds: i.e. that if $\phi \in C_c(G) \cap P(G)$, then $\phi = f * f^* = (f, f)$ for some $f \in E^2$. (The discrete group version of this is given in [**7**, Lemma VII.2.7].) It follows that $A(G)$ is an ideal in $B(G)$. I do not know if these results are true for locally compact groupoids in general.

For $f, g \in C_c(G)$, the coefficient $(f, g) = g * f^*$ itself belongs to $C_c(G)$. We define the Fourier algebra $A(G)$ to be *the closure in* $B(G)$ *of the subalgebra generated by the set of such elements* $g * f^*$. (Another possible definition of $A(G)$ is given by

Oty ([**62**]).) Of course, $A(G)$ is a commutative Banach algebra. Further, as in the group case, $A(G) \subset C_0(G)$. In the group case, $VN(G)$ is the dual of $A(G)$. Using the G-Hilbert module E^2, there is a version of this for the groupoid case, though it is more involved. For the technicalities, see [**66**].

The last part of [**66**] proves a duality theorem for $A(G)$ generalizing that of Eymard's in the group case. For the latter, recall (§1) that the character space of $A(G)$ is just G, the elements of G being characters under point evaluation. I do not know if this is still true in the groupoid case. The duality theorem of [**66**] is formulated in terms of bisections and certain multiplicative module maps from $A(G)$ into D. In one direction, the group Γ of bisections of G (§2) determines a pair of multiplicative maps on $A(G)$ as follows. If $a \in \Gamma$, then we define $\alpha^a : A(G) \to D$, $\beta_a : A(G) \to D$ by setting: $\alpha^a(\phi)(u) = \phi(a^u)$, $\beta_a(\phi)(u) = \phi(a_u)$. The pair (α^a, β_a) satisfy a number of interesting properties. For example, α^a is a D-module homomorphism for G, while β_a is the same for $G(r)$, the groupoid G with multiplication reversed. Crucial is the fact that if J is the homeomorphism $u \to v$ of G^0, where $a^u = a_v$, then for all $\phi \in A(G)$, we have $\beta_a(\phi) \circ J = \alpha^a(\phi)$.

We then consider the set $\Phi_{A(G)}$ of pairs of maps (α, β) satisfying these properties abstractly. We give Γ the topology of pointwise convergence on G^0, regarding each $a \in \Gamma$ as the (single) map $u \to (a_u, a^u) \in G^2$. Next, regarding each $(\alpha, \beta) \in \Phi_{A(G)}$ as the single map $(\phi, u) \to (\alpha(\phi)(u), \beta(\phi)(u)) \in \mathbb{C}^2$ on $A(G) \times G^0$, we also give $\Phi_{A(G)}$ the topology of pointwise convergence. The duality theorem of [**66**] then is that for a large class of locally compact groupoids G, *the map* $a \to (\alpha^a, \beta_a)$ *is a homeomorphism from* Γ *onto* $\Phi_{A(G)}$. In the group case, each bisection a is just a group element x and the maps α^a, β_a coincide on $A(G)$, being just the point evaluation at x. Further, $\Phi_{A(G)}$ is the space of characters of $A(G)$, and the above duality theorem reduces to Eymard's duality theorem.

A list of open questions in the continuous theory is given in [**66**]. This includes the natural questions as to whether $\|.\| = \|.\|_{cb}$ on $B(G)$ and whether the character space of $A(G)$ is identifiable with G.

References

[1] J. Baker, A. T. Lau and J. Pym, *Module homomorphisms and topological centers associated with weakly sequentially complete Banach algebras*, J. Functional Anal. 158(1998), 186-208.

[2] D. P. Blecher, P. S. Muhly and V. I. Paulsen, *Categories of operator modules - Morita equivalence and projective modules*, Mem. Amer. Math. Soc., Vol. 143, No. 681, 2000.

[3] A. Connes, *Sur la théorie non commutative de l'intégration*, Lecture Notes in Mathematics, 725(1979), 19-143.

[4] A. Connes, *Noncommutative Geometry*, Academic Press, Inc., New York, 1994.

[5] C-H. Chu and A. T. Lau, *Algebraic structures of harmonic functions on groups and Fourier algebras*, monograph, to appear, 2001.

[6] M. G. Cowling, *An application of Littlewood-Paley theory in harmonic analysis*, Math. Ann. 241(1979), 83-96.

[7] K. Davidson, *C^*-algebras by Example*, Fields Institute Monographs, American Mathematical Society, Providence, R.I., 1996.

[8] J. Dixmier, *C^*-algebras*, North-Holland Publishing Company, Amsterdam, 1977.

[9] E. G. Efros and Z-J. Ruan, *Operator Spaces*, London Mathematical Society Monographs New Series, Vol. 23, Clarendon Press, Oxford, 2000.

[10] E. G. Effros and Z-J. Ruan, Operator space tensor products and Hopf convolution algebras, preprint, 2001.

[11] P. Eymard, *L'algèbre de Fourier d'un groupe localement compact*, Bull. Soc. Math. France 192(1964), 181-236.

[12] B. E. Forrest, *Amenability and derivations of the Fourier algebra*, Proc. Amer. Math. Soc. 104(1988), 437-442.

[13] B. E. Forrest, *Amenability and bounded approximate identities in ideals of $A(G)$*, Illinois J. Math. 34(1990), 1-25.

[14] B. E. Forrest, *Arens regularity and discrete groups*, Pacific J. Math. 151(1991), 217-227.

[15] B. E. Forrest, *Some Banach algebras without discontinuous derivations*, Proc. Amer. Math. Soc. 114(1992), 965-970.

[16] B. E. Forrest, *Complemented ideals in the Fourier algebra and the Radon Nikodým property*, Trans. Amer. Math. Soc. 333(1992), 689-700.

[17] B. E. Forrest, *Amenability and ideals in $A(G)$*, J. Austral. Math. Soc. 53(1992), 143-155.

[18] B. E. Forrest, *Arens regularity and the $A_p(G)$ algebras*, Proc. Amer. Math. Soc. 119(1993), 595-598.

[19] B. E. Forrest and M. Skantharajah, *A note on a type of approximate identity in the Fourier algebra*, Proc. Amer. Math. Soc. 120(1994), 651-652.

[20] B. E. Forrest, *Amenability and the structure of the algebras $A_p(G)$*, Trans. Amer. Math. Soc. 343(1994), 233-243.

[21] B. E. Forrest, *Weak amenability and the second dual of the Fourier algebra*, Proc. Amer. Math. Soc. 125(1997), 2373-2378.

[22] B. E. Forrest, *Fourier analysis on coset spaces*, Rocky Mountain J. Math. 28(1998), 173-190.

[23] R. Godement, *Les fonctions de type positif et la théorie des groupes*, Trans. Amer. Math. Soc. 63(1948), 1-84.

[24] E. E. Granirer, *On some properties of the Banach algebras $A_p(G)$ for locally compact groups*, Proc. Amer. Math. Soc. 95(1985), 375-381.

[25] E. E. Granirer, *On some spaces of linear functionals on the algebras $A_p(G)$ for locally compact groups*, Colloq. Math. 52(1987), 119-132.

[26] E. E. Granirer, *A survey on some functional analytic properties of the Fourier algebra $A(G)$ of a locally compact group*, Southeast Asian Bull. Math. 20(1996), 1-12.

[27] E. E. Granirer, *Day points for quotients of the Fourier algebra $A(G)$, extreme non-ergodicity of their duals and extreme non-Arens regularity*, Illinois J. Math. 40(1996), 402-419.

[28] E. E. Granirer, *When quotients of the Fourier algebra $A(G)$ are ideals in their bidual and when $A(G)$ has WCHP*, Math. Japon. 46(1997), 69-72.

[29] E. E. Granirer, *Amenability and semisimplicity for second duals of quotients of the Fourier algebra $A(G)$*, J. Austral. Math. Soc. 63(1997), 289-296.

[30] E. E. Granirer, *The Schur property and the WRNP for submodules of the dual of the Fourier algebra $A(G)$*, C. R. Math. Rep. Acad. Sci. Canada 19(1997), 15-20.

[31] A. Grothendieck, *Résumé de la théorie métrique des produits tensoriels topologiques*, Boll. Soc. Mat. Sao-Paulo 8(1956), 1-79.

[32] P. Hahn, *Haar measure for measure groupoids*, Trans. Amer. Math. Soc. 242(1978), 1-33.

[33] C. Herz, *The theory of p-spaces with an application to convolution operators*, Trans. Amer. Math. Soc. 154(1971), 69-82.

[34] C. Herz, *Une généralization de la notion de transformée de Fourier-Stieltjes*, Ann. Inst. Fourier (Grenoble) 23(1974), 145-157.

[35] E. Hewitt and K. A. Ross, *Abstract Harmonic Analysis I*, Springer-Verlag, Berlin, 1963.

[36] E. Hewitt and K. A. Ross, *Abstract Harmonic Analysis II*, Springer-Verlag, New York, 1970.

[37] Z. Hu, *On the Set of Topologically Invariant Means on the Von Neumann Algebra VN(G)*, Illinois J. Math. 39(1995), 463-490.

[38] Z. Hu, *Extreme Non Arens Regularity of Quotients of the Fourier Algebra A(G)*, Colloq. Math. 72(1997), 237-249.

[39] Z. Hu, *The Von Neumann Algebra VN(G) of a Locally Compact Group and Quotients of Its Subspaces*, Canad. J. Math. 49(1997), 1117-1139.

[40] Z. Hu, *Spectrum of Commutative Banach Algebras and Isomorphism of C^*-algebras Related to Locally Compact Groups*, Studia Math. 129(1998), 207-223.

[41] Z. Hu, *Open Subgroups of G and Almost Periodic Functionals on A(G)*, Proc. Amer. Math. Soc. 128(2000), 2473-2478.

[42] B. E. Johnson, *Cohomology in Banach algebras*, Mem. Amer. Math. Soc. 127(1972).

[43] B. E. Johnson, *Non-amenability of the Fourier algebra of a compact group*, J. London Math. Soc. 50(1994), 361-374.

[44] E. Kaniuth and A. T. Lau, *A separation property of positive definite functions on locally compact groups and applications to Fourier algebras*, J. Functional Anal. 175(2000), 89-110.

[45] E. Kaniuth and A. T. Lau, *Spectral synthesis for $A(G)$ and subspaces of $VN(G)$*, Proc. Amer. Math. Soc. 129(2001), 3253-3263.

[46] M. Khoshkam and G. Skandalis, *Regular Representation of Groupoid C^*-algebras and Applications to Inverse Semigroups*, preprint, 2001.

[47] J. Kraus and Z-J. Ruan, *Approximation properties for Kac algebras*, Indiana Univ. Math. J. 48(1999), 469-535.

[48] A. Kumjian, D. Pask, I. Raeburn and J. Renault, *Graphs, groupoids and Cuntz-Krieger algebras*, J. Functional Anal. 144(1997), 505-541.

[49] E. C. Lance, *Hilbert C^*-modules*, London Mathematical Society Lecture Note Series 210, Cambridge University Press, 1995.

[50] N. P. Landsman, *Mathematical topics between classical and quantum mechanics*, Springer Monographs in Mathematics, Springer-Verlag, NY, 1998.

[51] A. T. Lau, *Fourier and Fourier-Stieltjes algebras of a locally compact group and amenability*, pp. 79-92 of: *Topological vector spaces and related areas, Hamilton, Ontario*, Pitman Research Notes in Mathematics Series 316, Longman Scientific and Technical, Harlow, 1994.

[52] A. T. Lau and V. Losert, *Ergodic sequences in the Fourier-Stieltjes algebra and measure algebra of a locally compact group*, Trans. Amer. Math. Soc. 351(1999), 417-428.

[53] A. T. Lau and R. J. Loy, *Weak amenability of Banach algebras on locally compact groups*, J. Functional Anal. 145(1997), 175-204.

[54] A. T. Lau, R. J. Loy and G. A. Willis, *Amenability of Banach and C^*-algebras on locally compact groups*, Studia Math. 119(1996), 161-178.

[55] A. T. Lau, *Some geometric properties on the Fourier and Fourier-Stieltjes algebras of locally compact groups, Arens regularity and related problems*, Trans. Amer. Math. Soc. 337(1993), 321-359.

[56] H. Leptin, *Sur l'algèbre de Fourier d'un groupe localement compact*, C. R. Acad. Sci. Paris Sér. A-B 266(1968), 489-494.

[57] V. Losert, *Properties of the Fourier algebra that are equivalent to amenability*, Proc. Amer. Math. Soc. 92(1984), 347-354.

[58] K. C. H. Mackenzie, *Lie groupoids and Lie algebroids in Differential Geometry*, London Mathematical Society Lecture Note Series, vol. 124, Cambridge University Press, Cambridge, 1987.

[59] K. McKennon, *Multipliers, positive functionals, positive definite functions and Fourier-Stieltjes transforms*, Mem. Amer. Math. Soc. 111(1971).

[60] P. S. Muhly, *Coordinates in Operator Algebra*, to appear, CBMS Regional Conference Series in Mathematics, American Mathematical Society, Providence, 180pp..

[61] P. S. Muhly, J. N. Renault and D. P. Williams, *Equivalence and isomorphism for groupoid C^*-algebras*, J. Operator Theory 17(1987), 3-22.

[62] K. Oty, *Fourier-Stieltjes algebras of r-discrete groupoids*, J. Operator Theory 41(1999), 175-197.

[63] A. L. T. Paterson, *Amenability*, Mathematical Surveys and Monographs, No. 29, American Mathematical Society, Providence, R. I., 1988.

[64] A. L. T. Paterson, *Groupoids, inverse semigroups and their operator algebras*, Progress in Mathematics, Vol. 170, Birkhäuser, Boston, 1999.

[65] A. L. T. Paterson, *Graph inverse semigroups, groupoids and their C^*-algebras*, to appear, J. Operator Theory.

[66] A. L. T. Paterson, *The Fourier algebra for locally compact groupoids*, preprint, 2001.

[67] V. I. Paulsen, *Completely bounded maps and dilations*, Pitman Research Notes in Mathematics Ser. 146, Longman and John Wiley and Sons, Inc., New York, 1986.

[68] J. -P. Pier, *Amenable locally compact groups*, Wiley, New York, 1984.

[69] A. Ramsay, *Topologies for measured groupoids*, J. Functional Anal. 47(1982), 314-343.

[70] A. Ramsay and M. E. Walter, *Fourier-Stieltjes algebras of locally compact groupoids*, J. Functional Anal. 148(1997), 314-367.

[71] H. Reiter, *Classical Harmonic Analysis and Locally Compact Groups*, Oxford Mathematical Monographs, Oxford University Press, Oxford, 1968.

[72] P. F. Renaud, *Centralizers of the Fourier algebra of an amenable group*, Proc. Amer. Math. Soc. 32(1972), 539-542.

[73] J. N. Renault, *A groupoid approach to C^*-algebras*, Lecture Notes in Mathematics, Vol. 793, Springer-Verlag, New York, 1980.

[74] J. N. Renault, *Répresentation de produits croisés d'algèbres de groupoïdes*, J. Operator Theory, 18(1987), 67-97.

[75] J. N. Renault, *The Fourier algebra of a measured groupoid and its multipliers*, J. Functional Anal. 145(1997), 455-490.

[76] Z-J. Ruan, *The operator amenability of $A(G)$*, Amer. J. Math. 117(1995), 1449-1474.

[77] Z-J. Ruan and G. Xu, *Splitting properties of operator bimodules and operator amenability of Kac algebras*, pp.193-216 in: *Operator theory, operator algebras and related topics*, (Timişoara, 1996), Theta Foundation, Bucharest, 1997.

[78] J. Vallin, *Bimodules de Hopf et poids operatoriels de Haar*, J. Operator Theory 35(1996), 39-65.

[79] J. Vallin, *Unitaire pseudo-multiplicatif associé à un groupoïde applications à la moyennabilité*, preprint.

[80] M. E. Walter, *W^*-algebras and nonabelian harmonic analysis*, J. Functional Anal. 11(1972), 17-38.

[81] A. Weinstein, *Groupoids: unifying internal and external symmetry. A tour through some examples*, Notices Amer. Math. Soc. 43(1996), 744-752.

[82] A. Weinstein, *Poisson Geometry*, Diff. Geom. Appl., 9(1998), 213-238.

[83] P. J. Woods, *Complemented ideals in the Fourier algebra of a locally compact group*, Proc. Amer. Math. Soc. 128(2000), 445-451.

Department of Mathematics, University of Mississippi, University, Mississippi 38677
E-mail address: mmap@olemiss.edu

Contemporary Mathematics
Volume **321**, 2003

Preserving the commutant under functional calculus

Gabriel T. Prăjitură

ABSTRACT. For an operator T and a function f, analitic in a neighborhood of the spectrum of T, we will discuss several conditions that will ensure that T and $f(T)$ have the same commutant or the same hyperinvariant subspaces.

1. Introduction

By an operator we mean a bounded linear operator on a complex Hilbert space. If T is an operator we will use $\sigma(T)$ to denote the spectrum of T. Also, $H(\sigma(T))$ will denote the set of all functions that are analytic in some neighborhood of $\sigma(T)$. We will denote by $\{T\}'$ the commutant of T; this is the set of all operators A such that $TA = AT$.

If T is an operator an f is a function such that $f(T)$ makes sense then it is always true that

$$\{T\}' \subseteq \{f(T)\}'.$$

For a proof, see Proposition 4.9 on page 204 in [**4**] for the case when $f \in H(\sigma(T))$ and Proposition 8.1 on page 285 in [**4**] for the case when T is normal and f is a Borel function. Most of the sufficient conditions for $\{T\}' = \{f(T)\}'$ are based on the possibility of getting $T = g(f(T))$ for some function g, which is connected to the existence of a kind of an inverse for the function f. Therefore it is not surprisingly that all these conditions will be in some neighborhood of "one - to - one".

2. The Conditions

For an operator T we will use $\sigma_p(T)$ to denote the set of all eigenvalue of T and $\sigma_c(T)$ for the compression spectrum of T. Recall that $\lambda \in \sigma_c(T)$ iff $\overline{\lambda} \in \sigma_p(T^*)$. Here and hereafter the bar stands for the complex conjugate.

If x and y are two norm - one vectors in a Hilbert space then $x \otimes y$ will denote the rank one operator that takes y into x; i. e. $x \otimes y(z) = \langle z, y \rangle x$.

Let $\alpha \in \sigma_p(T)$ and $\beta \in \sigma_c(T)$. Let x be a norm - one eigenvector corresponding to the eigenvalue α of T and y be a norm - one eigenvector coresponding to the eigenvalue $\overline{\beta}$ of T^*. We have that:

$$T\,(x \otimes y) = (Tx) \otimes y = (\alpha x) \otimes y = \alpha(x \otimes y)$$

2000 *Mathematics Subject Classification.* Primary: 47 A60; Secondary: 47 A10, 47 B15.
Key words and phrases. operator,functional calculus, spectrum, commutant, hyperinvariant.

while

$$(x \otimes y)\,T = x \otimes (T^* y) = x \otimes (\overline{\beta} y) = \beta(x \otimes y).$$

Also, for $f \in H(\sigma(T))$ we have that:

$$f(T)\,(x \otimes y) = (f(T)x) \otimes y = (f(\alpha)x) \otimes y = f(\alpha)(x \otimes y)$$

and

$$(x \otimes y)\,f(T) = x \otimes (f(T)^* y) = x \otimes (\overline{f(\beta)} y) = f(\beta)(x \otimes y).$$

From the above relations it is easy to see the following necessary condition.

2.1. PROPOSITION. *If T is an operator and f is in $H(\sigma(T))$ such that $\{T\}' = \{f(T)\}'$ then for every $\alpha \in \sigma_p(T)$ and $\beta \in \sigma_c(T)$, $\alpha \neq \beta$ implies $f(\alpha) \neq f(\beta)$.*

The oldest sufficient condition for the permanence of the commutant is the following theorem from [**7**].

2.2. THEOREM. *If T is an operator and f is in $H(\sigma(T))$ such that f is one-to-one on $\sigma(T)$ and f' does not have any zero in $\sigma(T)$ then T and $f(T)$ have the same commutant.*

PROOF. Since f is one-to-one on $\sigma(T)$ and f' does not have zeros on $\sigma(T)$, there is a neighborhood of $\sigma(T)$, W, such that f is one-to-one in W and f' does not have zeros in W.Therefore f is invertible in a neighborhood of $\sigma(T)$ and has an analytic inverse $g \in H(\sigma(f(T)))$. It is easy to see that $g(f(T)) = T$. □

The next result is an obvious consequence.

2.3. COROLLARY. *Let T_1, T_2 be operators and $f_j \in H(\sigma(T_j))$, such that f_j is one-to-one on $\sigma(T_j)$ and f_j' does not have any zero in $\sigma(T_j)$, for $j = 1, 2$. Then $f_1(T_1)f_2(T_2) = f_2(T_2)f_1(T_1)$ implies $T_1T_2 = T_2T_1$.*

For $f_1(z) = f_2(z) = e^z$ we obtain the results in [**13**] and [**11**].

The condition in Theorem 2.2 is hardly necessary, as can be seen in the following example from [**14**]. Let S be the unilateral shift and $T = \frac{1}{n} + S^*$ for some $n \geq 2$. The function z^2 is really far from being one-to-one on the spectrum of T and f' has a zero in the interior of the spectrum. Nevertheless, T and T^2 have the same commutant. To see this, let A be an operator such that $T^2A = AT^2$ and let λ be a complex number in the interior of the spectrum of T such that $|\lambda| > 1 - \frac{1}{n}$. Then λ is an eigenvalue for T while $-\lambda$ is not in the spectrum of T. Let $x \in \ker(T - \lambda)$. We have that $(T^2 - \lambda^2)A = A(T^2 - \lambda^2)$ or $(T + \lambda)(T - \lambda)A = A(T + \lambda)(T - \lambda)$. Therefore $(T + \lambda)(T - \lambda)Ax = 0$ and since $(T + \lambda)$ is invertible $(T - \lambda)Ax = 0$. Thus $(T - \lambda)Ax = A(T - \lambda)x$ or $TAx = ATx$ for every x in such a kernel. Hence $AT = TA$ because all these kernels span the space.

Before we will present a result a little bit stronger we need to discuss about the commutant of a direct sum of operators and some more definitions. A simple algebraic computation shows that the operator

$$\begin{pmatrix} X & Y \\ U & V \end{pmatrix}$$

commutes with $A \oplus B$ iff $AX = XA, BV = VB, AY = YB, BU = UA$. By Rosenblum's theorem (see [**10**]), if $\sigma(A) \cap \sigma(B) = \emptyset$ then there are no nonzero intertwining operators Y and U. Therefore, if $\sigma(A) \cap \sigma(B) = \emptyset$ then $\{A \oplus B\}' = \{A\}' \oplus \{B\}'$.

If C is a component of $\sigma(T)$ then the dimension of the subspace corresponding to the Riesz idempotent of C is called the (spectral) multiplicity of C. We will denote by $\sigma_{p0}(T)$ the set of all isolated eigenvalues of T having finite multiplicity (also called normal eigenvalues).

2.4. THEOREM. *If T is an operator and f is in $H(\sigma(T))$ such that:*
(a) f is one-to-one on $\sigma(T)$;
(b) f' does not have any zero in $\sigma(T) \setminus \sigma_{p0}(T)$;
(c) for every $\alpha \in \sigma_{p0}(T)$ with $f'(\alpha) = 0$ the spectral multiplicity is 1
then T and $f(T)$ have the same commutant.

PROOF. There is an open set W containing $\sigma(T) \setminus \sigma_{p0}(T)$ where f is invertible and f' is zero on $\sigma(T) \setminus W$. Since $\sigma(T) \setminus W$ is finite, say $\{\lambda_1, \ldots, \lambda_n\}$ T is similar to $A = A_0 \oplus A_1 \oplus \cdots \oplus A_n$, where $\sigma(A_0) = \sigma(T) \cap W$ and $\sigma(A_j) = \lambda_j$ for every $1 \leq j \leq n$.

Since the property "X and $f(X)$ have the same commutant" is invariant under similarities it will sufice to show that A has it. By 2.2, $\{A_0\}' = \{f(A_0)\}'$. Also, as by hypothesis for $1 \leq j \leq n$ A_j is an operator on a one - dimensional space, we have that $\{A_j\}' = \{f(A_j)\}'$.

The remark about the commutant of a direct sum finishes the proof. □

Next we will turn our attention to the particular case of normal operators. In this case we will consider continuous functions on the spectrum. Recall that if f is a continuous function, X a Borel subset of the domain of f such that f is one-to-one on X then $f(X)$ is a Borel set(see Theorem 11.12 on page 488 in [**2**]). For spectral measures and the functional calculus for normal operators see Chapter 2 in [**5**].

2.5. THEOREM. *Let N be a normal operator and f a continuous function on $\sigma(N)$. If there are Borel sets X and Y such that $\sigma(N) = X \cup Y$, f is one-to-one on X, $f(X) = f(\sigma(N))$ and Y has spectral measure 0 then N and $f(N)$ have the same commutant.*

PROOF. For a normal operator N and ϕ a Borel function on the spectrum of N, the commutant of N is always included in the commutant of $\phi(N)$ (see Proposition 8.1 on page 285 in [**4**]). In particular, this implies that the commutant of N is included in the commutant of $f(N)$. But, since $f(N)$ is itself a normal operator (see Proposition 1.12 on page 258 in [**4**]), it also implies that in order to show that the commutant of $f(N)$ is included in the commutant of N it suffices to find a Borel function g on $\sigma(f(N))$ such that $N = g(f(N))$.

Let $g : f(\sigma(N)) \longrightarrow \sigma(N)$ be the inverse of f that takes values in X. If K is a closed subset of $\sigma(N)$ then $g^{-1}(K) = f(X \cap K)$ which is a Borel set, being one-to-one continuous image of a Borel set. Thus g is a bounded measurable function on the spectrum of the normal operator $f(N)$ and so we can define $g(f(N))$. Since the spectral measure of Y is 0, it is not difficult to see that $N = g(f(N))$ (see Theorem 11.5 on page 53 and Proposition 10.4 on page 49 in [**5**]), which concludes the proof. □

2.6. COROLLARY. *Let N_1, N_2 be normal operators and for $j = 1, 2$, f_j a continuous function on $\sigma(T_j)$, such that there are Borel sets X_j and Y_j with the following properties: $\sigma(N_j) = X_j \cup Y_j$, f_j is one-to-one on X_j, $f_j(X_j) = f_j(\sigma(N_j))$ and Y_j has spectral measure 0. Then $f_1(N_1)f_2(N_2) = f_2(N_2)f_1(N_1)$ implies $N_1N_2 = N_2N_1$.*

In [**12**], in the statement of the theorem for exponentials the condition is $\sigma(N_j) \cap \sigma(N_j + 2\pi i)$ has spectral measure 0 for every $n \geq 1$. Taking $Y_j = \cup_{n \geq 1}(\sigma(N_j) \cap \sigma(N_j + 2\pi i))$ (which is a compact set, since only a finite number of the intersections are nonempty), $X_j = \sigma(N_j) \setminus Y_j$ and $f_j(z) = e^z$ we can apply Theorem 2.5 to obtain the result.

Although by far more general then the main result in [**12**], 2.5 is a particular case of one direction of the main theorem in [**6**]. As shown there, the result is actually true for invertible Borel functions whose inverse is also a Borel function and it is in fact an equivalence. Recall that 2.2 was really far from being necessary.

We want to conclude this section by mentioning the result of [**3**]. It is about a particular function, namely z^n and shows that $\{T\}' = \{T^n\}'$ iff for any $\lambda \neq 1$, n-th root of one, there is no nonzero operator λ commuting with T.

3. The class

Two properties closely related to having the same commutant are generating the same algebras and having the same hyperinvariant subspaces. But we need some more definitions. For an operator T we will denote by $\mathcal{U}(T)$ the uniform closed algebra (that is closed in the norm topology) generated by the rational fanctions in T. By generating the same algebra we mean when is $\mathcal{U}(T) = \mathcal{U}(f(T))$. In is not difficult to see that this property is equivalent to the existence of a sequence of rational functions $\{R_n\}_n$ such that $T = \lim_n R_n(f(T))$. From here we can conclude easily that generating the same algebras implies having the same commutant. For a fixed f we will denote by

$$C_1 = \{T; \mathcal{U}(T) = \mathcal{U}(f(T))\}.$$

This class of operators was studied in [**8**]. The spectral description of this class of operators is the subject of the next two results.

First we will characterize the closure of the class.

3.1. THEOREM. *For an operator T the following are equivalent.*
(a) $T \in \operatorname{cl} C_1$.
(c) *f is one-to-one on $\mathcal{P}_\pm(T)$ and if K is a component of $\sigma_{lre}(T) \cup \sigma_{p0}(T)$ then $f(K) \not\subset f(\mathcal{P}_\pm(T))$.*

Here $\mathcal{P}_\pm(T)$ is the subset of the semi-Fredholm domain of T with nonzero index and $\sigma_{lre}(T)$ is the left and right essential spectrum, i.e. the part of the spectrum where there is no left inverse and no right inverse modulo the compact operators.

Next we will characterize the interior. Let Z be the set of all zeros of f'.

3.2. THEOREM. *For an operator T the following are equivalent.*
(a) $T \in \operatorname{int} C_1$.
(b) *f is one-to-one on an open set containing $\sigma(T)$ or f is one-to-one on an open set containing $\sigma(T) \setminus \sigma_{p0}(T)$ and if $\lambda \in Z \cap \sigma_{p0}(T)$ then λ has multiplicity one.*

A subspace (closed linear manifold) $\mathcal{M}$ of a Hilbert space is called invariant for an operator T if $T(\mathcal{M}) \subset \mathcal{M}$. The set of all invariant subspaces of T is denoted by $\operatorname{Lat}(T)$. A hyperinvariant subspace is one that is invariant under each operator commuting with T. The set of all such subspaces is denoted by $\operatorname{HyperLat}(T)$. So

$$\operatorname{HyperLat}(T) = \cap_{A \in \{T\}'} \operatorname{Lat} A.$$

This last relation shows that having the same commutant implies having the same hyperinvariant subspaces.

As a consequence, all the suficient conditions for having the same commutant are also suficient for having the same hyperinvariant subspaces. It is not clear at all if the condition in 2.1 is still necessary. It can be shown that it is provided that at least one of α or β is isolated, but we don't know what happens in general. This property is by far more elusive than the other two. Only recently we finished to characterize it in finite dimensional spaces. The following result is from [**9**]

3.3. THEOREM. *If T is an operator on $\mathbb{C}^n$ then* $\operatorname{HyperLat} T = \operatorname{HyperLat} f(T)$ *if and only if*
a) f is one-to-one on $\sigma(T)$
and, if there is $\lambda \in \sigma(T)$ such that $f'(\lambda) = 0$ then either
b_1) $f''(\lambda) = 0$ and all λ Jordan blocks of T have dimension 1
or
b_2) $f''(\lambda) \neq 0$ and all λ Jordan blocks of T have the same odd dimension.

Let

$$C_2 = \{T; \{T\}' = \{f(T)\}'\}$$

and

$$C_3 = \{T; \operatorname{HyperLat} T = \operatorname{HyperLat} f(T)\}.$$

As we pointed before, $C_1 \subset C_2 \subset C_3$.

We believe that C_2 and C_3 have the same closure. As the example in the previos section shows for the function z^2, the closure of C_2 includes strictly the closure of C_1.

We have the following characterization of the interior.

3.4. THEOREM. *For an operator T the following are equivalent.*
(a) $T \in \operatorname{int} C_2$.
(b) $T \in \operatorname{int} C_3$.
(c) *f is one-to-one on an open set containing $\sigma(T)$ or f is one-to-one on an open set containing $\sigma(T) \setminus \sigma_{p0}(T)$ and if $\lambda \in Z \cap \sigma_{p0}(T)$ then λ has multiplicity one.*

PROOF. It is obvious that (a) implies (b).

(c) implies (a) follows by 3.2

Sketch of (b) implies (c)

If T does not satisfy one of the conditions in (c), we can construct into any ε from T an Apostol - Morrel model that is not in the class because of the necessary condition. Indeed, the Apostol - Morrel models give us the freedom to find isolated α and β. □

References

[1] I. N. Baker, James A. Deddens, J. L. Ullman, A Theorem on entire functions with applications to Toeplitz operators, Duke Math. J. **41** (1974), 739-745.

[2] Andrew M. Bruckner, Judith B. Bruckner, Brian S. Thomson, *Real Analysis*, Prentice-Hall, Upper Saddle River, New Jersey, 1997.

[3] Carl C. Cowen, Commutants and the operator equations $AX = \lambda XA$. Pacific J. Math. 80 (1979), no. 2, 337–340.

[4] John B. Conway, *A Course in Functional Analysis*, Springer-Verlag, New York Berlin Heidelberg, 1990.

[5] John B. Conway, *A Course in Operator Theory*, American Mathematical Society, Providence, Rhode Island, 1999.

[6] John B. Conway, T. A. Gillespie, Is a selfadjoint operator determined by its invariant subspace lattice? J. Funct. Anal. 64 (1985), no. 2, 178–189.

[7] M. Finkelstein, A. Lebow, A note on "nth roots of operators" Proc. Amer. Math. Soc. **21** (1969), 250

[8] Gabriel Prăjitură, *Permanence of invariant subspaces under functional calculus and related topics*, Thesis, University of Tennessee, Knoxville, 1999

[9] Gabriel Prăjitură, When is Hyperlat T = Hyperlat f(T) in finite dimension, to appear in Linear Algebra and its Applications
[10] Marvin Rosenblum, On the operator equation $BX - XA = Q$. Duke Math. J. **23** (1956), 263–269.
[11] Christoph Schmoeger, Remarks on commuting exponentials in Banach algebras, Proc. Amer. Math. Soc. **127** (1999), 1337-1338.
[12] Christoph Schmoeger, On normal operator exponentials, Proc. Amer. Math. Soc. **130** (2002), 697 - 702.
[13] E. M. E. Wermuth, A remark on commuting operator exponentials, Proc. Amer. Math. Soc. **125** (1997), 1685-1688.
[14] Warren R. Wogen, Private communication, 1999.

Department of Mathematics, SUNY Brockport, Brockport, NY, 14420
E-mail address: gprajitu@brockport.edu

Contemporary Mathematics
Volume **321**, 2003

L_p-spaces associated with a von Neumann algebra without trace: a gentle introduction via complex interpolation

Yves Raynaud

ABSTRACT. We expound the construction and the main structural properties of non commutative L_p spaces associated with type III von Neumann-algebras (i. e. von Neumann algebras having no non trivial normal semifinite trace). This exposition presupposes a minimal knowledge of von Neumann algebra theory; its starting point is based on Complex interpolation tools which are familiar to Functional Analysts; the introduction of more involved material from Tomita-Takesaki theory is postponed to the stage where it appears unavoidable. For the sake of simplicity we consider a σ-finite algebra (i. e. having a faithful normal state). The exposition is almost self-contained.

Introduction

Since quite a long time the noncommutative L_p-spaces are familiar to mathematicians working in the field of Functional Analysis, at least under their simplest form, the Schatten classes ([**MC**]). These are the non-commutative analogues of the complex ℓ_p-spaces; the measure space underlying to the usual ℓ_p-spaces, i. e. the set $\mathbb{N}$ of natural numbers equipped with the counting measure, is "replaced" (in the case of Schatten spaces) by the space $B(\ell_2)$ of bounded linear operators on ℓ_2 equipped with the trace application Tr (defined on the positive cone $B(\ell_2)_+$ with values in $[0, +\infty]$). In fact it is more correct to say that the *-algebra $B(\ell_2)$ and its trace play the role of the complex algebra ℓ_∞ (with the pointwise product and conjugation operation) and the sum map $(\ell_\infty)_+ \to [0, +\infty]$, $(x_n) \mapsto \sum\limits_n x_n$.

A new step in the construction of non-commutative L_p space was made with the theory of non-commutative integration (Dixmier, Segal, Nelson). The non-commutative algebras underlying to this construction are now the von Neumann algebras, i.e. the *-subalgebras of $B(H)$ (where H is a Hilbert space) containing the identity I which are closed in weak operator topology. In this theory one assumes that the von Neumann algebra $\mathcal{A}$ under consideration is equipped with a trace $\tau : \mathcal{A}_+ \to [0, +\infty]$ which has the basic properties of the trace of $B(H)$ (it

2000 *Mathematics Subject Classification.* Primary 46L52; Secondary 46B70, 46B25.
Key words and phrases. Non-commutative L_p-spaces, Von Neumann algebras, Complex interpolation, Modular theory.

is normal, faithful, semi-finite). The "measurable sets" of this theory are now the (orthogonal) projections belonging to $\mathcal{A}$, their "measure" is their trace. (It would be more exact to assimilate the set $\mathcal{P}(\mathcal{A})$ of projections in $\mathcal{A}$ to the complete Boole algebra obtained in measure theory by quotienting the σ-algebra Σ of μ-measurable sets by the ideal $\mathcal{I}_\mu$ of negligible sets- see [**Fr**] for the theory of L_p-spaces over complete measure algebras). The space $L_p(\mathcal{A}, \tau)$ is the completion of the space $\mathfrak{m}_p(\tau)$ of operators in $x \in \mathcal{A}$ such that $\tau(|x|^p) < \infty$, equipped with the norm $\|x\|_p = \tau(|x|^p)^{1/p} < \infty$. The fact that $\mathfrak{m}_p(\tau)$ is a linear subspace (in fact a two-sided ideal) of $\mathcal{A}$ and $\| \ \|_p$ a norm is intimately connected with the trace property $(\tau(x^*x) = \tau(xx^*))$.

Since the fundamental work on the theory of Von Neumann algebras by Murray and Von Neumann it is known that not every von Neumann algebra can be equipped with a n.s.f. trace. In this case ("type III von Neumann algebras") one is led to leave out the tracial property and to work with a normal semi-finite faithful weight. However the Dixmier-Segal approach fails completely for a weight and the non-commutative L_p spaces were only defined under several (equivalent) forms in the late 70's (Haagerup [**H**], Araki-Masuda [**AM**], Hilsum [**Hi**], Kosaki [**Ko1**], Leinert [**L**]), after the development of the Tomita-Takesaki theory for the study of type III von Neumann algebras.

The aim of this expository paper is to present the construction of these L_p-space for a type III von Neumann algebra which is σ-finite (i. e. admits a normal faithful positive linear functional φ_0). It contains only well known material, but the presentation is significantly different of those given in the papers cited above, in that we assume only a basic knowledge on von Neumann algebras and we define the objects of the study very quickly by complex interpolation (this point of view was initiated by Kosaki [**Ko1**], but with heavy use of Haagerup's construction; see also [**Te2**]). We refer to [**BL**] for the general theory of complex interpolation.

The two properties of the tracial $L_p(\mathcal{A}, \tau)$ spaces which are at the root of this presentation are the following: firstly $L_1(\mathcal{A}; \tau)$ is naturally identifiable with a predual of $\mathcal{A}$; secondly the spaces $L_1(\mathcal{A}; \tau)$ and $\mathcal{A}$ form naturally a compatible couple of Banach spaces in the sense of interpolation theory and $L_p(\mathcal{A}, \tau)$ can be isometrically identified with the complex interpolation space $(L_1(\mathcal{A}; \tau), \mathcal{A})_\theta$ (where $\theta = 1 - \frac{1}{p}$).

Since every von Neumann algebra $\mathcal{A}$ has a unique predual $\mathcal{A}_*$ (see [**S**]) it is natural to decide in the general case that $L_1(\mathcal{A})$ is this predual. The structure of $L_1(\mathcal{A})$ is well known ($\mathcal{A}$-bimodule structure, order, polar decomposition) and described in the first chapters of the classical books ([**D**], [**KaR**], [**S**], [**T**]). Then we define $L_p(\mathcal{A})$, $1 < p < \infty$ by complex interpolation (in Section 1). This way has the considerable advantage to present quasi immediately these objects in an form which is understandable with a limited knowledge on von Neumann algebras by Functional analysts outside of the subject; it has the disavantage to hide several structural properties of these objects.

The subsequent sections aim to reveal these structural properties. More deeper we progress in the exposition of these properties, more intensively we are led to invoke the Tomita-Takesaki theory (which is not a prerequisite for the reading of the present paper).

In the frame we choose there is not one, but two natural embeddings of $\mathcal{A}$ into its predual (the left embedding and the right one). This gives rise to two scales of interpolation spaces $L_p^\ell(\mathcal{A})$, $L_p^r(\mathcal{A})$, $1 < p < \infty$, which are easily seen to be anti-isometrically identifiable (i. e. by an antilinear isomorphism). The left L_p spaces inherit of a natural left-$\mathcal{A}$-module structure, while the right ones have are right $\mathcal{A}$-modules. In the case $p = 2$ general interpolation facts show that the L_2 spaces coincide with the completions H_{φ_0}, K_{φ_0} of the prehilbertian structures on $\mathcal{A}$ associated with the scalar products $\langle x, y\rangle = \varphi_0(y^*x)$, resp. $\langle x, y\rangle = \varphi_0(xy^*)$. In section 2 we show that the left and right spaces are in fact "naturally" linearly isometric. In the case $p = 2$ this is connected with the bases themselves of Tomita-Takesaki theory. For $p \geq 2$ the right and left L_p-spaces are imbedded in the corresponding L_2 spaces, and this permits to describe the linear isomorphism (which is however *not* inherited from that or the L_2-spaces). For $p \leq 2$ the isomorphism is defined by duality. After identifying the right and left L_p-spaces we obtain an object which has a structure of $\mathcal{A}$-bimodule (i. e. the left and right action of $\mathcal{A}$ commute).

In section 3 we consider a first connection between the L_p-spaces for different $p's$; namely the external product $L_p(\mathcal{A}) \times L_q(\mathcal{A}) \to L_r(\mathcal{A})$, $\frac{1}{r} = \frac{1}{p} + \frac{1}{q}$, which gives rise to a Hölder inequality. This product is easily defined by interpolation; however to derive certain properties requires more techniques from Tomita-Takesaki theory. For proving the associativity of the product we introduce the sub*-algebra of analytic elements of $\mathcal{A}$. For proving the basic norm equality $\|h^*.h\|_{p/2} = \|h\|_p^2$ we introduce a class of homogeneous operators on $L_2(\mathcal{A})$ (linked to that used in the construction of [**Hi**]; the arguments here are closer to those of [**AM**] or [**Ko2**]).

In section 4 we define the order on $L_p(\mathcal{A})$. The positive cone is the set of products $h^*.h$, where $h \in L_{2p}(\mathcal{A})$ (in a first phase one considers the closure of this set, which is ultimately shown to be closed). Every element in $L_p(\mathcal{A})$ is shown to be a linear combination of four positive elements, and a polar decomposition is shown to hold. The positive cones of $L_p(\mathcal{A})_+$ for different $p's$ are connected by a power operation $h \mapsto h^\alpha$, $\alpha = p_1/p_2$. These results rely on a further analysis of $1/p$-homogeneous operators (following ideas of [**AM**]).

In section 5 we study the effect of a "change of weight": if φ_0, ψ_0 are two faithful normal states on $\mathcal{A}$, we show that the resulting scales of $L_p(\mathcal{A})$ spaces are isomorphic in a very strong sense: the isomorphisms are isometric and preserve all the structural properties given in the preceding sections. Like to the commutative case, these isomorphisms are simply defined by a "change of density", i. e. of the image of the unit of $\mathcal{A}$ in $L_p(\mathcal{A})$.

1. Two complex interpolation scales between the algebra and its predual.

1.1. The interpolation couple $(\mathcal{A}, \mathcal{A}_*)$.

Let $\mathcal{A}$ be a Von Neumann algebra with a normal faithful state φ_0. We have two natural norm one linear embeddings of $\mathcal{A}$ into its predual $\mathcal{A}_*$ which are associated with φ_0: the left embedding $i_\ell : x \mapsto x.\varphi_0$ and the right embedding $i_r\ x \mapsto \varphi_0.x$. Recall that

$$\forall y \in \mathcal{A}, \quad x.\varphi_0(y) = \varphi_0(yx) \text{ and } \varphi_0.x(y) = \varphi_0(xy)$$

These embeddings are injective: if $x.\varphi_0 = 0$, consider the polar decomposition $x = u\,|x|$; then $0 = x.\varphi_0(u^*) = \varphi_0(u^*x) = \varphi_0(|x|)$; hence $|x| = 0$ since φ_0 is faithful, and finally $x = 0$. The ranges $i_\ell(\mathcal{A})$ and $i_r(\mathcal{A})$ are dense in $\mathcal{A}_*$ since if $y \in \mathcal{A}$ is in the annihilator of $i_\ell(\mathcal{A})$, then considering the polar decomposition $y^* = u\,|y^*|$ we have $0 = u.\varphi_0(y) = \varphi_0(yu) = \varphi_0(|y^*|)$, hence $|y^*| = 0$ and $y = 0$.

We can define two scales of complex interpolation spaces associated with these embeddings. For $1 < p < \infty$ we set $\theta = 1/p$ and

$$L_p^\ell(\mathcal{A};\varphi_0) = (i_\ell(\mathcal{A}), \mathcal{A}_*)_\theta \text{ and } L_p^r(\mathcal{A};\varphi_0) = (i_r(\mathcal{A}), \mathcal{A}_*)_\theta$$

(we shall write simply $L_p^\ell(\mathcal{A})$ and $L_p^r(\mathcal{A})$ when there is no ambiguity on the normal state φ_0). We have $i_\ell(\mathcal{A}) \subset L_p^\ell(\mathcal{A}) \subset \mathcal{A}_*$ and $i_r(\mathcal{A}) \subset L_p^r(\mathcal{A}) \subset \mathcal{A}_*$ (norm one inclusion maps). Moreover $i_\ell(\mathcal{A})$ (resp. $i_r(\mathcal{A})$) is dense in $L_p^\ell(\mathcal{A})$ (resp. $L_p^r(\mathcal{A})$).

1.2. The left and right actions of $\mathcal{A}$.

Note that $i_\ell(\mathcal{A})$, resp. $i_r(\mathcal{A})$, is a left, resp. a right ideal in $\mathcal{A}_*$; in other words $x.i_\ell(\mathcal{A}) \subset i_\ell(\mathcal{A})$ and $i_r(\mathcal{A}).x \subset i_r(\mathcal{A})$ for every $x \in \mathcal{A}$. In fact it is clear that $i_\ell(xy) = x.i_\ell(y)$ and $i_r(yx) = i_r(y).x$ for every $x, y \in \mathcal{A}$. If $x \in \mathcal{A}$, let $\pi_\infty^\ell(x)$ (resp $\pi_\infty^r(x)$) be the map $\mathcal{A} \to \mathcal{A}$, $y \mapsto xy$ (resp $x \mapsto yx$); similarly let $\pi_1^\ell(x)$ (resp. $\pi_1^r(x)$) be the map $\mathcal{A}_* \to \mathcal{A}_*$, $\psi \mapsto x.\psi$, resp. $\psi \mapsto \psi.x$ (we have $\pi_\infty^\ell(x) = \pi_1^r(x)'$ and $\pi_\infty^r(x) = \pi_1^\ell(x)'$). Then, for every $x \in \mathcal{A}$, we have clearly $i_\ell \circ \pi_\infty^\ell(x) = \pi_1^\ell(x) \circ i_\ell$ and $i_r \circ \pi_\infty^r(x) = \pi_1^r(x) \circ i_r$.

$$\begin{array}{ccc} \mathcal{A} & \xrightarrow{\pi_\infty^\ell(x)} & \mathcal{A} \\ {\scriptstyle i_\ell}\downarrow & & \downarrow{\scriptstyle i_\ell} \\ \mathcal{A}_* & \xrightarrow{\pi_1^\ell(x)} & \mathcal{A}_* \end{array} \qquad\qquad \begin{array}{ccc} \mathcal{A} & \xrightarrow{\pi_\infty^r(x)} & \mathcal{A} \\ {\scriptstyle i_r}\downarrow & & \downarrow{\scriptstyle i_r} \\ \mathcal{A}_* & \xrightarrow{\pi_1^r(x)} & \mathcal{A}_* \end{array}$$

By interpolation we deduce that the map $\pi_1^\ell(x)$ restricts to a bounded linear map $\pi_p^\ell(x) : L_p^\ell(\mathcal{A}) \to L_p^\ell(\mathcal{A})$, and similarly $\pi_1^r(x)$ restricts to a bounded linear map $\pi_p^r(x) : L_p^r(\mathcal{A}) \to L_p^r(\mathcal{A})$. In particular $L_p^\ell(\mathcal{A})$ is a left (non closed) ideal of $\mathcal{A}_*$ and $L_p^r(\mathcal{A})$ is a right ideal. We have clearly $\|\pi_\infty^\ell(x)\| = \|\pi_\infty^r(x)\| = \|x\|$ and $\|\pi_1^\ell(x)\| = \|\pi_1^r(x)\| = \|x\|$, so by interpolation $\|\pi_p^\ell(x)\| \le \|x\|$ and $\|\pi_p^r(x)\| \le \|x\|$. Note also that

$$\forall x, y \in \mathcal{A}, \qquad \pi_p^\ell(xy) = \pi_p^\ell(x)\pi_p^\ell(y) \text{ and } \pi_p^r(xy) = \pi_p^r(y)\pi_p^r(x)$$

(since this is true in the case $p = 1$). If $x = u\,|x|$ is the polar decomposition of $x \in \mathcal{A}$, we have $\|\pi_p^\ell(x)\| \le \|\pi_p^\ell(u)\|\,\|\pi_p^\ell(|x|)\| \le \|\pi_p^\ell(|x|)\|$ and the converse inequality follows similarly from the equality $|x| = u^*x$; so $\|\pi_p^\ell(x)\| = \|\pi_p^\ell(|x|)\|$. If $\varepsilon > 0$ and e is a spectral projector of $|x|$ such that $|x| \ge (1-\varepsilon)\,\|x\|\,e$ we have $(1-\varepsilon)\,\|x\|\,e = y\,|x|\,e$ with $y \in \mathcal{A}$, $\|y\| \le 1$ (take $y = f(|x|)$ with $f(t) = (1-\varepsilon)^{-1}\,\|x\|^{-1}\,t$ if $t > (1-\varepsilon)\,\|x\|$, $f(t) = 0$ if $t \le (1-\varepsilon)\,\|x\|$); hence $(1-\varepsilon)\,\|x\|\,\|e.\varphi_0\|_p \le \|\,|x|\,e.\varphi_0\|_p$, and consequently $\|\pi_p^\ell(|x|)\| \ge (1-\varepsilon)\,\|x\|$. So letting $\varepsilon \to 0$ we obtain $\|\pi_p^\ell(x)\| = \|x\|$ and similarly $\|\pi_p^r(x)\| = \|x\|$. So we have obtained that

FACT 1.1. *The map π_p^ℓ (resp. π_p^r) is an isometric homomorphism of unital Banach algebras from $\mathcal{A}$ (resp. $\mathcal{A}^{\mathrm{op}}$) into $B(L_p^\ell(\mathcal{A}))$ (the Banach algebra of bounded linear operators on $L_p^\ell(\mathcal{A})$). Here we denote by $\mathcal{A}^{\mathrm{op}}$ the "opposite algebra" of $\mathcal{A}$ (i. e. the same linear space with opposite product $x \cdot y = yx$).*

REMARK 1.2. *The left action π_p^ℓ is continuous on bounded sets of $\mathcal{A}$ for the σ-strong topology on $\mathcal{A}$ and the strong operator topology in $B(L_p^\ell(\mathcal{A}))$.*

PROOF. For $p = 1$ we have by the Cauchy-Schwartz inequality for every $\psi \in \mathcal{A}_*^+$:

$$\forall x, y \in \mathcal{A}, |x.\psi(y)| = |\psi(yx)| \leq \psi(yy^*)^{1/2}\psi(x^*x)^{1/2}$$

whence

$$\|x.\psi\|_{\mathcal{A}_*} \leq \|\psi\|_{\mathcal{A}_*}^{1/2} \psi(x^*x)^{1/2}$$

so $x.\psi$ converges in $\mathcal{A}_*$ norm to zero when x converges σ-strongly to zero. Then for every $a \in \mathcal{A}$ we have:

$$\|x.i_\ell(a)\|_p = \|i_\ell(xa)\|_p \leq \|xa\|_{\mathcal{A}}^{1-1/p} \|x.i_\ell(a)\|_{\mathcal{A}_*}^{1/p}$$

hence $xa\varphi_0$ converges to zero when $x \to 0$ σ-strongly in a bounded set; by density this remains true for every $h \in L_p(\mathcal{A})$ in place of $i_\ell(a)$. □

PROPOSITION 1.3. *For every $h \in L_p^\ell(\mathcal{A})$ the set of projections s in $\mathcal{A}$ such that $s.h = h$ has a least element. We denote it by $s_\ell(h)$ and call it the* left support *of h.*

PROOF. Let $\mathcal{S}_h$ be the set of the projections s such that $s.h = h$. First $\mathcal{S}_h$ is a downwards directed net: if $s_1, s_2 \in \mathcal{N}_h$ then $\frac{1}{2}(s_1 + s_2).h = h$, hence $(\frac{1}{2}(s_1 + s_2))^n.h = h$ for every $n \geq 1$. Since the sequence $(\frac{1}{2}(s_1 + s_2))^n)$ is σ-convergent to $s_1 \wedge s_2$ and is included in the unit ball of $\mathcal{A}$, we obtain by the preceding Remark 1.2 that $(s_1 \wedge s_2).h = h$. Then $\mathcal{S}_h$ has a greatest lower bound s_{min} which is the σ-strong limit of the net; again by the Remark 1.2, s_{min} belongs to $\mathcal{S}_h$ and is the least element of $\mathcal{S}_h$. □

1.3. The external product between $L_p^\ell(\mathcal{A})$ and $L_q^r(\mathcal{A})$ (for conjugate exponents).

Consider now the product maps

$$P_\ell : \quad \mathcal{A} \times \mathcal{A}_* \xrightarrow{P_\ell} \mathcal{A}_* \quad (x, \psi) \mapsto x.\psi$$
$$P_r : \quad \mathcal{A}_* \times \mathcal{A} \xrightarrow{P_r} \mathcal{A}_* \quad (\psi, x) \mapsto \psi.x$$

They coincide on the product $i_\ell(\mathcal{A}) \times i_r(\mathcal{A})$ with the map $(x.\varphi_0, \varphi_0.y) \mapsto x.\varphi_0.y$. By bilinear interpolation ([**BL**], th. 4.4.1) this last one extends uniquely to a bounded bilinear map from $(i_\ell(\mathcal{A}), \mathcal{A}_*)_\theta \times (\mathcal{A}_*, i_r(\mathcal{A}))_\theta = L_p^\ell(\mathcal{A}) \times L_q^r(\mathcal{A})$ to $\mathcal{A}_*$. Here $\frac{1}{p} = \theta$, $\frac{1}{q} = 1 - \theta$, so p, q is an arbitrary pair of conjugate exponents $(1 < p, q < \infty)$. We denote simply by $(h, k) \mapsto h.k$ this external product. It is clear that $h.(\varphi_0.y) = P_r(h, y)$ and $(x.\varphi_0).k = P_\ell(x, k)$ for every $x, y \in \mathcal{A}$, $h \in L_p^\ell(\mathcal{A}) \subset \mathcal{A}_*$, $k \in L_q^r(\mathcal{A}) \subset \mathcal{A}_*$. The interpolation inequality gives rise to Hölder inequality

$$\|h.k\| \leq \|h\|_p \|k\|_q$$

We have moreover

$$\forall x, y \in \mathcal{A}, \forall h \in L_p^\ell(\mathcal{A}), \forall k \in L_q^r(\mathcal{A}), \quad (x.h).(k.y) = x.(hk).y$$

by density since this is true when $h \in i_\ell(\mathcal{A}), k \in i_r(\mathcal{A})$.

1.4. The conjugation maps.

Let $C : \mathcal{A} \to \mathcal{A}$, $x \mapsto x^*$ be the conjugation map. We have the dual conjugation map $C_* : \mathcal{A}_* \to \mathcal{A}_*$, $\psi \mapsto \psi^*$ defined as usual by $\langle \psi^*, x\rangle = \overline{\langle \psi, x^*\rangle}$. Since $(x.\varphi_0)^* = \varphi_0.(x^*)$, we have a commutative diagram:

$$\begin{array}{ccc} \mathcal{A} & \xrightarrow{C} & \mathcal{A} \\ i_\ell \downarrow & & \downarrow i_r \\ \mathcal{A}_* & \xrightarrow{C_*} & \mathcal{A}_* \end{array}$$

and by interpolation we deduce a conjugation map $C_p^\ell : L_p^\ell(\mathcal{A}) \to L_p^r(\mathcal{A})$, $h \mapsto h^*$, wich is the restriction of C_* to $L_p^\ell(\mathcal{A})$; and similarly an inverse conjugation map $C_p^r : L_p^r(\mathcal{A}) \to L_p^\ell(\mathcal{A})$, $k \mapsto k^*$, wich is the restriction of C_* to $L_p^r\mathcal{A})$. These maps are isometric and antilinear. It is an easy exercise to see that

$$\forall x \in \mathcal{A}, \forall h \in L_p^\ell(\mathcal{A}), \forall k \in L_q^r(\mathcal{A}), \quad (x.h)^* = h^*.x^* \text{ and } (h.k)^* = k^*.h^*$$

1.5. The spaces $L_2^\ell(\mathcal{A})$ and $L_2^r(\mathcal{A})$.

It is a very classical construction (Guelfand-Naimark-Segal) to equip $\mathcal{A}$ with the complex scalar product

$$(x, y)_{\varphi_0} = \varphi_0(y^*x)$$

(note that $(x, x) = \varphi(x^*x) > 0$ when $x \neq 0$ since φ_0 is faithful) and to consider the completion $H_{\varphi_0}^\ell$ of the space $\mathcal{A}$ for the associated prehilbertian norm. Let $\lambda_\ell : \mathcal{A} \hookrightarrow H_{\varphi_0}$ be the corresponding inclusion. Similarly we can consider the scalar product $(x, y)_{\varphi_0}^r = \varphi_0(xy^*)$ and the associated Hilbert space K_{φ_0}, with the embedding $\lambda_r : \mathcal{A} \hookrightarrow K_{\varphi_0}$. It is clear that the conjugation $C : \mathcal{A} \to \mathcal{A}$, $x \mapsto x^*$ extends to an isometric antilinear isomorphism $C : \xi \mapsto \xi^*$ from H_{φ_0} onto K_{φ_0}. Hence K_{φ_0} is linearly isometric to the dual space of H_{φ_0}, the duality bracket is simply given by

$$\langle \xi, \eta\rangle_{H_{\varphi_0} \times K_{\varphi_0}} = (\xi, \eta^*)_{\varphi_0}$$

The following Proposition is a particular case of a more general result (formalized in [**W**] or [**CS**]):

PROPOSITION 1.4. *The space $L_2^\ell(\mathcal{A}; \varphi_0)$ is linearly isometric to the space H_{φ_0}, and similarly $L_2^r(\mathcal{A}; \varphi_0)$ is linearly isometric to K_{φ_0}. More precisely the norms of $L_2^\ell(\mathcal{A}; \varphi_0)$ and of H_{φ_0} coincide on their common linear subspace $\mathcal{A}$ (and similarly for $L_2^r(\mathcal{A}; \varphi_0)$ and K_{φ_0}).*

PROOF. For every $x \in \mathcal{A}$ we have:

$$\begin{aligned} \|\lambda_\ell(x)\|_{H_{\varphi_0}}^2 &= \varphi_0(x^*x) = \langle x.\varphi_0.x^*, I\rangle \\ &\leq \|x.\varphi_0.x^*\|_{\mathcal{A}_*} = \|i_\ell(x).i_r(x^*)\|_{\mathcal{A}_*} \\ &\leq \|i_\ell(x)\|_{L_2^\ell(\mathcal{A})} \|i_r(x^*)\|_{L_2^r(\mathcal{A})} = \|i_\ell(x)\|_{L_2^\ell(\mathcal{A})}^2 \end{aligned}$$

Consider now the dual of the complex interpolation space $L_2^r(\mathcal{A}) = (i_r(\mathcal{A}), \mathcal{A}_*)_{1/2}$. Since $i_r(\mathcal{A})$ is dense in $\mathcal{A}_*$, the conjugate map i_r' is a continuous injective embedding of $(\mathcal{A}_*)^* = \mathcal{A}$ into $\mathcal{A}^*$. By the duality theorem ([**BL**], th. 4.5.1), the dual space of $(i_r(\mathcal{A}), \mathcal{A}_*)_{1/2}$ is isometrically identified with the interpolation space $(\mathcal{A}^*, i_r'(\mathcal{A}))^{1/2}$

(for Calderón's second method of interpolation). Note that in fact $i'_r(\mathcal{A}) = i_\ell(\mathcal{A})$ up to the canonical embedding of $\mathcal{A}_*$ into its bidual $\mathcal{A}^*$, since:

$$\forall x, y \in \mathcal{A}, \langle i'_r(x), y\rangle = \langle x, i_r(y)\rangle = \varphi_0(yx) = \langle i_\ell(x), y\rangle$$

By a theorem of Bergh [**Be**] the norm of the upper interpolation space $(\mathcal{A}^*, i_\ell(\mathcal{A}))^{1/2}$ coincides with that of the lower interpolation space $(\mathcal{A}^*, i_\ell(\mathcal{A}))_{1/2}$ over the intersection $i^*_r(\mathcal{A}) = i_\ell(\mathcal{A})$. Moreover$(\mathcal{A}^*, i_\ell(\mathcal{A}))_{1/2} = (\mathcal{A}_*, i_\ell(\mathcal{A}))_{1/2}$ since $\mathcal{A}_*$ is the closure in $\mathcal{A}^*$ of the intersection space $i_\ell(\mathcal{A})$. Then for every $x \in \mathcal{A}$:

$$\begin{aligned}\|i_\ell(x)\|_{L^\ell_2(\mathcal{A})} &= \|i_\ell(x)\|_{(\mathcal{A}^*, i_\ell(\mathcal{A}))^{1/2}} \\ &= \|i_\ell(x)\|_{L^r_2(\mathcal{A})^*} = \sup\{\langle i_\ell(x), \xi\rangle \mid \xi \in L^r_2(\mathcal{A}), \|\xi\| \le 1\}\end{aligned}$$

By density we can restrict this supremum to the $\xi = i_r(y^*)$, $y \in \mathcal{A}$; then $\langle i_\ell(x), \xi\rangle = x.\varphi_0(y^*) = \varphi_0(y^*x) = (\lambda_\ell(x), \lambda_\ell(y))_{H_{\varphi_0}}$ and $\|i_r(y^*)\|_{L^r_2(\mathcal{A})} = \|i_\ell(y)\|_{L^\ell_2(\mathcal{A})}$. Thus:

$$\begin{aligned}\|i_\ell(x)\|_{L^\ell_2(\mathcal{A})} &= \sup\{(\lambda_\ell(x), \lambda_\ell(y))_{H_{\varphi_0}} \mid y \in \mathcal{A}, \|i_\ell(y)\|_{L^\ell_2(\mathcal{A})} \le 1\} \\ &\le \sup\{(\lambda_\ell(x), \lambda_\ell(y))_{H_{\varphi_0}} \mid y \in \mathcal{A}, \|\lambda_\ell(y)\|_{H_{\varphi_0}} \le 1\} \\ &= \|\lambda_\ell(y)\|_{H_{\varphi_0}} \qquad \square\end{aligned}$$

1.6. Duality of the spaces $L^\ell_p(\mathcal{A})$ and $L^r_q(\mathcal{A})$.

PROPOSITION 1.5. *For $1 < p < \infty$ the space $L^\ell_p(\mathcal{A})$ and $L^r_p(\mathcal{A})$ are uniformly convex and uniformly smooth. For conjugate indices p, q with $1 \le p < \infty$ the space $L^r_q(\mathcal{A})$ (resp. $L^\ell_q(\mathcal{A})$) is linearly isometrically isomorphic to the dual space of $L^\ell_p(\mathcal{A})$ (resp. $L^r_p(\mathcal{A})$). For conjugate indices p, q the duality between $L^r_q(\mathcal{A})$ and $L^\ell_p(\mathcal{A})$ is given by*

$$\forall h \in L^\ell_p(\mathcal{A}), k \in L^r_q(\mathcal{A}) \qquad \langle h, k\rangle = \langle h.k, I\rangle$$

where at the right handside I is the identity of $\mathcal{A}$ and the duality bracket is relative to the duality between $\mathcal{A}_*$ and $\mathcal{A}$.

PROOF. By the reiteration theorem ([**BL**], th. 4.6.1) the space $L^\ell_p(\mathcal{A})$, $1 < p < 2$ is a complex interpolation space between $\mathcal{A}_*$ and $L^\ell_2(\mathcal{A})$. Since the latter is an Hilbert space by Prop. 1.4, it results from [**CR**] that $L^\ell_p(\mathcal{A})$ is uniformly convex. The same argument works for $2 < p < \infty$, since then $L^\ell_p(\mathcal{A})$ is a complex interpolation space between $L^\ell_2(\mathcal{A})$ and $\mathcal{A}$, and for the right spaces $L^r_p(\mathcal{A})$ too.

As in the case $p = 2$ we have

$$L^\ell_p(\mathcal{A})^* = (\mathcal{A}_*, i_\ell(\mathcal{A}))^*_\theta = (i^*_\ell(\mathcal{A}), \mathcal{A}^*)^\theta = (i_r(\mathcal{A}), \mathcal{A}^*)^\theta$$

but $i_r(\mathcal{A})$ is w*-dense in $L^\ell_p(\mathcal{A})^*$ (if $h \in L^\ell_p(\mathcal{A})$ verifies $\langle h, i_r(\mathcal{A})\rangle = 0$ then h is zero as an element of $\mathcal{A}_*$); since $L^\ell_p(\mathcal{A})$ is reflexive the space $i_r(\mathcal{A})$ is norm-dense in $L^\ell_p(\mathcal{A})^*$; then by Bergh's theorem $(i_r(\mathcal{A}), \mathcal{A}^*)^\theta = (i_r(\mathcal{A}), \mathcal{A}^*)_\theta = (i_r(\mathcal{A}), \mathcal{A}_*)_\theta = L^r_q(\mathcal{A})$. When $h = x.\varphi_0 \in i_\ell(\mathcal{A}) \subset L^\ell_p(\mathcal{A})$ and $k \in L^r_q(\mathcal{A}) \subset \mathcal{A}_*$ we have $\langle h, k\rangle_{L^\ell_p(\mathcal{A}) \times L^r_q(\mathcal{A})} = \langle x, k\rangle_{\mathcal{A} \times \mathcal{A}_*} = \langle x.k, I\rangle_{\mathcal{A}_* \times \mathcal{A}}$. By density this remains true when $h \in L^\ell_p(\mathcal{A})$. $\square$

2. Linear identification of the two scales and Tomita-Takesaki theory.

2.1. Linear identification of $L_2^\ell(\mathcal{A})$ and $L_2^r(\mathcal{A})$.

As was pointed out before, the conjugation C provides only an antilinear identification of the right and left spaces $L_2^\ell(\mathcal{A},\varphi_0) = H_{\varphi_0}$ and $L_2^r(\mathcal{A},\varphi_0) = K_{\varphi_0}$.

The identity operator $\mathcal{A} \to \mathcal{A}, x \mapsto x$ defines an unbounded linear operator $\mathcal{I} : H_{\varphi_0} \to K_{\varphi_0}$ with dense domain $\mathcal{A}$. This operator $\mathcal{I}$ is closable: to see this note that the element $\xi_0 = I$ of H_{φ_0} is cyclic for the commutant $\pi_\ell^2(\mathcal{A})'$, i. e. $\{x'\xi_0 \mid x' \in \pi_\ell^2(\mathcal{A})'\}$ is dense in H_{φ_0} (this is an easy standard consequence of the fact that ξ_0 is separating for $\pi_\ell^2(\mathcal{A})$). On the other hand the antilinear isomorphism $B(H_{\varphi_0}) \to B(K_{\varphi_0})$, $T \mapsto C^*TC$ maps $\pi_\ell^2(\mathcal{A})$ onto $\pi_r^2(\mathcal{A})$ and $\pi_\ell^2(\mathcal{A})'$ onto $\pi_r^2(\mathcal{A})'$. For every $x \in \mathcal{A}$ and $x' \in \pi_\ell^2(\mathcal{A})'$ we have then:

$$\begin{aligned}(\mathcal{I}x\xi_0, Cx'\xi_0)_{K_{\varphi_0}} &= (x'\xi_0, C^*\mathcal{I}x\xi_0)_{H_{\varphi_0}} = (x'\xi_0, x^*\xi_0)_{H_{\varphi_0}} \\ &= (xx'\xi_0, \xi_0)_{H_{\varphi_0}} = (x'x\xi_0, \xi_0)_{H_{\varphi_0}} \\ &= (x\xi_0, x'^*\xi_0)_{H_{\varphi_0}}\end{aligned}$$

So if $x_n\xi_0 \to 0$ weakly in H_{φ_0} and the sequence $(\mathcal{I}x_n\xi_0)$ is bounded in K_{φ_0} then $\mathcal{I}x_n\xi_0 \to 0$ weakly in K_{φ_0}: so $\mathcal{I}$ is closable; we denote still by $\mathcal{I}$ its closure. From the preceding it is clear that $\mathcal{I}^*Cx'.\xi_0 = x'^*.\xi_0$ for all $x' \in \pi_\ell^2(\mathcal{A})'$; in fact it is true but far from being immediate that $\mathcal{I}^*$ is the closure of the unbounded operator of domain $\pi_r^2(\mathcal{A})'.\xi_0$, defined by $Cx'.\xi_0 \mapsto x'^*.\xi_0$.

Set $\Delta = \mathcal{I}^*\mathcal{I}$ and let $\mathcal{I} = U\Delta^{1/2}$ be the polar decomposition of the closed operator $\mathcal{I}$. Since $\mathcal{I}$ and $\mathcal{I}^*$ have dense range, the partial isometry U is in fact a unitary operator $H_{\varphi_0} \to K_{\varphi_0}$, which provides the announced linear identification of the two Hilbert spaces. The antilinear operator $C^*U : H_{\varphi_0} \to H_{\varphi_0}$ is traditionally denoted by J (or J_{φ_0}) in the literature.

Note that if instead of $\mathcal{I} : H_{\varphi_0} \to K_{\varphi_0}$ we start with $\mathcal{I}^{-1} : K_{\varphi_0} \to H_{\varphi_0}$ we obtain exactly the inverse linear identification map $U^* : K_{\varphi_0} \to H_{\varphi_0}$. For, we have $U\mathcal{I}^{-1} = U\Delta^{-1/2}U^* \geq 0$, so by unicity of the polar decomposition, $|\mathcal{I}^{-1}| = U\Delta^{-1/2}U^*$ and U^* is the "phase " of $\mathcal{I}^{-1}$. On the other hand it is straightforward to verify the equation $\mathcal{I}^{-1} = C^*\mathcal{I}C^*$, and so the phase of $\mathcal{I}^{-1}$ coincides with C^*UC^*; consequently, $C^*UC^* = U^*$, and we obtain the well known equality $J^2 = I_{H_{\varphi_0}}$. We have thus the following commutative diagram where the horizontal arrows are linear while the vertical ones are antilinear:

$$\begin{array}{ccc} H_{\varphi_0} & \xrightarrow{U} & K_{\varphi_0} \\ {\scriptstyle C}\downarrow & & \downarrow{\scriptstyle C^*} \\ K_{\varphi_0} & \xrightarrow{U^*} & H_{\varphi_0} \end{array}$$

The images $U\xi$ are generally not explicit, even for $\xi = x.\xi_0, x \in \mathcal{A}$. The only simple case is when $\xi = z.\xi_0$ with z belonging to the center $\mathcal{Z}(\mathcal{A})$ of $\mathcal{A}$: in this case $U(z.\xi_0) = z.\xi_0$. For, we have simultaneously $\mathcal{I}(z.\xi_0) = z.\xi_0$ and $\mathcal{I}^*(z.\xi_0) = z.\xi_0$, so $\Delta(z.\xi_0) = \mathcal{I}^*\mathcal{I}(z.\xi_0) = z.\xi_0$. As a consequence $\Delta^{1/2}(z.\xi_0) = z.\xi_0$, and the relation $\mathcal{I}(z.\xi_0) = z.\xi_0$ implies $U(z.\xi_0) = z.\xi_0$.

For further use we state some simple properties of the modular operators Δ and J:

i) $J^* = J = J^2$ hence $(J\xi, J\eta) = (\eta, \xi)$ for every $\xi, \eta \in H_{\varphi_0}$.
ii) $J\Delta^{1/2} = \Delta^{-1/2}J$ and more generally $J\Delta^z = \Delta^{-\bar{z}}J$ for every $z \in \mathbb{C}$.
iii) $J\Delta^{1/2}x\xi_0 = x^*\xi_0$ for every $x \in \mathcal{A}$.

2.2 The $\mathcal{A}$-bimodule structure of $L_2(\mathcal{A})$ and Tomita's theorem.
We can now transfer the right action π_2^r of $\mathcal{A}$ onto $L_2^r(\mathcal{A}) = K_{\varphi_0}$ to a right action of $\mathcal{A}$ onto $L_2^\ell(\mathcal{A}) = H_{\varphi_0}$ via the isometry U. Recall that $\pi_2^r(x)\lambda_2^r(y) = \lambda_2^r(yx)$ for every $x, y \in \mathcal{A}$. Since clearly $\pi_2^r(x) = C\pi_2^\ell(x^*)C^*$ for every $x \in \mathcal{A}$, the transferred right action of $x \in \mathcal{A}$ onto H_{φ_0} is the operator $U^*C\pi_2^\ell(x^*)C^*U = J\pi_2^\ell(x^*)J$ (simply denoted by Jx^*J). So we put $\xi.x = Jx^*J\xi$ for every $x \in \mathcal{A}$ and $\xi \in H_{\varphi_0}$. The space H_{φ_0} equipped with the right and left actions of $\mathcal{A}$ will be denoted by $L_2(\mathcal{A}; \varphi_0)$ (or simply $L_2(\mathcal{A})$).

THEOREM 2.1. *The right and left actions of $\mathcal{A}$ define a structure of $\mathcal{A}$-bimodule onto $L_2(\mathcal{A})$.*

This theorem tells essentially that the right and left actions commute, i. e. $J\pi_2^\ell(\mathcal{A})J \subset \pi_2^\ell(\mathcal{A})'$. In fact the equality $J\pi_2^\ell(\mathcal{A})J = \pi_2^\ell(\mathcal{A})'$ holds, and this is simply the particular case $t = 0$ of the following fundamental theorem of Tomita:

THEOREM 2.2. [Tomita] *For every $t \in \mathbb{R}$, $J\Delta^{it}\pi_2^\ell(\mathcal{A})\Delta^{-it}J = \pi_2^\ell(\mathcal{A})'$.*

SKETCH OF PROOF. It is equivalent to prove that $J\Delta^{it}\pi_2^\ell(\mathcal{A})'\Delta^{-it}J = \pi_2^\ell(\mathcal{A})$ and the hard task is to prove the inclusion $J\Delta^{it}\pi_2^\ell(\mathcal{A})'\Delta^{-it}J \subset \pi_2^\ell(\mathcal{A})$. The existing proofs are indirect and modelled on the following pattern. One shows that for every $x' \in \pi_2^\ell(\mathcal{A})'$ the strongly convergent integral

$$x_f = \int_{\mathbb{R}} f(t)J\Delta^{it}x'\Delta^{-it}Jdt \tag{2.1}$$

defines an element of $\pi_2^\ell(\mathcal{A})$, for each f in a sufficiently rich class $\mathcal{F} \subset L_1(\mathbb{R})$, i. e. a class which separates the bounded continuous functions $\mathbb{R} \to \mathbb{R}$. Then since for every $y' \in \pi_2^\ell(\mathcal{A})'$ we have

$$0 = [x_f, y'] = \int_{\mathbb{R}} f(t)[J\Delta^{it}x'\Delta^{-it}J, y']dt$$

the commutators $[J\Delta^{it}x'\Delta^{-it}J, y']$ all vanish, wich means that for every $t \in \mathbb{R}$,

$$J\Delta^{it}x'\Delta^{-it}J \in \pi_2^\ell(\mathcal{A})'' = \pi_2^\ell(\mathcal{A})$$

Following [**P**], choose for $\mathcal{F}$ the class of functions $f_\theta(t) = \dfrac{e^{-\theta t}}{2\cosh \pi t}$, $|\theta| < \pi$. Set $x_\theta = x_{f_\theta}$. If $\xi, \eta \in \mathcal{D}(\Delta^{1/2}) \cap \mathcal{D}(\Delta^{-1/2})$, the formula $g(z) = e^{i\theta z}(Jx'J\Delta^z\xi, \Delta^{-\bar{z}}\eta)$ defines a continuous function on the closed strip $\{z \in \mathbb{C} \mid -1/2 \leq \Re e\, z \leq 1/2\}$, which is analytic on the open strip. Then a simple calculation shows that:

$$\begin{aligned}
&e^{-i\theta/2}(x_\theta\Delta^{-1/2}\xi, \Delta^{1/2}\eta) + e^{i\theta/2}(x_\theta\Delta^{1/2}\xi, \Delta^{-1/2}\eta) = \\
&= \int_{\mathbb{R}} (g(-\frac{1}{2} + it) + g(\frac{1}{2} + it))\frac{dt}{2\cosh \pi t} \\
&= g(0) = (Jx'J\xi, \eta)
\end{aligned}$$

(since $\dfrac{1}{2\cosh \pi t}$ is the Poisson kernel of the strip at point 0). So we have formally

$$e^{-i\theta/2}\Delta^{1/2}x_\theta\Delta^{-1/2}+e^{i\theta/2}\Delta^{-1/2}x_\theta\Delta^{1/2}=Jx'J \tag{2.2}$$

Conversely if eq. (2.2) is verified by some $x_\theta \in B(H_{\varphi_0})$ then $x_\theta = x_{f_\theta}$. For, if $\xi,\eta\in\mathcal{D}(\Delta^{1/2})\cap\mathcal{D}(\Delta^{-1/2})$, we set

$$h(z)=e^{i\theta z}(x_\theta\Delta^z\xi,\Delta^{-\bar{z}}\eta)$$

then

$$\begin{aligned}\int_{\mathbb{R}} e^{-\theta t}(J\Delta^{it}x'\Delta^{-it}J\xi,\eta)\frac{dt}{2\cosh t} &= \int_{\mathbb{R}}(h(-\frac{1}{2}+it)+h(\frac{1}{2}+it))\frac{dt}{2\cosh t}\\ &= h(0)=(x_\theta\xi,\eta)\end{aligned}$$

So $x_\theta = x_{f_\theta}$ is the unique solution of (2.2) (in $B(H_\varphi)$). It is convenient to replace the formal equation (2.2) by an equation involving only bounded operators. For this aim set $A=(I+\Delta)^{-1}$ and $B=\Delta^{1/2}A$. Then A and B are positive bounded injective operators and $\Delta^{1/2}B=I-A$, $\Delta^{-1/2}B=A$ and $JB=BJ$. Then eq (2.2) is equivalent to

$$e^{-i\theta/2}(I-A)x_\theta A+e^{i\theta/2}Ax_\theta(I-A)=BJx'JB \tag{2.3}$$

If (2.3) is verified for some $x_\theta\in\mathcal{A}$, then applying to ξ_0 we have (since $\Delta\xi_0=\xi_0$)

$$e^{-i\theta/2}(I-A)x_\theta\xi_0+e^{i\theta/2}Ax_\theta\xi_0=BJx'\xi_0 \tag{2.4}$$

Since both $BJx'\xi_0=\Delta^{1/2}AJx'\xi_0$ and $(I-A)x_\theta\xi_0=\Delta^{1/2}Bx_\theta\xi_0$ belong to the domain $\mathcal{D}(\Delta^{-1/2})$ of $\Delta^{-1/2}$, so does $Ax_\theta\xi_0$, and hence $x_\theta\xi\in\mathcal{D}(\Delta^{-1/2})$. Consequently by injectivity of B:

$$e^{-i\theta/2}\Delta^{-1/2}x_\theta\xi_0+e^{i\theta/2}\Delta^{1/2}x_\theta\xi_0=Jx'\xi_0 \tag{2.5}$$

Note that for $\theta=0$ eq. (2.5) writes simply $x_0\xi_0=JBx'\xi_0$.

Under the condition $x_\theta\in\mathcal{A}$ one can give an equivalent reformulation of (2.5) where the modular operators Δ and J do not appear. If $x_\theta\in\mathcal{A}$ verifies (2.5), then for every $y\in\mathcal{A}$,

$$\begin{aligned}(y\xi_0,x'\xi_0)=(Jx'\xi_0,Jy\xi_0)&=e^{-i\theta/2}(\Delta^{-1/2}x_\theta\xi_0,Jy\xi_0)+e^{i\theta/2}(\Delta^{1/2}x_\theta\xi_0,Jy\xi_0)\\ &=e^{-i\theta/2}(x_\theta\xi_0,\Delta^{-1/2}Jy\xi_0)+e^{i\theta/2}(y\xi_0,J\Delta^{1/2}x_\theta\xi_0)\\ &=e^{-i\theta/2}(x_\theta\xi_0,y^*\xi_0)+e^{i\theta/2}(y\xi_0,x_\theta^*\xi_0)\\ &=((e^{-i\theta/2}yx_\theta+e^{i\theta/2}x_\theta y)\xi_0,\xi_0)\end{aligned}$$

Hence:

$$\forall y\in\mathcal{A},\quad (y\xi_0,x'\xi_0)=\varphi_0(e^{-i\theta/2}yx_\theta+e^{i\theta/2}x_\theta y) \tag{2.6}$$

Conversely if $x_\theta\in\mathcal{A}$ verifies (2.6) then $|(x_\theta\xi_0,y^*\xi_0)|=|(yx_\theta\xi_0,\xi_0)|\le C\,\|y\xi_0\|$ for every $y\in\mathcal{A}$. This implies that $x_\theta\xi_0\in\mathcal{D}((J\Delta^{1/2})^*)=\mathcal{D}(\Delta^{-1/2})$, and (2.5) is verified and consequently (2.4) too.

The point is that such a $x_\theta\in\mathcal{A}$ (verifying (2.6)) does exist for every $x'\in\pi_2^\ell(\mathcal{A})$. For, it is sufficient to prove it for self-adjoint $y\in\mathcal{A}$, $x'\in\pi_2^\ell(\mathcal{A})$. Then a simple Hahn-Banach argument shows that the selfadjoint normal form $\psi: y\mapsto(y\xi_0,x'\xi_0)$ belongs to the weakly compact convex subset $K=\{e^{-i\theta/2}x.\varphi_0+e^{i\theta/2}\varphi_0.x \mid x\in\mathcal{A}_{sa},\|x\|\le\frac{1}{2\cos\theta/2}\|x'\|\}$ of the real Banach space $(\mathcal{A}_*)_{sa}=(\mathcal{A}_{sa})_*$.

Note that for $\theta = 0$ this shows that $JBx'\xi_0 \in \mathcal{A}\xi_0$ for every $x' \in \pi_2^\ell(\mathcal{A})'$.

Now we can prove that eq. (2.6) implies (2.3). Let $y', z' \in \pi_2^\ell(\mathcal{A})'$. We have $JBy'\xi_0 = y_0\xi_0$, $JBz'\xi_0 = z_0\xi_0$ where by the preceding remark $y_0, z_0 \in \mathcal{A}$. Note that $Ay'\xi_0 = \Delta^{-1/2}By'\xi_0 = \Delta^{-1/2}Jy_0\xi_0 = y_0^*\xi_0$, while $(I-A)y'\xi_0 = \Delta^{1/2}By'\xi_0 = \Delta^{1/2}Jy_0\xi_0 = ((J\Delta^{1/2})^*y_0\xi_0$. Then if $x_\theta \in \mathcal{A}$ verifies (2.6) a little calculation shows that:

$$(BJx'JBy'\xi_0, z'\xi_0) = (y_0^*z_0\xi_0, x'\xi_0) = \varphi_0(e^{-i\theta/2}y_0^*z_0x_\theta + e^{i\theta/2}x_\theta y_0^*z_0) \tag{2.7}$$

and:

$$\begin{cases} ((I-A)x_\theta Ay'\xi_0, z'\xi_0) = (z_0\xi_0, y_0x_\theta^*\xi_0) = \varphi_0(x_\theta y_0^*z_0) \\ (Ax_\theta(I-A)y'\xi_0, z'\xi_0) = (z_0x_\theta\xi_0, y_0\xi_0) = \varphi_0(y_0^*z_0x_\theta) \end{cases} \tag{2.8}$$

From (2.7) and (2.8), we deduce (2.3) since $\pi_2(\mathcal{A})'\xi_0$ is dense in H_{φ_0}. Finally x_θ, which belongs to $\mathcal{A}$, is a solution of (2.3), hence it coincides with the unique solution of (2.3) in $B(H_{\varphi_0})$.

The preceding shows in particular (with $t = 0$) that $J\pi_2^\ell(\mathcal{A})'J \subset \pi_2^\ell(\mathcal{A})$. The converse inclusion results then from simple algebraic manipulations; we give it for the sake of completeness. It suffices to prove that $JxJy = yJxJ$ for every $x, y \in \mathcal{A}$. First we notice that for every $u \in \mathcal{A}$:

$$\begin{cases} (JuJy\xi_0, \xi_0) = (y, Ju^*J\xi_0) = (y, \Delta^{1/2}u\xi_0) \\ (yJuJ\xi_0, \xi_0) = (Ju\xi_0, y^*\xi_0) = (Jy^*\xi_0, u\xi_0) = (\Delta^{1/2}y, u\xi_0) \end{cases}$$

and since $\Delta^{1/2}$ is self-adjoint these two expressions are equal. Now let v', w' be arbitrary elements of $\pi_2^\ell(\mathcal{A})'$. Then $v = Jv'J$ and $w = Jw'J$ belong to $\pi_2^\ell(\mathcal{A})$, and it is easy to see that:

$$v'JxJyw' = J(vxw)Jy \text{ and } v'yJxJw' = yJ(vxw)J$$

consequently $(v'JxJyw'\xi_0, \xi_0) = (v'yJxJw'\xi_0, \xi_0)$ by the preceding; then since ξ_0 is cyclic for $\pi_2^\ell(\mathcal{A})'$ we obtain $JxJy = yJxJ$. □

Remark 2.3. The *-automorphisms $\sigma_t : \mathcal{A} \mapsto \mathcal{A}, x \mapsto \Delta^{it}x\Delta^{-it}$ form a 1-parameter strongly continuous group $(\sigma_t)_{t\in\mathbb{R}}$ called the *modular automorphism group* (associated with φ_0).

2.3. Linear identification of $L_p^\ell(\mathcal{A})$ and $L_p^r(\mathcal{A})$.

We consider first the case $2 < p < \infty$. On the subspace $i_\ell(\mathcal{A}) = \mathcal{A}.\xi_0$ of $L_2(\mathcal{A})$, we consider the analytic family of linear operators $(\Delta^{\frac{z}{2}})_{z\in S}$, where $S = \{z \in \mathbb{C} \mid 0 \le \Re e\, z \le 1\}$. Since $0 \le \Re e\,\frac{z}{2} \le \frac{1}{2}$, we have $\mathcal{D}(\Delta^{\frac{z}{2}}) \supset \mathcal{D}(\Delta^{1/2}) \supset \mathcal{A}\xi_0$. For every $x \in \mathcal{A}$, the map $S \to L_2(\mathcal{A})$, $z \mapsto \Delta^{\frac{z}{2}}x\xi_0$ is continuous and its restriction to the open strip is analytic. Moreover it is bounded, since (setting $\alpha = \Re e\, z/2$):

$$\|\Delta^{\frac{z}{2}}x\xi_0\|_2 = \|\Delta^\alpha x\xi_0\| \le \|x\xi_0\| + \|\Delta^{1/2}\xi_0\| = \|x\xi_0\|_2 + \|x^*\xi_0\|_2$$

For $z = it$ we have $\left\|\Delta^{it/2}x\xi_0\right\|_2 = \|x\xi_0\|_2$, and the map $t \mapsto \Delta^{it/2}x\xi_0$ is continuous in the $L_2(\mathcal{A})$-norm.

For $z = 1 + it$, we have

$$\Delta^{(1+it)/2}x\xi_0 = \Delta^{it/2}Jx^*\xi_0 = J\Delta^{it/2}x^*\Delta^{-it/2}\xi_0 = \xi_0.(\Delta^{-it/2}x\Delta^{it/2})$$

which belongs to $i_r(\mathcal{A})$ by Tomita's theorem. Moreover $\|\Delta^{(1+it)/2}x\xi_0\|_{i_r(\mathcal{A})} = \|\Delta^{-it}x\Delta^{it}\|_{\mathcal{A}} = \|x\|_{\mathcal{A}} = \|x.\xi_0\|_{i_\ell(\mathcal{A})}$. The map $t \mapsto \Delta^{(1+it)/2}x\xi_0$ is continuous for the norm of $L_2(\mathcal{A})$ but generally not for the norm of $i_r(\mathcal{A})$. These conditions insure that we can apply a variant of the interpolation theorem (due essentially to Stein, see [**S**] where it is stated for L_p-spaces; see also [**CJ**], th. 2 for a general version) and conclude that for every $0 < \theta < 1$ we have $\Delta^{\theta/2}x\xi_0 \in (L_2(\mathcal{A}), i_r(\mathcal{A})))^\theta = (L_2(\mathcal{A}), i_r(\mathcal{A})))_\theta = L_p^r(\mathcal{A})$, where $\frac{1}{p} = \frac{1-\theta}{2}$ and moreover $\|\Delta^{\theta/2}x\xi_0\|_p \leq \|x\xi_0\|_{(L_2(\mathcal{A}), i_\ell(\mathcal{A})))_\theta} = \|x\xi_0\|_p$.

Set $\frac{1}{q} = \frac{1}{2} - \frac{1}{p} = \frac{\theta}{2}$. The linear operator $x\xi_0 \mapsto \Delta^{1/q}x\xi_0$ can be extended by density to a norm-one operator U_p from $L_p^\ell(\mathcal{A})$ into $L_p^r(\mathcal{A})$. Note that since $\Delta^{1/q}$ is closed, we have necessarily $L_p^\ell(\mathcal{A}) \subset \mathcal{D}(\Delta^{1/q})$ and U_p is precisely the restriction of $\Delta^{1/q}$ to $L_p^\ell(\mathcal{A})$. Similarly one shows that the linear operator $\xi_0.x \mapsto \Delta^{-1/q}\xi_0.x$ extends to a norm-one operator V_p from $L_p^r(\mathcal{A})$ into $L_p^\ell(\mathcal{A})$, that $L_p^r(\mathcal{A}) \subset \mathcal{D}(\Delta^{-1/q})$ and that V_p is the restriction of $\Delta^{-1/q}$ to $L_p^r(\mathcal{A})$. As a consequence we have $V_pU_p = \mathrm{Id}_{L_p^\ell(\mathcal{A})}$, $U_pV_p = \mathrm{Id}_{L_p^r(\mathcal{A})}$ and U_p is an isometry from $L_p^\ell(\mathcal{A})$ onto $L_p^r(\mathcal{A})$.

In the case $1 < p < 2$ we define simply U_p by duality. Since the dual of $L_p^\ell(\mathcal{A})$ is linearly identified with $L_{p'}^r(\mathcal{A})$, where p' is the conjugate exponent of p, and that of $L_p^r(\mathcal{A})$ with $L_{p'}^\ell(\mathcal{A})$, we set $U_p = (U_{p'})'$.

The maps U_p preserve the conjugation, i. e. we have a commutative diagram:

$$\begin{array}{ccc} L_p^\ell(\mathcal{A}) & \xrightarrow{U_p} & L_p^r(\mathcal{A}) \\ C\downarrow & & \downarrow C^{-1} \\ L_p^r(\mathcal{A}) & \xrightarrow{U_p^{-1}} & L_p^\ell(\mathcal{A}) \end{array}$$

For on $L_2(\mathcal{A})$ the conjugation is simply the involution J; and if $p \geq 2$ then for $x \in \mathcal{A}$ we have $J\Delta^{1/q}x\xi_0 = \Delta^{-1/q}Jx\xi_0$. For $1 < p < 2$ we reason by duality.

Note that $J_p := C^{-1}U_p = U_p^{-1}C$ is an isometric antilinear involution of $L_p^\ell(\mathcal{A})$.

The identification of $L_p^\ell(\mathcal{A})$ and $L_p^r(\mathcal{A})$ by the linear isometry U_p permits to define a Banach space $L_p(\mathcal{A})$ equipped with both a left- and a right action of $\mathcal{A}$, and an isometric antilinear involution $J_p : h \mapsto h^*$. They are linked by the relations

$$\forall x \in \mathcal{A}, \forall h \in L_p(\mathcal{A}), \quad (x.h)^* = h^*.x^*$$

We have a left representation $\pi_p^\ell : \mathcal{A} \hookrightarrow B(L_p(\mathcal{A}))$ and a right representation $\pi_p^r : \mathcal{A}^{op} \hookrightarrow B(L_p(\mathcal{A}))$ (isometric Banach algebra homomorphisms) defined by $\pi_p^\ell(x)h = x.h, \pi_p^r(x)h = h.x$, and clearly we have the rule $\pi_p^r(x) = J_p\pi_p^\ell(x^*)J_p$.

PROPOSITION 2.4. *The space $L_p(\mathcal{A})$ is an $\mathcal{A}$-bimodule.*

PROOF. Assume first that $2 < p < \infty$. Set $\frac{1}{q} = \frac{1}{2} - \frac{1}{p}$. We have to prove that $\pi_p^r(\mathcal{A}) \subset \pi_p^\ell(\mathcal{A})'$, i. e. $xJ_pyJ_ph = J_pyJ_pxh$ for every $x, y \in \mathcal{A}$ and $h \in L_p(\mathcal{A})$; equivalently $x\Delta^{-1/q}JyJ\Delta^{1/q}h = \Delta^{-1/q}JyJ\Delta^{1/q}xh$ for every $x, y \in \mathcal{A}$ and $h \in$

$L_p^\ell(\mathcal{A})$ (equation in $L_2(\mathcal{A})$). Consider first the case $h = \xi_0$. For every $w \in \mathcal{A}$ the maps:

$$z \mapsto F(z) = (x\Delta^{-z/2}JyJ\Delta^{z/2}\xi_0, \xi_0.w) = (x\Delta^{-z/2}(\xi_0.y^*), \xi_0.w)$$

$$\text{and} \qquad z \mapsto G(z) = (\Delta^{-z/2}JyJ\Delta^{z/2}x\xi_0, \xi_0.w) = (JyJ\Delta^{z/2}x\xi_0, \Delta^{-\bar{z}/2}(\xi_0.w))$$

are bounded and continuous on the closed strip S, and analytic in the interior, since $\mathcal{A}\xi_0 \subset \mathcal{D}(\Delta^{1/2})$ and $\xi_0.\mathcal{A} = J\mathcal{A}\xi_0 \subset \mathcal{D}(\Delta^{1/2}J) = \mathcal{D}(\Delta^{-1/2})$. Since $\Delta^{it}JyJ\Delta^{-it} \in \pi_2^\ell(\mathcal{A})'$ by Tomita's theorem, we see that $F(it) = G(it)$ for every $t \in \mathbb{R}$, and consequently $F(z) = G(z)$ for every $z \in S$. In particular for $\theta = 2/q$ we obtain $(x\Delta^{-1/q}JyJ\Delta^{1/q}\xi_0, \xi_0.w) = (\Delta^{-1/q}JyJ\Delta^{1/q}x\xi_0, \xi_0.w)$, for every $w \in \mathcal{A}$ and consequently $x\Delta^{-1/q}JyJ\Delta^{1/q}\xi_0 = \Delta^{-1/q}JyJ\Delta^{1/q}x\xi_0$ by density of $\xi_0.\mathcal{A}$ in $L_2(\mathcal{A})$. Then for every $v \in \mathcal{A}$ we have:

$$x\Delta^{-1/q}JyJ\Delta^{1/q}v\xi_0 = xv\Delta^{-1/q}JyJ\Delta^{1/q}\xi_0 = \Delta^{-1/q}JyJ\Delta^{1/q}xv\xi_0$$

and by density of $\mathcal{A}.\xi_0$ we obtain the desired equality

$$x\Delta^{-1/q}JyJ\Delta^{1/q}h = \Delta^{-1/q}JyJ\Delta^{1/q}xh$$

for all $h \in L_p(\mathcal{A})$.

The case $1 < p < 2$ is treated by duality. If $h \in L_p^\ell(\mathcal{A}), k \in L_{p'}^r(\mathcal{A}), x \in \mathcal{A}$ we have:

$$\langle x.h, k\rangle = \langle x.h.k, I\rangle = \langle h.k, x\rangle = \langle h.k.x, I\rangle = \langle h, k.x\rangle$$

hence $\pi_p^\ell(x) = \pi_{p'}^r(x)'$ (the conjugate operator). On the other hand by the construction itself $J_p = (J_{p'})'$, hence $\pi_p^r(x) = J_p\pi_p^\ell(x^*)J_p = J_{p'}'\pi_{p'}^r(x^*)'J_{p'}' = \pi_{p'}^\ell(x)'$.□

Remark 2.5. We can now define the right support $s_r(h)$ of an element $L_p(\mathcal{A})$ by $s_r(h) = s_\ell(h^*)$. This is the least projection e in $\mathcal{A}$ such that $h.e = h$.

3. Relation between the $L_p(\mathcal{A})$-spaces for different p's: the external product.

3.1. The external product between $L_p(\mathcal{A})$ and $L_q(\mathcal{A})$: the general case.

Let $1 \leq r < \infty$. Consider now the product maps:

$$P_\ell: \quad \mathcal{A} \times L_r(\mathcal{A}) \xrightarrow{P_\ell} L_r(\mathcal{A}) \qquad (x,h) \mapsto x.h$$

$$P_r: \quad L_r(\mathcal{A}) \times \mathcal{A} \xrightarrow{P_r} L_r(\mathcal{A}) \qquad (h,x) \mapsto h.x$$

They coincide on the product $\mathcal{A}.\varphi_0 \times \varphi_0.\mathcal{A}$ with the map $(x.\varphi_0, \varphi_0.y) \mapsto x.\varphi_0.y$. which by bilinear interpolation extends uniquely to a bounded bilinear map $(h,k) \mapsto h.k$ from $(i_\ell(\mathcal{A}), L_2(\mathcal{A}))_\theta \times (L_2(\mathcal{A}, i_r(\mathcal{A}))_\theta = L_p^\ell(\mathcal{A}) \times L_q^r(\mathcal{A})$ to $L_r(\mathcal{A})$, with now $\frac{1}{p} = \frac{\theta}{r}$, $\frac{1}{q} = \frac{1-\theta}{r}$, so p, q is an arbitrary pair of exponents such that $\frac{1}{p} + \frac{1}{q} = \frac{1}{r}$. Again the interpolation inequality is interpreted as a Hölder inequality:

$$\|h.k\|_r \leq \|h\|_p \|k\|_q$$

There is a distinguished linear form on the space $L_1(\mathcal{A}) = \mathcal{A}_*$, corresponding to the unit element I of $\mathcal{A}$. This form is called the trace and denoted by Tr. The reason of this denomination is given by the following proposition:

PROPOSITION 3.1. *Let p, p' be a pair of conjugate exponents. For every $h \in L_p(\mathcal{A})$, $k \in L_{p'}(\mathcal{A})$ we have* $\mathrm{Tr}(h.k) = \mathrm{Tr}(k.h)$.

PROOF. This proposition means that the family of identification maps $U_p : L_p^\ell(\mathcal{A}) \to L_p^r(\mathcal{A})$ is compatible with the duality of $L_p(\mathcal{A})$ with $L_q(\mathcal{A})$, that is $\langle h, U_{p'}k\rangle = \langle k, U_p h\rangle$ for every $h \in L_p^\ell(\mathcal{A}), k \in L_{p'}^\ell(\mathcal{A})$. In the case $p = 2$ we have $\langle h, Uk\rangle = (h, CUk)_{H_{\varphi_0}} = (h, Jk)_{H_{\varphi_0}} = (k, Jh)_{H_{\varphi_0}} = \langle k, Uh\rangle$. In the case $p \neq 2$ we have $U_{p'} = U_p'$ by the construction itself. □

3.2. Associativity of the external product.

NOTATION. From now on we adopt the following notation: we denote by $\mathbf{1}_p$ the element φ_0 considered as belonging to $L_p(\mathcal{A})$. Note that $\mathbf{1}_p.\mathbf{1}_q = \mathbf{1}_r$ for every p, q, r with $\frac{1}{p} + \frac{1}{q} = \frac{1}{r}$, and $\mathbf{1}_\infty$ is the identity of $\mathcal{A}$.

LEMMA 3.2. *Two elements x, y of $\mathcal{A}$ verify the equality $\mathbf{1}_p.x = y.\mathbf{1}_p$ iff $y.\xi_0 = \Delta^{1/p}x.\xi_0$.*

PROOF. Assume first $p \geq 2$. The element $\mathbf{1}_p.x \in L_p(\mathcal{A})$ is identified with the element $\xi_0.x$ of $L_2(\mathcal{A}) = H_{\varphi_0}$ via the right embedding $L_p^r(\mathcal{A}) \hookrightarrow L_2^r(\mathcal{A})$. Similarly the element $y.\mathbf{1}_p \in L_p(\mathcal{A})$ is identified with the element $y.\xi_0$ of $L_2(\mathcal{A}) = H_{\varphi_0}$ via the left embedding $L_p^\ell(\mathcal{A}) \hookrightarrow L_2^\ell(\mathcal{A})$. These elements of H_{φ_0} are identified as elements of $L_p(\mathcal{A})$ iff $\xi_0.x = \Delta^{(1/2-1/p)}y\xi_0$. Since $\xi_0.x = Jx^*\xi_0 = \Delta^{1/2}x\xi_0$, this condition is equivalent to the equation $y.\xi_0 = \Delta^{1/p}x.\xi_0$.

Assume now that $1 \leq p \leq 2$. The element $\xi_0.x$ is identified with the element $\mathbf{1}_p.x$ via the right embedding $L_2^r(\mathcal{A}) \hookrightarrow L_p^r(\mathcal{A})$, while the element $y.\xi_0$ is identified with the element $y.\mathbf{1}_p$ via the left embedding $L_2^\ell(\mathcal{A}) \hookrightarrow L_p^\ell(\mathcal{A})$. If h is an element of $L_{p'}(\mathcal{A})$, let $\xi, \eta \in H_{\varphi_0}$ be its images in $L_2(\mathcal{A})$ by the right, resp. left embeddings. Then $\eta = \Delta^{1/2-1/p'}\xi$ and:

$$\langle h, \mathbf{1}_p.x\rangle = \langle \xi, \xi_0.x\rangle = (\xi, x^*\xi_0)_{H_{\varphi_0}}$$

$$\langle y.\mathbf{1}_p, h\rangle = \langle y\xi_0, \eta\rangle = (y\xi_0, J\eta)_{H_{\varphi_0}}$$

Hence $\mathbf{1}_p.x = y.\mathbf{1}_p$ iff $(\xi, x^*\xi_0)_{H_{\varphi_0}} = (y\xi_0, J\Delta^{1/2-1/p'}\xi)_{H_{\varphi_0}}$ for every $\xi \in L_p^\ell(\mathcal{A}) \subset H_{\varphi_0}$. Equivalently, $(\Delta^{1/2}x\xi_0, J\xi)_{H_{\varphi_0}} = (y\xi_0, \Delta^{1/p'-1/2}J\xi)_{H_{\varphi_0}}$ which by a density argument is equivalent to the fact that $y\xi_0 \in \mathcal{D}(\Delta^{1/p'-1/2})$ and $\Delta^{1/p'-1/2}y\xi_0 = \Delta^{1/2}x\xi_0$, i. e. $y\xi_0 = \Delta^{1/p}x\xi_0$. □

LEMMA 3.3. *Let $1 \leq p, q \leq \infty$ with $\frac{1}{p} + \frac{1}{q} \leq 1$. For every $h \in L_p(\mathcal{A}), k \in L_q(\mathcal{A}), x, y \in \mathcal{A}$ we have:*

$$x.(h.k).y = (x.h).(k.y) \text{ and } (h.x).k = h.(x.k)$$

PROOF. The first assertion follows by density from the case $h = a.\mathbf{1}_p$, $k = \mathbf{1}_q.b$, where $a, b \in \mathcal{A}$ since then $h.k = a.\mathbf{1}_r.b$. (with $\frac{1}{p} + \frac{1}{q} = \frac{1}{r}$).

The proof of the second assertion is more delicate. It suffices to prove it when $h = \mathbf{1}_p$, $k = \mathbf{1}_q$ since then by the preceding:

$$\forall a, b \in \mathcal{A}\ (a.\mathbf{1}_p.x).(\mathbf{1}_q.b) = a.((\mathbf{1}_p.x).\mathbf{1}_q).b = a.(\mathbf{1}_p.(x.\mathbf{1}_q)).b = (a.\mathbf{1}_p).(x.\mathbf{1}_q.b)$$

and the general case follows by density.

Assume that $x \in \mathcal{A}$ is such that $x\xi_0 \in \mathcal{D}(\Delta^{-1/p}) \cap \mathcal{D}(\Delta^{1/q})$ and

$$\exists\, y, w \in \mathcal{A} \text{ such that: } y\xi_0 = \Delta^{-1/p}x\xi_0 \text{ and } w\xi_0 = \Delta^{1/q}x\xi_0 \tag{A}$$

Then $\mathbf{1}_p.x = y.\mathbf{1}_p$, $x.\mathbf{1}_q = \mathbf{1}_q.w$ and $y.\mathbf{1}_r = \mathbf{1}_r.w$ (since $w\xi_0 = \Delta^{1/r}y\xi_0$). So

$$(\mathbf{1}_p.x).\mathbf{1}_q = (y.\mathbf{1}_p).\mathbf{1}_q = y.\mathbf{1}_r = \mathbf{1}_r.w = \mathbf{1}_p.(\mathbf{1}_q.w) = \mathbf{1}_p.(x.\mathbf{1}_q)$$

The set of $x \in \mathcal{A}$ verifying (A) is strongly dense in $\mathcal{A}$. In fact the elements

$$x_n = \frac{1}{\sqrt{2\pi n}} \int_{\mathbb{R}} e^{-\frac{1}{2}nt^2} \Delta^{it} x \Delta^{-it} dt$$

belong to $\mathcal{A}$, are such that $x_n\xi_0 \in \mathcal{D}(\Delta^z)$ and $\Delta^z x_n \xi_0 \in \mathcal{A}.\xi_0$ for every $z \in \mathbb{C}$, and $\|x_n\| \le \|x\|$, $x_n \to x$ (strongly*) when $n \to \infty$ (see [**P**]). Then $x_n.\mathbf{1}_q \to x.\mathbf{1}_q$ (for $q = 2$ it results from strong convergence; for $2 < p < \infty$ it follows by interpolation; for $1 \le p < 2$ by left embedding in $L_p(\mathcal{A})$) and similarly $\mathbf{1}_p.x_n \to \mathbf{1}_p.x$. The equality $(\mathbf{1}_p.x).\mathbf{1}_q = \mathbf{1}_p.(x.\mathbf{1}_q)$ follows. □

REMARK 3.4. Let $\mathcal{A}$ be the set of the elements $x \in \mathcal{A}$ such that the map $\mathbb{R} \to \mathcal{A}$, $t \mapsto \sigma_t(x) = \Delta^{it}x\Delta^{-it}$ extends to an entire function $\mathbb{C} \to \mathcal{A}$, $z \mapsto \sigma_z(x)$. These elements are called *analytic*. Then $\mathcal{A}$ is a *-subalgebra of $\mathcal{A}$, and the preceding construction shows that $\mathcal{A}$ is w*-dense in $\mathcal{A}$. Note also that $\sigma_z(\mathcal{A}) = \mathcal{A}$ for every $z \in \mathbb{C}$ and $\sigma_{z'}(\sigma_z(a)) = \sigma_{(z+z')}(a)$ for all $a \in \mathcal{A}$ and $z, z' \in \mathbb{C}$.

PROPOSITION 3.5. $1 \le p, q, r \le \infty$ *with* $\frac{1}{p} + \frac{1}{q} + \frac{1}{r} \le 1$. *Then for every* $h \in L_p(\mathcal{A})$, $k \in L_q(\mathcal{A})$ *and* $\ell \in L_r(\mathcal{A})$ *we have* $(h.k).\ell = h.(k.\ell)$.

PROOF. Assume first that $h = x.\mathbf{1}_p$, $k = \mathbf{1}_q$, $\ell = \mathbf{1}_r.y$, with $x, y \in \mathcal{A}$. Then by the preceding lemma 3.3:

$$((x.\mathbf{1}_p).\mathbf{1}_q).(\mathbf{1}_r.y) = (x.\mathbf{1}_\alpha).(\mathbf{1}_r.y) = x.\mathbf{1}_s.y = (x\mathbf{1}_p).(\mathbf{1}_\beta.y) = (x.\mathbf{1}_p).(\mathbf{1}_q.(\mathbf{1}_r.y))$$

where $\frac{1}{\alpha} = \frac{1}{p} + \frac{1}{q}$, $\frac{1}{\beta} = \frac{1}{q} + \frac{1}{r}$ and $\frac{1}{s} = \frac{1}{p} + \frac{1}{q} + \frac{1}{r}$. By density we obtain

$$(h.\mathbf{1}_q).\ell = h.(\mathbf{1}_q.\ell)$$

for every $h \in L_p(\mathcal{A})$ and $\ell \in L_r(\mathcal{A})$. Take now the case $k = x.\mathbf{1}_q$, $x \in \mathcal{A}$. Then using the preceding and Lemma 3.3:

$$(h.(x\mathbf{1}_q)).\ell = ((h.x).\mathbf{1}_q).\ell = (h.x).(\mathbf{1}_q.\ell) = h.(x.(1_q.\ell)) = h.((x.\mathbf{1}_q).\ell)$$

The case of a general $k \in L_q(\mathcal{A})$ follows by density. □

3.3. Norm of the product of two adjoint elements.

THEOREM 3.6. *For every* $h \in L_{2p}(\mathcal{A})$ *we have* $\|h^*.h\|_p = \|h\|_{2p}^2$.

The inequality $\|h^*.h\|_p \le \|h^*\|_{2p}\|h\|_{2p} = \|h\|_{2p}^2$ is trivial; the converse inequality requires some preliminary work. A useful tool will be the following notion of homogeneous operators on H_{φ_0}. Recall that the *-algebra $\mathcal{A}$ of analytic operators was introduced in Remark 3.4. Note that $J\mathcal{A}\xi_0 = \mathcal{A}\xi_0$ since $Ja\xi_0 = \Delta^{1/2}a^*\xi_0 = \sigma_{-i/2}(a^*)\xi_0$ when $a \in \mathcal{A}$.

DEFINITION 3.7. Let $\alpha \geq 0$. A closed operator T with dense domain is called *α-homogeneous* if $J\sigma_{i\alpha}(x)JT \subseteq TJxJ$ for every $x \in \mathcal{A}$; in other words[1]:

$$\forall \xi \in \mathcal{D}(T), \eta \in \mathcal{D}(T^*), \forall x \in \mathcal{A}, (J\sigma_{i\alpha}(x)JT\xi, \eta) = (JxJ\xi, T^*\eta) \qquad (3.1)$$

The 0-homogeneous operators are simply the operators affiliated with $\mathcal{a}$. It is easy to see that T is $1/p$-homogeneous iff T^* is, and that if T is $1/p$-homogeneous then T^*T is $2/p$-homogeneous.

We say that an element ξ of H_{φ_0} *represents* an element h of $L_p(\mathcal{a})$ if $\xi.\xi_0 = h.\mathbf{1}_{p'}$, i. e. they coincide modulo the left embeddings into $\mathcal{a}_*$; in other words if $(x.\xi, \xi_0) = \mathrm{Tr}(x.h.\mathbf{1}_{p'})$ for every $x \in \mathcal{a}$. If T is a $1/p$-homogeneous operator containing ξ_0 in its domain, we say that T represents an element $h \in L_p(\mathcal{a})$ iff $T\xi_0$ represents h.

LEMMA 3.8. *If $2 \leq p < \infty$ every element in $L_p(\mathcal{a})$ is representable by a $1/p$-homogeneous operator. If $1 \leq p < 2$, the set of the elements of $L_p(\mathcal{a})$ representable by a $1/p$-homogeneous operator is dense in $L_p(\mathcal{a})$; in fact it contains the dense subspace $\mathcal{A}.\mathbf{1}_p$.*

PROOF. a) Assume $2 \leq p < \infty$. Recall that if $h \in L_p(\mathcal{a})$, its image $\xi = h.\mathbf{1}_r$ by the left embedding of $L_p(\mathcal{a})$ into $L_2(\mathcal{a}) = H_{\varphi_0}$ is in the domain of Δ^r (where $\frac{1}{r} = \frac{1}{2} - \frac{1}{p}$). Define an unbounded operator T_0 on $\mathcal{A}.\xi_0 = J\mathcal{A}.J.\xi_0$ by $TJxJ\xi_0 = J\sigma_{i/p}(x)J\xi$. This definition is unambiguous since ξ_0 is separating for $\mathcal{a}'$. We claim that T_0 is closable, i. e. T_0^* has dense range. In fact we verify that $\mathcal{a}.\xi_0$ is contained in $\mathcal{D}(T_0^*)$. If $y \in \mathcal{a}$, we have:

$$\begin{aligned}(T_0JxJ\xi_0, y\xi_0) &= (J\sigma_{i/p}(x)J\xi, y\xi_0)\\ &= (\xi, J\sigma_{-i/p}(x^*)Jy\xi_0) = (\xi, yJ\sigma_{-i/p}(x^*)J\xi_0)\end{aligned}$$

and since

$$\begin{aligned}J\sigma_{-i/p}(x^*)J\xi_0 &= J\Delta^{1/p}x^*\Delta^{-1/p}J\xi_0\\ &= J\Delta^{1/p}x^*\xi_0 = J\Delta^{1/p}J\Delta^{1/2}x\xi_0\\ &= \Delta^{1/2-1/p}x\xi_0\end{aligned}$$

we obtain

$$\begin{aligned}(T_0JxJ\xi_0, y\xi_0) &= (\xi, y\Delta^{1/r}x\xi_0) = (\Delta^{1/r}y^*\xi, x\xi_0)\\ &= (JxJ\xi_0, J\Delta^{1/r}y^*\xi)\end{aligned}$$

hence $y\xi_0 \in \mathcal{D}(T_0^*)$ with $T_0^*y\xi_0 = J\Delta^{1/r}y^*\xi$. So T_0 is closable; let T be its closure; by the definition of T_0 it is clear that T is $1/p$-homogeneous.

b) Assume only $1 \leq p < \infty$. Consider an element $h = x.\mathbf{1}_p$, where $x \in \mathcal{A}$; then h is represented by the element $x.\xi_0 = x\Delta^{1/p}.\xi_0$ of H_{φ_0}. The operator $T_0 = x.\Delta^{1/p}$ has a dense domain $\mathcal{D}(T_0) = \mathcal{D}(\Delta^{1/p})$ and its adjoint is $T_0^* = \Delta^{1/p}x^*$. Since $x^* \in \mathcal{A}$, it is easy to see that $\mathcal{D}(T_0^*) \supset \mathcal{D}(\Delta^{1/p})$ (and $\Delta^{1/p}x^*.\xi = \sigma_{-i/p}(x^*)\Delta^{1/p}.\xi$ for every $\xi \in \mathcal{D}(\Delta^{1/p})$); hence T_0^* has dense domain and T_0 is closable. Let T be the closure of T_0. We claim that T is $1/p$-homogeneous. In fact for every $y \in \mathcal{A}$, $\xi \in \mathcal{D}(T^*)$

[1] It would be called $(-\alpha)$-homogeneous in the terminology of [Hi].

and $\eta \in \mathcal{D}(T_0) = \mathcal{D}(\Delta^{1/p})$ we have:

$$\begin{aligned}(JyJ\xi, T\eta) &= (JyJ\xi, x.\Delta^{1/p}\eta)\\ &= (x^*JyJ\xi, \Delta^{1/p}\eta) = (JyJx^*\xi, \Delta^{1/p}\eta)\\ &= (JyJ\Delta^{-1/p}T^*\xi, \Delta^{1/p}\eta) = (J\Delta^{1/p}\eta, yJ\Delta^{-1/p}JT^*\xi)\\ &= (\Delta^{-1/p}J\eta, y\Delta^{1/p}JT^*\xi) = (J\eta, \sigma_{it/p}(y)JT^*\xi)\\ &= (J\sigma_{it/p}(y)JT^*\xi, \eta)\end{aligned}$$

Since $\mathcal{D}(T_0)$ is a core for T, this remains true for $\eta \in \mathcal{D}(T)$. □

PROPOSITION 3.9. *If T is a α-homogeneous self-adjoint positive operator then for every $z \in \mathbb{C}$ with $\Re e\, z \geq 0$:*

$$\forall \xi, \eta \in \mathcal{D}(T^z), \forall x \in \mathcal{A}, (J\sigma_{i\alpha\bar{z}}(x)JT^z\xi, \eta) = (JxJ\xi, T^{\bar{z}}\eta) \tag{2}$$

In particular for every $t \in \mathbb{R}$ we have $T^{it} \in \Delta^{i\alpha t}.\mathcal{a}$; and for every $p \geq 0$, the operator T^p is $\alpha.p$-homogeneous.

PROOF. It is sufficient to prove (2) for ξ, η belonging to a core of T^z, so we may suppose that both ξ, η belong to some $e.H_{\varphi_0}$, where e is a spectral projection of T of the form $e = \chi_{[a,b]}(T)$, $0 < a < b < \infty$. Set

$$f_1(z) = (JxJ\xi, T^{\bar{z}}\eta) \quad \text{and} \quad f_2(z) = (J\sigma_{i\alpha\bar{z}}(x)JT^z\xi, \eta)$$

Then both f_1 and f_2 are entire functions of exponential type which are bounded on the imaginary axis. Moreover $f_1(n) = f_2(n)$ for every $n \in \mathbb{N}$. This is seen by repeated use of (3.1) using the relation $\sigma_\zeta(\sigma_z(x)) = \sigma_{\zeta+z}(x)$:

$$\begin{aligned}f_1(n) = (JxJ\xi, T^n\eta) &= (J\sigma_{i\alpha}(x)JT\xi, T^{n-1}\eta) = (J\sigma_{2i\alpha}(x)JT^2\xi, T^{n-2}\eta) = \ldots\\ &= (J\sigma_{ni\alpha}(x)JT^n\xi, \eta) = f_2(n)\end{aligned}$$

By Carlson's uniqueness theorem (see [**Bo**] or [**Lv**], sect. 8.3) we have $f_1(z) = f_2(z)$ for every $z \in \mathbb{C}$.

For $z = it$, $t \in \mathbb{R}$ we obtain that $T^{it}JxJ = J\sigma_{\alpha t}(x)JT^{it} = J\Delta^{i\alpha t}x\Delta^{-i\alpha t}JT^{it} = \Delta^{i\alpha t}JxJ\Delta^{-i\alpha t}T^{it}$ for every $x \in \mathcal{A}$, hence for every $x \in \mathcal{a}$; in other words $\Delta^{-i\alpha t}T^{it}$ commutes with $J\mathcal{a}J = \mathcal{a}'$, and thus belongs to $\mathcal{a}$. □

COROLLARY 3.10. *Let T be a closed operator with dense range and $T = U\,|T|$ its polar decomposition. Then T is α-homogeneous iff $|T|$ is α-homogeneous and $U \in \mathcal{a}$.*

PROOF. If T is α-homogeneous then T^* is also α-homogeneous; then it is easy to see that T^*T is 2α-homogeneous, and consequently by the Prop. 3.9, $|T| = (T^*T)^{1/2}$ is α-homogeneous. Then for every $x \in \mathcal{A}$, $\xi \in \mathcal{D}(T)$, $\eta \in \mathcal{D}(T^*)$ we have:

$$\begin{aligned}(JxJU\,|T|\,\xi, \eta) &= (JxJT\xi, \eta)\\ &= (J\sigma_{-i\alpha}(x)J\xi, T^*\eta) = (J\sigma_{-i\alpha}(x)J\xi, |T|\,U^*\eta)\\ &= (JxJ\,|T|\,\xi, U^*\eta) = (UJxJ\,|T|\,\xi, \eta)\end{aligned}$$

By density we infer that $(JxJUP\xi,\eta) = (UJxJP\xi,\eta)$ for every $\xi,\eta \in H_{\varphi_0}$, where P is the support projection of $|T|$. Note that $P = T^0$ belongs to $\mathcal{a}$ since it is 0-homogeneous by the Prop. 3.9. Since $U = UP$ we have then:

$$JxJU = JxJUP = UJxJP = UPJxJ = UJxJ$$

hence U commutes with $\mathcal{a}'$ and thus belongs to $\mathcal{a}$. The converse implication is easy. □

LEMMA 3.11. *Let $1 \leq p < \infty$ and T be a $1/2p$-homogeneous operator on H_{φ_0}; assume that $\xi_0 \in \mathcal{D}(T^*T)$ and that $T\xi_0$ represents an element k of $L_{2p}(\mathcal{a})$. Then $T^*T\xi_0$ represents the element k^*k of $L_p(\mathcal{a})$.*

PROOF. Let $\frac{1}{r} = \frac{1}{2} - \frac{1}{2p}$, so $r = 2p'$. Then for every $x \in \mathcal{a}$:

$$\begin{aligned}\mathrm{Tr}(xk^*k\mathbf{1}_{p'}) &= \mathrm{Tr}(\mathbf{1}_r xk^*k\mathbf{1}_r) = \mathrm{Tr}((kx^*\mathbf{1}_r)^*k\mathbf{1}_r)\\ &= (k.\mathbf{1}_r, (k.x^*).\mathbf{1}_r)_{L_2(\mathcal{a})}\\ &= (T\xi_0, J_{2p}xJ_{2p}T\xi_0)_{H_{\varphi_0}}\\ &= (T\xi_0, \Delta^{-1/r}JxJ\Delta^{1/r}T\xi_0)_{H_{\varphi_0}}\\ &= (T\xi_0, J\Delta^{1/r}x\Delta^{-1/r}JT\xi_0)_{H_{\varphi_0}}\end{aligned}$$

Assume now that $x \in \mathcal{A}$. Then we obtain, using the fact that T is $1/p$-homogeneous:

$$\begin{aligned}\mathrm{Tr}(xk^*k\mathbf{1}_{p'}) &= (T\xi_0, J\sigma_{-i/r}(x)JT\xi_0)\\ &= (T^*T\xi_0, J\sigma_{-i/2p}(\sigma_{-i/r}(x))J\xi_0)\\ &= (T^*T\xi_0, J\sigma_{-i/2}(x)J\xi_0)\\ &= (T^*T\xi_0, J\Delta^{1/2}xJ\xi_0)\\ &= (T^*T\xi_0, x^*\xi_0)\\ &= \mathrm{Tr}(xT^*T\xi_0.\xi_0)\end{aligned}$$

hence (by w*-density of $\mathcal{A}$): $T^*T\xi_0.\xi_0 = k^*.k.\mathbf{1}_{p'}$. □

LEMMA 3.12. *Let T be a $1/p$-homogeneous positive self-adjoint operator on H_{φ_0}; assume that $\xi_0 \in \mathcal{D}(T)$ and that $T\xi_0$ represents an element h of $L_p(\mathcal{a})$. Then $T^{1/2}\xi_0$ represents an element k of $L_{2p}(\mathcal{a})$ such that $k^*.k = h$ and $\|k\|_{2p}^2 = \|h\|_p$.*

PROOF. By the preceding Lemma 3.11 it is sufficient to prove that $T^{1/2}\xi_0$ represents an element k of $L_{2p}(\mathcal{a})$ such that $\|k\|_{2p}^2 \leq \|h\|_p$. Consider the map $F : \bar{S} \to H_{\varphi_0}, z \mapsto T^z\xi_0$ where $\bar{S}$ is as usual the closed strip $\{z \in \mathbb{C} \mid 0 \leq \Re e\, z \leq 1\}$; it is norm continuous and analytic on the interior of the strip. By the Proposition 3.9, for every $t \in \mathbb{R}$, we have $T^{it} = \Delta^{it/p}u_t$, where $u_t \in \mathcal{a}$; actually (u_t) is a strongly continuous family of partial isometries in $\mathcal{a}$. Hence $T^{it}\xi_0 = \Delta^{it/p}u_t\xi_0 = \Delta^{it/p}u_t\Delta^{-it/p}\xi_0$ belongs to $\mathcal{a}.\xi_0$, with norm $\|T^{it}\xi_0\|_\infty \leq 1$. On the other hand, $T^{1+it}\xi_0 = T^{it}T\xi_0 = \Delta^{it/p}u_tT\xi_0$. This represents an element of $L_p(\mathcal{a})$ with norm $\leq \|h\|_p$. For, $u_tT\xi_0$ represents $u_t.h$, u_t is a partial isomertry of $\mathcal{a}$, while $\Delta^{it/p}$ is the restriction to $L_2(\mathcal{a})$ of the map $\sigma'_{-t/p} : \mathcal{a}_* \to \mathcal{a}_*, \psi \mapsto \psi\circ\sigma_{-t/p}$ which is an onto isometry of both spaces $\mathcal{a}_*$ and $i_\ell(\mathcal{a})$, hence of all $L_q(\mathcal{a})$, $1 \leq q \leq \infty$.

Arguing by duality, it is not hard to see that F is L_p-valued and $\sup_{z\in\bar{S}} \|F(z)\|_p \leq \|h\|_p$; then $F(\theta) \in (i_\ell(\mathcal{Q}), L_p^\ell(\mathcal{Q}))^\theta = L_{p/\theta}(\mathcal{Q})$ for every $0 < \theta < 1$. In particular

$$\|k\|_{2p} = \|T^{1/2}\xi_0\|_{2p} \leq \|T\xi_0\|_p^{1/2} = \|h\|_{2p}^{1/2} \qquad \square$$

PROOF OF THEOREM 3.6:. By density of $\mathcal{A}.\mathbf{1}_{2p}$ in $L_{2p}(\mathcal{Q})$, we may suppose that $h = x.\mathbf{1}_{2p}$, with $x \in \mathcal{A}$. By Lemma 3.8, h is represented by the $2p$-homogeneous operator $T = \overline{x\Delta^{1/2p}}$. The domain of the adjoint operator $T^* = \Delta^{1/2p}.x^*$ contains $T\xi_0 = x.\xi_0$; hence $\xi_0 \in \mathcal{D}(T^*T)$. So by Lemma 3.11, T^*T represents h^*h and by Lemma 3.12, $|T| = (T^*T)^{1/2}$ represents an element $k \in L_{2p}(\mathcal{Q})$ such that $\|k\|_{2p}^2 = \|h^*.h\|_p$. Consider the polar decomposition $T = u\,|T|$ of T: by Corollary 3.10 the partial isometry u belongs to $\mathcal{Q}$. Then $h = u.k$, $k = u^*.h$ and thus $\|h\|_{2p} = \|k\|_{2p}$. $\square$

4. Positive cones in the $L_p(\mathcal{Q})$ spaces and their relations.

4.1. $L_p(\mathcal{Q})$ as ordered topological vector space.

NOTATION 4.1. Let Γ_p be the subset $\{k^*.k \mid k \in L_{2p}(\mathcal{Q})\}$ in $L_p(\mathcal{Q})$. We denote by $L_p(\mathcal{Q})_+$ the closure of Γ_p in $L_p(\mathcal{Q})$.

Note that if $h \in L_q(\mathcal{Q})$ with $\frac{2}{q} + \frac{1}{p} = \frac{1}{r} \leq 1$ then we have trivially $h^*.\Gamma_p.h \subset \Gamma_r$ and $h^*.L_p(\mathcal{Q})_+.h \subset L_r(\mathcal{Q})_+$.

PROPOSITION 4.2. *The set $L_p(\mathcal{Q})_+$ is a closed convex cone containing no line. Consequently it defines an order on the space $L_p(\mathcal{Q})$ which is compatible with the structure of topological vector space. For $p = 1$ the cone $L_1(\mathcal{Q})_+$ coincides with Γ_1 and with the positive cone of $\mathcal{Q}_*$.*

PROOF. Since $\mathcal{Q}.\mathbf{1}_{2p}$ is dense in $L_{2p}(\mathcal{Q})$, the set $\Gamma_p^0 = \{\mathbf{1}_{2p}.x^*.x.\mathbf{1}_{2p} \mid x \in \mathcal{Q}\}$ is dense in Γ_p. But Γ_p^0 is a convex cone since it is equal to $\{\mathbf{1}_{2p}.a.\mathbf{1}_{2p} \mid a \in \mathcal{Q}_+\}$; consequently $L_p(\mathcal{Q})_+$ is a closed convex cone. For $p = 1$ we have $\Gamma_1 \subset \mathcal{Q}_{*+}$ since for every $h \in L_2(\mathcal{Q})$ and $x \in \mathcal{Q}$:

$$\mathrm{Tr}((h^*.h).(x.x^*)) = \mathrm{Tr}((x^*.h^*).(h.x)) = \|h.x\|_2^2$$

In fact it is well known that (since the vector ξ_0 is cyclic and separating for $\mathcal{Q}$) every positive element ψ of $\mathcal{Q}_*$ takes the form ω_ξ for some element $\xi \in H_{\varphi_0}$, where $\omega_\xi(x) = (x\xi, \xi)_{\varphi_0}$. Identifying ξ with an element h of $L_2(\mathcal{Q})$ we have $\psi(x) = \mathrm{Tr}(h^*.(x.h)) = \mathrm{Tr}((h.h^*).x)$, i. e. ψ identifies with the element $h.h^*$ of Γ_1.

For $p \geq 1$, let u_p be the bounded linear mapping $L_p(\mathcal{Q}) \to L_1(\mathcal{Q})$, $h \mapsto \mathbf{1}_{2p'}.h.\mathbf{1}_{2p'}$. It is straightforward that u_p maps Γ_p into Γ_1, and consequently $L_p(\mathcal{Q})_+$ into $L_1(\mathcal{Q})_+$. Since u_p is injective (as the composition of the injective maps $h \mapsto h.\mathbf{1}_{2p'}$, $L_p(\mathcal{Q}) \to L_{(2p')'}(\mathcal{Q})$, and $k \mapsto \mathbf{1}_{2p'}.k$, $L_{(2p')'}(\mathcal{Q}) \to L_1(\mathcal{Q})$) and $L_1(\mathcal{Q})_+ = \mathcal{Q}_{*+}$ contains no line, so does $L_p(\mathcal{Q})_+$. $\square$

An element h of $L_p(\mathcal{Q})$ is said to be *self-adjoint* if $h^* = h$. Every element h in $L_p(\mathcal{Q})$ can be written uniquely $h = h_1 + ih_2$ where h_1, h_2 are self-adjoint; then $\|h_1\| \vee \|h_2\| \leq \|h\|$. The set $L_p(\mathcal{Q})_{sa}$ of self-adjoint elements in $L_p(\mathcal{Q})$ is a closed real subspace of $L_p(\mathcal{Q})$. The elements of Γ_p are trivially self-adjoint; hence positive elements are self-adjoint. The right and left supports (see prop. 1.3 and rem. 2.5 for definitions) of a self-adjoint element h coincide trivially, we denote them simply by $s(h)$.

Two self-adjoint elements h, k of $L_p(\mathcal{A})$ are called *disjoint* if there exist a projection $e \in \mathcal{A}$ such that $h = e.h$, $k = e^{\perp}.k$. Note that we have then $h = h.e = e.h.e$, $k = k.e^{\perp} = e^{\perp}.k.e^{\perp}$.

Our next goal is to prove the following polar decomposition results for in $L_p(\mathcal{A})$.

THEOREM 4.3. *Every element h in $L_p(\mathcal{A})$ has a unique decomposition $h = u\,|h|$, where $|h| \in L_p(\mathcal{A})_+$ and $u \in \mathcal{A}$ is a partial isometry such that $uu^* = s_\ell(h)$, $u^*u = s(|h|)$. Moreover the polar decomposition of h^* is $h^* = u^*\,|h^*|$.*

THEOREM 4.4. *Every self-adjoint element in $L_p(\mathcal{A})$ is the difference of two disjoint positive elements in $L_p(\mathcal{A})$. This decomposition is unique.*

COROLLARY 4.5. *Every element in the unit ball of $L_p(\mathcal{A})$ is a linear combination of four positive elements in the unit ball of $L_p(\mathcal{A})$.*

The proof of the Theorems 4.3 and 4.4 is based on a deeper analysis of the representation of elements in $L_p(\mathcal{A})$ by $1/p$-homogeneous operators. The crucial point is the following result on the domain of $1/p$-homogeneous operators; we borrow the argument to [**M**] and [**AM**].

FACT 4.6. *If $p \geq 2$ then $\mathcal{A}.\xi_0 = J\mathcal{A}.\xi_0$ is a core for every $1/p$-homogeneous operator T on H_{φ_0} such that $\xi_0 \in \mathcal{D}(T)$ and $T\xi_0 \in L_p^\ell(\mathcal{A})$. Moreover if $\frac{1}{r} = \frac{1}{2} - \frac{1}{p}$ the domain of such an operator contains the left image of $L_r(\mathcal{A})$ in $H_{\varphi_0} = L_2(\mathcal{A})$; in fact T is the closure of the operator $L_r^\ell(\mathcal{A}) \mapsto L_2(\mathcal{A})$, $h.\mathbf{1}_p \mapsto k.h$ (where $k \in L_p(\mathcal{A})$ is defined by: $k.\mathbf{1}_r = T\xi_0$) .*

We shall sketch the proof of this basic Fact later (see the Appendix). For the moment we derive simple corollaries of this Fact.

COROLLARY 4.7. *Let $p \geq 2$. Every symmetric $1/p$-homogeneous operator T on H_{φ_0} which represents an element of $L_p(\mathcal{A})$ is self-adjoint.*

PROOF. By the hypothesis we have $T \subseteq T^*$. Since $\xi_0 \in \mathcal{D}(T) \subset \mathcal{D}(T^*)$ and $T\xi_0 \in L_p(\mathcal{A})$, then by Fact 4.6 the operators T and T^* have a common core on which they agree; hence they are equal. □

COROLLARY 4.8. *Let $p \geq 2$.*
a) If an element h of $L_p(\mathcal{A})$ is represented by a $1/p$-homogeneous operator on H_{φ_0}, this operator is unique.
b) If h, h^* are represented by $1/p$-homogeneous operators T, S then $S = T^*$.
c) In particular if a self-adjoint element ξ of $L_p(\mathcal{A})$ is represented by a $1/p$-homogeneous operator on H_{φ_0}, this operator is self-adjoint.

PROOF. a) If $\xi = T_1\xi_0 = T_2\xi_0$ where T_1, T_2 are $1/p$-homogeneous, then T_1 and T_2 coincide on $J\mathcal{A}\xi_0$ since $T_1JxJ\xi_0 = J\sigma_{i/p}(x)JT_1\xi_0 = J\sigma_{i/p}(x)J\xi = T_2JxJ\xi_0$ for every $x \in \mathcal{A}$. Since T_1 and T_2 coincide on a common core they are equal.
b) By the preceding it suffices to prove that $\xi_0 \in \mathcal{D}(T^*)$ and that $T^*\xi_0$ represents h^*. For every $x \in \mathcal{A}$ we have:

$$\begin{aligned}(\xi_0, TJxJ\xi_0) &= (\xi_0, J\sigma_{i/p}(x)JT\xi_0) = (J\sigma_{-i/p}(x^*)J\xi_0, T\xi_0) \\ &= (J\Delta^{1/p}J\Delta^{1/2}x\xi_0, T\xi_0) = (\Delta^{1/r}x\xi_0, T\xi_0)\end{aligned}$$

where $\frac{1}{r} = \frac{1}{2} - \frac{1}{p}$.

Since $p \geq 2$ then $T\xi_0 \in \mathcal{D}(\Delta^{1/r})$ (since $T\xi_0 \in L_p(\mathcal{A}).\mathbf{1}_r$), so we have

$$(\xi_0, TJxJ\xi_0) = (x\xi_0, \Delta^{1/r}T\xi_0) = (J\Delta^{1/r}T\xi_0, Jx\xi_0)$$

Since $J\mathcal{A}\xi_0$ is a core for T, this proves that $\xi_0 \in \mathcal{D}(T^*)$ and $T^*\xi_0 = J\Delta^{1/r}T\xi_0$; since $T\xi_0$ represents h we know that $J\Delta^{1/r}T\xi_0$ represents h^*. □

PROOF OF THEOREMS 4.3, 4.4 : CASE $p \geq 2$. In this case every element h of $L_p(\mathcal{A})$ is represented by a $1/p$-homogeneous operator T by Lemma 3.8; if h is self-adjoint, then T is self-adjoint by Cor. 4.8. Morover it is easy to see that the left support of h coincides with that of T (the projection on the closure of the range of T). Let $T = u|T|$ be the polar decomposition of T: we know that $|T|$ is a $1/p$-homogeneous operator, $\mathcal{D}(|T|) = \mathcal{D}(T)$ contains ξ_0 and $u \in \mathcal{A}$. Let $|h| = u^*h$, then $|h|$ is represented by $|T|$, hence is positive by Lemma 3.12; we have then $uu^* = s_\ell(T) = s_\ell(h)$ and $u^*u = s(|T|) = s(h)$.

Conversely if $h = vk$ with $k \geq 0$, $vv^* = s_\ell(h)$, $v^*v = s(k)$, let S be the $1/p$-homogeneous operator which represents k. Since k is self-adjoint so is S. We claim that S is positive. In fact let $S = S_+ - S_-$ its decomposition into positive and negative parts and e_+, e_- the corresponding spectral projections. Then $h_+ = e_+h$ and $h_- = e_-h$ are positive (since they are represented by S_+, S_- respectively). We have $h_+ = h + h_- \geq h$, hence $0 \leq h_- = e_-he_- \leq e_-h_+e_- = 0$; thus $h_- = 0$ and consequently $S_- = 0$. Since vS represents h we have $T = vS$, so $vv^* = s_\ell(T)$, $v^*v = s(k) = s(T)$ and $T = vS$ is the polar decomposition of T. Hence $v = u$, $S = |T|$ and $k = |h|$.

Since h^* is represented by T^* and $T^* = u^*|T^*|$ is the polar decomposition of T^*, that of h^* is $h^* = u^*|h^*|$.

If now h is self-adjoint, so is T; then $u \in \mathcal{A}$, is a self-adjoint partial isometry commuting with T; we have $u = e_+ - e_-$ where e_+ e_- are disjoint projection of $\mathcal{A}$ which are spectral projections of T. Then $T_+ = e_+T$, $T_- = -e_-T$ are positive self-adjoint $1/p$-homogeneous operators; they represent $h_+ = e_+.h$, $h_- = -e_-.h$, which are thus positive elements of $L_p(\mathcal{A})$ by Lemma 3.12. Then $h = h_+ - h_-$ is the desired decomposition of h.

Conversely if $h = h_1 - h_2$ is such a decomposition, with h_1, h_2 positive and disjoint, set $e_1 = s(h_1)$, $e_2 = s(h_2)$, $v = e_1 + e_2$, $k = h_1 + h_2$, then $h = uk$ is the polar decomposition of h; so $k = |h|$, $h_1 = \frac{1}{2}(h + |h|) = h_+$, $h_2 = \frac{1}{2}(|h| - h) = h_-$.□

Note that by Lemma 3.12 and the preceding reasoning we can state:

REMARK 4.9. Assume $p \geq 2$. Then $L_p(\mathcal{A})_+ = \Gamma_p$ consists exactly of elements representable by positive self-adjoint $1/p$-homogeneous operators.

LEMMA 4.10. *If $h \in L_p(\mathcal{A})$, $h' \in L_{p'}(\mathcal{A})$ are positive then* $\mathrm{Tr}(h.h') \geq 0$.

PROOF. By density we may assume that $h \in \Gamma_p$ and $h' \in \Gamma_{p'}$; so $h = k^*.k$ and $h' = k'^*.k'$ for some $h' \in L_{2p}(\mathcal{A})$, $k' \in L_{2p'}(\mathcal{A})$. Then

$$\mathrm{Tr}(h.h') = \mathrm{Tr}(k^*.k.k'^*.k') = \mathrm{Tr}(k'.k^*.k.k'^*) = \|k.k'^*\|_2^2 \geq 0 \qquad \square$$

Now we define the powers h^α, $0 \leq \alpha \leq p/2$ of a positive element of $L_p(\mathcal{A})$, $p \geq 2$.

PROPOSITION 4.11. *Let $p \geq 2$ and $h \in L_p(\mathcal{A})_+$. Let T be the unique $1/p$-homogeneous positive, self-adjoint operator which represents h. Then $\xi_0 \in \mathcal{D}(T^{p/2})$. For every $0 < \alpha \leq p/2$ we set $\xi_\alpha = T^\alpha \xi_0$. Then ξ_α represents an element $h^\alpha \in L_{p/\alpha}(\mathcal{A})_+$ with norm $\|h^\alpha\|_{p/\alpha} = \|h\|_p^\alpha$. Moreover for every $\alpha, \beta > 0$ we have $h^{\alpha+\beta} = h^\alpha . h^\beta$ (if $\alpha + \beta \leq p/2$) and $(h^\alpha)^\beta = h^{\alpha\beta}$ (if $\alpha\beta \leq p/2$).*

PROOF. First we define $\xi_\alpha = T^\alpha \xi_0$ for $0 < \alpha \leq 1$. By the reasoning of Lemma 3.12, ξ_α represents an element $h^\alpha \in L_{p/\alpha}(\mathcal{A})_+$ with norm $\|h^\alpha\|_{p/\alpha} \leq \|h\|_p^\alpha$. Clearly $(h^\alpha)^\beta$ and $h^{\alpha\beta}$ are represented by the same element $(T^\alpha)^\beta \xi_0 = T^{\alpha\beta}\xi_0$ for every $\alpha, \beta \in (0,1]$. Also $h^{\alpha+\beta}$ is represented by $T^{\alpha+\beta}\xi_0 = T^\alpha T^\beta \xi_0$, which represents $h^\alpha . h^\beta$ by the same reasoning as was used in Lemma 3.11. In particular $\|h\|_p = \|(h^{1/n})^n\|_p \leq \|h^{1/n}\|_{np}^n \leq \|h\|_p$ for every positive integer n. If $0 < \alpha < 1$, choose $n \geq 1/\alpha$, then $\|h\|_p^{1/n} = \|h^{1/n}\|_{np} = \|(h^\alpha)^{1/n\alpha}\|_{np} \leq \|h^\alpha\|_{p/\alpha}^{1/n\alpha}$, and thus $\|h^\alpha\|_{p/\alpha} = \|h\|_p^\alpha$.

Now if $\frac{1}{r} = \frac{1}{2} - \frac{1}{p} \geq \frac{1}{p}$, we have by Fact 4.6: $\mathcal{D}(T) \supseteq L_r(\mathcal{A}).\mathbf{1}_p \supseteq L_p(\mathcal{A}).\mathbf{1}_r$; so $T\xi_0 \in \mathcal{D}(T)$, which means that $\xi_0 \in \mathcal{D}(T^2)$. Moreover $T^2\xi_0$ represents $h^2 \in L_{p/2}(\mathcal{A})$. Iterating, we see that $\xi_0 \in \mathcal{D}(T^m)$, $m = \left[\frac{p}{2}\right]$, and $T^m\xi_0$ represents $h^m \in L_{p/m}(\mathcal{A})$. If $\gamma = \frac{p}{2} - m > 0$, then $T^\gamma \xi_0 \in L_{p/\gamma} \subseteq \mathcal{D}(T^m)$ since $\frac{\gamma}{p} = \frac{1}{2} - \frac{m}{p}$. Thus $\xi_0 \in \mathcal{D}(T^{m+\gamma}) = \mathcal{D}(T^{p/2})$. Then $T^{p/2}\xi_0$ represents $h^m . h^\gamma \in L_\rho(\mathcal{A})$ where $\frac{1}{\rho} = \frac{m}{p} + \frac{\gamma}{p} = \frac{1}{2}$.

The properties of h^α, $\alpha \in (0, \frac{p}{2}]$ are now easily derived since $h^\alpha = k^{2\alpha/p}$ where $k = h^{p/2} \in L_2(\mathcal{A})_+$ is represented by $T^{p/2}\xi_0$. □

If X is a Banach space, and $x \in X$, a *norming functional* for X is element $x' \in X^*$ such that $\|x'\| = 1$ and $\|x\| = \langle x, x' \rangle$. Recall that if X^* is a strictly convex Banach space every non-zero element of X has a unique norming functional.

PROPOSITION 4.12. *Let $1 \leq p < \infty$ every positive element in $L_p(\mathcal{A})$ has a a positive norming functional (as element of $L_{p'}(\mathcal{A})$). In particular if $p > 1$ the unique norming functional of a positive element of $L_p(\mathcal{A})$ is positive.*

PROOF. The case $p = 1$ is standard since then $L_p(\mathcal{A}) = \mathcal{A}_*$. We assume then $p > 1$. Let p' be the conjugate exponent.

a) The case $1 < p < 2$. If h is a general element of $L_p(\mathcal{A})$ and $h' \in L_{p'}(\mathcal{A})$ is a norming functional for h, then h'^* is a norming functional for h^*. Consequently every self-adjoint $h \in L_p(\mathcal{A})$ has a self-adjoint norming functional. If h is positive, let h' be a self-adjoint norming functional for h. Let $h' = h'_+ - h'_-$ its decomposition into positive and negative part (here we apply precisely the case of Th. 4.4 which was proved above since $p' > 2$). Then $\mathrm{Tr}(h.h'_+) \geq \mathrm{Tr}(h.h') = \|h\|$ and $\|h'_+\| \leq \|h'\| \leq 1$. So h'_+ is a positive norming functional for h.

b) The case $2 \leq p < \infty$. Consider $h \in L_p(\mathcal{A})_+$ and set $h' = h^{p/2-1}.h^{p/2}$ Then $\|h'\|_{p'} \leq \|h\|_p^{p-1}$ and $\mathrm{Tr}(h.h') = \mathrm{Tr}(h^{p/2}.h^{p/2}) = \|h^{p/2}\|_2^2 = \|h\|_p^p$. So $\dfrac{h'}{\|h'\|}$ is a norming functional for h. Finally $h' = h^{p/2-1}.h.h^{p/2-1}$ is positive. □

PROOF OF THEOREMS 4.3, 4.4 IN THE CASE $1 < p < 2$. Let $h \in L_p(\mathcal{A})$ with $\|h\| = 1$ and h' be its unique norming functional in $L_{p'}(\mathcal{A})$. Since $p' \geq 2$ we have the polar decomposition $h' = u\,|h'|$. Then $h'^* = u^*\,|h'^*|$ and $\mathrm{Tr}(uh\,|h'^*|) =$

$\mathrm{Tr}(h\,|h'^*|\,u) = \mathrm{Tr}(hh') = 1$ and $\|uh\| \le \|h\| = 1$, hence uh is the unique norming functional for $|h'^*|$; then $uh \ge 0$ by Prop. 4.12 . Note that $s_\ell(h) = s_r(h')$, $s_r(h) = s_\ell(h')$: for example, $\mathrm{Tr}(s_r(h').h.h') = \mathrm{Tr}(h.h'.s_r(h')) = \mathrm{Tr}(h.h') = 1$, $\|s_r(h').h\|_p \le 1$ imply that $s_r(h').h = h$, so $s_r(h') \ge s_\ell(h)$. Let $k = uh$; since $u^*u = s_\ell(h'^*) = s_r(h') = s_\ell(h)$, we have $u^*k = h$. These two equations imply $s_r(k) = s_r(h)$; then $uu^* = s_\ell(h') = s_r(h) = s_r(k)$. Thus $h = u^*.k$ is a polar decomposition for h. Conversely from a polar decomposition $h = v.\ell$ for h we deduce a polar decomposition $h'^* = vk'$ for h'; by unicity of this last one, we have $v = u^*$ and $\ell = v^*.h = u.h = k$.

If h is self-adjoint, so is h'. Then u is self-adjoint so $h = u\,|h| = h^* = |h|\,u$; hence u commutes with $|h|$, and similarly with h; so do the projections $e_+ = (u^2+u)/2$ and $e_- = (u^2-u)/2$. Then $h = h_+ - h_-$ with $h_+ = e_+\,|h| = e_+\,|h|\,e_+ \ge 0$ and $h_- = e_-\,|h| = e_-\,|h|\,e_- \ge 0$. This gives the decomposition of h into disjoint positive and negative parts. Its unicity stems from that of the polar decomposition. □

PROPOSITION 4.13. *The cones $L_p(\mathfrak{A})_+$ and $L_{p'}(\mathfrak{A})_+$ are polar each of the other.*

This means that $L_p(\mathfrak{A})_+ = \{h \in L_p(\mathfrak{A}) \mid \mathrm{Tr}(h.h') \ge 0,\ \forall h' \in L_{p'}(\mathfrak{A})_+\}$.

PROOF. The fact that $L_p(\mathfrak{A})_+$ is included in the polar cone of $L_{p'}(\mathfrak{A})_+$ was proved in Lemma 4.10. We prove now the reverse inclusion.

Assume that $\mathrm{Tr}(h.h') \ge 0$ for every $h' \in L_{p'}(\mathfrak{A})_+$. We have $h = h_1 + ih_2$ where h_1, h_2 are self-adjoint. For every $h' \in L_{p'}(\mathfrak{A})_+$ the scalar products $\mathrm{Tr}(h_i.h')$, $i = 1,2$ are real; so the inequality $\mathrm{Tr}(h.h') \ge 0$ forces the scalar product $\mathrm{Tr}(h_2.h')$ to vanish. Using Cor. 4.5, this implies that $\mathrm{Tr}(h_2.h') = 0$ for every $h' \in L_{p'}(\mathfrak{A})$; consequently $h_2 = 0$, and h is self-adjoint. Let now $h = h_+ - h_-$ be the decomposition of h into positive and negative part given by Theorem 4.4, and e a projection of $\mathfrak{A}$ such that $h_+ = e.h = h.e$. If $h_- \neq 0$ there exists by Prop. 4.12 an element $h' \in L_{p'}(\mathfrak{A})+$ such that $\mathrm{Tr}(h_-.h) \neq 0$, hence $\mathrm{Tr}(h_-.h) > 0$. Then $h'_1 = e^\perp.h'.e^\perp \ge 0$ and $\mathrm{Tr}(h.h'_1) = \mathrm{Tr}(-h_-.h') < 0$, a contradiction. □

COROLLARY 4.14. *If $0 \le h \le k$ in $L_p(\mathfrak{A})$ then $\|h\|_p \le \|k\|_p$. More generally if h, k are self-adjoint elements of $L_p(\mathfrak{A})$ such that $-k \le h \le k$ then $\|h\|_p \le \|k\|_p$.*

PROOF. Assume first $0 \le h \le k$, and let h' be a positive norming functional for h. Since $k - h \ge 0$ we have $\|k\|_p \ge \mathrm{Tr}(k.h') = \mathrm{Tr}((k-h).h') + \mathrm{Tr}(h.h') \ge \mathrm{Tr}(h.h') = \|h\|_p$.

Assume now $-k \le h \le k$ and let $h = h_+ - h_-$ a decomposition into positive and negative part given by the Theorem 4.4, and e a projection such that $h_+ = e.h = h.e$. Then $h_+ = e.h.e \le e.k.e$, while $h_- = -e^\perp.h.e^\perp \le e^\perp.k.e^\perp$. So $0 \le h_+ + h_- \le e.k.e + e^\perp.k.e^\perp$ and consequently $\|h_+ + h_-\| \le \|e.k.e + e^\perp.k.e^\perp\|$. Let $u = e - e^\perp$; since $h_+ + h_- = u.h$, $e.k.e + e^\perp.k.e^\perp = \frac{1}{2}(k + u.k.u)$ and u is unitary we have then $\|h\| = \|h_+ + h_-\| \le \frac{1}{2}\|k + u.k.u\| \le \|k\|$. □

REMARK 4.15. The L_p-norm is in fact strictly monotone ($0 \le h \le k$, $h \neq k$ implies $\|h\|_p < \|k\|_p$); when $1 < p < \infty$ this is an easy consequence of the strict convexity.

4.2. Square roots of positive elements.

The key result of this subsection is the following:

THEOREM 4.16. *Let $1 \leq p < \infty$. For every positive $h_1, h_2 \in L_{2p}(\mathcal{A})$ we have*

$$\|h_1 - h_2\|_{2p} \leq \|h_1^2 - h_2^2\|_p^{1/2}$$

We prove first the following Lemma:

LEMMA 4.17. *Assume that $1 < p < \infty$. Let $h \in L_{2p}(\mathcal{A})$ be a self-adjoint element, with decomposition $h = h_+ - h_-$ into disjoint positive and negative parts, and $k \in L_{p'}(\mathcal{A})$ be the norming functional for h^2. Then $h_+.k$ and $h_-.k$ are positive.*

PROOF. Assume first that h is positive. If $h \neq 0$ then $\dfrac{h.k}{\|h.k\|_{(2p)'}}$ is the normal functional for h, hence it is positive. In the general case we have $h_+ = e.h = h.e$ for some projection in $\mathcal{A}$, $h = e.h.e - e^\perp.h.e^\perp$ and $h^2 = e.h^2.e + e^\perp.h^2.e^\perp$. Then $k_0 = e.k.e + e^\perp.k.e^\perp$ is also a norming functional for h^2. Since $p > 1$ we conclude that $k = k_0$ by unicity of the norming functional. Let $|h| = h_+ + h_-$ then since k is a norming functional for $|h|^2$ we have $|h|.k \geq 0$. Then $h_+.k = e.|h|.k.e$ and $h_-.k = e^\perp.|h|.k.e^\perp$ are positive too. □

PROOF OF THEOREM 4.16. It is inspired from the case $p = 1$ treated in [**A**]. Let $h = h_1 - h_2$ and k be the norming functional for h^2 if $p > 1$ ($k = I$ if $p = 1$). Then:

$$\begin{aligned}
\|h\|_{2p}^2 = \|h^2\|_p = \operatorname{Tr}(h^2.k) &= \operatorname{Tr}((h_1 - h_2).(h_+ - h_-).k)\\
&= \operatorname{Tr}((h_1.h_+ + h_2h_-).k) - \operatorname{Tr}(h_1.h_-.k) - \operatorname{Tr}(h_2.h_+.k)\\
&\leq \operatorname{Tr}((h_1.h_+ + h_2h_-).k) + \operatorname{Tr}(h_1.h_-.k) + \operatorname{Tr}(h_2.h_+.k)\\
&= \operatorname{Tr}((h_1 + h_2).(h_+ + h_-).k)\\
&= \operatorname{Tr}((h_1 + h_2).(h_+ - h_-).u.k)
\end{aligned}$$

where $u = e - e^\perp$ is the self-adjoint operator of $\mathcal{A}$ such that $h = u|h|$. We have used the fact that $\operatorname{Tr}(h_1.h_-.k) \geq 0$, $\operatorname{Tr}(h_2.h_+.k) \geq 0$ since $h_1, h_2, h_-.k$ and $h_+.k$ are all positive. Note that $u.k = k.u$ by the proof of the Lemma 4.17. Then:

$$\begin{aligned}
\|h\|_{2p}^2 &\leq \operatorname{Tr}((h_1 + h_2)(h_1 - h_2).u.k)\\
&= \operatorname{Tr}(h_1^2 - h_2^2).u.k) + \operatorname{Tr}((h_1.h_2 - h_2.h_1).u.k)
\end{aligned}$$

But in the last line the first term is real, the second one is purely imaginary (since $u.k$ and $h_1^2 - h_2^2$ are self-adjoint, and $h_1.h_2 - h_2.h_1$ is anti-self-adjoint). Necessarily the second term vanishes and we obtain

$$\|h\|_{2p}^2 \leq \operatorname{Tr}(h_1^2 - h_2^2).u.k) \leq \|h_1^2 - h_2^2\|_p$$

since $\|k\|_{p'} = 1$. □

COROLLARY 4.18. *For every $h \in L_p(\mathcal{A})_+$ there exists a unique element in $L_{2p}(\mathcal{A})_+$, denoted by $h^{1/2}$, such that $h = (h^{1/2})^2$. In particular $L_p(\mathcal{A})_+$ coincides with the cone Γ_p.*

PROOF. The uniqueness is clear. It results from Lemma 3.8 and 3.11 that the set of positive elements of $L_p(\mathcal{A})_+$ representable by a positive self-adjoint $1/p$-homogeneous operator is dense in $L_p(\mathcal{A})_+$ (it contains the elements $\mathbf{1}_{2p}a^*a.\mathbf{1}_{2p}$, $a \in \mathcal{A}$, see the proof of Th. 3.6); every element in this set has a square root by

Lemma 3.12. By the preceding theorem 4.16 the set of positive elements in $L_p(\mathcal{A})$ having a square root is closed, so it is equal to $L_p(\mathcal{A})_+$. □

We are now in position to extend the definition of the powers of positive elements $h \in L_p(\mathcal{A})_+$ to the case $1 \le p \le 2$.

DEFINITION 4.19. Let $1 \le p < \infty$. For every $h \in L_p(\mathcal{A})_+$ and every $\alpha \in (0,p]$ we set $h^\alpha = ((h^{1/2})^\alpha)^2$.

The map $h \mapsto h^\alpha$ is a bijection betwen the positive cones of $L_p(\mathcal{A})$ and of $L_{p/\alpha}(\mathcal{A})$ verifying the norm equation $\|h^\alpha\|_{p/\alpha} = \|h\|_p^\alpha$. It satisfies the ordinary calculation rules for the exponents.

Lozanovskii's decompositions in $L_p(\mathcal{A})$.

We have seen in §1 that if $\frac{1}{r} = \frac{1}{p} + \frac{1}{q} \ge 1$ then $L_p(\mathcal{A}).L_q(\mathcal{A}) \subseteq L_r(\mathcal{A})$ with the corresponding Hölder inequality. This embedding can now be shown to be an equality.

THEOREM 4.20. *Let $\frac{1}{r} = \frac{1}{p} + \frac{1}{q} \ge 1$; then $L_p(\mathcal{A}).L_q(\mathcal{A}) = L_r(\mathcal{A})$. Every $h \in L_r(\mathcal{A})$ has a decomposition $h = k.\ell$ with $k \in L_p(\mathcal{A}), \ell \in L_q(\mathcal{A})$ and $\|h\|_r = \|k\|_p \cdot \|\ell\|_q$.*

PROOF. It is clear by polar decomposition that we may assume that $h \ge 0$. Then set $k = h^{r/p}$ and $\ell = h^{r/q}$. □

5. Change of weight.

In the preceeding sections we constructed the scale of $L_p(\mathcal{A};\varphi_0)$ spaces associated with a fixed normal faithful state φ_0. We show now that this construction is essentially independant from this reference state. This generalizes the well known fact that two equivalent measures on the same measure space give raise to isometric, lattice isomorphic L_p-spaces.

THEOREM 5.1. *Let φ_0, ψ_0 be two normal faithful states on the v. N. a. $\mathcal{A}$. There exists a scale $(V_p)_{1\le p\le\infty}$ of linear isometric $\mathcal{A}$-bimodule isomorphisms from $L_p(\mathcal{A};\psi_0)$ onto $L_p(\mathcal{A};\varphi_0)$ wich preserve the conjugation and the external products.*

PROOF. Let us denote by $\mathbf{1}_p^{\varphi_0}$, resp. $\mathbf{1}_p^{\psi_0}$ the states φ_0, resp. ψ_0 viewed as elements of $L_p(\mathcal{A};\varphi_0)$, resp $L_p(\mathcal{A};\psi_0)$.

We define first V_2. Since the state ψ_0 belongs to the positive cone of $\mathcal{A}_* = L_1(\mathcal{A};\varphi_0)$ it has by Cor. 4.18 a unique square root $k_0 \in L_2(\mathcal{A};\varphi_0)_+$. We have then for every $x \in \mathcal{A}$:

$$\|x.k_0\|_2^2 = \operatorname{Tr}(k_0^* x^* x k_0) = \operatorname{Tr}(x^* x k_0^2) = \psi_0(x^* x) = \|x.\mathbf{1}_2^{\psi_0}\|_2^2$$

hence the map $x.\mathbf{1}_2^{\psi_0} \mapsto x.k_0$ extends to linear isometry V_2 from $L_2(\mathcal{A};\varphi_0)$ into $L_2(\mathcal{A};\psi_0)$. In fact this isometry is onto because k_0 is cyclic for the left action of $\mathcal{A}$, i. e. $\mathcal{A}.k_0$ and is dense in $L_2(\mathcal{A};\varphi_0)$: to see this, note that the projection on $\overline{\mathcal{A}.k_0}$ commutes with the left action of $\mathcal{A}$, hence by Tomita's theorem is given by the right action of some projection $e \in \mathcal{A}$; then $k_0.e^\perp = 0$, and $\psi_0(e^\perp) = \|k_0.e^\perp\|_2^2 = 0$ wich implies $e^\perp = 0$.

The operator $S^0_{\psi_0}: \mathfrak{a}.\mathbf{1}_2^{\psi_0} \mapsto \mathfrak{a}.\mathbf{1}_2^{\psi_0}, x.\mathbf{1}_2^{\psi_0} \mapsto x^*.\mathbf{1}_2^{\psi_0}$ is transferred by the isometry V_2 to the operator $S^0_{k_0}: \mathfrak{a}.k_0 \mapsto \mathfrak{a}.k_0, x.k_0 \mapsto x^*.k_0$ and the same is true for their closures S_{ψ_0}, S_{k_0}, i. e. $S_{k_0} = V_2 S_{\psi_0} V_2^*$. The polar decomposition $S_{\psi_0} = J_{\psi_0}\Delta_{\psi_0}^{1/2}$ is then transferred to the polar decomposition $S_{k_0} = J_{k_0}\Delta_{k_0}^{1/2}$. Let $A = J_{\varphi_0} S_{k_0}$, where J_{φ_0} is the involution $h \mapsto h^*$ in $L_2(\mathfrak{a};\varphi_0)$. Then the operator A is positive, since

$$\forall x \in \mathfrak{a}, (Axk_0, xk_0) = (k_0.x, xk_0) = (k_0, x.k_0.x^*) \geq 0$$

because $k_0 \geq 0$ and $x.k_0.x^* \geq 0$. Hence A is symmetric, i. e. $A \subseteq A^*$. Now $\mathfrak{a}.k_0$ is a core for S_{k_0}, hence for A, while $k_0.\mathfrak{a}$ is a core for $S^*_{k_0}$, so $J_{\varphi_0}k_0.\mathfrak{a} = \mathfrak{a}.k_0$ is a core for $A^* = S^*_{k_0}J_{\varphi_0}$; since A and A^* coincide on a common core, they are equal. So A is a positive selfadjoint operator, and $S_{k_0} = J_{\varphi_0}A$ is necessary the polar decomposition of S_{k_0}. In particular $J_{k_0} = J_{\varphi_0}$ and $A = \Delta_{k_0}^{1/2} = V_2\Delta_{\psi_0}^{1/2}V_2^*$.

As a consequence we have $V_2 h^* = (V_2 h)^*$ for every $h \in L_2(\mathfrak{a};\psi_0)$. Since V_2 preserves clearly the left actions of $\mathfrak{a}$, it preserves the right actions too.

Now we define V_p. For this aim, let us show that the function $[0,1] \to L_1(\mathfrak{a};\varphi_0), \alpha \mapsto k_0^{2\alpha}.\mathbf{1}^{\varphi_0}_{1/1-\alpha}$ can be extended to a map $F: \bar{S} \to L_1(\mathfrak{a};\varphi_0)$ which is analytic in open strip S and bounded, continuous on the closed strip $\bar{S}$. Recall that $k_0 = T_0\xi_0$ where T_0 is 1/2-homogeneous (relatively to the modular group of φ_0)) and $\xi_0 = \mathbf{1}_2^{\varphi_0}$; then $k_0^{\alpha}.\mathbf{1}^{\varphi_0}_{2/1-\alpha} = T_0^{\alpha}\xi_0$ for all $0 \leq \alpha \leq 1$; moreover $u_t := T_0^{it}\Delta_{\varphi_0}^{-it}$ belongs to $\mathfrak{a}$, and (u_t) is a strongly continuous family of unitaries of $\mathfrak{a}$.

Note that by fact 4.6 T_0 is nothing but the closure of the operator $x.\xi_0 \mapsto k_0.x$ (T_0 is usually denoted by $\Delta^{1/2}_{k_0,\xi_0}$); hence we have in particular

$$T_0^{it}k_0 = T_0^{1+it}\xi_0 = T_0(T_0^{it}\xi_0) = T_0 u_t\xi_0 = k_0.u_t \tag{5.1}$$

Set $G(z) = T_0^z\xi_0$ and $F(z) = G(2z).\xi_0$ if $0 \leq \Re e\, z \leq 1/2$, $F(z) = k_0.G(2z-1)$ if $1/2 \leq \Re e\, z \leq 1$. This definition is coherent if the two expressions of $F(z)$ coincide when $\Re e\, z = 1/2$; but if $z = \frac{1}{2} + it$ we have by (5.1):

$$G(2z).\xi_0 = (T_0^{1+2it}\xi_0).\xi_0 = (k_0.u_{2t}).\xi_0 = k_0.(u_{2t}\xi_0) = k_0.(T_0^{2it}\xi_0) = k_0.G(2z-1)$$

Now G is clearly analytic $S \to L_2(\mathfrak{a};\varphi_0)$, continuous on the closed strip; so F has the desired properties of analyticity and continuity. Furthermore on the imaginary axis it is $\mathfrak{a}$-valued and strongly continuous.

Moreover for every $t \in \mathbb{R}$:

$$\begin{cases} F(it) = (T_0^{2it}\xi_0)\xi_0 = (u_{2t}\xi_0).\xi_0 = u_{2t}.\varphi_0 \\ F(1+it) = k_0.T_0^{1+2it}\xi_0 = k_0.k_0.u_{2t} = \psi_0.u_{2t} \end{cases}$$

As a consequence we have an analytic family of bounded linear operators $W_z: i_\ell^{\psi}(\mathfrak{a}) \to \mathfrak{a}_*$, $x.\psi_0 \mapsto xF(z)$, such that each W_{it}, $t \in \mathbb{R}$, is an isometry from $i_\ell^{\psi}(\mathfrak{a})$ into $i_\ell^{\varphi_0}(\mathfrak{a})$, while each W_{1+it}, $t \in \mathbb{R}$ is an isometry for the norm of $\mathfrak{a}_*$. Consequently $W_{1/p}$ is bounded (with norm less than 1) for the norms of the interpolation spaces and $L_p(\mathfrak{a};\psi_0)$, $L_p(\mathfrak{a};\varphi_0)$, i. e. $\|x.k_0^{2/p}\|_{L_p(\mathfrak{a};\varphi_0)} \leq \|x.\mathbf{1}_p^{\varphi_0}\|_{L_p(\mathfrak{a};\psi_0)}$ for all $x \in \mathfrak{a}$. Similarly one can show (using the map $z \mapsto F(\bar{z})^*$) that $\|k_0^{2/p}.x\|_{L_p(\mathfrak{a};\varphi_0)} \leq \|\mathbf{1}_p^{\varphi_0}.x\|_{L_p(\mathfrak{a};\psi_0)}$, and by duality we obtain that $\|x.k_0^{2/p}\|_{L_p(\mathfrak{a};\varphi_0)} = \|x.\mathbf{1}_p^{\varphi_0}\|_{L_p(\mathfrak{a};\psi_0)}$ (note that $\mathrm{Tr}_{\varphi_0}(xk_0^{2/p}.k^{2/q}y) = \mathrm{Tr}_{\psi_0}(xy)$ for every $x, y \in \mathfrak{a}$). Since it is easy to see that $\mathfrak{a}.k^{1/p}$ is dense in $L_p(\mathfrak{a})$, it follows that $W_{1/p}$ extends to an isometry of

$L_p(\mathcal{A};\psi_0)$ onto $L_p(\mathcal{A};\varphi_0)$. We set $V_p = W_{1/p}$. This isometry clearly preserves the left action of $\mathcal{A}$.

Note that for conjugate exponents $1 < p, p' < \infty$ the isometry $(V'_{p'})^{-1}$ is precisely the closure of the isometry $\mathbf{1}_p^{\varphi_0}.x \mapsto k_0^{2/p}.x$. This isometry preserves the right actions of $\mathcal{A}$; it is easy to see also that $J_pV_p = J_p(V'_{p'})^{-1}$. We prove now that $V_p = (V'_{p'})^{-1}$. By duality it suffices to reason in the case $p > 2$.

Let $\frac{1}{p} + \frac{1}{r} = \frac{1}{2}$. Note that trivially: $V_p(h).k_0^{2/r} = V_2(h.\mathbf{1}_r^{\psi_0})$ for every $h \in L_p(\mathcal{A};\psi_0)$. If now $h = \mathbf{1}_p^{\psi_0}.y = J_p y^* \mathbf{1}_p^{\psi_0}$ with $y \in \mathcal{A}$ then $h.\mathbf{1}_r^{\psi_0} = \Delta_{\psi_0}^{1/p-1/2} J y^* \mathbf{1}_2^{\psi_0}$ $= \Delta_{\psi_0}^{1/p} y \mathbf{1}_2^{\psi_0}$ (see sect. 2.3), hence $V_p(h).k_0^{2/r} = \Delta_{k_0}^{1/p} y k_0$. The equality $V_p(\mathbf{1}_p^{\psi_0}.y) = k_0^{2/p}.y$ which we want to prove is then equivalent to:

$$k_0^{2/p}.y.k_0^{2/r} = \Delta_{k_0}^{1/p} y k_0 \tag{5.2}$$

We prove this again by an argument of analytic functions. We prove in fact that:

$$T_0^z y T_0^{1-z} \xi_0 = \Delta_{k_0}^{z/2} y k_0 \tag{5.3}$$

for all z in the closed strip $\bar{S}$. Remark that $JT_0J(\xi_0.x) = k_0.x$ for every $x \in \mathcal{A}$; hence $\widehat{T}_0 := (JT_0J)^{-1}$ coincide with the closure Δ_{ξ_0,k_0} of the operator $x.k_0 \mapsto \xi_0.x$. Hence by reason of symmetry we have $\widehat{T}_0^{it} = v_t \Delta_{k_0}^{it/2}$ for some family (v_t) of unitaries of $\mathcal{A}$, and $T_0^{it} = Jv_tJ\Delta_{k_0}^{it}$; from this one deduces easily that $T_0^{it} y T_0^{-it} = \Delta_{k_0}^{it/2} y \Delta_{k_0}^{-it/2}$ for every $y \in \mathcal{A}$. Coming back to (5.3) we obtain for $z = \alpha + it$

$$T_0^z y T_0^{1-z} \xi_0 = T_0^\alpha T_0^{it} y T_0^{1-it} T_0^{1-\alpha} \xi_0 = T_0^\alpha \sigma_t^{k_0}(y) k_0^{1-\alpha}$$

which is defined by Fact 4.6 since $\sigma_t^{k_0}(y)k_0^{1-\alpha} \in L_{2/(1-\alpha)} \in \mathcal{D}(T_0^\alpha)$. The left member of (5.3) is analytic on S (and weakly continuous on $\bar{S}$) since for every $x \in \mathcal{A}$:

$$(T_0^z y T_0^{1-z} \xi_0, x\xi_0) = (y T_0^{1-z} \xi_0, T_0^{-\bar{z}} x \xi_0)$$

and $\mathcal{A}\xi_0 \subset \mathcal{D}(T_0)$, so $z \mapsto T_0^z$ is analytic and $z \mapsto T_0^{-\bar{z}} x\xi_0$ is anti-analytic. So it is sufficient to prove (5.3) for $\Re e\, z = 0$. If $z = it$ we have:

$$T_0^{it} y T_0^{1-it} \xi_0 = T_0^{it} y T_0^{-it} k_0 = \Delta_{k_0}^{it/2} y \Delta_{k_0}^{-it/2} k_0 = \Delta_{k_0}^{it/2} y k_0$$

Hence (5.3) is verified. For $z = 2/p$ we have by Fact 4.6:

$$T_0^{2/p} y T_0^{1-2/p} \xi_0 = T_0^{2/p} y k_0^{2/r} = k_0^{2/p} y k_0^{2/r}$$

which with (5.3) proves (5.2) and finally the equality $V_p = (V'_{p'})^{-1}$.

Since the maps V_p preserve the $\mathcal{A}$-bimodule structure of the spaces $L_p(\mathcal{A})$ and the identification maps between their left- and right embeddings in $\mathcal{A}_*$, they also preserve the external products between these spaces (which were defined by bilinear complex interpolation in section 3.1). □

REMARK 5.2. It is clear that the identification maps V_p preserve the order, the polar decomposition and the square roots. In fact they the power operations, i.e. $V_{p/\alpha}(h^\alpha) = (V_p(h))^\alpha$ when $0 \le \alpha \le p$ and $h \in L_p(\mathcal{A};\psi_0)_+$: this can be seen easily for a dyadic α; then note that the map $\alpha \mapsto h^\alpha \mathbf{1}_{p/(1-\alpha)}$ is continuous.

Appendix.

Here we provide a sketch of the proof of Fact 4.6 (see mainly [AM], Lemma 3.4).

Note that by definition of α-homogeneity, we have $\mathcal{A}.\xi_0 = J\mathcal{A}.\xi_0 \subset \mathcal{D}(T)$. Moreover for $x \in \mathcal{A}$ we have:

$$\begin{aligned} Tx\xi_0 &= TJ\Delta^{1/2}x^*\xi_0 = TJ\sigma_{-i/2}(x^*)JT\xi_0 \\ &= J\sigma_{i/p}(\sigma_{-i/2}(x^*))JT\xi_0 = J\sigma_{-i/r}(x^*)JT\xi_0 \\ &= J\Delta^{1/r}x^*\Delta^{-1/r}JT\xi_0 \end{aligned}$$

If $T\xi_0$ represents an element h of $L_p(\mathcal{a})$ the last expression represents the element $h.x$. In other words

$$T(x.\mathbf{1}_2) = h.x.\mathbf{1}_r$$

Since $\mathcal{A}.\mathbf{1}_r$ is dense in $L_r(\mathcal{a})$ and T is closed, it follows easily that $L_r(\mathcal{a}).\mathbf{1}_p \subseteq \mathcal{D}(T)$ and that for every $k \in L_r(\mathcal{a})$ we have $T(k.\mathbf{1}_p) = h.k$.

Now we show that $\mathcal{A}.\xi_0$ is a core of T. Due to Cor. 3.10 it is sufficient to consider the case of a self-adjoint α-homogeneous operator T. It follows from Prop. 3.9 that $u_t = T^{itp}\Delta^{-it} \in \mathcal{a}$ for every $t \in \mathbb{R}$. Let $p = s(T) = T^0$ be the support of T (which belongs to $\mathcal{a}$ too). It is elementary to check the following relations:

$$\forall s,t \in \mathbb{R}, \qquad \begin{cases} u_{s+t} = u_s\sigma_s(u_t) \\ u_t^*u_t = \sigma_t(p) \\ u_tu_t^* = p \end{cases}$$

(a one-parameter family (u_t) verifying the first condition is called a *cocycle* relative to the automorphism group (σ_s)). Then there exists a unique normal semifinite weight w on $\mathcal{a}$ (with support p) such that $u_t = \Delta^{it}_{w,\varphi_0}\Delta^{-it}$. (Here the modular operator Δ_{w,φ_0} is defined like Δ as follows. Let H_w be the GNS Hilbert space associated with the weight w (i. e. the completion of the left ideal $\mathfrak{n}_w = \{x \in \mathcal{a} \mid w(x^*x) < \infty\}$ for the norm $\|x\|_w = w(x^*x)^{1/2}$). Let C_{w,φ_0} be the (closure of) the conjugaison operator $H_{\varphi_0} \to H_w$, $x \mapsto x^*$: then $\Delta^{1/2}_{w,\varphi_0} = |C_{w,\varphi_0}|$). This result is due to A. Connes in the case $p = I$; the proof is adapted to the case of a general p in [AM], th. B1.

We have then $T^{itp} = \Delta^{it}_{w,\varphi_0}$ and by the uniqueness part in Stone's representation theorem it results that $T = \Delta^{1/p}_{w,\varphi_0}$. Since $\mathfrak{n}_w$ is a core for $\Delta^{1/2}_{w,\varphi_0}$ and $0 < 1/p \leq 1/2$ it is also a core for $\Delta^{1/p}_{w,\varphi_0}$; since $\mathfrak{n}_w \subset \mathcal{a}.\xi_0 \subset \mathcal{D}(T)$, the space $\mathcal{a}.\xi_0$ is a core for T.

Since $\mathcal{A}$ is σ-strongly dense in $\mathcal{a}$, it results that $\mathcal{A}.\xi_0$ is also a core for T (since the closure of the restriction of T to $\mathcal{A}.\xi_0$ contains in fact $L_r(\mathcal{a}).\mathbf{1}_p$).

References

[A] H. Araki, *Some properties of modular conjugation operator of von Neumann algebras and a non-commutative Radon-Nikodym theorem with a chain rule*, Pacific J. Math. **50** (1974), 309-354.

[AM] H. Araki, T. Masuda, *Positive cones and L_p-spaces for von Neumann algebras*, Publ. RIMS, Kyoto Univ. **18** (1982), 339-411.

[Be] J. Bergh, *On the relation between the two complex methods of interpolation*, Indiana Univ. Math. J. **28** (1979), 775-778.

[BL] J. Bergh, J. Löfström, *Interpolation Spaces: An Introduction*, Springer Verlag, 1976.

[Bo] R. P. Boas, *Entire functions*, Academic Press New York, 1954.

[CJ] M. Cwikel, S. Janson, *Interpolation of analytic families of operators*, Studia Math. **43** (1984), 61-71.

[CR] M. Cwikel, S. Reisner, *Interpolation of uniformly convex Banach spaces*, Proc. Amer. Math. Soc. **84** (1982), 555-559.

[CS] F. Cobos, T. Schonbek, *On a theorem by Lions and Peetre about interpolation between a Banach space and its dual*, Houston J. Math. **24** (1998), 325–344.

[D] J. Dixmier, *Les algèbres d'opérateurs dans l'espace hilbertien*, 2d ed., Gauthiers Villars, 1969.

[F] F. Fidaleo, *Canonical operator space structures on non-commutative L_p spaces*, J. Funct. Anal. **169** (1999), 226-250.

[Fr] D. H. Fremlin, *Topological Riesz spaces and measure theory*, Cambridge University Press, 1974.

[H] U. Haagerup, *L^p-spaces associated with an arbitrary von Neumann algebra*, Colloques Internationaux C.N.R.S. **274** (1979), 175-184.

[Hi] M. Hilsum, *Les espaces L^p d'une algèbre de von Neumann*, J. Functional Analysis **40** (1981), 151-169.

[I] H. Izumi, *Construction of noncommutative L_p-spaces with a complex parameter arising from modular actions*, Internat. J. Math. **8** (1997), 1029-1066.

[KaR] R.V. Kadison, J.R. Ringrose, *Fundamentals of the theory of operator algebras I and II*, Academic Press, 1983 and 1986.

[Ko1] H. Kosaki, *Applications of the complex interpolation method to a von Neumann Algebra: Non-commutative L^p-Spaces*, J. Funct. Anal. **56** (1984), 29-78.

[Ko2] H. Kosaki, *Positive Cones and L^p-spaces associated with a von Neumann algebra*, J. Operator Theory **6** (1981), 13-23.

[Ko3] H. Kosaki, *Positive Cones associated with a von Neumann algebra*, Math. Scand. **47** (1980), 295-307.

[Ko3] H. Kosaki, *Positive Cones associated with a von Neumann algebra*, Math. Scand. **47** (1980), 295-307.

[L] M. Leinert, *Integration with respect to a weight*, Inter. J. Math. **2** (1991), 177-182.

[Lv] B. Ya. Levin, *Lectures on entire functions*, Translations of Mathematical Monographs 150, Amer. Math. Soc., 1996.

[MC] C. A. McCarthy, c_p, Israel J. Math. **5** (1967), 249–271.

[N] E. Nelson, *Notes on non-commutative integration*, J. Funct. Anal. **15** (1974), 103-116.

[P] G. K. Pedersen, *C^*-algebras and their automorphism group*, Academic Press, 1979.

[S] S. Sakai, *C^*-Algebras and W^*-Algebras*, Springer Verlag, 1971.

[Se] I. E. Segal, *A non-commutative extension of abstract integration*, Ann. of Math. **57** (1953), 401-457.

[T] M. Takesaki, *Theory of operator algebras I*, Springer-Verlag, 1979.

[Te1] M. Terp, *L^p-spaces associated with von Neumann algebras*, Notes, Math. Institute, Copenhagen Univ., 1981.

[Te2] M. Terp, *Interpolation spaces between a von Neumann algebra and its predual*, J. Operator Th. **8** (1982), 327–360.

[Y] F. J. Yeadon, *Non-commutative L_p-spaces*, Math. Proc. Cambridge Philos. Soc. **77** (1975), 91-102.

[W] F. Wattbled, *Interpolation complexe d'un espace de Banach et de son antidual*, C. R. Acad. Sci. Paris, Série I **321** (1995), 1437-1440.

Equipe d'Analyse Fonctionnelle (CNRS), Institut de Mathématiques de Jussieu, Case 186, 4, place Jussieu, 75252 Paris Cedex 05, France.

E-mail address: yr@ccr.jussieu.fr

Contemporary Mathematics
Volume **321**, 2003

Banach and operator space structure of C^*-algebras

Haskell P. Rosenthal*

Introduction

A C^*-algebra is often thought of as the non-commutative generalization of a $C(K)$-space, i.e. the space of continuous functions on some locally compact Hausdorff space, vanishing at infinity. We go one step further, for we seek to compare the Banach space properties of C^*-algebras *and* their naturally complemented subspaces, with those of $C(K)$-spaces. (For a recent survey on $C(K)$ spaces or Banach spaces, see [**Ro3**].) This involves the recent theory of operator spaces, or quantized Banach spaces. We briefly review this concept at the beginning of section 1; the reader is referred to [**ER**] and [**Pi**] for in depth coverage. For the definition of complete boundedness of maps, complete isomorphisms, etc., see section 1.

Our presentation here is expository; only simple deductions are given, often from rather deep principles.

Section 1 shows how C^*-algebras share certain Banach space properties of $C(K)$-spaces. For example, we state Pfitzner's theorem that C^*-algebras have Pełczyński's property (V) as Theorem 1.1, and then deduce that non-reflexive completely complemented subspaces of C^*-algebras contain complete isomorphic copies of c_0 in Corollary 1.2. We then use an old result

2000 *Mathematics Subject Classification.* 46L05, 47C15.
Key words and phrases. (nuclear) C^*-algebras, von Neumann algebras, $C(K)$-spaces.
*Research partially supported by NSF grant DMS-0075047.

of the author's to deduce that non-reflexive completely complemented subspaces of von Neumann algebras contain complete isomorphic copies of ℓ^∞, in Corollary 1.3.

We next discuss the important class of *nuclear* C^*-algebras; in many ways, these are the closest, in Banach space structure, to $C(K)$-spaces. These include C^*-algebras with separable duals, and more generally, type I C^*-algebras. These are described via Definition 1.5, after which we give a description of the CAR, or Fermion algebra, a fundamental nuclear non-type I (separable) C^*-algebra. Section one concludes with remarkable results of Glimm and Kirchberg. In particular, Kirchberg proved the non-commutative analogue of Milutin's theorem: *Every non-type I nuclear separable C^*-algebra is completely isomorphic to the CAR algebra.*

Section two deals with quantized versions of the separable extension property (SEP) for Banach spaces. These were introduced in [**Ro2**] and developed further in [**OR**] and [**AR**]. Recall that a separable Banach space has the SEP provided it is complemented in every separable superspace. Sobczyk established that c_0 *has* the SEP, and Zippin proved that c_0 is the *only* separable space with this property, up to isomorphism. We give a proof of Sobczyk's theorem following Theorem 2.7, which motivates the approach taken in [**AR**] (and was also given there). This proof uses the Borsuk extension theorem (Theorem 2.7).

Our first quantized result of the SEP: *A separable operator space has the CSEP provided it is completely complemented in every separable operator superspace.* (This is equivalent to the definition given in Section 2, Definition 2.2, in virtue of the injectivity of $B(H)$, Theorem 2.3). The spaces ROW and COLUMN, denoted R and C, are defined following the above mentioned proof, and then the author's discovery from [**Ro2**] (as refined in [**AR**]) is given as Theorem 2.12: $c_0(M_{n,\infty} \oplus M_{\infty,n})$ *has the 2-CSEP for all* n. Of course this implies that $c_0(R \oplus C)$ has the CSEP. There follows some discussion of a possible converse to this, namely the conjecture: a separable operator space with the CSEP completely embeds in $c_0(R \oplus C)$. This conjecture is at a far deeper level than the results obtained so far.

Unfortunately, $\mathbf{K}$ (the space of all compact operators) *fails* the CSEP, as discovered by Kirchberg. We give some discussion of an alternate proof, from [**OR**], showing that there is a separable operator superspace Y of K_0 with Y/K_0 completely isometric to c_0 and K_0 completely uncomplemented in Y (Theorem 2.9). (K_0 denotes the c_0-sum of M_n's.) T. Oikhberg and the author succeeded in proving that $\mathbf{K}$ has our second quantized version of the SEP, the CSCP (Theorem 2.12). In virtue of a discovery in [**OR**], this property may be formulated: *a locally reflexive separable operator space has the CSCP provided it is completely complemented in every locally reflexive separable operator superspace.* This suggests the obviously quite deep problem: if an operator has the CSCP, does it completely embed in $\mathbf{K}$?

The last part of Section 2 deals with the approach taken in [**AR**], namely that of fundamental properties of complete M-ideals (see Definition 2.15).

We do not discuss here the basic new tool used in [**AR**] for dealing with these, *M-complete approximate identities.* Rather, we just show the basic principles developed in [**AR**] which lead to Theorems 2.8 and 2.12. Among these are the lifting result from [**AR**], stated as Theorem 2.21 here, which says qualitatively that *if $\mathcal{J} \subset Y \subset \mathcal{A}$ with $\mathcal{J}$ a nuclear ideal in a C^*-algebra and Y locally reflexive with $Y/\mathcal{J}$ separable, then $\mathcal{J}$ is completely complemented in Y.*

The proof of Theorem 2.21 in [**AR**] uses ideas from work of Effros-Haagerup [**ER**], where this result is established for $Y = \mathcal{A}$ itself. An alternative proof of a pure operator space isometric extension of the latter is given in the appendix to [**AR**], and formulated here as Theorem 2.16. The article concludes with a sketch of the proof that $\mathbf{K}$ has the CSCP. Several open problems are also discussed, in both Sections 1 and 2.

1. C^*-algebras from a Banach space perspective

What do C^*-algebras look like, as Banach spaces, up to linear homeomorphism? What do their naturally complemented subspaces look like? The appropriate way to approach these questions is through the theory of *quantized* Banach spaces, or operator spaces, whose natural morphisms are completely bounded maps. We first briefly recall this concept; for fundamental background and references, see [**ER**] and [**Pi**].

An *operator space* X is a complex Banach space which is a closed linear subspace of $B(H)$, the bounded linear operators on some Hilbert space H, endowed with its natural tensor product structure with $\mathbf{K} = K(\ell_2)$ (where $K(H)$ denotes the space of compact operators on H). We let $\mathbf{K} \otimes_{\text{op}} X$ denote the closed linear span in $B(\ell_2 \otimes_2 H)$ of the operators $A \otimes T$ where $A \in \mathbf{K}$, $T \in X$, and $\ell_2 \otimes_2 H$ is the Hilbert space tensor product of ℓ_2 and H. A linear operator $T : X \to Y$ between operator spaces X and Y is called *completely bounded* if $I_{\mathbf{K}} \otimes T$ is a bounded linear operator from $\mathbf{K} \otimes X$ to $\mathbf{K} \otimes Y$, endowed with their natural norms. Of course, $I_{\mathbf{K}} \otimes T$ then uniquely extends to a bounded linear operator from $\mathbf{K} \otimes_{\text{op}} X$ to $\mathbf{K} \otimes_{\text{op}} Y$, which we also denote by $I_{\mathbf{K}} \otimes T$; then we *define* $\|T\|_{\text{cb}} = \|I_{\mathbf{K}} \otimes T\|$. It now follows easily that if X_i, Y_i are operator spaces and $T_i : X_i \to Y_i$ are completely bounded maps, then also $T_1 \otimes T_2$ is (i.e., extends to) a completely bounded map from $X_1 \otimes_{\text{op}} X_2$ to $Y_1 \otimes_{\text{op}} Y_2$, with $\|T_1 \otimes T_2\|_{\text{cb}} \leq \|T_1\|_{\text{cb}} \|T_2\|_{\text{cb}}$. We should point out that when X_1 and X_2 are C^*-algebras, $X_1 \otimes_{\text{op}} X_2$ is also called the *spatial tensor product*, and also the *minimal* tensor product, for it is the least tensor norm on $X_1 \otimes X_2$ whose completion (i.e. $X_1 \otimes_{\text{op}} X_2$) is a C^*-algebra (cf. [**Mu**] for a proof of this theorem).

Now many natural Banach space concepts naturally extend to the context of operator spaces. Thus operator spaces X and Y are called *completely isomorphic* if there exists an invertible linear operator $T : X \to Y$ with T and T^{-1} completely bounded. If $\|T\|_{\text{cb}} \|T^{-1}\|_{\text{cb}} \leq \lambda$, we say X and Y are *λ-completely isomorphic*; then we set $d_{\text{cb}}(X, Y) = \inf\{\lambda \geq 1 :$

X is λ-completely isomorphic to $Y\}$. If $X \subset Y$ with Y an operator space and X a closed linear subspace, then X is regarded as an operator space via the natural structure $\mathbf{K} \otimes_{\mathrm{op}} X \subset \mathbf{K} \otimes_{\mathrm{op}} Y$. X is then called *completely complemented in* Y if there is completely bounded (linear) projection P mapping Y onto X. If $\|P\|_{\mathrm{cb}} \leq \lambda$, we say X is λ*-completely complemented in* Y; if $\|I - P\|_{\mathrm{cb}} \leq \lambda$, we say X is λ*-completely co-complemented in* Y.

Now of course $\mathbf{K}$ may be identified with those infinite matrices representing compact operators on ℓ_2 with respect to its natural basis. For an operator space X, $\mathbf{K} \otimes_{\mathrm{op}} X$ may also be identified with a Banach space of infinite matrices with elements in X. Now let M_{00} denote all infinite matrices of scalars with only finitely many non-zero entries. Then if M_n denotes the space of $n \times n$ matrices of complex numbers, we may regard $M_0 \subset M_1 \subset \cdots \subset M_n \subset M_{n+1} \subset \cdots \subset M_{00} \subset \mathbf{K} = \overline{M}_{00}$. It follows easily that if $P_n : \mathbf{K} \to M_n$ in the canonical projection, then

$$P_n \otimes I_X \to I_{\mathbf{K}} \otimes I_X \quad \text{in the SOT on } K \otimes_{\mathrm{op}} X \tag{1.1}$$

(SOT denotes the Strong Operator Topology). For operator spaces X and Y and $T : X \to Y$ a bounded linear operator, we define $\|T\|_n$ by

$$\|T\|_n = \|P_n \otimes T\| \ . \tag{1.2}$$

(Equivalently, $\|T\|_n = \|I_n \otimes T\|$, where I_n denotes the identity operator on $B(\ell_2) = M_n$). It then follows easily that T is completely bounded iff $(\|T\|_n)_{n=1}^{\infty}$ is bounded, and then

$$\|T\|_{\mathrm{cb}} = \sup_n \|T\|_n \ . \tag{1.3}$$

(This easy equivalence is often taken as the definition of complete boundedness, c.f. [**ER**]). Identifying $\mathbf{K} \otimes_{\mathrm{op}} X$ with infinite matrices, we then have that a bounded linear operator $T : X \to Y$ is completely bounded when (Tx_{ij}) belongs to $\mathbf{K} \otimes_{\mathrm{op}} Y$ for all (x_{ij}) in $\mathbf{K} \otimes_{\mathrm{op}} X$, and then $(I_{\mathbf{K}} \otimes T)(x_{ij}) = (Tx_{ij})$.

Evidently, the concept of an operator space X is captured by the Banach space $K \otimes_{\mathrm{op}} X$. Remarkable axioms of Z. J. Ruan abstractly characterize this tensor product, without reference to the ambient Hilbert space. For this and the "correct" notion of duality, as well as various other tensor products on operator spaces, see [**ER**], [**Pi**], and also [**BP**] for the latter.

Next we recall some basic concepts concerning C^*-algebras (see also standard references such as [**Mu**] and [**Ped**]). A (concrete) C^*-algebra $\mathcal{A}$ is defined to be a norm closed subalgebra of $B(H)$ (for some H) so that $T^* \in \mathcal{A}$ for all $T \in \mathcal{A}$. Ruan's axioms have their deep historical precedent in the abstract axioms given by Gelfand and Naimark, and we make no real distinction between abstract and concrete C^*-algebras. A von-Neumann algebra $\mathcal{N}$ is defined to be a unital adjoint-closed subalgebra of $B(H)$ which is *closed* in the SOT. Since $B(H)$ is naturally a dual space, with predual C_1 (the space of trace class operators on H), it follows that von Neumann algebras are also dual spaces, for they are weak* closed in $B(H)$. A remarkable result of Sakai asserts that conversely, any C^*-algebra $\mathcal{A}$ which is isometric to a dual space (just as a Banach space) is $*$-isomorphic to a von Neumann algebra, and moreover if X and Y are Banach spaces with X^* and Y^* isometric to $\mathcal{A}$, then X and Y are isometric; thus von Neumann algebras have unique preduals, up to isometry.

Standard results show that C^*-algebras have a unique operator space structure. In fact, if $\mathcal{A}_1, \mathcal{A}_2$ are C^*-algebras and $T : \mathcal{A}_1 \to \mathcal{A}_2$ is just an algebraic surjective $*$-isomorphism, then T is already a complete isometry. However every infinite dimensional Banach space X has many possible operator space structures. Two of these are distinguished as being the smallest, denoted MIN, and the largest, denoted MAX. These are functorially presented as follows; for any operator spaces X and Y and bounded linear map $T : Y \to (X, \mathrm{MIN})$, T is completely bounded and $\|T\| = \|T\|_{\mathrm{cb}}$; for any bounded linear map $T : (X, \mathrm{MAX}) \to Y$, T is completely bounded and $\|T\| = \|T\|_{\mathrm{cb}}$.

We now deal with some fundamental Banach space results for C^*-algebras. The first one is due to H. Pfitzner [**Pf**] (c_0 denotes the space of scalar sequences tending to 0; ℓ^∞ the space of all bounded sequences of scalars).

THEOREM 1.1. *Let $\mathcal{A}$ be a C^*-algebra, X a Banach space and $T : \mathcal{A} \to X$ a non weakly compact operator. There exists a commutative C^*-subalgebra $\mathcal{B}$ of $\mathcal{A}$ and a subspace Y of $\mathcal{B}$ isometric to c_0 so that $T|Y$ is an isomorphism.*

REMARK. The commutative version of this result (i.e. when $\mathcal{A}$ itself is commutative) is due to A. Pełczyński [**Pe2**].

COROLLARY 1.2. *A non-reflexive completely complemented subspace of a C^*-algebra contains a subspace completely isomorphic to c_0.*

PROOF. Let $P : \mathcal{A} \to X$ be a completely bounded projection onto X, with $\mathcal{A}$ a C^*-algebra, X a non-reflexive subspace. Then of course, P is non-weakly compact. Now choose $\mathcal{B}$ and Y as in 1.1, it follows that $\mathcal{B}$ must be endowed with the MIN operator structure. Since $P|Y$ is completely bounded, $P(Y)$ must *also* have an operator space structure equivalent to MIN, whence $P|Y$ is a complete isomorphism and so $P(Y)$ is completely isomorphic to c_0. □

For the next result, recall that a Banach space is isometrically injective if it is contractively complemented in every superspace. Equivalent formulations and the operator space version will be given in the next section. For any measure μ on a measurable space, $(L1(\mu))^*$ is isometrically injective, and in fact every commutative von Neumann algebra is isometric to such a space (and μ and the measurable space may be chosen with $(L1(\mu))^* = L^\infty(\mu)$). We now apply this, Theorem 1.1, and a result of the author's to obtain the von-Neumann algebra version of the previous result.

COROLLARY 1.3. *A non-reflexive completely complemented subspace of a von Neumann algebra contains a subspace completely isomorphic to ℓ^∞.*

PROOF. Let P, $\mathcal{A}$, X, $\mathcal{B}$ and Y be as in the proof of 1.2, with $\mathcal{A}$ a von Neumann algebra. Let then $\mathcal{N} = \mathcal{B}''$, the double commutant of $\mathcal{B}$, which by a standard theorem due to von Neumann, equals the von Neumann algebra generated by $\mathcal{B}$. $\mathcal{N}$ is commutative, and hence is isometric to $(L1(\mu))^*$ for some $L1(\mu)$-space. So $\mathcal{N}$ is isometrically injective. Since $P|\mathcal{N}$ is not weakly compact, it follows by a result of the author's [**Ro1**] that there exists a subspace Z of $\mathcal{N}$ with Z isomorphic to ℓ^∞ and $P|Z$ an isomorphism. Again, since $\mathcal{N}$ has MIN as its operator space structure, $P|Z$ is a complete isomorphism and so $P(Z)$ is completely isomorphic to ℓ^∞. □

REMARKS. 1. Corollaries 1.2 and 1.3 are obtained in [**Ro2**] as Proposition 2.19, with a somewhat different argument for 1.3. As noted there, it follows also that 1.2 and 1.3 also hold if one just deletes the term "completely" in their statements.

2. Unlike the commutative case, a complemented reflexive subspace of a C^*-algebra may be infinite dimensional. It is a result of G. Pisier that any such space must be isomorphic to a Hilbert space, (c.f. [**R**], Theorem 13). This suggests the following problem in the operator space category: *Characterize the Hilbertian operator spaces which are completely isomorphic to completely complemented subspaces of C^*-algebras.*

The classical Gelfand-Naimark Theorem asserts that every commutative C^*-algebra is a $C(K)$ space, i.e. $*$-isomorphic to the space of continuous functions vanishing at infinity on a locally compact Hausdorff space K. However, in terms of their Banach space structure, many C^*-algebras seem far removed from $C(K)$ spaces; for example, if H is infinite dimensional, $B(H)$ fails the approximation property [**Sz**], but $C(K)$ spaces he the metric approximation property in a natural way. From this perspective, the following class of C^*-algebras might be viewed as the "correct" non-commutative version of $C(K)$ spaces.

DEFINITION 1.4. *A C^*-algebra $\mathcal{A}$ is nuclear provided for every finite dimensional subspace F of $\mathcal{A}$ and $\varepsilon > 0$, there exists a finite rank $T : \mathcal{A} \to A$ so that $\|Tf - f\| \le \varepsilon\|f\|$ for all $f \in F$, such that there exist n and compete contractions $U : \mathcal{A} \to M_n$ and $V : M_n \to A$ with $T = VU$.*

We have taken the operator space definition (in fact, this is precisely how one *defines* nuclear operator spaces; c.f [**EOR**]). 1.4 is equivalent to the original formulation: *$\mathcal{A}$ is nuclear provided there is exactly one (pre) C^*-norm on $\mathcal{A} \otimes \mathcal{B}$ for all C^*-algebras $\mathcal{B}$.* (This equivalence is due to Choi-Effros and Kirchberg (independently) in the completely positive setting; the refinement to completely contractive maps as above is due to Smith. For detailed references, see [**W**].) Nuclear C^*-algebras include the following family.

DEFINITION 1.5. *A C^*-algebra $\mathcal{A}$ is called type I if every irreducible $*$-representative φ of $\mathcal{A}$ on a Hilbert space H satisfies: $K(H) \subset \varphi(\mathcal{A})$.*

(Recall that $\varphi : \mathcal{A} \to B(H)$ is a $*$-representation if φ is an algebraic $*$-homeomorphism. φ is called irreducible if $\varphi(\mathcal{A})$ has no invariant (closed linear) subspaces other than $\{0\}$ and H.)

It should be pointed out that infinite dimensional type I C^*-algebras are never type I von Neumann algebras. Of course von Neumann algebras are also C^*-algebras, but their topology is really the weak*-topology. In fact, an older definition, later proved equivalent: A C^* algebra $\mathcal{A}$ is type I if and only if every factor representation of $\mathcal{A}$ induces a type I von Neumann algebra (which must then simply be $B(H)$ for some H). It is also a theorem that a type I von Neumann algebra is an ℓ^∞-direct sum of algebras of the form $L^\infty(\mu, B(H))$; where μ is a measure with $(L_1(\mu))^* = L^\infty(\mu)$ and $L^\infty(\mu, B(H))$ denotes the bounded μ-measurable $B(H)$-valued (equivalence classes of) functions on the ambient measure space. Now if $\mathcal{A}$ is a C^*-algebra, $\mathcal{A}^{**}$ is a von Neumann algebra; it is then the case that $\mathcal{A}$ *is a type I C^*-algebra if and only if $\mathcal{A}^{**}$ is a type I von Neumann algebra.*

This is, in turn, a special witness to the deep results of Connes [**C**] and Choi-Effros [**CE2**] that a C^*-algebra $\mathcal{A}$ is nuclear if and only if $\mathcal{A}^{**}$ is injective (as defined in Section 2).

REMARKS. It is a standard result that every C^*-subalgebra of a type I C^*-algebra is also type I. However C^*-subalgebras of nuclear C^*-algebras need not be nuclear. Profound work of Kirchberg yields that these coincide with the class of exact C^*-algebras, also introduced by him [**Ki2**]. It follows from the definition that nuclear C^*-algebras have the Banach metric approximation property; in fact, their duals *also* have this property. there are known natural examples of exact non-nuclear C^*-algebras with the metric approximation property, but it is unknown if there exist such algebras which *fail* the metric or even the general (unbounded) approximation property for Banach spaces. The following question is also open: *Suppose $\mathcal{A}$ is a C^*-algebra whose dual has the approximation property. Is $\mathcal{A}$ nuclear?*

The CAR, or Fermion algebra is a fundamental example of a non type I nuclear C^*-algebra. This may be defined as the "infinite" tensor product $\overset{\infty}{\underset{1}{\otimes}} M_2$ of the 2×2 matrices. We prefer the following intuitive description.

We identify $B(\ell_2)$ with infinite matrices and define CAR_d to be all $T \in B(\ell_2)$ so that there exist an $n \geq 0$ and an $A \in M_{2^n}$ with

$$T = \begin{bmatrix} A & & & \\ & A & & \\ & & A & \\ & & & \ddots \end{bmatrix} \tag{1.4}$$

It follows that T is indeed bounded; in fact $\|T\| = \|A\|$. We also define $\mathcal{A}_n$ to be all T as in (1.4) with $A \in M_{2^n}$. Then it follows that CAR_d is a unital $*$-subalgebra $*$-isomorphic to M_{2^n}, with

$$\mathrm{CAR}_d = \cup_{n=0}^{\infty} \mathcal{A}_n \ . \tag{1.5}$$

We now define the CAR algebra $\mathbf{A}$ to be the norm-closure of CAR_d. It then follows that $\mathbf{A}$ acts irreducibly on ℓ_2, and moreover, $\mathbf{A}$ has no non-zero compact operators, so $\mathbf{A}$ is clearly not type I. To see that $\mathbf{A}$ is nuclear, for each n, let $i_n : M_{2^n} \to \mathcal{A}_n$ be the $*$-isomorphism given by (1.4), and define a map $\pi_n : \mathcal{A}_{n+1} \to \mathcal{A}_n$ by

$$(1.6) \qquad \begin{cases} \pi_n T = i_n \begin{pmatrix} \frac{A+D}{2} & 0 \\ 0 & \frac{A+D}{2} \end{pmatrix} & \text{if} \\ \\ T = i_{n+1} \begin{pmatrix} A & B \\ C & D \end{pmatrix} \end{cases}$$

where A, B, C, D are in M_{2^n}.

It follows easily that π_n is a completely contractive projection mapping $\mathcal{A}_{n+1}$ onto $\mathcal{A}_n$. But then by (1.5), there exists a unique completely contractive surjective projection $P_n : \mathbf{A} \to \mathcal{A}_n$ so that

$$(1.7) \qquad P_n(T) = \pi_n \cdots \pi_{m-1} \pi_m T$$

whenever $T \in \mathcal{A}_m$ with $m > n$. Of course then $P_n x \to x$ for all $x \in \mathbf{A}$, showing $\mathbf{A}$ is indeed nuclear.

The CAR algebra, $\mathbf{A}$ is quite different from type I C^*-algebras. In fact it follows from the results of [**HRS**] that if $\mathcal{B}$ is a type I C^*-algebra, then $\mathbf{A}^*$ is not Banach isomorphic (i.e. linearly homeomorphic) to a subspace of $\mathcal{B}^*$, hence $\mathcal{A}$ and $\mathcal{B}$ are not Banach isomorphic. (Here $\mathcal{B}^*$ denotes the Banach space dual of $\mathcal{B}$.) A remarkable result of J. Glimm yields that $\mathcal{A}$ is a universal witness to a C^*-algebra being non-type I.

THEOREM 1.6. [**G**]*; see also Theorem 6.73 of* [**Ped**]*) Let $\mathcal{A}$ be a non-type I C^*-algebra. Then the CAR algebra is $*$-isomorphic to a C^*-subquotient of $\mathcal{A}$. That is, there is a C^*-subalgebra $\mathcal{B}$ of $\mathcal{A}$ such that* $\mathbf{A}$ *is $*$-isomorphic to a quotient algebra of $\mathcal{B}$.*

Now actually, if one applies the refined formulation given in [**Ped**] and a lifting theorem due to Choi-Effros [**CE1**], one obtains (as observed in [**Ki1**])

COROLLARY 1.7. *Let $\mathcal{A}$ be a separable non-type I C^*-algebra. Then the CAR algebra is completely isometric to a completely contractively complemented subspace of $\mathcal{A}$.*

PROOF. According to Theorem 6.73 of [**Ped**], one may choose q a projection in $\mathcal{A}^{**}$ and $\mathcal{B}$ a C^*-subalgebra of $\mathcal{A}$ so that q commutes with $\mathcal{B}$, $q\mathcal{B}$ is $*$-isomorphic to $\mathbf{A}$ (the CAR algebra), and

$$(1.8) \qquad q\mathcal{B} = q\mathcal{A}q \ .$$

Since $q\mathcal{B}$ is thus nuclear, by the lifting result in [**CE1**], there exists a complete contraction $L : q\mathcal{B} \to \mathcal{B}$ so that

$$(1.9) \qquad \pi L = I_{q\mathcal{B}}$$

where $\pi \to q\mathcal{B}$ is the quotient map given by $\pi b = qb$ for all $b \in \mathcal{B}$. Then define $U : \mathcal{A} \to q\mathcal{B}$ by

$$Ua = qaq \quad \text{for all } a \in \mathcal{A}. \tag{1.10}$$

Finally, *define* W by

$$W = L(q\mathcal{B}) \ . \tag{1.11}$$

Then for all $w \in W$,

$$\begin{aligned} LU(w) &= L(qwq) \\ &= L(qw) \quad \text{because } q \text{ commutes with } \mathcal{B} \\ &= w \quad \text{by (1.9)}. \end{aligned} \tag{1.12}$$

It then follows, since of course U is a complete contraction, that W is completely isometric to $\mathbf{A}$ *and* W is completely contractively complemented in $\mathcal{A}$ via the projection $P = LU$. □

It is a famous discovery in Banach space theory, due to A. Milutin [**M**], that for any compact metric space K, $C(K)$ is isometric to a contractively complemented subspace of $C(\mathbf{D})$, where $\mathbf{D}$ denotes the Cantor discontinuum. Milutin then deduces that if K is uncountable, $C(K)$ is isomorphic to $C(\mathbf{D})$. E. Kirchberg has established the quantized version of this result, proving the following converse.

THEOREM 1.8. [**Ki1**] *Let $\mathcal{A}$ be a separable nuclear C^*-algebra. Then $\mathcal{A}$ is completely isometric to a completely contractively complemented subspace of the CAR algebra.*

A simple application of the operator space version of the Pełczyński decomposition method ([**Pe1**]) then yields

COROLLARY 1.9. [**Ki1**] *Any separable non-type I nuclear C^*-algebra is completely isomorphic to the CAR-algebra.*

PROOF. For operator spaces X and Y, let $X \overset{cc}{\hookrightarrow} Y$ denote: X is completely isometric to a completely contractively complemented subspace of Y. Now let $\mathcal{D} = (\mathbf{A} \oplus \mathbf{A} \oplus \cdots)_{c_0}$. Of course $\mathcal{D}$ is a nuclear C^*-algebra. Now if $\mathcal{B}$ is a separable non-type I nuclear C^*-algebra, then by Corollary 1.7

$$\mathcal{B} \overset{cc}{\hookrightarrow} \mathcal{D} \tag{1.13}$$

(since $\mathcal{B} \overset{cc}{\hookrightarrow} \mathbf{A}$ and trivially $\mathbf{A} \overset{cc}{\hookrightarrow} \mathcal{D}$). But by Corollary 1.7 and Theorem 1.8,

$$\mathcal{D} \overset{cc}{\hookrightarrow} \mathbf{A} \overset{cc}{\hookrightarrow} \mathcal{B} \ . \tag{1.14}$$

It then follows by the Pełczyński decomposition method that $\mathcal{B}$ is completely isomorphic to $\mathcal{D}$, whence also $\mathcal{D}$ is completely isomorphic to $\mathbf{A}$. □

This leaves the following open question:

Problem. Classify the separable (infinite-dimensional) type I C^*-algebras to complete isomorphism.

I believe the classification should be the *same* as the Banach classification. However I also believe the answer is far more intricate than the known commutative case; see [**Ro2**] for an exposition of the latter.

2. The CSEP and the CSCP

We are concerned here with quantized versions of the Separable Extension Property for Banach spaces, which is defined as follows.

Definition 2.1. *A Banach space Z has the Separable Extension Property (SEP) provided for all separable Banach spaces $X \subset Y$ and bounded linear operators $T : X \to Z$, there exists a bounded linear operator $\tilde{T} : Y \to Z$ extending T. That is, we have the diagram*

$$\begin{array}{ccc} Y & \overset{\tilde{T}}{\searrow} & \\ \cup & & \\ X & \xrightarrow{T} & Z \end{array} \tag{2.1}$$

If $\lambda \geq 1$ is such that $\tilde{T}$ can always be chosen with $\|\tilde{T}\| \leq \lambda \|T\|$, we say that Z has the λ-SEP.

Our first quantized version of the SEP goes as follows.

Definition 2.2. *An operator space Z has the Complete Separable Extension Property (the CSEP) provided for all separable operator spaces $X \subset Y$ and completely bounded $T : X \to Z$, there exists a completely bounded $\tilde{T} : Y \to Z$ extending T. That is. we have that (2.1) holds for completely bounded maps. Again, if $\tilde{T}$ can always be chosen with $\|\tilde{T}\|_{cb} \leq \lambda \|T\|_{cb}$, we say that Z has the λ-CSEP.*

It turns out that if Z has the CSEP, then Z has the λ-CSEP for some λ, and of course a similar statement holds for the SEP itself. We are mainly interested in the case of separable Z. Sobczyk proved in 1941 that c_0 has the 2-SEP and 2 is best possible here [**Sob**]; in fact, if Z is infinite-dimensional separable with the λ-SEP, then $\lambda \geq 2$. Zippin proved in 1977 the far deeper converse; *every infinite dimensional separable Banach space with the SEP is isomorphic to c_0* [**Z**].

We first give a proof of Sobczyk's theorem which motivates the approach to the quantized versions of the SEP given in [**AR**]. We recall that a Banach space Z is called *isomorphically injective* if for all Banach spaces $X \subset Y$ and operators $T : X \to Z$, there exists a $\tilde{T}$ satisfying (2.1). If we require also that we can choose $\tilde{T}$ with $\|\tilde{T}\| = \|T\|$, we say that Z is *isometrically injective.* Similarly, an operator space Z is called *isomorphically injective* if for all operator spaces $X \subset Y$ and completely bounded $T : X \to Y$, there is a completely bounded $\tilde{T}$ satisfying (2.1); if again we require that $\tilde{T}$ can be chosen with $\|\tilde{T}\|_{\mathrm{cb}} = \|T\|_{\mathrm{cb}}$, Z is called *isometrically injective.* (In the literature, isometrically injective operator spaces are just termed injective.) It is easily seen, from the Hahn-Banach theorem, that ℓ^∞ is isometrically injective. The non-commutative version of this is true (throughout, we use complex scalars).

THEOREM 2.3. *$B(H)$ is an isometrically injective operator space, for any Hilbert space H.*

This was proved for the case of completely positive maps and self-adjoint operator spaces in the domain by Arveson [**A**], and later in general by Paulsen (cf. [**Pa**]) and Wittstock [**Wi**]. It follows from 2.3 that *a separable operator space has the CSEP if and only if it is completely complemented in every separable operator superspace.* Similarly, an operator space is isomorphically injective if and only if it is completely complemented in every operator superspace.

The following result is an immediate consequence of Corollaries 1.2 and 1.3.

COROLLARY 2.4. *Let X be a non-reflexive operator space. If X is separable with the CSEP, X contains a subspace completely isomorphic to c_0. If X is isomorphically injective, X contains a subspace completely isomorphic to ℓ^∞.*

It is known that a Banach space is isometrically injective if and only if it is isometric to $C(\Omega)$ for some extremely disconnected compact Hausdorff space Ω. It is a famous open problem if every isomorphically injective Banach space is isomorphic to an isometrically injective one. Similarly, we have the quantized version: *Is every isomorphically injective operator space completely isomorphic to an isometrically injective operator space?*

We next deal with the proof of Sobczyk's theorem, which motivates the approach to the quantized versions given in [**AR**]. We first formulate complementation in terms of lifts.

DEFINITION 2.5. *Let $X \subset Y$ be Banach (resp. operator) spaces and let $\pi : Y \to Y/X$ be the quotient map. A bounded linear map $L : Y/X \to Y$ is a lift of $I_{Y/X}$ if $I_{Y/X} = \pi L$. That is, the following diagram holds:*

$$\begin{array}{ccc} & & Y \\ & \overset{L}{\nearrow} & \downarrow \pi \\ Y/X & \xrightarrow{I} & Y/X \end{array} \tag{2.2}$$

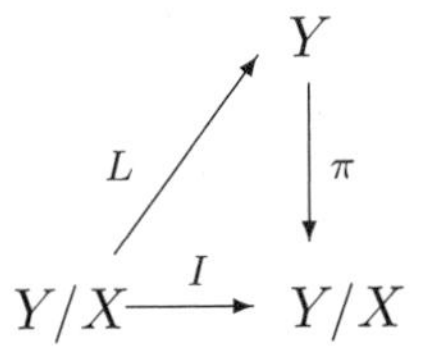

Then it is easily seen that X is complemented in Y if and only if $I_{Y/X}$ admits a lift. In fact, we have the following simple result.

PROPOSITION 2.6. *Let $X \subset Y$ be Banach (resp. operator) spaces, and let $\lambda \geq 1$. Then X is λ-co-complemented in Y (resp. λ-completely co-complemented in Y) if and only if $I_{Y/X}$ admits a lift L with*

$$\|L\| \leq \lambda \quad (\textit{resp. } \|L\|_{cb} \leq \lambda). \tag{2.3}$$

PROOF. Suppose L satisfies (2.2). Then setting $Q = L\pi$, we easily verify that Q is a projection on Y with kernel equal to X, and of course $\|Q\| \leq \|L\|$ (resp. $\|Q\|_{\text{cb}} \leq \|L\|_{\text{cb}}$). Conversely, if $Q : Y \to Y$ is a bounded linear projection with $X = \operatorname{Ker} Q$, let $W = Q(Y)$. It follows easily that $\pi|W$ maps W onto Y/X and

$$\|\pi w\| \geq \frac{1}{\|Q\|}\|w\| \qquad \text{for all } w \in W. \tag{2.4}$$

Then $L = (\pi|W)^{-1}$ is the desired lift; moreover in the completely bounded setting, $\|L\|_{\text{cb}} \leq \|Q\|_{\text{cb}}$. □

Next we recall a classical theorem of Borsuk [**B**], which asserts that if K is a compact metrizable subset of a compact Hausdorff space Ω, then there is a *linear* operator $L : C(K) \to C(\Omega)$ of norm one so that $(Lf)|K = f$ for all $f \in K$. This may be reformulated in the language of C^*-algebras as follows.

THEOREM 2.7. *Let $\mathcal{A}$ be a unital commutative C^*-algebra and $\mathcal{J}$ be a (closed) ideal in $\mathcal{A}$ so that $\mathcal{A}|\mathcal{J}$ is separable. Then there exists a contractive lift $\mathcal{A}|\mathcal{J} \to \mathcal{A}$ of $I_{\mathcal{A}|\mathcal{J}}$.*

PROOF. By the Gelfand-Naimark theorem, $\mathcal{A}$ is $*$-isometric to $C(\Omega)$ for some compact Hausdorff space Ω, and moreover, if we just assume $\mathcal{A} = C(\Omega)$, then for some closed subset K of Ω, $\mathcal{J} = \{f \in \mathcal{A} : f(k) = 0 \text{ for all } k \in K\}$. $\mathcal{A}|\mathcal{J}$ is separable if and only if K is metrizable; thus 2.6 is just a reformulation of Borsuk's theorem. □

Now we give a

PROOF OF SOBCZYK'S THEOREM. We first note that c_0 is an *ideal* in ℓ^∞, which can be regarded as a C^*-algebra. Let then $X \subset Y$ be separable Banach spaces and $T : X \to c_0$ be a given operator. Since ℓ^∞ is isometrically injective, we may choose $\tilde{T} : Y \to \ell^\infty$ extending T with $\|\tilde{T}\| = \|T\|$. Let $\mathcal{A}$ be the (commutative) C^*-subalgebra of ℓ^∞ generated by c_0 and $T(Y)$. Then of course c_0 is an ideal in $\mathcal{A}$ also, and hence is co-contractively complemented in $\mathcal{A}$ by Theorem 2.7 and Proposition 2.6; that is, we may choose a projection $P : \mathcal{A} \to c_0$ with $\|I - P\| = 1$; hence $\|P\| \leq 2$. But then letting $S = P\tilde{T}$, S is an extension of T to Y, and $\|S\| \leq 2$. Thus c_0 has the 2-SEP. □

The remainder of this section deals with *quantized* versions of this argument. Before going into this, we note that it is an open question if the conclusion of Theorem 2.7 holds if we drop the assumption that $\mathcal{A}$ is commutative. In fact, it is open, if every ideal of a separable C^*-algebra $\mathcal{A}$

is *complemented* in $\mathcal{A}$; i.e. if there exists a bounded linear lift L of $I_{\mathcal{A}/\mathcal{J}}$. Deep work of Ando [**A**] yields that the answer is affirmative, however if $\mathcal{A}/\mathcal{J}$ has the bounded approximation property; in particular, if $\mathcal{A}/\mathcal{J}$ has the metric approximation property then $I_{\mathcal{A}/\mathcal{J}}$ admits a contractive lift. An important special case of this problem: *let Y be a separable subspace of $B(\ell_2)$ containing* $\mathbf{K}$. *Is* $\mathbf{K}$ *complemented in Y?* Trivially we may assume Y is a C^*-algebra, by just replacing Y by its generated C^*-algebra. An approach to a possible positive resolution to this is given in [**AR**]. The author now guesses, however, this last problem has a negative answer. (A further complement to this problem is given by Proposition 2.20 below.)

We now summarize the main known results on separable spaces with the CSEP. Let R (resp. C) denote the ROW (resp. COLUMN) operator space. Identifying $B(\ell_2)$ with infinite matrices, R is simply all such matrices with non-zero entries only in the first row. Similarly, C is all matrices with non-zero entries only in the first column. It is easily seen that R and C are isometric to ℓ_2; their matrix representation inside $B(\ell_2)$ determines their operator space structure. More generally, if $1 \leq j, k, \leq \infty$, $M_{j,k}$ denotes all operators in $B(\ell_2)$ whose matrices have non-zero entries only in their first j rows and k columns. Thus $M_{\infty,\infty} = B(\ell_2)$, $M_{1,\infty} = R$, $M_{\infty,1} = C$.

If X, Y are operator spaces, $X \oplus Y$ denotes their ℓ^∞ direct sum. If $X_1, X_2, \ldots$ are given operator spaces, $(X_1 \oplus X_2 \oplus \cdots)_{c_0}$ denotes their c_0-direct sum; i.e. the Banach space of all sequences (x_n) with $x_n \in X_n$ for all n and $\|x_n\| \to 0$; $(X_1 \oplus X_2 \oplus \cdots)_{\ell^\infty}$ denotes their ℓ^∞-direct sum. Both of these spaces are just endowed with the corresponding ℓ^∞-direct sum operator structures. Finally, we denote $(X \oplus X \oplus \cdots)_{c_0}$ by $c_0(X)$ and $(X \oplus X \oplus \cdots)_{\ell^\infty}$ by $\ell^\infty(X)$. The following is the main "positive" result on the CSEP.

THEOREM 2.8. *For all $n \geq 1$, $c_0(M_{n,\infty} \oplus M_{\infty,n})$ has the 2-CSEP.*

(This is established in [**Ro2**] with "$2+\varepsilon$"in place of "2", $\varepsilon > 0$ arbitrary. The refinement eliminating $\varepsilon > 0$ is given in [**AR**].)

Now for each $n < \infty$, $c_0(M_{n,\infty})$ is completely isomorphic to $c_0(R)$ and $c_0(M_{\infty,n})$ is completely isomorphic to $c_0(C)$; however, just the Banach-Mazur distance itself of $c_0(R)$ to $c_0(M_{n,\infty})$ leads to infinity; i.e. $d(c_0(\ell_2), c_0(M_{n,\infty})) \to \infty$ as $n \to \infty$. Thus it appears surprising that the λ-CSEP constant in 2.8 is best possible, namely $\lambda = 2$.

CONJECTURE. *Let X be a separable operator space with the CSEP. Then X is completely isomorphic to a subspace of $c_0(R \oplus C)$.*

Of course any completely complemented subspace of $c_0(R \oplus C)$ has the CSEP; naturally this includes c_0 itself. See Conjecture 4.6 of [**Ro2**] for a list of 21 completely complemented operator subspaces of $c_0(R \oplus C)$, which could conceivably be the entire family of separable infinite-dimensional ones with the CSEP.

Now $R \oplus C$ itself is isometrically injective (as an operator space), so has the CSEP. As shown in Proposition 22 of [**Ro2**], any reflexive operator space with the CSEP is isomorphically injective. A problem thus related to the above conjecture: *is every separable (infinite-dimensional) isomorphically injective operator space completely isomorphic to R, C or $R \oplus C$?* The problem has been answered affirmatively by Robertson in case the operator space is actually isometrically injective [**R**].

Of course, $M_{n,\infty} \oplus M_{\infty,n}$ is completely isometric to a subspace of $\mathbf{K}$ for all n. However $\mathbf{K}$ itself *fails* the CSEP. This result is due to Kirchberg [**Ki2**]. In fact, he obtains that $\mathbf{K}_0$ fails the CSEP, where

$$\mathbf{K}_0 = (M_1 \oplus M_2 \oplus \cdots \oplus M_n \oplus \cdots)_{c_0} . \tag{2.5}$$

Here, we regard $\mathbf{K}_0$ as a C^*-algebra; notice that $\mathbf{K}_0$ is an ideal in the "finite" von Neumann algebra $\mathcal{N} = (M_1 \oplus M_2 \oplus \cdots)_\infty$.

A new proof that $\mathbf{K}_0$ fails the CSEP is given in [**OR**], via the following result (See Corollary 4.9 of [**OR**]).

THEOREM 2.9. *There exists an operator space Y containing $\mathbf{K}_0$ so that $Y/\mathbf{K}_0$ is completely isometric to c_0 and $\mathbf{K}_0$ is completely uncomplemented in Y.*

The particular construction in [**OR**] yields that $\mathbf{K}_0$ is Banach cocontractively complemented in Y The same holds for Kirchberg's construction mentioned above; hence these results cannot resolve the complemented ideal problem mentioned earlier. (Actually we show later that this is a consequence of a general principle; see Proposition 2.20.)

Is there a quantized version of the SEP which $\mathbf{K}$ satisfies? For $\mathbf{K}$ is often thought of as quantized c_0. There is indeed; the "culprit" in the counter example Y of 2.9: Y fails to be *locally reflexive* as an operator space.

DEFINITION 2.10. *An operator space X is called* locally reflexive *if there is a $\lambda \ge 1$ that for all $\varepsilon > 0$ and finite dimensional subspaces F and G of X^* and X^{**} respectively, there exists a linear operator $T : G \to X$ satisfying*

$$\langle Tg, f\rangle = \langle g, f\rangle \quad \textit{for all } g \textit{ in } G \textit{ and } f \in F \tag{2.6}$$

and

$$\|T\|_{cb} < \lambda + \varepsilon . \tag{2.7}$$

If λ works, X is called λ-locally reflexive.

If X is any Banach space, then X is 1-locally reflexive and hence (X, MIN) is 1-locally reflexive. Remarkable permanence properties yield that if X is λ-locally reflexive, so is any subspace (cf. [**ER**], [**Pi**]). A C^*-algebra is either 1-locally reflexive or non-locally reflexive [**EH**]. Nuclear C^*-algebras are locally reflexive, but for example $B(H)$ is *not* (for H infinite-dimensional).

As noted above, a separable operator space has the CSEP provided it is completely complemented in every separable operator superspace.

DEFINITION 2.11. *A separable locally reflexive operator space Z has the Complete Separable Complementation Property (the CSCP) provided every* *complete isomorph of Z is completely complemented in every separable locally reflexive operator superspace. Equivalently, given $X \subset Y$ separable locally reflexive operator spaces and $T : X \to Z$ a complete surjective homomorphism, there exists a completely bounded $\tilde{T}$ satisfying (2.1).*

This concept was introduced in [**Ro2**], where it was established that $\mathbf{K}_0$ has the CSCP. Subsequently, T. Oikhberg and the author jointly established the following result.

THEOREM 2.12. [**OR**] **K** *has the CSCP.*

Now it follows that then every completely complemented subspace of **K** has the CSCP. An affirmative answer to the following would extend Zippman's result [**Z**].

CONJECTURE. *Every operator space with the CSCP is completely isomorphic to a subspace of* **K**.

For a possible list of all *primary* completely complemented subspaces of **K**, see Conjecture 4.10 of [**Ro2**]. There are 11 such candidates. Taking direct sums of these yields a finite (up to complete isomorphism) family of operator spaces which could conceivably *be* the list of all operator spaces with the CSCP. Such a result, if true, would involve many deep new ideas.

The following equivalences for the CSCP are established in [**OR**].

THEOREM 2.13. *Let X be a separable locally reflexive operator space. Then the following are equivalent*

(a) *X has the CSCP.*
(b) *X is completely complemented in every separable locally reflexive superspace.*
(c) *Assuming $X \subset B(H)$, then X is completely complemented in Y for all separable locally reflexive Y with $X \subset Y \subset B(H)$.*

Let us note that it is *false* that spaces Z with the CSCP enjoy the extension property given in (2.1) for completely bounded maps, even if Y is separable locally reflexive. Indeed it is established in [**OR**] that *if $Y = C_1$ (the trace class operators) and for all $X \subset Y$ and completely bounded $T : X \to Z$, there is a completely bounded $\tilde{T} : Y \to Z$ satisfying (2.1), then Z has the CSEP.* However, spaces with the CSCP do satisfy a form of this principle (Theorem 1.6(d) of [**OR**]).

PROPOSITION 2.14. *Let Z have the CSCP and $X \subset Y$ be separable locally reflexive such that X is locally complemented in Y. Then for every completely bounded $T : X \to Z$, there is a completely bounded $\tilde{T} : Y \to Z$ extending T (i.e., so that (2.1) holds).*

Recall that if X and Y are operator spaces, with $X \subset Y$, then X is called *locally complemented in* Y if there is a $C < \infty$ so that X is C-completely complemented in Z for all $X \subset Z \subset Y$ with Z/X finite dimensional. It is proved in [**Ro2**] that if Y is locally reflexive, then X is locally complemented in Y if and only if X^{**} is completely complemented in Y^{**}.

As noted above, there exist completely bounded operators from subspaces of C_1 into $\mathbf{K}$ with no completely bounded extensions. However the following problem *may* have an affirmative answer.

PROBLEM. Let $\mathcal{A}$ be a separable nuclear C^*-algebra, X a subspace, and $T : X \to \mathbf{K}$ be a completely bounded map. Does there exist a completely bounded extension $\tilde{T} : \mathcal{A} \to \mathbf{K}$?

We finally sketch a route throough Theorems 2.8 and 2.12, following the approach in [**AR**]. The following concept is fundamental.

DEFINITION 2.15. *Let $X \subset Y$ be Banach/operator spaces.*

(a) *X is called an M-summand in Y if there exists a closed linear subspace Z of Y with $X \oplus Z = Y$ so that*

$$\|x + z\| = \max\{\|x\|, \|z\|\} \quad \text{for all } x \in X \text{ and } z \in Z. \tag{2.8}$$

In the operator space case, X is called a complete M-summand if Z can also be chosen so that

$$\|(x_{ij} + z_{ij})\| = \max\{\|(x_{ij})\|, \|(z_{ij})\|\} \tag{2.9}$$

for all n and $n \times n$ matrices (x_{ij}) and (z_{ij}), of elements of X and Z respectively.

(b) *X is called an M-ideal (resp. complete M-ideal) in Y if $X^{**} = X^{\perp\perp}$ is an M-summand (resp. complete M-summand) in Y^{**}.*

It is a remarkable theorem that *if $\mathcal{A}$ is a C^*-algebra and $X \subset \mathcal{A}$, then X is an M-ideal iff X is an algebraic (closed 2-sided) ideal iff X is a complete M-ideal.* This is due to Alfsen-Effros [**AE**] and Smith-Ward [**SW**]. The entire idea of M-ideals appears in the seminal work in [**AE**]. The following isometric lifting result is established in the appendix to [**AR**], extending a lifting result in [**EH**] to the pure operator spaces setting. (An operator space X is called *nuclear* if it satisfies Definition 1.4, replacing "$\mathcal{A}$" by "X" in 1.4.)

THEOREM 2.16. *Let $X \subset Y$ be operator spaces with X a nuclear complete M-ideal in Y, Y locally reflexive, and Y/X separable. Then $I_{Y/X}$ admits a completely contractively life $L : Y/X \to Y$.*

We need one more concept, which appears fundamental for the study of the CSEP.

DEFINITION 2.17. *An operator space X is said to be of finite matrix type if there is a $C \geq 1$ so* that for any finite dimensional operator space G, there is an integer n with

$$\|T\|_{cb} \leq C\|T\|_n \quad \text{for all linear operators } T : G \to X. \tag{2.10}$$

If C works, we say X is C-finite.

This concept was introduced in [**Ro2**]. In [**OR**], it is proved that *if X is separable and $c_0(X)$ has the CSEP, then X is of finite matrix type.* This suggests the

CONJECTURE. *If X is separable with the CSEP, then X is of finite matrix type.*

It also shows that $\mathbf{K}_0$ fails the CSEP, since it is easily seen that $\mathbf{K}_0$ in *not* of finite matrix type. (In fact, Theorem 2.9 is proved in the course establishing this result). The following yields the needed remaining ingredients for Theorem 2.8.

PROPOSITION 2.18.

(a) *If an operator space X is C-finite, so is $\ell^\infty(X)$.*
(b) *If X is C-finite, X is C-locally reflexive.*
(c) *$M_{n,\infty} \oplus M_{\infty,n}$ is 1-finite, for all n.*
(d) *$c_0(M_{n,\infty} \oplus M_{\infty,n})$ is a complete M-ideal in $\ell^\infty(M_{n,\infty} \oplus M_{\infty,n})$, for all n.*

(a),(b) are proved in [**AR**] and (c) is established in [**Ro2**]. Also, (d) is proved in [**AR**], as a simple consequence of the criterion developed there for X to be a complete M-ideal in Y, namely that existence of a *complete M-approximate identity.*

PROOF OF THEOREM 2.8. Let $n \geq 1$, let $X \subset Y$ be separable operator spaces, let $Z = c_0(M_{n,\infty} \oplus M_{\infty,n})$, and let $T : X \to Z$ be a completely bounded map. It follows easily that setting $W = \ell^\infty(M_{n,\infty} \oplus M_{\infty,n})$, then W is completely contractively complemented in $B(\ell_2)$ and hence W is isometrically injective, by Theorem 2.3. Then there exists an extension $\tilde{T} : Y \to W$ of T with

$$\|\tilde{T}\|_{\mathrm{cb}} = \|T\|_{\mathrm{cb}} \ . \tag{2.11}$$

Now by Proposition 2.18, W is 1-finite and hence 1-locally reflexive. It is easily seen directly that Z is nuclear; of course this also follows since Z is completely isometric to a completely contractively complemented subspace of $\mathbf{K}$. Furthermore, since Z is a complete M-ideal in W, Z is a complete M-ideal in $\tilde{W} \overset{\mathrm{def}}{=} \overline{Z + \tilde{T}(Y)}$. Again, $\tilde{W}$ is 1-locally reflexive, and so by Theorem 2.16, Z is completely co-contractively complemented in $\tilde{W}$. But then there exists a projection P from $\tilde{W}$ onto Z with

$$\|P\|_{\mathrm{cb}} \leq 2 \ . \tag{2.12}$$

Finally, letting $S = P\tilde{T}$, then S is a cb extension of T with $\|S\|_{\mathrm{cb}} \leq 2\|T\|_{\mathrm{cb}}$. □

The following is one of the crucial ingredients needed to establish Theorem 2.12.

THEOREM 2.19. [**OR**] *Let $X \subset Y$ be separable operator spaces, $\tilde{X} \subset B(H)$, and $T : X \to \tilde{X}$ a complete isomorphism from X onto $\tilde{X}$. Then there exists a $\tilde{Y}$ with $\tilde{X} \subset \tilde{Y} \subset B(H)$ and a complete surjective isomorphism $\tilde{T} : Y \to \tilde{Y}$ extending T.*

This result bears on the "complemented ideal" problem discussed following the proof of Sobczyk's theorem above.

PROPOSITION 2.20. *Let $X \subset Y$ be separable operator spaces with X completely isomorphic to $\mathbf{K}$ or $\mathbf{K}_0$. If Y/X has the bounded approximation property, then X is complemented in Y.*

PROOF. Suppose X is completely isomorphic to $\mathbf{K}$. Let $T : X \to \mathbf{K}$ be a complete surjective isomorphism, and let $\mathbf{K} \subset \tilde{Y} \subset B(\ell_2)$ and $\tilde{T} : Y \to \tilde{Y}$ be a complete isomorphism extending T. Then $\tilde{Y}/\mathbf{K}$ is isomorphic to Y/X and so has the bounded approximation property. $\mathbf{K}$ is moreover an M-ideal in $\tilde{Y}$, since it is an ideal in $B(\ell_2)$. By a result of Ando mentioned above [**A**], $\mathbf{K}$ is complemented in $\tilde{Y}$ and hence X is complemented in Y. Now if X is completely isomorphic to $\mathbf{K}_0$, set $W = (M_1 \oplus M_2 \oplus \cdots)_\infty$. It is known that W is completely isomorphic to $B(\ell_2)$. (This is a simple application of the decomposition method for operator spaces). It follows, since Theorem 2.19 is invariant under complete isomorphisms, that one may replace $B(\ell_2)$ by W in its statement. Thus we now let $T : X \to \mathbf{K}_0$ be a complete isomorphism and choose $\mathbf{K}_0 \subset \tilde{Y} \subset W$ and $\tilde{T} : Y \to \tilde{Y}$ a complete isomorphism extending T. Again, $\mathbf{K}_0$ is an M-ideal in $\tilde{Y}$ and $\tilde{Y}/\mathbf{K}_0$ is isomorphic to Y/X and so has the bounded approximation property, so again by [**A**], $\mathbf{K}_0$ is complemented in $\tilde{Y}$ and hence X is complemented in Y. □

REMARKS. A different proof is given in [**AR**] of the special case of Ando's result used above, via the concept of extendable *local liftings* (ell's) introduced there. Thus, given Banach spaces $X \subset Y$, (X, Y) is said to admit ell's if there is a $C \geq 1$ so that for all finite-dimensional $E \subset Y/X$, there exists a $T : Y/X \to Y^{**}$ with $\|T\| \leq C$ so that $T(E) \subset Y$ and $e = \pi T(e)$ where $\pi : Y \to Y/X$ is the quotient map. It is proved in [**AR**] that *if $\mathcal{J}$ is a nuclear ideal in a C^*-algebra $\mathcal{A}$ so that $\mathcal{A}/\mathcal{J}$ is separable, then $\mathcal{J}$ is complemeted in $\mathcal{A}$ provided $(\mathcal{J}, \mathcal{A})$ has extendable local liftings.* (This is easily seen to be the case if $\mathcal{A}/\mathcal{J}$ has the bounded approximation property.) It follows that *the conclusion of 2.20 holds if one replaces the assumption that Y/X has the bounded approximation property by the more general one that (X, Y) has ell's.*

The following result is our last needed ingredient for 2.12.

THEOREM 2.21. [**AR**] *Let $\mathcal{J} \subset Y \subset \mathcal{A}$ with $\mathcal{J}$ a nuclear ideal in a C^*-algebra $\mathcal{A}$ and Y a λ-locally reflexive operator space with $Y/\mathcal{J}$ separable. Then for every $\varepsilon > 0$, there exists a completely bounded lift $L : Y/\mathcal{J} \to Y$ of $I_{Y/\mathcal{J}}$ with $\|L\|_{cb} < \lambda + \varepsilon$.*

This result is generalized to a pure operator space theorem in [**AR**, Theorem 2,4]. The hypotheses then require more than the assumption that $\mathcal{J}$ is a complete M-ideal, for the proof makes use of the special approximate identities that exist in (non-unital) C^*-algebras. Theorem 2.21 generalizes a result of Effros-Haagerup [**EH**], which establishes 2.21 when $Y = \mathcal{A}$ itself. The fact that Y need not be 1-locally reflexive causes difficulties, however, and the proof in [**AR**] is somewhat delicate, although based in part on the techniques in [**EH**]. Note also that the result of [**EH**] itself follows from Theorem 2.16, whose proof in [**AR**] follows Ando's construction rather than the argument in [**EH**]. In fact, we don't know if the conclusion of Theorem 2.16 holds, if "1" and "contractive" are deleted in its statement.

We finally conclude with the

PROOF OF THEOREM 2.12. Let $X \subset Y$ be separable operator spaces with Y locally reflexive and let $T : X \to \mathbf{K}$ be a complete surjective isomorphism. By Theorem 2.19, there exists a $\tilde{Y}$ with $\mathbf{K} \subset \tilde{Y} \subset B(\ell_2)$ and a complete isomorphism $\tilde{T} : Y \to \tilde{Y}$ extending T. But then $\tilde{Y}$ is separable locally reflexive, and of course $\mathbf{K}$ is a nuclear ideal in $\mathcal{A} = B(\ell_2)$. Theorem 2.20 yields a completely bounded projection P from $\tilde{Y}$ onto $\mathbf{K}$, and hence $S = P\tilde{T}$ is the desired extension from Y to $\mathbf{K}$. □

References

[AE] E. Alfsen and E. Effros, *Structure in real Banach space*, Ann. of Math. **96** (1972), 98–173.

[A] T. Ando, *A theorem on non-empty intersection of convex sets and its applications*, J. Approx. Th. **13** (1975), 158–166.

[AR] A. Arias and H.P. Rosenthal, *M-complete approximate identities in operator spaces*, Studia Math. **141** (2000), 143–200.

[Arv] W.B. Arveson, *Subalgebras of C^*-algebras*, Acta Math. **123** (1969), 141–224.

[B] K. Borsuk, *Über isomorphic der funktionalräume*, Bull. Int. Acad. Polon. Sci. A**113** (1933), 1–10.

[BP] D.P. Blecher and V. Paulsen, *Tensor products of operator spaces*, J. Func. Anal. **99** (1991), 262–292.

[CE1] M.-D. Choi and E. Effros, *The completely positive lifting problem for C^*-algebras*, Ann. of Math. **104** (1976), 585–609.

[CE2] M.-D. Choi and E. Effros, *Nuclear C^*-algebras and the approximation property*, Amer. J. Math **100** (1978), 61–79.

[C] A. Connes, *Classification of injective factors*, Ann. of Math. **104** (1976), 73–115.

[EH] E. Effros and U. Haagerup, *Lifting problems and local reflexivity for C^*-algebras*, Duke Math. J. **52** (1985), 103–128.

[EOR] E. Effros, N. Ozawa and Z.-J. Ruan, *On injectivity and nuclearity for operator spaces*, Duke Math. J., to appear.

[ER] E. Effros and Z.-J. Ruan, *Operator spaces*, Oxford University Press, Oxford, 2000.

[G] J. Glimm, *Type I C^*-algebras*, Ann. of Math. **73** (1961), 572–612.

[HWW] P. Harmand, D. Werner and W. Werner, *M-ideals in Banach spaces and Banach algebras*, Lecture Notes in Math. 1547, Springer-Verlag, 1993

[HRS] U. Haagerup, H.P. Rosenthal and F.A. Sukochev, *Banach embedding properties of non-commutative L^p-spaces*, Memoirs of the Amer. Math. Soc., to appear.

[Ki1] E. Kirchberg, *On subalgebras of the CAR-algebra*, J. Funct. Anal. **129** (1995), 35–63.

[Ki2] E. Kirchberg, *On non-semisplit extensions, tensor products and exactness of group C^*-algebras*, Invent. Math. **112** (1993), 449–489.

[M] A.A. Milutin, *Isomorphisms of spaces of continuous functions on compacts of power continuum.* Tieoria Func. (Kharkov) **2** (1966), 150–156 (Russian).

[Mu] G.J. Murphy, *C^*-algebras and operator theory*, Academic Press, San Diego, 1990.

[OR] T. Oikhberg and H.P. Rosenthal, *Extension properties for the space of compact operators*, J. Funct. Anal. **179** (2001), 251–308.

[Pa] V. Paulsen, *Completely bounded maps and dilations*, Wiley, New York, 1986.

[Ped] G.K. Pedersen, *C^*-algebras and their automorphism groups*, Academic Press, London, 1989.

[Pe1] A. Pełczyński, *Projections in certain Banach spaces*, Studia Math. **19** (1960), 209–228.

[Pe2] A. Pełczyński, *Banach spaces on which every unconditionally converging operator is weakly compact*, Bull. Acad. Pol. Sci. ser. Math. Astr. Phys. **10** (1962), 265–270.

[Pf] H. Pfitzner, *Weak compactness in the dual of a C^*-algebra is determined commutatively*, Math. Ann. **298** (1994), 349–371.

[Pi] G. Pisier, *An introduction to the theory of operator spaces*, Cambridge Univ. Press, to appear.

[R] A.G. Robertson, *Injective matricial Hilbert spaces*, Math. Proc. Cambridge Phil. Soc. **110** (1991), 183–190.

[Ro1] H.P. Rosenthal, *On complemented and quasi-complemented subspaces of quotients of $C(S)$ for σ-Stonian S*, Proc. Nat. Acad. Sci. USA (1968), 1165–1169.

[Ro2] H.P. Rosenthal, *The complete separable extension property*, J. Operator Theory **43** (2000), 324–374.

[Ro3] H.P. Rosenthal, *The Banach spaces $C(K)$*, Handbook of the Geometry of Banach spaces, vol.2, W.B. Johnson and J. Lindenstrauss, eds., Elsevier, Amsterdam, to appear.

[S] R.R. Smith, *Completely bounded maps between C^*-algebras*, J. London Math. Soc. **27** (1983), 157–166.

[SW] R. Smith and J. Ward, *M-ideal structure in Banach algebras*, J. Funct. Anal. **27** (1978), 337–349.

[Sob] A. Sobczyk, *Projection of the space (m) on its subspace (c_0)*, Bull. Amer. Math. Soc. **47** (1941), 938–947.

[Sz] A. Szankowski, *$B(H)$ does not have the approximation property*, Acta. Math. **147** (1981), 89–108.

[W] S. Wassermann, *Exact C^*-algebras and related topics*, Lecture Notes Series, Seoul National University, 1994.

[Wi] G. Wittstock, *Ein operatorwertigen Hahn-Banach Satz*, J. Funct. Anal. **40** (1981), 127–150.

[Z] M. Zippin, *The separable extension problem*, Israel J. Math. **26** (1977), 372–387.

The University of Texas at Austin, Department of Mathematics, 1 University Station C1200, Austin, TX 78712-0257, USA

E-mail address: rosenthl@math.utexas.edu

Contemporary Mathematics
Volume **321**, 2003

How many operators exist on a Banach space?

Th. Schlumprecht

Abstract. We present some results concerning the following question: Given an infinite dimensional Banach space X. Are there two normalized basic sequences (x_n) and (y_n), with $\liminf_{n\to\infty} \|x_n - y_n\| > 0$, so that the map $x_n \mapsto y_n$ extends to a linear bounded operator between the linear span (x_n) and the linear span (y_n)?

1. Introduction

Over the last years the following question caught the interest of an increasing number of researchers in the area of infinite dimensional Banach space theory and became for many of them one of the central problems.

(Q1) Assume X is an infinite dimensional Banach space. Does there exist a linear bounded operator T on X which is not of the form $T = \lambda\mathrm{Id} + C$, where λ is a scalar, C is a compact operator on X, and Id denotes the identity on X?

Clearly, if $E \subset X$ is a finite dimensional subspace and $S : E \to X$ is any linear operator, then S can be extended to a linear operator $T : X \to X$ of finite rank. Thus, there must be "many" finite rank operators on X, and thus "many" elements in the closure of the finite rank operators, which is, assuming X has the approximation property, the ideal of compact operators. Of course, the identity must also be a linear bounded operator. Therefore question (Q1) asks wheather or not there are always some other linear bounded operators defined on a Banach space, beyond the class of operators which must exist by elementary reasons.

Note that a counterexample to (Q1) would be the first example of a Banach space X for which the invariant subspace problem has a positive answer: Does every operator T on X have a non trivial subspace Y ($Y \neq \{0\}$ and $Y \neq X$) so that $T(Y) \subset Y$? Indeed, due to a result of V. I. Lomonosov [**Lo**], any operator on a Banach space which commutes with a compact operator $C \neq 0$ must have a none trivial invariant subspace. A counterexample to (Q1) might also be a prime candidate for a Banach space on which all Lipschitz function admit a point of differentiability [**Li**].

2000 *Mathematics Subject Classification.* Primary 46B03, 46B20, 47L10.

The research was supported by NSF.

Recall ([**GM1**] and [**GM2**]) that an infinite dimensional Banach space X is called a *hereditarily indecomposable space* (HI) if no infinite dimensional closed subspace of X is the complemented sum of two further infinite dimensional closed subspaces. The first known example of an HI space, we denote it by GM, was constructed by T. W. Gowers and B. Maurey in 1993 [**GM1**]. Since then HI spaces with additional properties were constructed (e.g. [**AD**], [**F1**], and [**OS**]).

V. Ferenczi [**F2**] showed that a Banach space over $\mathbb{C}$ is (HI) if and only if every operator $T : Y \to X$, with Y being a subspace of X, is of the form $\text{In}_{(Y,X)} + S$, where λ is a scalar, $\text{In}_{(Y,X)}$ denotes the inclusion map from Y into X and S is a *strictly singular operator*, i.e. an operator which on no infinite dimensional subspace is an isomorphism into X.

Although a counterexample to (Q1) does not need to be an HI space, these spaces are nevertheless natural candidates for counterexamples. In [**AS**] it was shown that the space constructed in [**GM1**] admits an operator (defined on all of GM) which is strictly singular but not compact. Already earlier it was shown in [**Go1**] that such an operator can be constructed on a subspace of GM.

Given the seemingly easy task to write down on a concrete space a nontrivial operator (i.e. not of the form $\lambda +$ *compact*) the proof in [**AS**] is quite involved. It would be interesting, but probably technically even harder, to establish the existence of non trivial operators defined on other known HI spaces. Since it seems to be hard to answer (Q1) already for specific spaces, the tools to give an answer (at least a positive one) in the general case are probably not developed yet. It also seems that (Q1) is the type of question which will not be answered directly, but by solving first some other, more structure theoretical questions.

Following more along the line of the concept of HI spaces we turn therefore to the following "easier question" (for a positive answer) and ask whether or not it is always possible to define a non trivial operator on a subspace.

(Q2) Assume X is an infinite dimensional Banach space. Is there a closed subspace Y of X and an operator $T : Y \to X$, so that T is not of the form $T = \lambda\text{In}_{(Y,X)} + C$, where λ is a scalar and $C : Y \to X$ is a compact operator?

It is easy to see, that (Q2) can be equivalently reformulated as follows

(Q3) Assume that X is an infinite dimensional Banach space. Does there exist two normalized basic sequences (x_n) and (y_n) so that (x_n) dominates (y_n) and so that $\inf_{n\in\mathbb{N}} \|x_n - y_n\| > 0$?

We want to go one step further and formulate a more structure theoretical approach to our problem. We call a Banach space X with a basis (e_i) *a space of Class 1* if:

(C1) Every block basis of (e_i) has a subsequence which is equivalent to some subsequence of (e_i).

REMARK. Note that the spaces ℓ_p, $1 \le p < \infty$, and c_0 are clearly (C1). Actually in ℓ_p, $1 \le p < \infty$, and c_0 all blockbasis are equivalent to each other, a property which characterizes the unit bases of ℓ_p, $1 \le p < \infty$, and c_0 [**Z**]. Moreover, in [**CJT**] it was shown that Tsirelson's space T (as described in [**FJ**]) as well as its dual T^* (the original *Tsirelson space* defined in [**T**]) are spaces of Class 1. Recently the result of [**CJT**] was generalized to all *finitely mixed Tsirelson spaces* in [**LT**]. The reader unfamiliar with the usual notations and concepts (like mixed Tsirelson spaces) is referred to the last paragraphs of this section, where all the notions of this paper are introduced.

It seems that until the early nineties all known Banach spaces had subspaces which were of Class 1. Then, in 1991, the author [**Sch**] of this work constructed a space, nowadays denoted by S, which fails (C1) in an extreme way: In all infinite dimensional subspaces, spanned by a block, one is not only able to find two normalized blocks (x_n) and (y_n) having no subsequences which are equivalent to each other. But one can even find in each subspace two normalized blocks (x_n) and (y_n), for which the map $x_n \mapsto y_n$ extends to a bounded linear and strictly singular operator between the span of (x_n) and the span of (y_n).

Let us therefore define the following second class of Banach spaces. Let us call a Banach space X with a basis (e_i) a space of Class 2 if:

(C2) Each block basis (z_n) has two further block bases (x_n) and (y_n) so that the map $x_n \mapsto y_n$ extends to a bounded linear and strictly singular operator between the span of (x_n) and the span of (y_n).

The main purpose of this paper is to establish criteria for a Banach space to be of Class 2 and thereby address the following problem.

(Q4) Does every infinite dimensional Banach space contain an infinite dimensional Banach space which is either of Class 1 or of Class 2?

At first sight, Question (Q4) may seem quite a daring one, so let us motivate it. If one believes (Q2) to have a positive answer, one could argue that experience teaches us that positive results in Banach space theory are often derived from dichotomy principles. Therefore (C2) could be a candidate for the second alternative in a dichotomy result in which (C1) is the first alternative.

Secondly, note that within the class of complex (HI) spaces (Q2) is equivalent to (Q4). Indeed, if (Q2) is true for all (HI) spaces and (x_i) is a block basis of an (HI) space X with a basis, then we can argue as follows. Since the closed span Z of (x_i) is an (HI) space and we assumed (Q2) to be true for all (HI) spaces we deduce that there exists a subspace Y of Z and a nontrivial operator $T : Y \to Z$. By subtracting a multiple of the inclusion of Y into Z, if necessary, we may assume that T is strictly singular but not compact. Using the usual perturbation arguments and taking differences, if need be, we conclude that T maps a normalized block basis (z_n) of (x_n) to a seminormalized sequence (y_n), which is also a block basis of (x_n). Thus, the map $z_n \mapsto y_n$ extends to a bounded-linear and strictly singular operator.

Thirdly, if one believes Question (Q2) to have a counterexample, one might try to look first for a counterexample of (Q4). Problem (Q4) could have, contrary to (Q2), a counterexample with an unconditional basis, which probably will be easier to define. Then, starting with a counterexample to (Q4), one might look for a modification thereof to obtain a counterexample to question (Q2). In this context, it is worthwhile to recall that the key argument towards defining the first known (HI) space was, in fact, the definition of a space of Class 2 having an unconditional basis.

Finally note that a negative answer of (Q3) is equivalent to the statement that the following class of Banach spaces is not empty. A Banach space X with a basis (e_i) is said to be *a space of Class 3* if:

(C3) For any two block bases (x_n) and (z_n) of (e_i), with $\inf_{n\in\mathbb{N}} \|x_n - z_n\| > 0$, neither the map $x_n \mapsto z_n$ nor the map $z_n \mapsto x_n$ extends to a linear bounded operator between the spans of (x_n) and (z_n).

In this paper we are interested in formulating criteria which imply that a given Banach space X is of Class 2. Therefore we want to find sufficient conditions for X to contain two seminormalized basic sequences (x_n) and (y_n) for which the map $x_n \mapsto y_n$ extends to a bounded linear and strictly singular operator.

In Section 2 we discuss the following problem: Given two normalized basic sequences (x_n) and (y_n), which conditions should we impose on the spreading models of (x_n) and (y_n), in order to insure the existence of subsequences $(\tilde{x}_n)$ and $(\tilde{y}_n)$ of (x_n) and (y_n) respectively, so that $\tilde{x}_n \mapsto \tilde{y}_n$ extends to a strictly singular and bounded linear operator? The properties of spreading models can be quite different form the properties of the underlying generating sequences. For example, the spreading model of the basis of Schreier's space (using the notations of Definition 1.6 this is the space $S(\mathcal{S}_1)$, with $\mathcal{S}_1 = \{E \in [\mathbb{N}]^{<\infty} : \#E \le \min E\}$) is isometrically equivalent to the ℓ_1-unit vector basis. On the other hand Schreier's space is hereditarily c_0. Therefore we expect to need to impose rather strong conditions on the spreading models of (x_n) and (y_n) in order to conclude that for some subsequences (m_k), (n_k) and $\mathbb{N}$ it follows that $x_{m_k} \mapsto y_{n_k}$ extends to bounded linear and strictly singular map $[x_{m_k} : k \in \mathbb{N}] \to [y_{n_k} : k \in \mathbb{N}]$.

The main result of Section 2 is the following answer to our question.

THEOREM 1.1. *Let(x_n) and (y_n) be two normalized weakly null sequences having spreading models (e_n) and (f_n) respectively.*

Assume that (e_n) is not equivalent to the c_0-unit vector basis and that the following condition holds.

(1) *There is a sequence (δ_n) of positive numbers decreasing to zero so that*

$$\Big\|\sum_{i=1}^{\infty} a_i f_i\Big\| \le \max_{\substack{n\in\mathbb{N}\\ i_1<i_2<\dots i_n}} \delta_n \Big\|\sum_{j=1}^{n} a_{i_j} e_i\Big\|.$$

Then there are subsequences $(\tilde{x}_n)$ and $(\tilde{y}_n)$ of (x_n) and (y_n) respectively, so that $\tilde{x}_n \mapsto \tilde{y}_n$ extends to a strictly singular and bounded linear operator.

In order to formulate our second result, we need to recall the Cantor Bendixson index of subsets of $[\mathbb{N}]^{<\infty}$, the set of finite subsets of $\mathbb{N}$.

A set $\mathcal{A} \subset [\mathbb{N}]^{<\infty}$ is called *hereditary* if for all $A \in \mathcal{A}$ and all $B \subset A$ it follows that $B \in \mathcal{A}$. We call $\mathcal{A}$ *spreading* if $A = \{a_1, a_2, \dots a_k\} \in \mathcal{A}$, with $a_1 < a_2 < \dots < a_k$, and if $b_1 < b_2 < \dots b_k$ are in $\mathbb{N}$, so that $b_i \ge a_i$ for $i = 1, \dots k$, then $\{b_1, \dots b_k\} \in \mathcal{A}$. We always consider the topology of pointwise convergence on $[\mathbb{N}]^{<\infty}$(identifying $A \subset \mathbb{N}$ with χ_A, the characteristic function of A).

For an hereditary and compact $\mathcal{A} \subset [\mathbb{N}]^{<\infty}$ we define

$$\begin{aligned}(2)\qquad \mathcal{A}^{(1)} &= \{A \in \mathcal{A} | A \text{ is accumulation point of } \mathcal{A}\}\\ &= \{A \in \mathcal{A} | \exists N \subset \mathbb{N}, N \text{ infinite } \forall n \in N \quad A \cup \{n\} \in \mathcal{A}\}.\end{aligned}$$

Since $\mathcal{A}$ is compact in $[\mathbb{N}]^{<\infty}$ each $A \in \mathcal{A}$ is subset of a maximal element of $\mathcal{A}$ and therefore we conclude that

$$(3)\qquad \mathcal{A}^{(1)} \subsetneqq \mathcal{A}, \text{ if } \mathcal{A} \ne \emptyset.$$

Secondly it follows that $\mathcal{A}^{(1)}$ is also compact and hereditary. We can therefore define $\mathcal{A}^{(\alpha)}$ for each $\alpha < \omega_1$ by transfinite induction. We let $\mathcal{A}^{(0)} = \mathcal{A}$ and assuming we defined $\mathcal{A}^{(\beta)}$ for all $\beta < \alpha$ we put

$$\mathcal{A}^{(\alpha)} = \big(\mathcal{A}^{(\beta)}\big)^{(1)} \text{ if } \alpha = \beta + 1, \text{ and } \mathcal{A}^{(\alpha)} = \bigcap_{\beta<\alpha} \mathcal{A}^{(\beta)} \text{ if } \alpha = \sup_{\beta<\alpha} \beta. \tag{4}$$

Since $[\mathbb{N}]^{<\infty}$ is countable it follows from (3) that there is an $\alpha < \omega_1$ for which $\mathcal{A}^{(\alpha)}$ is empty and we define the Cantor Bendixson index of $\mathcal{A}$ by

$$\mathrm{CB}(\mathcal{A}) = \min\{\alpha < \omega_1 : \mathcal{A}^{(\alpha)} = \emptyset\}. \tag{5}$$

We note that $\mathrm{CB}(\mathcal{A})$ is a successor ordinal. Indeed, assume that $\alpha < \omega_1$ is a limit ordinal and that $\mathcal{A}^{(\beta)} \neq \emptyset$ for all $\beta < \alpha$. Then it follows from the fact that $\mathcal{A}^{(\beta)}$ is hereditary that $\emptyset \in \mathcal{A}^{(\beta)}$ for all $\beta < \alpha$, and, thus, by (4) that $\emptyset \in \mathcal{A}^{(\alpha)}$ which implies that $\mathrm{CB}(\mathcal{A}) > \alpha$.

The *strong Cantor Bendixson index* of $\mathcal{A}$ is defined as follows (cf. [**AMT**]):

$$\mathrm{CBS}(\mathcal{A}) = \sup_{N\subset\mathbb{N},\#N=\infty}\ \inf_{M\subset N,\#M=\infty} \mathrm{CB}\big(\mathcal{A}\cap[M]^{<\infty}\big) \tag{6}$$

Remark. Note that $\mathrm{CBS}(\mathcal{A})$ could be a limit ordinal. Indeed, let $(N_n)_{n\in\mathbb{N}}$ be a sequence of infinite pairwise disjoint subsets of $\mathbb{N}$ whose union is $\mathbb{N}$. Take $\mathcal{A} = \bigcup_{n\in\mathbb{N}} \mathcal{A}_n$, with $\mathcal{A} := \{A \subset N_n \text{ finite } : \#A = n\}$. Then it follows that $\mathrm{CBS}(\mathcal{A}) = \omega$.

On the other hand it is clear that for a successor ordinal γ and an hereditary $\mathcal{A} \subset [\mathbb{N}]^{<\infty}$ we have that

$$\mathrm{CBS}(\mathcal{A}) \geq \gamma \iff \exists N \subset \mathbb{N}, \#N = \infty \forall M \subset N, \#M = \infty \quad \mathrm{CB}(\mathcal{A}\cap[M]^{<\infty}) \geq \gamma.$$

Using the Cantor Bendixson index we can characterize spaces which have subspaces of Class 1 and quantify the property of not having a subspace of Class 1. More generally, we will use the Cantor Bendixson index to measure "how far away" two basic sequences are to each other, up to passing to subsequences".

Definition 1.2. Assume that $\overline{x} = (x_n)$ and $\overline{z} = (y_n)$ are two seminormalized basic sequences. For $c \geq 1$ we define

$$\mathcal{C}(\overline{x},\overline{z},c) = \{A \in [\mathbb{N}]^{<\infty} : \exists B \in [\mathbb{N}]^{<\infty}, \#B = \#A, \quad (x_n)_{n\in A} \sim_c (z_n)_{n\in B}\}, \tag{7}$$

by "$(x_n)_{n\in A} \sim_c (z_n)_{n\in B}$" we mean that $(x_n)_{n\in A}$ and $(z_n)_{n\in B}$ are c-equivalent (see end of section), where on A and B we consider the order given by $\mathbb{N}$. For $\gamma < \omega_1$ let (put $\inf\emptyset = \infty$)

$$\begin{aligned} c(\overline{x},\overline{z},\gamma) &= \inf\{c \geq 1 : \mathrm{CBS}\big(\mathcal{C}(\overline{x},\overline{z},c)\big) \geq \gamma + 1\} \text{ (with } \inf\emptyset = \infty) \\ &= \inf\{c \geq 1 : \exists N \overset{\infty}{\subset} \mathbb{N} \forall M \overset{\infty}{\subset} N \quad \mathrm{CB}\big(\mathcal{C}(\overline{x},\overline{z},c)\cap[M]^{<\infty}\big) \geq \gamma + 1\} \end{aligned} \tag{8}$$

$$\gamma_0(\overline{x},\overline{z}) = \sup\{\gamma < \omega_1 : c(\overline{x},\overline{z},\gamma) < \infty\}. \tag{9}$$

In the case that $\overline{z}$ is the unit basis of ℓ_1 we define $\mathcal{B}(\overline{x},c) = \mathcal{C}(\overline{x},\overline{z},c)$, $b(\overline{x},\gamma) = c(\overline{x},\overline{z},\gamma)$, and $\beta_0(\overline{x}) = \gamma_0(\overline{x},\overline{z})$ and for a Banach space X and $\beta < \omega_1$ we put

(10)

$$\begin{aligned} B(\beta) = B(\beta, X) &= \sup\{b(\overline{z},\beta) : \overline{z} = (z_n) \subset B_X \text{ semi normalized, weakly null}\} \\ &= \sup\left\{ b > 0 : \begin{array}{c} \exists (z_n) \subset B_X \quad \text{semi normalized and weakly null} \\ \mathrm{CBS}(\mathcal{B}(\overline{z},b))\} \geq \beta + 1 \end{array} \right\}. \end{aligned}$$

$$(11)\quad \beta_0(X) = \sup\{\beta < \omega_1 : B(\beta) > 0\}$$
$$= \sup\{\beta_0(\overline{z}) : \overline{z} = (z_n) \subset B_X \quad \text{semi normalized and weakly null}\}$$

It is clear that $\beta_0(X)$ must be a limit ordinal.

PROPOSITION 1.3. *(see proof after Corollary 4.9 in Section 4). Assume X is a Banach space with a basis (x_n). If (z_n) is another basic sequence for which $\gamma_0((z_n),(x_n)) = \omega_1$ then a subsequence of (z_n) is ismorphically equivalent to a subsequence of (x_n) or to the spreading model of a subsequence of (x_n).*

Moreover, X is of Class 1 if and only if for all blockbases (z_n) of (x_n) it follows that $\gamma_0((z_n),(x_n)) = \gamma_0((x_n),(z_n)) = \omega_1$.

In Section 6 we will prove the following result, implying a condition for a space to be in Class 2.

THEOREM 1.4. *Let X be a Banach space with a basis (e_i) not containing c_0. Assume that there is an ordinal $\gamma \in [0, \beta_0(X)]$ so that:*

a) *There is a semi normalized block basis $\overline{y} = (y_n) \subset B_X$ with $\beta_0(\overline{y}) \le \gamma$.*
b) $\inf_{\beta<\gamma} B(\beta, X) > 0$.

Then there is a seminormalized block basis (x_n) in X and a subsequence $(\tilde{y}_n)$ of (y_n) so that the map $x_n \mapsto \tilde{y}_n$ extends to a strictly singular operator $T : [x_n : n \in \mathbb{N}] \to [\tilde{y}_n : n \in \mathbb{N}]$.

COROLLARY 1.5. *Assume that Z is a reflexive Banach space with a basis and each block subspace of Z satisfies the condition of Theorem 1.4.*

Then Z is of Class 2.

REMARK. Note that in the statement of Theorem 1.4 either $\gamma = \beta_0(X)$ in which case it is clear that (a) is satisfied for all block bases. Or $\gamma < \beta_0(X)$ in which case it is clear that (b) is satisfied.

Also note that the assumption in Theorem 1.4 says the following:

On the one hand there is a block basis $\overline{y}$ which is "far away from the ℓ_1- unit vector basis, in the sense that given any $\varepsilon > 0$ there is an $\alpha < \gamma$ so that $b(\overline{y}, \alpha) < \varepsilon$, i.e. on sets of strong Cantor Bendixson index α the best equivalence constant between the ℓ_1 basis and $\overline{y}$ is at most ε.

On the other hand if we let $c = \inf_{\beta<\gamma} B(\beta, X)$, we can choose for any $\alpha < \gamma$ and any $\varepsilon > 0$ a block sequence $\overline{z}$ for which $b(\overline{z}, \alpha) > c - \varepsilon$.

In other words the assumption in Theorem 1.4 states that we require the existence of blockbases which have "different ℓ_1-behavior".

But if one wants to attack Problem (Q4) one could start by assuming that our given Banach space X has no infinite dimensional subspace of Class 1 and therefore by Proposition 1.3 every block basis (y_n) must admit a further block basis (z_n) so that $\gamma_0((y_n),(z_n)) < \omega_1$ or $\gamma_0((z_n),(y_n)) < \omega_1$. So, instead of going the quite "uneconomical route" (as in Theorem 1.4) and comparing two blockbases to the ℓ_1-unit vector basis one should compare them to each other directly.

Here exactly lies one major hurdle which one has to overcome before reaching further results. The proof of Theorem 1.1 as well as the proof of Theorem 1.4 make use of certain "partial unconditionality results" namely a result by E. Odell [**O2**] (see Theorem 2.2) and a result by S. A. Argyros, S. Mercourakis and A. Tsarpalias [**AMT**] (see Lemma 6.1 in Section 6). Both results give partial answers to the question, under which conditions a family of finite rank projections on certain subsets of a basis are uniformly bounded.

In order to make advances on (Q4) we would need to address the following question which asks for generalizations of the results in [**O2**] and [**AMT**].

(Q5) Assume that X is a Banach space with a basis (e_n), $\gamma \in [0, \omega_1)$, and assume that for all normalized block bases (x_n) of (e_n) it follows that $c((x_n),(e_n),\gamma) < \infty$ $c((e_n),(x_n),\gamma) < \infty$.

Secondly, consider for a blockbasis (x_n) and an $r > 0$ the set

$$\mathcal{U}((x_n), r) = \{A \in [\mathbb{N}]^{<\infty} : \|P_{[x_n:n\in A]}\| \leq r\},$$

where $P_{[x_n:n\in A]} : [x_n : n \in \mathbb{N}] \to [x_n : n \in A]$ is the defined to be the usual projection onto $[x_n : n \in A]$.

Now, does it follow that any block basis (x_n) has a subsequence (z_n) for which $\mathrm{CBS}(\mathcal{U}((z_n), r)) \geq \gamma$ for some $r > 0$?

The result cited from [**O2**] gives a positive answer if $\gamma = \omega$ and the result cited from [**AMT**] gives a positive answer if we replaced in the assumption $c((e_n),(x_n),\gamma)$ by $b((x_n),\gamma)$.

To be able to prove Theorem 1.4 we will introduce in Section 3 a family $(\mathcal{F}_\alpha)_{\alpha<\omega_1}$ of subsets of $[\mathbb{N}]< \infty$, having the property that for each $\alpha < \omega_1$ the strong Cantor Bendixson index is $\alpha + 1$ (see Corollary4.9). The family $(\mathcal{F}_\alpha)$ is defined in a similar fashion as the *Schreier families* $(\mathcal{S}_\alpha)_{\alpha<\omega_1}$ [**AA**], with one crucial difference: one obtains $\mathcal{F}_{\alpha+1}$ from $\mathcal{F}_\alpha$ by adding to each element of $A \in \mathcal{F}_\alpha$ at most one new element, therefore the step from $\mathcal{F}_\alpha$ to $\mathcal{F}_{\alpha+1}$ is much smaller than the step from $\mathcal{S}_\alpha$ to $\mathcal{S}_{\alpha+1}$.

Feeling that this family $(\mathcal{F}_\alpha)$ could be an important tool to analyse the combinatorics of blockbases we extensively discuss the properties of this family in the Sections 3, 4 and 5, in particular we prove a result which, roughly speaking, says that any hereditary set $\mathcal{A} \subset [\mathbb{N}]^{<\infty}$ restricted to some appropriate $M \overset{\infty}{\subset} \mathbb{N}$ is equal to "a version of $\mathcal{F}_\alpha$" (see Theorem 4.10). Based on this result we will be able to prove Theorem 1.4 in Section 6.

We will need the following notations and conventions.

For simplicity all our Banach spaces are considered to be over the field $\mathbb{R}$. If X is a Banach space B_X and S_X denotes the unit ball and the unit sphere respectively.

c_{00} denotes the vectorspace of sequences in $\mathbb{R}$ which eventually vanish and (e_i) denotes the usual vector basis of c_{00} and when we consider a Banach space X with a normalized basis, we think of X being the completion of some norm on c_{00} with (e_i) being that basis. For $x = \sum x_i e_i \in c_{00}$ we let $\mathrm{supp}(x) = \{i \in \mathbb{N} : x_i \neq 0\}$ be the *support of* x and if $E \subset \mathbb{N}$ we let $E(x) = \sum_{i\in E} x_i e_i$.

We say that a basic sequence (x_n) dominates another basic sequence (y_n) if the map $x_n \mapsto y_n$ extends to a linear bounded map between the span of (x_n) and the span of (y_n), i.e. if the there is a $c > 0$ so that for all $(a_i) \in c_{00}$ $\|\sum_{i=1}^n a_i y_i\| \leq c\|\sum_{i=1}^n a_i x_i\|$.

We say that (x_n) and (y_n) are c-equivalent, $c \geq 1$ and we write $(x_n) \sim_c (y_n)$ if for any $(a_i) \in c_{00}$ it follows that $\frac{1}{c}\|\sum_{i=1}^n a_i x_i\| \leq \|\sum_{i=1}^n a_i y_i\| \leq c\|\sum_{i=1}^n a_i x_i\|$.

The closed linear span of a subsequence $(x_n)_{n\in\mathbb{N}}$ of a Banach space is denoted by $[x_n : n\in\mathbb{N}]$.

Let E be a Banach space with a 1-spreading basis (e_i), i.e.

$$\Big\|\sum_{i=1}^m a_i e_i\Big\| = \Big\|\sum_{i=1}^m a_i e_{n_i}\Big\|,$$

whenever $m \in \mathbb{N}$, $(a_i)_{i=1}^m \subset \mathbb{R}$ and $n_1 < n_2 < \dots n_m$ are in $\mathbb{N}$,

and let (x_n) be a seminormalized basic sequence. We say that (e_i) is the spreading model of (x_n) if for any $k \in \mathbb{N}$ and any $(a_i)_{i=1}^k \subset \mathbb{R}$ it follows that

$$\lim_{n_1\to\infty} \lim_{n_2\to\infty} \dots \lim_{n_k\to\infty} \Big\|\sum_{i=1}^k a_i x_{n_i}\Big\| = \Big\|\sum_{i=1}^k a_i e_i\Big\|.$$

Recall [**BL**] that any seminormalized basic sequence has a subsequence with spreading model and that, using our notations in Definition 1.2, a seminormalized sequence (x_n) has a subsequence whose spreading model is isomorphic to (e_i) if and only if $c((x_n),(e_i),\omega) < \infty$).

For a set M the set of all finite subsets of M is denoted by $[M]^{<\infty}$ and the set of all infinite subsets is denoted by $[M]^\infty$. We write $M \overset{\infty}{\subset} N$ if M is an infinite subset of N.

DEFINITION 1.6. (Mixed Schreier and Tsirelson spaces)

For an hereditary, spreading and compact subset $\mathcal{F}$ of $[\mathbb{N}]^{<\infty}$ containing all singletons of $[\mathbb{N}]^{<\infty}$ we define *the $\mathcal{F}$-Schreier space $S(\mathcal{F})$* to be the completion of c_{00} under the norm defined by

$$(S(\mathcal{F})) \qquad \|x\| = \sup_{E\in\mathcal{F}} \sum_{i\in E} |x_i|, \text{ whenever } x = (x_i) \in c_{00}.$$

Let $(\mathcal{F}_n)_{n=1}^\infty$ be an increasing sequence of hereditary, spreading and compact subsets of $[\mathbb{N}]^{<\infty}$ and let $(\Theta_n) \subset (0,1]$ be a non increasing sequence. The *mixed Schreier space* $S((\Theta_n),(\mathcal{F}_n))$ is the completion of c_{00} under the norm defined by

$$(S((\Theta_n),(\mathcal{F}_n))) \qquad \|x\| = \sup_{n\in\mathbb{N}} \sup_{E\in\mathcal{F}_n} \Theta_n \sum_{i\in E} |x_i|, \text{ whenever } x = (x_i) \in c_{00}.$$

If in addition $(\ell_n)_{n\in\mathbb{N}}$ is a sequence in $\mathbb{N}$ increasing to ∞, we define the *mixed Schreier space with the additional admissibility condition given by $(\ell_n)_{n\in\mathbb{N}}$* to be the space $S((\Theta_n),(\mathcal{F}_n),(\ell_n))$ with the following norm

$$(S((\Theta_n),(\mathcal{F}_n),(\ell_n))) \qquad \|x\| = \sup_{n\in\mathbb{N}} \sup_{\ell_n \le \min E, E\in\mathcal{F}_n} \Theta_n \sum_{i\in E} |x_i|,$$

whenever $x = (x_i) \in c_{00}$.

If $\mathcal{F}$ of $[\mathbb{N}]^{<\infty}$ is hereditary, spreading and compact, and if $E_1 < E_2 < \dots E_\ell$ are in $\mathcal{F}$, we say that $(E_i)_{i=1}^\ell$ is *$\mathcal{F}$-admissible* if $\{\min E_i : i = 1,2\dots\ell\} \in \mathcal{F}$. The *mixed Tsirelson space* $T((\Theta_n),(\mathcal{F}_n))$ is the completion of c_{00} under the norm which is implicitly defined by

$(T((\Theta_n),(\mathcal{F}_n)))$

$$\|x\| = \max\Big\{ \sup_{n\in\mathbb{N}} |x_n|, \sup_{n\in\mathbb{N}} \sup_{(E_i)_{i=1}^n \ \mathcal{F}_n\text{-adm.}} \Theta_n \sum_{i=1}^n \|E_i(x)\| \Big\} \text{ for } x=(x_i)\in c_{00}.$$

A *finitely mixed Tsirelson space* is a space $T((\Theta_n),(\mathcal{F}_n))$ where $(\Theta_n) \subset (0,1]$ and $(\mathcal{F}_n)$ are finite sequences.

2. Conditions on the spreading models implying the existence of nontrivial operators

DEFINITION 2.1. For two normalized basic sequences (x_n) and (y_n) and an $\varepsilon > 0$ we define

$$\Delta_{((x_n),(y_n))}(\varepsilon) = \sup\left\{\left\|\sum a_i y_i\right\| : (a_i) \in c_{00},\ \max_{i\in\mathbb{N}} |a_i| \le \varepsilon, \text{ and } \left\|\sum a_i x_i\right\| \le 1\right\}.$$

We say that (x_n) *strongly dominates* (y_n) if

$$\lim_{n\to\infty} \inf_{A\subset\mathbb{N},\#A=n} \|\sum_{i\in A} x_i\| = \infty \text{ and} \tag{12}$$

$$\lim_{\varepsilon\searrow 0} \Delta_{((x_n),(y_n))}(\varepsilon) = 0. \tag{13}$$

REMARK. Assume (x_n) and (y_n) are two normalized basic sequences.

a) If $\Delta_{((x_n),(y_n))}(\varepsilon) < \infty$ for some $\varepsilon > 0$ (and thus for all $\varepsilon > 0$) then (x_n) dominates (y_n), i.e. the map $x_n \mapsto y_n$ extends to a bounded linear operator.
b) Note that in the case that (x_n) is subsymmetric condition (12) means that (x_n) is not equivalent to the c_0 unit vector basis.
c) No normalized basic sequence strongly dominates the unit vector basis in ℓ_1.

We will need the following result by E. Odell on *Schreier unconditionality.*

THEOREM 2.2. **[O2]** *Let (x_n) be a normalized weakly null sequence in a Banach space and $\eta > 0$. Then (x_n) contains a subsequence $(\tilde{x}_n)$ so that*

$$\forall(\alpha_i) \in c_{00} \forall F \subset \mathbb{N}, \min F \ge \#F : \quad \left\|\sum_{i\in F} \alpha_i \tilde{x}_i\right\| \le (2+\eta)\left\|\sum_{i=1}^{\infty} \alpha_i \tilde{x}_i\right\|.$$

If (x_n) is a bimonotone sequence above factor $(2+\eta)$ could be replaced by $(1+\eta)$.

REMARK. Assume that (x_n) and (y_n) are normalized basic sequence, that (x_n) strongly dominates (y_n) and that either (x_n) is weakly null or 1-subsymmetric.

By passing to subsequences (of (x_n) and (y_n) simultaneously) we can assume that (x_n) has a spreading model (e_i) and that the conclusion of Theorem 2.2 holds, i.e. that

$$\|\sum_{i\in F} \alpha_i e_i\| \sim_2 \|\sum_{i\in F} \alpha_i x_i\| \le 3\|\sum_{i=1}^{\infty} \alpha_i x_i\| \tag{15}$$

for all $(\alpha_i) \in c_{00}$ and finite $F \subset \mathbb{N}$, with $\min F \ge \#F$.

Since (x_n) satisfies (12) (e_n) cannot be equivalent to the c_0-unit basis and we deduce therefore that

$$c_n = \|\sum_{i=1}^{n} e_i\| \nearrow \infty. \tag{16}$$

We define now the following basic sequence $(\tilde{y})$ which dominates (y_n) by

$$\|\sum_{i=1}^{\infty} \alpha_i \tilde{y}_i\| = \|\sum_{i=1}^{\infty} \alpha_i y_i\| + \max_n \frac{1}{\sqrt{c_n}} \max_{k_1<\ldots k_n} \|\sum_{i=1}^{n} \alpha_{k_i} e_i\| \text{ whenever } (\alpha_i) \in c_{00}. \tag{17}$$

PROPOSITION 2.3. *Using the assumptions and notations as in above Remark it follows that*

a) *(x_n) strongly dominates $(\tilde{y}_n)$.*
b) *For $\eta > 0$ it follows that*

$$\ell_\eta = \{\#\{i \in \mathbb{N} : |\alpha_i| \geq \eta\} : (\alpha_i) \in c_{00}, \|\sum \alpha_i \tilde{y}_i\| \leq 1\} < \infty.$$

PROOF. In order to show (a) let $\eta > 0$. Since (x_n) strongly dominates (y_n) we can choose an $\varepsilon' > 0$ so that $\Delta_{((x_n),(y_n))}(\varepsilon') < \eta/2$. Secondly we choose $n_0 \in \mathbb{N}$ so that $1/\sqrt{c_{n_0}} < \eta/12$ and we let $\varepsilon = \min(\varepsilon', \eta/2n_0)$.

If $(\alpha_i) \in c_{00}$ with $\|\sum \alpha_i x_i\| \leq 1$ and $\max |\alpha_i| \leq \varepsilon$ we deduce for $n \in \mathbb{N}$, with $n \leq n_0$, that

$$\frac{1}{\sqrt{c_n}} \max_{n \leq k_1 < \ldots k_n} \|\sum_{i=1}^{n} \alpha_i e_i\| \leq n\varepsilon \leq \eta/2.$$

If $n \in \mathbb{N}$, with $n \geq n_0$, we deduce from (15) that

$$\frac{1}{\sqrt{c_n}} \max_{n \leq k_1 < \ldots k_n} \left\|\sum_{i=1}^{n} \alpha_{k_i} e_i\right\| \leq \frac{2}{\sqrt{c_n}} \max_{n \leq k_1 < \ldots k_n} \left\|\sum_{i=1}^{n} \alpha_{k_i} x_i\right\| \leq \frac{6}{\sqrt{c_n}} \left\|\sum_{i=1}^{\infty} \alpha_i x_i\right\| \leq \eta/2,$$

which together with the choice of ε' implies that $\|\sum \alpha_i \tilde{y}_i\| \leq eta$ and finishes the proof of the claim in (a).

To show (b), let $(\alpha_i) \in c_{00}$ so that $\|\sum \alpha_i \tilde{y}_i\| \leq 1$ and let $\eta > 0$. We put $I = \{i \in \mathbb{N} : \|\alpha_i\| > \eta\}$ and let I' the set of the $\lceil \#I/2 \rceil$ largest elements of I. Note that $\#I' \leq \min I$ which implies by our definition of $(\tilde{y}_n)$ and the fact that (e_i) suppression 1-unconditional (thus 2-unconditional) that

$$1 \geq \left\|\sum \alpha_i \tilde{y}_i\right\| \geq \frac{1}{\sqrt{c_{\#I'}}} \left\|\sum_{i \in I'} \alpha_i e_i\right\| \geq \frac{1}{2\sqrt{c_{\#I'}}} \left\|\sum_{i \in I'} \eta e_i\right\| \geq \eta \sqrt{c_{\#I'}}/2,$$

which implies that $\ell_\eta \leq 2 \max\{n \in \mathbb{N} : \sqrt{c_n} \geq 2/\eta\}$ and finishes the claim (b).

Finally, (c) follows easily from (b). □

LEMMA 2.4. *Let (e_i) be a suppression 1-unconditional, 1-subsymmetric and normalized basis.*

For a normalized, basic and weakly null sequence (y_n) the following statements are equivalent.

a) *(y_n) has a subsequence (z_n) which is strongly dominated by (e_i).*
b) *(y_n) has a subsequence (z_n) having a spreading model (f_i) which is strongly dominated by (e_i).*
c) *(y_n) has a subsequence (z_n) for which there is a sequence $(\delta_n) \subset (0, \infty)$, with $\delta_n \searrow 0$, if $n \nearrow \infty$, so that*

$$\left\|\sum_{i=1}^{\infty} \alpha_i z_i\right\| \leq \max_{n \in \mathbb{N}} \delta_n \max_{j_1 < j_2 < \ldots j_n} \left\|\sum_{i=1}^{n} \alpha_{j_i} e_i\right\|, \quad \textit{whenever } (\alpha_j) \in c_{00}.$$

d) *(y_n) has a subsequence (z_n) for which there is a sequence $(\varepsilon_n) \subset (0, \infty)$, with $\varepsilon_n \searrow 0$, if $n \nearrow \infty$, and a subsequence (ℓ_n) of $\mathbb{N}$ so that*

$$\left\|\sum_{i=1}^{\infty} \alpha_i z_i\right\| \leq \max_{n \in \mathbb{N}} \varepsilon_n \max_{n \leq j_1 < j_2 < \ldots j_{\ell_n}} \left\|\sum_{i=1}^{\ell_n} \alpha_{j_i} e_i\right\|, \quad \textit{whenever } (\alpha_j) \in c_{00}.$$

PROOF. (a)⇒(b) Assume (z_n) is a subsequence of (y_n) which is strongly dominated by (e_i). By passing to a further subsequence we can assume that (z_n) has a spreading model (f_i), being subsymmetric. If $\varepsilon > 0$ and $(\alpha_i)_{i=1}^k \subset \mathbb{R}$, with $\|(\alpha_i)\|_\infty \leq \varepsilon$, and $\|\sum_{i=1}^k \alpha_i e_i\| \leq 1$ it follows that

$$\left\|\sum_{i=1}^k \alpha_i f_i\right\| = \lim_{n_1\to\infty} \ldots \lim_{n_k\to\infty} \left\|\sum_{i=1}^k \alpha_i y_{n_i}\right\| \leq \Delta_{((e_i),(y_i))}(\varepsilon),$$

which implies (b).
(b)⇒(a) Let (f_i) be the spreading model of a subsequence (u_i) of (y_i). We first note that it is enough to show that for a fixed $\eta > 0$ we need to find a subsequence (v_i) of (u_i) and an $\varepsilon = \varepsilon(\eta)$ so that $\Delta_{((e_i),(y_i))}(\varepsilon) < \eta$. Then a standard diagonal argument using a sequence (η_n) decreasing to 0 will show that a subsequence of (u_i) is strongly dominated by (e_i).

For a fixed $\eta > 0$ we first choose $(\varepsilon_n) \subset (0,1)$ so that $\sum_{n\in\mathbb{N}} \varepsilon_n n < \eta/2$ and so that $\sum_{n\in\mathbb{N}} \Delta_{(e_i),(f_i)}(\varepsilon_n) < \varepsilon/4$. Since (e_n) satisfies condition (12) it follows that $\ell_n = \min\{\ell \in \mathbb{N} : \|\sum_{i=1}^\ell e_i\| \geq 1/\varepsilon_{n+1}\}$ is finite for every $n \in \mathbb{N}$. By passing again to a subsequence of we can assume that

$$\left\|\sum_{i=1}^{\ell_n} \alpha_i f_i\right\| \sim_2 \left\|\sum_{i=1}^{\ell_n} \alpha_i u_{k_i}\right\|, \tag{18}$$

whenever $n \leq k_1 < k_2 < \ldots k_{\ell_n}$ and $(\alpha_i)_{i=1}^{\ell_n} \subset \mathbb{R}$.

If $(\alpha_i) \in c_{00}$ with $\|\sum \alpha_i e_i\| \leq 1$ and $\max_{i\in\mathbb{N}} |\alpha_i| \leq \varepsilon_1$ we deduce that

$$\begin{aligned}\left\|\sum \alpha_i u_i\right\| &\leq \sum_{n=1}^\infty \left\|\sum_{i\in\mathbb{N},\varepsilon_{n+1}<|\alpha_i|\leq\varepsilon_n} \alpha_i u_i\right\| \leq \sum_{n=1}^\infty n\varepsilon_n + \left\|\sum_{n\leq i,\varepsilon_{n+1}<|\alpha_i|\leq\varepsilon_n} \alpha_i u_i\right\| \\ &\leq \sum_{n=1}^\infty n\varepsilon_n + 2\left\|\sum_{n\leq i,\varepsilon_{n+1}<|\alpha_i|\leq\varepsilon_n} \alpha_i f_i\right\| \leq \sum_{n=1}^\infty n\varepsilon_n + 2\Delta_{((e_i)(f_i))}(\varepsilon_n) < \eta,\end{aligned}$$

(for the third inequality note that since $\|\sum \alpha_i e_i\| \leq 1$ it follows that $\#\{i : |\alpha_i| \geq \varepsilon_{n+1}\} \leq \ell_n$) which finishes the proof of (b)⇒(a).
(a)⇒(c) Assume that (y_i) is strongly dominated by (e_i). Since we could replace (y_n) by $(\tilde{y}_n)$ as in the remark after Theorem 2.2 and apply Proposition 2.3 we can assume that (y_n) satisfies the properties (a) and (b) of Proposition 2.3. After a renorming we can also assume that (e_i) dominates (y_i) with constant 1, i.e. that we have $\|\sum \alpha_i y_i\| \leq \|\sum \alpha_i e_i\|$ for all $(\alpha_i) \in c_{00}$. Choose a strictly to 0 decreasing sequence $(\varepsilon_i)_{i=0}^\infty$ with $\varepsilon_0 = 1$ and $\Delta_{((e_i),(y_i))}(\varepsilon_i 2^{i+1}) \leq 2^{-i-1}/i$ and define for $i \in \mathbb{N}_0$

$$K_i = \ell_{\varepsilon_{i+1}} = \max\left\{\#\{i : |\alpha_i| > \varepsilon_{i+1}\} : (\alpha_i) \in c_{00}, \|\sum \alpha_i y_i\| \leq 1\right\}.$$

Finally, choose for $i \in \mathbb{N}$ $\delta_i = 2\Delta_{((e_i),(y_i))}(1)$ if $i \in \{1, 2, \ldots K_0\}$ and $\delta_i = 1/n$ if $i \in \{K_{n-1}+1, K_{n-1}+2, \ldots, K_n\}$ for some $n \in \mathbb{N}$.

Assume now that $(\alpha_i) \in c_{00}$ with $\|\sum_{i=1}^\infty \alpha_i y_i\| = 1$. Therefore there must be an $n \in \mathbb{N}_0$ for which

$$\left\|\sum_{i\in\mathbb{N},\varepsilon_{n+1}<|\alpha_i|\leq\varepsilon_n} \alpha_i y_i\right\| \geq 2^{-n-1} = 2^{-n-1}\left\|\sum_{i=1}^\infty \alpha_i y_i\right\|.$$

We deduce that

$$\begin{aligned}
\left\|\sum_{i=1}^{\infty}\alpha_i y_i\right\| &\le 2^{n+1}\left\|\sum_{i\in\mathbb{N},\varepsilon_{n+1}<|\alpha_i|\le\varepsilon_n}\alpha_i y_i\right\|\\
&= 2^{n+1}\left\|\sum_{i\in\mathbb{N},\varepsilon_{n+1}<|\alpha_i|\le\varepsilon_n}\alpha_i e_i\right\|\left\|\sum_{i\in\mathbb{N},\varepsilon_{n+1}<|\alpha_i|\le\varepsilon_n}\frac{\alpha_i}{c}y_i\right\|\\
&\left[\text{with } c=\|\sum_{i\in\mathbb{N},\varepsilon_{n+1}<|\alpha_i|\le\varepsilon_n}\alpha_i e_i\|\right]\\
&\le 2^{n+1}\left\|\sum_{i\in\mathbb{N},\varepsilon_{n+1}<|\alpha_i|\le\varepsilon_n}\alpha_i e_i\right\|\Delta_{((e_i)(y_i))}(\varepsilon_n 2^{n+1})\\
&\left[\text{Note: } c=\|\sum_{i\in\mathbb{N},\varepsilon_{n+1}<|\alpha_i|\le\varepsilon_n}\alpha_i e_i\|\ge\|\sum_{i\in\mathbb{N},\varepsilon_{n+1}<|\alpha_i|\le\varepsilon_n}\alpha_i y_i\|\ge 2^{-n-1}\right]\\
&\le \max_{j_1<j_2<\dots j_{K_n}}\left\|\sum_{i=1}^{K_n}\alpha_{j_i}e_i\right\|\cdot\begin{cases}2\Delta_{((e_i),(y_i))}(1) & \text{if } n=0\\ 1/n & \text{if } n\ge 1\end{cases}\\
&[\text{by choice of } K_n]\\
&= \delta_{K_n}\max_{j_1<j_2<\dots<j_{K_n}}\left\|\sum_{i=1}^{K_n}\alpha j_i e_i\right\|\le\max_{k\in\mathbb{N}}\delta_k\max_{j_1<j_2<\dots<j_k}\left\|\sum_{i=1}^{k}\alpha j_i e_i\right\|.
\end{aligned}$$

which concludes this part of the proof.
(c)⇒(d) Let (z_n) and (δ_n) be given as in (c). For $k\in\mathbb{N}$, we let $\tilde{\ell}_k=\max\{\ell:\delta_\ell>2^{-k^2}\}$.

We can assume that $\tilde{\ell}_k\ge k$, by passing to a slower decreasing sequence $\tilde{\delta}_k$, if necessary. We also can assume that (x_i) is a monoton basis.

Then it follows from (c) for $(\alpha_i)\in c_{00}$ that there is an $\ell\in\mathbb{N}$ so that

$$\begin{aligned}
\left\|\sum_{i=1}^{\infty}\alpha_i z_i\right\| &\le \delta_\ell\max_{j_1<j_2<\dots j_\ell}\left\|\sum_{i=1}^{\ell}\alpha_{j_i}e_i\right\|\\
&\le\begin{cases}\delta_1\max_{j_1<j_2<\dots j_{\tilde{\ell}_1}}\left\|\sum_{i=1}^{\tilde{\ell}_1}\alpha_{j_i}e_i\right\| & \text{if } \ell\le\tilde{\ell}_1\\[2ex]
2^{-k^2}\max_{j_1<j_2<\dots j_{\tilde{\ell}_{k+1}}}\left\|\sum_{i=1}^{\tilde{\ell}_{k+1}}\alpha_{j_i}e_i\right\| & \text{if } \ell\in[\tilde{\ell}_k+1,\tilde{\ell}_{k+1}]\\
& \text{for some } k\in\mathbb{N}.\end{cases}
\end{aligned}$$

Therefore the claim follows in the case that $\ell\le\tilde{\ell}_1$ if we put $\varepsilon_1=\delta_1$ and $\ell_1=\tilde{\ell}_1$. If $\ell>\tilde{\ell}_1$ we can find a $k\in\mathbb{N}$ so that

$$\left\|\sum_{i=1}^{\infty}\alpha_i z_i\right\|\le 2^{-k^2}\max_{j_1<j_2<\dots j_{\tilde{\ell}_{k+1}}}\left\|\sum_{i=1}^{\tilde{\ell}_{k+1}}\alpha_{j_i}e_i\right\| \text{ and }\left\|\sum_{i=1}^{\infty}\alpha_i y_i\right\|>2^{-(k-1)^2}\left\|\sum_{i=1}^{\tilde{\ell}_k}\alpha_{j_i}e_i\right\|.$$

Thus we conclude

$$\begin{aligned}\Big\|\sum_{i=1}^{\infty}\alpha_i z_i\Big\| &\le 2^{-k^2}\Bigg[\max_{j_1<j_2<\dots j_{\tilde\ell_{k+1}}}\Big\|\sum_{i=k}^{\tilde\ell_{k+1}}\alpha_{j_i}e_i\Big\|+\Big\|\sum_{i=1}^{k-1}\alpha_{j_i}e_i\Big\|\Bigg]\\ &\le 2^{-k^2}\Bigg[\max_{j_1<j_2<\dots j_{\tilde\ell_{k+1}}}\Big\|\sum_{i=k}^{\tilde\ell_{k+1}}\alpha_{j_i}e_i\Big\|+\Big\|\sum_{i=1}^{\tilde\ell_k}\alpha_{j_i}e_i\Big\|\Bigg]\\ &\le 2^{-k^2}\max_{k\le j_1<j_2<\dots j_{\tilde\ell_{k+1}}}\Big\|\sum_{i=1}^{\tilde\ell_{k+1}}\alpha_{j_i}e_i\Big\|+2^{-k^2}2^{(k-1)^2}\Big\|\sum_{i=1}^{\infty}\alpha_i z_i\Big\|\end{aligned}$$

which implies that

$$\big[1-2^{-k^2}2^{(k-1)^2}\big]\Big\|\sum_{i=1}^{\infty}\alpha_i z_i\Big\|\le 2^{-k^2}\max_{k\le j_1<j_2<\dots j_{\tilde\ell_{k+1}}}\Big\|\sum_{i=1}^{\tilde\ell_{k+1}}\alpha_{j_i}e_i\Big\|$$

and finishes the proof of our claimed implication if we choose $\varepsilon_k = 2^{-k^2}/[1-2^{-k^2}2^{(k-1)^2}]$ and $\ell_k=\tilde\ell_{k+1}$.

(d)$\Rightarrow$(a) Assume (z_n), (ε_n) and (ℓ_n) are chosen as required in (d). Let $\eta>0$, choose $n_0\in\mathbb{N}$ so that $\varepsilon_{n_0}\le\eta/2$ and choose $\varepsilon<\eta/2n_0$.

If now $(\alpha_i)\in c_{00}$, $\max_i|\alpha_i|<\varepsilon$, and $\|\sum\alpha_i e_i\|\le 1$ then it follows from condition (d) that

$$\Big\|\sum_{i=1}^{\infty}\alpha_i z_i\Big\|\le\max_{n\in\mathbb{N},n\le j_1<\dots j_{\ell_n}}\varepsilon_n\Big\|\sum_{i=1}^{\ell_n}\alpha_{j_i}e_i\Big\|\le\eta n_0+\max_{n\ge n_0,n\le j_1<\dots j_{\ell_n}}\varepsilon_n\Big\|\sum_{i=1}^{\ell_n}\alpha_{j_i}e_i\Big\|<\eta,$$

which implies that $\Delta_{(e_i),(z_i)}(\varepsilon)<\eta$ and proves the claim and finishes the proof of the Lemma. □

Proof of Theorem 1.1. Using Lemma 2.4 (c)$\Rightarrow$(a) for the sequences (f_i) and (e_i) and then Lemma 2.4 (b)$\Rightarrow$(a) for the sequences (y_i) and (e_i) we can assume, by passing to a subsequence of (y_n) if necessary, that (e_i) strongly dominates (y_i) and by passing to the sequence $(\tilde y_n)$ as defined in the remark after Theorem 2.2 we can assume that (a) and (b) of Proposition 2.3 are satisfied.

Using Lemma 2.4 we can again pass to a subsequence, still denoted by (y_n) for which we find sequences $(\tilde\delta_n)\subset(0,\infty)$ and $\ell_n\subset\mathbb{N}$, with $\delta_n\searrow 0$ and $\ell_n\nearrow\infty$ for $n\nearrow\infty$, so that

$$\Big\|\sum\alpha_i y_i\Big\|\le\frac{1}{2}\max_k\tilde\delta_k\max_{k\le n_1<n_2<\dots n_k}\Big\|\sum_{i=1}^{k}\alpha_{n_i}e_i\Big\|,\ \text{whenever } (\alpha_i)\in c_{00}.$$

By using Theorem 2.2 and passing to a subsequence of (x_n), if necessary, we can assume that

$$\text{(19)}\qquad\Big\|\sum\alpha_i y_i\Big\|\le\max_k\delta_k\max_{k\le n_1<n_2<\dots n_k}\Big\|\sum_{i=1}^{k}\alpha_{n_i}x_{n_i}\Big\|\le 3\Big\|\sum\alpha_i x_i\Big\|,$$

whenever $(\alpha_i)\in c_{00}$. Thus (x_n) dominates (y_n) and in order to show that the formal identity $I:x_n\mapsto y_n$ extends to a strictly singular operator, let (u_n) be a seminormalized block of (x_n), write u_n, for $n \in \mathbb{N}$ as

$$u_n = \sum_{i=k_{n-1}+1}^{k_n} \alpha_i^{(n)} x_i, \text{ and let } v_n = I(u_n) \sum_{i=k_{n-1}+1}^{k_n} \alpha_i^{(n)} y_i.$$

Using Theorem 2.2 and the fact that (e_n) has to satisfy (12), we can assume that $\lim_{n\to\infty} \max_{k_{n-1}<i\le k_n} |\alpha_i^{(n)}| = 0$, otherwise we could pass to an appropriate seminormalized block of (u_n). From (19) we can easily deduce that $\lim_{n\to\infty} \|I(u_n)\| = 0$, which proves that I cannot be an isomorphism. □

Theorem 1.1 gives a necessary and sufficient condition, for the property that a basic sequence strongly dominates an other one. Of course strong domination is a much stronger condition then domination. Nevertheless, if our goal is to state a condition on the spreading models of the sequences (x_n) and (y_n) which forces that (x_n) dominates (y_n) then strong domination of the spreading models is needed as the following remark shows.

REMARK. Assume that (e_i) and (f_i) are two normalized 1-subsymmetric (1-spreading and 1-unconditional) basic sequences, so that (e_i) dominates (f_i) but does not strongly dominate it. Moreover assume that $F = [f_i : i \in \mathbb{N}]$ does not contain a subspace isomorphic to c_0.

We can therefore find for $n \in \mathbb{N}$ an element $a^{(n)} = (a_i^{(n)}) \in c_{00}$, so that $\max_i |a_i^{(n)}| \to 0$ if $n \to \infty$, and $c', c > 0$ so that

$$1 = \Big\|\sum_{i\in\mathbb{N}} a_i^{(n)} f_i\Big\| \ge c\Big\|\sum_{i\in\mathbb{N}} a_i^{(n)} e_i\Big\| \ge c'. \tag{20}$$

Now we let (x_n) be the basis for the *Schreier space associated to* (e_i), i.e. the norm defined by

$$\Big\|\sum a_i x_i\Big\| = \max_{\substack{n\in\mathbb{N}\\ n\le i_1<\dots i_n}} \Big\|\sum_{j=1}^{n} a_{i_j} e_j\Big\| \text{ whenever } (a_i) \in c_{00}. \tag{21}$$

As in the original Schreier space (where (e_i) is set to be the ℓ_1-basis) it is easy to see that (e_i) is a spreading model of (x_i). Let $(\tilde{x}_n)$ be a subsequence of (x_n). For $n \in \mathbb{N}$ define $m_n = \max \operatorname{supp}(a_i^{(n)})$ and $u_n = \sum_{i=1}^{m_n} a_i^{(n)} \tilde{x}_{m_n+i}$. Then $\|u_n\| \ge c'/c$ and, again, as in the original Schreier space, we can show that a subsequence of u_n is equivalent to the c_0-unit basis. Since F does not contain a copy of c_0 we deduce that the map $\tilde{x}_n \mapsto f_n$ can not be extended to a linear bounded operator.

In general the condition that a subsymmetric and normalized basis (e_i) strongly dominates another basis (f_i) is much stronger than the condition that (e_i) dominates (f_i) without (e_i) being equivalent to (f_i). But in the case of $E = \ell_1$, we have the following.

PROPOSITION 2.5. *Assume (y_n) is a normalized basic weakly null sequence.*

(y_n) has a subsequence which is strongly dominated by ℓ_1 if and only if (y_n) has a subsequence whose spreading model is not equivalent to the unit vector basis of ℓ_1.

PROOF. By Lemma 2.4 it follows immediately that (y_n) is not strongly dominated by ℓ_1 if it has a subsequence whose spreading model is equivalent to the unit vector basis of ℓ_1.

To show the converse we need to show by Lemma2.4 that if a 1-subsymmetric basis (f_i) is not strongly dominated by ℓ_1 then (f_i) is equivalent to the unit vector basis of ℓ_1.

Let f_n^*, $n \in \mathbb{N}$, be the coordinate functionals on F. Then (f_n^*) is also a 1-subsymmetric basis (of its closed linear span).

By assumption there is a $\delta_0 > 0$ and for each $n \in \mathbb{N}$ a $y_n = \sum_{i=1}^\infty a_i^{(n)} f_i \in [f_i : i \in \mathbb{N}]$ with $0 \le a_i^{(n)} < 1/n$, for $i \in \mathbb{N}$, $\sum_{i=1}^\infty a_i^{(n)} = 1$, and $\|\sum_{i=1}^\infty a_i^{(n)} f_i\| \ge \delta_0$.

For each $n \in \mathbb{N}$ choose $y_n^* = \sum_{i=1}^\infty \beta_i^{(n)} f_i^* \in [f_i^* : i \in \mathbb{N}]$ with $0 \le \beta_i^{(n)}$, for $i \in \mathbb{N}$, $\|\sum_{i=1}^\infty \beta_i^{(n)} f_i^*\| = 1$, and $y_n^*(y_n) = \sum_{i=1}^\infty a_i^{(n)} \beta_i^{(n)} \ge \delta_0/2$.

Letting $c_n = \#\{i : \beta_i^{(n)} \ge \delta_0/4\}$, $n = 1, 2 \ldots$ it follows from the conditions on $(a_i^{(n)})$ that

$$\delta_0/2 \le \sum_{i=1}^\infty a_i^{(n)} \beta_i^{(n)} \le c_n \frac{1}{n} + \delta_0/4,$$

thus $c_n \ge \frac{n\delta_0}{4}$. Since (f^*n) is 1-subsymmetric it follows for all $k \in \mathbb{N}$ that $\|\frac{\delta_0}{4} \sum_{i=1}^k f_i^*\| \le 1$ and thus that f_i^* is equivalent to unit basis of c_0, from which we finally deduce that (f_i) is equivalent to the ℓ_1-basis. □

The following proposition describes another situation in which Theorem 1.1 applies. Its proof can be compiled from the techniques in [**AOST**], Section 3. Nevertheless, the proof is still quite technical, and since Theorem 1.4 provides a generalization, we will not give a proof here. Before we can state the result we need the following Definition from [**AOST**].

DEFINITION 2.6. Let (x_i) be a 1-subsymmetric basic sequence. The *Krivine set* of (x_i) is the set of p's ($1 \le p \le \infty$) with the following property: For all $\varepsilon > 0$ and $n \in \mathbb{N}$ there exists $m \in \mathbb{N}$ and $(\lambda_k)_{k=1}^m \subset \mathbb{R}$, such that for all $(a_i)_1^n \subseteq \mathbb{R}$,

$$\frac{1}{1+\varepsilon} \|(a_i)_{i=1}^n\|_p \le \left\| \sum_{i=1}^n a_i y_i \right\| \le (1+\varepsilon)\|(a_i)_{i=1}^n\|_p, \text{ where}$$

$$y_i = \sum_{k=1}^m \lambda_k x_{(i-1)m+k}, \quad \text{for} \quad i = 1, \ldots, n,$$

and $\|\cdot\|_p$ denotes the norm of the space ℓ_p.

The proof of Krivine's theorem [**K**] as modified by Lemberg [**Le**] (see also [**Gu**], remark II.5.14), shows that for every 1-subsymmetric basic sequence (x_i) the Krivine set of (x_i) is non-empty. It is important to note that our definition of a Krivine p requires not merely that ℓ_p be block finitely representable in $[x_i]$ but each ℓ_p^n unit vector basis is obtainable by means of an identically distributed block basis.

PROPOSITION 2.7. *Assume X is a Banach space containing a normalized basic sequence (x_n) which has a spreading model (e_i) which is not equivalent to the ℓ_1-unit vector basis, but whose Krivine set contains the number 1.*

Then there is a normalized basic sequence (z_n) in X which strongly dominates (x_n).

3. The transfinite family $(\mathcal{F}_\alpha)$ and some of its basic properties

In this section we discuss a well ordered family $(\mathcal{F}_\alpha)_{\alpha<\omega}$ of subsets of the finite subsets of $\mathbb{N}$. Its definition is similar to the definition of the Schreier family $(S_\alpha)_{\alpha<\omega}$ [**AA**]. The Schreier set of order α, S_α, corresponds to our set $\mathcal{F}_{\omega^\alpha}$, in the sense that they have the same Cantor Bendixson index.

For every limit ordinal $\alpha<\omega_1$ we consider a sequence of sets $(A_n(\alpha))_{n\in\mathbb{N}}$ so that for each $n\in\mathbb{N}$ $A_n(\alpha)$ is a finite subset of $[0,\alpha)$ and

$$A_n(\alpha)\subset A_{n+1}(\alpha) \text{ for, } n\in\mathbb{N}, \text{ and } \lim_{n\to\infty}\max A_n(\alpha)=\alpha. \tag{22}$$

We call $(A_n(\alpha))_{n\in\mathbb{N}}$ the *sequence approximating* α. If for every limit ordinal $\alpha<\omega$ (we write $\alpha\in\mathrm{Lim}(\omega_1)$) $(A_n(\alpha))_{n\in\mathbb{N}}$ is a sequence approximating α, we call the family $(A_n(\alpha))_{n\in\mathbb{N},\alpha\in\mathrm{Lim}(\omega_1)}$ *an approximating family.*

Given an approximating family $(A_n(\alpha))_{n\in\mathbb{N},\alpha\in\mathrm{Lim}(\omega_1)}$ the sets $\mathcal{F}_\alpha\subset[\mathbb{N}]^{<\infty}$, $\alpha<\omega_1$, are defined by transfinite recursion as follows

$$\mathcal{F}_0=\{\emptyset\}. \tag{23}$$

Assuming for some $0<\alpha<\omega_1$ the sets $\mathcal{F}_\beta\subset[\mathbb{N}]^{<\infty}$, with $\beta<\alpha$, are already defined we proceed as follows.

$$\mathcal{F}_\alpha=\big\{\{n\}\cup E: n\in\mathbb{N}, E\in\mathcal{F}_\beta\big\}\cup\{\emptyset\} \text{ if } \alpha=\beta+1 \text{ and} \tag{24}$$

$$\mathcal{F}_\alpha=\Big\{\ E\in[\mathbb{N}]^{<\infty}: E\in\bigcup_{\beta\in A_{\min E}(\alpha)}\mathcal{F}_\beta\Big\} \text{ if } \alpha\in\mathrm{Lim}(\omega_1) \tag{25}$$

We say that the *transfinite family* $(\mathcal{F}_\alpha)_{\alpha<\omega}$ *is defined by the approximating family*
$(A_n(\alpha))_{n\in\mathbb{N},\alpha\in\mathrm{Lim}(\omega_1)}$.

We first state some elementary properties of our family $(\mathcal{F}_\alpha)$.

PROPOSITION 3.1. *Assume that $(\mathcal{F}_\alpha)_{\alpha<\omega_1}$ is the transfinite family associated to an approximating family $(A_n(\alpha))_{n\in\mathbb{N},\alpha\in Lim(\omega_1)}$.*

a) *For $\alpha<\omega_1$, $\mathcal{F}_\alpha$ is hereditary, spreading and compact in $[\mathbb{N}]^{<\infty}$.*

b) *For $\alpha<\omega_1$ it follows that*

$$\begin{aligned}\mathcal{F}_{\alpha+1}&=\big\{\{n\}\cup E: n\in\mathbb{N}, n<\min E, \text{ and } E\in\mathcal{F}_\alpha\big\}\cup\{\emptyset\}\\&=\big\{E\in[\mathbb{N}]^{<\infty}: E\neq\emptyset, E\setminus\{\min E\}\in\mathcal{F}_\alpha\big\}\cup\{\emptyset\}.\end{aligned}$$

c) *For $\alpha\le\beta<\omega_1$ there is an $m\in\mathbb{N}$ so that*

$$\mathcal{F}_\alpha\cap\big[\{m,m+1,\ldots\}\big]^{<\infty}\subset\mathcal{F}_\beta\cap\big[\{m,m+1,\ldots\}\big]^{<\infty}.$$

PROOF. We can prove (a) by transfinite induction for all $\alpha<\omega_1$ while (b) follows from the fact that $\mathcal{F}_\alpha$ is hereditary and spreading.

To show (c) we fix α and prove the claim by transfinite induction for all $\beta>\alpha$. Assuming the claim is true for all $\gamma<\beta$. If $\beta=\gamma+1$ the claim follows since $\mathcal{F}_\gamma\subset\mathcal{F}_{\gamma+1}$. If β is a limit ordinal and if $(A_n(\beta))$ is its approximating sequence we proceed as follows.

First we choose $n\in\mathbb{N}$ so that $\beta^{(n)}=\max A_n(\beta))>\alpha$, and, using the induction hypothesis we can find an $\ell\in\mathbb{N}$ so that

$$\mathcal{F}_\alpha\cap[\{\ell,\ell+1,\ldots\}]^{<\infty}\subset\mathcal{F}_{\beta^{(n)}}\cap[\{\ell,\ell+1\ldots\}]^{<\infty}.$$

Secondly we observe from the definition of $\mathcal{F}_\beta$ it follows that

$$\mathcal{F}_{\beta(n)} \cap [\{n, n+1\ldots\}]^{<\infty} \subset \mathcal{F}_\beta \cap [\{n, n+1\ldots\}]^{<\infty}.$$

Therefore the claim follows by choosing $m = \max\{\ell, n\}$. □

In the definition of approximating families we allow the sets $A_n(\alpha)$ to have more than one element, contrary to the definition of the Schreier families (see [**AA**]), because we want to ensure that the transfinite families are directed.

PROPOSITION 3.2. *Assume for every $k \in \mathbb{N}$ $(A_n^{(k)}(\alpha))_{n\in\mathbb{N},\alpha\in Lim(\omega_1)}$ is an approximating family defining the transfinite family $(\mathcal{F}_\alpha^{(k)})_{\alpha<\omega_1}$.*

For each $k \in \mathbb{N}$, $n \in \mathbb{N}$ and $\alpha \in Lim(\omega_1)$ define $B_n^{(k)}(\alpha) = \bigcup_{i=1}^k A_n^{(i)}(\alpha)$ and let $\mathcal{G}_\alpha^{(k)}$, $\alpha < \omega_1$ be defined using the approximating family $(B_n^{(k)}(\alpha))$.

Further more define for $n \in \mathbb{N}$ and $\alpha \in Lim(\omega_1)$ $B_n(\alpha) = \bigcup_{i=1}^n A_n^{(i)}(\alpha)$ and let $(\mathcal{G}_\alpha)_{\alpha<\omega_1}$ be the transfinite family defined by $(B_n(\alpha))$.

Then it follows for all $\alpha < \omega_1$

a) *$\bigcup_{i=1}^k \mathcal{F}_\alpha^{(i)} \subset \mathcal{G}_\alpha^{(k)}$ for $k \in \mathbb{N}$.*
b) *For all $k \in \mathbb{N}$ it follows that $\mathcal{F}_\alpha^{(k)} \cap [\{k, k+1, \ldots\}]^{<\infty} \subset \mathcal{G}_\alpha$.*

PROOF. By transfinite induction. □

DEFINITION 3.3. Let $\mathcal{A} \subset [\mathbb{N}]^{<\infty}$ and $N = \{n_i : i \in \mathbb{N}\} \overset{\infty}{\subset} \mathbb{N}$, $n_i \nearrow \infty$, if $i \nearrow \infty$.

a) We call the set $\mathcal{A} \cap [N]^{<\infty}$ *the restriction of $\mathcal{A}$ onto N*.
b) We call the family $\mathcal{A}^N = \big\{\{n_i : i \in E\} : E \in \mathcal{A}\big\}$ the *spreading of $\mathcal{A}$ onto N*.

Using the definition of $(\mathcal{F}_\alpha)$, we obtain the following recursive description of $\mathcal{F}_\alpha^N$.

PROPOSITION 3.4. *Assume $N \overset{\infty}{\subset} \mathbb{N}$, $N = (n_i)$, $n_i \nearrow \infty$ and $\alpha < \omega_1$. Then*

a) *If $\alpha = \beta + 1$*

$$\begin{aligned}\mathcal{F}_\alpha^N &= \big\{\{n\} \cup F : n \in N \text{ and } F \in \mathcal{F}_\beta^N\big\} \cup \{\emptyset\}\\ &= \big\{F \in [N]^{<\infty} \setminus \{\emptyset\} : F \setminus \{\min F\} \in \mathcal{F}_\beta\big\} \cup \{\emptyset\}.\end{aligned}$$

b) *If $\alpha \in Lim(\omega_1)$ and $(A_n(\alpha))$ is the sequence approximating α, then*

$$\mathcal{F}_\alpha^N = \big\{F \in [N]^{<\infty} : F \in \bigcup_{\beta\in A_{\min\{i:n_i\in F\}}(\alpha)} \mathcal{F}_\beta^N\big\}.$$

PROPOSITION 3.5. *Assume $M, N \overset{\infty}{\subset} \mathbb{N}$ and $m_0 \in \mathbb{N}$ so that*

a) *$\#(M \cap [1, m_0]) \le \#(N \cap [1, m_0])$*
b) *$M \cap [m_0, \infty) \subset N \cap [m_0, \infty)$*

Then it follows for $\alpha < \omega_1$ that $\mathcal{F}_\alpha^M \cap [\{m_0, m_0+1, \ldots\}]^{<\infty} \subset \mathcal{F}_\alpha^N \cap [\{m_0, m_0 + 1, \ldots\}]^{<\infty}$.

PROOF. We prove the claim by transfinite induction on $\alpha < \omega_1$, using at each induction step Proposition 3.4 (a) or (b). □

PROPOSITION 3.6. *If $N \overset{\infty}{\subset} \mathbb{N}$, $N = (n_i)$, $n_i \nearrow \infty$, and $\beta < \alpha < \omega_1$ then there is an $\ell \in \mathbb{N}$ so that*

$$\mathcal{F}_\beta^{\{n_i : i \geq \ell\}} \subset \mathcal{F}_\alpha^N.$$

PROOF. Using Proposition 3.1 (c) we can choose $\ell \in \mathbb{N}$ so that $\mathcal{F}_\beta \cap [\{\ell, \ell + 1, \ldots\}]^{<\infty} \subset \mathcal{F}_\alpha \cap [\{\ell, \ell+1, \ldots\}]^{<\infty}$. Thus it follows (for the first "$\subset$" recall that $\mathcal{F}_\beta$ is spreading)

$$\begin{aligned}\mathcal{F}_\beta^{\{n_i : i \geq \ell\}} &= \big\{\{n_{i+\ell-1} : i \in E\} : E \in \mathcal{F}_\beta\big\} \subset \big\{\{n_j : j \in E\} : E \in \mathcal{F}_\beta, \text{ and } E \geq \ell\big\} \\ &\subset \big\{\{n_j : j \in E\} : E \in \mathcal{F}_\alpha, \text{ and } E \geq \ell\big\} \subset \mathcal{F}_\alpha^N,\end{aligned}$$

which finishes the proof. □

PROPOSITION 3.7. *Let $(A_n(\alpha))_{n \in \mathbb{N}, \alpha \in Lim(\omega_1)}$ and $(B_n(\alpha))_{n \in \mathbb{N}, \alpha \in Lim(\omega_1)}$ be two approximating families defining the transfinite families $(\mathcal{F}_\alpha)_{\alpha < \omega_1}$ and $(\mathcal{G}_\alpha)_{\alpha<\omega_1}$ respectively. For $\alpha < \omega_1$ and $N \overset{\infty}{\subset} \mathbb{N}$ there is an $M \overset{\infty}{\subset} N$ so that $\mathcal{G}_\alpha^M \subset \mathcal{F}_\alpha$.*

PROOF. We proof the claim by transfinite induction on $\alpha < \omega_1$. Assume that the claim is true for all $\beta < \alpha$. If α is a successor it follows immediately that the claim is true for α.

Assume $\alpha = \sup_{\gamma < \alpha} \gamma = \sup_{n \in \mathbb{N}} \max B_n(\alpha) = \sup_{n \in \mathbb{N}} \max A_n(\alpha)$.

Using Proposition 3.1(c) we can choose an $m_i \in \mathbb{N}$ for each $i \in \mathbb{N}$, so that for all $\gamma \in B_i(\alpha)$ it follows that $\mathcal{F}_\gamma \cap [\{m_i, m_i + 1, \ldots\}]^{<\infty} \subset \mathcal{F}_\alpha$.

Secondly, using the induction hypothesis, we can find $N \overset{\infty}{\supset} M_1 \overset{\infty}{\supset} M_2 \ldots$ so that for all $k \in \mathbb{N}$ and all $\gamma \in B_k(\alpha)$ it follows that $\mathcal{G}_\gamma^{M_k} \subset \mathcal{F}_\gamma$. Since we can make sure that $\min M_k \geq m_k$, for $k \in \mathbb{N}$, it follows that $\mathcal{G}_\gamma^{M_k} \subset \mathcal{F}_\alpha$ for all $k \in \mathbb{N}$ and $\gamma \in B_k(\alpha)$. If we finally let M be a diagonal sequence of the M_i's we deduce the claim from Proposition 3.5. □

From the property that $\mathcal{F}_\alpha$ is spreading it is easy to see that for any $L \overset{\infty}{\subset} \mathbb{N}$ it follows that $\mathcal{F}_\alpha^L \subset \mathcal{F}_\alpha \cap [L]^{<\infty}$. If one is willing to change the approximating family the converse becomes true.

PROPOSITION 3.8. *Let $(\mathcal{F}_\alpha)_{\alpha<\omega_1}$ be a transfinite family which is defined by an approximating family $(A_n(\alpha))_{n \in \mathbb{N}, \alpha \in Lim(\omega_1)}$, and let $L \overset{\infty}{\subset} \mathbb{N}$.*

Then there is an approximating family $(B_n(\alpha))_{n \in \mathbb{N}, \alpha \in Lim(\omega_1)}$ defining a transfinite family $(\mathcal{G}_\alpha)$ for which it follows that for any $\alpha < \omega_1$

$$\mathcal{F}_\alpha \cap [L]^{<\infty} \subset \mathcal{G}_\alpha^L. \tag{26}$$

PROOF. We write L as $L = \{\ell_1, \ell_2, \ldots\ldots\}$ with $\ell_1 < \ell_2, < \ldots$. For a limit ordinal $\alpha < \omega_1$ and an $i \in \mathbb{N}$ we put $B_i(\alpha) = A_{\ell_i}(\alpha)$ and show by transfinite induction that $\mathcal{F}_\alpha \cap [L]^{<\infty} \subset \mathcal{G}_\alpha^L$, where the family $(\mathcal{G}_\alpha)$ is defined based on the approximating family $(B_i(\alpha))_{n \in \mathbb{N}, \alpha \in \mathrm{Lim}(\omega_1)}$.

Assuming the claim to be true for all $\beta < \alpha$ the claim follows immediately from Proposition 3.4 (a) for α if α is a successor. If α is a limit ordinal we observe that for an $F \in \mathcal{F}_\alpha \cap [L]^{<\infty}$ we have $F \in \bigcup_{\beta \in A_n(\alpha)} \mathcal{F}_\beta \cap [L]^{<\infty}$ with $n = \min F$. Choosing $i \in \mathbb{N}_0$ so that $\ell_i = n$ we deduce from the induction hypothesis that

$$F \in \bigcup_{\beta \in A_n(\alpha)} \mathcal{F}_\beta \cap [L]^{<\infty} \subset \bigcup_{\beta \in A_n(\alpha)} \mathcal{G}_\beta^L = \bigcup_{\beta \in B_i(\alpha)} \mathcal{G}_\beta^L$$

which implies by Proposition 3.4 (b) the claim. □

4. The transfinite family $(\mathcal{F}_\alpha)$ is universal

The main goal in this section is to prove that the family $(\mathcal{F}_\alpha)$ which was introduced in section 3 is universal (see Theorem 4.10 for the precise statement).

In the following Defintion we are using the transfinite family (F_α) to measure the complexity of hereditary sets $\mathcal{A} \subset [\mathbb{N}]^{<\infty}$. As we will see later this measure is equivalent to the Cantor Bendixson index introduced in Section 1.

DEFINITION 4.1. Consider an approximating family $(A_n(\alpha))_{n\in\mathbb{N},\alpha\in\mathrm{Lim}(\omega_1)}$ defining a transfinite family $(\mathcal{A}_\alpha)_{\alpha<\omega_1}$.

Let $P \overset{\infty}{\subset} \mathbb{N}$, $\alpha < \omega_1$, and $\mathcal{A} \subset [\mathbb{N}]^{<\infty}$ be hereditary. We say that $\mathcal{A}$ *is* α*-large on* P, if for all $M \overset{\infty}{\subset} P$ there is an $N \overset{\infty}{\subset} M$ so that $\mathcal{F}_\alpha^N \subset \mathcal{A}$.

Moreover, we call $\mathbb{I}(\mathcal{A}, P) = \sup\{\alpha : \mathcal{A} \text{ is } \alpha\text{-large on } P\}$ *the complexity of* $\mathcal{A}$ *on* P.

Remark. Since the notion α-large depends on $(\mathcal{F}_\alpha)$ which depends on the choice of the approximating family, we should have rather used the notion $\mathcal{F}_\alpha$-large instead of α-large. But in Corollary 4.4 we will show that the property of being α-large is independent to the underlying approximating family. For the results up to Corollary 4.4 we consider the approximating family and its transfinite family to be fixed.

PROPOSITION 4.2. *(Stabilization of* $\mathbb{I}(\mathcal{A}, P)$ *with respect to* P*).*
Assume $\mathcal{A} \subset [\mathbb{N}]^{<\infty}$ *is hereditary,* $P \overset{\infty}{\subset} \mathbb{N}$ *and let* $\alpha_0 = \mathbb{I}(\mathcal{A}, P)$*. Then there is a* $Q \overset{\infty}{\subset} P$ *so that for all* $L \overset{\infty}{\subset} Q$ *it follows that* $\mathbb{I}(\mathcal{A}, L) = \alpha_0$*.*

PROOF. Since $\mathbb{I}(\mathcal{A}, P) < \alpha_0 + 1$ we deduce that there is a $Q \overset{\infty}{\subset} P$ so that for all $N \overset{\infty}{\subset} Q$ it follows that $\mathcal{F}_{\alpha_0+1}^N \not\subset \mathcal{A}$. This implies that for all $N \overset{\infty}{\subset} Q$ we have $\mathbb{I}(\mathcal{A}, N) < \alpha_0 + 1$, and thus $\mathbb{I}(\mathcal{A}, N) \le \alpha_0$. On the other hand it is clear that $\mathbb{I}(\mathcal{A}, N) \ge I(\mathcal{A}, P)$ for all $N \overset{\infty}{\subset} P$, which implies the claim. □

PROPOSITION 4.3. *Let* $P \overset{\infty}{\subset} \mathbb{N}$*,* $\alpha < \omega_1$*, and* $\mathcal{A} \subset [\mathbb{N}]^{<\infty}$ *hereditary.*

a) *If* $\beta < \alpha$ *and if* $\mathcal{A}$ *is* α*-large on* P*, then* $\mathcal{A}$ *is* β*-large on* P*.*
b) *If* $\alpha = \beta + 1$*, then*

$$\mathcal{A} \textit{ is } \alpha\textit{-large on } P \iff \forall Q \overset{\infty}{\subset} P \quad \exists L \overset{\infty}{\subset} Q \textit{ so that}$$

$$(*)\begin{cases} \forall \ell \in L \quad \forall M \overset{\infty}{\subset} L \quad \exists N \overset{\infty}{\subset} M \\ \qquad \mathcal{F}_\beta^N \subset \mathcal{A}|_\ell := \{E : \{\ell\} \cup E \in \mathcal{A}, \ell < E\} \\ \qquad [\textit{i.e. } \mathcal{A}|_\ell \textit{ is } \beta\textit{-large on } L] \end{cases}$$

c) *If* $\alpha \in \mathit{Lim}(\omega_1)$ *and* $(A_n(\alpha))$ *is the approximating sequence for* α *it follows that*

$$\mathcal{A} \textit{ is } \alpha\textit{-large on } P \iff \forall n \in \mathbb{N} \quad \mathcal{A} \textit{ is } \max A_n(\alpha)\textit{-large on } P$$
$$[\iff (\textit{by part (a)})\, \forall \beta < \alpha \quad \mathcal{A} \textit{ is } \beta\textit{-large on } P]$$

Together with (a), (c) implies that if $\mathbb{I}(\mathcal{A}, P) < \omega_1$ *it follows that the set of all* $\alpha < \omega_1$ *for which* $\mathcal{A}$ *is* α*-large on* P *is the closed interval* $[0, \mathbb{I}(\mathcal{A}, P)]$*.*

PROOF. (a) follows immediately from Proposition 3.6.

For (b)"$\Rightarrow$" let $\mathcal{A}$ be α-large and $Q \overset{\infty}{\subset} P$. Then there is an $L \overset{\infty}{\subset} Q$ with $\mathcal{F}_\alpha^L \subset \mathcal{A}$.

Since $\alpha = \beta + 1$ it follows from Proposition 3.4 that

$$\mathcal{F}_\alpha^L = \{\{\ell\} \cup E : E \in \mathcal{F}_\beta^L, \ell \in L \text{ and } \ell < E\} \cup \{\emptyset\} \subset \mathcal{A}.$$

Therefore it follows for all $\ell \in L$ that $\mathcal{F}_\beta^L \subset \mathcal{A}|_\ell$ and therefore it follows for any $M \overset{\infty}{\subset} L$ that $\mathcal{F}_\beta^M \subset \mathcal{A}_\ell$.

In order to prove "$\Leftarrow$" of (b) assume that $M \overset{\infty}{\subset} P$. We need to find $N \overset{\infty}{\subset} M$ so that $\mathcal{F}_\alpha^N \subset \mathcal{A}$. By assumption we find an $L = (\ell_i) \subset M$ satisfying $(*)$.

By induction we can choose $n_1 < n_2 < \ldots$ and $N_1 \supset N_2 \ldots$ so that for all $k \in \mathbb{N}$:

$$n_{i+1} \in N_i \text{ for } i = 1, \ldots k-1 \tag{27}$$

$$\#(N_i \cap [0, n_{i+1}]) \geq i+1 \text{ for } i = 1, \ldots k-1 \tag{28}$$

$$\mathcal{F}_\beta^{N_i} \subset \mathcal{A}|_{n_i}, \text{ for } i = 1, \ldots k. \tag{29}$$

Indeed, choose $n_1 = \ell_1$ and, using the property $(*)$ (with $\ell = n_1$) we find $N_1 \overset{\infty}{\subset} L$ so that $\mathcal{F}_\beta^{N_1} \subset \mathcal{A}|_{n_1}$. Then we choose an $n_2 \in N_1 \cap \{n_1+1, n_1+2, \ldots\}$, large enough in order to satisfy (28), and apply $(*)$ again (with $\ell = n_2$) to find an $N_2 \overset{\infty}{\subset} N_1$ with $\mathcal{F}_\beta^{N_2} \subset \mathcal{A}|_{n_1}$. We continue in that way.

Now we claim that $\mathcal{F}_\alpha^{\{n_i : i \in \mathbb{N}\}} \subset \mathcal{A}$, which would finish this part of the proof. Indeed, if $E = \{n_{i_1}, n_{i_2}, \ldots n_{i_k}\} \in \mathcal{F}_\alpha^{\{n_i : i \in \mathbb{N}\}}$, with $n_{i_1} < n_{i_2} < \ldots < n_{i_k}$, then, by Proposition 3.4 it follows that $\{n_{i_2}, n_{i_3} \ldots n_{i_k}\} \in \mathcal{F}_\beta^{\{n_i : i \in \mathbb{N}\}}$. Also note that by choice of n_{i_2}, $\#(N_{i_2 - 1} \cap [0, n_{i_2}]) \geq i_2 = \#\{n_i : i \in \mathbb{N}\} \cap [0, n_{i_2}]$. Therefore, by Proposition 3.5 $\{n_{i_2}, n_{i_3} \ldots n_{i_k}\} \in \mathcal{F}_\beta^{N_{i_2 - 1}}$, and thus, since $\mathcal{F}_\beta^{N_{i_2-1}} \subset \mathcal{F}_\beta^{N_{i_1}}$, it follows that $\{n_{i_2}, n_{i_3} \ldots n_{i_k}\} \in \mathcal{F}_\beta^{N_{i_1}}$, which implies by (29) that $\{n_{i_2}, n_{i_3} \ldots n_{i_k}\} \in \mathcal{A}|_{n_{i_1}}$ and thus $\{n_{i_1}, n_{i_2} \ldots n_{i_k}\} \in \mathcal{A}$.

The claim (c)"$\Rightarrow$" follows from (a). In order to show (c)"$\Leftarrow$" let $L \overset{\infty}{\subset} P$. Using the assumption and part (a) we find $L \overset{\infty}{\supset} N_1 \overset{\infty}{\supset} N_2 \overset{\infty}{\supset} \ldots$ so that $\mathcal{F}_\gamma^{N_i} \subset \mathcal{A}$ for all $i \in \mathbb{N}$ and $\gamma \in A_i(\alpha)$. Then choose $N = (n_i)$ to be a diagonal sequence of $(N_i)_{i=1}^\infty$ in such a way that for $i \in \mathbb{N}$ n_i is in N_i and is at least as big as the i-th element of N_i.

It follows that $\mathcal{F}_\alpha^N \subset \mathcal{A}$. Indeed, let $E \in \mathcal{F}_\alpha^N$ and, thus, $E \in \mathcal{F}_{\gamma_0}^N$ for some $\gamma_0 \in A_{i_0}(\alpha)$ where $n_{i_0} = \min E$. Note that by the choice of N we have that $i_0 = \#(N \cap [1, n_{i_0}]) \leq \#(N_{i_0} \cap [1, n_{i_0}])$ and $N \cap [n_{i_0}, \infty) \subset N_{i_0} \cap [n_{i_0}, \infty)$, and we deduce from Proposition 3.5 that $E \in \mathcal{F}_{\gamma_0}^N \cap [\{n_{i_0}, n_{i_0}+1, \ldots\}]^{<\infty} \subset \mathcal{F}_{\gamma_0}^{N_{i_0}} \cap [\{n_{i_0}, n_{i_0}+1, \ldots\}]^{<\infty} \subset \mathcal{A}$, which finishes the proof. □

Now we can conclude that the property of being α-large for a set $\mathcal{A}$ does not depend on the choice of the approximating sequences.

Corollary 4.4. *For an $\mathcal{A} \subset [\mathbb{N}]^{<\infty}$, $P \overset{\infty}{\subset} \mathbb{N}$ and $\alpha < \omega_1$, the property of being α-large does not depend on the choice of approximating family one has chosen for defining the sets $\mathcal{F}_\beta$.*

Proof. Assume that $(\mathcal{G}_\alpha)$ is a transfinite family defined by another approximating family. We will show by transfinite induction that if $\mathcal{F}_\alpha^N \subset \mathcal{A}$ for some

$N \overset{\infty}{\subset} \mathbb{N}$ then there is an $M \overset{\infty}{\subset} N$ so that $\mathcal{G}_\alpha^M \subset \mathcal{A}$. Assume that our claim is true for all $\beta < \alpha$ for some $\alpha < \omega_1$ and let $N \overset{\infty}{\subset} \mathbb{N}$ be such that $\mathcal{F}_\alpha^N \subset \mathcal{A}$.

If $\alpha = \beta + 1$ the claim follows from the induction hypothesis and Proposition 4.3 part (b). If $\alpha = \sup_{\beta<\alpha} \beta$ the claim follows from applying Proposition 4.3 part (a) and (c). □

COROLLARY 4.5. *For $\mathcal{A} \subset [\mathbb{N}]^{<\infty}$ and $P \overset{\infty}{\subset} \mathbb{N}$ it follows that*

$$\mathbb{I}(\mathcal{A}, P) \leq \sup_{p\in P} \sup_{L \overset{\infty}{\subset} P} \mathbb{I}(\mathcal{A}|_p, L) + 1. \tag{30}$$

Moreover if $\mathbb{I}(\mathcal{A}, P)$ is a limit ordinal then it even follows that

$$\mathbb{I}(\mathcal{A}, P) \leq \sup_{p\in P} \sup_{L \overset{\infty}{\subset} P} \mathbb{I}(\mathcal{A}|_p, L). \tag{31}$$

PROOF. Put $\alpha_0 = \mathbb{I}(\mathcal{A}, P)$. If α_0 is a successor our claim follows directly from Proposition 4.3 part (b). If α_0 is a limit ordinal and $\beta < \alpha_0$ arbitrary (and thus $\beta + 1 < \alpha_0$) we conclude from Proposition 4.3 part (b) that there is a $p \in P$ and an $L \overset{\infty}{\subset} P$ so that $\mathbb{I}(\mathcal{A}|_p, L) \geq \beta$.

Since $\beta < \alpha_0$ was arbitrary it follows that $\alpha_0 \leq \sup_{p\in P} \sup_{L \overset{\infty}{\subset} P} \mathbb{I}(\mathcal{A}|_p, L)$. □

COROLLARY 4.6. *Assume that for $\mathcal{A} \subset [\mathbb{N}]^{<\infty}$ and $P \overset{\infty}{\subset} \mathbb{N}$ we have that*

$$\forall \alpha < \omega_1 \exists N_\alpha \overset{\infty}{\subset} P \quad \mathcal{F}_\alpha^{N_\alpha} \subset \mathcal{A}. \tag{32}$$

Then there is an $L \overset{\infty}{\subset} P$ so that $[L]^{<\infty} \subset \mathcal{A}$.

Therefore, if an hereditary set $\mathcal{A} \subset [\mathbb{N}]^{<\infty}$ has the property, that for no $L \overset{\infty}{\subset} P$ it follows that $[L]^{<\infty} \subset \mathcal{A}$, the complexity of $\mathcal{A}$ must be some countable ordinal α_0.

We will say that the complexity of $\mathcal{A}$ on P is ω_1 and write $\mathbb{I}(\mathcal{A}, P) = \omega_1$ if (32) is satisfied.

PROOF. We first show that there is an $n \in P$, so that (32) holds for $\mathcal{A}|_n$ (instead of $\mathcal{A}$). Indeed, otherwise we could find for each $n \in P$ an α_n so that $\mathcal{F}_{\alpha_n}^N \not\subset \mathcal{A}|_n$ for all $N \overset{\infty}{\subset} P$. Letting $\alpha = \sup a_n$ we deduce that $\mathcal{F}_\alpha^N \not\subset \mathcal{A}|_n$ for any $n \in P$ and any $N \overset{\infty}{\subset} P$. By Proposition 4.3 (b) this would contradict (32) for $\alpha + 1$. Note that n could have been chosen out of any given cofinite subset of P.

We can iterate this argument and produce a strictly increasing sequence $(n_i) \subset P$ so that for every $k \in \mathbb{N}$ (32) holds for $\mathcal{A}|_{n_1,n_2,,\ldots n_k} = \{A \subset \mathbb{N} : \{n_1, \ldots n_k\} \cup A \in \mathcal{A}\}$ holds. This implies that that $[\{n_i : i \in \mathbb{N}\}]^{<\infty} \subset \mathcal{A}$. □

COROLLARY 4.7. *For $\alpha < \omega_1$ and $L \overset{\infty}{\subset} \mathbb{N}$ it follows that $\mathbb{I}(\mathcal{F}_\alpha, L) = \alpha$.*

PROOF. Since $\mathcal{F}_\alpha^L \subset \mathcal{F}_\alpha \cap [L]^{<\infty}$ it is clear that $\mathbb{I}(\mathcal{F}_\alpha, L) \geq \alpha$. Assume that for some $\beta > \alpha$ and some $N = \{n_1^{(1)}, n_2^{(1)} \ldots\} \overset{\infty}{\subset} \mathbb{N}$ it follows that $\mathcal{F}_\beta^N \subset \mathcal{F}_\alpha$.

By Propostion 4.3 (a) we can assume that $\beta = \alpha + 1$ and we claim that it would follow that there is a family $(N_\beta)_{\alpha<\beta<\omega_1}$ of infinite subsets of $\mathbb{N}$, with $N_\gamma \setminus N_\beta$ being finite, if $\gamma < \beta$, so that

$$\mathcal{F}_\beta^{N_\beta} \subset \mathcal{F}_\alpha. \tag{33}$$

Using Corollary 4.6 this would imply that for some $L \overset{\infty}{\subset} \mathbb{N}$ so that $[L]^{<\omega} \subset \mathcal{F}_\alpha$, contradicting the compactness of $\mathcal{F}_\alpha$.

We will show the existence of N_β by transfinite induction of $\beta > \alpha$. For $\beta = \alpha + 1$ N_β exists by assumption. If $\beta = \gamma + 1$ and if $N_\gamma = \{n_1^{(\gamma)}, n_2^{(\gamma)}, \dots\} \overset{\infty}{\subset} \mathbb{N}$ is as in (33) we choose $N_{\gamma+1} = \{n^{(\alpha+1)}_{n_i^{(\gamma)}} : i \in \mathbb{N}\}$ and observe that (note that $\mathcal{F}^{N_\gamma}_{\gamma+1} \subset \mathcal{F}_{\alpha+1}$ since $\mathcal{F}^{N_\gamma}_{\gamma} \subset \mathcal{F}_\alpha$)

$$\begin{aligned}\mathcal{F}^{N_{\gamma+1}}_{\gamma+1} &= \Big\{\{n^{(\alpha+1)}_{n_i^{(\gamma)}} : i \in E\} : E \in \mathcal{F}_{\gamma+1}\Big\} = \Big\{\{n^{(\alpha+1)}_m : m \in F\} : F \in \mathcal{F}^{N_\gamma}_{\gamma+1}\Big\} \\ &\subset \Big\{\{n^{(\alpha+1)}_m : m \in F\} : F \in \mathcal{F}_{\alpha+1}\Big\} = \mathcal{F}^{N_{\alpha+1}}_{\alpha+1} \subset \mathcal{F}_\alpha.\end{aligned}$$

If $\beta \in \mathrm{Lim}(\omega_1)$ and if $(A_i(\beta))$ is the sequence approximating β we first note that we can choose infinite subsets $\mathbb{N} \overset{\infty}{\supset} M_1 \overset{\infty}{\supset} M_2 \dots$ so that for all $i \in \mathbb{N}$ it follows that $\bigcup_{\gamma \in A_i(\beta)} \mathcal{F}^{N_i}_\gamma \subset \mathcal{F}_\alpha$. Then we choose $N_\beta = \{n_i^{(\beta)} : i \in \mathbb{N}\}$ with $n_i^{(\beta)}$ being the i-th element of M_i, for $i = 1, 2 \dots$. We deduce then the claim in this case from Proposition 3.5. □

Using our results on the family $(\mathcal{F}_\alpha)_{\alpha<\omega}$ we can now show the relation ship between $\mathbb{I}$ and the Cantor Bendixson index.

LEMMA 4.8. *Assume that $\mathcal{A} \subset [\mathbb{N}]^{<\infty}$ is hereditary and compact.*

a) *Define the* expansion of $\mathcal{A}$ *by $Ep(\mathcal{A}) = \big\{\{n\} \cup A : A \in \mathcal{A}\big\} \cup \{\emptyset\}$. (Note that $\mathcal{F}_{\alpha+1} = Ep(\mathcal{F}_\alpha)$ for $\alpha < \omega_1$ and that $F \in Ep(\mathcal{A})$ if and only if $F = \emptyset$ or $F \setminus \{\min F\} \in \mathcal{A}$.)*

For $\alpha \le CB(\mathcal{A})$ it follows that $Ep\big(\mathcal{A}^{(\alpha)}\big) = \big(Ep(\mathcal{A})\big)^{(\alpha)}$.

b) *Assume that there is a sequence $(\ell_n) \subset \mathbb{N}$ with $\lim_{n\to\infty} \ell_n = \infty$ and for $n \in \mathbb{N}$ hereditary and compact sets $\mathcal{A}_n \subset [\{\ell_n, \ell_n + 1, \ell_n + 2, \dots\}]^{<\infty}$ so that $\alpha_n = CB(\mathcal{A}_n)$ strictly increases to some $\alpha < \omega_1$, and assume that $\mathcal{A} = \bigcup \mathcal{A}_n$. Then it follows that $CB(\mathcal{A}) = \alpha + 1$.*

c) *For any $n \in \mathbb{N}$ and any $\alpha < \omega_1$ it follows that $\big(\mathcal{A}|_n\big)^{(\alpha)} = \big(\mathcal{A}^{(\alpha)}\big)|_n$*

PROOF. (a) First assume that $\alpha = 1$ and, thus, that $\mathrm{CB}(\mathcal{A}) \ge 1$. Clearly, $\mathrm{Ep}\big(\mathcal{A}^{(1)}\big)$ as well as $\big(\mathrm{Ep}(\mathcal{A})\big)^{(1)}$ contain $\emptyset$ as an element. For $F \in [\mathbb{N}]^{<\infty}$, $F \ne \emptyset$, we observe that

$$\begin{aligned} &F \in \big(\mathrm{Ep}(\mathcal{A})\big)^{(1)} \\ &\iff \exists N \overset{\infty}{\subset} \mathbb{N}, \min N > \max F\, \forall n \in N \quad F \cup \{n\} \in \mathrm{Ep}(\mathcal{A}) \\ &\iff \exists N \overset{\infty}{\subset} \mathbb{N}, \min N > \max F\, \forall n \in N \quad (F \cup \{n\}) \setminus \{\min F \cup \{n\}\} \in \mathcal{A} \\ &\iff \exists N \overset{\infty}{\subset} \mathbb{N}, \min N > \max F\, \forall n \in N \quad (F \setminus \{\min F\}) \cup \{n\} \in \mathcal{A} \\ &\iff F \setminus \{\min F\} \in \mathcal{A}^{(1)} \iff F \in \mathrm{Ep}\big(\mathcal{A}^{(1)}\big). \end{aligned}$$

For general $\alpha \le \mathrm{CB}(\mathcal{A})$ the claim now follows easily by transfinite induction.
(b) Note that for $\beta < \alpha$ we can choose $m \in \mathbb{N}$ so that $\alpha_m > \beta$ and thus $\emptyset \in \mathcal{A}_m^{(\beta)} \subset \mathcal{A}^{(\beta)}$. Since α is a limit ordinal we deduce that $\emptyset \in \mathcal{A}^{(\alpha)}$ and, thus, that $\mathrm{CB}(\mathcal{A}) \ge \alpha + 1$.

On the other hand note that for any $m \in \mathbb{N}$ it follows that

$$\mathcal{A}^{(\alpha_m)} = \Big(\bigcup_{i=1}^{m} \mathcal{A}_i\Big)^{(\alpha_m)} \cup \Big(\bigcup_{i=m+1}^{\infty} \mathcal{A}_i\Big)^{(\alpha_m)} \subset [\{\ell_m, \ell_m + 1, \dots\}]^{<\infty}$$

and, thus, that $\mathcal{A}^{(\alpha)} \subset \bigcap_{m\in\mathbb{N}}[\{\ell_m, \ell_m + 1, \dots\}]^{<\infty} = \{\emptyset\}$, which implies that $\mathrm{CB}(\mathcal{A}) \le \alpha + 1$.
To prove (c) let first $\alpha = 1$ and $n \in \mathbb{N}$. For $F \in [\mathbb{N}]^{<\infty} \setminus \{\emptyset\}$, $\min F > n$, it follows that

$$\begin{aligned} F \in \big(\mathcal{A}|_n\big)^{(1)} &\iff \exists M \overset{\infty}{\subset} \mathbb{N}, \min M > n \forall m \in M \quad \{n\} \cup F \cup \{m\} \in \mathcal{A} \\ &\iff \{n\} \cup F \in \mathcal{A}^{(1)} \iff F \in \big(\mathcal{A}^{(1)}\big)|_n \end{aligned}$$

For general α we conclude the claim by transfinite induction. □

COROLLARY 4.9. *For $\alpha < \omega_1$, a hereditary and compact $\mathcal{A} \subset [\mathbb{N}]^{<\infty}$, and an $P \overset{\infty}{\subset} \mathbb{N}$ it follows that*

$$\text{(34)} \qquad \mathcal{A} \textit{ is } \alpha\textit{-large on } P \iff \forall Q \overset{\infty}{\subset} P \quad CB(\mathcal{A} \cap [Q]^{<\infty}) \ge \alpha + 1.$$

PROOF. In order to show "$\Rightarrow$" it is enough to observe that $\mathrm{CB}(\mathcal{F}_\alpha^N) = \alpha + 1$ for any $N \overset{\infty}{\subset}$ which follows by transfinite induction on α using (a) (in the successor case) and (b) (in the case of limit ordinals) of Lemma 4.8.

We also show "$\Leftarrow$" by transfinite induction on $\alpha < \omega_1$ and assume that the implication is true for all $\tilde{\alpha} < \alpha$.

Assume that $\mathcal{A} \subset [\mathbb{N}]^{<\infty}$ is compact and hereditary so that for all $Q \overset{\infty}{\subset} P$ it follows that $\mathrm{CB}(\mathcal{A} \cap [Q]^{<\infty}) \ge \alpha + 1$.

If $\alpha = \beta + 1$, it is by the induction hypothesis and by Proposition 4.3 enough to show:
Claim. $\forall Q \overset{\infty}{\subset} P \exists L \overset{\infty}{\subset} Q \forall \ell \in L \forall K \overset{\infty}{\subset} L, \ell < \min K \quad \mathrm{CB}(\mathcal{A} \cap [K]^{<\infty}) \ge \beta + 1.$

Assume the claim is not true and choose a $Q \overset{\infty}{\subset} P$ so that for all $L \overset{\infty}{\subset} Q$ there is an $\ell \in L$ and a $K \overset{\infty}{\subset} L$ so that $\mathrm{CB}(\mathcal{A} \cap [K]^{<\infty}) \le \beta$.

We first put $L_1 = Q$ and then choose $\ell_1 \in L_1$ and $K_1 \overset{\infty}{\subset} L_1$, with $\min K_1 > \ell_1$ so that $\mathrm{CB}(\mathcal{A}|_{\ell_1} \cap [K_1]^{<\infty}) \le \beta$. Then we let $L_2 = K_1$ and choose an $\ell_2 \in L_2$ and a $K_1 \overset{\infty}{\subset} L_2$ with $\min K_2 > \ell_2$ so that $\mathrm{CB}(\mathcal{A}|_{\ell_2} \cap [K_2]^{<\infty}) \le \beta$. We can continue in this way and eventually get a strictly increasing sequence $L = (\ell_i)$ and (L_i) with $Q = L_1 \overset{\infty}{\supset} L_2 \overset{\infty}{\supset} \dots$ so that $\mathrm{CB}(\mathcal{A}|_{\ell_i} \cap [L_{i+1}]^{<\infty}) \le \beta$. Thus it follows for each $i \in \mathbb{N}$ that

$$\big(\mathcal{A}|_{\ell_i} \cap [L]^{<\infty}\big)^{(\beta)} = \big(\mathcal{A}|_{\ell_i} \cap [\{\ell_{i+1}, \ell_{i+1}+1, \dots\}]^{<\infty}\big)^{(\beta)} \subset \big(\mathcal{A}|_{\ell_i} \cap [L_{i+1}]^{<\infty}\big)^{(\beta)} = \emptyset,$$

which implies by Lemma 4.8 (c) that $\mathcal{A}^{(\beta)} \cap [L]^{<\infty}$ must be finite and, thus, that $\mathrm{CB}(\mathcal{A} \cap [L]^{<\infty}) \le \beta + 1$, contradicting the assumption. This proves the claim and the induction step in the case that α is a successor.

If α is a limit ordinal and $\mathrm{CB}(\mathcal{A} \cap [Q]^{<\infty}) \ge \alpha + 1$ for all $Q \overset{\infty}{\subset} P$ it follows that for all $\beta < \alpha$ we have $\mathrm{CB}(\mathcal{A} \cap [Q]^{<\infty}) \ge \beta$, and, thus, by our induction hypothesis that $\mathcal{A}$ is β-large on P, for all $\beta < \alpha$, which implies, by Proposition 4.3 (c), that $\mathcal{A}$ is α-large on P. □

Using the equivalence of the strong Cantor Bendixson index and the concept of α-largeness and using Corollary 4.6 we can prove Proposition 1.3 of Section 1.

Proof of Proposition 1.3. Assume that X is a Banach space with a seminormalized basis (x_i) and let (z_n) another seminormalized basic sequence with $\gamma_0((z_n),(x_{k_n}))=\omega_1$.

By Corollary 4.9 we deduce for $1\le c<\infty$ and $\gamma<\omega_1$ that

$$c((z_n),(x_n),\gamma)=\inf\Big\{c\ge 1:\exists N\overset{\infty}{\subset}\mathbb{N}\quad \mathcal{C}((z_n),(x_n),c)\text{ is }\gamma\text{-large on }\mathbb{N}\Big\}<\infty.$$

Since $c((z_n),(x_n),\gamma)$ is non decreasing in γ we deduce from the uncountability of $[0,\omega_1)$ that there is a $c_0>0$ so that $c((z_n),(x_n),\gamma)\ge c_0$ for all $\gamma<\omega_1$. But this means that for any $\gamma<\omega_1$ the set $\mathcal{C}((z_n),(x_n),c_0/2)$ is γ large on some set $N_\gamma\overset{\infty}{\subset}\mathbb{N}$. From Corollary 4.6 we deduce therefore that there is an $L=(\ell_i)\overset{\infty}{\subset}\mathbb{N}$ so that $[L]^{<\infty}\subset\mathcal{C}((z_n),(x_n),c_0/2)$. By passing to a subsequence of (z_n) we might simply assume that $L=\mathbb{N}$.

Therefore we can choose for any $n\in\mathbb{N}$ a sequence $m_1^{(n)}<m_2^{(n)}<\dots m_n^{(n)}$ in N so that $(z_i)_{i=1}^n$ is $c_0/2$ equivalent to $(x_{m_i^{(n)}})_{i=1}^n$. Passing possibly to a subsequence of (z_n) and having possibly to redefine the $m_i^{(n)}$'s (note for a fixed sequence $(n_k)\subset\mathbb{N}$ we could change the choice of $m_i^{(n)}$ in such a way that $m_i^{(n)}=m_i^{(n_k)}$ if $i\le n$ and $n\in[n_{k-1}+1,n_k]$) we can assume one of the following two cases happens.

Case 1. There is a sequence (m_n) so that $m_i^{(n)}=m_i$ for all $n\in\mathbb{N}$ and $i\le n$.

In this case it follows that (z_i) is isomorphically equivalent to (x_{m_i}).

Case 2. For any $i\in\mathbb{N}$ it follows that $\lim_{n\to\infty}m_i^{(n)}=\infty$.

In the second case it follows that (z_i) is equivalent to a spreading model of a subsequence of (x_n). This proves the first part of Proposition 1.3.

In order to deduce the second part we assume that for all block bases (z_n) of (x_n) it follows $\gamma_0((z_n),(x_n))=\gamma_0((x_n),(z_n))=\omega_1$ and fix a block basis (z_n). By the first part of the proof a subsequence of (z_n) could be equivalent to a subsequence of (x_n), then we are done. Otherwise (z_n) is equivalent to a spreading model of a subsequence of (x_n) which means in particular that (z_n) is subsymmetric. Now we change the roles of (x_n) and (z_n), use the assumption $\gamma_0((x_n),(z_n))=\omega_1$, and go again through the arguments of the first part of the proof and observe that since (z_i) is subsymmetric both cases collaps to one and that a subsequence of (x_n) is isomorphically equivalent to a subsequence of (z_i). This proves that X is a space of Class 1. The other direction of the stated equivalence is trivial. □

We are now in the position to state and prove Theorem 4.10 concerning the universality of the transfinite families.

Theorem 4.10. *Let $\mathcal{A}\subset[\mathbb{N}]^{<\infty}$ be not empty and hereditary, and assume that $\alpha_0=\mathbb{I}(\mathcal{A},P)<\omega_1$. Then there is an approximating family $(B_n(\alpha))_{n\in\mathbb{N},\alpha\in Lim(\omega_1)}$ defining the transfinite family $(\mathcal{G}_\alpha)_{\alpha<\omega_1}$, and there is an $L\overset{\infty}{\subset}P$ so that*

$$\mathcal{G}^L_{\alpha_0}\subset\mathcal{A}\cap[L]^{<\infty}\subset\mathcal{G}_{\alpha_0}\cap[L]^{<\infty}. \tag{35}$$

Proof. Let $(\mathcal{F}_\alpha)$ be a transfinite family being chosen a priori. We will prove the claim by transfinite induction for all $\alpha_0<\omega_1$.

If $\alpha_0=0$, we deduce that $L=\{\ell\in P:\{\ell\}\notin\mathcal{A}\}$ is infinite and, thus, since $\mathcal{A}$ is hereditary and not empty it follows that

$$\mathcal{A}\cap[L]^{<\infty}=\{\emptyset\}=\mathcal{F}_0.$$

Assume the claim to be true for all hereditary $\tilde{\mathcal{A}} \subset [\mathbb{N}]^{<\infty}$ with $\mathbb{I}(\tilde{\mathcal{A}}, P) < \alpha_0$, where $\alpha_0 \geq 1$. Let $\mathcal{A} \subset [\mathbb{N}]^{<\infty}$ be hereditary with $\mathbb{I}(\mathcal{A}, P) = \alpha_0$. By passing to a subsequence of P, if necessary, we can assume that $\mathbb{I}(\mathcal{A}, L) = \alpha_0$ for all $L \overset{\infty}{\subset} P$ (we are using Proposition 4.2). Since $\mathcal{A}$ is not $\alpha_0 + 1$-large on P we deduce from Proposition 4.3 part (b) that there is a $Q \overset{\infty}{\subset} P$ so that

$$\forall M \overset{\infty}{\subset} Q \; \exists m \in M \quad \mathcal{A}|_m \text{ is not } \alpha_0\text{-large on } M. \tag{36}$$

We start by applying (36) to $M_1 = Q$ and find an $m_1 \in M_1$ for which $\beta_{m_1} = \mathbb{I}(\mathcal{A}|_{m_1}, M_1) < \alpha_0$ (recall that by Proposition 4.3 the set of ordinals α for which $\mathcal{A}|_{m_1}$ is α-large is a closed interval). By the induction hypothesis we can find an approximating family $(B_n^{(1)}(\gamma))_{n\in\mathbb{N},\gamma\in\mathrm{Lim}(\omega_1)}$ which defines a transfinite family $(\mathcal{G}_\gamma^{(1)})_{\gamma<\omega_1}$ and an $M_2 \subset M_1$ so that $\mathcal{A}|_{m_1} \cap [M_2]^{<\infty} \subset \mathcal{G}^{(1)}_{\beta_{m_1}} \cap [M_2]^{<\infty}$.

Since $\mathcal{A}|_{m_1}$ is β_{m_1}-large on M_1 (which does not depend on the choice of the approximating family) we also can require that $\big(\mathcal{G}^{(1)}_{\beta_{m_1}}\big)^{M_2} \subset \mathcal{A}|_{m_1}$.

By repeating this argument we find an increasing sequence $(m_i)_{i\in\mathbb{N}}$, sets $M_i \overset{\infty}{\subset} \mathbb{N}$, for $i \in \mathbb{N}$, a sequence of ordinals $(\beta_{m_i})_{i\in\mathbb{N}} \subset [0, \alpha_0)$, and approximating families $(B_n^{(i)}(\gamma))_{n\in\mathbb{N},\gamma\in\mathrm{Lim}(\omega_1)}$ defining transfinite families $(\mathcal{G}^{(i)})_{\gamma<\omega_1}$, for $i \in \mathbb{N}$, so that for all $i \in \mathbb{N}$

$$m_i \in M_i, M_{i+1} \overset{\infty}{\subset} M_i, \text{ and } m_i < \min M_{i+1}, \tag{37}$$

$$\big(\mathcal{G}^{(i)}_{\beta_{m_i}}\big)^{M_{i+1}} \subset \mathcal{A}|_{m_i} \cap [M_{i+1}]^{<\infty} \subset \mathcal{G}^{(i)}_{\beta_{m_i}} \cap [M_{i+1}]^{<\infty}, \tag{38}$$

Putting $M = \{m_1, m_2, \ldots\}$ we deduce from (37) and (38) that for all $i \in \mathbb{N}$

$$\big(\mathcal{G}^{(i)}_{\beta_{m_i}}\big)^{\{m_{i+1},m_{i+2},\ldots\}} \subset \mathcal{A}|_{m_i} \cap [M]^{<\infty} \subset \mathcal{G}^{(i)}_{\beta_{m_i}} \cap [M]^{<\infty}. \tag{39}$$

(for the second "$\subset$" recall that we defined $\mathcal{A}|_m$ in such a way that $\mathcal{A}|_m \subset [\{m+1, m+2, \ldots\}]^{<\infty}$) which implies that for any $m \in M$ and any $\tilde{M} \overset{\infty}{\subset} M$ we have $\mathbb{I}(\mathcal{A}|_m, \tilde{M}) = \beta_m$ and, thus, by Corollary 4.5, it follows that

$$\sup_{m\in M, m\geq k} \beta_m + 1 = \alpha_0, \qquad \text{for all } k \in \mathbb{N}. \tag{40}$$

To finish the proof we distinguish between the case that α_0 is a successor and the case that α_0 is a limit ordinal.

If $\alpha_0 = \gamma+1$ we deduce from (40) that the set $\tilde{L} = \{m \in M : \beta_m = \gamma\}$ is infinite. Using Proposition 3.2 (b) we can find an approximating family $(B_n(\alpha))_{n\in\mathbb{N},\alpha<\omega_1}$ so that for any $\alpha < \omega_1$ and any $i \in \mathbb{N}$ it follows that $\mathcal{G}^{(i)}_\alpha \cap [\{i, i+1 \ldots\}]^{<\infty} \subset \mathcal{G}_\alpha$, where $(\mathcal{G}_\alpha)$ is defined by $(B_n(\alpha))_{n\in\mathbb{N},\alpha<\omega_1}$. We therefore deduce that

$$\begin{aligned}
\mathcal{A} \cap [\tilde{L}]^{<\infty} &= \big\{\{\ell\} \cup E : \ell \in \tilde{L}, \ell < E \text{ and } E \in \mathcal{A}|_\ell\big\} \cap [\tilde{L}]^{<\infty} \cup \{\emptyset\} \\
&\subset \big\{\{\ell\} \cup E : \exists i \in \mathbb{N} \quad \ell = m_i \in \tilde{L}, \ell < E, E \in \mathcal{G}^{(i)}_\gamma\big\} \cap [\tilde{L}]^{<\infty} \cup \{\emptyset\} \\
&\quad \text{(by (39))} \\
&\subset \big\{\{\ell\} \cup E : \ell \in \tilde{L}, \ell < E \text{ and } E \in \mathcal{G}_\gamma\big\} \cap [\tilde{L}]^{<\infty} \cup \{\emptyset\} = \mathcal{G}_{\alpha_0} \cap [\tilde{L}]^{<\infty} \\
&\quad (\text{since } i \leq m_i).
\end{aligned}$$

If $\alpha_0 \in \operatorname{Lim}(\omega_1)$ we also define the approximating family $(B_n(\alpha))_{n\in\mathbb{N},\alpha<\omega_1}$ as in Proposition 3.2 (b), but add in the case of $\alpha = \alpha_0$ the ordinals $\beta_{m_1}+1, \beta_{m_2}+1, \dots \beta_{m_n}+1$ to the set $B_n(\alpha_0)$ (still denoting it $B_n(\alpha_0)$). The transfinite family defined by $(B_n(\alpha))_{n\in\mathbb{N},\alpha<\omega_1}$ is denoted by $(\mathcal{G}_\alpha)$. We put $\tilde{L} = M = \{m_1, m_2, \dots\}$.

Now if $E \in \mathcal{A} \cap [\tilde{L}]^{<\infty}$, $E \neq \emptyset$, we write $E = \{m\} \cup F$ with $m = m_n = \min E$ and $F \in \mathcal{A}|_{m_n} \subset \mathcal{G}^{(n)}_{\beta_{m_n}}$ (by (39)). From the definition of the family $(\mathcal{G}_\alpha)_{\alpha<\omega_1}$ we conclude that $F \in \mathcal{G}_{\beta_{m_n}}$ and thus $E \in \mathcal{G}_{\beta_{m_n}+1}$. Since $n \leq m_n \leq E$ and since $\beta_{m_n}+1 \in B_n(\alpha_0)$ we deduce that $E \in \mathcal{G}_{\alpha_0} \cap [\tilde{L}]^{<\infty}$.

Therefore we derive in both cases (α_0 being a successor and α_0 being a limitordinal) that

$$\mathcal{A} \cap [\tilde{L}]^{<\infty} \subset \mathcal{G}_{\alpha_0} \cap [\tilde{L}]^{<\infty}. \tag{41}$$

On the other hand it follows from the definition of α_0 and from Proposition 4.3 that $\mathcal{A}$ is α_0-large (which by Corollary 4.4 does not depend on the transfinite family). We can therefore chose an $L \overset{\infty}{\subset} \tilde{L}$ so that $\mathcal{G}^L_{\alpha_0} \subset \mathcal{A}$ which implies that

$$\mathcal{G}^L_{\alpha_0} \subset \mathcal{A} \cap [L]^{<\infty} \subset \mathcal{G}_{\alpha_0} \cap [L]^{<\infty}$$

and finishes the proof of Theorem 4.10. □

We will have to apply Theorem 4.10 not only for one $\mathcal{A} \subset [\mathbb{N}]^{<\infty}$ but for a sequence $(\mathcal{A}_n)$ simultaneously. Therefore we need the following reformulation.

COROLLARY 4.11. *Assume we are given a* $P \overset{\infty}{\subset} N$*, a sequence* $(\mathcal{A}_\ell)_{\ell\in\mathbb{N}}$ *of nonempty and hereditary subsets of* $[\mathbb{N}]^{<\infty}$*, an increasing sequence* $(\ell_k) \subset \mathbb{N}$*, and a sequence of ordinals* (α_ℓ) *so that*

$$\alpha_\ell \geq \mathbb{I}(\mathcal{A}_\ell, Q), \text{ whenever } \ell \in \mathbb{N} \text{ and } Q \overset{\infty}{\subset} P. \tag{42}$$

Then there is a transfinite family $(\mathcal{G}_\alpha)_{\alpha<\omega_1}$ *and* $K \overset{\infty}{\subset} P$, $K = \{k_1, k_2, \dots\}$, (k_i) *strictly increasing, so that for all* $n \in \mathbb{N}$ *and* $\ell \leq \ell_n$

$$\mathcal{A}_\ell \cap [\{k_n, k_{n+1}, \dots\}]^{<\infty} \subset G^K_{\alpha_\ell}. \tag{43}$$

PROOF. We first use Proposition 4.2 and an easy diagonalization argument to assume that $\mathbb{I}(\mathcal{A}_\ell, Q) = \mathbb{I}(\mathcal{A}_\ell, P)$ for all $Q \overset{\infty}{\subset} P$ (note that (42) stays valid if we pass to subsequences).

For $i \in \mathbb{N}$ we then apply Theorem 4.10 to each of $\mathcal{A}_\ell$, $\ell \leq \ell_i$, in order to get an approximating family $(B^{(i)}_n(\gamma))_{n\in\mathbb{N},\gamma\in\operatorname{Lim}(\omega_1)}$ with associated transfinite families $(\mathcal{G}^{(i)}_\alpha)_{\alpha<\omega_1}$ and an $L_i \overset{\infty}{\subset} P$ so that for any $\ell \leq \ell_i$ it follows that $L_i \overset{\infty}{\subset} L_{i-1}$ (with $L_0 = P$) and $\mathcal{A}_\ell \cap [L_i]^{<\infty} \subset \mathcal{G}_{\alpha_\ell}$.

Now define for $n \in \mathbb{N}$ and $\gamma \in \operatorname{Lim}(\omega_1)$ as in Proposition 3.2, i.e. $B_n(\gamma) = \bigcup_{j=1}^n B^{(j)}_n(\gamma)$, let $(\mathcal{H}_\alpha)$ be the transfinite family associated to $B_n(\gamma)$, and Let $K = \{k_1, k_2 \dots\}$ be a diagonal sequence of the L_i's. Then we deduce that for any $m \in \mathbb{N}$ and any $\ell \leq \ell_m$ it follows that

$$\begin{aligned}\mathcal{A}_\ell \cap [\{k_m, k_{m+1} \dots\}]^{<\infty} &= \mathcal{A}_\ell \cap [L_m]^{<\infty} \cap [\{k_m, k_{m+1}, \dots\}]^{<\infty} \\ &\subset \mathcal{G}^\ell_{\alpha_\ell} \cap [\{k_m, k_{m+1}, \dots\}]^{<\infty} \subset \mathcal{H}_{\alpha_\ell} \cap [K]^\infty.\end{aligned}$$

By Proposition 3.8 we then choose $(G_\alpha)_{\alpha<\omega_1}$ so that $\mathcal{H}_\alpha \cap [K]^{<\infty} \mathcal{G}^K_\alpha$, for $\alpha < \omega_1$. □

5. Some Consequences of Theorem 4.10

Using Theorem 4.10 we deduce the following generalization of Ramsey's theorem for finite sets.

COROLLARY 5.1. *Let* $\mathcal{F} \subset [\mathbb{N}]^{<\infty}$ *be hereditary and let* $P \overset{\infty}{\subset} \mathbb{N}$.

If $\mathcal{F} = \mathcal{A} \cup \mathcal{B}$ *with* $\mathcal{A}$ *and* $\mathcal{B}$ *also being hereditary then there is a* $Q \overset{\infty}{\subset} P$ *so that*

$$\max(\mathbb{I}(\mathcal{A}, Q), \mathbb{I}(\mathcal{B}, Q)) = \mathbb{I}(\mathcal{F}, Q) \tag{44}$$

PROOF. By passing to an infinite subsequence of P and using Proposition 4.2, if necessary, we can assume that there are ordinals α_0, β_0 and γ_0 so that for all $\tilde{P} \overset{\infty}{\subset} P$

$$\alpha_0 = \mathbb{I}(\mathcal{A}, \tilde{P}), \quad \beta_0 = \mathbb{I}(\mathcal{B}, \tilde{P}), \text{ and } \gamma_0 = \mathbb{I}(\mathcal{F}, \tilde{P}).$$

We need to show that $\max(\alpha_0, \beta_0) = \gamma_0$.

Assume that this is not true and, thus, assume that $\max(\alpha_0, \beta_0) < \gamma_0$.

By Theorem 4.10 and the fact that $\mathbb{I}(\mathcal{A}, P)$ and $\mathbb{I}(\mathcal{B}, P)$ are stabilized in the sense of Proposition 4.2 we find a transfinite family $(\mathcal{G}_\alpha)_{\alpha<\omega_1}$ and an $L \overset{\infty}{\subset} P$ so that $\mathcal{A} \cap [L]^{<\infty} \subset \mathcal{G}_{\alpha_0}$ and $\mathcal{B} \cap [\tilde{L}]^{<\infty} \subset \mathcal{G}_{\beta_0}$ and therefore, it would follow from Corollary 4.7 that $\mathbb{I}(\mathcal{A}, L) \leq \mathbb{I}(\mathcal{G}_{\max(\alpha_0,\beta_0)}, L) = \max(\alpha_0, \beta_0) < \gamma_0$ which is a contradiction. □

We introduce the following "addition" of subsets of $[\mathbb{N}]^{<\infty}$

DEFINITION 5.2. For $\mathcal{A}, \mathcal{B} \subset [\mathbb{N}]^{<}$ we define $\mathcal{A} \sqcup \mathcal{B} = \{A \cup B : A \in \mathcal{A}, B \in \mathcal{B}\}$

REMARK. At first sight one might believe that $\mathbb{I}(\mathcal{F}_\alpha \sqcup \mathcal{F}_\beta, N) = \alpha + \beta$. But this cannot be true since on one hand addition on the ordinal numbers is not commutative, on the other hand it is clear that $\mathcal{A} \sqcup \mathcal{B} = \mathcal{B} \sqcup \mathcal{A}$, for any $\mathcal{A}, \mathcal{B} \subset [\mathbb{N}]^{<\infty}$. For example it is easy to see that $\mathcal{F}_{\omega+1} \sqcup \mathcal{F}_\omega = 2\omega + 1 \neq 2\omega = \omega + 1 + \omega$.

One could define the following "commutative addition of ordinal numbers":

$$\alpha \sqcup \beta = \mathbb{I}(\mathcal{F}_\alpha \sqcup \mathcal{F}_\beta, \mathbb{N}).$$

It might be interesting to determine the properties of this binary operation and compare it with the addition of ordinal numbers. Nevertheless it is easy to prove by transfinite induction on β that for all $\alpha, \beta < \omega_1$ it follows that

$$\mathcal{F}_{\alpha+\beta} \subset \mathcal{F}_\alpha \sqcup \mathcal{F}_\beta. \tag{45}$$

PROPOSITION 5.3. *Assume that* $\alpha < \beta < \omega_1$ *and that* $\gamma < \omega_1$

a) *There is an* $m \in \mathbb{N}$ *so that* $\mathcal{F}_\alpha \cap [\{m, m+1, \dots\}]^{<\infty} \sqcup \mathcal{F}_\gamma \subset \mathcal{F}_\beta \cap [\{m, m+1, \dots\}]^{<\infty} \sqcup \mathcal{F}_\gamma$

b) *For any* $N \overset{\infty}{\subset} \mathbb{N}$ *it follows that* $\mathbb{I}(\mathcal{F}_\alpha \sqcup \mathcal{F}_\gamma, N) < \mathbb{I}(\mathcal{F}_\beta \sqcup \mathcal{F}_\gamma, N)$

PROOF. (a) follows immediately from Proposition 3.1(c). To prove (b) define $\delta =$ $\mathbb{I}(\mathcal{F}_\alpha \sqcup \mathcal{F}_\gamma, N)$. Let $M \overset{\infty}{\subset} N$. By Proposition 4.3 (last part) there is an $L \overset{\infty}{\subset} M$ so that $\mathcal{F}_\delta^L \subset \mathcal{F}_\alpha \sqcup \mathcal{F}_\gamma$. Then note that

$$\begin{aligned} \mathcal{F}_{\delta+1}^L &= \{\{\ell\} \cup D : D \in \mathcal{F}_\delta^L \text{ and } \ell \in L\} \\ &\subset \{\{\ell\} \cup A \cup G : A \in \mathcal{F}_\alpha \cap [L]^{<\infty}, G \in \mathcal{F}_\gamma \cap [L]^{<\infty} \ell \in L\} \\ &\subset \{\tilde{A} \cup G : \tilde{A} \in \mathcal{F}_{\alpha+1} \cap [L]^{<\infty}, G \in \mathcal{F}_\gamma\} \subset \mathcal{F}_{\alpha+1} \sqcup \mathcal{F}_\gamma. \end{aligned}$$

Thus, by (a), $\delta = \mathbb{I}(\mathcal{F}_\alpha \sqcup \mathcal{F}_\gamma, N) < \delta + 1 \le \mathbb{I}(\mathcal{F}_{\alpha+1} \sqcup \mathcal{F}_\gamma, N) \le \mathbb{I}(\mathcal{F}_\beta \sqcup \mathcal{F}_\gamma, N)$ which finishes the proof. □

PROPOSITION 5.4. *Assume that $(\mathcal{F}_\alpha)_{\alpha<\omega_1}$ and $(\mathcal{G}_\alpha)_{\alpha<\omega_1}$ are two transfinite families and let $N \overset{\infty}{\subset} \mathbb{N}$ and $\alpha < \beta$ be such that $\mathbb{I}(\mathcal{F}_\alpha \sqcup \mathcal{F}_\beta, N)$ and $\mathbb{I}(\mathcal{G}_\alpha \sqcup \mathcal{G}_\beta, N)$ are stabilized in the sense of Proposition 4.2, i.e. $\mathbb{I}(\mathcal{F}_\alpha \sqcup \mathcal{F}_\beta, N) = \mathbb{I}(\mathcal{F}_\alpha \sqcup \mathcal{F}_\beta, \tilde{N})$ and $\mathbb{I}(\mathcal{G}_\alpha \sqcup \mathcal{G}_\beta, N) = \mathbb{I}(\mathcal{G}_\alpha \sqcup \mathcal{G}_\beta, \tilde{N})$ for all $\tilde{N} \subset N$.*

Then it follows that $\mathbb{I}(\mathcal{F}_\alpha \sqcup \mathcal{F}_\beta, N) = \mathbb{I}(\mathcal{G}_\alpha \sqcup \mathcal{G}_\beta, N)$.

PROOF. Write N as $N = \{n_1, n_2, \ldots\}$, with $n_i \nearrow \infty$. By Proposition 3.7 we can find an $M = \{m_1, m_2, \ldots\} \subset N$, $m_i \nearrow \infty$, so that $\mathcal{G}_\alpha^M \subset \mathcal{F}_\alpha$ and $\mathcal{G}_\beta^M \subset \mathcal{F}_\beta$. Define $L = \{m_{n_i} : i \in \mathbb{N}\}$. Then it follows that $\mathbb{I}(\mathcal{G}_\alpha \sqcup \mathcal{G}_\beta, N) = \mathbb{I}(\mathcal{G}_\alpha^M \sqcup \mathcal{G}_\beta^M, L) \le \mathbb{I}(\mathcal{F}_\alpha \sqcup \mathcal{F}_\beta, L) = \mathbb{I}(\mathcal{F}_\alpha \sqcup \mathcal{F}_\beta, N)$, which finishes the proof by symmetry. □

PROPOSITION 5.5. *(Cancellation Lemma)*

Let $\alpha, \beta < \omega_1$ and consider a map $\Psi : \mathcal{F}_\alpha \sqcup \mathcal{F}_\beta \to [\mathbb{N}]^{<\infty}$ with the following property:

There exists a hereditary $\mathcal{B} \subset [\mathbb{N}]^{<\infty}$ and an $N \overset{\infty}{\subset} \mathbb{N}$ so that $\Psi(\mathcal{F}_\alpha \sqcup \mathcal{F}_\beta) \subset \mathcal{B}$ and $\mathbb{I}(\mathcal{B}, N) \le \beta$.

If $\mathcal{C} \subset [\mathbb{N}]^{<\infty}$ is hereditary and contains the set $\{A \setminus \Psi(A) : A \in \mathcal{F}_\alpha \sqcup \mathcal{F}_\beta\}$ then there is an $M \overset{\infty}{\subset} N$ so that $\mathbb{I}(\mathcal{C}, M) \ge \alpha$.

PROOF. Assume that our claim is not true and that for all $M \overset{\infty}{\subset} N$ it follows that $\mathbb{I}(\mathcal{C}, M) < \alpha$. First, by applying Proposition 4.2 and passing to a subsequence of N, we can assume that $\mathbb{I}(\mathcal{B}, \tilde{N}) = \mathbb{I}(\mathcal{B}, N) \le \beta$ for all $\tilde{N} \overset{\infty}{\subset} N$. By applying Proposition 4.2 a second time we can also assume that $\alpha_0 = \mathbb{I}(\mathcal{C}, \tilde{N}) = \mathbb{I}(\mathcal{C}, N) < \alpha$, for all $\tilde{N} \overset{\infty}{\subset} N$, and define $\beta_0 = \mathbb{I}(\mathcal{B}, N) \le \beta$.

Using Theorem 4.10 (since we are in the stabilized situation we can apply it simultaneously to $\mathcal{C}$ and $\mathcal{B}$ as in the proof of Corollary 4.11) we obtain a transfinite family $(\mathcal{G}_\gamma)$ and a $M \overset{\infty}{\subset} N$ so that $\mathcal{B} \cap [M]^{<\infty} \subset \mathcal{G}_{\beta_0}$ and $\mathcal{C} \cap [M]^{<\infty} \subset \mathcal{G}_{\alpha_0}$ and thus $(\mathcal{B} \sqcup \mathcal{C}) \cap [M]^{<\infty} \subset \mathcal{G}_{\beta_0} \sqcup \mathcal{G}_{\alpha_0}$.

On the other hand it is clear that $\mathcal{F}_\alpha \sqcup \mathcal{F}_\beta \subset \mathcal{C} \sqcup \mathcal{B}$. Now, using Proposition 4.2 we first find a subset $L \overset{\infty}{\subset} M$ so that $\mathbb{I}(\mathcal{G}_{\beta_0} \sqcup \mathcal{G}_{\alpha_0}, \tilde{L}) = I(\mathcal{G}_{\beta_0} \sqcup \mathcal{G}_{\alpha_0}, L) = \mathbb{I}(\mathcal{G}_{\beta_0} \sqcup \mathcal{G}_{\alpha_0}, M)$ for all $\tilde{L} \subset L$. Then we pass to a subset $K \overset{\infty}{\subset} L$ so that and $\mathbb{I}(\mathcal{F}_\alpha \sqcup \mathcal{F}_\beta, \tilde{K}) = \mathbb{I}(\mathcal{F}_\alpha \sqcup \mathcal{F}_\beta, K) = \mathbb{I}(\mathcal{F}_\alpha \sqcup \mathcal{F}_\beta, L)$, for all $\tilde{K} \overset{\infty}{\subset} K$.

Finally we deduce the following chain of inequalities

$$\mathbb{I}(\mathcal{F}_\alpha \sqcup \mathcal{F}_\beta, K) \le \mathbb{I}(\mathcal{C} \sqcup \mathcal{B}, K) \le \mathbb{I}(\mathcal{G}_{\alpha_0} \sqcup \mathcal{G}_{\beta_0}, K) = \mathbb{I}(\mathcal{F}_{\alpha_0} \sqcup \mathcal{F}_{\beta_0}, K)$$

(the last inequality follows from Proposition 5.4). But on the other it follows from Proposition 5.3 that $\mathbb{I}(\mathcal{F}_\alpha \sqcup \mathcal{F}_\beta, K) > \mathbb{I}(\mathcal{F}_{\alpha_0} \sqcup \mathcal{F}_\beta, K) \ge \mathbb{I}(\mathcal{F}_{\alpha_0} \sqcup \mathcal{F}_{\beta_0}, K)$ which is a contradiction. □

6. Proof of Theorem 1.4

We now turn to the proof of Theorem 1.4. We will first restate it in an equivalent form using the equivalence of the Cantor Bendixson index and the concept of α-largeness. Recall the definitions of $\mathcal{B}(\overline{z}, b)$, $B(X, \beta)$, $\beta_0(\overline{z})$ and $\beta_0(X)$ in Definition

1.2 for X being a Banach space, $\overline{z} = (z_n) \subset X$ seminormalized, $b > 0$ and $\beta < \omega_1$. Using Corollary 4.9 we get

$$b(\overline{z}, \beta) = \inf\Big\{b \geq 1 : \exists N \overset{\infty}{\subset} \mathbb{N} \quad \mathcal{B}(\overline{z}, b) \text{ is } \beta\text{-large on } N\Big\}, \tag{46}$$

where for $b \geq 1$ and $\mathcal{B}(\overline{z}, b) = \{A \in [\mathbb{N}]^{<\infty} : (z_n)_{n\in A} \sim_b \ell_1^A - \text{unit vector basis}\}$.

We secondly want to replace in the definition of $b(\overline{z}, \beta)$ the set $\mathcal{B}(\overline{z}, b)$ by a somewhat more convenient set. We will need the following special case of a result from [**AMT**] (see also [**AG**] Lemma 3.2).

LEMMA 6.1. *Assume (x_n) is a weakly null and semi-normalized sequence in a Banach space X. Let $\delta > 0$, $N \overset{\infty}{\subset} \mathbb{N}$ and $\varepsilon_n > 0$, for $n \in \mathbb{N}$. Then there exists an $M \overset{\infty}{\subset} \mathbb{N}$, $M = \{m_1, m_2, \ldots\}$ so that for all finite $F \subset M$ the following implication is true:*

(47)

If there is an $x^* \in B_{X^*}$, with $x^*(x_m) \geq \delta$ for all $m \in F$, then there is a $y^* \in B_{X^*}$, with $y^*(x_m) \geq \delta$ for all $m \in F$ and $|y^*(x_{m_i})| < \varepsilon_i$ for all $i \in \mathbb{N}$, with $m_i \notin F$.

For the sake of being self contained and reader-friendly we present a proof of this special case of the above cited result of [**AG**].

PROOF. By recursion we choose for every $k \in \mathbb{N}$, $m_k \in \mathbb{N}$ and $L_k \overset{\infty}{\subset} N$ with $m_1 < m_2 < \ldots m_k$, $L_k \overset{\infty}{\subset} L_{k-1} \ldots \overset{\infty}{\subset} L_1 \overset{\infty}{\subset} L_0 = N$, and $m_k = \min L_{k-1}$ so that for all $F \subset \{1, \ldots, k\}$ all $L \overset{\infty}{\subset} L_k$, $L = \{\ell_1, \ell_2, \ldots\}$, and all $n \in \mathbb{N}$ the following implication holds:

(48)

$$[\exists x* \in B_{X^*} \quad x^*(x_{m_i}) \geq \delta \text{ for all } i \in F \text{ and } x^*(x_{\ell_i}) \geq \delta \text{ for all } i = 2, 3, \ldots n]$$

$$\Rightarrow \Bigg[\exists y^* \in B_{X^*} \quad y^*(x_{m_i}) \geq \delta \text{ for all } i \in F, \quad y^*(x_{\ell_i}) \geq \delta \text{ for all } i = 2, 3, \ldots n$$

$$|y^*(x_{m_i})| < \varepsilon_i \text{ for all } i \in \{1, \ldots k\} \setminus F \text{ and } |y^*(x_{\ell_1})| < \varepsilon_{k+1}\Bigg].$$

Clearly the claim of the Lemma follows if we can accomplish such a choice of m_k's and L_k's.

Assume for some $k \geq 1$ we have chosen L_{k-1} (recall: $L_0 = N$) and $m_1 < \ldots m_{k-1}$.

We define $m_k = \min L_{k-1}$ and

$$\mathcal{L} = \{L \subset L_{k-1} : \forall F \subset \{1, \ldots k\} \quad \forall n \in \mathbb{N} \quad (L, F, n) \text{ satisfies (48) }\}.$$

It is easy to see that $\mathcal{L}$ is closed in the pointwise topology and we can apply Ramsey's theorem.

In the case that there is an $L \in \mathcal{L}$ so that $[L]^\infty \subset \mathcal{L}$ we are done. We have to show that the alternative in Ramsey's theorem leads to a contradiction.

Assume that there is an $\tilde{L} \overset{\infty}{\subset} L_{k-1}$ so that $[\tilde{L}]^\infty \cap \mathcal{L} = \emptyset$. Thus for any $L = \{\ell_1, \ell_2 \ldots\} \overset{\infty}{\subset} \tilde{L}$ there is an $F = F_L \subset \{1, 2 \ldots k\}$ and an $n = n_L \in \mathbb{N}$ so that there exists an $x^* = x_L^* \in B_{X^*}$ with

$$x^*(x_{m_i}) > \delta \text{ for all } i \in F, \text{ and } x^*(x_{\ell_i}) > \delta, \text{ for all } i = 2, 3, \ldots n, \tag{49}$$

but for any $y^* \in B_{X^*}$ satisfying (49) there must be either an $i \in \{1, \dots k\} \subset F$ with $|y^*(x_{m_i})| > \varepsilon_i$ or $|y^*(x_{\ell_1})| \geq \varepsilon_{k+1}$.

We first use again Ramsey's theorem to assume without loss of generality that the sets F_L do not depend on L. Thus $F_L = F$ for all $L \overset{\infty}{\subset} \tilde{L}$.

Fixing for a moment such an $L = \{\ell_1, \ell_2 \dots\} \overset{\infty}{\subset} \tilde{L}$ we let $j_0 = \max\{j \in [0, k] : j \notin F\}$. If $j_0 = 0$ (meaning $F = \{1, 2 \dots k\}$) we put $z_L^* = x_L^*$ and observe that we must have $|z_L^*(x_{\ell_1})| \geq \varepsilon_{k+1}$. If $j_0 \geq 1$ we apply the fact that our induction hypothesis is true for $j_0 - 1$ and are able to find a $z_L^* \in B_{X^*}$ satisfying (49) and secondly

$$|z_L^*(x_{m_i})| < \varepsilon_i \text{ if } i \in \{1, 2, \dots k\} \setminus F \tag{50}$$

and, thus, we also must have $|z_L^*(x_{\ell_1})| > \varepsilon_{k+1}$ (apply the induction hypothesis to the set
$L = \{m_{j_0}, \dots m_k, \ell_2, \ell_3, \dots\} \subset L_{j_0 - 1}$, $F \cap [1, j_0 - 1]$ and $n + k - j_0 + 1$).

We write $\tilde{L} = \{\tilde{\ell}_1, \tilde{\ell}_2, \dots\}$ and claim that for any $m \in \mathbb{N}$ there is a x_m^* so that $|x_m^*(x_{\tilde{\ell}_i})| \geq \varepsilon_{k+1}$, for $i = 1, 2, \dots m$. This would be a contradiction to the assumption that (x_i) is weakly null.

For each $j = 1, 2, \dots m$ we find an n_j so that the triple $(\tilde{L}_j, F, n_j)$ does not satisfy (48) with $\tilde{L}_j = \{\tilde{\ell}_j, \tilde{\ell}_{m+1}, \tilde{\ell}_{m+2} \dots\}$. Choose j_0 so that n_{j_0} is the maximum of $(n_j)_{j=1}^m$. Then choose $x_m^* = z_{L_{j_0}}^*$ (where z_L is defined as above).

We observe that x_m^* satisfies (49) with respect to all of the L_j's. Secondly it satisfies (50) and thus it must follow that $|x_m^*(x_{\tilde{\ell}_j})| > \varepsilon_{k+1}$ for all $j =, 2, \dots m$. This finishes the proof of the claim and, thus the proof of the Lemma. □

We will slightly reformulate Lemma 6.1.

Corollary 6.2. *Assume (x_n) is a weakly null and semi-normalized sequence in a Banach space X. Let $\delta > 0$, $N \overset{\infty}{\subset} \mathbb{N}$ and $\varepsilon > 0$.*

Then there exists an $M \overset{\infty}{\subset} \mathbb{N}$, $M = \{m_1, m_2, \dots\}$ so that for all finite $F \subset M$ the following implication is true:

$$\begin{aligned} &\text{If there is an } x^* \in B_{X^*}, \text{ with } x^*(x_m) \geq \delta \text{ for all } m \in F, \text{ then there is a} \\ &z^* \in B_{X^*}, \text{ with } z^*(x_m) \geq \delta - \varepsilon \text{ for all } m \in F \text{ and } z^*(x_{m_i}) = 0, \\ &\text{for all } i \in \mathbb{N}, \text{ with } m_i \notin F. \end{aligned} \tag{51}$$

Proof. After passing to a subsequence of x_n we can assume that there is for each $n \in \mathbb{N}$ an $x_n^* \in (2/\|x_n\|) B_{X^*}$, with $x_n^*(x_m) = \delta_{(n,m)}$, whenever $n, m \in \mathbb{N}$. Choose for $n \in \mathbb{N}$ $\varepsilon_n = 2^{-n}\varepsilon/(2 + \sup_{n \in \mathbb{N}} \|x_n\|)$ and apply Lemma 6.1 in order to obtain $M = (m_i) \overset{\infty}{\subset} \mathbb{N}$.

If $F \subset M$ is finite and $x^* \in B_{X^*}$ so that $x^*(x_i) \geq \delta$, whenever $i \in F$, then we let $y^* \in B_{X^*}$ be as prescribed in (47) and let $z^* = \tilde{z}^*/\|\tilde{z}^*\|$ where $\tilde{z}^* = y^* - \sum_{m \in M \setminus F} y^*(x_m) x_m^*$. □

We introduce notations similar to $\mathcal{B}(\overline{z}, b)$, $b(\overline{z}, \beta)$, $B(\beta, X)$, $\beta_0(\overline{z})$ and $\beta_0(X)$.

Definition 6.3. Let $\overline{x} = (x_n)$ be a seminormalized sequence in a Banach space X. For $a > 0$ we put

$$\mathcal{A}(\overline{x}, a) = \big\{A \in [\mathbb{N}]^{<\infty} : \exists x^* \in B_{X^*} \forall i \in A \quad x^*(x_i) \geq a\big\}. \tag{52}$$

For $\alpha < \omega_1$ we let

$$
\begin{aligned}
(53)\qquad a(\overline{x},\alpha) &= \sup\{a \geq 0 : \exists N \overset{\infty}{\subset} \mathbb{N} \quad \mathcal{A}(\overline{x},a) \text{ is } \alpha\text{-large on } N\} \\
&= \sup\{a \geq 0 : \exists M \overset{\infty}{\subset} \mathbb{N} \quad \mathcal{F}_\alpha^M \subset \mathcal{A}(\overline{x},a)\},
\end{aligned}
$$

$$(54)\qquad \alpha_0(\overline{x}) = \sup\{\alpha < \omega_1 : a(\overline{x},\alpha) > 0\},$$

$$(55)\qquad A(\alpha,X) = \sup\{a(\overline{z},\alpha) : \overline{z} \subset B_X \text{ seminormalized and weakly null}\}, \text{ and}$$

$$(56)\qquad \alpha_0(X) = \sup\{\alpha{<}\omega_1 : A(\alpha,X) > 0\} = \sup\{\alpha_0(\overline{z}) : \overline{z}{\subset} B_X \text{ seminorm., weakly null}\}.$$

LEMMA 6.4. *Let $\overline{x}$ be a seminormalized sequence in a Banach space X with $\alpha_0(\overline{x}) < \omega_1$.*

Then there is subsequence $\overline{y} = (y_n)$ of $\overline{x}$ with the following properties.

a) *For all $\alpha < \omega_1$ and all $a' < a(\overline{y},\alpha)$ the set $\mathcal{A}(\overline{y},a')$ is α-large on $\mathbb{N}$.*
b) *For all $\alpha < \omega_1$ and all subsequences $\overline{z}$ of $\overline{y}$ it follows that $a(\overline{y},\alpha) = a(\overline{z},\alpha)$.*
c) *The map $[0,\omega_1) \ni \alpha \mapsto a(\overline{y},\alpha)$ is decreasing and continuous.*

Moreover if $\beta_0 < \alpha_0(\overline{x})$ (and, thus $a(\overline{x},\beta_0) > 0$) and if $0 < \eta < a(\overline{x},\beta_0)$ the subsequence $\overline{y}$ can be chosen so that $a(\overline{y},\beta_0) > a(\overline{x},\beta_0) - \eta$.

PROOF. We first note that if $(\overline{u}) = (u_n)$ is almost a subsequence of $\overline{v} = (v_n)$ (i.e. for some $n_0 \in \mathbb{N}$ it follows that $(u_{n_0+i})_{i\in\mathbb{N}}$ is a subsequence of $\overline{v}$) then $a(\overline{u},\alpha) \leq a(\overline{v},\alpha)$. In particular this means that $\alpha_0(\overline{u}) \leq \alpha_0(\overline{v})$ and therefore it is enough to find a subsequence of $\overline{x}$ which satisfies (a), (b) and (c) for all $\alpha < \alpha_0(\overline{x})$.
Claim. Let $\overline{y} = (y_n)$ be a subsequence of $\overline{x}$ and let $\alpha < \alpha_0(\overline{x})$. Then there is a subsequence $\overline{z}$ of $\overline{y}$ so that for all $a' < a(\overline{z},\alpha)$ it follows that $\mathcal{A}(\overline{z},a')$ is α-large on $\mathbb{N}$.

Because of the observation at the beginning of the proof the sequence $\overline{z}$ in the claim has the property that $a(\overline{u},\alpha) = a(\overline{z},\alpha)$ for any sequence $\overline{u}$ which is almost a subsequence of $\overline{z}$.

In order to show the claim we let $\varepsilon_i \searrow 0$, put $a_0 = a(\overline{y},\alpha)$ and choose an $N_1 = (n_i^{(1)}) \overset{\infty}{\subset} \mathbb{N}$ so that $\mathcal{F}_\alpha^{N_1} \subset \mathcal{A}(\overline{y},a_0-\varepsilon_1)$. Letting now $\overline{z}^{(1)} = (y_{n_i^{(1)}})$ we deduce that $\mathcal{F}_\alpha \subset \mathcal{A}(\overline{z}^{(1)},a_0-\varepsilon_1)$ and, thus, $\mathcal{F}_\alpha \subset \mathcal{A}(\overline{z},a_0-\varepsilon_1)$ for any subsequence $\overline{z}$ of $\overline{z}^{(1)}$ (recall that $\mathcal{F}_\alpha$ is spreading), which finally implies that $\mathcal{A}(\overline{z},a_0-\varepsilon_1)$ is α-large for any sequence $\overline{z}$ which is almost a subsequence of $\overline{z}^{(1)}$.

Now we let $a_1 = a(\overline{z}^{(1)},\alpha)$ and continue this way, eventually finding $\mathbb{N} \overset{\infty}{\supset} N_1 = (n_i^{(1)}) \overset{\infty}{\supset} N_2 = (n_i^{(2)}) \overset{\infty}{\supset} \ldots$, so that if we put $\overline{z}^{(k)} = (y_{n_i^{(k)}})$ and $a_k = a(\overline{z}^{(k)},\alpha)$ it follows that $\mathcal{F}_\alpha \subset \mathcal{A}(\overline{z}^{(k)},a_{k-1}-\varepsilon_k)$. We deduce that $a_{k-1}-\varepsilon_k \leq a_k \leq a_{k-1}$. Letting $\overline{z}$ be a diagonal sequence of the $\overline{z}^{(k)}$'s it follows that $a(\overline{z},\alpha) = a = \inf a_k$ and, since every subsequence of $\overline{z}$ is almost a subsequence of each $\overline{z}^{(k)}$ it follows for each $k \in \mathbb{N}$ that $\mathcal{A}(\overline{z},a_{k-1}-\varepsilon_k)$ is α-large, which implies the claim.

Note also, that if we had assumed that none of the ε_i's would exceed a value η then it follows that $a(\overline{z},\alpha) > a(\overline{y},\alpha) - \eta$ (this proves the part of our claim starting with "moreover" if we let $\alpha = \beta_0$). Writing now the interval $[0,\alpha_0(\overline{x}))$ as a sequence (α_n) and applying successively the above claim to each α_n, we obtain by diagonalization a subsequence $\overline{y}$ of $\overline{x}$ so that (a) and (b) of our statement are satisfied. It is also clear that $a(\overline{y},\alpha)$ is decreasing in α. Let α be a limit ordinal and $a' < a = \lim_{\beta\to\alpha} a(\overline{y},\beta) = \inf_{\beta<\alpha} a(\overline{y},\beta)$, then it follows that for every $\beta < \alpha$ that

$\mathcal{A}(\overline{y}, a')$ is β-large on $\mathbb{N}$. By Proposition 4.3 this implies that $\mathcal{A}(\overline{y}, a')$ is α-large on $\mathbb{N}$ for all $a' < a$, which implies the claimed continuity. $\square$

PROPOSITION 6.5. *Let $\overline{x} = (x_n)$ be a weakly null, and normalized sequence in a Banach space X and $c > \eta > 0$. Then there is a subsequence $\overline{y}$ of $\overline{x}$ so that*

$$\mathcal{A}(\overline{y}, 2c + \eta) \subset \mathcal{B}(\overline{y}, \frac{1}{c}) \subset \mathcal{A}(\overline{y}, c - \eta) \tag{57}$$

and, thus it follows that

$$\frac{1}{2}a(\overline{x}, \alpha) \le \frac{1}{b(\overline{x}, \alpha)} \le a(\overline{x}, \alpha) \text{ and } \frac{1}{2}A(X, \alpha) \le \frac{1}{B(X, \alpha)} \le A(X, \alpha), \tag{58}$$

$$\alpha_0(\overline{x}) = \beta_0(\overline{x}), \text{ and } \alpha_0(X) = \beta_0(X). \tag{59}$$

Proposition 6.5 will follow from Corollary 6.2 and the following simple observation.

LEMMA 6.6. *Assume $E = (\mathbb{R}^n, \|\cdot\|)$, $n \in \mathbb{N}$ is an n-dimensional normed space for which the unit vector basis $(e_i)_{i=1}^n$ of $\mathbb{R}^n$ is a normalized basis. Define:*

$$c_1 = \max\left\{c \ge 0 : \forall A \subset \{1, 2 \dots n\} : \exists x^* \in B_{E^*} \quad \begin{matrix} x^*(e_i) \ge c \text{ if } i \in A \\ x^*(e_i) = 0 \text{ if } i \notin A \end{matrix}\right\}$$

$$c_2 = \max\{c \ge 0 : cB_{\ell_\infty^n} \subset B_{E^*}\} \text{ and } c_3 = \min\Big\{\|\sum_{i=1}^n a_i e_i\| : \sum_{i=1}^n |a_i| = 1\Big\}$$

Then it follows that $c_1 \ge c_2 = c_3 \ge \frac{1}{2}c_1$.

PROOF. It is clear that $c_1 \ge c_2$. To show that $c_2 \ge c_3$ we first observe that by the maximality of c_2 we can find an $x^* \in S_{X^*}$ of the form $x^* = \sum_{i=1}^n x_i^* e_i^* \in S_{E^*}$ (e_i^* being the i-th coordinate functional) so that $|x_i^*| = c_2$ for $i = 1, \dots n$. Then choose $x = (x_i) \in S_E$ so that $x^*(x) = \sum x_i x_i^* = 1$. Thus $1 = \sum x_i x_i^* \le c_2 \sum |x_i| \le c_2/c_3$, which implies the claimed inequality.

In order to show $c_3 \ge c_2$ and $c_3 \ge c_1/2$ let $(a_i)_{i=1}^n \in \mathbb{R}^n$ with $\sum_{i=1}^n |a_i| = 1$. First we can choose an $x^* = \sum_{i=1}^n x_i^* e_i^* \in B_{E^*}$ with $|x_i^*| = c_2$ and $\text{sign}(x_i^*) = \text{sign}(a_i)$, for $1 \le i \le n$. This proves that $\|\sum_{i=1}^n a_i e_i\| \ge x^*(\sum_{i=1}^n a_i e_i) \ge c_2$, which implies $c_3 \ge c_2$. Secondly, we can assume that $\sum_{i=1}^n a_i^+ \ge 1/2$ and a simliar argument implies that $\|\sum_{i=1}^n a_i e_i\| \ge c_1/2$, and thus $c_3 \ge c_1/2$. $\square$

PROOF OF PROPOSITION 6.5. Assume $c > \eta > 0$ to be given. Using Corollary 6.2 we can find a subsequence $\overline{y} = (y_n)$ of $\overline{x}$ so that

$\mathcal{A}(\overline{y}, 2c + \eta) \subset \tilde{\mathcal{A}}(\overline{y}, 2c)$, and (trivially) $\tilde{\mathcal{A}}(\overline{y}, c) \subset \mathcal{A}(\overline{y}, c)$, where we put for $r > 0$

$\tilde{\mathcal{A}}(\overline{y}, r) = \{A \in [\mathbb{N}]^{<\infty} : \forall B \subset A \exists x^* \in B_{X^*} \quad x^*(y_i) \ge r$ if $i \in B$ and $x^*(y_i) = 0$ if $i \notin B\}$.

Secondly we deduce from Lemma 6.6 that

$$\tilde{\mathcal{A}}(\overline{y}, 2c) \subset \{A \in [N]^\infty : \forall (a_i)_{i \in A} \subset \mathbb{R} \quad \|\sum_{i \in A} a_i y_i\| \ge c \sum_{i \in A} |a_i|\} \subset \tilde{\mathcal{A}}(\overline{y}, c),$$

proving the claim (note that $1/c_3$, where c_3 as defined in Lemma 6.6 is the smallest c so that E is c-equivalent to ℓ_1^n). $\square$

We now can restate Theorem 1.4 as follows.

THEOREM 6.7. *Let X be a Banach space with a basis not containing c_0. Assume that there is an ordinal $\beta_0 \in [0, \alpha_0(X)]$ so that the following two conditions hold.*

a) *There is a seminormalized weakly null sequence $\overline{y} \subset B_X$ with $a(\overline{y}, \beta_0) = 0$.*
b) $\inf_{\gamma<\beta_0} B(X, \gamma) > 0$.

Then there is a seminormalized block basis (x_n) in X and a subsequence $(\tilde{y}_n)$ of (y_n) so that the map $x_n \mapsto \tilde{y}_n$ extends to a strictly singular and linear bounded operator

$$T : [x_n : n \in \mathbb{N}] \to [\tilde{y}_n : n \in \mathbb{N}].$$

In order to prove Theorem 6.7 we will need several Lemmas.

LEMMA 6.8. *Let $(\mathcal{F}_\alpha)$ be a transfinite family and $\overline{x} = (x_n)$ be a weakly null and semi-normalized sequence in a Banach space X satisfying the conclusions of Lemma 6.4. Let $1 \le \alpha < \omega_1$ and assume that $a(\overline{x}, \alpha) > 0$. For any $\eta > 0$ there is a subsequence $\overline{z}$ of $\overline{x}$ so that:*

For $A \in \mathcal{F}_\alpha$, or $A \in \mathcal{F}_\alpha \sqcup \mathcal{F}_\alpha$, there is a $z^*_A \in B_{X^*}$ so that $z^*_A(x_m) = 0$ if $m \in \mathbb{N} \setminus A$ and $z^*_A(x_m) \ge a(\overline{x}, \alpha) - \eta$, or $z^*_A(x_m) \ge (a(\overline{x}, \alpha) - \eta)/2$, if $m \in A$, respectively.

PROOF. Let $1 \le \alpha < \omega_1$, with $a(\overline{x}, \alpha) > 0$, and $\eta > 0$. Put $a = a(\overline{x}, \alpha)$. From the definition of $a(\cdot, \cdot)$ it follows that there is an $M_1 \subset \mathbb{N}$ so that $\mathcal{F}^{M_1}_\alpha \subset \mathcal{A}(\overline{x}, a - \eta/4)$ and applying Corollary 6.2 we find an $M_2 \overset{\infty}{\subset} M_1$ so that (51) holds for $\delta = a - \eta/4$ and $\varepsilon = \eta/4$. In particular this means that our claim holds for all $A \in \mathcal{F}^{M_2}_\alpha$. Secondly we deduce that $\mathcal{F}^{M_2}_\alpha \sqcup \mathcal{F}^{M_2}_\alpha \subset \mathcal{A}(\overline{x}, (a - \eta/2)/2)$. Since each $A \in \mathcal{F}_\alpha \sqcup \mathcal{F}_\alpha$ is the disjoint union of two elements of $\mathcal{F}_\alpha$, it also follows that for any $A \in \mathcal{F}^{M_2}_\alpha \sqcup \mathcal{F}^{M_2}_\alpha$ we find a $z^*_A \in B_{X^*}$ so that $z^*_A(x_m) = 0$ if $m \in M_2 \setminus A$ and $z^*_A(x_m) \ge (a(\overline{x}, \alpha) - \eta)/2$, if $m \in A$.

Choosing finally $\overline{z}$ to be the subsequence defined by M_3, will finish the proof. □

Using Lemma 6.8 successively for different α's and the appropriate choices of η we conclude from a simple diagonalization argument the following Corollary.

COROLLARY 6.9. *Let $\overline{x} = (x_n)$ be a weakly null and semi-normalized sequence in a Banach space X satisfying the conclusions of Lemma 6.4, let $(\ell_k) \subset \mathbb{N}$ be strictly increasing, and let $(\alpha_k) \subset [0, \alpha_0(\overline{x}))$. Then there is a subsequence $\overline{z}$ so that for any $k \in \mathbb{N}$ and any $\ell \le \ell_k$ it follows that:*

For $A \in \mathcal{F}_{\alpha_\ell} \cap [\{k, k+1, \dots\}]^{<\infty}$, or $A \in [\mathcal{F}_{\alpha_\ell} \sqcup \mathcal{F}_{\alpha_\ell}] \cap [\{k, k+1, \dots\}]^{<\infty}$, there is a $z^*_A \in B_{X^*}$ so that $z^*_A(x_m) = 0$ if $m \in \mathbb{N} \setminus A$ and $z^*_A(x_m) \ge a(\overline{z}, \alpha_\ell)/2$, or $z^*_A(x_m) \ge a(\overline{z}, \alpha_\ell)/4$, if $m \in A$, respectively.

LEMMA 6.10. *Let $\overline{x} = (x_n)$ be a weakly null and semi-normalized sequence in a Banach space X satisfying the conclusions of Lemma 6.4. Let (δ_k) be decreasing sequence in $(0, 1)$, with $\delta_k \le 1/(k+1)$, for any $k \in \mathbb{N}$.*

Then there is a subsequence $\overline{z}$ of $\overline{x}$ and a sequence of ordinals α_k which increases to $\alpha_0(\overline{x})$ so that for each $k \in \mathbb{N}$

a) $\frac{1}{2}\delta_k \le a(\overline{z}, \alpha_k) \le \delta_k$
b) *α_k is of the form $\alpha_k = \tilde{\alpha}_k + k$ (here we identify positive integers with finite ordinals of the same cardinality).*

PROOF. We define for $k \in \mathbb{N}$ $\beta_k = \min\{\beta : a(\overline{x}, \beta) \le \delta_k\}$. Since $\delta_k \le 1/(k+1)$ it is easy to see that $\beta_k \ge (k+1)$. If β_k, is a successor, say $\beta_k = \gamma_k + 1$, then

$a(\overline{x}, \gamma_k) \geq \delta_k$ If β_k is a limit ordinal we deduce from the continuity that $a(\overline{x}, \beta_k) = \delta_k$, and we let $\gamma_k = \beta_k$.

It follows from (45) that $a(\overline{x}, 2\gamma_k) \geq \frac{1}{2}a(\overline{x}, \gamma_k) \geq \frac{1}{2}\delta_k$. Since $\gamma_k \geq k$ we will be able to find an α_k between β_k and $2\gamma_k$ which is of the form as required in (b). $\square$

LEMMA 6.11. *Assume that $\overline{x} = (x_n)$ is a weakly null and seminormalized sequence in a Banach space X satisfying the conclusion of Lemma 6.4. Let α_k be ordinals increasing to $\alpha_0(\overline{x})$ so that for each $k \in \mathbb{N}$ α_k can be written as $\alpha_k = \tilde{\alpha}_k + k$ for some $\tilde{\alpha}_k < \omega_1$. For $k \in \mathbb{N}$ let $\varepsilon_k = a(\overline{z}, \alpha_k)$.*

a) *There exists a transfinite family $(\mathcal{G}_\alpha)$ and a subsequence $\overline{z} = (z_n)$ of $\overline{x}$ so that for all $(a_i) \in c_{00}$ it follows that*

$$\left\|\sum a_i z_i\right\| \leq 8 \sum_{k=1}^{\infty} \varepsilon_{k-1} \max_{A \in \mathcal{G}_{\alpha_k}} \sum_{i \in A} |a_i|.$$

b) *Let $(\mathcal{F}_\alpha)$ be a transfinite family. Then there is a subsequence $\overline{z} = (z_n)$ of $\overline{x}$ so that for all $(a_i) \in c_{00}$ it follows that*

$$\left\|\sum a_i z_i\right\| \geq \frac{1}{8} \max_{k \in \mathbb{N}, A \in \mathcal{F}_{\alpha_k}} \varepsilon_k \sum_{i \in A} |a_i|.$$

PROOF. From our assumption on $\overline{x}$ it follows that for all $Q \overset{\infty}{\subset} \mathbb{N}$ and all $k \in \mathbb{N}$ it follows that $\mathbb{I}(\mathcal{A}(\overline{x}, 2\varepsilon_k), Q) \leq \alpha_k$. Therefore we can apply Corollary 4.11 to obtain a transfinite family $(\mathcal{G}_\alpha)_{\alpha<\omega_1}$ and a $K = \{k_1, k_2, \ldots\} \overset{\infty}{\subset} \mathbb{N}$, $k_i \nearrow \infty$, if $i \nearrow \infty$, so that for all ℓ

$$\mathcal{A}(\overline{x}, 2\varepsilon_\ell) \cap [\{k_\ell, k_{\ell+1}, \ldots\}]^{<\infty} \subset \mathcal{G}^K_{\alpha_\ell}. \tag{60}$$

Putting $\overline{z} = (x_{k_i})$ this implies that for all ℓ

$$\mathcal{A}(\overline{z}, 2\varepsilon_\ell) \cap [\{\ell, \ell+1, \ldots\}]^{<\infty} \subset \mathcal{G}_{\alpha_\ell}. \tag{61}$$

Let $(a_i) \in c_{00}$. We find a $z^* \in B_{X^*}$ which norms $\sum a_i z_i$ and ($\varepsilon_0 = \sup \|z_i\|$) and deduce that

$$\begin{aligned} z^*\Big(\sum_{i=1}^{\infty} a_i z_i\Big) &= \sum_{k=1} \sum_{2\varepsilon_k < z^*(z_i) \leq 2\varepsilon_{k-1}} z^*(z_i) a_i + \sum_{2\varepsilon_k < -z^*(z_i) \leq 2\varepsilon_{k-1}} z^*(z_i) a_i \\ &\leq \sum_{k=1}^{\infty} 2\varepsilon_{k-1} \left[\sum_{2\varepsilon_k < z^*(z_i) \leq 2\varepsilon_{k-1}} |a_i| + \sum_{2\varepsilon_k < -z^*(z_i) \leq 2\varepsilon_{k-1}} |a_i| \right]. \end{aligned} \tag{62}$$

Now note that the set $\{i \in \mathbb{N} : 2\varepsilon_k < z^*(z_i)\}$ is the union of two sets in $\mathcal{G}_{\alpha_k}$ namely $\{i \geq k : 2\varepsilon_k < z^*(z_i)\}$ and $\{i \leq k : 2\varepsilon_k < z^*(z_i)\}$ (where the second set lies in $\mathcal{G}_{\alpha_k}$ because α_k is the k-th successor of some $\tilde{\alpha}_k$). Similarly we can proceed with $\{i \in \mathbb{N} : 2\varepsilon_k < -z^*(z_i)\}$ and we therefore derive from (62) that

$$z^*\Big(\sum_{i=1}^{\infty} a_i z_i\Big) \leq \sum_{k=1}^{\infty} 2\varepsilon_{k-1} 4 \max_{A \in \mathcal{G}_{\alpha_k}} \sum_{i \in A} |a_i|, \tag{63}$$

which finishes the proof of part (a).

In order to prove part (b) we first choose $\ell_0 = 0$ and $(\ell_k)_{k\in\mathbb{N}} \subset \mathbb{N}$ fast enough increasing so that for all $(a_i) \in c_{00}$ it follows that

$$\Big\|\sum a_i z_i\Big\| \geq \sup_{A\subset\mathbb{N},\#A=k} \frac{\varepsilon_{\ell_k}}{4} \sum_{i\in A} |a_i|. \tag{64}$$

After passing to a subsequence we can assume that the conclusion of Corollary 6.9 is satisfied.

Let $(a_i) \in c_{00}$, $k \in \mathbb{N}$, $\ell \leq \ell_k$ and $A \in \mathcal{F}_{\alpha_\ell} \cap [\{k, k+1, \dots\}]^{<\infty}$. Letting $A^+ = \{i \in A : a_i \geq 0\}$ and $A^- = A \setminus A^+$ we can choose $z^*_{A^+}$ and $z^*_{A^-}$ as prescribed in Corollary 6.9 and deduce that

$$\Big\|\sum_{i=k}^{\infty} a_i z_i\Big\| \geq \frac{1}{2} z^*_{A^+}\Big(\sum_{i=k}^{\infty} a_i z_i\Big) - \frac{1}{2} z^*_{A^-}\Big(\sum_{i=k}^{\infty} a_i z_i\Big) \geq \frac{1}{4}\varepsilon_\ell \sum_{i\in A} |a_i|. \tag{65}$$

Therefore we obtain

$$\begin{aligned}
\Big\|\sum_{i=1}^{\infty} a_i z_i\Big\| &\geq \frac{1}{2} \max_{k\in\mathbb{N}} \Big[\Big\|\sum_{i=k}^{\infty} a_i z_i\Big\| + \Big\|\sum_{i=1}^{k-1} a_i z_i\Big\|\Big] \\
&\geq \frac{1}{8} \max_{k\in\mathbb{N}} \Big[\max_{\substack{\ell\in(\ell_{k-1},\ell_k],\\ A\in\mathcal{F}_{\alpha_\ell}, A\geq k}} \varepsilon_\ell \sum_{i\in A} |a_i| + \varepsilon_{\ell_{k-1}} \sum_{i=1}^{k-1} |a_i|\Big] \\
&\geq \frac{1}{8} \max_{k\in\mathbb{N}} \max_{\substack{\ell\in(\ell_{k-1},\ell_k]\\ A\in\mathcal{F}_{\alpha_\ell}}} \varepsilon_\ell \sum_{i\in A} |a_i| = \frac{1}{8} \max_{\ell\in\mathbb{N}, A\in\mathcal{F}_{\alpha_\ell}} \varepsilon_\ell \sum_{i\in A} |a_i|,
\end{aligned}$$

proving part (b). □

LEMMA 6.12. *Assume that $\overline{u} = (u_n)$ and $\overline{v} = (v_n)$ are two weakly null and seminormalized sequences in a Banach space X satisfying the conclusions of Lemma 6.4 such that*

$$a(\overline{u}, \alpha) < \frac{1}{5} a(\overline{v}, \alpha). \tag{66}$$

Let $\overline{x} = \overline{u} + \overline{v} = (u_n + v_n)$. Then there is a subsequence $\overline{z}$ of $\overline{x}$ so that

$$\mathcal{F}_\alpha \subset \mathcal{A}\big(\overline{z}, a(\overline{v}, \alpha)/4\big), \text{ in particular } a(\overline{y}, \alpha) \geq a(\overline{v}, \alpha)/4 \tag{67}$$

for all subsequences $\overline{y}$ of $\overline{z}$.

PROOF. By passing to a subsequence we can assume that the conclusions of Lemma 6.8 are satisfied for $\overline{v}$, α, and $\eta = \frac{1}{10} a(\overline{v}, \alpha)$. In particular we find for each $A \in \mathcal{F}_\alpha \sqcup \mathcal{F}_\alpha$, a $v^*_A \in B_{X^*}$ which has the property that $v^*_A(x_m) \geq (a(\overline{v}, \alpha) - \eta)/2 = \frac{9}{20} a(\overline{v}, \alpha)$, whenever $m \in A$, and $v^*_A(x_m) = 0$, if $m \notin A$.

For $A \in \mathcal{F}_\alpha \sqcup \mathcal{F}_\alpha$ we define $\psi(A) = \{i \in A : v^*_A(u_i) < -\frac{1}{5} a(\overline{v}, \alpha)\}$, and note that $\Psi(\mathcal{F}_\alpha \sqcup \mathcal{F}_\alpha) \subset \mathcal{A}(\overline{u}, \frac{1}{5} a(\overline{v}, \alpha))$ and $\mathbb{I}(\mathcal{A}(\overline{u}, \frac{1}{5} a(\overline{v}, \alpha)), \mathbb{N}) \leq \alpha$ (using (66)). Secondly we note that $\{A \setminus \Psi(A) : A \in \mathcal{F}_\alpha \sqcup \mathcal{F}_\alpha\}$ contains $\mathcal{A}(\overline{v} + \overline{u}, \frac{1}{4} a(\overline{v}, \alpha))$. This means that we can apply Proposition 5.5 and derive that for some $N \overset{\infty}{\subset} \mathbb{N}$ we have $\mathbb{I}(\mathcal{A}(\overline{v} + \overline{u}, \frac{1}{4} a(\overline{v}, \alpha)), N) \geq \alpha$ and thus that there is an $M \overset{\infty}{\subset} N$ so that $\mathcal{F}^M_\alpha \subset \mathcal{A}(\overline{v} + \overline{u}, \frac{1}{4} a(\overline{v}, \alpha))$ which implies the claim for the subsequence of $\overline{v} + \overline{u}$ defined by M. □

LEMMA 6.13. *For $\alpha < \omega_1$, $(\alpha_n) \subset [0, \omega_1)$ strictly increasing, $(\Theta_n) \subset [0,1]$ being non decreasing, and $(\ell_n) \subset \mathbb{N}$, with $\lim_{n\to\infty} \ell_n = \infty$ it follows that the Schreier spaces $S(\mathcal{F}_\alpha)$, $S((\Theta_n), (\mathcal{F}_{\alpha_n}))$ and $S((\Theta_n), (\mathcal{F}_n), (\ell_n))$ (recall Definition 1.6) are hereditarily c_0, assuming that in the case of the space $S((\Theta_n), (\mathcal{F}_{\alpha_n}))$ the sequence (Θ_n) decreases to 0.*

PROOF. We will proof by transfinite induction on $\alpha < \omega_1$ that the spaces $S(\mathcal{F}_\alpha)$, $S((\Theta_n), (\mathcal{F}_{\alpha_n}))$ and $S((\Theta_n), (\mathcal{F}_n), (\ell_n))$, where $\alpha_n \nearrow \alpha$, is hereditarily c_0 (of course the claim for the second and third space is vacuous if α is not a limit ordinal).

Assume the claim is true for all $\beta < \alpha$. If α is a successor, say $\alpha = \beta + 1$ it is clear that the norm on $S(\mathcal{F}_\beta)$ and $S(\mathcal{F}_{\beta+1})$ are equivalent, and, thus, the claim follows from the induction hypothesis.

If α is a limit ordinal then $S(\mathcal{F}_\alpha)$ is a special case of the space $((\Theta_n), (\mathcal{F}_{\alpha_n}), (\ell_n))$. Indeed, we let $\Theta_n = 1$, for $n \in \mathbb{N}$, and if $A_n(\alpha)$ is the approximating sequence we write $\bigcup A_n(\alpha)$ as a strictly increasing sequence $\{\alpha_n : n \in \mathbb{N}\}$ and put $\ell_n = \min\{\ell : \alpha_n \in A_\ell(\alpha)\}$.

Therefore we only need to consider the norms

$$\|x\| = \begin{cases} \sup_{n\in N} \sup_{F\in\mathcal{F}_{\alpha_n}, \ell_n \le F} \Theta_n \sum_{i\in F} |x_i| & \text{or} \\ \sup_{n\in N} \sup_{F\in\mathcal{F}_{\alpha_n}} \Theta_n \sum_{i\in F} |x_i| \end{cases}.$$

Let X be the completion of c_{00} under $\|\cdot\|$, and let, without of generality, Y be the closed subspace spanned by a normalized block (y_n). Either there is a further block (z_n) and an $n_0 \in \mathbb{N}$ so that the norm $\|\cdot\|$ is on the closed subspace spanned by (z_n) equivalent to the norm $\|\cdot\|_{\alpha_{n_0}}$ on $S(\mathcal{F}_{\alpha_{n_0}})$. Then our claim follows from the induction hypothesis. Or we can choose a normalized block (z_n) in Y and increasing sequences $(k_n), (m_n)$ in $\mathbb{N}$, with $k_n < m_n < k_{n+1}$ so that for each $n \in \mathbb{N}$ it follows that (we denote for $z = \sum a_i e_i \in c_{00}$ and $k \in \mathbb{N}$ $z|_{[\ell_k,\infty)} = \sum_{i\ge k} a_i e_i$)

$$\|z_n\|_{\alpha_k} < 2^{-n}, \text{ whenever } k \le k_n \tag{68}$$

$$\max\operatorname{supp}(z_n) \ge \ell_n \text{ (assuming admissibility) and } 1 = \|z_n\| = \Theta_{m_n}\|z_n\|_{\alpha_{m_n}} \tag{69}$$

$$\left.\begin{matrix} \Theta_k\|z_n\|_{\alpha_k} \\ \text{respectively} \\ \Theta_k\|z_n|_{[k,\infty)}\|_{\alpha_k} \end{matrix}\right\} < 2^{-n}, \text{ whenever } k \ge k_{n+1}. \tag{70}$$

(for (70) we are using either the (ℓ_n)-admissibility condition or the fact that $\Theta_n \searrow 0$, if $n \nearrow \infty$).

Finally it is easy to see that (68), (70)and (69) imply that (z_{m_n}) is equivalent to the c_0-unit basis. $\square$

We are now ready for the proof of Theorem 6.7.

PROOF OF THEOREM 6.7. We choose a seminormalized and weakly null sequence $\overline{y} \subset B_X$ with the property that $a(\overline{y}, \beta_0) = 0$ and (by passing to a subsequence) assume that $\overline{y}$ satisfies the conclusion of Lemma 6.4. We let

$$c = \min\Big\{\frac{1}{5}, \frac{1}{2} \inf_{\gamma<\beta_0} B(X,\gamma)\Big\}. \tag{71}$$

Using first Lemma 6.10 we find an increasing sequence of ordinals $(\alpha_k) \subset [0, \alpha_0(\overline{y}))$ so that, after passing to a subsequence, we can assume that $\frac{1}{2}(c/2)^{3k} \le a(\overline{y}, \alpha_k) \le$

$(c/2)^{3k}$ and that α_k is the k-th successor of another ordinal, for all $k \in \mathbb{N}$. Using secondly Lemma 6.11(a) we can assume, after passing again to a subsequence, that there is a transfinite family $(\mathcal{G}_\alpha)_{\alpha<\omega_1}$ so that for any $(a_i) \in c_{00}$ it follows that

$$\Big\|\sum a_i y_i\Big\| \le 8\sum_{k=1}^{\infty}(c/2)^{3k-3}\max_{A\in\mathcal{G}_{\alpha_k}}\sum_{i\in A}|a_i| \le \frac{16}{c^3}\max_{k\in\mathbb{N},A\in\mathcal{G}_{\alpha_k}} c^{3k}\sum_{i\in A}|a_i| \tag{72}$$

(where the second inequality is an easy application of the inequality of Holder).

We now choose (infinitely or finitely many) sequences $\overline{x}^{(1)}$, $\overline{x}^{(2)}$ in X as follows.

We first choose $\overline{x}^{(1)} \subset B_X$ weakly null, seminormalized, so that $a(\overline{x}^{(1)}, \alpha_1) > c$. By Lemma 6.4 (using also the part which starts with "moreover" for α_1) we can assume that the conclusions of Lemma 6.4 are satisfied.

Now we consider the following two cases:
Case 1: For all $k \in \mathbb{N}$ it follows that $a(\overline{x}^{(1)}, \alpha_k) \ge c^{3k}$. In that case we are done for the moment and do not continue to choose other sequences in X.
Case 2: There is a smallest k_1 (necessarily bigger than 1) for which $a(\overline{x}^{(1)}, \alpha_{k_1}) < c^{3k_1}$.

In this case we choose a seminormalized and weakly null sequence (recall (71))

$$\overline{y}^{(1)} \subset \frac{5}{c}c^{3k_1}B_X \text{ with } a(\overline{y}^{(1)}, \alpha_{k_1}) \ge 5c^{3k_1} \tag{73}$$

and define $(x^{(2)}(n)) = \overline{x}^{(2)} = \overline{x}^{(1)} + \overline{y}^{(1)}$. Using now Lemma 6.12 we find a subsequence $(x(2, n_i^{(2)}))_{i\in\mathbb{N}}$ of $\overline{x}^{(2)} = (x(2,n))_{n\in\mathbb{N}}$ so that

$$a((x(2, n_i^{(2)}))_{i\in\mathbb{N}}, \alpha_{k_1}) \ge c^{3k_1}. \tag{74}$$

Once again we can consider two cases, namely
Case 1: For all $k \in \mathbb{N}$, $k \ge k_1$, it follows that $a((x(2, n_i^{(2)})), \alpha_k) \ge c^{3k}$.
Case 2: There is a smallest k_2 (must be bigger than k_1) for which $a(\overline{x}(2, n_i^{(2)}), \alpha_{k_2})| < c^{3k_2}$.

As before we stop in Case 1 and define in Case 2 sequences $\overline{y}^{(2)}$, $\overline{x}^{(3)}$ and $n_i^{(3)}$ as before.

We proceed in that way and eventually find an $\ell \in \mathbb{N}\cup\{\infty\}$, a strictly increasing sequence $(k_j)_{1\le j<\ell+1}$ and sequences $\overline{y}^{(j)} \subset \frac{5}{c}c^{3k_j}B_X$, and $\overline{x}^{(j)} = \overline{x}(1) + \overline{y}^{(1)} + \dots \overline{y}^{(j-1)} \subset X$, and subsequences of $N = \mathbb{N} = (n_i^{(1)}) \supset (n_i^{(2)}) \dots$, for $1 \le j < \ell+1$ so that for any $j \in \mathbb{N}$, $1 \le j < \ell + 1$, it follows that

$$(\overline{y}^{(j)}) \subset \frac{5}{c}c^{3k_j}B_X \tag{75}$$

satisfying the conclusions of Lemma 6.4 and

$$a((x(j, n_i^{(j)})), \alpha_k) \ge c^{3k}, \text{ whenever } k_{j-1} \le k < k_j \tag{76}$$

(where $k_0 = 1$ and $k_\ell = \infty$ if $\ell < \infty$). Moreover (76) holds for all subsequences of $(x^{(j)}, n_i^{(j)}))_{i\in\mathbb{N}}$.

Now we choose $\overline{x}$ to be the diagonal sequence of $x(j, n_i^{(j)})$, if $\ell = \infty$ (i.e. $x_i = (x(i, n_i^{(i)}))_{i\in\mathbb{N}}$) or we choose $\overline{x} = x(\ell, n^{(\ell)_i})$ if $\ell < \infty$.

For any $k \in \mathbb{N}$, there is a $j \in \mathbb{N}$, $1 \le j < \ell+1$, so that $k_{j-1} \le k < k_j$ and we notice from (75) that $\overline{x}$ is up to some finitely many elements a subsequence of $x^{(j-1)}$

and a sequence whose elements are of norm not exceeding $\frac{5}{c}\sum_{j\ge k_j} c^{3j}$. Thus, $\overline{x}$ is seminormalized and we conclude from (76) that

$$a(\overline{x},\alpha_k) \ge c^{3k} - \frac{5}{c}\sum_{i\ge k_j} c^{3i} \ge \frac{1}{2}c^{3k}, \text{ whenever } j\in\mathbb{N} \text{ and } k_{j-1}\le k<k_j. \tag{77}$$

Using Lemma 6.11 again we can assume (by passing to subsequence) that for any $(a_i)\in c_{00}$

$$\Big\|\sum a_i x_i\Big\| \ge \frac{1}{16}\max_{k\in\mathbb{N},A\in\mathcal{G}_{\alpha_k}} c^{3k}\sum_{i\in A}|a_i|. \tag{78}$$

Combining (72) and (78) we showed that for any $(a_i)\in c_{00}$ it follows that

$$\frac{c^3}{16}\Big\|\sum a_i y_i\Big\| \le \max_{k\in\mathbb{N},A\in\mathcal{G}_{\alpha_k}} c^{3k}\sum_{i\in A}|a_i| \le 16\Big\|\sum a_i x_i\Big\|.$$

Thus the mapping $x_i\mapsto y_i$ extends to a linear bounded operator T which factors through a space which is hereditarily c_0 (by Lemma 6.13). On the other hand we assumed that X does not contain c_0, therefore T must be strictly singular. □

References

[AA] D. Alspach and S.A. Argyros, Complexity of weakly null sequences, *Dissertationes Mathematicae*, **321** (1992), 1–44.

[AD] S. A. Argyros and I.Deliyanni, Examples of asymptotic l_1 Banach spaces. *Trans. Amer. Math. Soc.* 349 (1997), no. 3, 973–995.

[AG] S. A. Argyros and I. Gasparis, Unconditional structures of weakly null sequences, *Transaction of the A.M.S.*, **353**, no 5, (2001), 2019 - 2058.

[AMT] S. A. Argyros, S. Mercourakis and A. Tsarpalias, Convex unconditionality and summability of weakly null sequences, *Israel Journal of Mathematics*, **107** (1998), 157–193.

[AO] G. Androulakis and E. Odell, *Distorting mixed Tsirelson spaces*, Israel J. Math.,**109** (1999), 125–149

[AOST] G. Androulakis, E. Odell, Th. Schlumprecht and N. Tomczak-Jaegermann, On the structure of the spreading models of a Banach space, preprint.

[AS] G. Androulakis and Th. Schlumprecht, Strictly singular, non-compact operators exist on the space of Gowers and Maurey, *J. London Math. Soc.*(2) 64 (2001), no. 3, 655–674.

[BL] B. Beauzamy and J.-T. Lapreste, Modeles etales des espaces de Banach, *Travaux en Cours*, Hermann, Paris, 1984.

[CJT] P. G. Casazza, W. B. Johnson and L. Tzafriri, On Tsirelson's space, *Israel Journal of Math.*, **47** (1984, 81 –98.

[F1] V. Ferenczi, A uniformly convex hereditarily indecomposable Banach space, *Israel J. Math*,**102** (1997), 199–225.

[F2] V. Ferenczi, Operators on subspaces of hereditarily indecomposable Banach spaces, *Bull. London Math. Soc. 29*, pp. 338-344, 1997.

[FJ] T. Figiel and W. B. Johnson, A uniformly convex Banach space which contains no l_p, *Compositio Math.* **29** (1974), 179–190.

[Go1] T. W Gowers,A new dichotomy for Banach spaces, *Geom. Funct. Anal.* **6** (1996), no. 6, 1083–1093.

[Go2] T. W Gowers, A remark about the scalar-plus-compact problem. Convex geometric analysis (Berkeley, CA, 1996), 111–115, *Math. Sci. Res. Inst. Publ.*, 34, Cambridge Univ. Press, Cambridge, 1999.

[Gu] S. Guerre-Delabrière, Classical Sequences in Banach Spaces, Marcel Dekker, Inc., New York (1992).

[GM1] T. Gowers and B. Maurey, The unconditional basis sequence problem , *Journal A.M.S.*, **6** (1993), 523-530.

[GM2] T. Gowers and B. Maurey, Banach spaces with small spaces of operators. *Math. Ann.*, **307** (1997), no. 4, 543–568.

[K] J.L. Krivine, Sous-espaces de dimension finie des espaces de Banach réticules, Ann. of Math. **104** (1976), 1–29.

[Le] H. Lemberg, Sur un theorémè de J.-L. Krivine sur la finie représentation de ℓ_p dans un espace de Banach. C. R. Acad. Sci. Paris Sér. I Math. 292 (1981), no. 14, 669–670.

[Li] J. Lindenstrauss, personal communication.

[Lo] V. I. Lomonosov, Invariant subspaces of the family of operators that commute with a completely continuous operator, *Funkcional. Anal. i Priložen.* **7** (1973), no. 3, 55–56.

[LT] D. Leung and W. Tang, The Bourgain ℓ_1-index of mixed Tsirelson space, preprint.

[O1] E. Odell, Applications of Ramsey theorems to Banach space theory, Notes in Banach spaces, ed. H.E. Lacey, Univ. of Texas Press, Austin, TX (1980), 379–404

[O2] E. Odell, On Schreier unconditional Sequences, *Contemporary Mathematics*, **144**, 1993, 197–201.

[OS] E. Odell and Th. Schlumprecht, A Banach space block finitely universal for monotone bases. *Trans. Amer. Math. Soc.***352** (2000), no. 4, 1859–1888.

[Sch] Th. Schlumprecht, An arbitrarily distortable Banach space, *Israel J. Math.* **76**(1991), no. 1-2, 81–95.

[T] B. S. Tsirelson, Not every Banach space contains an embedding of ℓ_p or c_0, *Israel Journal of Math.*, **32** (1979), 32–38.

[Z] M. Zippin, On perfectly homogeneous bases in Banach spaces, *Israel J. Math,***4** 1966 265–272.

Department of Mathematics, Texas A&M University, College Station, TX 77843, USA

E-mail address: schlump@math.tamu.edu

Contemporary Mathematics
Volume **321**, 2003

Maximal algebra norms

Geoffrey V Wood

ABSTRACT. A norm on a Banach space X is *maximal* if there is no equivalent norm which has a strictly larger group of isometries. This property can be viewed as one of the operator norm on $B(X)$. This suggests a notion of maximality for Banach algebras, which was first introduced by E. R. Cowie in 1981. This work was largely unpublished, so details of the work are given here and how it relates to recent work of M. Hansen and R. Kadison. Some open questions are posed.

1. Introduction

To motivate this work, we begin with a brief survey of results about maximality for Banach spaces. For a more comprehensive survey, see [**3**].

Let X be a complex Banach space, $B(X)$ the bounded linear operators on X, and $G(X)$ the group of invertible isometries on X.

DEFINITION 1. The norm on X is *maximal* if no equivalent norm has a strictly larger group of isometries.

Thus a norm being maximal means that there is 'as much symmetry as the topology allows.'

LEMMA 1. *The following are equivalent:*
i) the norm on X is maximal,
ii) $G(X)$ is a maximal bounded group in $B(X)$.

There are some stronger notions of symmetry.

DEFINITION 2. The norm on X is *transitive* if, for every $x, y \in X$, $\|x\| = \|y\| = 1$, there exists an isometry U on X such that $Ux = y$. (Equivalently: for each $x \in X$, $\|x\| = 1$, the orbit $\{Ux : U \in G(X)\}$ is the unit sphere of X)

DEFINITION 3. The norm on X is *almost transitive* if, given $\varepsilon > 0$ and $x, y \in X$, $\|x\| = \|y\| = 1$, there exists an isometry U on X such that $\|Ux - y\| < \epsilon$. (Equivalently: for each $x \in X$, $\|x\| = 1$, the orbit $\{Ux : U \in G(X)\}$ is dense in the unit sphere of X)

DEFINITION 4. The norm on X is *convex transitive* if, for each $x \in X$, $\|x\| = 1$, the convex hull of the orbit $\{Ux : U \in G(X)\}$ is dense in the unit ball of X.

2000 *Mathematics Subject Classification.* Primary 46B04, 46B10, 46B20.
Key words and phrases. Transitive norm, maximality, convex transitive, Hilbert space.

We have the implications:

transitive ⇒ almost transitive ⇒ convex transitive ⇒ maximal.

The first two implications are immediate from the definitions; the third is straightforward - the details are in [**24**].

Hilbert spaces are transitive and a finite-dimensional maximal space is necessarily a Hilbert space ([**24**]). Thus all four properties are equivalent for finite dimensional spaces.

It is a long-standing open question (Mazur's problem) whether every separable transitive Banach space is a Hilbert space. (Examples of a non-complete separable transitive space which is not a pre-Hilbert space and a non-separable complete transitive space which is not a Hilbert space are given in [**24**].)

$C(K)$ spaces are never transitive or almost transitive ([**26**]), but can be convex-transitive when there are sufficiently many homeomorphisms on X. A characterisation is given in [**26**].

There has been some interest in whether there is a locally compact space K with $C_0(K)$ transitive or almost transitive. It was conjectured in [**26**] that no such space exist. See [**14**], [**16**], [**1**], [**2**] and [**8**] for progress on this conjecture, and [**3**] for a more encyclopaedic set of references.

$L^p[0,1]$ is almost transitive for $1 \leq p < \infty$, but is only transitive when $p = 2$. ([**24**].)

$L^\infty[0,1]$ and $C(\mathbb{T})$ are convex transitive but not almost transitive. ([**10**] and [**24**].) Here $\mathbb{T}$ denotes the unit circle in $\mathbb{C}$.

$l_p (p \neq 2)$ and $C[0,1]$ are maximal but not convex-transitive. ([**24**], [**13**] and [**18**].) Sufficient conditions for $C_0(K)$ to be maximal are given in [**18**], but see [**9**] and [**20**].

In [**11**], E. R. Cowie introduced a stronger notion than maximal.

DEFINITION 5. The norm on X is *uniquely maximal* if it is maximal and no equivalent norm has the same group of isometries.

Cowie proves in [**11**] that for Banach spaces uniquely maximal is equivalent to convex transitive.

In this paper we are concerned only with complex Banach spaces and algebras. There are corresponding definitions for real spaces. This case is different and in some ways more difficult. For example, Partington showed in [**23**] that $C_{\mathbb{R}}[0,1]$ is not maximal; it is still not known whether $C_{\mathbb{R}}(\mathbb{T})$ is maximal.

2. $B(X)$ spaces

Lemma 1 shows that the property of maximality for X can be given as a property of the algebra $B(X)$. We will give the definitions for an arbitrary unital Banach algebra A.

DEFINITION 6. $u \in A$ is *unitary* if $\|u\| = \|u^{-1}\| = 1$. We denote by U_A the set of unitary elements in A.

DEFINITION 7. A is *(algebra) maximal* if U_A is a maximal bounded group in A.

Then Lemma 1 states that $B(X)$ is maximal if and only if X is maximal.

THEOREM 1. **[10]** *The following are equivalent:*

i) A is maximal,

ii) there is no equivalent algebra norm on A which has a larger group of unitary elements,

iii) there is no x in A, $\|x\| > 1$*, such that the group generated by* U_A *and x is bounded.*

PROOF. Clearly i) and iii) are equivalent. i) implies ii) since if an equivalent norm $\|.\|_1$ has a larger group of isometries this will be bounded for the original norm.

For ii) $\Rightarrow$ i), suppose that A is not maximal, and let H be a bounded group containing U_A. Put

$$\|x\|_1 = \sup\{\|sx\| : s \in H\}.$$

Then $\|.\|_1$ is an equivalent space norm for A, and

$$\|a\|_2 = \sup\{\|ax\|_1 : \|x\|_1 \leq 1\}$$

defines an equivalent algebra norm. It is clear that $\|s\|_2 \leq 1$ for all $s \in H$, and so the elements of S are unitary for $\|.\|_2$. This contradicts ii).

□

There are analogues of convex transitive and uniquely maximal for algebras, though as we shall see, these properties for $B(X)$ are not equivalent to the corresponding properties for X. However in the same way that convex transitive and uniquely maximal are equivalent for Banach spaces, the corresponding notions turn out to be equivalent for $B(X)$ algebras. These were first introduced in [**10**]; corresponding ideas appear in [**17**].

DEFINITION 8. A is *(algebra) uniquely maximal* if U_A is a maximal bounded group in A, and no equivalent algebra norm has the same group of unitary elements.

THEOREM 2. **[10]** *If A in uniquely maximal then there is no* x_0 *in A,* $\|x_0\| > 1$*, such that the semigroup generated by* U_A *and* x_0 *is bounded.*

PROOF. Suppose that A is uniquely maximal and that there is an $x_0 \in A$, $\|x_0\| > 1$ such that the semigroup S generated by U_A and x_0 is bounded. Then

$$\|x\|_1 = \sup\{\|sx\| : s \in S\}$$

defines an equivalent space norm, and

$$\|a\|_2 = \sup\{\|ax\|_1 : \|x\|_1 \leq 1\}$$

defines an equivalent algebra norm. It is clear that $\|u\|_2 = 1$ for all $u \in U_A$. Since A is uniquely maximal, $\|a\|_2 = \|a\|$ for all $a \in A$. But $\|x_0\|_2 = 1 < \|x_0\|$, which is a contradiction. □

We shall denote by $\overline{\text{co}}(S)$ the norm-closure of the convex hull of S.

DEFINITION 9. A is *unitary* (or *algebra convex transitive*) if $\overline{\text{co}}(U_A)$ is the unit ball of A.

THEOREM 3. **[10]** *If* $B(X)$ *is unitary then* X *and* X^* *are convex transitive as Banach spaces. In fact X has the stronger property:*

(*) *For each* $f \in X^*$, $\|f\| = 1$*, the unit ball of* X^* *is the closed convex hull of the set* $\{U^*f : U \in G(X)\}$.

PROOF. Suppose $B(X)$ is unitary, and let $x \in X, \|x\| = 1$. Choose $f \in X^*$ such that

$$1 = f(x) = \|x\| = \|f\|.$$

Now for $y \in X, \|y\| \leq 1$, the rank one operator $f \otimes y$ is contained in $\overline{\text{co}}\{U : U \in G(X)\}$. Therefore we have

$$y = (f \otimes y)(x) \in \overline{\text{co}}\{Ux : U \in G(X)\}.$$

Thus the norm on X is convex transitive.

To prove property (*), let $f, g \in X$, with $\|f\| = \|g\| = 1$. Fix $\delta > 0$. Choose $x_0 \in X, \|x_0\| = 1$ such that $|f(x_0)| > 1-\delta$. Since $B(X)$ is unitary, given $\epsilon > 0$, there exists $T \in \overline{\text{co}}\{U : U \in G(X)\}$ with $\|g \otimes x_0 - T\| < \epsilon$. Therefore $\|(g \otimes x_0 - T)^* f\| < \epsilon$, or $\|f(x_0)g - T^* f\| < \epsilon$. Thus $f(x_0)g \in \overline{\text{co}}\{U^* f : U \in G(X)\}$. Since δ is arbitrary, $g \in \overline{\text{co}}\{U^* f : U \in G(X)\}$, and the result is proved. □

In [**4**], this has been generalised as follows:

THEOREM 4. *For a Banach space X, consider the following conditions:*

(1) *$B(X)$ is unitary.*

(2) *For every α in X^{**}, $\|\alpha\| = 1$ we have*

$$\overline{co}\{T^{**}(\alpha) : T \in G(X)\} \supseteq \text{ the unit ball of } X\,.$$

(3) *For every f in X^*, $\|f\| = 1$ we have*

$$\overline{co}\{T^*(f) : T \in G(X)\} = \text{ the unit ball of } X^*\,.$$

(4) *For every x in X, $\|x\| = 1$ we have*

$$\overline{co}\{T(x) : T \in G(X)\} = \text{ the unit ball of } X\,.$$

Then $1 \Rightarrow 2 \Rightarrow 3 \Rightarrow 4$.

We have shown that if $B(X)$ is convex transitive as an algebra then X is convex transitive as a space. The converse is not true as the next example shows.

EXAMPLE 1. Let $X = C(\mathbb{T})$. Then X is convex transitive as a Banach space and hence uniquely maximal. X does not have the property (*) since if $\mu \in M(\mathbb{T})$ is a discrete measure, then the set $\text{co}\{U^*\mu : U \in G(X)\}$ will consist only of discrete measures and cannot be dense in the unit ball of X^*. It follows from Theorem 2.7 that $B(X)$ is not algebra convex transitive. We will see that $B(X)$ is not algebra uniquely maximal either.

However if we use the weak operator topology in $B(X)$, then we do get an equivalence.

THEOREM 5. [**11**] *X is convex transitive if and only if the convex hull of the unitary elements of $B(X)$ is weak-operator dense in the unit ball.*

PROOF. Suppose that X is convex transitive, but that $co\{U : U \in G(X)\}$ is not weak operator dense in the unit ball. Then there exists $T \in B(X)$, $\|T\| = 1$, $x \in X$, $f \in X^*$ with $\|x\| = \|f\| = 1$, $f(Tx) = 1$ but $|f(Ux)| \leq \delta < 1$ for all $U \in G(X)$. But $co\{Ux : U \in G(X)\}$ is dense in the unit ball, so $\|f\| \leq \delta$, which is a contradiction.

Conversely, suppose that the convex hull of $\{U : U \in G(X)\}$ is weak-operator dense in the unit ball. Let $x, y \in X$, $\|x\| = 1$, $\|y\| = 1$. Choose $f \in X^*$ such that $\|f\| = 1$ and $f(x) = 1$. Now the 1-dimensional operator $f \otimes y$ has norm 1 and so is contained in $\overline{\text{co}}^{weak-op}\{U : U \in G(X)\}$. Thus there exists T_n in $co\{U : U \in G(X)\}$ such that $\|T_n - f \otimes y\| \to 0$. Therefore $|g(T_n x) - g(y)f(x)| \to 0$ or $g(T_n x) \to g(y)$ for all $g \in X^*$. This means that $y \in \overline{\text{co}}^{weak}\{Ux : U \in G(X)\} = \overline{\text{co}}\{Ux : U \in G(X)\}$. Since x and y are arbitrary, X is convex transitive. □

We now consider whether a stronger hypothesis than convex transitivity for X would be sufficient for $B(X)$ to be unitary. We first show that even if X is almost transitive, then $B(X)$ need not be unitary.

EXAMPLE 2. [**10**] For $X = L^p[0,1], 1 \leq p < \infty$, $B(X)$ is not unitary.

We cannot use Theorem 3 this time, since in these cases we *do* have the stronger property (*).

We give a proof when $p = 1$. The other cases are similar.

PROOF. We will show that the one-dimensional operator $m \otimes f \notin \overline{\text{co}}(G(X))$, where f is the identically one function and m is Lebesgue measure. The isometries on $L^1[0,1]$ are characterised in [**19**] as follows:

Let γ be a regular set isomorphism on the Lebesgue measurable sets on $[0,1]$. That is a map from the σ-ring of measurable sets of $[0,1]$ defined modulo null sets, which preserves countable unions, complements and null sets. Let $\hat{\gamma}$ be the corresponding map of $L^1[0,1]$ to itself generated by $\hat{\gamma}(\chi_A) = \chi_{\gamma A}$. Define a measure on $[0,1]$ by $\mu(A) = m(\gamma^{-1}A)$. Then

$$Uf(t) = h(t)f(\gamma t) \quad a.e.,$$

where $|h(t)| = \dfrac{d\mu}{dm} a.e.$, defines an isometry on $L^1[0,1]$. Furthermore all isometries arise in this way.

Suppose that $m \otimes f \in \overline{\text{co}}(G(X))$. Then for $\epsilon > 0$, there are isometries U_i such that

$$\|m \otimes f - \sum_1^N c_i U_i\| < \epsilon,$$

where $c_i > 0$ and $\sum_1^N c_i = 1$.

For a measurable set A, we have

$$\|(m \otimes f)\chi_A - \sum_1^N c_i U_i \chi_A\| < \epsilon \|\chi_A\|,$$

or

$$\int_{[0,1]} |m(A) - \sum_1^N c_i U_i \chi_A| dt < \epsilon \|\chi_A\|.$$

Now for each i, $U_i\chi_A = h_i \chi_{\gamma_i A}$, where γ_i is a regular set isomorphism. If $B = (\cup_{i=1}^N \gamma_i A)^c$, then

$$\int_B |m(A) - \sum_1^N c_i U_i \chi_A| dt < \epsilon \|\chi_A\|,$$

or $m(A)m(B) < \epsilon m(A)$. Thus $m(B) < \epsilon$ and so $m(\cup_{i=1}^N \gamma_i A) > 1 - \epsilon$.

We will show that this cannot be true for *all* measurable sets A. Suppose that it is and let A_n be a decreasing sequence of subsets with $\cap A_n = \varnothing$. For each n, there exists $i, 1 \leq i \leq N$ such that $m(\gamma_i A_n) > \dfrac{1-\epsilon}{N}$. By passing to a subsequence if necessary, we can assume that the same i is chosen for all n, that is

$$m(\gamma_i A_n) > \frac{1-\epsilon}{N}, \qquad \text{for all } n.$$

But $\cap_n \gamma_i(A_n) \subset \gamma_i(\cap_n A_n)$, and so

$$m(\cap_n \gamma_i(A_n)) \leq m(\gamma_i(\cap_n A_n)) = 0,$$

since $\cap_n A_n = \varnothing$ and γ_i preserves null sets.

But, since the sets A_n are decreasing, so are the sets $\gamma_i A_n$. Therefore $m(\cap_n \gamma_i(A_n)) = \lim m(\gamma_i(A_n)) \geq \dfrac{1-\epsilon}{N}$. This contradiction proves that $B(X)$ is not unitary. □

By the Russo-Dye Theorem ([**25**]), $B(X)$ unitary when X is a Hilbert space, and this is the only known example. Therefore we have the conjecture:

If $B(X)$ is unitary, then X is a Hilbert space.

Whether $B(X)$ is unitary when X is transitive also seems to be open. It is conceivable - but unlikely - that for separable spaces this is equivalent to Mazur's problem.

In [**4**] the conjecture is proved under the additional hypothesis that the unit balls of one of X, X^*, or X^{**} admits holomorphic functions that are not linear.

For Banach spaces X, uniquely maximal and convex transitive are equivalent. Even though neither of these properties imply the corresponding property of $B(X)$, we now show that for $B(X)$, algebra uniquely maximal and algebra convex transitive (i.e. unitary) are still equivalent.

THEOREM 6. [**10**] *For a Banach space X, $B(X)$ is unitary if and only if $B(X)$ is algebra uniquely maximal.*

PROOF. Suppose that $B(X)$ is unitary and that $\|.\|_1$ is an equivalent algebra norm on $B(X)$ with isometries G_1 with $G(X) \subseteq G_1$. Then

$$\{T : \|T\| \leq 1\} = \overline{\text{co}}G(X) \subseteq \overline{\text{co}}G_1 \subseteq \{T : \|T\|_1 \leq 1\}.$$

Hence $\|T\|_1 \leq \|T\|$ for all $T \in B(X)$.

But the norm in $B(X)$ is minimal ([**6**] Theorem 8, page 161), so the norms are equal and $B(X)$ is algebra uniquely maximal.

Conversely, suppose that $B(X)$ is uniquely maximal, and that $B = \overline{\text{co}}G(X)$ is not the unit ball of $B(X)$. Then there exists $S_0 \notin B$ with $\|S_0\| = 1$. Choose $f \in B(X)^*$ with $|f(T)| \leq 1$ for all $T \in B$ and $f(S_0) = r > 1$. Define a new Banach space norm on $B(X)$ by

$$\|T\|_1 = \max\{\|T\|, \sup\{|f(UT)| : U \in G(X)\}\}.$$

Then $\|VT\|_1 = \|T\|_1$ for all $V \in G(X)$, and it easy to check that

$$\|S\|_2 = \sup\{\|ST\|_1 : \|T\|_1 \leq 1\}$$

defines an equivalent algebra norm for which the elements of the group $G(X)$ are unitary. Since $B(X)$ is algebra uniquely maximal, this is necessarily equal to the uniform norm. But $\|S_0\| = 1$, yet

$$\|S_0\|_2 \geq \|S_0\|_1 \geq \sup|f(UT)| : U \in G(X)\} \geq |f(S_0)| = r > 1.$$

This is a contradiction, and so $B(X)$ is unitary. □

3. Unital Banach Algebras

For arbitrary unital Banach algebras, uniquely maximal and unitary are no longer equivalent. The second part of the proof of Theorem 5 is valid for any Banach algebra, so we have the implication in one direction:

THEOREM 7. *If A is uniquely maximal then it is unitary.*

The converse of this is not true for all algebras. We show in the next section that, if G is a discrete group with a finite subgroup, then $l_1(G)$ is unitary but not uniquely maximal. However if we have, in addition to being unitary, the semigroup property used in Theorem 2, then the algebra *is* uniquely maximal.

THEOREM 8. [**12**] *If A is unitary and there is no a in A, $\|a\| > 1$ such that the semigroup generated by U_A and a is bounded, then A is uniquely maximal.*

PROOF. Suppose that $\|.\|_1$ is an equivalent norm such that

$$u \in U_A \Rightarrow \|u\|_1 = \|u^{-1}\|_1 = 1.$$

Since $\|.\|$ is a unitary norm,

$$\{a \in A : \|a\| \leq 1\} = \overline{\text{co}}(U_A) \subseteq \{a \in A : \|a\|_1 \leq 1\}.$$

Thus $\|a\|_1 \leq \|a\|$ for all $a \in A$. If the norms are not the same, then there exists $x \in A$ such that $\|x\|_1 = 1 < \|x\|$. The semigroup S generated by U_A and x is contained in $\{a \in A : \|a\|_1 \leq 1\}$ and so will be bounded in the original norm. This contradicts our hypothesis, so the norms must be the same. □

We have seen that uniquely maximal implies unitary but the converse is not true. It is clear that we also have that uniquely maximal implies maximal and again the converse is not true. In [**17**] these two properties maximal and unitary are referred to as *maximal unitary*. We shall see in the next section that maximal unitary does not in general imply uniquely maximal, but does in certain cases.

The Russo-Dye theorem ([**25**] or [**7**], Theorem 2 page 98) gives the following:

THEOREM 9. *C^*-algebras are unitary.*

Palmer proved in [**21**] the stronger result that the unit ball is the closed convex hull of the exponential elements in U_A. In fact, in [**22**], he showed that the open unit ball is contained in the convex hull of the exponential elements in U_A.

It is discussed in both [**10**] and [**17**] how far maximal, uniquely maximal and unitary are characteristic of C^*-algebras. Characterisations of C^*-algebras can be obtained in certain circumstances. The following results were proved in [**10**] and [**17**].

We begin with the finite dimensional case. We noted earlier that a maximal finite dimensional space is necessarily a Hilbert space, and this has analogues for algebras.

THEOREM 10. [**10**] *A finite dimensional semisimple maximal algebra is a C^*-algebra.*

PROOF. If A is a semisimple finite dimensional algebra, then it is the direct sum of full matrix algebras $M_1, M_2, \dots M_n$. This algebra can be given an equivalent C^*-norm $\|.\|_1$ by using the operator norm on each M_i (as operators on a finite dimensional Hilbert space) and by taking the supremum norm over the components. If $u \in A$ is unitary for the original norm, then $u = \sum_1^n u_i$, with u_i a unitary matrix.

Clearly all these elements are unitary for the $\|.\|_1$ norm, and since the norm is maximal, we must have $\|.\| = \|.\|_1$. □

The condition of semisimplicity is necessary here as the following example shows.

EXAMPLE 3. [**10**] The algebra

$$A = \{\begin{pmatrix} a & b \\ 0 & a \end{pmatrix} : a, b \in \mathbb{C}\}$$

is not semi-simple.

Now $U_A = \{\begin{pmatrix} \lambda & 0 \\ 0 & \lambda \end{pmatrix} : |\lambda| = 1\}$, and these are the only elements in A with bounded powers. Thus A is necessarily maximal but not a C^*- algebra.

In [**10**] Cowie proves that a uniquely maximal algebra is always semisimple. This follows from the following interesting result.

THEOREM 11. [**10**] *If A is a uniquely maximal Banach algebra and I is a non-zero ideal, then, for $a \in I$, $\|a\| = \sup\{\|ab\| : b \in I, \|b\| \leq 1\}$.*

PROOF. For $x \in A$, denote by $\|x + I\|$ the norm of the image of x under the quotient map $A \to A/I$.

Then $\|x\|_1 = \|x\| + \|x + I\|$ defines an equivalent Banach space norm on A, and $\|x\|_2 = \sup\{\|xy\|_1 : y \in A, \|y\|_1 \leq 1\}$ defines an equivalent algebra norm on A. The unitary elements of A will still be unitary for $\|.\|_2$, so by the uniquely maximality of A, $\|x\| = \|x\|_2$, for all $x \in A$.

Let $a \in I$, $\|a\| = 1$. We have $\|a\| = \sup\{\|ay\|_1 : \|y\|_1 = 1\}$. Given $\epsilon > 0$, there is $y \in A$ such that $1 \geq \|ay\|_1 > 1 - \epsilon$ and $\|y\|_1 = 1$. Since $ay \in I$, $\|ay\|_1 = \|ay\|$, so we have:

$$1 \geq \|ay\| > 1 - \epsilon \text{ and } \|y\| + \|y + I\| = 1.$$

Since $\|a\| = 1$, $\|y\| > 1 - \epsilon$ and so $\|y + I\| < \epsilon$.

There exists b in I such that $\|y + b\| < \|y + I\| + \epsilon < 2\epsilon$, and so $\|a(y + b)\| < 2\epsilon$. Now $\|ab\| = \|a(y + b) - ay\| \geq \|ay\| - 2\epsilon > 1 - 3\epsilon$. We also have

$$\|b\| \leq \|y + b\| + \|y\| \leq \|y + I\| + \epsilon + \|y\| = 1 + \epsilon.$$

Therefore, if $c = \dfrac{b}{1 + \epsilon}$, then $\|c\| \leq 1$ and $\|ac\| \geq \dfrac{1 - 3\epsilon}{1 + \epsilon}$. Since ϵ is arbitrary, we have the result. □

It is clear that a finite dimensional Jacobson radical R cannot have this property, since $R^n = (0)$ for some n. It follows that a finite dimensional uniquely maximal algebra is necessarily semisimple. Thus we have the following:

THEOREM 12. [**10**] *A finite-dimensional uniquely maximal algebra is a C^*-algebra.*

In [**17**] the authors prove the corresponding result when A is maximal unitary. Thus we have:

THEOREM 13. *([**10**], [**17**]). Let A be a finite dimensional unital algebra. Then the following are equivalent:*

i) A is uniquely maximal,

ii) A is maximal unitary,

iii) A is a C^*-algebra.

However a finite dimensional unitary algebra need not be a C^*-algebra. An example is given in the next section.

The same result is true for commutative unital Banach algebras.

THEOREM 14. [**10**] *A commutative uniquely maximal algebra is a C^*-algebra.*

PROOF. If A is a commutative uniquely maximal algebra then, for each $x \in A$, the spectral radius $\rho(x) = \|x\|$. For if not, there is $x \in A$ with $\rho(x) < 1 < \|x\|$. Then x has bounded positive powers and so, since A is commutative, the semigroup generated by U_A and x will be bounded.

Thus, if Ω is the maximal ideal space of A, the Gelfand map $A \to C(\Omega)$ is an isometry. Since A is unitary, its image in $C(\Omega)$ will be a $*$-subalgebra. Thus A is isometrically isomorphic to $C(\Omega)$. □

The corresponding result for maximal unitary algebras is proved in [**17**], so we have:

THEOREM 15. *([**10**], [**17**]). Let A be a commutative unital Banach algebra. Then the following are equivalent:*
i) A is uniquely maximal,
ii) A is maximal unitary,
iii) A is a C^*-algebra.

However a commutative unitary algebra need not be a C^*-algebra. In [**17**] examples of unitary norms on $\mathbb{C}^n$ are given. These are generated by subgroups of $\mathbb{T}^n$ of the form $\mathbb{T}(\{1\} \times \mathbb{T}_{m_2} \times \mathbb{T}_{m_3} \ldots \times \mathbb{T}_{m_n})$, where $\mathbb{T}_m$ denotes the group of complex m-th roots of unity. Clearly these are not maximal bounded groups and the algebras generated are not C^*-algebras.

Finally, we have that unitary is characteristic of C^*-algebras in the case of subalgebras of $B(H)$

THEOREM 16. [**10**],[**17**] *A closed subalgebra A of a C^*-algebra for which the norm is unitary is a C^*-algebra.*

PROOF. It is sufficient to prove the result when A is a closed subalgebra of $B(H)$. The unitary elements in A will be unitary operators on H, and so $U^* = U^{-1}$. It follows that A is a $*-$subalgebra of $B(H)$ and hence it will be a C^*-algebra. □

4. Group algebras

There is another class of unitary algebras - the group algebras. Let G be a discrete group and $l_1(G)$ be the group algebra

$$l_1(G) = \{\sum a_x : x \in G, \sum |a_x| < \infty\}.$$

Then the group of unitaries for this algebra is the set

$$U = \{\lambda x : |\lambda| = 1, x \in G\}.$$

It is clear from the definition that the unit ball of $l_1(G)$ is the closed convex hull of U, and so $l_1(G)$ is unitary.

It is surprising that C^*-algebras and $l_1(G)$ algebras share this property, as in other ways, the two classes are very different. For example, in a C^*-algebra, the hermitian elements generate the algebra, while $l_1(G)$ has no non-trivial hermitian elements. (see [**7**].)

The following characterisation of groups G for which $l_1(G)$ is uniquely maximal is given in [**12**].

Definition 10. A group is *ICC* (has *infinite conjugacy classes*) if $\{e\}$ is the only finite conjugacy class.

Theorem 17. [**12**] *Let G be a group. Then the following are equivalent*
i) $l_1(G)$ *is uniquely maximal,*
ii) G is an ICC group.

We can characterise those abelian groups G for which $l_1(G)$ is maximal as follows

Theorem 18. *For an abelian group G, the following are equivalent:*
i) $l_1(G)$ *is maximal*
ii) G is torsion-free (i.e. no elements of finite order).

Proof. If G has a finite subgroup H of order n and $f = \frac{1}{n}\sum\{x : x \in H\}$, then f is an idempotent ($f^2 = f$) and $v = e - 2f$ satisfies $v^2 = e$. If U is the set of unitary elements of $l_1(G)$ then U and v generate a bounded group. By Theorem 1, this means that $l_1(G)$ is not maximal. Conversely, if G is torsion-free then the dual group $\hat{G}$ is connected, and the Beurling-Helson Theorem ([**5**]) shows that in this case, the only elements of bounded powers are the elements of U. Again by Theorem 1, $l_1(G)$ is maximal. □

In particular, $l_1(Z)$ does not have non-trivial elements of bounded powers, and so is an example of a maximal unitary algebra which is not uniquely maximal.

The corresponding problem for non-abelian groups seems to be open.

References

[1] Becerra Guerrero J. and Rodriguez Palacios A., The geometry of convex transitive Banach spaces, *Bull. London Math. Soc.* 31 (1999) no. 3, 323-331.

[2] ———, Transitivity of the norm in Banach spaces having a Jordan structure, *Manuscripta Math.* 102 (2000), 111-127.

[3] ———, Transitivity of the norm in Banach spaces, *Extracta Math.* 17 (2002), 1-58.

[4] Becerra Guerrero J., Rodriguez Palacios A. and Wood G. V., Banach spaces whose algebras of operators are unitary: a holomorphic approach (to appear *Bulletin Lond. Math. Soc.*).

[5] Beurling A. and Helson H., Fourier-Stieltjes transforms with bounded powers, *Math. Scand.* 1 (1953) 120-126.

[6] Bonsall F. F., Minimal property of the norm in some Banach algebras, *Journal London Math. Soc.* 29 (1954) 156-164.

[7] Bonsall F. F. and Duncan J., *Numerical Ranges II*, London Math. Soc. Lecture Note Series 10, Cambridge Univ. Press (1973).

[8] Cabello Sanchez F., Transitivity of M-spaces and Wood's conjecture, *Math. Proc. Camb. Phil. Soc.* 124 (1998) 513-520.

[9] Cabello Sanchez A. and Cabello Sanchez F., Maximal norms on Banach spaces of continuous functions, *Math. Proc. Camb. Phil. Soc.* 129 (2000) 325-330.

[10] Cowie E. R., *Isometries in Banach algebras*, Ph.D. Thesis, Swansea 1981.

[11] Cowie E. R., A note on uniquely maximal Banach spaces, *Proc. Edinburgh Math. Soc.* 26 (1983) 85-87.

[12] Cowie E. R., An analytic characterisation of groups with no finite conjugacy classes, *Proc. Amer. Math. Soc.* (1983) 7-10.

[13] Gordon Y. and Lewis D. R., Isometries of diagonally symmetric Banach spaces, *Israel J. Math.* 28 (1977) 45-67.

[14] Greim, P., Jamison J. E. and Kamińska A., Almost transitivity of some function spaces. *Math. Proc. Cambridge Philos. Soc.* 116 (1994) no. 3, 475-488.

[15] ———, Corrigendum: "Almost transitivity of some function spaces", *Math. Proc. Cambridge Philos. Soc.* 121 (1997) no. 1, 191-192.
[16] Greim P. and Rajagopalan M., Almost transitivity in C_0L. *Math. Proc. Cambridge Philos. Soc.* 121 (1997) no. 1, 75-80.
[17] Hansen M. L. and Kadison R. V., Banach algebras with unitary norms, *Pacific J Math.* 175 (1996) 535-552.
[18] Kalton, N. J. and Wood, G. V., Orthonormal systems in Banach spaces and their applications. *Math. Proc. Camb. Phil. Soc.* 79 (1976) 493-510.
[19] Lamperti J., On the isometries of certain function spaces, *Pacific J Math.* 8 (1958) 459-466.
[20] Lin, P.-K., Maximal norms on $C(S)$, *Bull. London Math. Soc.* 29 (1997) no. 3 345-349.
[21] Palmer T. W., Characterizations of C^* algebras, *Bull. Amer. Math. Soc.* 74 (1968) 538-540.
[22] ———, Characterizations of C^* algebras II, *Trans. Amer. Math. Soc.* 148 (1970) 577-588.
[23] Partington, J. R., Maximal norms on Banach spaces, *Bull. London Math. Soc.* 17 (1985) no. 1, 55–56.
[24] Rolewicz, S., *Metric Linear Spaces.* Monografie Matematyczne, Warszawa (1972) (Second Edition 1984).
[25] Russo B. and Dye H. A., A note on unitary operators in C* algebras, *Duke Math. J.* 33 (1966) 413-416.
[26] Wood, G. V. Maximal symmetry in Banach spaces. *Proc. Roy. Irish Acad.* 82A (1982) 177-186.

UNIVERSITY OF WALES SWANSEA, SWANSEA, WALES, UK SA2 8PP
E-mail address: G.V.Wood@swansea.ac.uk

Contemporary Mathematics
Volume **321**, 2003

On Banach Spaces with Small Spaces of Operators

András Zsák

ABSTRACT. If X is a Banach space with a basis, then for every block subspace Y of X we define a seminorm $|||\cdot|||$ on the space of operators $T\colon Y\to X$ in a similar way to a seminorm already defined by Gowers and Maurey. We prove the following theorem, which for practical purposes is an extension of the theorem of Gowers and Maurey. Let $\mathcal{S}$ be a countable semigroup of operators on c_{00} such that every $S\in\mathcal{S}$ satisfies $\|S\|_{\ell_1\to\ell_1}\le 1$, $\|S\|_{c_0\to c_0}\le 1$, and the matrix of S with respect to the usual basis of c_{00} has finite rows and columns. Then there exists a Banach space X (the completion of c_{00} under a certain norm) such that for every block subspace Y of X, every bounded operator $T\colon Y\to X$ is in the $|||\cdot|||$-closure of the set of restrictions to Y of the elements of the algebra $\mathcal{A}$ generated by $\mathcal{S}$.

1. Introduction

In this paper we prove a generalization of a theorem of Tim Gowers and Bernard Maurey from [**GM2**]. Their result roughly speaking says the following: given a semigroup $\mathcal{S}$ of operators, there exists a Banach space X whose algebra of operators $L(X)$ is in some sense close to the algebra generated by $\mathcal{S}$. In order to explain how our generalization works, we need to describe the type of semigroups to which the Gowers-Maurey result applies. As usual, we write c_{00} for the space of finite (real or complex) sequences. The usual (algebraic) basis of c_{00} will be denoted by $(e_n)_{n\in\mathbb{N}}$. For a bijection $\rho\colon A\to B$ between subsets of $\mathbb{N}$, we define a linear map $S_\rho\colon c_{00}\to c_{00}$ by $S_\rho(e_n)=e_{\rho(n)}$ when $n\in A$, and $S_\rho(e_n)=0$ otherwise. We call S_ρ a *spread* from A to B. If ρ is the (unique) order-preserving bijection from A to B, then we write $S_{A,B}$ for S_ρ, and call it the *order-preserving spread* from A to B. The result of Gowers and Maurey applies to semigroups $\mathcal{S}$ of order-preserving spreads that satisfy the following technical condition:

(1) for pairs $(i,j)\neq(k,l)$ of positive integers there are only finitely many elements $S\in\mathcal{S}$ such that $S(e_i)=e_j$ and $S(e_k)=e_l$.

Our result allows semigroups of arbitrary spreads (i.e., not just order-preserving ones), and we replace the technical condition with the only requirement that the semigroup should be countable (note that this is implied by the technical condition). The conclusion of our theorem is that given such a semigroup $\mathcal{S}$, there is a Banach

2000 *Mathematics Subject Classification.* Primary 46B03, 46B20, 47L10.

space X such that $\mathcal{S} \subset L(X)$, and $L(X)$ is the closure of the algebra generated by $\mathcal{S}$ in some seminorm. There is one point at which our result is weaker than that of Gowers and Maurey. Their theorem also states that elements of the semigroup have norm at most one in $L(X)$, whereas in our case the semigroup will in general contain elements of arbitrarily large norm. This is necessary for reasons to be explained later on.

Our construction follows very closely that of Gowers and Maurey. The notions of (m, f)-forms, rapidly increasing sequences, special functionals from [**GM2**] will be used with slight modifications, and many of the lemmas will be used here with appropriate changes. Two new ideas will be introduced: a notion of admissible sequences of functionals, and a weight defined on the given semigroup to determine the norm of its elements on the space to be constructed. We shall explain the role of these when they are defined.

It turns out that our generalization extends (with hardly any changes) to countable semigroups consisting of operators satisfying what we shall call the minimum support property (MSP). This is the result we shall present in Theorem 2 in the next section. When dealing with a semigroup of MSP operators, we need to impose further restrictions on the semigroup (apart from countability). However, these conditions are automatically satisfied by semigroups of spreads.

The paper is organized as follows. In the next section we state the main theorem after defining its ingredients. The third section contains the proof of the theorem, which requires a large number of technical results. Along the way we shall point out which are the main lemmas in the proof. In section four we establish some general properties that the spaces given by our main result have. These are properties that may be useful in applications. In the final section we discuss a specific problem to which our result may be applied, which in fact led to the generalization of the Gowers-Maurey theorem.

2. Definitions and Statement of the Main Theorem

Recall that c_{00} is the space of finite (real or complex) sequences with basis $(e_n)_{n\in\mathbb{N}}$. Given $x = (x_n)_{n\in\mathbb{N}} \in c_{00}$, the *support* of x, $\operatorname{supp}(x)$, is the set of $n \in \mathbb{N}$ for which $x_n \neq 0$, and the *range* of x is the smallest interval containing $\operatorname{supp}(x)$ and is denoted by $\operatorname{ran}(x)$. (An interval is a subset of $\mathbb{N}$ of the form $\{m, m+1, \dots, n\}$.) For subsets E, F of $\mathbb{N}$ we write $E < F$ if $\max E < \min F$ (by convention $\max \emptyset = 0$, $\min \emptyset = \infty$). For $x, y \in c_{00}$ we shall write $x < y$ if $\operatorname{supp}(x) < \operatorname{supp}(y)$. A sequence $x_1 < x_2 < x_3 < \dots$ in c_{00} (finite or infinite) is called *successive*. A *block subspace* of c_{00} is one generated by a successive sequence of non-zero vectors; this sequence is then a basis, called a *block basis*, of the subspace.

For a subset $A \subset \mathbb{N}$ and a vector $x = \sum_{n=1}^{\infty} x_n e_n$ we let $Ax = \sum_{n\in A} x_n e_n$ — in other words we identify A with the projection onto it. The spaces we shall later construct belong to the class $\mathcal{X}$ of normed spaces $(c_{00}, \|\cdot\|)$ for which $(e_n)_{n\in\mathbb{N}}$ is a normalized, bimonotone basis; that is, $\|e_n\| = 1$ for all $n \in \mathbb{N}$, and for all $x \in c_{00}$ and for every interval $E \subset \mathbb{N}$ we have $\|Ex\| \leq \|x\|$. Note that in such a space $\|x\|_\infty \leq \|x\| \leq \|x\|_1$.

Completing an element of $\mathcal{X}$ yields a Banach space $\overline{X}$ for which the sequence $(e_n)_{n\in\mathbb{N}}$ is a normalized, bimonotone Schauder basis. So every $x \in \overline{X}$ can be written as a norm convergent sum $\sum_{n=1}^{\infty} a_n e_n$ for a unique sequence of scalars $(a_n)_{n\in\mathbb{N}}$. The above definitions of support, range, successive vectors, etc. extend in an obvious way to $\overline{X}$ (or indeed to any Banach space with a basis). The only difference is that subspaces of $\overline{X}$ are always assumed to be closed (unless stated otherwise), so, for example, a block subspace is the *closed* linear span of a block basis $(u_n)_{n\in\mathbb{N}}$ and is denoted by $[u_n]_{n=1}^{\infty}$.

The first ingredient of our main result is a seminorm on certain spaces of continuous operators. Let $X = (c_{00}, \|\cdot\|)$ be a normed space from the class $\mathcal{X}$. For $n \in \mathbb{N}$ we define an equivalent norm on X by

$$\|x\|_{(n)} = \sup \sum_{i=1}^{n} \|E_i x\|,$$

where the supremum is taken over all successive sequences $E_1 < \ldots < E_n$ of intervals. Notice that $\|x\| \leq \|x\|_{(n)} \leq n\|x\|$ for every $x \in c_{00}$, and that $(c_{00}, \|\cdot\|_{(n)})$ also belongs to $\mathcal{X}$. Now let Y be a block subspace of X. Define $\mathcal{L}(Y)$ to be the set of successive sequences $\mathbf{x} = (x_n)_{n\in\mathbb{N}}$ of vectors in Y satisfying $\|x_n\|_{(n)} \leq 1$ for every $n \in \mathbb{N}$, and $\lim_{n\to\infty} \|x_n\|_\infty = 0$. Then for $T \in L(Y, X)$ we define $|||T||| = \sup_{\mathbf{x}\in\mathcal{L}(Y)} \limsup_n \|T(x_n)\|$. It is routine to check that this is a seminorm on $L(Y, X)$. This definition is essentially the same as the one in [**GM2**] except for the extra ℓ_∞-norm condition on elements of $\mathcal{L}(Y)$. It will later become clear why the extra condition was necessary.

The next ingredient of the main theorem is a class of operators on c_{00}. Let $T\colon c_{00} \to c_{00}$ be a linear map. The matrix $(t_{ij})_{i=1,j=1}^{\infty}$ of T defined by

$$T(e_i) = \sum_{j=1}^{\infty} t_{ji} e_j$$

has finite columns. If the rows of (t_{ij}) are also finite we say that T satisfies the *minimum support property (MSP)*. This is equivalent to

$$\min \operatorname{supp}(T(e_n)) \to \infty \quad \text{as } n \to \infty.$$

Now consider the adjoint T^* of T defined on the algebraic dual of c_{00}. This is formally represented by the transpose (t^*_{ij}) of (t_{ij}). Thus for a sequence $x = (x_n)_{n\in\mathbb{N}}$ of scalars and $m \in \mathbb{N}$ we have

$$(T^*(x))_m = \sum_{n=1}^{\infty} t^*_{mn} x_n = \sum_{n=1}^{\infty} t_{nm} x_n.$$

Note that T^* has finite rows, so T^* always has MSP for any linear map T on c_{00}. If, however, T has MSP, then T^* has finite columns as well as finite rows, so it gives rise to an operator on c_{00} with MSP — we call it the formal adjoint of T and we also denote it by T^*. Thus the collection of operators with MSP is a vector space that is closed under taking adjoints. It is simple to verify that it is also closed under composition — i.e., it is a $*$-algebra. Indeed, assume that S and T are operators with MSP, and K is a positive integer. Then first we find an integer $m \in \mathbb{N}$ such that $\min \operatorname{supp}(Te_i) > K$ for all $i > m$ (using MSP for T), and then we find $n \in \mathbb{N}$ such that $\min \operatorname{supp}(Se_i) > m$ whenever $i > n$ (using MSP for S). It follows that if $i > n$, then $\min \operatorname{supp}(TSe_i) > K$, so the composite TS also has MSP.

The final ingredient is a class of weight functions associated to countable semigroups of operators with MSP. It is mentioned in the introduction as one of the new tools added to the Gowers-Maurey construction. Let $\mathcal{I}$ be the set of operators consisting of all interval projections and the identity operator (recall that for every

$A \subset \mathbb{N}$ we have a projection defined by $A(x) = \sum_{n\in A} x_n e_n$; when A is an interval we call this an interval projection). Let us now fix a countable semigroup $\mathcal{S}$ of operators with MSP, and let us define $\mathcal{U}$ to be the semigroup generated by $\mathcal{S} \cup \mathcal{I}$. Thus $\mathcal{U}$ is the set of all operators of the form

$$E_{n+1}S_nE_n \dots E_2S_1E_1,$$

where each $E_i \in \mathcal{I}$ and each $S_i \in \mathcal{S}$. Note that all elements of $\mathcal{U}$ have MSP.

LEMMA 1. *Let $\mathcal{U}$ be defined as above. Then there exists a function*

$$\omega\colon \mathcal{U} \to \mathbb{N} \cup \{0\}$$

with the following properties:

(i) *$\omega(E) = 0$ for all $E \in \mathcal{I}$;*
(ii) *$\omega(VU) \le \omega(U) + \omega(V)$ for all $U, V \in \mathcal{U}$;*
(iii) *for each $r \in \mathbb{N}$ there exists a finite set $\mathcal{F}_r$ of finite sequences in $\mathcal{S}$ such that for any $U \in \mathcal{U}$ with $\omega(U) \le r$ there exist $(S_1, \dots, S_n) \in \mathcal{F}_r$ and $E_1, \dots, E_{n+1} \in \mathcal{I}$ such that $U = E_{n+1}S_nE_n \dots E_2S_1E_1$.*

PROOF. Enumerate $\mathcal{S}$ as $S_1, S_2, \dots$. Define $\omega\colon \mathcal{U} \to \mathbb{N} \cup \{0\}$ by letting

$$\omega(U) = \min \sum_{j=1}^{n} i_j,$$

the minimum being taken over all sequences $i_1, \dots, i_n$ ($n \in \mathbb{N}$) for which there exist $E_1, \dots, E_{n+1}$ in $\mathcal{I}$ such that $U = E_{n+1}S_{i_n}E_n \dots E_2S_{i_1}E_1$. Properties (i) and (ii) follow immediately from the definition. For (iii) note that if $U \in \mathcal{U}$ has $\omega(U) \le r$, then there exist $i_1, \dots, i_n$ in $\mathbb{N}$ with $\sum_{j=1}^n i_j \le r$, and elements $E_1, \dots, E_{n+1}$ of $\mathcal{I}$ such that $U = E_{n+1}S_{i_n}E_n \dots E_2S_{i_1}E_1$. Thus $\mathcal{F}_r = \{(S_{i_1}, \dots, S_{i_n}) : \sum_{j=1}^n i_j \le r\}$ will do. $\square$

From now on a pair $(\mathcal{S}, \omega)$ will stand for a countable semigroup $\mathcal{S}$ of operators with MSP together with a function ω as given by the above result. Implicitely associated to a pair is the definition of the semigroup $\mathcal{U}$ and the sequence $\mathcal{F}_r$ ($r = 1, 2, \dots$) from the above proof.

We are now ready to state our main theorem.

THEOREM 2. *Let $(\mathcal{S}, \omega)$ be a pair such that $\max\left(\|S\|_{\ell_1\to\ell_1}, \|S\|_{c_0\to c_0}\right) \le 1$ for all $S \in \mathcal{S}$. Then there exists a Banach space X (the completion of a space $(c_{00}, \|\cdot\|) \in \mathcal{X}$) with the following properties:*

(i) *for every $S \in \mathcal{S}$ we have $\|S\|_{L(X)} \le 2^{\omega(S)}$;*
(ii) *if Y is a block subspace of X, then every element of $L(Y, X)$ is in the $|||\cdot|||$-norm closure of the set of restrictions to Y of the elements of the algebra generated by $\mathcal{S}$;*
(iii) *the seminorm $|||\cdot|||$ satisfies the algebra inequality $|||UV||| \le |||U|||\,|||V|||$.*

We conclude this section with a few remarks on the theorem. Note that when $\mathcal{S}$ is a semigroup of spreads, then the requirement that $\max\left(\|S\|_{\ell_1\to\ell_1}, \|S\|_{c_0\to c_0}\right) \le 1$ for all $S \in \mathcal{S}$ is automatically satisfied. So in this case we only need $\mathcal{S}$ to be countable.

The most important property of X is the $|||\cdot|||$-density in $L(X)$ of the algebra $\mathcal{A}$ generated by $\mathcal{S}$. So for any $T \in L(X)$ and for any $\epsilon > 0$, there exist $U \in \mathcal{A}$ and $S \in L(X)$ such that $T = U + S$ and $|||S||| < \epsilon$. Later we will see that an operator with small $|||\cdot|||$-norm is in some sense close to a compact operator. For example the perturbation of a Fredholm operator by an operator with small $|||\cdot|||$-norm remains Fredholm with the same index — a result useful in applications.

As we remarked earlier, the construction of Gowers and Maurey yields somewhat stronger conclusions for a semigroup of order-preserving spreads satisfying the technical condition 1: instead of (i) above, one has $\|S\|_{L(X)} \leq 1$ for all S in the semigroup (properties (ii) and (iii) are the same). We now explain why this cannot hold for more general semigroups. Let $\mathcal{S}$ be a semigroup of operators with MSP, and let X be a Banach space such that $\mathcal{S}$ is uniformly bounded in $L(X)$. If a linear map T on c_{00} is the pointwise limit on c_{00} of a sequence in $\mathcal{S}$, then it defines a bounded operator on X. The collection of all such pointwise limits, however, may make $L(X)$ too large, and property (ii) above (i.e., the $|||\cdot|||$-density in $L(X)$ of the algebra generated by $\mathcal{S}$) may fail. A good example is given by taking $\mathcal{S}$ to consist of all projections onto finite subsets of $\mathbb{N}$. Then the collection of pointwise limits of sequnces in $\mathcal{S}$ is the collection of all projections onto (not necessarily finite) subsets of $\mathbb{N}$. So if X (the completion of an element of $\mathcal{X}$) is such that $\mathcal{S}$ is uniformly bounded in $L(X)$, then $(e_n)_{n\in\mathbb{N}}$ is an unconditional basis of X. It is not difficult to verify that in this case $L(X)$ is not separable even in the $|||\cdot|||$-norm, and so the algebra generated by $\mathcal{S}$ cannot be dense in $L(X)$. In [**GM2**] the technical condition 1 implies that $\mathcal{S}$ is closed under pointwise limits. In our result it is the function ω used in the construction of the space X that ensures that a sequence in $\mathcal{S}$ that converges pointwise on c_{00} to an operator T that is not in the $|||\cdot|||$-closure of the algebra generated by $\mathcal{S}$ cannot be uniformly bounded (and so T will not define a continuous operator on X).

3. Proof of the Main Theorem

In order to prove the theorem we need to define, for a given pair $(\mathcal{S},\omega)$, a norm on c_{00} such that the corresponding normed space is in $\mathcal{X}$, and then to show that its completion satisfies properties (i) to (iii) of the theorem. The definition of the norm is done in the same way as in [**GM2**] using (m,f)-forms and special functionals. However, the definition of these norming functionals will be different here. We introduce the notion of admissibility that was already mentioned in the introduction as one of the new tools added to the Gowers-Maurey construction. Then we define our version of (m,f)-forms and special functionals, and then we will be ready to describe the required norm on c_{00}.

To motivate the following definitions, note that if S is an order-preserving spread, then for successive vectors $x_1 < x_2 < \ldots < x_n$ the sequence $Sx_1, \ldots, Sx_n$ is also successive. This property was crucial in [**GM2**], but it clearly does not hold in general for MSP operators or even for spreads.

For a subset $A \subset \mathbb{N}$ and a linear map $T : c_{00} \to c_{00}$, we define

$$T(A) = \bigcup_{n\in A} \operatorname{supp}(Te_n).$$

Then $T(A)$ has the property that for all $x \in c_{00}$ with $\operatorname{supp}(x) \subset A$ we have $\operatorname{supp}(Tx) \subset T(A)$. Indeed, $T(A)$ is the smallest such set.

Given a pair $(\mathcal{S},\omega)$, we say that a successive sequence $E_1 < \ldots < E_k$ of intervals is $(\mathcal{S},\omega)$-*admissible* if for all $U \in \mathcal{U}$ there exists $1 \leq i \leq k$ such that

$$|\operatorname{ran}(\textstyle\bigcup_{j=1}^{i-1} E_j)| < 2^{\omega(U)},$$

and the sets $U^*(E_i), \ldots, U^*(E_k)$ are successive in some order. If the pair $(\mathcal{S}, \omega)$ is clear from the context, then we just say that the sequence $E_1, \ldots, E_k$ is admissible. A successive sequence $x_1 < \ldots < x_k$ of vectors is $(\mathcal{S}, \omega)$-admissible if $\operatorname{ran}(x_1), \ldots, \operatorname{ran}(x_k)$ is.

Remark. The idea of admissibility is as follows. Let $x_1^* < \ldots < x_k^*$ be an admissible sequence of functionals. Given $U \in \mathcal{U}$, let i be the integer given by the definition of admissibility. Then the functional $U^*(x_1^* + \ldots + x_k^*) = y^* + z^*$, where $y^* = U^*(x_1^* + \ldots + x_{i-1}^*)$ and $z^* = U^*(x_i^*) + \ldots + U^*(x_k^*)$. We shall see later that y^* is in some sense "small", so it has essentially no effect, whereas z^* is the sum of successive functionals, with which we can deal. It may seem strange that we did not require that $U^*(x_i^*) < \ldots < U^*(x_k^*)$ in the definition above but that they are successive in *some* order. This actually makes no difference to Theorem 2, but it may be important in applications.

The definition of (m, f)-forms uses functions $f : [1, \infty) \to [1, \infty)$, which have the following properties:
(i) $f(1) = 1$ and $\forall x > 1$ we have $f(x) < x$;
(ii) f is strictly increasing and tends to infinity;
(iii) $\lim_{x \to \infty} f(x) x^{-q} = 0$ for every $q > 0$;
(iv) the function F defined by $F(x) = x$ for $x \in [0, 1]$ and $F(x) = x/f(x)$ for $x \in [1, \infty)$ is concave and increasing;
(v) $f(xy) \leq f(x) f(y)$ for every $x, y \in [1, \infty)$;
(vi) the right derivative of f at 1 is positive.
Note that $\log_2(1 + x)$ and $\sqrt{\log_2(1 + x)}$ have these properties. The family $\mathcal{F}$ of all functions satisfying these conditions was introduced by T. Schlumprecht (see [**S**]) in his construction of an arbitrarily distortable Banach space (except for condition (vi) which was added in [**GM2**]). Given $X \in \mathcal{X}$, $f \in \mathcal{F}$ and a positive integer m, we let $A_m(X, f)$ be the collection of all functionals on X that can be written as $f(m)^{-1} \sum_{i=1}^m x_i^*$ for an admissible sequence $x_1^* < \ldots < x_m^*$ of functionals all of which have norm at most 1. An *(m, f)-form* on X is a functional $x^* \in A_m(X, f)$ with $\|x^*\| \leq 1$. If for all $m \in \mathbb{N}$ all elements of $A_m(X, f)$ have norm at most 1, then we say that X *satisfies the lower-f-estimate*. This is equivalent to the assertion that for every m, for every admissible sequence $E_1 < \ldots < E_m$ of intervals, and for every $x \in c_{00}$ we have

$$\|x\| \geq f(m)^{-1} \sum_{i=1}^{m} \|E_i x\|.$$

We next define special functionals. First we fix $f \in \mathcal{F}$ to be the function $f(x) = \log_2(1 + x)$ for the rest of this paper. Given $X \in \mathcal{X}$, we shall write $A_m(X)$ for $A_m(X, f)$. Let Q be the set of vectors of c_{00} each having rational coordinates and ℓ_∞-norm at most 1. Fix a subset $J = \{j_1, j_2, \ldots\}$ of $\mathbb{N}$ such that $f(j_1) > 256$ and whenever $m < n$ we have $2 j_m < \log \log \log j_n$. Let $K, L \subset J$ be the sets $\{j_1, j_3, j_5, \ldots\}$ and $\{j_2, j_4, j_6, \ldots\}$, respectively. Let σ be an injection from the set of all finite, successive sequences of elements of Q to L. We define a *special functional (of length k)* on X to be a functional y^* on X of the form $y^* = f(k)^{-1/2} \sum_{i=1}^k y_i^*$, where $k \in K$, $y_1^* < \ldots < y_k^*$ is an admissible sequence of elements of Q, $y_1^* \in A_{j_{2k}}(X)$, and $y_i^* \in A_{\sigma(y_1^*, \ldots, y_{i-1}^*)}(X)$ for $i = 2, \ldots, k$. We shall further require that, putting $m_1 = j_{2k}$ and $m_i = \sigma(y_1^*, \ldots, y_{i-1}^*)$ for $i = 2, \ldots, k$, we have $m_1 < \ldots < m_k$. Note that if X satisfies the lower-f-estimate, then every special functional belongs to $A_k(X, \sqrt{f})$ for some $k \in K$. As mentioned before, our definitions of (m, f)-forms and special functionals are the same as the ones in [**GM2**] except for the admissibility requirement.

We are now ready to start the proof of Theorem 2. Let $(\mathcal{S}, \omega)$ be a pair as in the statement of the theorem. We define $\|\cdot\|$ to be the smallest norm on c_{00} such that every x in $X = (c_{00}, \|\cdot\|)$ satisfies

$$\begin{aligned}\|x\| = \|x\|_\infty &\vee \sup\{\|E(x)\| : E \text{ is an interval}\}\\ &\vee \sup\{2^{-\omega(U)}\|U(x)\| : U \in \mathcal{U}\}\\ &\vee \sup\{|x^*(x)| : m \geq 2,\ x^* \in A_m(X)\}\\ &\vee \sup\{|x^*(x)| : x^* \text{ is a special functional}\}\end{aligned} \tag{2}$$

In a moment we verify that such a norm does indeed exist. Our task is then to show that X satisfies the conclusions of Theorem 2. (It is actually the completion of X we need but, as we only ever deal with finitely supported vectors in the proof, it is more convenient to let X denote the normed space $(c_{00}, \|\cdot\|)$ and leave the completion to the end.) For now note that the first two terms (and the minimality of $\|\cdot\|$) ensure that $X \in \mathcal{X}$. The third term guarantees that property (i) in the main theorem holds, whereas the fourth one implies that X satisfies the lower-f-estimate. The second term is, in fact, already contained in the third one, but it is included for emphasis.

LEMMA 3. *The norm defined above exists and for every x in X we have*

$$\begin{aligned}\|x\| = \|x\|_\infty &\vee \sup\{2^{-\omega(U)}|x^*(Ux)| : m \geq 2,\ x^* \in A_m(X),\ U \in \mathcal{U}\}\\ &\vee \sup\{2^{-\omega(U)}|x^*(Ux)| : x^* \textit{ is a special functional},\ U \in \mathcal{U}\}.\end{aligned} \tag{3}$$

PROOF. For each countable ordinal α we define a space $X_\alpha = (c_{00}, \|\cdot\|_{(\alpha)})$ by recursion. For $x \in c_{00}$ we let $\|x\|_{(0)} = \|x\|_\infty$. For a successor $\alpha + 1$ and for $x \in c_{00}$ we let

$$\begin{aligned}\|x\|_{(\alpha+1)} = \|x\|_{(\alpha)} &\vee \sup\{\|E(x)\|_{(\alpha)} : E \text{ is an interval}\}\\ &\vee \sup\{2^{-\omega(U)}\|U(x)\|_{(\alpha)} : U \in \mathcal{U}\}\\ &\vee \sup\{|x^*(x)| : m \geq 2,\ x^* \in A_m(X_\alpha)\}\\ &\vee \sup\{|x^*(x)| : x^* \text{ is a special functional on } X_\alpha\}.\end{aligned}$$

Finally, for a limit ordinal λ we define

$$\|x\|_{(\lambda)} = \sup_{\alpha<\lambda} \|x\|_{(\alpha)}.$$

We now define $\|x\| = \sup_{\alpha<\omega_1} \|x\|_{(\alpha)}$, and claim that this has the required properties. By ordinal induction, the sequence $(\|x\|_{(\alpha)})_{\alpha<\omega_1}$ is non-decreasing and bounded above by the ℓ_1-norm of x. Hence $\|\cdot\|$ is well-defined and satisfies $\|x\|_\infty \leq \|x\| \leq \|x\|_1$. As $\mathbb{R}$ has no uncountable subset well-ordered by the usual ordering, for each $x \in c_{00}$ there is an ordinal $\alpha < \omega_1$ with $\|x\| = \|x\|_{(\alpha)}$. Taking the union of these ordinals over the countable set of elements of c_{00} with rational coordinates, we obtain a countable ordinal λ with $\|x\| = \|x\|_{(\lambda)}$ for all $x \in c_{00}$. Now for $x \in c_{00}$ and for an interval E we have

$$\|Ex\| = \|Ex\|_{(\lambda)} \leq \|x\|_{(\lambda+1)} = \|x\|.$$

Hence $(e_n)_{n\in\mathbb{N}}$ is a bimonotone basis of $X = (c_{00}, \|\cdot\|)$. Similarly, for every $x \in c_{00}$ and $U \in \mathcal{U}$, we have $2^{-\omega(U)}\|Ux\| \leq \|x\|$. Thus $\|U\|_{L(X)} \leq 2^{\omega(U)}$ for every $U \in \mathcal{U}$.

Now let $x^* \in A_m(X)$ $(m \geq 2)$. Then $x^* \in A_m(X_\lambda)$, so we have

$$|x^*(x)| \leq \|x\|_{(\lambda+1)} = \|x\|.$$

Hence elements of $A_m(X)$ have norm at most 1, so X satisfies the lower-f-estimate. It is similar to show that the unit ball of X^* is also closed under taking special functionals.

We thus have proved that the right-hand side of 2 and that of 3 are at most the left-hand side. This proves equality in 2 (just take $E = \operatorname{ran}(x)$ in the second term of the right-hand side). To show that $\|\cdot\|$ is the smallest norm satisfying 2, assume that $\|\cdot\|'$ is another such norm. Then an easy ordinal induction shows that $\|x\|_{(\alpha)} \leq \|x\|'$ for all $\alpha < \omega_1$. Hence $\|x\| \leq \|x\|'$.

It remains to show that the left-hand side of 3 is at most the right-hand side. Assume that $\|x\| > \|x\|_\infty$, and let $\epsilon > 0$ such that $\|x\| > \|x\|_\infty + \epsilon$. Let β be the least ordinal for which there exists $U \in \mathcal{U}$ with $\|x\| - \epsilon < 2^{-\omega(U)}\|U(x)\|_{(\beta)}$. Then $\beta > 0$ and, by minimality, it must be a successor $\alpha+1$, say. Now choose $\delta > 0$ such that

$$\|x\| - \epsilon + \delta < \frac{1}{2^{\omega(U)}}\|U(x)\|_{(\alpha+1)}.$$

Consider the definition of $\|Ux\|_{(\alpha+1)}$. If there is a $V \in \mathcal{U}$ with $\|Ux\|_{(\alpha+1)} - \delta < 2^{-\omega(V)}\|VUx\|_{(\alpha)}$ then, by the subadditivity of ω, we have

$$\|x\| - \epsilon < \frac{1}{2^{\omega(U)}}\|U(x)\|_{(\alpha+1)} - \delta < \frac{1}{2^{\omega(U)}}\frac{1}{2^{\omega(V)}}\|VU(x)\|_{(\alpha)} \leq \frac{1}{2^{\omega(VU)}}\|VU(x)\|_{(\alpha)}.$$

But this contradicts the minimality of $\beta = \alpha+1$. So there is a functional x^*, either an element of some $A_m(X_\alpha)$ $(m \geq 2)$ or a special functional on X_α, such that

$$\|Ux\|_{(\alpha+1)} - \delta < |x^*(Ux)|.$$

So we have

$$\|x\| - \epsilon < \frac{1}{2^{\omega(U)}}|x^*(Ux)|.$$

Since the norms of X_α are increasing, the dual norms are decreasing; so $A_m(X_\alpha) \subset A_m(X)$, and special functionals on X_α are also special functionals on X. Moreover, ϵ can be chosen arbitrarily small, so the result follows. □

REMARK. The ordinal induction in the above proof is not absolutely necessary. It is possible to define a norm by ordinary induction. In this case, however, the unit ball of the dual space X^* may not be closed under taking special functionals, so subsequent proofs would have to be modified slightly.

So far we have constructed a space $X = (c_{00}, \|\cdot\|)$ with the following properties: X belongs to $\mathcal{X}$, it satisfies the lower-f-estimate (recall that $f(x) = \log_2(1+x)$), and $L(X)$ contains $\mathcal{U}$ with $\|U\|_{L(X)} \leq 2^{\omega(U)}$ for all $U \in \mathcal{U}$ (i.e., property (i) of Theorem 2 is satisfied). Lemma 3 above also tells us how elements of X are normed: every $x \in X$ either has the ℓ_∞-norm, or is normed by a functional $2^{-\omega(U)}U^*(x^*)$, where $U \in \mathcal{U}$, and x^* is either an (m, f)-form or a special functional. Note also that a special functional on X is a $(k, \sqrt{f})$-form for some $k \in K$.

Our task now is to show that X satisfies properties (ii) and (iii) of Theorem 2. The proof of property (ii) breaks down into a number of steps. We first introduce

RIS vectors. These are sums of successive elements of c_{00}, and are the natural objects to be normed by (m, f)-forms and special functionals. We then prove a series of technical lemmas culminating in Lemma 7 (and its consequence Lemma 9) giving bounds on norms of RIS vectors. After a couple of further technical lemmas on admissibility, we turn to Lemma 13, the main result towards proving property (ii). The proof of property (ii) is finally completed in Lemma 14. The proof of property (iii) is somewhat easier and is done in the last three lemmas of this section. This will conclude the proof of Theorem 2.

We define RIS vectors the same way as Gowers and Maurey. Let $N \in \mathbb{N}$ and $\epsilon > 0$. We say that the successive sequence $x_1 < \ldots < x_N$ of vectors *satisfies an RIS(ϵ) condition* if there are integers $n_1, \ldots, n_N$ satisfying $\frac{2N}{f'(1)} f^{-1}(\frac{4N^2}{\epsilon^2}) < n_1 < \ldots < n_N$ such that $\|x_i\|_{(n_i)} \leq 1$ for each i and

$$\epsilon\sqrt{f(n_i)} > |\operatorname{ran}(x_1 + \ldots + x_{i-1})| \quad \text{for } i = 2, \ldots, N.$$

(Note that f is still the function $\log_2(1 + x)$; the norms $\|\cdot\|_{(n)}$ $(n \in \mathbb{N})$ are those defined on page 3 just before the definition of the seminorm $|||\cdot|||$.) An $RIS(\epsilon)$ *vector* is the sum of a sequence satisfying an $RIS(\epsilon)$ condition.

The next two lemmas show how a (k, g)-form (for some $g \in \mathcal{F}$) acts on an arbitrary vector and on an RIS vector. These are versions of corresponding lemmas in [**GM2**]: the difference is that operators with MSP do not preserve successive vectors, and so admissibility will play an important role (cf. remarks following the definition of admissibility).

LEMMA 4. *Let $g \in \mathcal{F}$, and let k be a positive integer. Let x^* be a (k, g)-form and $U \in \mathcal{U}$. Then there are intervals $E_{i_0}, \ldots, E_k$ successive in some order such that for every $x \in X$ we have*

$$2^{-\omega(U)}|x^*(Ux)| \leq (i_0 - 1)g(k)^{-1}\|x\|_\infty + g(k)^{-1}\sum_{i=i_0}^{k} \|E_i x\|,$$

where $i_0 = 1$ or 2. In particular, $2^{-\omega(U)}|x^*(Ux)| \leq 2g(k)^{-1}\|x\|_{(k)}$.

PROOF. We may write x^* as a sum $x_1^* + \ldots + x_k^*$ of an admissible sequence of functionals all of which have norm at most $g(k)^{-1}$. By definition, there exists $1 \leq i \leq k$ such that

$$|\operatorname{ran}(x_1^* + \ldots + x_{i-1}^*)| < 2^{\omega(U)},$$

and $U^*(x_i^*), \ldots, U^*(x_k^*)$ are successive in some order. Then

$$\begin{aligned} 2^{-\omega(U)}|U^*(x_1^* + \ldots + x_{i-1}^*)(x)| &\leq 2^{-\omega(U)}\|U\|_{c_0 \to c_0}\|x_1^* + \ldots + x_{i-1}^*\|_1\|x\|_\infty \\ &\leq g(k)^{-1}\|x\|_\infty. \end{aligned}$$

On the other hand, the intervals $E_j = \operatorname{ran}(U^*(x_j^*))$ for $j = i, \ldots, k$ are successive in some order, and

$$2^{-\omega(U)}|U^*(x_i^* + \ldots + x_k^*)(x)| \leq g(k)^{-1}\sum_{j=i}^{k} \|E_j x\|.$$

Since $x^*(Ux) = U^*(x^*)(x)$, the result follows. □

LEMMA 5. *Let $g \in \mathcal{F}$ such that $\sqrt{f} \le g \le f$. Let $x_1, \dots, x_N$ satisfy an $RIS(\epsilon)$ condition. Let x^* be a (k, g)-form and $U \in \mathcal{U}$. Then, putting $x = x_1 + \dots + x_N$, we have*

$$2^{-\omega(U)}|x^*(Ux)| \le 1 + \epsilon + \frac{2N}{\sqrt{f(k)}}.$$

In particular, if $k > f^{-1}(4N^2/\epsilon^2)$ then $2^{-\omega(U)}|x^*(Ux)| \le 1 + 2\epsilon$.

PROOF. Let $n_1, \dots, n_N$ be the integers appearing in the RIS condition. Choose i such that $n_i < k \le n_{i+1}$. Then

$$\begin{aligned} |x^*(U(x_1 + \dots + x_{i-1}))| &\le \|x^*\|_\infty \|U\|_{\ell_1 \to \ell_1} |\text{ran}(x_1 + \dots + x_{i-1})| \\ &< g(k)^{-1} \epsilon \sqrt{f(n_i)} \le \epsilon. \end{aligned}$$

Since $\|U\|_{L(X)} \le 2^{\omega(U)}$, we have $2^{-\omega(U)}|x^*(Ux_i)| \le 1$. Finally, Lemma 4 implies that for $j > i$ we have

$$2^{-\omega(U)}|x^*(Ux_j)| \le 2g(k)^{-1}\|x_j\|_{(k)} \le 2f(k)^{-1/2}.$$

Putting all these facts together, we obtain the result. □

We are now ready to prove a norm estimate on RIS vectors in Lemma 7 below. The proof follows that of the corresponding result of [**GM2**]. In particular, we need the following technical lemma, which we quote without proof.

LEMMA 6. *Let $n \in \mathbb{N}$, and let $x \in c_{00}$ with $\|x\|_{(n)} \le 1$. Then there exists a measure w on $A = \text{ran}(x)$ such that $w(A) = 1$ and $w(E) \ge \|Ex\|$ for every interval E with $\|Ex\| \ge n^{-1}$.*

LEMMA 7. *Let $g \in \mathcal{F}$ with $\sqrt{f} \le g \le f$. Let $x_1, \dots, x_N$ satisfy an $RIS(\epsilon)$ condition, where $0 < \epsilon \le 1$; let n_1 be the first integer appearing in the condition. Set $x = x_1 + \dots + x_N$. If $\|x\|_\infty \le n_1^{-1}$ and for every interval E*

$$\|Ex\| \le 1 \vee \sup\{2^{-\omega(U)}|x^*(UEx)| : k \ge 2,\ x^* \text{ is a } (k,g)-\text{form},\ U \in \mathcal{U}\}, \tag{4}$$

then $\|x\| \le (1 + 2\epsilon)Ng(N)^{-1}$.

PROOF. Since $\|x\|_{(n_1)} \le N$, Lemma 6 implies that there exists a measure w on $\text{ran}(x)$ such that $w(\text{ran}(x)) = N$ and $\|Ex\| \le w(E)$ for every interval $E \subset \text{ran}(x)$ with $\|Ex\| \ge Nn_1^{-1}$. Let G be the function defined from g as in property (iv) of elements of $\mathcal{F}$ (see page 6). We will prove the following assertion:

$$\|Ex\| \le (1 + 2\epsilon)G(w(E)) \text{ for every interval } E \subset \text{ran}(x) \text{ with } \|Ex\| \ge Nn_1^{-1}. \tag{5}$$

This implies the result by putting $E = \text{ran}(x)$.

First note that $Nn_1^{-1} < 1$ by the RIS condition, and the assertion clearly holds when $Nn_1^{-1} \le \|Ex\| \le 1$. Indeed, we then have

$$\|Ex\| = G(\|Ex\|) \le G(w(E)) < (1 + 2\epsilon)G(w(E)).$$

Now suppose the assertion is false. Let E be a minimal counterexample. Then, by the above, we must have $\|Ex\| > 1$, and so $\|Ex\| > (1 + 2\epsilon)G(w(E)) > 1 + 2\epsilon$. Hence, by assumption 4, there exists a (k, g)-form x^* with $k \ge 2$, and an element $U \in \mathcal{U}$ such that

$$2^{-\omega(U)}|x^*(UEx)| > (1 + 2\epsilon)G(w(E)) \ \vee\ g(2)^{-1}(1 + n_1^{-1})\|Ex\|. \tag{6}$$

(Note that $g(2)^{-1}(1+n_1^{-1}) \le f(2)^{-1/2}(1+n_1^{-1}) < 1$.) By Lemma 5 we must also have $k \le f^{-1}(4N^2/\epsilon^2)$, and hence $kNn_1^{-1} < f'(1)/2$. By Lemma 4, for $i_0 = 1$ or 2 there are intervals $E_{i_0}, \dots, E_k$ that are successive in some order such that

$$(7) \qquad 2^{-\omega(U)}|x^*(UEx)| \le (i_0-1)g(k)^{-1}\|Ex\|_\infty + g(k)^{-1}\sum_{i=i_0}^{k}\|E_i x\|.$$

We may clearly assume that each E_i is contained in E. In fact, since $k \ge 2$, each E_i is properly contained in E, otherwise from 7 we get

$$\begin{aligned} 2^{-\omega(U)}|x^*(UEx)| \le& (i_0-1)g(k)^{-1}n_1^{-1} + g(k)^{-1}\|Ex\| \\ \le& g(2)^{-1}(n_1^{-1}+1)\|Ex\|, \end{aligned}$$

which contradicts 6. Hence, by minimality of E, for each i we have either $\|E_i x\| < Nn_1^{-1}$ or $\|E_i x\| \le (1+2\epsilon)G(w(E_i))$. Let A be the set of i satisfying the first of these possibilities and B the set of those satisfying the latter. If we let $s = |A| + (i_0 - 1)$, then $|B| = k - s$. Note that $k - s > 0$ otherwise from 7 we get

$$\begin{aligned} 2^{-\omega(U)}|x^*(UEx)| \le& (i_0-1)g(k)^{-1}n_1^{-1} + g(k)^{-1}|A|Nn_1^{-1} \\ \le& g(k)^{-1}kNn_1^{-1} < g(k)^{-1}f'(1)/2 < 1, \end{aligned}$$

which again is a contradiction. Now write w_i for $w(E_i)$ and w for $w(E)$. We then have, by concavity of G, that

$$\sum_{i\in B}\|E_i x\| \le (1+2\epsilon)\sum_{i\in B}G(w_i) \le (1+2\epsilon)(k-s)G(w/(k-s)).$$

We also have that $\sum_{i\in A}\|E_i x\| < |A|Nn_1^{-1}$ and $\|Ex\|_\infty \le Nn_1^{-1}$. Putting these inequalities together and using 7, we obtain

$$\begin{aligned} 2^{-\omega(U)}|x^*(UEx)| \le& (1+2\epsilon)g(k)^{-1}(k-s)G(\frac{w}{k-s}) + g(k)^{-1}sNn_1^{-1} \\ \le& (1+2\epsilon)\left((1-\frac{s}{k})G(\frac{w}{k-s})G(k) + \frac{s}{k}f'(1)/2\right). \end{aligned}$$

Now $f'(1)/2 \le g'(1) = G(1) - G'(1)$. Hence, using the submultiplicativity of g and the concavity of G, we finally get

$$\begin{aligned} 2^{-\omega(U)}|x^*(UEx)| \le& (1+2\epsilon)\left((1-\frac{s}{k})G(w/(1-\frac{s}{k})) + \frac{s}{k}(G(1)-G'(1))\right) \\ \le& (1+2\epsilon)(G(w+\frac{s}{k}) - \frac{s}{k}G'(1)) \le (1+2\epsilon)G(w). \end{aligned}$$

This contradiction with 6 completes the proof. □

Before continuing we verify that there is a $g \in \mathcal{F}$ for which assumption 4 of the above lemma holds. For this we quote a technical result form [**GM2**].

LEMMA 8. *Let $K_0 \subset K$. There exists a function $g \in \mathcal{F}$ such that $\sqrt{f} \le g \le f$, $g(k) = \sqrt{f(k)}$ for every $k \in K_0$ and $g(n) = f(n)$ whenever $N \in J\backslash K_0$ and $n \in [\log N, \exp N]$.*

Let g be the element of $\mathcal{F}$ given by the above result when $K_0 = K$. Since $g \le f$, for every positive integer m, every (m,f)-form on X is in particular an (m,g)-form. Also, since $g(k) = \sqrt{f(k)}$ for every $k \in K$, every special functional on X is a (k,g)-form for some $k \in K$. Hence, by Lemma 3, for every $x \in X$

$$\|x\| = \|x\|_\infty \vee \{2^{-\omega(U)}|x^*(Ux)| : k \ge 2,\ x^* \text{ is a } (k,g)\text{-form},\ U \in \mathcal{U}\}.$$

We now use this observation and Lemma 7 to estimate the $\|\cdot\|_{(n)}$-norm of an RIS vector.

LEMMA 9. *Let $0 < \epsilon \leq 1$, and $0 \leq \delta$. Let M, N and n be integers such that $M \in L$, $N/n \in [\log M, \exp M]$ and $f(N) \leq (1+\delta)f(N/n)$. Let $x_1, \ldots, x_N$ satisfy an $RIS(\epsilon)$ condition, and let $x = x_1 + \ldots + x_N$. Further assume that $\|x\|_\infty \leq n_1^{-1}$, where n_1 is the first integer appearing in the RIS condition. Then $\|x\|_{(n)} \leq (1+3\epsilon)(1+\delta)Nf(N)^{-1}$.*

PROOF. Let g be the function as in the observation preceding the statement of this lemma. Let $E_1, \ldots, E_n$ be successive intervals each contained in $\mathrm{ran}(x)$. Following the notation in the proof of Lemma 7, assertion 5 in the proof implies that

$$\begin{aligned}
\sum_{i=1}^n \|E_i x\| &\leq (1+2\epsilon)\Big(\sum_{i=1}^n G(w(E_i)) + nNn_1^{-1}\Big) \\
&\leq (1+2\epsilon)(N/g(N/n) + nNn_1^{-1}) \\
&= (1+2\epsilon)(N/f(N/n) + nNn_1^{-1}) \\
&\leq (1+2\epsilon)((1+\delta)N/f(N) + nNn_1^{-1}) \\
&\leq (1+3\epsilon)(1+\delta)Nf(N)^{-1}.
\end{aligned}$$

The last line uses the fact that $n_1 > \frac{2N}{f'(1)} f^{-1}(4N^2/\epsilon^2)$. This completes the proof. $\square$

We now prove three lemmas on admissibility. The first two show that any block subspace of X contains arbitrarily long admissible sequences. For this note the following: if T is a linear operator with MSP then for all m there is an $n > m$ such that for subsets E and F of $\mathbb{N}$ with $E < \{m\}$ and $\{n\} < F$ we have $T(E) < T(F)$. Indeed, let $k = \max T(\{1, \ldots, m\})$. By MSP, there exists $n > m$ such that $\min \mathrm{supp}(Te_i) > k$ for all $i > n$. This n will do. In general it is, of course, not possible to find n satisfying this for all T in some collection (unless the collection is finite). But this can be done in some special cases. For $r \in \mathbb{N}$ let $\mathcal{U}_r = \{U \in \mathcal{U} : \omega(U) \leq r\}$. Recall that implicit in the definition of the pair $(\mathcal{S}, \omega)$ is the sequence $\mathcal{F}_r$ $(r = 1, 2, \ldots)$ of finite sets of finite sequences of elements of $\mathcal{S}$ defined in the proof of Lemma 1.

LEMMA 10. *Let $r \in \mathbb{N}$. Then for all $m \in \mathbb{N}$ there exists $n > m$ such that for all subsets E, F of $\mathbb{N}$ with $E < \{m\} < \{n\} < F$ and for all $U \in \mathcal{U}_r$ we have $U^*(E) < U^*(F)$.*

PROOF. For positive integers $p < q$ we use the notation $[p, q]$ for the interval $\{p, p+1, \ldots, q\}$, and $[p, \infty)$ for the infinite interval $\{p, p+1, p+2, \ldots\}$. When $p = 1$ we write $[q]$ instead of $[1, q]$.

Let $m \in \mathbb{N}$. Choose $k > m$ such that for any sequence $(S_1, \ldots, S_i)$ in $\mathcal{F}_r$ we have

$$\max S_1^*(\ldots(S_{i-1}^*(S_i^*([m])))\ldots) < k.$$

(Note that if A and B are operators on c_{00} then for $E \subset \mathbb{N}$ the set $A(B(E))$ is in general strictly bigger than (AB)(E).) If $T_1, \ldots, T_i$ are operators with MSP then

$$\min T_i(\ldots(T_2(T_1([n,\infty))))\ldots) \to \infty \text{ as } n \to \infty.$$

Thus there exists $n > k$ such that

$$\min S_1^*(\dots(S_{i-1}^*(S_i^*([n,\infty))))\dots) > k$$

for all $(S_1,\dots,S_i) \in \mathcal{F}_r$. We claim that this n will do.

Let $U \in \mathcal{U}_r$. Then $U = E_{i+1}S_iE_i\dots E_2S_1E_1$ for some $(S_1,\dots,S_i) \in \mathcal{F}_r$ and for some $E_1,\dots,E_{i+1}$ in $\mathcal{I}$. We then have $U^* = E_1S_1^*E_2\dots E_iS_i^*E_{i+1}$. Now if $l \in [m]$ then $\operatorname{supp}(U^*(e_l)) \subset S_1^*(\dots(S_{i-1}^*(S_i^*([m])))\dots)$. Hence for $E \subset [m]$ we have $U^*(E) \subset [k-1]$. If $l \in [n,\infty)$ then $\operatorname{supp}(U^*(e_l)) \subset S_1^*(\dots(S_{i-1}^*(S_i^*([n,\infty))))\dots)$. Hence for a subset $F \subset [n,\infty)$ we have $U^*(F) \subset [k+1,\infty)$. □

LEMMA 11. *Let $E_1 < E_2 < \dots$ be an infinite sequence of intervals. Then there exists a subsequence $(E_{i(n)})_{n\in\mathbb{N}}$ such that the sequence $E_{i(1)},\dots,E_{i(k)}$ is admissible for every k. In particular, every block subspace of c_{00} contains arbitrarily long, admissible sequences of vectors.*

PROOF. We find the subsequence inductively. Set $i(1) = 1$ and let $k_0 = 0$. Suppose we have defined $i(1) < \dots < i(m-1)$ for some $m \geq 2$ together with numbers $k_0 < \dots < k_{m-2}$. Choose $k_{m-1} > k_{m-2}$ with $|\operatorname{ran}(\bigcup_{j=1}^{m-1} E_{i(j)})| < 2^{k_{m-1}}$. By the previous lemma, we can now choose $i(m) > i(m-1)$ such that $U^*(E_{i(m-1)}) < U^*(E_{i(m)})$ for all $U \in \mathcal{U}_{k_{m-1}}$. The subsequence $(E_{i(n)})_{n\in\mathbb{N}}$ defined this way has the required property. Indeed, given $U \in \mathcal{U}$, let $m \in \mathbb{N}$ be minimal with $U \in \mathcal{U}_{k_m}$. Then $\omega(U) \geq k_{m-1}$, and so

$$|\operatorname{ran}(\textstyle\bigcup_{j=1}^{m-1} E_{i(j)})| < 2^{k_{m-1}} \leq 2^{\omega(U)};$$

and, since $U \in \mathcal{U}_{k_m} \subset \mathcal{U}_{k_{m+1}} \subset \dots$, we have

$$U^*(E_{i(m)}) < U^*(E_{i(m+1)}) < U^*(E_{i(m+2)}) < \dots.$$

□

The next result will be used in the proof of the main lemma, Lemma 13. For $r \in \mathbb{N}$ we have already defined collections $\mathcal{F}_r$ and $\mathcal{U}_r$. We now define

$$\mathcal{S}_r = \{S_nS_{n-1}\dots S_1 : (S_1,\dots,S_n) \in \mathcal{F}_r\}.$$

Note that each $\mathcal{S}_r$ is finite and $\mathcal{S} = \bigcup_r \mathcal{S}_r$.

LEMMA 12. *Let $x_1,\dots,x_n$ be a successive sequence of vectors, and let $E_i = \operatorname{ran}(x_i)$ for each i. Suppose that*

$$S_m(\dots(S_1(E_1))\dots) < \dots < S_m(\dots(S_1(E_n))\dots)$$

for every $(S_1,\dots,S_m) \in \mathcal{F}_r$. Then for each $U \in \mathcal{U}_r$ there exists $S \in \mathcal{S}_r$ such that there are at most two values of i with $0 \neq U(x_i) \neq S(x_i)$.

PROOF. Given $U \in \mathcal{U}_r$, there exist $(S_1,\dots,S_m) \in \mathcal{F}_r$ and $F_1,\dots,F_{m+1} \in \mathcal{I}$ such that $U = F_{m+1}S_mF_m\dots F_2S_1F_1$. For $j = 0,1,\dots,m$ define

$$T_j = S_jS_{j-1}\dots S_1,$$
$$R_j = F_{j+1}S_jF_j\dots F_2S_1F_1.$$

We claim that for each $j = 0,1,\dots,m$ there exist k and l such that

$$R_j(x_i) = \begin{cases} T_j(x_i) & \text{if } k < i < l, \\ 0 & \text{if } i < k \text{ or } l < i. \end{cases}$$

The result follows from the case $j = m$ by putting $S = T_m$. The proof of the claim is by induction on j. For $j = 0$ we have $T_j = I$, $R_j = F_1$, so the result is clear. Let $j \geq 1$ and assume the result for $j - 1$. Then there exist k' and l' satisfying the claim for $j - 1$. As $R_j = F_{j+1}S_jR_{j-1}$ we have

$$R_j(x_i) = \begin{cases} F_{j+1}T_j(x_i) & \text{if } k' < i < l', \\ 0 & \text{if } i < k' \text{ or } l' < i. \end{cases}$$

Now $(S_1, \dots, S_j) \in \mathcal{F}_r$, so by assumption we have $T_j(x_1) < \dots < T_j(x_n)$. Hence there exist k and l with $k' \leq k \leq l \leq l'$ such that $F_{j+1}T_j(x_i) = T_j(x_i)$ for $k < i < l$ and $R_j(x_i) = 0$ for $i < k$ or $l < i$. So the proof of the inductive step is complete. □

The next result is the main lemma in showing property (ii) in Theorem 2. Its proof follows that of the corresponding result of [**GM2**]. The extra complication caused by admissibility is taken care of by Lemma 12. The notation $x > \{m\}$ below, of course, means that $\min \operatorname{supp}(x) > m$.

LEMMA 13. *Let Y be a block subspace of X, and let $T \in L(Y, X)$. Then for every $\epsilon > 0$ there is an integer m such that if $x \in Y$ with $x > \{m\}$, $\|x\|_{(m)} \leq 1$ and $\|x\|_\infty \leq 1/m$ then*

$$d(Tx, m \operatorname{conv}\{\lambda Sx : |\lambda| = 1,\, S \in \mathcal{S}_m\}) \leq \epsilon.$$

PROOF. Without loss of generality $\|T\| \leq 1$. After perturbing T in the operator norm we may also assume that the matrix of T relative to the natural bases of Y and X has finitely many non-zero entries in every row and column. (Note that to make the rows of T finite, we rely on the fact that $(e_n)_{n\in\mathbb{N}}$ is a shrinking basis of X; we shall prove this at the beginning of the next section using only results we have already proved.)

Assume the result is false. Then there exists $\epsilon > 0$ such that for all n there exists $y_n \in Y$ with $y_n > \{n\}$, $\|y_n\|_{(n)} \leq 1$, $\|y_n\|_\infty \leq 1/n$ and

$$d(Ty_n, n \operatorname{conv}\{\lambda Sy_n : |\lambda| = 1,\, S \in \mathcal{S}_n\}) > \epsilon.$$

Let F_n be the smallest interval containing y_n, $T(y_n)$ and every $S(y_n)$ for $S \in \mathcal{S}_n$. After passing to a subsequence, we may assume, using Lemma 11, that for every k the sequence $F_1, \dots, F_k$ is admissible. Write C_n for the compact, convex set $n \operatorname{conv}\{\lambda Sy_n : |\lambda| = 1,\, S \in \mathcal{S}_n\}$. By the Hahn-Banach theorem there is a norm-1 functional y_n^* such that

$$\sup\{|y_n^*(x)| : x \in C_n + \epsilon B_X\} < y_n^*(Ty_n).$$

We then have $y_n^*(Ty_n) > \epsilon$ and $|y_n^*(Sy_n)| < 1/n$ for all $S \in \mathcal{S}_n$. We may assume, replacing y_n^* by $F_n y_n^*$ if necessary, that $\operatorname{ran}(y_n^*) \subset F_n$.

Let $N \in L$. We define an *N-pair* (x, x^*) as follows. Choose $n_1, \dots, n_N$ so that the sequence $y_{n_1}, \dots, y_{n_N}$ satisfies an $RIS(1/3)$ condition. Set

$$x = \frac{f(N)}{N}(y_{n_1} + \dots + y_{n_N}) \quad \text{and} \quad x^* = \frac{1}{f(N)}(y_{n_1}^* + \dots + y_{n_N}^*).$$

We then have $x^*(Tx) > \epsilon$ and $|x^*(Sx)| < 1/N^2$ for all $S \in \mathcal{S}_N$ (since $n_1 > N^2$). By Lemma 9, we also have $\|x\| \leq 2$ and $\|x\|_{(\sqrt{N})} \leq 6$. Further observe that x^* is an (N, f)-form by the admissibility of the sequence $(F_n)_{n\in\mathbb{N}}$.

Fix $k \in K$. We now define pairs $(x_1, x_1^*), \dots, (x_k, x_k^*)$. The ultimate aim is to construct a vector y for which $\|Ty\| > \|y\|$. This contradiction will then prove the result.

Let $N_1 = j_{2k}$. Construct an N_1-pair (x_1, x_1^*) as described above. After small perturbation we may assume that $x_1^* \in Q$ and that $N_2 = \sigma(x_1^*)$ satisfies $N_2 > N_1$ and $|\mathrm{ran}(x_1)| < \frac{1}{4}\sqrt{f(N_2)} < 2^{N_2}$. By making the perturbation small enough, we still have the properties $x_1^*(Tx_1) > \epsilon$ and $|x_1^*(Sx_1)| < 1/N_1^2$ for all $S \in \mathcal{S}_{N_1}$ of an N_1-pair. The properties $\|x_1\| \leq 2$ and $\|x_1\|_{(\sqrt{N_1})} \leq 6$ also hold.

Next choose an N_2-pair (x_2, x_2^*) so that

$$S_n(\dots(S_1(\mathrm{ran}(x_1)))\dots) < S_n(\dots(S_1(\mathrm{ran}(x_2)))\dots)$$

for all $(S_1, \dots, S_n) \in \mathcal{F}_{N_2}$. After small perturbation we have that $x_2^* \in Q$ and, putting $N_3 = \sigma(x_1^*, x_2^*)$, we have $N_3 > N_2$ and $|\mathrm{ran}(x_1 + x_2)| < \frac{1}{4}\sqrt{f(N_3)} < 2^{N_3}$, while retaining the properties $x_2^*(Tx_2) > \epsilon$ and $|x_2^*(Sx_2)| < 1/N_2^2$ for all $S \in \mathcal{S}_{N_2}$ of an N_2-pair. Remember also that $\|x_2\| \leq 2$ and $\|x_2\|_{(\sqrt{N_2})} \leq 6$.

Having chosen pairs $(x_1, x_1^*), \dots, (x_{i-1}, x_{i-1}^*)$, we set $N_i = \sigma(x_1^*, \dots, x_{i-1}^*)$ and choose an N_i-pair (x_i, x_i^*) such that

$$S_n(\dots(S_1(\mathrm{ran}(x_{i-1})))\dots) < S_n(\dots(S_1(\mathrm{ran}(x_i)))\dots)$$

for all $(S_1, \dots, S_n) \in \mathcal{F}_{N_i}$. We then perturb x_i^* so that it belongs to Q and, setting $N_{i+1} = \sigma(x_1^*, \dots, x_i^*)$, we have $N_i < N_{i+1}$ and $|\mathrm{ran}(x_1 + \dots + x_i)| < \frac{1}{4}\sqrt{f(N_{i+1})} < 2^{N_{i+1}}$. By making the perturbation small, we retain the properties of an N_i-pair: $x_i^*(Tx_i) > \epsilon$, $|x_i^*(Sx_i)| < 1/N_i^2$ for all $S \in \mathcal{S}_{N_i}$. In addition we have $\|x_i\| \leq 2$ and $\|x_i\|_{(\sqrt{N_i})} \leq 6$.

Continue in this way until k pairs have been chosen. It is easy to ensure that the sequences $x_1, \dots, x_k$ and $x_1^*, \dots, x_k^*$ are admissible. Let $y = x_1 + \dots + x_k$ and $y^* = f(k)^{-1/2}(x_1^* + \dots + x_k^*)$. Then y^* is a special functional of length k, so $\|Ty\| \geq y^*(Ty) > \epsilon k f(k)^{-1/2}$.

We now derive an upper bound on the norm of y, and thereby obtain the required contradiction. Let $z^* = f(k)^{-1/2} \sum_{i=1}^k z_i^*$ be a special functional of length k and let $U \in \mathcal{U}$. We will show that $2^{-\omega(U)}|z^*(Uy)| \leq 6$. Let $m_1 = j_{2k}$ and $m_i = \sigma(z_1^*, \dots, z_{i-1}^*)$ for $i = 2, \dots, k$. Thus z_i^* is an (m_i, f)-form for each i. Let $\Omega = \omega(U)$ and choose i with $N_i < \Omega \leq N_{i+1}$. Then

$$2^{-\Omega}|z^*(U(x_1 + \dots + x_{i-1}))| \leq 2^{-\Omega}\|z^*\|_\infty \|U(x_1 + \dots + x_{i-1})\|_1 \leq f(k)^{-1/2}.$$

Now, by construction, for all $(S_1, \dots, S_n) \in \mathcal{F}_{N_{i+1}}$ we have

$$S_n(\dots(S_1(\mathrm{ran}(x_i)))\dots) < \dots < S_n(\dots(S_1(\mathrm{ran}(x_k)))\dots).$$

Hence, by Lemma 12, there exists $S \in \mathcal{S}_{N_{i+1}}$ such that there are at most two values of $j \geq i$ with $0 \neq Ux_j \neq Sx_j$. For such a j we still have $2^{-\Omega}|z^*(Ux_j)| \leq \|x_j\| \leq 2$. Thus, letting $A = \{j > i : Ux_j = Sx_j\}$, we have

$$2^{-\Omega}|z^*(Uy)| \leq f(k)^{-1/2} + 4 + f(k)^{-1/2} \sum_{l=1}^{k} \sum_{j \in A} 2^{-\Omega}|z_l^*(Sx_j)|.$$

We now bound the terms $|z_l^*(Sx_j)|$. Let t be maximal such that $z_l^* = x_l^*$ for $1 \leq l < t$. Then for $l \in A$ with $l < t$ we have $|z_l^*(Sx_l)| = |x_l^*(Sx_l)| < 1/N_l^2 < 1/k^2$, since $S \in \mathcal{S}_{N_{i+1}} \subset \mathcal{S}_{N_l}$ and $N_l \geq N_1 = j_{2k}$. If $t \in A$ then we have $2^{-\Omega}|z_t^*(Sx_t)| =$

$2^{-\Omega}|z_t^*(Ux_t)| \le \|x_t\| \le 2$. If $l \ne j$ or $l = j > t$ then $m_l \ne N_j$ by injectivity of σ. If $m_l < N_j$ then in fact $m_l < \sqrt{N_j}$, so, by Lemma 4, we have

$$2^{-\Omega}|z_l^*(Sx_j)| = 2^{-\Omega}|z_l^*(Ux_j)| \le 2f(m_l)^{-1}\|x_j\|_{(m_l)} \le 12f(m_l)^{-1} < k^{-2},$$

using also $m_l \ge m_1 = j_{2k}$. If $m_l > N_j$ then we actually have $m_l > f^{-1}(36N_j^2)$, so, by Lemma 5, we have $2^{-\Omega}|z_l^*(Sx_j)| = 2^{-\Omega}|z_l^*(Ux_j)| \le 2f(N_j)/N_j < k^{-2}$. Putting all these facts together we obtain

$$2^{-\Omega}|z^*(Uy)| \le 4f(k)^{-1/2} + 4 < 6.$$

We now obtain a norm estimate for y using Lemma 7. Let $y' = \frac{1}{6}y = \frac{1}{6}x_1 + \ldots + \frac{1}{6}x_k$. Let g be the function given by Lemma 8 when $K_0 = K - \{k\}$. Then every (m,f)-form is in particular an (m,g)-form, and every special functional of length $k' \in K_0$ is a (k',g)-form. So what we have proved above shows that for every interval E

$$\|Ey'\| \le 1 \vee \sup\{2^{-\omega(U)}|z^*(UEy')| : k' \ge 2,\ z^* \text{ is a } (k',g)\text{-form},\ U \in \mathcal{U}\}.$$

(Note that if $U \in \mathcal{U}$ and E is an interval then UE is also in $\mathcal{U}$ with $\omega(UE) \le \omega(U)$.) By construction, $\frac{1}{6}x_1, \ldots, \frac{1}{6}x_k$ satisfies an $RIS(1/2)$ condition. The first integer in this RIS condition is $\sqrt{N_1} = \sqrt{j_{2k}}$, and it is easy to verify that $\|y'\|_\infty \le \sqrt{N_1}^{-1}$. Hence, by Lemma 7, y has norm at most $12kf(k)^{-1}$. This implies that $\|Ty\| > \frac{\epsilon}{12}f(k)^{1/2}\|y\|$. This gives the required contradiction when k is sufficiently large. □

We can now complete the proof of property (ii) in exactly the same way as done in [**GM2**].

LEMMA 14. *Let Y be a block subspace of X and $T \in L(Y,X)$. For $\epsilon > 0$ let m be the integer given by the previous lemma. Let $\mathcal{A}_m = m\,\mathrm{conv}\{\lambda S : |\lambda| = 1, S \in \mathcal{S}_m\}$. Then there exists $U \in \mathcal{A}_m$ with $|||T - U||| \le 13\epsilon$.*

PROOF. As before, we shall assume that $\|T\| \le 1$ and that the matrix of T has finitely many non-zero entries in every row and column.

Suppose the result is false. Then for all $U \in \mathcal{A}_m$ there is a sequence $\mathbf{x} = \mathbf{x}(U) \in \mathcal{L}(Y)$ such that $\limsup_n \|(T-U)x_n\| > 13\epsilon$. Let $\mathcal{O}_1, \ldots, \mathcal{O}_k$ be a covering of $\mathcal{A}_m$ by open sets of diameter less than ϵ. Fix $U_j \in \mathcal{O}_j$, and let $\mathbf{x}_j = \mathbf{x}(U_j)$ for $j = 1, \ldots, k$. Then $\limsup_n \|(T-U)x_{jn}\| > 12\epsilon$ for each j and $U \in \mathcal{O}_j$.

Choose $N \in L$ with $N > k$ and $N > m^2$. For each $j = 1, \ldots, k$ choose $n_1 < \ldots < n_N$ so that the sequence $x_{jn_1}, \ldots, x_{jn_N}$ satisfies an $RIS(1/3)$ condition, and $\|(T-U_j)x_{jn_l}\| > 13\epsilon$ for each l. We may also assume that each x_{jn_l} has ℓ_∞-norm at most n_1^{-1}. We then have $\|(T-U)x_{jn_l}\| > 12\epsilon$ for each l and for each $U \in \mathcal{O}_j$. We can easily arrange that if F_{jl} is the smallest interval containing the support of x_{jn_l}, Tx_{jn_l} and Sx_{jn_l} for all $S \in \mathcal{S}_m$, then $\{m\} < F_{11} < \ldots < F_{kN}$, and the sequence $F_{11}, \ldots, F_{kN}$ of intervals is admissible. For $j = 1, \ldots, k$ set $y_j = \frac{f(N)}{N}(x_{jn_1} + \ldots + x_{jn_N})$.

Now choose a partition of unity $\phi_1, \ldots, \phi_k$ of the compact set $\mathcal{A}_m$ with respect to the open cover $\{\mathcal{O}_j\}$. Define

$$y(U) = \sum_{j=1}^{k} \phi_j(U) y_j.$$

By Lemma 9, $\|y_j\|_{(m)} \le 6$ for each j, and so $\|y(U)\|_{(m)} \le 6$ for every $U \in \mathcal{A}_m$. We also have that $y(U) > \{m\}$ and $\|y(U)\|_\infty < 1/m$. We now show that

$\|(T-U)y(U)\| > 6\epsilon$ for every $U \in \mathcal{A}_m$. Let z_{jl}^* be a support functional at $(T-U)x_{jn_l}$ such that $\mathrm{supp}(z_{jl}^*) \subset F_{jl}$. Then $z^* = f(kN)^{-1}\sum_{j=1}^{k}\sum_{l=1}^{N} z_{jl}^*$ is a (kN,f)-form, and hence $\|(T-U)y(U)\| \geq |z^*((T-U)y(U))|$. Now if $U \in \mathcal{O}_j$ then

$$z^*((T-U)y_j) \geq \frac{f(N)}{f(kN)}12\epsilon,$$

and if $U \notin \mathcal{O}_j$ then we have $\phi_j(U) = 0$. This implies that

$$\|(T-U)y(U)\| \geq z^*((T-U)y(U)) = \sum_{j:U\in\mathcal{O}_j} \phi_j(U)z^*((T-U)y_j) \geq \frac{f(N)}{f(kN)}12\epsilon > 6\epsilon.$$

The last inequality follows from $N > k$.

We now complete the argument by applying a fixed-point theorem. For each $U \in \mathcal{A}_m$ let $\Gamma(U) = \{V \in \mathcal{A}_m : \|(T-V)y(U)\| \leq 6\epsilon\}$. By Lemma 13, $\Gamma(U)$ is a non-empty, compact, convex subset of $\mathcal{A}_m$. Since $U \mapsto y(U)$ is continuous, the map $U \mapsto \Gamma(U)$ is upper-semicontinuous (i.e., the set $\{U \in \mathcal{A}_m : \Gamma(U) \subset \mathcal{W}\}$ is open for every open subset $\mathcal{W}$ of $\mathcal{A}_m$). Hence, by a fixed-point theorem (see e.g. [**DG**], §5 Theorem 11.3), there is a U with $U \in \Gamma(U)$ — a contradiction. □

We shall complete the proof of Theorem 2 by showing the submultiplicativity of $|||\cdot|||$. This is preceded by two quick results.

LEMMA 15. *Let $x = x_1 + \ldots + x_n$ be the sum of successive vectors. Assume that $\|x_i\| \leq C$ for all i. Then for $m \in \mathbb{N}$ we have*

$$\|x\|_{(m)} \leq C(n+2m).$$

PROOF. Let $E_1 < \ldots < E_m$ be successive intervals. For $1 \leq i \leq m$ set

$$A_i = \{j \in [n] : \mathrm{supp}(x_j) \subset E_i\}.$$

Then for each i we have

$$\|E_i x\| \leq C(|A_i| + 2),$$

using the bimonotonicity of the basis. Hence,

$$\sum_{i=1}^{m} \|E_i x\| \leq C(\sum_{i=1}^{m} |A_i| + 2m) \leq C(n+2m).$$

The result follows. □

LEMMA 16. *Let $U \in L(X)$. Then $\forall \epsilon > 0\ \exists k \in \mathbb{N}$ such that $\forall n \geq k\ \exists l_n \in \mathbb{N}$ such that for every vector x satisfying $\{l_n\} < x$, $\|x\|_\infty < n^{-1}$ and $\|x\|_{(n)} \leq 1$ we have $\|Ux\| \leq |||U||| + \epsilon$.*

PROOF. If the claim is false, then we can find $\epsilon > 0$, an increasing sequence $n_1 < n_2 < \ldots$ in $\mathbb{N}$ and a successive sequence $x_1 < x_2 < \ldots$ in c_{00} such that

$$\|x_i\|_\infty \leq n_i^{-1},\ \|x_i\|_{(n_i)} \leq 1 \text{ and } \|Ux_i\| > |||U||| + \epsilon.$$

But this implies the absurd inequality that $|||U||| > |||U||| + \epsilon$. □

LEMMA 17. *For $U, V \in L(X)$ we have*

$$|||UV||| \leq |||U|||\ |||V|||.$$

PROOF. After perturbing U and V in the operator norm, we may assume that the matrices of U and V have only finitely many non-zero entries in every row and column. Assume that $|||UV||| > \alpha$. We will show that $|||U|||\ |||V||| \geq \alpha$. The result then follows.

By our assumption, there is a sequence $(x_n)_{n\in\mathbb{N}}$ in $\mathcal{L}(X)$ satisfying

$$\|UVx_n\| > \alpha \text{ for all } n.$$

After passing to a subsequence, we may assume that if F_n is the smallest interval containing x_n, Ux_n, Vx_n and $UV(x_n)$, then for every k, the sequence $F_1, F_2, \dots F_k$ is admissible. We may also arrange that $\|x_n\|_\infty \leq n^{-1}$ for all n. Fix $0 < \epsilon \leq 1/2$. Let k and $(l_n)_{n\geq k}$ be the numbers given by Lemma 16 applied to V and ϵ. For each $n \geq k$ fix a very large element N_n of L. We now build a sequence $(y_n)_{n\geq k}$ of RIS vectors from the x_i by induction. Having defined $y_k < y_{k+1} < \dots < y_{n-1}$, we define y_n as follows. Set $N = N_n$, choose $i_1 < i_2 < \dots < i_N$ such that $x_{i_1}, x_{i_2}, \dots, x_{i_N}$ satisfies an $RIS(\epsilon/4)$ condition. Now set

$$y_n = \frac{f(N)}{N}(x_{i_1} + x_{i_2} + \dots + x_{i_N});$$

by choosing i_1 sufficiently large, we can ensure that $\{l_n\} < y_n$ and $y_{n-1} < y_n$ hold. This defines the sequence $(y_n)_{n\geq k}$. We now estimate some norms. Let $n \geq k$ and $N = N_n$. For $1 \leq j \leq N^\epsilon$ let

$$z_j = \frac{f(N^{1-\epsilon})}{N^{1-\epsilon}} \sum_{l=(j-1)N^{1-\epsilon}+1}^{jN^{1-\epsilon}} x_{i_l}.$$

Note that each z_j is a scalar multiple of an RIS vector, and that

$$y_n = \frac{f(N)}{f(N^{1-\epsilon})} \frac{1}{N^\epsilon} \sum_{j=1}^{N^\epsilon} z_j.$$

Now, by Lemma 9, we have

$$\|z_j\|_{(n)} \leq 1 + \epsilon \qquad \text{for each } j$$

(if N is sufficiently large); moreover, $\|z_j\|_\infty \leq n^{-1}$ certainly holds. Since $\{l_n\} < z_j$ for each j, we have $\|Vz_j\| \leq (1+\epsilon)(|||V||| + \epsilon)$. Then, by Lemma 15 with $C = (1+\epsilon)(|||V|||+\epsilon)$, we have

$$\|Vy_n\|_{(n)} \leq \frac{f(N)}{f(N^{1-\epsilon})} \frac{1}{N^\epsilon} C(N^\epsilon + 2n) < (1+\epsilon)(1+2\epsilon)(|||V||| + \epsilon)$$

if N is sufficiently larger than n. Thus, letting $c = (1+\epsilon)(1+2\epsilon)(|||V|||+\epsilon)$, we have that $(V(\frac{y_n}{c}))_{n\geq k}$ belongs to $\mathcal{L}(X)$. Hence, for all large n, we have

$$\|UVy_n\| \leq (|||U||| + \epsilon)c.$$

Finally, by the lower-f-estimate, we have

$$\|UVy_n\| > \alpha \text{ for all } n.$$

Since ϵ was arbitrary, the last two inequalities imply that

$$\alpha \leq |||U|||\ |||V|||,$$

as required. □

4. Immediate Consequences and Remarks

Let $(\mathcal{S}, \omega)$ be a pair such that $\max(\|S\|_{\ell_1 \to \ell_1}, \|S\|_{c_0 \to c_0}) \le 1$ for all $S \in \mathcal{S}$. Let $X = X(\mathcal{S})$ be the space given by Theorem 2. We first show that X is reflexive, using (as in [**GM1**]) a characterization of reflexivity due to R. C. James (see e.g. [**LT**] Theorem 1.b.5). It is necessary to show that the basis $(e_n)_{n\in\mathbb{N}}$ of X is shrinking and boundedly complete. For the latter assume that the sequence $(\sum_{i=1}^n x_i e_i)_{n\in\mathbb{N}}$ does not converge in X. Then there is an $\epsilon > 0$ and a sequence $p_1 < q_1 < p_2 < q_2 < \dots$ such that $\|\sum_{i=p_n}^{q_n} x_i e_i\| > \epsilon$ and the sequence $([p_n, q_n])_{n\in\mathbb{N}}$ of intervals is admissible. Then, by the lower-f-estimate (recall that f is the function $\log_2(1+x)$ from $\mathcal{F}$), we have

$$\|\textstyle\sum_{i=1}^{q_n} x_i e_i\| > \epsilon n f(n)^{-1} \qquad \text{for all } n.$$

Hence the sequence $(\|\sum_{i=1}^n x_i e_i\|)_{n\in\mathbb{N}}$ is not bounded, which shows that the basis of X is boundedly complete. To show that it is also shrinking we need the notion of ℓ_{1+}^n-vectors and some of their properties from [**GM1**].

We say that $x \in X$ is an *ℓ_{1+}^n-vector with constant C* if it can be written as $x_1 + \dots + x_n$ for a successive sequence $x_1 < \dots < x_n$ of vectors satisfying $\|x_j\| \le \frac{C}{n}\|x\|$ for each j. Note that if $E_j = \operatorname{ran}(x_j)$, then $\sum_{j=1}^n \|E_j x\| \le C\|x\|$. It turns out that if we take arbitrary intervals, then the same inequality holds up to a small constant. More precisely we have

LEMMA 18. *Let $x \in X$ be an ℓ_{1+}^n-vector. Then for intervals $E_1 < \dots < E_m$ we have*

$$\sum_{i=1}^m \|E_i x\| \le C(1 + 2m/n)\|x\|.$$

Hence, $\|x\|_{(m)} \le C(1+2m/n)\|x\|$.

PROOF. Let us write $x = x_1 + \dots + x_n$ as in the definition of an ℓ_{1+}^n-vector. Then for any $j \in [n]$ and for any interval E we have $\|Ex_j\| \le \|x_j\| \le \frac{C}{n}\|x\|$. Let us define $A_i = \{j \in [n] : \operatorname{supp}(x_j) \subset E_i\}$ for $i = 1, \dots, m$. Then

$$\sum_{i=1}^m \|E_i x\| \le \sum_{i=1}^m \sum_{j=1}^n \|E_i x_j\| \le \sum_{i=1}^m (|A_i| + 2)\frac{C}{n}\|x\| \le C(1 + 2m/n)\|x\|,$$

as claimed. □

The next two results show that ℓ_{1+}^n-vectors exist in abundance.

LEMMA 19. *Let n, k be positive integers and $C > 1$. Let $x_1 < \dots < x_{n^k}$ be norm-1 vectors, and let $x = \sum_{i=1}^{n^k} x_i$. If $\|x\| \ge C^{-k} n^k$, then there is a subset A of $\{1, 2, \dots, n^k\}$ (with $|A| \ge n$) such that $\sum_{i\in A} x_i$ is an ℓ_{1+}^n-vector with constant C.*

PROOF. By induction on k. If $k = 1$ then x is an ℓ_{1+}^n-vector with constant C, so we can take $A = \{1, 2, \dots, n\}$. Let $k > 1$, and assume the result for $k-1$. For $i = 1, \dots, n$ define $y_i = \sum_{j=(i-1)n^{k-1}+1}^{in^{k-1}} x_j$. If for all i we have $\|y_i\| \le \frac{C}{n}\|x\|$ then $x = y_1 + \dots + y_n$ is an ℓ_{1+}^n-vector with constant C, so we are done by letting $A = \{1, 2, \dots, n^k\}$. Otherwise there exists an i with $\|y_i\| > C^{-(k-1)} n^{k-1}$. In this case the result follows by the induction hypothesis. □

LEMMA 20. *Let $x_1 < x_2 < \dots$ be an infinite sequence of norm-1 vectors in X. Then for every $n \in \mathbb{N}$ and for every $C > 1$, there exists a finite subset A of $\mathbb{N}$ (with $|A| \ge n$) such that $\sum_{i\in A} x_i$ is an ℓ_{1+}^n-vector with constant C.*

PROOF. Choose a sufficiently large integer k with $f(n^k)^{1/k} < C$. By Lemma 11, there is a subset B of $\mathbb{N}$ of size n^k such that $(x_i)_{i\in B}$ is an admissible sequence of vectors. Let $x = \sum_{i\in B} x_i$. By the lower-f-estimate we have $\|x\| \geq n^k/f(n^k)$. Hence, by the previous lemma, there is a subset A of B such that $\sum_{i\in A} x_i$ is an ℓ_{1+}^n-vector with constant $f(n^k)^{1/k}$. This completes the proof. □

We can now prove that the basis $(e_n)_{n\in\mathbb{N}}$ of X is shrinking: we argue by contradiction. Assume that there is an $\epsilon > 0$, a norm-1 functional x^* and a sequence $x_1 < x_2 < \ldots$ of norm-1 vectors such that $x^*(x_i) > \epsilon$ for all $i \in \mathbb{N}$. By Lemma 20, for any $n \in \mathbb{N}$ there is a finite set A of $\mathbb{N}$ such that $|A| \geq n$ and $y' = \sum_{i\in A} x_i$ is an ℓ_{1+}^n-vector with constant 2. By Lemma 18, we have $\|y'\|_{(n)} \leq 6\|y'\| \leq 6|A|$. So $y = \frac{1}{6|A|}y'$ satisfies $\|y\|_{(n)} \leq 1$, $\|y\|_\infty < n^{-1}$ and $x^*(y) > \epsilon/6$. Given $N \in L$, it is now clear how to construct a sequence $y_1 < y_2 < \ldots < y_N$ satisfying an RIS(1/3) condition such that $x^*(y_i) > \epsilon/6$ for all i. Also putting $z = \sum_{i=1}^N y_i$, we have $\|z\|_\infty \leq n_1^{-1}$, where n_1 is the first integer in the RIS condition. Hence, using Lemma 9, we have

$$\frac{\epsilon N}{6} \leq x^*(z) \leq \|z\| \leq 2Nf(N)^{-1}.$$

This is a contradiction if N is sufficiently large.

We shall now turn to the algebra of operators on X. Let us write $\mathcal{A}$ for the algebra generated by $\mathcal{S}$. Recall that $\mathcal{A}$ is $|||\cdot|||$-dense in $L(X)$ (cf. property (ii) in Theorem 2). Let us define $\mathcal{N} = \{T \in L(X) : |||T||| = 0\}$. Then $\mathcal{N}$ is a closed ideal of $L(X)$: the ideal property follows from property (iii) in Theorem 2. It clearly contains all finite-rank operators, and it is not hard to show that it is contained in the ideal of strictly singular operators. Recall that an operator between Banach spaces is *strictly singular* if it is not an isomorphism on any infinite-dimensional subspace.

LEMMA 21. *For any block subspace Y of X and for any $0 < \eta < 1$ there exists an element $(y_n)_{n\in\mathbb{N}} \in \mathcal{L}(Y)$ such that $\|y_n\| > 1 - \eta$ for all n. Hence for $T \in L(Y,X)$ and for $\alpha > |||T|||$ every block subspace Z of Y contains an element z with $\|Tz\| < \alpha\|z\|$.*

PROOF. By Lemma 20 every block subspace of X contains ℓ_{1+}^n-vectors with constant C for every $n \in \mathbb{N}$ and $C > 1$. Fix $0 < \delta < 1$ such that $(1+\delta)^{-2} > 1 - \eta$. Let $n \in \mathbb{N}$, and choose $N \in \mathbb{N}$ so that $2n/N < \delta$. Let $x \in Y$ be an ℓ_{1+}^N-vector with constant $1+\delta$. Then, by Lemma 18, we have $\|x\|_{(n)} \leq (1+\delta)^2\|x\|$. Also, writing x as a sum $x_1 + \ldots + x_N$ according to the definition of an ℓ_{1+}^N-vector, we have $\|x_i\|_\infty \leq \|x_i\| \leq (1+\delta)N^{-1}\|x\|$ for each i. Hence $\|x\|_\infty \leq (1+\delta)N^{-1}\|x\| \leq (1+\delta)N^{-1}\|x\|_{(n)}$. Now let $y = x/\|x\|_{(n)}$. Then y satisfies $\|y\|_{(n)} \leq 1$, $\|y\| \geq (1+\delta)^{-2} > 1 - \eta$ and $\|y\|_\infty \leq n^{-1}$. It is now simple to construct inductively a sequence $(y_n)_{n\in\mathbb{N}}$ with the required properties.

For the last part fix an arbitrary β with $|||T||| < \beta < \alpha$, and choose $0 < \eta < 1$ with $\beta(1-\eta)^{-1} < \alpha$. Take an element $(z_n)_{n\in\mathbb{N}} \in \mathcal{L}(Z)$ with $\|z_n\| > 1 - \eta$ for all n. Then

$$\limsup_n \|Tz_n\| \leq |||T||| < \beta,$$

so for sufficiently large n we have

$$\|Tz_n\| < \beta \leq \beta\|z_n\|(1-\eta)^{-1} < \alpha\|z_n\|.$$

□

The second part of the lemma implies immediately that if $|||T||| = 0$ then T is strictly singular. Thus $\mathcal{N}$ is contained in the ideal of strictly singular operators as claimed. Now let $\mathcal{N}_\mathcal{A} = \mathcal{N} \cap \mathcal{A}$ be the ideal of $\mathcal{A}$ consisting of the $|||\cdot|||$-zero elements. Let $\mathcal{G}$ be the Banach algebra defined to be the completion of the algebra $(\mathcal{A}/\mathcal{N}_\mathcal{A}, |||\cdot|||)$. We then define a map

$$\phi\colon L(X) \longrightarrow \mathcal{G}$$

as follows. For $T \in L(X)$ there is, by property (ii) of Theorem 2, a sequence $(U_n)_{n\in\mathbb{N}}$ in $\mathcal{A}$ such that $|||T - U_n||| \to 0$ as $n \to \infty$. Then $(U_n)_{n\in\mathbb{N}}$ is Cauchy in $\mathcal{G}$, so it converges to some $\phi(T) \in \mathcal{G}$. It is clear that $\phi(T)$ does not depend on the particular choice of the sequence $(U_n)_{n\in\mathbb{N}}$. This defines the map ϕ which is clearly linear. It is also multiplicative by property (iii) in our main theorem. Moreover,

$$|||\phi(T)||| = \lim_{n\to\infty} |||U_n||| = |||T||| \leq \|T\|_{L(X)}.$$

Thus ϕ is a norm-decreasing algebra homomorphism whose restriction to $\mathcal{A}$ is the "identity", and the kernel of ϕ is the ideal $\mathcal{N}$ of $L(X)$. Note that the Banach algebra $\mathcal{G}$ is separable, since $\mathcal{S}$ is countable, and as we have seen above the kernel of ϕ consists of strictly singular operators. So in some sense we can say that X has a "small" space of operators. In applications it is sometimes possible to embed $\mathcal{G}$ into some well-known algebra (like the Wiener algebra or the algebra of operators on ℓ_1). Composing by ϕ, we obtain a continuous algebra homomorphism of $L(X)$ into this algebra, which in turn provides information about operators on X. This is indeed used in applications of the main result in [**GM2**].

We close this section with a quantitative version of the observation that $|||\cdot|||$-norm zero operators are strictly singular. If $T\colon Y \to Z$ is a bounded linear map between infinite-dimensional Banach spaces, we define

$$s(T) = \sup_{Y_0} \inf_{x\in S_{Y_0}} \|Tx\|,$$

where the supremum is taken over all infinite-dimensional subspaces of Y, and S_{Y_0} denotes the unit sphere of Y_0. It is clear that T is strictly singular if and only if $s(T) = 0$. It is well-known that a strictly singular perturbation of a Fredholm operator is also Fredholm with the same index. A quantitative version of this states (see e.g. [**LT**] Proposition 2.c.9) that, given a Fredholm operator $U\colon Y \to Z$, there exists an $\epsilon > 0$ such that if $s(T) < \epsilon$, then $U + T$ is also Fredholm and has the same index as U. The last part of Lemma 21 implies that if Y is a block subspace of our space X, then for $T\colon Y \to X$ we have $s(T) \leq |||T|||$. Thus a small $|||\cdot|||$-perturbation of a Fredholm operator in $L(Y, X)$ is also Fredholm with the same index. In particular a Fredholm element of $L(Y, X)$ can be arbitrarily closely approximated in $|||\cdot|||$-norm by a Fredholm element of the algebra $\mathcal{A}$. This fact is also often useful in applications.

5. The Square-cube Problem

In this final section we mention a specific problem whose study led to the main result of this paper. The problem asks if there is a Banach space X such that $X^2 \cong X^3$ and $X \not\cong X^2$. We shall refer to this as the square-cube problem. It arises in relation to the following question: given a general (infinite-dimensional) Banach space, for which positive integers m, n are the powers X^m and X^n of X isomorphic?

If X is one of the classical sequence-, or function-spaces, then any two of its powers are isomorphic, or equivalently X is isomorphic to its square. This led Banach to raise the question whether every (infinite-dimensional) Banach space is isomorphic to its square. The first counterexample was constructed by R. C. James. His space J has the property that its embedding into the second dual J^{**} has codimension 1. It follows that $J^m \cong J^n$ if and only if $m = n$. If a space X has distinct powers that are isomorphic, then in fact the following has to be true: there are integers $a \geq 1$ and $d \geq 1$ such that $X^m \cong X^n$ if and only if either $m = n$ or $\min(m,n) \geq a$ and $m \equiv n \pmod{d}$. Whether or not all values of a and d can occur this way is the question we would like to answer. Note that the case $a = 1, d = 1$ is the classical situation. The possibility of $a = 1, d = 2$, i.e., the existence of a Banach space X with the properties $X \cong X^3$ and $X \not\cong X^2$ was first shown by Tim Gowers in [**G**] as a solution in the negative of the Schroeder-Bernstein problem for Banach spaces. Such a space was also constructed by Gowers and Maurey in [**GM2**] as one of the applications of their theorem. Both constructions generalize to give examples for the cases $a = 1, d \geq 2$. The cases $a \geq 2, d \geq 1$ are still open. In particular it is not yet known if there is a positive answer to the square-cube problem. However we shall now describe (without too many details) a promising candidate constructed by the main theorem of this paper, and we also explain why the stronger version of the Gowers-Maurey theorem was necessary.

Let $\{A_1, A_2\}$ and $\{B_1, B_2, B_3\}$ be two partitions of $\mathbb{N}$ into infinite sets. Let $\mathcal{S}$ be the semigroup generated by the order-preserving spreads S_{A_i,B_j} and their adjoints. Applying the Gowers-Maurey theorem to $\mathcal{S}$ we obtain a space X whose algebra of operators $L(X)$ contains the algebra $\mathcal{A}$ generated by $\mathcal{S}$. Actually, in this generality $\mathcal{S}$ may not satisfy the technical condition (1) mentioned in the introduction; however, this condition is not required for defining X, but rather for proving the crucial property that $\mathcal{A}$ is $|||\cdot|||$-dense in $L(X)$. At any rate we have a space with a subspace $W = \overline{\operatorname{span}}\{e_n : n \in A_1\}$ and enough operators to guarantee that $X \cong W^2 \cong W^3$. Now if $\mathcal{S}$ *does* satisfy the technical condition, then $\mathcal{A}$ is $|||\cdot|||$-dense in $L(X)$, in other words $L(X)$ contains the bare minimum to guarantee the property $W^2 \cong W^3$. One would then hope that $W \not\cong W^2$, and so we would have a positive solution for the square-cube problem. (We note that this way of obtaining a candidate by a Gowers-Maurey type theorem is essentially the only one.) In order to verify the technical condition we need a good description of elements of $\mathcal{S}$. This is easiest when the two partitions of $\mathbb{N}$ into infinite sets possess many symmetries. The natural choice is the partitioning of $\mathbb{N}$ into congruence classes mod2 and into classes mod3. In this case $\mathcal{S}$ consists of all spreads between congruence classes $\bmod 2^a3^b$ and $\bmod 2^c3^d$ with $a+b = c+d \geq 1$. The technical condition follows immediately; however it is also easy to see that W *is* isomorphic to $X = W^2$. The isomorphism is of a very simple kind: it is a sum $U = S_{C_1,D_1} + \ldots + S_{C_n,D_n}$ of (order-preserving) spreads, where $\{C_1, \ldots, C_n\}$ is a partition of A_1, and $\{D_1, \ldots, D_n\}$ is a partition of $\mathbb{N}$. Its inverse is the formal adjoint $U^* = S_{D_1,C_1} + \ldots + S_{D_n,C_n}$. We call an isomorphism between W and X of this kind a *trivial isomorphism.*

It seems very likely that whenever the partitions $\{A_1, A_2\}$ and $\{B_1, B_2, B_3\}$ consist of unions of congruence classes the algebra $\mathcal{A}$ contains a trivial isomorphism between W and X. One explanation for this is the large number of relations satisfied by the generators of $\mathcal{S}$. It turns out that it is possible to identify exactly those relations that are always satisfied independently of the partitions and to construct the two partitions of $\mathbb{N}$ in such a way to ensure that only these relations are satisfied in $\mathcal{S}$. However, these constructions do not allow one to verify the technical condition. This is why it was necessary to replace that condition by something weaker. According to our main theorem countability of the semigroup is sufficient, so we can apply the theorem to $\mathcal{S}$ independently of how the two partitions of $\mathbb{N}$ are chosen. In particular we obtain a space X such that the algebra $\mathcal{A}$ generated by $\mathcal{S}$ is dense in $L(X)$, and the set of relations of $\mathcal{S}$ is as small as possible. We now have a better chance that the subspace W (which of course still has the property that $W^2 \cong W^3$) is not isomorphic to its square. We have no proof of this yet, indeed the task of showing that there are no trivial isomorphisms between W and X seems to be rather difficult too. However, we have several partial results in [**Zs**] that strongly suggest that our candidate solves the square-cube problem.

Acknowledgements

The results of this paper were obtained during the author's Ph.D. I am greatly indebted to my research supervisor Tim Gowers for his help during my Ph.D., and for his suggestions in the preparation of this article.

References

[DG] J. Dugundji and A. Granas: Fixed point theory, Volume I, Warszawa 1982.

[G] W. T. Gowers: A solution to the Schroeder-Bernstein problem for Banach spaces. Bull. London Math. Soc. 28 (1996), no. 3, 297–304.

[GM1] W. T. Gowers, B. Maurey: The unconditional basic sequence problem, Jour. A.M.S. **6** (1993), 851 - 874.

[GM2] W. T. Gowers, B. Maurey: Banach spaces with small spaces of operators, Math. Ann. **307**, 543–568 (1997).

[J] R. C. James: A non reflexive Banach space isometric to its second conjugate, Proc. Nat. Acad. Sc. U.S.A. (1951), 174 - 177.

[LT] Lindenstrauss and Tzafriri: Classical Banach spaces I and II, Springer-Verlag, reprint of the 1977, 1979 editions.

[M] B. Maurey: Banach spaces with small spaces of operators, Lecture notes of a mini-course given at the Summer School in Spetses, August 1994.

[S] T. Schlumprecht: An arbitrarily distortable Banach space, Israel J. Math. **76** (1991), 81 - 95

[Zs] A. Zsák: Algebras of Operators on Banach spaces. Ph. D. Thesis, Cambridge University, 2000.

Department of Mathematics, Texas A&M University, College Station, TX 77843, USA

E-mail address: azsak@math.tamu.edu

Contemporary Mathematics
Volume **321**, 2003

A remark on p-summing norms of operators

Artem Zvavitch

Abstract. In this paper we improve a result of W. B. Johnson and G. Schechtman by proving that the p-summing norm of any operator with n-dimensional domain can be well-approximated using $C(p)n\log n(\log\log n)^2$ vectors if $1 < p < 2$, and using $C(p)n^{p/2}\log n$ if $2 < p < \infty$.

1. p-summing norms

Throughout this paper we will follow notations of N. Tomczak-Jaegermann [**To**].

Definition 1.1. *Let X and Y be Banach spaces. An operator $u : X \longrightarrow Y$ is called* **p-summing** *if there exists a constant c such that for all finite sequences $\{x_j\}$ in X one has*

$$\left(\sum_j ||ux_j||^p\right)^{1/p} \leq c \sup_{x^*\in X^* \text{ and } ||x^*||\leq 1} \left(\sum_j |(x_j, x^*)|^p\right)^{1/p}.$$

The infimum of constants c satisfying this inequality is denoted by $\pi_p(u)$ and is called the p-summing norm of u.

Here we would like to discuss the following natural question: Given an operator u with n-dimensional domain, how many vectors do we need to approximate the p-summing norm of u?

To present a more precise version of this question we will first give the following definition:

Definition 1.2. *For positive integer n, let $\pi_p^{(n)}(u)$ is the smallest constant c such that for arbitrary vectors $x_1, \dots, x_n \in X$ one has*

$$\left(\sum_{j=1}^n ||ux_j||^p\right)^{1/p} \leq c \sup_{x^*\in X^* \text{ and } ||x^*||\leq 1} \left(\sum_{j=1}^n |(x_j, x^*)|^p\right)^{1/p}.$$

One can see that

$$||u|| = \pi_p^{(1)}(u) \leq \pi_p^{(2)}(u) \leq \cdots \leq \pi_p^{(n)}(u) \leq \pi_p(u),$$

2000 *Mathematics Subject Classification.* Primary 46E30, 46B07; Secondary 46B09, 60G15.
Key words and phrases. p-summing norm, L_p, ℓ_l^n.

and

$$\lim_{n\to\infty} \pi_p^{(n)}(u) = \pi_p(u).$$

Now we are ready to state a precise version of the question above: Given a operator u with n-dimensional domain what is the smallest number N such that $\pi_p(u) \le C\pi_p^{(N)}(u)$.

The first case to study is $p = 2$. In this case a useful result of N. Tomczak-Jaegermann (see [**To**], page 143) states that if u is 2-summing of rank n than $\pi_2(u) \le \sqrt{2}\pi_2^{(n)}(u)$.

For $p = 1$, S. Szarek [**Sz**] proved that $\pi_1(u) \le C\pi_1^{(Cn\log n)}(u)$. Finally W. B. Johnson and G. Schechtman proved in [**J-S**], that $\pi_p(u) \le C\pi_p^{(Cn\log^3 n)}(u)$ for $1 < p < 2$ and $\pi_p(u) \le C\pi_p^{(Cn^{p/2}\log^3 n)}(u)$, for $2 < p < \infty$.

Here we would like to improve a result of W. B. Johnson and G. Schechtman by reducing the power of $\log n$. We will follow an approach of [**J-S**] but we will use different methods for approximation of the expectation (or tails) of a given random process. Our goal is to prove the following theorem:

THEOREM 1.1. *Suppose that dim$(X)=n$, $u : X \to Y$ is a linear operator and $\varepsilon > 0$. Then*

$$\pi_p(u) \le (1+\varepsilon)\pi_p^{(N)},$$

as long as

(i): $1 < p < 2$ *and* $N \ge C(p)\varepsilon^{-2}n(\log n)(\log\log(\varepsilon^{-2}n) + \log\varepsilon^{-1})^2$,
(ii): $2 < p < \infty$ *and* $N \ge C(p)\varepsilon^{-5}n^{\frac{p}{2}}\log(\varepsilon^{-5}n)$.

Given a linear operator $u : X \to Y$ of finite rank, $1 < q < \infty$ and positive integers n, k define

$$\nu_q^{(n,k)}(u) = \inf\left\{\sum_{i=1}^{k}\nu_q^{(n)}(u_i) : u = \sum_{i=1}^{k} u_i\right\},$$

where $\nu_q^{(n)}(u)$ is the infimum of

$$||A||\ |||w||\ ||B||$$

where $u = BwA$ and

$$A : X \to \ell_\infty^n,\ w : \ell_\infty^n \to \ell_q^n \text{ diagonal},\ B : \ell_q^n \to Y.$$

The next theorem (theorem 24.2 [**To**]) shows a connection between $\pi_p^n(u)$ and $\nu_q^{(n,k)}(u)$:

THEOREM 1.2. *Let* $\frac{1}{p} + \frac{1}{q} = 1$ *and*

$$\hat{\nu}_q^{(n)}(u) = \lim_{k\to\infty}\nu_q^{(n,k)}(u),$$

then the ideal norms $\pi_p^{(n)}$ *and* $\hat{\nu}_q^{(n)}$ *are in trace duality.*

The next theorem is a key step in the proof of theorem 1.1. After proving it, we will finish our proof of theorem 1.1 using an iterational procedure, and theorem 1.2.

THEOREM 1.3. *Let $n \le M$ be positive integers; $u : X \to Y$ a linear operator with X finite dimensional and dim$(Y) \le n$. Then, (for $\frac{1}{p} + \frac{1}{q} = 1$),*

(i): *For* $1 < p < 2$,

$$\nu_q^{(\frac{7}{8}M,2)}(u) \le \left(1 + C_p\left(\frac{n}{M}\log n\left[\log\frac{M}{n} + \log\log M\right]^2\right)^{\frac{1}{2}}\right)\nu_q^{(M,1)}(u).$$

(ii): *For* $2 < p < \infty$,

$$\nu_q^{(\frac{7}{8}M,2)}(u) \le \left(1 + c(p)\left(\frac{n^{p/2}}{M}\log M\right)^{1/5}\right)\nu_q^{(M,1)}(u).$$

2. Probabilistic lemma

Before proving theorem 1.3 we need to prove the following probabilistic lemma, which will improve the result of proposition 2.1, [**J-S**].

The assumptions of this lemma is the "Lewis lemma" (see [**L**], or [**S-Z**]): in [**S-Z**] it was shown that for any n-dimensional subspace X of $L_p(\Omega,\mu)$, $0 < p < \infty$, one can find a probability measure τ on Ω and a subspace $\tilde{X}$ of $L_p(\Omega,\tau)$ isometric to X which admits a basis $h_1,\dots,h_n$ *orthonormal* in $L_2(\Omega,\tau)$, such that

$$\sum_{i=1}^{n} h_i^2 \equiv n.$$

In addition we note that if X is a subspace of ℓ_p^M and τ is the probability measure on $\{1,\dots,M\}$ given by the theorem above then, as observed in [**J-S**], we can split each atom of τ of mass large than $4/M$ into pieces each of size between $2/M$ and $4/M$. This will enlarge the number of atoms by at most $M/2$. The new measure λ is such that

$$\lambda\{i\} \le 4/M \text{ for all } i,$$

and it is supported on $\bar{K} = \{1,\dots,K\}$, where $K \le 3M/2$. Finally, $\tilde{X}$ is still isometric (we denote this isometry by J) to a subspace $J\tilde{X}$ of $L_p(\bar{K},\lambda)$, and $J\tilde{X}$ admits an orthonormal basis $h_1,\dots,h_n$ whose sum of squares is a constant (clearly, X is also isometric to $J\tilde{X}$).

LEMMA 2.1. *Let X be an n dimensional subspace of $L_p(\bar{K},\lambda)$ with $2 < p < \infty$, and assume that X has characteristics from the Lewis lemma (plus the remark after it, concerning the splitting of big atoms). Let B_p be the unit ball of X and $\{\varepsilon_j\}_{j=1}^K$ be independent random variables taking the values $+1$ or -1 with equal probability, then*

$$\mathbb{P}\left\{\sup_{x\in B_p}\left|\sum_{j=1}^{K}\lambda\{j\}\varepsilon_j|x(j)|^p\right| > c(p)\left(\frac{n^{p/2}}{K}\log K\right)^{1/5}\right\} \le \frac{1}{2}.$$

Proof: To prove this lemma we will use a method of proof due to J. Bourgain, J. Lindenstrauss and V. Milman [**B-L-M**].

Fix $0 < t < 1/2$ to be chosen later ($t = c(p)\left(\frac{n^{p/2}}{K}\log K\right)^{1/5}$).

Let $\mathcal{F}$ be a $\frac{1}{2}t$-net (with respect of the metric $d(x,y) = \sum\lambda\{j\}\,||x_i|^p - |y_i|^p|$) on the boundary of B_p then,

$$d(x,y) = \sum\lambda\{j\}\,||x_i|^p - |y_i|^p| \le p\sum\lambda\{j\}\max\{|x_i|,|y_i|\}^{p-1}\,|x_i - y_i| \le$$

$$p\left(\sum\lambda\{j\}\max\{|x_i|,|y_i|\}^p\right)^{(p-1)/p}\left(\sum\lambda\{j\}\,|x_i-y_i|^p\right)^{1/p} \le c_p||x-y||_p,$$

so by a standard volume argument, we get that $\log|\mathcal{F}| \le c_p n \log t^{-1}$.

It is easy to see that:

$$\mathbb{P}\left\{\sup_{x\in B_p}\left|\sum_{j=1}^{K}\lambda\{j\}\varepsilon_j|x(j)|^p\right| > \frac{1}{2}t\right\} \le \mathbb{P}\left\{\sup_{x\in \mathcal{F}}\left|\sum_{j=1}^{K}\lambda\{j\}\varepsilon_j|x(j)|^p\right| > t\right\}.$$

Using the Lewis lemma (see **[J-S]**, Proposition 2.1) we get that $||x||_\infty \le n^{1/2}||x||_p$, for $x \in X$, so that the l_∞-diameter of B_p is less or equal to $2n^{1/2}$.

For $k = 1, 2, \ldots, l = [\log_{(1+t)} n^{1/2}] + 1$ let $A_k \subset B_p$ be such that

$$B_p \subset \bigcup_{g\in A_k}(g + \frac{t(1+t)^k}{3}B_\infty).$$

X has characteristics from the Lewis lemma, so applying Proposition 2.1 from **[J-S]** $\left[\ \log N(B_p, B_\infty, t) \le c(p)n\log Kt^{-2}\ \right]$ we get:

$$\log|A_k| \le c(p)n\log Kt^{-2}(1+t)^{-2k}.$$

For every $x \in \mathcal{F}$ and $1 \le k \le l$ let $f^{k,x} \in A_k$ satisfy $||x - f^{k,x}||_\infty \le t(1+t)^k/3$. Put

$$C_{k,x} = \{i : |f_i^{k,x}| \ge (1+t)^{k-1}\},$$

$$D_{k,x} = C_{k,x} \setminus \bigcup_{h>k} C_{h,x},\ D_{0,x} = \{1, \ldots, K\} \setminus \bigcup_{k=1}^{l} C_{k,x},$$

and

$$\hat{x} = xI_{D_{0,x}} + \sum_{k=1}^{l}(1+t)^k I_{D_{k,x}}.$$

Note that if $i \in C_{k,x}$, $k \ge 1$, then

$$|x_i| \ge (1+t)^{k-1} - t(1+t)^k/3 \ge (1+t)^{k-2},$$

while if $i \notin C_{k,x}$

$$|x_i| \le (1+t)^{k-1} + t(1+t)^k/3 \le (1+t)^{k+1}.$$

Hence for every $i \in D_{k,x}$, $k \ge 1$,

$$(1+t)^{k-2} \le |x_i| \le (1+t)^{k+2},$$

while for $i \in D_{0,x}$, $|x_i| \le (1+t)^2$. It follows that

$$(1+t)^{-2} \le \frac{|\hat{x}_i|}{|x_i|} \le (1+t)^2,$$

then

$$1 - 4pt \le \frac{|\hat{x}_i|^p}{|x_i|^p} \le 1 + 4pt.$$

Then

$$\left|\sum_{j=1}^{K}\lambda\{j\}\varepsilon_j|\hat{x}(j)|^p - \sum_{j=1}^{K}\lambda\{j\}\varepsilon_j|x(j)|^p\right| \le \sum_{j=1}^{K}\lambda\{j\}\,||\hat{x}(j)|^p - |x(j)|^p| \le c_p t.$$

So we may consider $\hat{x}$ instead of x.

To prove our lemma we first estimate the probability of the following event:

$$\left|\sum_{j=1}^{K} \lambda\{j\}\varepsilon_j|\hat{x}(j)|^p I_{D_{0,x}}\right| \le t_o,\ x \in \mathcal{F},$$

$$\left|\sum_{j=1}^{K} \lambda\{j\}\varepsilon_j(1+t)^{pk} I_{D_{k,x}}\right| \le t_k,\ 1 \le k \le l, x \in \mathcal{F},$$

where

$$\sum t_k \le t.$$

Let B_k $1 \le k \le l$ be the collection of all sets of form $D_{k,x}$. It follows from the definition that

$$\log|B_k| \le \sum_{h=k}^{l} \log|A_h| \le c_p n \log K t^{-3}(1+t)^{-2k}.$$

Now we can apply Bernstein's inequality (see, for example, [**B**], page 39), which states that for given $\{a_i\}_{i=1}^n \in \mathbb{R}^n$, with $\sum_{i=1}^{n} a_i^2 = 1$ we get

$$\mathbb{P}(\left|\sum a_i\varepsilon_i\right| > u) \le 2\exp\left(\frac{-u^2}{2}\right).$$

So that if

$$a_j = \frac{\lambda\{j\}|\hat{x}(j)|^p}{\sum_{j\in D_{0,x}} \lambda^2\{j\}|\hat{x}(j)|^{2p}},\ u = \frac{t_0}{\sum_{j\in D_{0,x}} \lambda^2\{j\}|\hat{x}(j)|^{2p}},$$

then

$$\mathbb{P}\left(\left|\sum_{j=1}^{K} \lambda\{j\}\varepsilon_j|\hat{x}(j)|^p I_{D_{0,x}}\right| > t_o\right) \le 2\exp\left(\frac{-t_0^2}{2\sum_{j\in D_{0,x}} \lambda^2\{j\}|\hat{x}(j)|^{2p}}\right).$$

Applying $\sum_{j\in D_{0,x}} \lambda\{j\}|\hat{x}(j)|^p \le 2$, $\lambda\{j\}|\hat{x}(j)|^p \le \frac{2}{K}(1+t)^{2p}$, we get

$$\mathbb{P}\left(\left|\sum_{j=1}^{K} \lambda\{j\}\varepsilon_j|\hat{x}(j)|^p I_{D_{0,x}}\right| > t_o\right) \le 2\exp(-c_p t_0^2 K(1+t)^{-2p}).$$

In the same manner we get:

$$\mathbb{P}\left(\left|\sum_{j=1}^{K} \lambda\{j\}\varepsilon_j|\hat{x}(j)|^p I_{D_{k,x}}\right| > t_k\right) < 2\exp(-c_p t_k^2 K(1+t)^{-pk}).$$

So, in order to prove our lemma it is enough to estimate

$$\begin{aligned} &2|\mathcal{F}|\exp(-c_p t_0^2 K) + 2\sum_{k=1}^{l} |B_k|\exp(-c_p t_k^2 K(1+t)^{-pk}) \\ = \quad & 2 \quad \exp(4n\log t^{-1} - c_p t_0^2 K) \\ + \quad & c_p \sum_{k=1}^{l} \exp(c_p n \log K t^{-3}(1+t)^{-2k} - c_p t_k^2 K(1+t)^{-pk}). \end{aligned}$$

Now we may substitute $t_o = c_1(p)t$, $t_l = c_2(p)t$, $t_k = c_2(p)(1+t)^{(2-p)(l-k)/2}t$ to get:

$$\begin{aligned} & 2 \quad \exp(4n \log t^{-1} - c_p t^2 K) \\ & + \quad c_p \sum_{k=1}^{l} \exp(c_p n \log K t^{-3}(1+t)^{-2k} - c_p K (1+t)^{(2-p)(l-k)-pk} t^2) \\ & = \quad 2 \quad \exp(4n \log t^{-1} - c_p t^2 K) \\ & \quad + \quad c_p \sum_{k=1}^{l} \exp(c_p t^{-3}(1+t)^{-2k}(n \log K - c_p K (1+t)^{2l-pl} t^5)) \end{aligned}$$

$$= 2 \exp(4n \log t^{-1} - c_p t^2 K) + c_p \sum_{k=1}^{l} \exp(c_p (1+t)^{-2k} n (\log K - c_p K n^{-p/2} t^5)).$$

To conclude the proof of our lemma we take $t = c_p \left(\frac{n^{p/2}}{K} \log K\right)^{1/5}$ (i.e. such that $\log K - c_p K n^{-p/2} t^5 < -c_p')$.

□

The next lemma shows that in some applications of lemma 2.1, we may "omit" requirements for "Lewis lemma" characteristics.

LEMMA 2.2. *Let X be an n-dimensional subspace of $L_p(\bar{M}, \mu)$ consider $JX \subset L_p(\bar{K}, \mu')$, where J is the isometry defined by the splitting of big atoms. Then there is a partition $K_1 \cup K_2$ of $\bar{K}$ into two sets of cardinality at most $\frac{7}{8}M$ such that for each $x \in JX$ and $j = 1, 2$:*

(i):

$$||1_{K_j} x||^p_{L_p(\bar{K}, \mu')} \leq \left(\frac{1}{2} + C_p \left(\frac{n}{M} \log n \left[\log \frac{M}{n} + \log \log M \right]^2 \right)^{\frac{1}{2}} \right) ||x||^p_{L_p(\bar{M}, \mu)},$$

when $1 < p < 2$;

(ii):

$$||1_{K_j} x||^p_{L_p(\bar{K}, \mu')} \leq \left(\frac{1}{2} + c(p) \left(\frac{n^{p/2}}{M} \log M \right)^{1/5} \right) ||x||^p_{L_p(\bar{K}, \mu)},$$

for $2 < p < \infty$.

Proof: To prove (i) we use a result of M. Talagrand from his paper on embedding n-dimensional subspaces of L_p into L_p^N with N not too large (see proof of Proposition 2.3 from [**T**]) and get that if $Z \in L_p(\bar{M}, \tau)$ is a isometric copy of X after a Lewis change of density, and if $JZ \subset L_p(\bar{K}, \lambda)$ then

$$\sup_{z \in B_p(JZ)} \left| \sum_{i=1}^{K} \lambda\{i\} \varepsilon_i |z(i)|^p \right| \leq C_p \left(\frac{n}{K} \log n \left[\log \frac{K}{n} + \log \log K \right]^2 \right)^{\frac{1}{2}} \tag{1}$$

for most choices of signs $\varepsilon_i = \pm 1$.

One can see that for a fixed J the left hand side is invariant under a change of density f, i.e. we can replace the subspace Y of $L_p(\bar{K}, \lambda)$ with its image under the natural isometry from $L_p(\bar{K},\lambda)$ to $L_p(\bar{K},\mu)$ defined by $Tx = x/f^{\frac{1}{p}}$. And using that $M < K < \frac{3}{2}M$ we get

$$\sup_{x\in B_p(JX)}\left|\sum_{i=1}^{K}\mu'\{i\}\varepsilon_i|z(i)|^p\right| \le C_p\left(\frac{n}{M}\log n\left[\log\frac{M}{n}+\log\log M\right]^2\right)^{\frac{1}{2}}$$

for most choices of signs $\varepsilon_i = \pm 1$. Since also for most choices of signs the difference between the number of plus signs and minus signs is less than $K/8$, (i) follows. (ii) follows similarly, using lemma 2.1.

□

3. Proof of the main result

Proof of Theorem 1.3: For some probability measure μ on $\bar{M} = \{1\dots M\}$, we consider $A : Y^* \to L_p(\bar{M},\mu)$, and $B : L_1(\bar{M},\mu) \to X^*$ so that $||A||\,||B|| = \nu_q^{(M)}(u)$ and $u^* = Bi_{p,1}A$, where $i_{p,1}$ is the formal identity mapping from $L_p(\bar{M},\mu)$ onto $L_1(\bar{M},\mu)$.

Next, using lemma 2.2, we get a partition $K_1 \cup K_2 = \bar{K}$ into two sets of cardinality at most $\frac{7}{8}M$ such that

$$||1_{K_j}JAy||^p_{L_p(\bar{K},\mu')} \le \left(\frac{1}{2}+c(p)\left(\frac{n^{p/2}}{M}\log M\right)^{1/5}\right)||Ay||^p_{L_p(\bar{M},\mu)}$$

for each $y \in Y^*$, $j = 1,2$ and $p > 2$.

Denote for $j = 1,2$ the injection from $L_p(K_j,\mu'_{|K_j})$ to $L_1(K_j,\mu'_{|K_j})$ by $i^j_{p,1}$ and let P be the conditional expectation projection from $L_1(\bar{K},\mu')$ onto $J[L_1(\bar{M},\mu)]$ following by J^{-1}. Thus

$$u^* = B\ P\ i^1_{p,1}\ 1_{K_1}JA + B\ P\ i^2_{p,1}\ 1_{K_2}JA$$

so, using definition of $\nu_q^{(n,k)}(u)$,

$$\nu_q^{(\frac{7}{8}M,2)}(u) \le \sum_{j=1}^{2}\nu_q^{(\frac{7}{8}M)}([\ P\ i^j_{p,1}\ 1_{K_j}JA]^*) \le$$

$$\le \sum_{j=1}^{2}||1_{K_j}JA||\ ||i^j_{p,1}||\ ||BP|| \le$$

$$\left(\frac{1}{2}+c(p)\left(\frac{n^{p/2}}{M}\log M\right)^{1/5}\right)^{\frac{1}{p}}||A||\ ||B||\sum_{j=1}^{2}\mu'(K_j)^{\frac{1}{q}}$$

$$\le ||A||\ ||B||\left(1+c(p)\left(\frac{n^{p/2}}{M}\log M\right)^{1/5}\right)^{\frac{1}{p}}$$

$$\le ||A||\ ||B||\left(1+c(p)\left(\frac{n^{p/2}}{M}\log M\right)^{1/5}\right).$$

This completes the proof when $p > 2$; the other case is similar.

Finally, we are ready to prove Theorem 1.1. Without loss of generality we may assume that $\dim(Y) \leq n$. Using Theorem 1.2 it is enough to prove that

$$\hat{\nu}_q^{(N)}(v) \leq (1+\varepsilon)\nu_q^M(v),$$

for all $v : Y \to X$ and all positive integers $M \geq n$. Iterating Theorem 1.3, we get that

$$\nu_q^{([\frac{7}{8}]^k M, 2^k)}(u) \leq \prod_{j=1}^{k} \left(1 + c(p)\left(\frac{n^{p/2}}{[\frac{7}{8}]^{j-1}M}\log([\frac{7}{8}]^{j-1}M)\right)^{1/5}\right) v_q^{(M)}(u),$$

for all k such that $(\frac{7}{8})^k M \geq n$, and for all $p > 2$. The product on the right hand side of the above inequality is smaller than $1+\varepsilon$ as long as

$$c(p)\left(\frac{n^{p/2}}{[\frac{7}{8}]^{j-1}M}\log([\frac{7}{8}]^{j-1}M)\right)^{1/5} \leq c\varepsilon.$$

Set $N = [\frac{7}{8}]^k M$, then if

$$N \geq C(p)\varepsilon^{-5} n^{p/2}\log(\varepsilon^{-5}n),$$

then

$$\hat{\nu}_q^{(N)}(u) \leq \nu_q^{(N,2^k)}(u) \leq \nu_q^{(N)}(u).$$

This completes the proof when $p > 2$; the case $1 < p < 2$ is similar.

□

References

[B] K. M. Ball *An Elementary Introduction to Modern Convex Geometry.* Flavors of geometry, 1-58, Math. Sci. Res. Inst. Publ., **31**, Cambridge Univ. Press, Cambridge, (1997).

[B-L-M] J. Bourgain, J. Lindenstrauss and V. Milman, *Approximation of zonoids by zonotopes*, Acta Math., **162**(1989) 73-141.

[J-S] W.B. Johnson and G. Schechtman, *Computing p-Summing Norms With Few Vectors*, Israel Journal of Math.(1994) **87**, 19-31.

[L] D. R. Lewis, *Finite dimensional subspaces of L_p*, Studia Math. **63** (1978), 207–212.

[S-Z] G. Schechtman, A. Zvavitch, *Embedding subspaces of L_p into ℓ_p^N*, $0 < p < 1$, Math. Nachr. **227** (2001) pp. 133-142

[Sz] S. J. Szarek *Computing summing norms and type constants on few vectors,* Studia Math. **98** (1990), 147-156.

[T] M. Talagrand, *Embedding subspaces of L_p in L_p^N*, Geometric aspects of functional analysis (Israel, 1992-1994), 311-325, Oper. Theory Adv. Appl., 77, Birkhuser, Basel, 1995.

[To] N. Tomczak-Jaegermann, *Banach-Mazur distances and finite-dimensional operator ideal,* Pitman Monographs and Surveys in Pure and Applied Mathematics **38**, Longman, London, 1989.

Mathematics Department, Mathematical Sciences Bldg, University of Missouri, Columbia, MO 65211 USA

E-mail address: zvavitch@math.missouri.edu

SOFTBALL

Steps to Success

Diane L. Potter, EdD
Gretchen A. Brockmeyer, EdD
Springfield College
Springfield, Massachusetts

Leisure Press
Champaign, Illinois

Library of Congress Cataloging-in-Publication Data

Potter, Diane L., 1935-
Softball : steps to success / Diane L. Potter, Gretchen A. Brockmeyer.
p. cm.—(Steps to success activity series)
ISBN 0-88011-358-8
1. Slow pitch softball—Training. I. Brockmeyer, Gretchen A., 1943- . II. Title. III. Series.
GV881.4.T72P68 1989 88-34421
796.357′8—dc19 CIP

ISBN: 0-88011-358-8

Developmental Editor: Judy Patterson Wright, PhD
Production Director: Ernie Noa
Copy Editor: Peter Nelson
Assistant Editors: Kathy Kane and Robert King
Proofreader: Laurie McGee
Typesetters: Sandra Meier, Angela Snyder, and Yvonne Winsor
Text Design: Keith Blomberg
Text Layout: Jayne Clampitt
Cover Design: Jack Davis
Cover Photo: Bill Morrow
Illustrations By: Tim Offenstein and Gretchen Walters
Printed By: United Graphics, Inc.

Printed in the United States of America

10 9 8 7 6 5 4 3

Leisure Press
A Division of Human Kinetics Publishers, Inc.
Box 5076, Champaign, IL 61825-5076
1-800-747-4457

Contents

Series Preface

The Steps to Success Activity Series is a breakthrough in skill instruction through the development of complete learning progressions—the *steps to success*. These *steps* help students quickly perform basic skills successfully and prepare them to acquire advanced skills readily. At each step, students are encouraged to learn at their own pace and to integrate their new skills into the total action of the activity, which motivates them to achieve.

The unique features of the Steps to Success Activity Series are the result of comprehensive development—through analyzing existing activity books, incorporating the latest research from the sport sciences and consulting with students, instructors, teacher educators, and administrators. This groundwork pointed up the need for three different types of books—for participants, instructors, and teacher educators—which we have created and together comprise the Steps to Success Activity Series.

The *participant book* for each activity is a self-paced, step-by-step guide; learners can use it as a primary resource for a beginning activity class or as a self-instructional guide. The unique features of each *step* in the participant book include

- sequential illustrations that clearly show proper technique for all basic skills,
- helpful suggestions for detecting and correcting errors,
- excellent drill progressions with accompanying *Success Goals* for measuring performance, and
- a complete checklist for each basic skill for a trained observer to rate the learner's technique.

A comprehensive *instructor guide* accompanies the participant's book for each activity, emphasizing how to individualize instruction. Each *step* of the instructor's guide promotes successful teaching and learning with

- teaching cues (*Keys to Success*) that emphasize fluidity, rhythm, and wholeness,
- criterion-referenced rating charts for evaluating a participant's initial skill level,
- suggestions for observing and correcting typical errors,
- tips for group management and safety,
- ideas for adapting every drill to increase or decrease the difficulty level,
- quantitative evaluations for all drills (*Success Goals*), and
- a complete test bank of written questions.

The series textbook, *Instructional Design for Teaching Physical Activities*, explains the *steps to success* model, which is the basis for the Steps to Success Activity Series. Teacher educators can use this text in their professional preparation classes to help future teachers and coaches learn how to design effective physical activity programs in school, recreation, or community teaching and coaching settings.

After identifying the need for participant, instructor, and teacher educator texts, we refined the *steps to success* instructional design model and developed prototypes for the participant and the instructor books. Once these prototypes were fine-tuned, we carefully selected authors for the activities who were not only thoroughly familiar with their sports but also had years of experience in teaching them. Each author had to be known as a gifted instructor who understands the teaching of sport so thoroughly that he or she could readily apply the *steps to success* model.

Next, all of the participant and instructor manuscripts were carefully developed to meet the guidelines of the *steps to success* model. Then our production team, along with outstanding artists, created a highly visual, user-friendly series of books.

The result: The Steps to Success Activity Series is the premier sports instructional series available today. The participant books are the best available for helping you to become a master player, the instructor guides will help you to become a master teacher, and the teacher educator's text prepares you to design your own programs.

This series would not have been possible without the contributions of the following:

- Dr. Joan Vickers, instructional design expert,
- Dr. Rainer Martens, Publisher,
- the staff of Human Kinetics Publishers, and
- the *many* students, teachers, coaches, consultants, teacher educators, specialists, and administrators who shared their ideas—and dreams.

Judy Patterson Wright
Series Editor

Preface

Over the years, it has been our observation that softball is one of two activities (the other being volleyball) in physical education curricula that are typically simply played, not taught, expecially at the secondary level. Softball can be a wonderful game to play for participants of all ages. However, in our opinion, it is a wonderful game only when one has developed an understanding of, an appreciation for, and an ability to master the game concepts as well as the physical skills. The teacher or coach who asks an unskilled player to apply yet-undeveloped skills and knowledge in a complex game setting is placing the participant in a no-win situation. The student who places himself or herself in that kind of situation is asking for frustration, failure, and a dislike of the game.

This book is designed to take you through a progression of skill-development practice. You will move from practicing individual skills to combining two, three, and four skills in gamelike drill settings, then to applying the skills in modified games. Finally, you will be given the opportunity to display and demonstrate your skills and knowledge in regulation game play.

The focus of this book is on the slow pitch game because that is the one most often used in physical education classes and recreational settings. References are made throughout this book to the other versions of softball, and the differences between them and slow pitch are explained.

The practice drills described in this book are designed so that you can practice inside on a gymnasium floor or outside on any grass area, as well as on a regulation softball diamond. You will find that you do not need a fully lined softball field, 10-player teams, coaches, and uniforms in order to practice the fundamental skills. A wall, lines on a gymnasium floor, a rebound net, or a blanket hung over a clothesline can all be used by individuals, pairs, or groups of various sizes for practicing softball skills.

The preparation of this book was made possible by the assistance of many people, not all of whom can be mentioned by name. Over the years many students and players we have taught and coached have challenged our views of the sport and how it should be taught. From them we have learned and thus further developed the approach presented in this book. To all of you, we are forever indebted.

Our sincere thanks go to four Springfield College students who were responsible for the pictures provided to the illustrator for this book: to David Blizard, for his excellent photography, which resulted in the pictures used throughout this book; and to Jodie Dobkowski, Shelly Quirk, and Christopher Mayhew, who were the subjects for the pictures. A special thanks goes to Bruce Oldershaw, director of the Audio Visual Aids (AVA) Department at Springfield College, for his assistance with film developing. Our sincere appreciation goes to Kenneth Dawley of the AVA Department and to Tammy Oswell, a former player and student, who developed all of the film and printed the photographs.

We are indebted to colleagues Lynn Johnson, teacher and softball coach at Springfield College, and Dr. Judy Patterson Wright, series editor, for encouragement, for serving as sounding boards for ideas, and for support throughout the preparation of the manuscript.

Diane L. Potter
Gretchen A. Brockmeyer

The Steps to Success Staircase

Get ready to climb a staircase—one that will lead you to become a skilled softball player. You cannot leap to the top; you get there by climbing one step at a time.

Each of the 25 steps you will take is an easy transition from the one before. The first few steps of the staircase provide a solid foundation—you will practice the techniques of basic, single skills. As you progress further, you will combine the single skills together in ways they are typically used in game situations. As you refine your physical skills, you will also learn game play concepts as you apply the skills and combinations in modified games.

The psychological concept of anticipation is introduced in the early steps and then reinforced throughout the book. Being able to anticipate, being ready, and becoming proficient at reading and reacting to a game situation enables you to more fully and actively participate in the game of softball. You will learn to anticipate in batting, running, fielding, and throwing so that you make the proper plays and fulfill your various offensive and defensive responsibilities. As you near the top of the staircase, the climb will ease, and you'll find that you have developed a sense of confidence about your softball abilities that makes further progress possible and playing the game a real joy.

To prepare to become a good climber, familiarize yourself with this section as well as the sections ''The Game of Softball'' and ''Preparing Your Body for Success'' for an orientation and in order to understand how to set up your practice sessions around the steps.

Follow the same sequence each step of the way:

1. Read the explanation of what is covered in the step, why the step is important, and how to execute or perform the step's focus, which may be a basic skill, concept, tactic, or combination of them.
2. Follow the numbered illustrations in the Keys to Success that show exactly how to position your body to execute each basic skill successfully. There are general parts to each skill: preparation (getting into a starting position), execution (performing the skill), and follow-through (finishing the action). The numbered items listed are the *major* points you should focus on for each part.
3. Look over the common errors that may occur and the recommendations of how to correct them.
4. Read the directions and the Success Goal for each drill. Practice accordingly and record your score. Compare your score with the Success Goal for the drill. All Success Goals assume that you use the correct form as described in the Keys to Success. Because the drills are arranged in an easy-to-difficult progression, you need to meet the Success Goal of each drill before moving on to practice the next one. This easy-to-difficult sequence is designed specifically to help you achieve continual success. The drills are set up to provide you with purposeful practice and enough repetition to help you improve your skills.
5. As soon as you can reach all the Success Goals for a step, you are ready for a qualified observer—such as your teacher, coach, or trained partner—to evaluate your basic skill technique against the Keys to Success Checklist. This is a subjective evaluation of the quality of your basic technique or form because using correct form can enhance your performance. In some steps the Checklist expands on the items listed in the three parts of the Keys to Success. These additional skill pointers will give you a more extensive evaluation of your performance. Your evaluator can tailor specific goals for you, if they are needed, by using the Individual Program sheet (see the Appendix).

6. Repeat these procedures for each of the 25 Steps to Success. Then rate yourself according to the directions for "Rating Your Total Progress."

Good luck on the step-by-step journey of developing your softball skills, building confidence, experiencing success, and having fun!

Key

Position	Symbol	Pos. No.
Pitcher	P	1
Catcher	C	2
First Baseman	1b	3
Second Baseman	2b	4
Third Baseman	3b	5
Shortstop	SS	6
Left Fielder	LF	7
Center Fielder	CF	8
Right Fielder	RF	9
Short Fielder (slow pitch only)	SF	10

X	player
——→	player movement
- - -→	thrown ball
-·-·-→	hit ball
∼∼→	rolled ball
ooo→	rebounding ball
H	hitter
F	fielder
BBR	batter-baserunner
B	baserunner
HP	home plate
DP	double play
OF	outfield
IF	infield
LE	less experienced
ME	more experienced
RH	right-handed player
LH	left-handed player
T	tosser or thrower
ST	sidearm thrower
R	relay person

The Game of Softball

The game we know today as softball was invented by George Hancock in 1887 at the Farragut Boat Club in Chicago. Hancock intended softball to be a game the rich members of the boat club could play indoors. Later, however, an outdoor version of the game, called *kittenball,* was developed by Lewis Rober, who introduced it to his fellow Minneapolis firemen. Today *softball* (as it was finally named at a 1926 YMCA convention) is played all over the world by literally millions of people from all walks of life. The skills needed to play the game are few; very simply, one must be able only to catch, throw, hit, and run bases with a moderate degree of skill.

The game of softball has several variations, each with certain unique rules that set it apart. There are official rules for coed slow pitch softball and for separate men's and women's games of fast pitch, modified, slow pitch, and 16-inch slow pitch. The rules of men's and women's games vary only slightly; however, the rules for fast pitch and slow pitch games make the games very distinct from each other.

The primary emphasis in this book is on slow pitch because it is more appropriate for class play.

PLAYING A GAME

Official games of softball are played on a field like that depicted in Diagram 1. The *playing field* is the area within which the ball may be legally played and fielded. The playing field usually has as its boundaries an outfield fence, and two side fences extending from the ends of the backstop to the outfield fence and running parallel to and 25 to 30 feet from the foul lines. The area outside the playing field is the *out of play/dead ball territory* (see Diagram 1). The playing field is made up of *fair territory,* which is that part of the playing field between and including the first and third base foul lines and the outfield fence, including the airspace above; and *foul territory,* that part of the playing field between the first and third base foul lines and the out of play/dead ball territory. The playing field is further divided into the

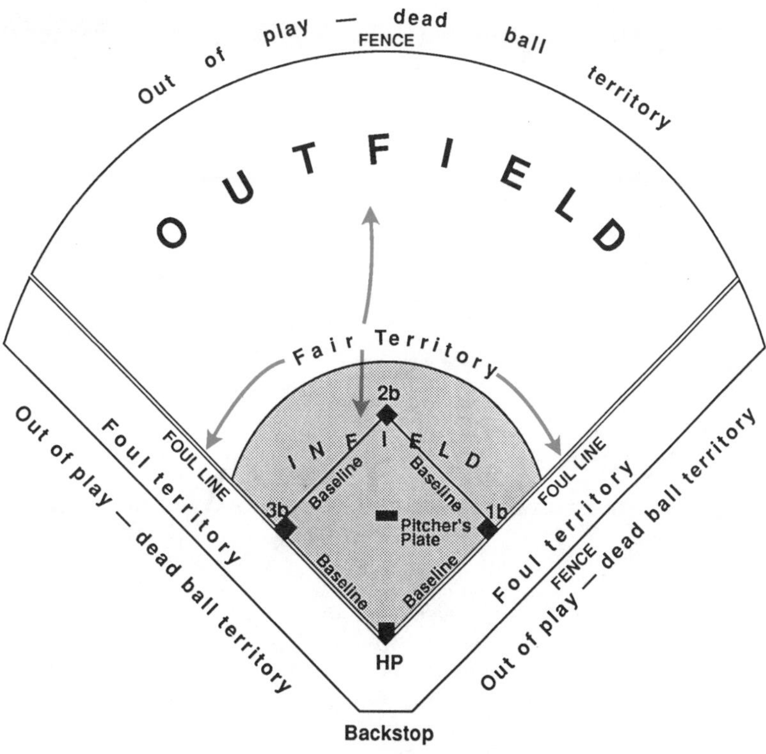

Diagram 1 Softball playing field.

infield, that portion of fair territory that includes areas normally covered by infielders; and the *outfield*, that portion of fair territory that is outside the diamond formed by the baselines, or the area not normally covered by an infielder between first and third bases and the outfield fence. Most softball playing fields have a dirt infield (see the shaded area on Diagram 1) and a grass outfield.

Distances between bases, pitching distances, and distances from home plate to the outfield fence vary, depending on the game being played. Unofficial games are played on all kinds of fields that have at least a home plate and three bases set out in a diamond figure.

Any of the games of softball involves two teams alternately playing offense and defense. A team is on offense when it is "at bat," attempting to score runs. The defensive team is "in the field," attempting to prevent the team at bat from scoring runs. Fast pitch and modified teams have 9 players in the field on defense, whereas all slow pitch versions have 10 defensive players. Defensive positions (see Diagram 2) are identified by the numbers 1 through 10 as follows: pitcher (1), catcher (2), first baseman (3), second baseman (4), third baseman, (5), shortstop (6), left fielder (7), center fielder (8), right fielder (9), and (slow pitch only) short fielder (10) (see Diagram 2). The left fielder, the center fielder, the right fielder, and the short fielder are called *outfielders*. *Infielders* are the first baseman, the second baseman, the third baseman, and the shortstop. The pitcher and catcher, although playing in the infield area and having some of the same responsibilities as infielders, are together usually called the *battery*, rather than infielders.

Games are played in *innings*; an inning is completed when each team has had a time at bat, making three outs, and has played defense in the field for the three outs of the other team. (An *out* occurs when an offensive player does not reach a base safely.) A *regulation game* consists of seven innings. In competitive play, the choice of first or last at bat in an inning is decided by a coin toss, unless stated differently in the rules of the organization governing the game. The *visiting team* is up to bat first in an inning; the *home team* bats last. Typically, in any kind of league play, the team upon whose field the game is being played is the home team.

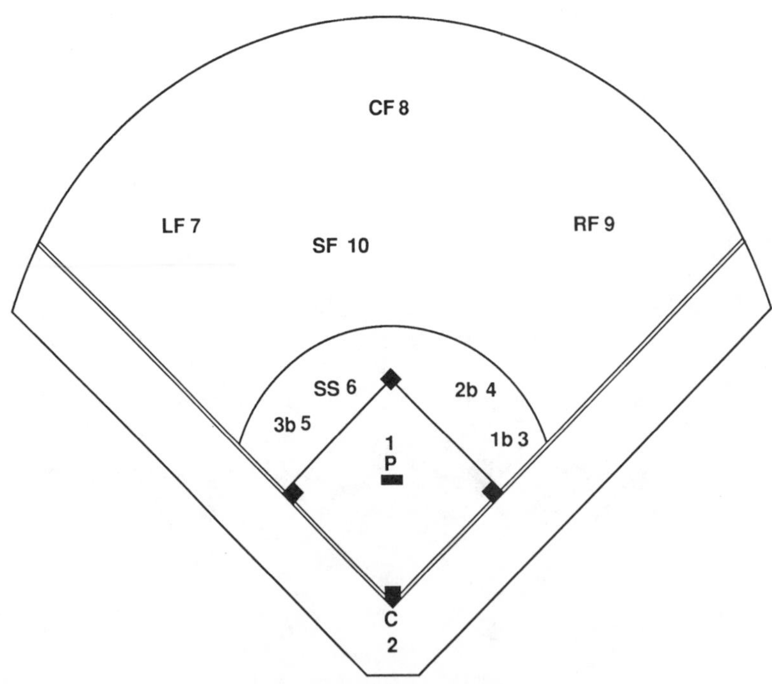

Diagram 2 Starting positions.

A *run* is scored each time a baserunner legally touches first base, second base, third base and home plate before the defensive team makes the third out of the inning. The winner of a game is the team that scores more runs.

In addition to the preceding basic rules of play, other rules are introduced and explained as they apply to specific skills and concepts taught in this book. Reference is made throughout the book to official and modified rules of play. *Official* rules are those used in an official game between two high schools, colleges, or summer league teams. High schools play under the rules of either the National Federation of State High School Associations or the National Association of Girls and Women in Sport (NAGWS). College women play under the rules of either the National Collegiate Athletic Association (NCAA) or the NAGWS. Recreational teams play under the rules of the Amateur Softball Association (ASA). International play is governed by the rules of the International Softball Federation. The majority of the official softball rules for high schools, colleges, and the ASA are the same because they follow the rules made by the International Softball Federation. However, the aforementioned organizations do have a few rules that apply only to their own competitions. For the purposes of this book, discussion of rules as they occur in the learning progression is based on the NAGWS rule book.

Modified rules, on the other hand, are rules that the teacher or coach makes up, usually to encourage students to focus on a particular skill or combination of skills. The player rotation rules used in the modified game ''scrub'' (Step 23), for example, are designed to ensure that every player experiences playing all the defensive field positions. Those rules are not official, though, and would not be used during an official game.

A glossary is provided at the end of the book. Turn to that section and become familiar with the definitions of softball terms *before* you begin your climb up the steps to success staircase. This glossary does not include *all* the terms used in softball, however. If you are interested in looking at a complete set of rules and definitions, check an official rule book at your local library or a sports store where softball equipment and clothing are sold.

EQUIPMENT SELECTION AND SAFETY CONCERNS

In order to play softball safely, you need to have reliable equipment and practice facilities free from danger. Here are some suggestions for selecting personal equipment and for safety checks prior to practicing softball skills.

When selecting personal equipment for softball, consider the following:

1. Fielder's glove:
   - All leather, including laces and bound edges of hand opening (be wary of plastic or synthetic materials in laces and edges, especially)
   - Easily adjustable strap over back of hand (Velcro® is nice)
   - Size large, but manageable; shorter finger length for second baseman, third baseman, and shortstop; longer for outfielder, first baseman, and catcher (if fielder's glove used instead of special mitt at first base or catcher position)
2. Bat:
   - Length and weight enabling you to swing the barrel of the bat into the contact zone easily
   - Grip size that feels comfortable in your hands

For safety's sake, do the following before practicing or playing softball:

1. Periodically check your personal equipment:
   - Glove—for broken laces, especially in the web area
   - Footwear—well fitting, no holes, strong arch support, soles that provide good traction (especially on wet or damp ground)
   - Shirt—loose fitting so that it does not restrict your movements, especially for throwing
   - Pants or shorts—again, loose fitting for free movement

- Sunglasses—if used, should have nonbreakable safety lenses

2. Check playing field for these:
   - Glass and other sharp objects
   - Holes in the field
   - Dangerous obstructions, such as football blocking sleds, lacrosse or field hockey goals, and so on
   - Loose equipment lying around, especially balls and bats
3. When practicing skills, remember these tips:
   - When working with a partner on throwing, fielding, and so on, line up so neither of you looks directly into the sun.
   - When inside, or outside near a building, be aware of windows, lights, and other people; do not practice with a window in the line of flight of the ball.
   - Be sympathetic to the skill ability of partners.
   - Do not throw the bat!
   - Do not hit rocks with the bat!

SOFTBALL TODAY

Participation opportunities abound for softball enthusiasts of all ages and abilities. The Amateur Softball Association (ASA) is the governing body for softball in the United States. Under the sponsorship of the ASA, national tournaments are held each year in a wide variety of classifications of fast pitch and slow pitch. There is also a full ASA program for junior teams.

Softball is played by millions of people in over 60 countries of the world. International competition opportunities are growing every year. The World Championship and the Pan American Games are well established. It is hoped that in the near future, softball will be included as an Olympic sport. With the inclusion of softball in the National Sports Festival (now the United States Olympic Festival), this country is well on its way to developing Olympic-level softball competitors.

Should you not aspire to Olympic-level competition, there are still literally hundreds of thousands of recreational teams providing opportunities for participation at every level of ability. Softball truly is a sport for everyone.

The NAGWS Softball Guide can be purchased from

The American Alliance for Health, Physical Education, Recreation and Dance
1900 Association Drive
Reston, VA 22091
ISBN 0-88314-389-5

The National Governing Body for Softball is

Amateur Softball Association
2801 N.E. 50th Street
RR #4, Box 385
Oklahoma City, OK 73111

Preparing Your Body for Success

There is a recommended softball workout sequence to follow. Prior to practicing, you need a 5- to 10-minute warm-up period to increase your heart rate and increase your flexibility. After finishing your practice, end with a 5-minute cool-down period. If you follow this sequence, you will not only help prepare your body and mind to play softball, but will also help prevent injuries.

Your first goal is to "get your blood moving." Starting at home plate, jog around the perimeter of the field and return to home plate. Next select and complete one of the aerobic exercises below. Either selection will not only increase your heart rate but will also utilize some movements used in the game.

WARM-UP SPRINTS

1. Stand at first base, facing home plate with your left foot against the inside edge of the base, knees bent, hands on knees. Staying low, pivot to your right and sprint to second base, taking your first step with your left foot (this is a *crossover step* start). Stop at second base (do not overrun the base). Immediately assume the same position at this base and, using the crossover step start, sprint to third (again, do not overrun). Repeat from third going to home plate (you *may* overrun home plate).

   Stand in a track-start position at home plate (crouch position facing first base, feet in forward stride with one foot on the plate). Sprint to first base, overrunning the base. Turn *to the left,* return to the base quickly, assume the track-start position, and sprint to second (as before, do not overrun). Continue as in the first round, only use the track start at each base until crossing home plate. These two complete laps complete the exercise. Run at top speed between bases. Set up and start to sprint to the next base quickly; little time should elapse between sprints.
2. Place two gloves 10 feet apart on the ground (or use some other appropriate flat markers—*do not use balls or bats!*). Run around the gloves in a figure-eight pattern crossing between the gloves. Do 10 repetitions starting around the first glove in a clockwise direction, then 10 traveling in the opposite direction (you need to unwind!).

FLEXIBILITY EXERCISES

Concentrate on your body as you perform the following exercises; if necessary, close your eyes so that you can more easily focus attention on your body. Do these exercises in a relaxed state of mind.

Move slowly into the stretch position. Hold the stretch position for 8 to 10 seconds. *Do not bounce in the stretch position*. During the 8- to 10-second stretch, relax. At the end of 10 seconds, gently try to increase the range of the stretch.

Do the exercises in order, working from your head to your toes. Select one exercise for each body part and follow the directions. Avoid comparing your stretch to anyone else's.

Head

1. *Neck rotation:* Stand with hands on hips, feet shoulder width apart, and head erect. Turn your head to the right, relax, and slowly count to 10. Turn your head to the left, relax, and slowly count to 10. Repeat both sides 3 times.

2. *Standing neck rolls:* With your head erect, relax and roll your head in a circular motion starting to your right shoulder, then forward, over to your left shoulder, and returning to an erect position, all while counting to 8. Do not drop your head back—this movement could put too much pressure on the bones of your neck. Do 3 right circles and 3 left circles.

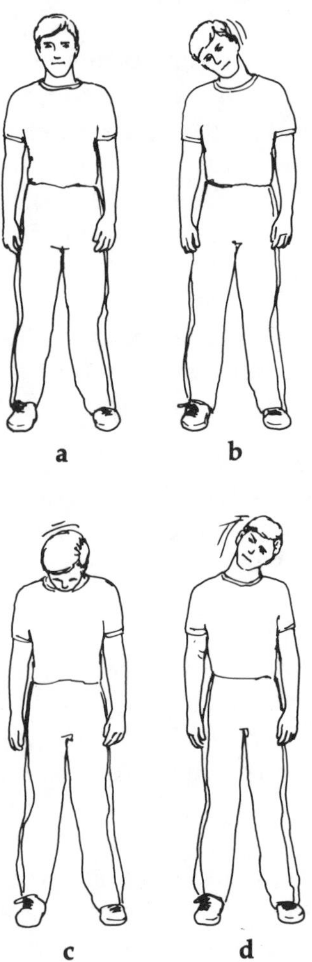

Shoulders

1. *Standing overhead shoulder stretch:* Bring your right hand to your upper back from above (scratch the middle of your back from above). Grasp your right elbow with your left hand, relax your arm, and stretch by gently pulling laterally on your elbow, holding for 8 to 10 seconds. Repeat with your left arm. Repeat each 3 times.

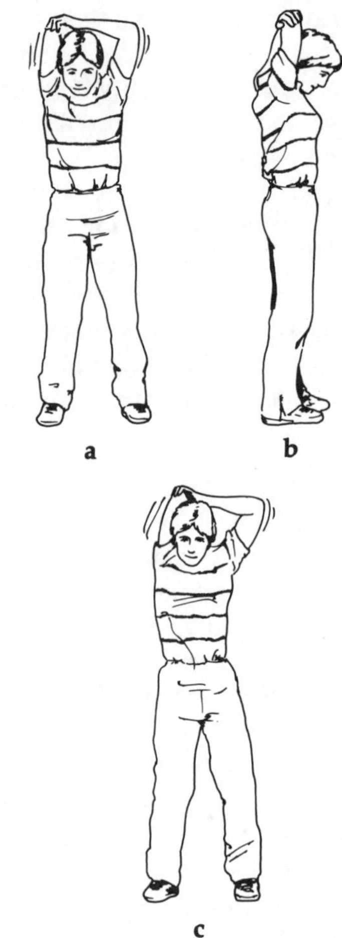

2. *Standing horizontal shoulder stretch:* Extend your right arm horizontally across your chest. Bend your left arm, hooking your right arm at the elbow. Gently pull on your right elbow with your left arm, increasing the right shoulder

stretch. Repeat for the left shoulder to complete one sequence. Repeat the sequence 3 times.

Arms

1. *Standing overhead arm stretch:* Extend your arms overhead, crossing them at the wrists and placing your palms together. Stretch and hold this position for 8 to 10 seconds. Interlock your fingers and bring your locked hands down, rolling them in toward your head, bringing them in under your chin with bent arms against your chest. Hold for 8 to 10 seconds, then roll your hands back up to the extended position over your head. Repeat 3 times.

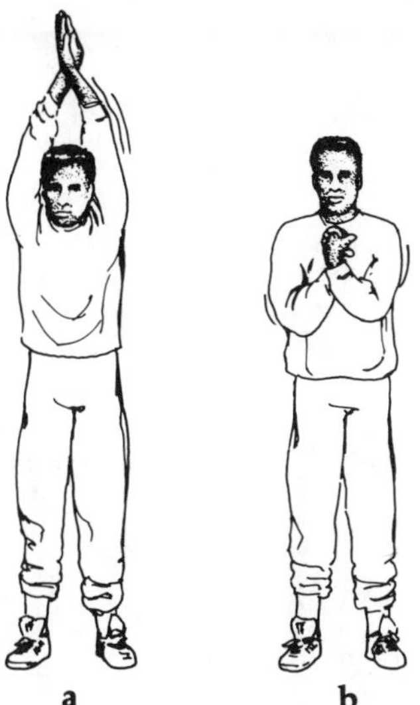

a b

2. *Standing arm circles:* With your arms stretched out at shoulder height, circle your arms in small, tight circles forward 10 times, then backward 10 times. Next slowly do 10 forward arm circles using small, then gradually larger, circles. Reverse direction and make 10 gradually larger circles backward. Repeat total sequence 3 times.

Wrists

1. *Kneeling wrist bends:* Place the palms of your hands on a flat, firm surface and stretch your wrists to at least 90 degrees in four hand positions: forward, backward, right, and left.

a b c d

2. *Standing wrist stretch:* Extend your left arm forward parallel to the ground. Bend your wrist so that your fingers point up and the palm of your hand is facing away from you. Grasp your left hand fingers with your right hand; gently pull back to increase the stretch in your left wrist. Hold the stretch for 8 to 10 seconds. Repeat with the right wrist. Repeat the exercise for both wrists with the fingers pointed down. Repeat the entire sequence 3 times.

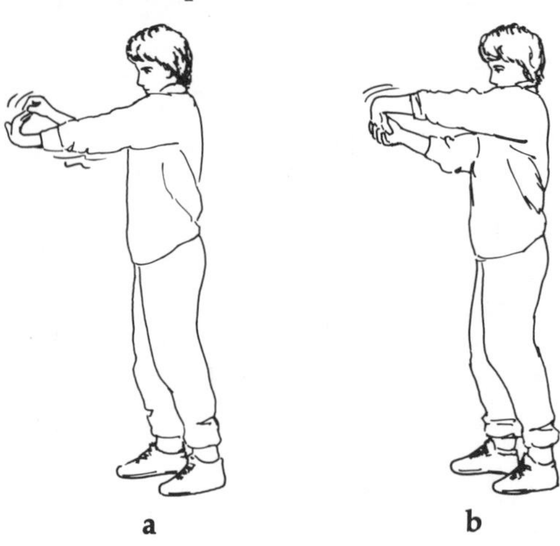

Torso

1. *Standing side bends:* With your feet shoulder width apart, join both hands overhead and bend your torso to the left, keeping your shoulders and hips square to the front. Hold for 8 to 10 seconds. Relax, then attempt to increase your stretch. Repeat to the opposite side. Repeat both sides 3 times.

2. *Standing trunk rotation:* With your feet shoulder width apart, extend both arms horizontal to the left. Rotate your torso, hips, and arms, first to the right, then to the left (this a, b, c, cycle equals 1 repetition). Repeat 8 times. Next do 8 repetitions rotating your arms and torso in opposition to your hips (see d and e).

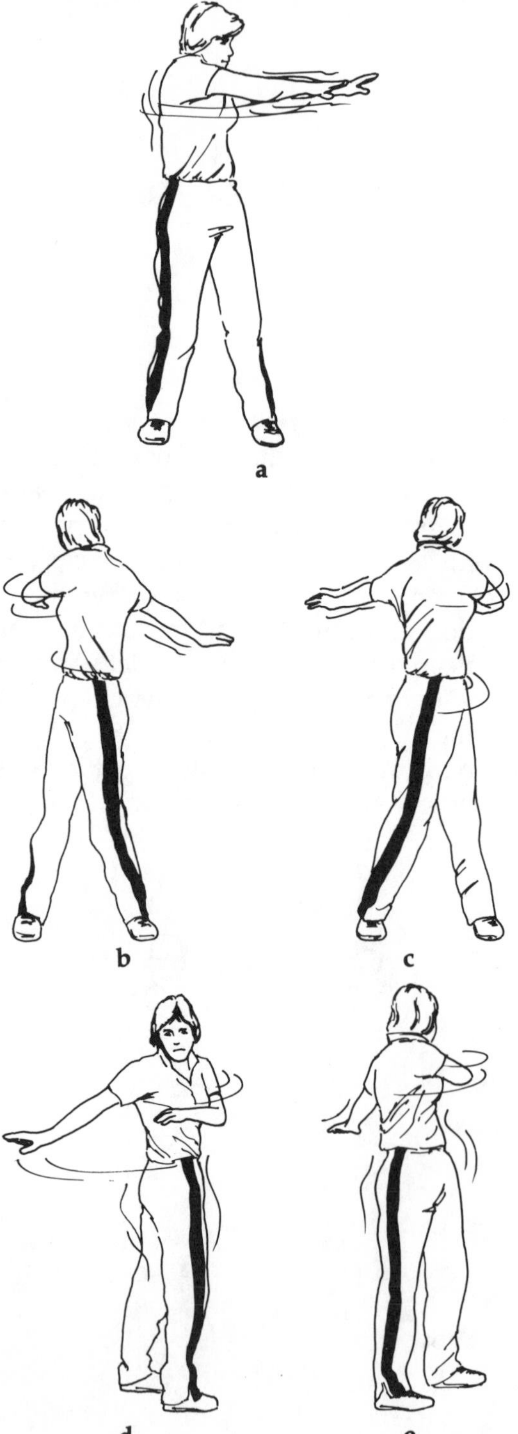

Lower Back

1. *Sitting lower back stretch:* Extend your legs forward and together on the ground. Reach toward your feet, trying to touch your chest to your knees. Hold 8 to 10 seconds. Relax, then try to increase your stretch. Repeat 3 times. Caution: To avoid injury, *do not bounce* at any time during the exercise.

2. *Sitting twister:* Extend your left leg, bend your right leg, and place your right foot on the ground to the left of your left knee. Twist your torso to the right, placing your right hand on the ground to the rear for support. Press your left arm against your right knee. Hold for 8 to 10 seconds. Switch arm and leg positions and repeat. Repeat total sequence 3 times.

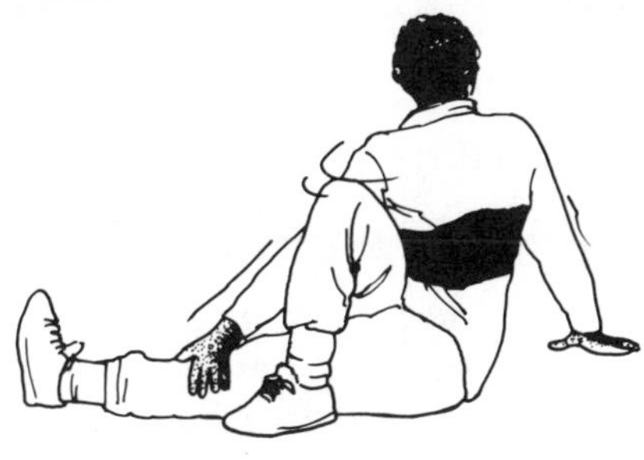

Legs

1. *Standing hamstring stretch:* With your feet shoulder width apart, slowly bend forward at the waist and attempt to touch your fingers to the ground while keeping your knees slightly bent. Relax your upper body, especially your arms and neck. Hold this position for 8 to 10 seconds. Gently increase your stretch and hold for 8 to 10 seconds. Remember, *do not bounce* at any time during your stretch. Slowly roll up one vertebrae at a time to standing. Repeat the sequence 8 times.

a

b

2. *Crossed-leg standing hamstring stretch:* Cross one leg in front of the other. With your feet touching, bend over at the waist, attempting to touch your toes with your fingers. Relax your upper body, especially your arms and neck. Hold this position for 8 to 10 seconds. Gently increase the stretch, trying to touch your fingers to your toes, and hold for 8 to 10 seconds. Bend your knees slightly and stand up. Repeat the sequence 8 times, bending your knees each time you return to the standing position.

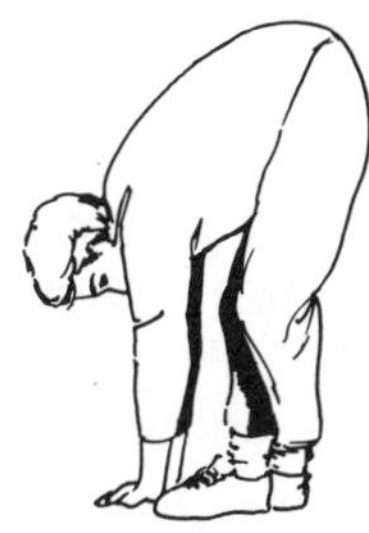

3. *Back-lying hamstring stretch with partner:* With one leg up and keeping the leg straight, have your partner gently move the leg toward your chest. Signal your partner to *stop* when you feel the stretch

in the back of your leg. Hold the stretch position for 8 to 10 seconds. Repeat with your other leg. Repeat the total sequence 3 times, then change roles with your partner.

4. *Front-lying quad stretch:* Bend your right leg, grasp your right foot with your right hand, and pull your heel toward your seat. Hold for 8 to 10 seconds. Relax, then try to increase the stretch (touch your heel to your seat, if possible) and hold for 8 to 10 seconds. Repeat with the other leg. Repeat the sequence 3 times.

Feet and Ankles

1. *Sitting ankle rolls:* Circle your right ankle clockwise 8 times, then counterclockwise 8 times. Repeat with your left ankle. Repeat entire sequence 3 times.

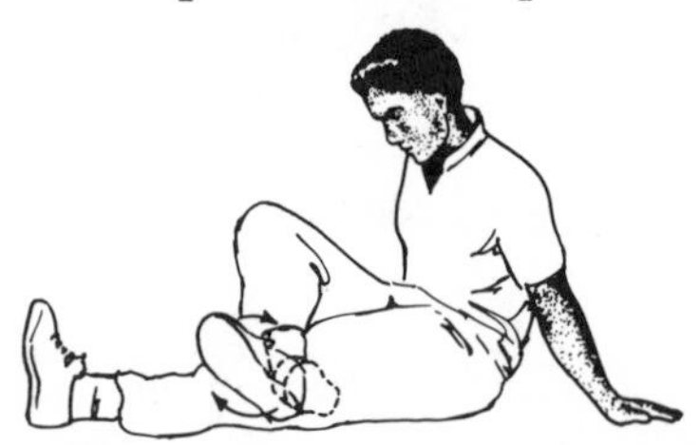

2. *Side-lying ankle flex:* While on your left side, place both hands, your left elbow, and your bent left leg on the ground for support and balance. Raise and hold your right leg 1 to 2 feet above the ground. Holding the foot at 90 degrees, flex and extend the outer side of your right foot 10 times. Repeat the exercise on your other side with your left foot. Repeat the sequence 3 times.

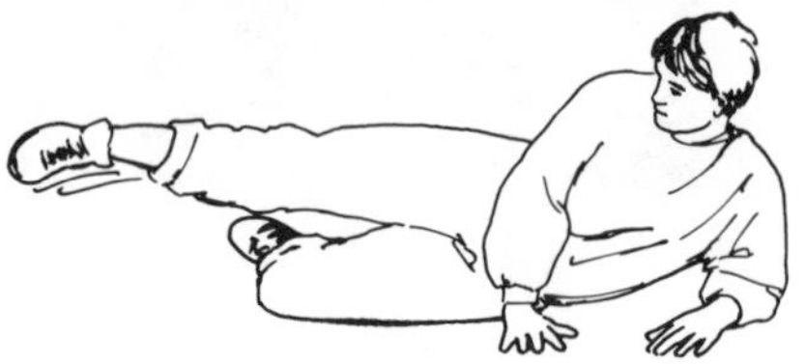

3. *Sitting Achilles tendon stretch:* Extend both legs together on the ground, while keeping your back upright and straight. Flex your ankles, attempting to bring your toes closer to your knees. Hold for 10 seconds. Relax, then repeat 3 times.

COOL-DOWN PERIOD

At the end of each practice session, take a few minutes to stretch out the muscles used the most in the practice session. Concentrate on your body as you properly stretch each part. A properly stretched muscle is a relaxed muscle. This is the best time to stretch, because your muscles are warm. It takes only 5 minutes to repeat *at least one each* of the head, shoulder, arm, torso, back, leg, and ankle warm-up exercises.

Step 1 Catching

Your first basic softball skill is catching the ball. Although throwing and catching go together, it is a good idea for you to know something about catching first, so that you do not get hurt when someone throws you the ball. The term *catching* also describes the special way the catcher receives the ball from the pitcher. However, it is important to first learn how to catch and bring the ball and your body into the throwing position—all in one continuous motion.

The techniques of the overhand throw are covered in Step 2. Even if you are a less experienced player, you will not need to spend a great deal of time on these first two steps. The two skills are presented separately because it is easier to focus your attention on one skill at a time before putting them together to play catch. If you are more experienced in the game of softball, use Steps 1 and 2 as a quick review of the correct techniques used in catching and throwing a ball. Then move on to Step 3 to combine the two skills in practice settings that are more gamelike. Later you will learn fielding skills that will help you catch ground balls and fly balls.

WHY IS BEING ABLE TO CATCH THE BALL IMPORTANT?

Can you imagine someone's throwing you the ball and your not knowing what to do to protect yourself? Catching correctly not only keeps you from getting hurt, but also makes it easier for you to get ready to throw the ball. Catching and throwing a ball are the two fundamental defensive softball skills. In a game your ability to catch and quickly throw the ball helps you throw out a baserunner attempting to advance to the next base.

HOW TO CATCH THE BALL

Initially, catching a ball involves tracking the ball—or watching it and determining the path it is taking—then moving your body and glove into that path in order to catch the ball. As the ball comes toward you, stand square to the ball with your glove-side foot slightly ahead. Reach out in front of your body to make contact with the ball and simultaneously shift your weight onto your front foot.

If the ball is arriving above your waist, point the fingers of your glove and of your throwing hand up, as shown in Figure 1.1a. If the ball is below your waist, your fingers point down (see Figure 1.1b). A ball coming directly at your waist is often the hardest to catch; to catch it, position your glove hand palm down, fingers parallel to the ground and thumb down, but your throwing hand palm up, as shown in Figure 1.1c.

As the ball comes into your glove, squeeze the ball with the thumb and ring finger of your glove hand, at the same time covering the ball with your throwing hand. Make a two-finger grip on the ball by placing your index and middle fingers on one seam and your thumb on a seam on the opposite side of the ball from your fingers (see Figure 1.2). Give with the ball, cushioning its impact, by drawing the ball and glove toward your throwing-side shoulder. At the same time, shift your weight onto your back foot and pivot (turn) your body so that your glove side is toward the throwing target. With your weight on your back foot, separate your hands, and bring the ball in your throwing hand to the throwing position. Your glove-side elbow should point at the throwing target. This follow-through position for catching a ball is the same as the preparatory position for the throw. This action makes it possible for you

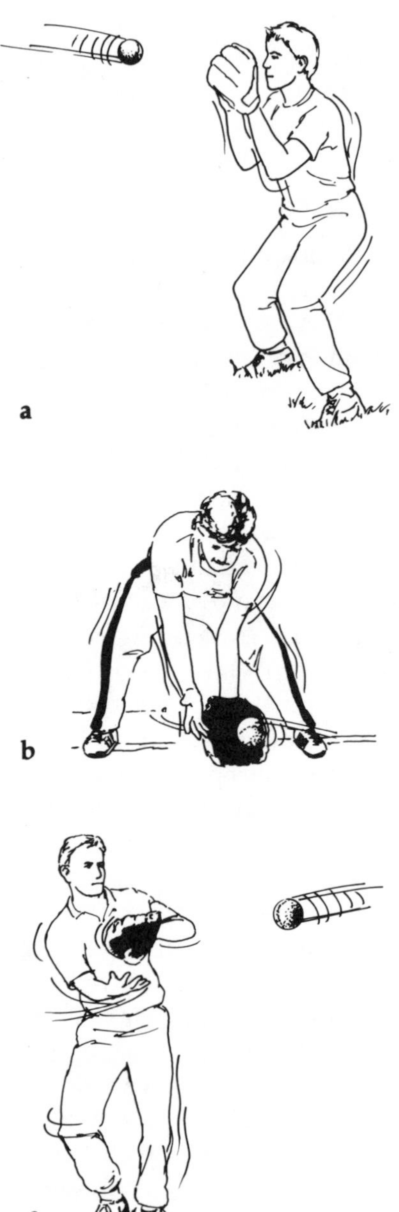

Figure 1.1 Catching form for ball above waist (a), for ball below waist (b), and ball at waist (c).

to make the transition between the catch and the throw one continuous motion when a quick release is needed. Figure 1.3 shows the catch-to-throw transition when the ball arrives above your waist.

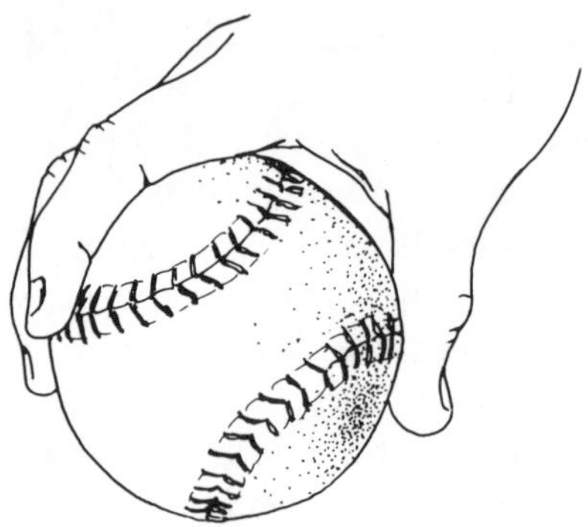

Figure 1.2 Two-fingered grip.

Using two hands to catch the ball not only makes for a surer catch, but also makes it easier for you to throw the ball quickly because you already have the ball in your throwing hand as soon as you catch the ball. The transition from catch to throw becomes uncoordinated and time consuming, however, with a one-handed, glove-only catching technique. One-handed catches should be used only when the ball is out of your two-handed reach.

Figure 1.3 Keys to Success: ***Catching***

Preparation Phase

1. Feet in forward stride, glove-side foot ahead
2. Face thrower, hips and shoulders square
3. Focus on ball in thrower's hand
4. Adjust glove to ball height
   - ball above waist, fingers up
   - ball below waist, fingers down
   - ball at waist, fingers parallel to ground
5. Focus on ball in flight

Execution Phase

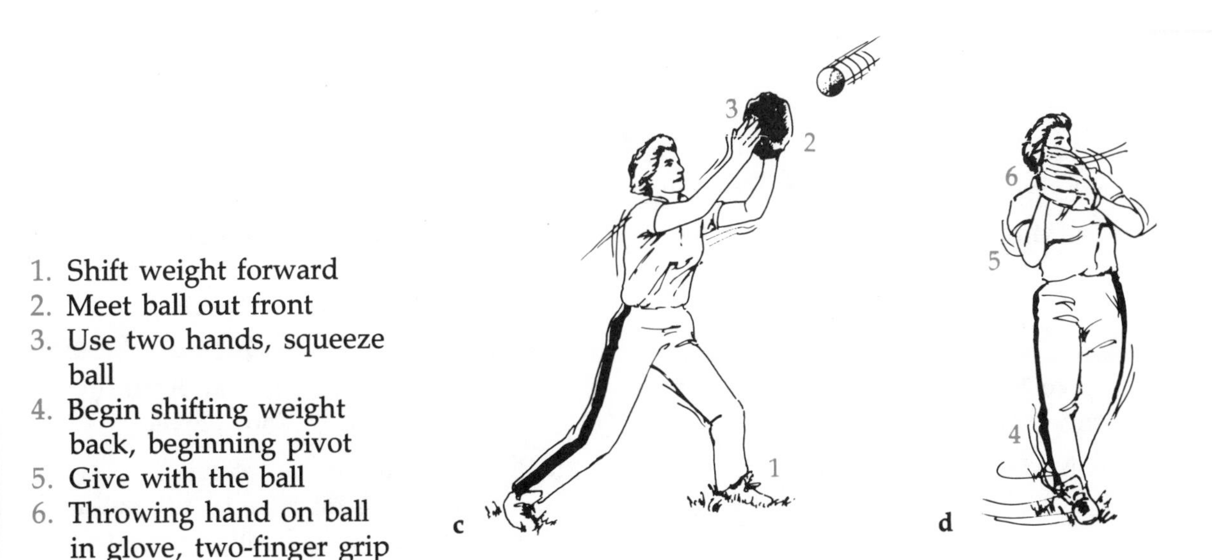

1. Shift weight forward
2. Meet ball out front
3. Use two hands, squeeze ball
4. Begin shifting weight back, beginning pivot
5. Give with the ball
6. Throwing hand on ball in glove, two-finger grip

Follow-Through Phase

1. Continue to shift weight back
2. Glove side toward target
3. Glove elbow points to target
4. Weight on back foot
5. Arms extend, move ball to throwing position
6. Throwing wrist extends

Detecting Catching Errors

Catching errors frequently occur because of failure to watch the ball go into your glove or to use two hands for the catch. The most common catching errors are listed below, along with suggestions on how to correct them.

ERROR	CORRECTION
1. The ball drops out of your glove a lot.	1. Use two hands. As soon as the ball enters your glove, cover it with your throwing hand. At the same time, squeeze the ball with your glove hand.
2. You have trouble making the ball go into the glove.	2. Watch the ball all the way from the thrower's hand to your glove.
3. The ball stings your hands when you catch it.	3. Give with the ball as you make contact; reach out in front to meet the ball and draw the glove and ball to your throwing shoulder.

Catching Drills

1. *Mimetic Drill*

Without a ball, put together all three phases of catching the ball. Practice doing the preparation, execution, and follow-through actions in sequence. Make your movements one continuous motion from the start to the throwing position; do not stop anywhere in the sequence until the end.

You may try this with your eyes closed, too. Try to picture the ball coming toward you, and you catching it using one fluid motion to bring it to the throwing position. This helps you focus your attention on how your body feels as you execute the skill.

Stand in front of a mirror and watch yourself go through the three phases of catching a ball mimetically. Check your own technique.

Success Goal = 30 total sequences with correct form (see Figure 1.3).

10 with eyes open

10 with eyes closed

10 in front of a mirror

Your Score =

(#) _____ sequences with eyes open

(#) _____ sequences with eyes closed

(#) _____ sequences with mirror

2. *Self-Toss Drill*

Without using a glove, gently toss a ball up into the air in front of your body so that it goes just above head height. Using both hands, reach up and out to catch the ball, draw it in to your throwing shoulder, drop your hands, and toss it again. Keep the action going—toss and catch, toss and catch, and so on. If you are a less experienced player, use a whiffle or nerf ball at first, if you like. Move to a real softball when you are successful at catching the whiffle or nerf ball.

Select a word phrase to repeat to yourself to help you focus on the catching action outlined in the execution phase of Figure 1.3. For instance, tell yourself, "Give with the ball," "cushion the ball," or "collapse the arms."

Toss the ball onto a high, slanted surface such as the roof of a shed or garage so that it will roll off for you to catch. Focus on the cue "reach and give" when catching the ball in this portion of the drill.

Success Goal = 20 total catches without dropping the ball

10 from a self-toss

10 from a roof toss

Your Score =

(#) _____ catches (self-toss)

(#) _____ catches (off roof)

3. Move Into Position Drill

From a starting point (e.g., pitcher's plate, spot on a foul line, line on the gymnasium floor, crack in a driveway), mark off distances of 5, 10, and 15 feet on each of three straight lines: one line to the front, one out to your glove side, and one out to your throwing side.

Stand on your starting point. Toss the ball up and forward so that it will fall in the vicinity of your first mark (5 feet ahead). Move under the ball and catch it. Turn around and toss the ball so that it will come down at your starting point. Move under the ball for the catch. Repeat the sequence, making a total of 4 catches at the 5-foot distance.

Now change the distance you must move to catch the ball: Repeat the drill sequence to both the 10- and the 15-foot forward marks. As the distance increases, your tosses will have to be higher to allow you time to get under the ball.

Now extend your skill in this drill by varying the direction you must move to make the catch. First toss to the glove-side marks. Repeat the sequence for a total of 4 catches at each distance. Then toss to the throwing-side marks, repeating until you have 4 catches at each distance. Remember, when tossing the ball to your glove side and your throwing side, face forward, toss the ball sideways, *then* turn to run under the ball. Otherwise, you won't get practice turning and reorienting your body.

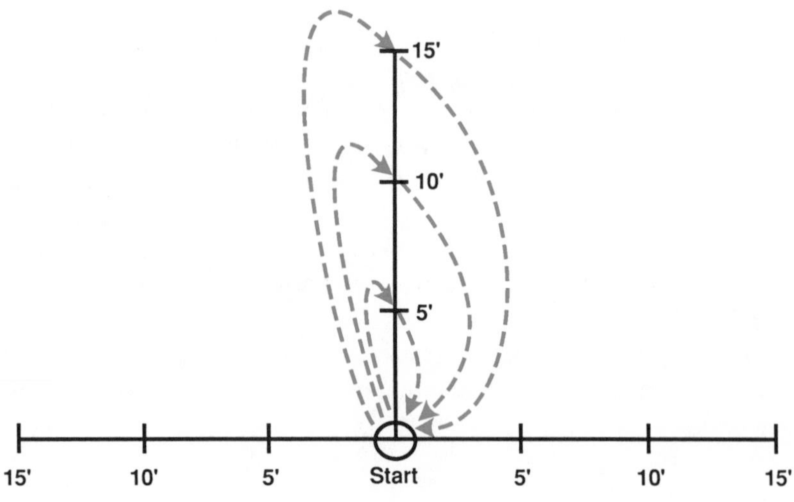

Success Goal = 36 total catches out of 36 self-tosses without dropping the ball

Forward direction	Glove side	Throwing side
4 at 5 feet	4 at 5 feet	4 at 5 feet
4 at 10 feet	4 at 10 feet	4 at 10 feet
4 at 15 feet	4 at 15 feet	4 at 15 feet

Your Score =

Forward direction	Glove side	Throwing side
(#) _____ at 5 feet	(#) _____ at 5 feet	(#) _____ at 5 feet
(#) _____ at 10 feet	(#) _____ at 10 feet	(#) _____ at 10 feet
(#) _____ at 15 feet	(#) _____ at 15 feet	(#) _____ at 15 feet

4. *Partner Toss Drill*

a. With a partner, stand 10 feet apart, facing one another. Using your gloves, one ball, and underhand tosses, toss and catch back and forth for 15 catches each. Repeat the sequence for a total of 30 catches each.

Be sure to stand in a forward stride position, your glove-side foot ahead. When executing the catch, bring the ball to the throwing position and stop. *Then* bring the ball to waist height so you can toss the ball underhand to your partner. *Remember,* you are working on the skill of catching; go through all three phases right to the end.

b. Stand 10 feet apart facing a partner. Have your partner make 10 tosses to each of the following levels: above the waist, below the waist, and at the waist.

Success Goals =

a. 30 total catches out of 30 tosses without dropping the ball

b. 30 total catches at varying heights without dropping the ball

10 above the waist

10 below the waist

10 at the waist

Your Score =

a. (#) _____ basic catches

b. (#) _____ above the waist

(#) _____ below the waist

(#) _____ at the waist

Catching
Keys to Success Checklist

Catching the ball involves tracking the ball, moving into position by getting into the path of the ball, and finally collecting the ball in both hands. Another important aspect of catching is making the follow-through phase of the catch the preparation phase for the throw. Ask your teacher, your coach, or a trained observer to evaluate your technique according to the following checklist. A checkmark should be placed in front of each item as the item is observed.

Preparation Phase

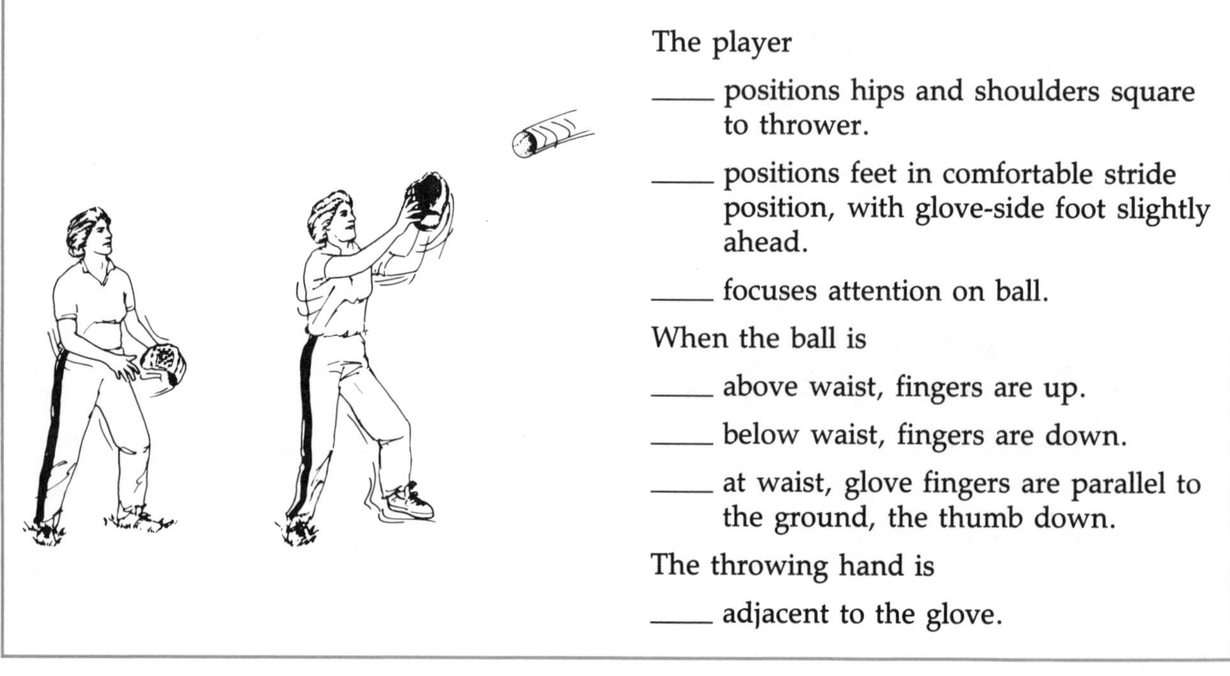

The player

____ positions hips and shoulders square to thrower.

____ positions feet in comfortable stride position, with glove-side foot slightly ahead.

____ focuses attention on ball.

When the ball is

____ above waist, fingers are up.

____ below waist, fingers are down.

____ at waist, glove fingers are parallel to the ground, the thumb down.

The throwing hand is

____ adjacent to the glove.

Execution Phase

The player

____ reaches out with both hands to contact ball in front of body.

____ shifts weight onto front foot.

____ catches ball in glove and immediately puts throwing hand on ball, making two-finger grip.

____ squeezes ball in glove with thumb and ring finger of glove hand.

____ cushions ball by collapsing arms and drawing ball in toward shoulder.

____ begins to shift weight to back foot and begins turning the glove side toward throwing target.

Follow-Through Phase

The player's

____ weight is shifted onto back foot.

____ glove side is turned toward throwing target.

____ glove-side elbow is pointed toward throwing target.

____ hands are separated and ball is in throwing hand in throwing position.

____ throwing-hand wrist is extended.

Step 2 Throwing Overhand

Once you catch the ball, you usually have to throw the ball to someone else. Catching and throwing thus go together and could be considered a combination skill because it is difficult to practice one without the other.

There are three general types of throws in softball: overhand, sidearm, and underhand. The full, hard underhand throw is commonly used only by the fast pitch pitcher. However, various underhand *tosses*, including the slow pitch pitch and the short *feed* to second base to start the double play, might technically be called underhand throws. The sidearm throw (Step 20) is used for relatively short throws where the ball must travel quickly and parallel to the ground. It is the least accurate of the throws and should be used sparingly.

The overhand throw is the kind of throw used most often to get the ball from one player to another. It is especially useful when the ball must travel a significant distance and when accuracy is a factor. Because of the major role the overhand throw plays in the game of softball, it is the first throw to *learn* if you are a less experienced player or to *review* if you are more experienced.

WHY IS THE OVERHAND THROW IMPORTANT?

Because the overhand throw is the strongest and most accurate throw, it is used by all fielders. Outfielders *must* use the overhand throw because of the great distance the ball must travel and because the ball needs to go in a straight line. If you were to throw the ball sidearm from the outfield to the catcher at home plate, the ball would curve. You must also use the overhand throw when throwing from shortstop or third base to first base because those distances are significant. A strong, accurate throw is needed to make the out on the batter running to first base.

HOW TO THROW OVERHAND

You covered the phases of catching the ball in Step 1. Those phases included taking a two-finger grip on the ball while you bring the ball and glove to your throwing-side shoulder. You then separate your hands, bringing the ball into the throwing position and pointing your glove toward the throwing target. This part of the follow-through phase of catching is the same as the beginning of the preparation phase for throwing. Think of these two phases as an overlapping transition that combines catching and throwing.

As you take the ball out of your glove and begin to bring it to your shoulder, be sure to grip the ball across the seams. Now say to yourself, "Turn, step, and throw." These cues will remind you to *turn* your glove side toward the target while extending your throwing hand back; to *step* in the direction of the target with your glove-side foot, shifting your weight onto that foot; and to *throw* the ball by bringing your arm forward, leading with the elbow. Keep your elbow high and your upper arm parallel to the ground. Rotate your forearm through the vertical, keeping the ball high as it goes by your head (see Figure 2.1).

Figure 2.1 Keys to Success: *Throwing Overhand*

Preparation Phase

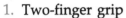

1. Two-finger grip
2. Staggered stride, weight on back foot
3. Glove side to target
4. Arms extended, glove to target
5. Wrist cocked, ball to rear

a

b

Execution Phase

c

d

e

1. Step toward target
2. Push off rear foot
3. Elbow leads throw, hand trails
4. Weight on front foot
5. Hips square
6. Forearm rotates through vertical
7. Ball high
8. Glove hand low
9. Snap wrist

Follow-Through Phase

1. Weight forward
2. Knees bent
3. Throwing hand low
4. Throwing shoulder forward
5. Assume balanced position

f

Detecting Overhand Throwing Errors

Overhand throwing errors typically are caused by the improper release of the ball or the failure to go through the throw's full range of motion. Less experienced players tend to *push* the ball rather than throw it, because the hand and ball move ahead of the elbow. However, the elbow must *lead* the forward motion of the throw. The most common overhand throwing errors are listed below, along with suggestions on how to correct them.

ERROR	CORRECTION
1. You ''push'' the ball rather than throw it.	1. Lead with your elbow as you bring the ball forward.
2. The ball's trajectory is too high.	2. Snap your wrist as you release the ball, or release the ball a bit later.
3. The ball curves when you throw it a long distance.	3. Check that you are using an overhand delivery, not a sidearm one. Be sure your elbow is up and your forearm passes your head vertically.
4. The ball goes to the right or the left of the target.	4. Check that you are using an overhand delivery, not a sidearm one. Step toward the target as you throw. Imagine that you are throwing your hand toward the target as you release the ball.

Overhand Throwing Drills

Note: These individual practice drills for the overhand throw are best done against a fence, a net, or a mat. Because the ball will not rebound from such a backstop, use a bucket of balls. Collect them after all are thrown.

1. Fence Drill

Stand 20 feet away from an unobstructed fence. Throw the ball directly against the fence 10 times.

Repeat the phrase ''turn, step, and throw'' to yourself as you go through the complete throwing action. Concentrate on your form. Be sure that your arm goes through the full range of motion. Make the entire throwing motion smooth.

Success Goal = 10 throws with correct form as shown in Figure 2.1

Your Score =

____ did your arm go through the full range of motion?

____ does your throwing action feel smooth?

2. Target Accuracy Drill

When you play softball, it is crucial to be able to throw the ball where you want. If you cannot throw the ball to the proper place, you will not be fully doing your part to help your team get baserunners out.

Mark a target about chest-high on a fence or a wall. Make the target a rectangle that is big enough for you to hit with a throw from 20 feet (if you have little experience, make the target at least 8 feet wide; if you are more experienced, make the target narrower).

a. Stand 20 feet from the target. Using the overhand throwing motion, throw 10 balls at the target. As you throw, say to yourself, ''Turn, step, and throw.''

   Remember, you want to use the *full range* of the throwing motion. Even though this drill uses a target, you should focus on the ''turn, step, and throw'' actions of the overhand motion.

b. Throw 10 more balls at the same target. This time, concentrate on the target throughout the throwing action.

Success Goals =

a. at least 5 throws hitting the target

b. 8 out of 10 overhand throws hitting the target

Your Score =

a. (#) _____ throws hit target

b. (#) _____ throws hit target

3. *Increase the Distance Drill*

You need to be accurate when throwing the ball different distances. You sometimes have to throw the ball a long distance (from the outfield), and other times your throw will be shorter (from the infield).

On the floor (or field) in front of the target used in the last drill, mark off distances 20, 30, 40, 50, and 60 feet.

a. Stand on the 20-foot mark. Using the overhand throw, deliver 10 consecutive balls at the target. Count the number of on-target hits.

Each time you meet part of the following Success Goal, move back to the next mark and repeat the drill. You may not move back to a greater distance unless you meet the goal for each target. See whether you can work your way back to a distance of 60 feet (the distance from home plate to first base).

Success Goal = 36 overhand throws hitting the target

8 of 10 at 20 feet

8 of 10 at 30 feet

7 of 10 at 40 feet

7 of 10 at 50 feet

6 of 10 at 60 feet

Your Score =

(#) _____ on-target hits at 20 feet

(#) _____ on-target hits at 30 feet

(#) _____ on-target hits at 40 feet

(#) _____ on-target hits at 50 feet

(#) _____ on-target hits at 60 feet

b. The players to whom you throw the ball in a softball game are much smaller than the target you have been throwing at. In fact, the target for a game throw is the glove of the person catching the ball.

Make your drill target narrower. Repeat the drill from the 20-, 30-, and 40-foot markers. Deliver 10 overhand throws from each distance and count the number of on-target hits. Remember, you cannot move back to the next distance until you meet the appropriate part of the Success Goal.

If you are a more experienced player, add two target hits to each Success Goal listed below. Or, you can try to get 8 out of 10 on-target hits from the 50- and 60-foot marks.

Success Goal = 21 on-target hits

8 of 10 at 20 feet

7 of 10 at 30 feet

6 of 10 at 40 feet

For more experienced players

10 of 10 at 20 feet

9 of 10 at 30 feet

8 of 10 at 40 feet

or

8 of 10 at 50 feet

8 of 10 at 60 feet

Your Score =

(#) _____ at 20 feet

(#) _____ at 30 feet

(#) _____ at 40 feet

or

(#) _____ at 50 feet

(#) _____ at 60 feet

4. H-O-R-S-E Drill

Use the floor markings from the Increase the Distance Drill (Drill 3). Less experienced players should use the larger target for this game; more experienced players, the smaller target.

Each of the five distance marks represents a letter in the word *horse* (20 feet = "H," 30 = "O," and so on). Using a maximum of 5 throws per trial per distance, try to spell H-O-R-S-E using the fewest possible total throws. Count the number of throws out of 5 that it takes you to hit the target at each distance. If you hit the target in 5 or fewer throws, you earn the letter that distance represents. Then move on to the next distance. When you do not earn a letter in the 5 allotted throws, you must move back to "H."

Try to become a thoroughbred, according to the following chart.

Success Goal = spell H-O-R-S-E at the Morgan level or better

Shetland pony = 21 to 25 throws

Quarter horse = 16 to 20 throws

Arabian = 11 to 15 throws

Morgan = 6 to 10 throws

Thoroughbred = only 5 throws

Your Score =

H, (#) _____ of throws at 20 feet

O, (#) _____ of throws at 30 feet

R, (#) _____ of throws at 40 feet

S, (#) _____ of throws at 50 feet

E, (#) _____ of throws at 60 feet

(#) _____ total number of throws

__________ (breed of horse)

5. Vary the Direction Drill

Because a fielder is expected to throw the ball to all the bases, it is important that you practice throwing in different directions. Remember, when throwing the ball, you must step in the direction of the target.

Put two 6-foot-square targets on a wall or fence. Place the targets chest-high and at least 10 feet apart. Stand on a spot between the targets and 30 feet from them (more experienced players should stand 40 or 50 feet away).

a. Using the overhand throw, deliver 10 balls to the target on your glove-hand side. Count the number that hit the target. Then deliver 10 balls to the target on your throwing-hand side. Count the number that hit this target.

Success Goal = 16 out of 20 balls hit target

8 of 10 hit glove-side target

8 of 10 hit throwing-side target

Your Score =

(#) _____ hit glove-side target

(#) _____ hit throwing-side target

(#) _____ total target hits

b. Using the same setup, deliver 20 balls to the targets, alternating glove-hand side and throwing-hand side: Deliver a ball to the glove-side target, then the next ball to the throwing-side target; repeat 10 times. Count the number of balls that hit each target.

Success Goal = 12 out of 20 balls that hit targets

6 of 10 hit glove-side targets

6 of 10 hit throwing-side target

Your Score =

(#) _____ hit glove-side target

(#) _____ hit throwing-side target

(#) _____ total target hits

c. Now compare the total number of target hits in the two target direction drill variations you just completed. Were you more successful when you delivered 10 balls in a row to each target, or when you delivered 10 balls to each target alternately?

____ separately ____ alternately

Why do you think you were more successful at one of the drill variations? Which drill is more like game play throwing? Maybe you should try the alternate target drill variation again and try to do better than you did before.

Overhand Throwing Keys to Success Checklist

The overhand throw is the most commonly used softball throw. You have developed your skills of throwing various distances and directions. Because technique is critical to accurate throwing, you need to check your form to make sure it's correct. Ask your teacher, your coach, or a trained observer to evaluate your technique qualitatively according to the following checklist. A check should be placed on the line in front of each *key to success* that you do correctly.

Preparation Phase

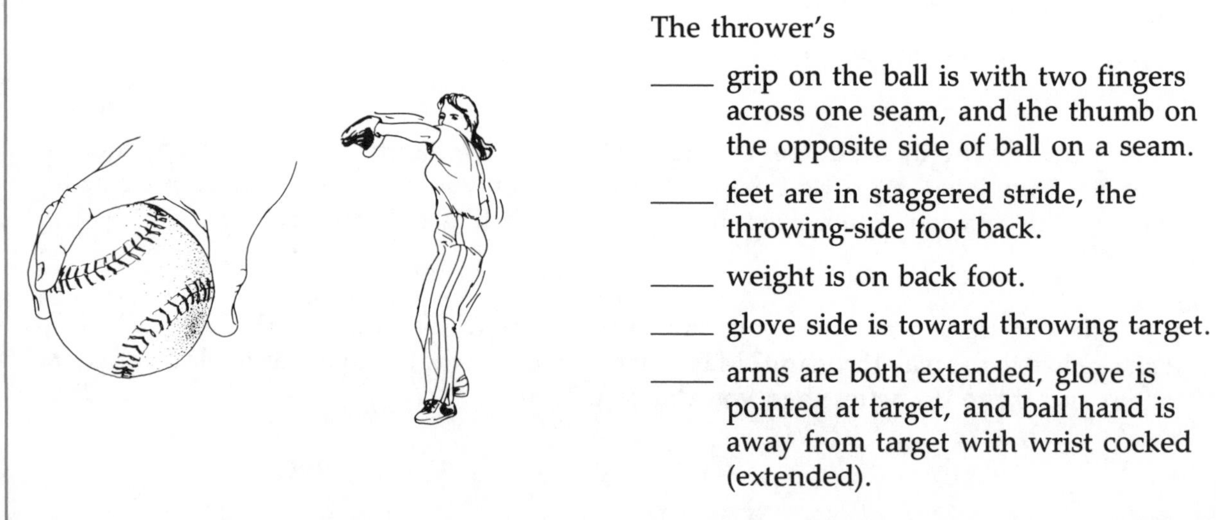

The thrower's

____ grip on the ball is with two fingers across one seam, and the thumb on the opposite side of ball on a seam.

____ feet are in staggered stride, the throwing-side foot back.

____ weight is on back foot.

____ glove side is toward throwing target.

____ arms are both extended, glove is pointed at target, and ball hand is away from target with wrist cocked (extended).

Execution Phase

The thrower

- ____ steps in direction of throwing target with glove-side foot.
- ____ starts ball forward by leading with throwing-side elbow.
- ____ makes sure ball hand trails elbow until shoulders are square to throwing target.
- ____ forcefully rotates throwing-side forearm forward toward target, passing through vertical.
- ____ keeps wrist cocked and the throwing hand and elbow high, the ball passing by the head.
- ____ forces body weight forward by driving off back foot.
- ____ has fingers directly behind ball, snaps wrist and releases ball toward target.
- ____ has glove hand low.

Follow-Through Phase

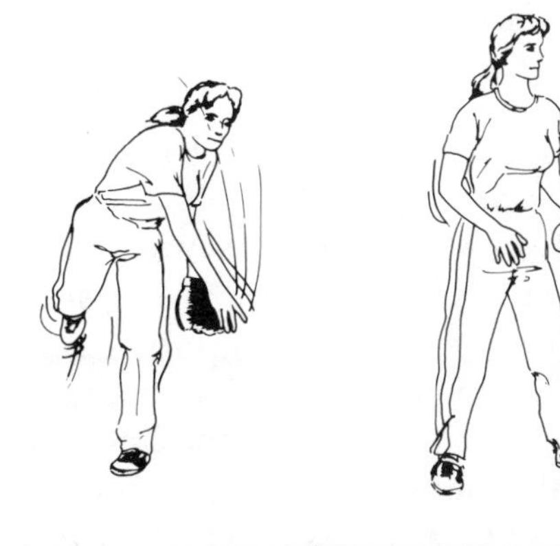

The thrower

- ____ takes full weight on glove-side foot.
- ____ drives throwing-side shoulder forward and down.
- ____ brings throwing hand down past glove-side knee.
- ____ brings throwing-side foot forward and shifts weight into balanced position.

Step 3 Combining Two Basics

At this point you know how to catch the ball and throw it overhand. You are now ready to combine these skills into the game of catch. ''Do you want to play catch?'' is the question you ask another player when you want to warm up your arm before entering a game or practice situation. What you are really doing is both throwing and catching.

WHY IS THIS TWO-SKILL COMBINATION IMPORTANT?

Throwing and catching form the cornerstone of defensive play in softball. Once the team at bat hits the ball, a defensive player must first catch the ball, then throw it to the proper location, which depends upon the game situation. In a game, the ball is seldom hit or thrown directly at you. That makes it difficult for you to use your basic catching technique in a stationary position. Nor will you always be called upon to throw from a stationary position in a game situation. You must be able to *adapt* the basic throwing and catching techniques to all the various circumstances that arise in a game: a ball coming directly at you in the air, to your glove side, on your throwing side, over your head, on the ground (covered later), and so forth.

You should always try to get your entire body directly in front of the ball when catching. Catching the ball outside of the midline of the body should be attempted only when it is impossible to get your body in front of the ball. For most people, the glove-side catch (either two-handed or one-handed) is easier than the catch on the throwing side, because the glove hand is on the same side of the body as the ball (see Figures 3.1a and b). To glove the ball one-handed, merely extend your arm and body toward the ball, making sure the pocket of your glove is open to receive the ball.

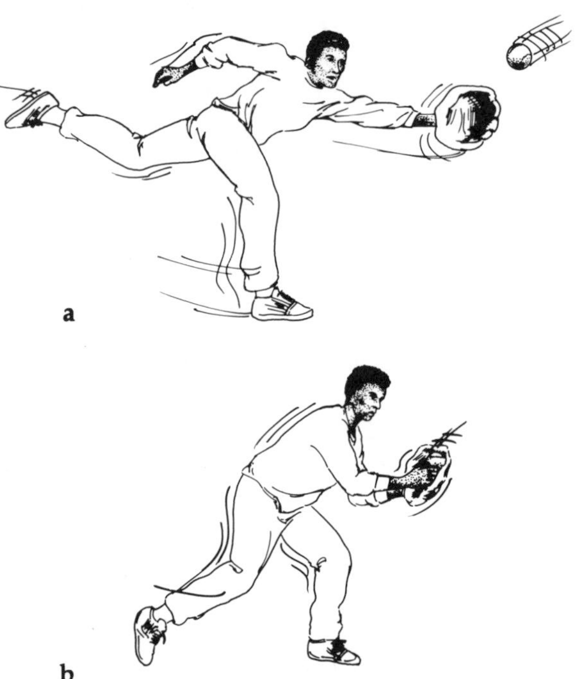

Figure 3.1 Gloving the ball with one hand (a), and with two hands (b).

Catching the ball on the throwing side involves *backhanding* the ball (see Figures 3.2a and b). This means moving your glove hand across your body to the throwing side and turning the glove over, its thumb side toward the ground, the fingers parallel to the ground, and the open pocket facing the line of direction of the throw. Because you must reach across your body to make this catch, you cannot extend as far as when going to the glove side. If necessary, you can increase your reach by turning your back on the ball's origin and stepping toward the ball with your glove-side foot. However, in order to make the catch, you have probably sacrificed the smooth transition into the throw. Thus, whenever you can, get your body in front of the ball; then you will be in a good position to follow up with a throw.

Figure 3.2 Backhanding the ball with two hands (a), and with one hand (b).

You will have ample opportunity to work on the concept of *anticipation* in this step. If you are a less experienced player, begin practicing moving to the ball by directing your partner to throw the ball to a specific side. You will then know what to expect and in what direction you will need to move. When you are practicing, work on reading and reacting to the path of the ball. See the ball leave the thrower's hand, move to it and catch it, and get into position to throw.

As you become more confident in your movement to the ball and your catching technique, ask your partner to decide where to throw the ball without telling you. Now you must truly anticipate. See the ball early, but read and react to its path, and go catch the ball! A more experienced player should start at this point with the drills.

Challenge one another by varying the direction and the speed of the throws. However, stay within your partner's range of movement for successful practice of the fundamental skills. Work to the limits of that range when you wish to practice the backhanded and glove-side catches. Now, *let's play catch.*

Overhand Throwing and Catching Drills

Note: A toss-back or rebound net could be substituted for the wall in Drills 1-5 when you are practicing outside by yourself.

Stand facing a wall at such a distance that your thrown ball rebounds back to you in the air. Throw the ball overhand against the wall so that it comes back to you about chest-high. Catch the ball and return a throw to the wall.

Work on your footwork. Move into position to catch the ball at chest height as soon as you see the ball coming off the wall. If the ball rebounds slightly to your side, step with the foot on that side to position your body directly in front of the ball. Do not reach out and catch the ball off to the side—get in front of the ball. Now is the time to begin practicing *correct* technique.

Make your catch and throw all one continuous motion. Remember that the follow-through phase of the catch becomes the preparation phase for the throw (see Figure 3.3).

Now you will practice the catch-and-throw combination in ways that they are used in game play. In order to be a good defensive player, you must be able to catch and throw the ball to the proper base as quickly and as accurately as possible. Making the catch and throw one continuous motion saves time. Moving into position to catch the ball makes it easier for you to have an accurate throw. Practice the following drills, concentrating on your technique.

Figure 3.3 Continuous motion from catching to throwing.

1. *Stationary Throw and Catch*

Stand at your spot at the wall. Throw the ball overhand straight to the wall and catch the rebound at about chest height. In a continuous motion, make a return throw to the wall. Repeat this "throw-catch-throw" series until you have completed 15 throws and catches with a continuous, rhythmic motion.

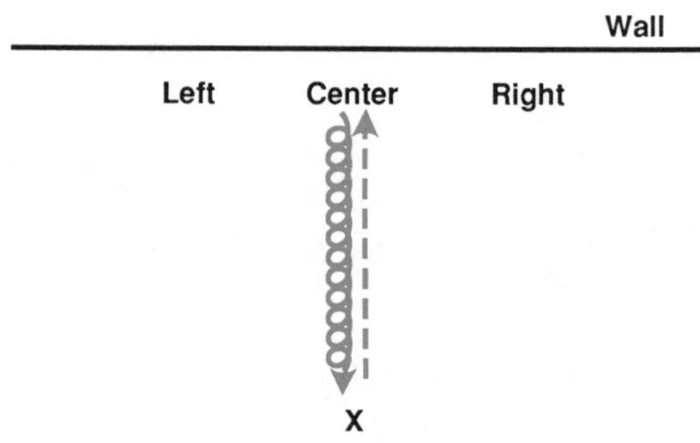

Success Goal = getting the rhythm of throw-catch-throw as a continuous motion

Your Score =

_____ yes, I established a rhythm

_____ no, I did not feel a rhythm

2. *Moving to Your Glove Side*

In a game the ball that comes to you at chest height (usually a line drive hit or a throw from a teammate) is often a bit off target. You still must be able to make the catch and throw. You must move to catch and throw the ball in the same motion if you want to make the play successfully.

Throw the ball slightly off center to your glove side so that you must move to your glove side to catch the rebound. Catch the ball and, in a continuous motion, prepare to throw it again at the wall. Hold the ball. Now move back to your starting point and repeat the throw-move-catch-prepare sequence.

Do this 10 times. Are you sure that you moved in front of the ball before catching it? Remember to move your feet so you can get into catching position. Count the number of catches you make with your glove in front of you.

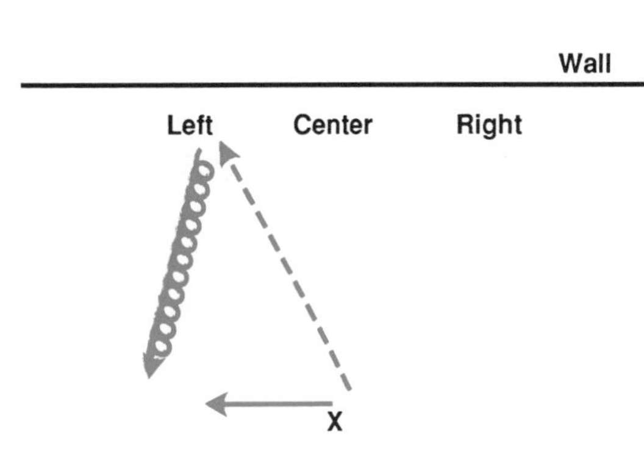

Success Goal = 8 of 10 balls caught in front of your body

Your Score = (#) _____ caught balls

3. *Moving to Your Throwing-Hand Side*

Throw the ball to the throwing-hand side. Move to that side to make the catch. Catch the ball and, in the same motion, prepare to throw. Hold the ball. Now go back to the starting point and repeat the throw-move-catch-prepare sequence until you have completed 10 throws and catches.

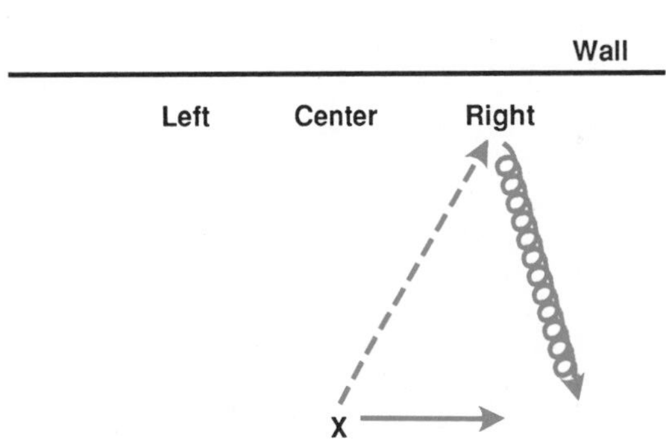

Success Goal = 7 out of 10 catches made directly in front of you

Your Score = (#) _____ caught in front of you

4. Moving Forward to Catch

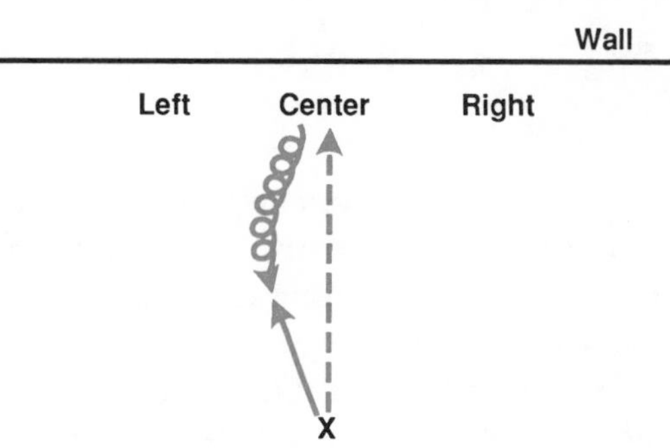

Throw the ball softly against the wall and move forward to catch the short rebound. Catch the ball and, in the same motion, prepare to throw. Hold the ball. Go back to the starting point and repeat the soft throw-move-catch-prepare sequence until you have completed 10 throws and catches.

Success Goal = 7 out of 10 catches made directly in front of you

Your Score = (#) _____ caught in front of you

5. Throw, Move to Catch, Throw to Target

Now you'll combine the skills you just completed so that you will throw, move, catch, and throw just as you might in a game. In a double play, you might have to throw the ball to a base, move back to the base you are covering, receive a throw, then throw the ball to a different base. This drill will help you develop the footwork and the catch-and-throw skills that you need to be successful in game play. It is essential that you execute the overhand throw correctly, using the full range of motion of the arm and hand.

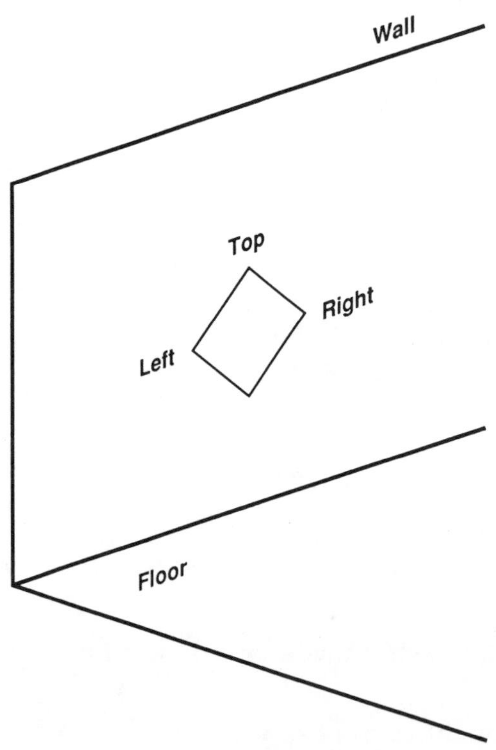

Aim Left

Put a target on the wall at chest height. Make the target a diamond that is 4 feet high and 4 feet wide. Stand at your original spot in front of the target. Throw the ball overhand to the left point of the diamond. Move to the left, catch the rebound, and throw the ball back to the center of the target. Move to catch the ball as it comes off the target. Repeat this until you can make 5 consecutive throw-move-catch-throw-catch sequences without dropping the ball or missing the target on your overhand throws. Count the number of times you try until you get 5 in a row.

Success Goal = 5 consecutive successful sequences to the left in fewer than 10 attempts

Your Score = (#) _____ attempts until you get 5 in a row

Aim Right

Using the same target, repeat the previous drill to the right side. Throw the ball overhand toward the right-hand point of the diamond target. Move to the right, catch the rebound, and throw the ball back to the center of the target. Repeat this sequence until you can complete 5 consecutive throw-move-catch-throw-move-catch sequences without dropping the ball or missing the target. Count the number of attempts you make until you successfully repeat the series 5 times in a row.

Success Goal = 5 consecutive successful sequences to the right in fewer than 10 attempts

Your Score = (#) _____ attempts

Aim Top

Use the same target and move back 5 feet for your throw. Throw the ball overhand to the target more softly so that you will have to move forward to catch it. After you catch the ball, throw it back to the target, trying to hit its top point. Move backward to catch the ball at chest height and throw the ball back to the middle of the target. Repeat this sequence until you make 7 catches and throws in a row.

Success Goal = 7 consecutive catches and throws in a row while moving up and back to make the catch

Your Score = (#) _____ consecutive catches

Which of the drills was hardest for you? Did you have more difficulty moving to the glove side, the throwing-hand side, forward, or back? Maybe for you the catching was easy but hitting the target on the throw after the catch was tough. These drills are ones that are good practice for you anytime and at any skill level. Work hardest on the drill that is the most difficult for you; make it become the easiest for you. Then you will be able to move to catch, throw, and move to catch a return throw whenever it happens in a game.

6. Partner Throw and Catch

You are going to play catch with a partner in Drills 6 through 11. Your practice here will help you anticipate just how a teammate will work with you as you catch and throw in a game.

Stand facing your partner from about 20 feet. If the sun is a factor in being able to see the ball, position yourselves so neither of you is looking into the sun.

Throw back and forth using the overhand throw. Throw with only the amount of force your partner is able to handle when catching. Make the ball travel a horizontal path at chest height.

Success Goal = 40 throws, developing a rhythmic pattern of catching and throwing in one continuous motion.

Your Score = (#) _____ throws you felt a rhythm

7. Quick-Release Drill

This drill is the same as the Partner Throw and Catch Drill (Drill 6) except that each partner starts the drill with a softball in hand. Each partner throws a ball to the other partner at the same time. Catch and immediately release (throw) the ball to your partner. Try to keep the two balls going continuously without a drop or a bad throw. *Be sure* to use correct form, especially the footwork of stepping and turning when throwing.

Success Goal = 20 successful sequences of the combination of catching and quick-releasing, neither partner making an error

Your Score = (#) _____ double combinations

8. Accuracy Drills

You and a partner stand 20 feet apart. The catching partner holds the glove as a target. The throwing person aims for the glove target. If you are a less experienced player, work on throws and catches slightly to the side or in front of the receiver. If you are a more experienced player, extend the range of movement needed to catch the ball.

Set Target Positions

Make 10 throws and catches, then switch roles. The thrower delivers the throws

- directly at the receiver,
- chest-high to the receiver's glove side,
- chest-high to the receiver's throwing side, and
- short, in front of the receiver.

Success Goal = 31 combined on-target throws for the thrower and successful catches for the receiver out of 40 attempts

9 out of 10 directly at the receiver

8 out of 10 to the receiver's glove side

7 out of 10 to the receiver's throwing side

7 out of 10 short of the receiver

Your Score =

(#) _____ at the receiver

(#) _____ to the receiver's glove side

(#) _____ to the receiver's throwing side

(#) _____ short of the receiver

Alternative Set Target Positions

Deliver 10 overhand throws to the target set by your partner. Your partner will catch the ball and get into position to throw the ball back to you. Repeat this until you have thrown 10 times to each target.

a. If you are a less experienced softball player not familiar with game situations yet, set these target positions:

- Throwing-side shoulder
- Waist
- Glove-side hip
- Throwing-side hip

Success Goal = 28 total on-target throws out of 40 attempts

7 out of 10 throws to the throwing-side shoulder

7 out of 10 throws to the waist

7 out of 10 throws to the glove-side hip

7 out of 10 throws to the throwing-side hip

Your Score =

(#) _____ to the throwing-side shoulder

(#) _____ to the waist

(#) _____ to the glove-side hip

(#) _____ to the throwing-side hip

b. If you are a more experienced softball player, select target positions that would simulate different game situations:

- Just below your knees, as in making a tag play
- High and out front with a stretch, as on a force play
- Head-high to your glove side, as on a stretch in a force play
- Knee-high to your throwing side for a backhand catch

Success Goal = 36 total on-target throws out of 40 attempts

9 out of 10 throws just below the knees

9 out of 10 throws waist-high in front

9 out of 10 throws head-high to the glove side

9 out of 10 throws knee-high to the throwing side

Your Score =

(#) _____ just below the knees

(#) _____ waist-high in front

(#) _____ head-high to the glove side

(#) _____ knee-high to the throwing side

Varying Target Positions

In all of these drills, you knew exactly where your partner was going to set the target. Likewise, the receiver knew where the ball was going to be coming for the catch. In a game, however, nothing is that easy. The next drill will help you learn to react to a thrown ball, catch it, and throw it back—a sequence that often happens in a game.

You and your partner stand 20 feet apart. You throw the ball back and forth to one another, keeping the catching and throwing one continuous motion. The catching partner sets the glove in any target position previously practiced as soon as he or she throws the ball. The object is to change the target each time there is a throw. Then you have to react to the ball as it approaches, catch it, then find the new target and throw to it.

Throw the ball back and forth 20 times each. Count the number of catches made without the receiver's having to move more than one step in any direction.

Success Goal = 15 out of 20 catches by receiver

Your Score = (#) _____ catches

9. *Line Drive Drill*

One of the hardest hit balls to catch is the line drive. It is batted sharply and directly into the playing field. Because it is coming so hard and so fast, it is difficult to get in front of it in time to catch it. You must also be certain to give with the ball as you catch it, to cushion the force.

With a partner, stand 60 feet apart, facing one another. Play catch using the overhand throw to simulate a line drive. Snap your wrist forcefully as you release the ball so that it travels with considerable speed in a horizontal path, as would a line drive.

Although most line drives are hit waist-high or above, some are below the waist. During your practice, be sure to set targets for your partner both above and below your waist and to the glove and throwing sides. (Use the backhand catching technique on all line drives to your throwing side.) When you catch the ball, move it directly into throwing position, but do not immediately throw it.

Success Goal = 36 total successful line drive throws out of 48 attempts, and 24 total successful catches out of 36 successful throws

a. 6 out of 8 throws above the waist
   in front
   to the glove side
   to the throwing side
b. 6 out of 8 throws below the waist
   in front
   to the glove side
   to the throwing side
c. 4 out of 6 catches above the waist
   in front
   to the glove side
   backhand to the throwing side
d. 4 out of 6 catches below the waist
   in front
   to the glove side
   backhand to the throwing side

Your Score =

a. throws above the waist
   (#) _____ in front
   (#) _____ to the glove side
   (#) _____ to the throwing side

b. throws below the waist
   (#) _____ in front
   (#) _____ to the glove side
   (#) _____ to the throwing side
c. catches above the waist
   (#) _____ in front
   (#) _____ to the glove side
   (#) _____ to the throwing side
d. catches below the waist
   (#) _____ in front
   (#) _____ to the glove side
   (#) _____ to the throwing side

10. Leaping Line Drive Drill

This is the same drill as the Line Drive Drill (Drill 9), except you now throw your line drive above your partner's head so that a leaping catch must be attempted. The receiver comes down with the ball in the throwing position and immediately throws it back to you.

Repeat this sequence for 15 leaping line drives. Count the number of successful line drive catches and throws back.

Success Goal = 8 out of 15 successful combined leaping line drive catches and throws back

Your Score = (#) _____ successful combined catches and throws

11. Continuous Line Drive Catch and Throw Drill

You and your partner stand 60 feet apart. Throw a line drive to your partner, who catches it and immediately throws a line drive back to you. You catch the line drive and, in one motion, throw a line drive back to your partner.

Continue this line drive catching and throwing for 5 minutes. Count the number of consecutive catches you two make together. If the ball is dropped by either partner, you must begin counting again at 1. In this drill, the footwork of turning and throwing is essential. Work on it.

Success Goal = 25 or more consecutive line drive catches between partners

Your Score = highest (#) _____ consecutive catches

Step 4 Fielding a Ground Ball

Now that you are comfortable throwing and catching with a partner, you need to be able to catch the ball as it would come to you in a real game situation. This is called *fielding*. First, you will learn how to field balls that come to you on the ground. Later you will learn how to field fly balls coming to you in the air.

WHY IS FIELDING GROUND BALLS IMPORTANT?

When the batter hits the ball on the ground (sometimes called a *grounder*), the defensive player must field the ball before making the throw. Proper fielding by the infielders is important because the play most often made by an infielder on a hit ball is fielding a grounder. If the ground ball is not fielded properly, the ball will go through into the outfield, thus allowing the batter to reach first base safely. Because baserunners must reach first base to be in a position to score runs, it is vital that infielders be very consistent in fielding ground balls and making throws to the appropriate bases. A ground ball that is not fielded even by the outfielders allows the baserunners to advance additional bases. This puts them into scoring position and increases the pressure on the defense while playing the game.

HOW TO FIELD GROUND BALLS

Whenever possible, you, the fielder, should move to a position directly in front of the ground ball before attempting to field it. Fielding the ball on the run makes both catching and throwing the ball very difficult. Being in a stationary position as you field the ball makes it very easy to get down low and watch the ball go directly into your glove, as shown in Figure 4.1. When fielding a ground ball and making a throw to a base to put a baserunner out, try to make the fielding and the throw one continuous motion. "Getting rid of the ball" quickly is important in order to put the baserunner out. Remember, however, that your first priority is to field the ball.

Figure 4.1 Keys to Success: ***Fielding Ground Balls***

Preparation Phase

1. Stagger stride
2. Knees bent, weight on balls of feet
3. Back flat
4. Focus on ball

Execution Phase

1. Hands low, glove open to ball
2. Meet ball out front
3. Use two hands
4. Watch ball into glove
5. Throwing hand on ball

b

c

Follow-Through Phase

1. Shift weight back
2. Glove side to target
3. Ball to overhand throwing position
4. Glove-side elbow to target

d e

Detecting Ground Ball Fielding Errors

Ground ball fielding errors usually occur because the fielder does not get in front of the ball or takes his or her eyes off the ball as it approaches. The most common errors committed when fielding a ground ball are listed below, along with suggestions on how to correct them.

ERROR	CORRECTION
1. You are late getting into position to field the ball.	1. Start to move into position as soon as the ball starts coming your way.
2. The ball goes "through" you, under your glove.	2. Remember the catching skills for a ball coming below your waist: Keep your glove fingers pointed down; don't squash the ball like a bug with your glove. Bend your knees, keep your head down, and get down to the ball.
3. The ball bounces off your glove.	3. Watch the ball go into the pocket of your glove and cover it at once with your throwing hand.

Ground Ball Fielding Drills

1. Wall Fielding Drill

Stand 15 feet away from a wall or rebound net if you are practicing by yourself outside. Tip the frame of the net slightly forward so the ball rebounds along the ground. Throw the ball low against the wall (1 to 2 feet above the ground) so that it rebounds back to you along the ground. Get down into fielding position directly in front of the ball. Field the ball and, in one continuous motion, come up into throwing position and throw low to the wall for a repeat grounder.

Success Goal = 10 consecutive fielding plays with correct form

Your Score = (#) ____ fielding plays

2. Footwork Drill

In this drill you get into position to field ground balls that are not coming directly toward you. You need to be able to move to balls coming to your right and your left when fielding grounders because that's where the hitters are trying to place their hits in game play. The hitter's job is to try to make it as difficult as possible for you to field the ball coming your way.

Move laterally into position to field a ball only a short distance away by using a *slide step* (see Figure 4.2): Remain facing the wall or net in a low fielding position; take a sideways step with the foot on the ground ball side, then close with the other foot. Repeat these sliding steps until you are directly in front of the ball.

a

b

c

Figure 4.2 Slide step.

For a ball rebounding at too great a distance to reach using the sliding step, use the *crossover* approach (see Figure 4.3): From a ready position (knees bent and hands at knees), pivot the ball-side foot toward the ball and cross over that foot with the other foot, taking a first step running to the ball.

Using a wall or rebound net as in the previous drill, direct your throw at a slight angle to the wall or net so that the rebound goes away from your starting position. Throwing the ball slightly off center to the right causes the rebound to go to your right. Throwing the ball to the left of center causes the ball to go to your left.

Figure 4.3 Crossover step.

Start by directing your throws barely off-center. Gradually increase the angle of the throws to increase the distance you must move to the ball. Use the appropriate footwork for the distance you must move to the ball. As the angle of the throw increases, move back away from the wall or net to allow yourself time to reach the ball. Vary your throws so that you practice going both to your right and to your left for the ball. Fielding the ball to your throwing-hand side is more difficult because you must travel further to get into fielding position in front of the ball.

Success Goal = 7 out of 10 total successful fielding plays

4 out of 5 going to your glove side

3 out of 5 going to your throwing hand side

Your Score =

(#) ____ fielding plays to glove side

(#) ____ fielding plays to throwing side

3. Play Ground Ball Catch

You and a partner stand 30 feet apart. The person with the ball throws ground balls to the fielding partner so that the ball travels *on the ground* at least two-thirds of the distance. The fielding person fields the ball and returns the ball in the air (as if throwing out a baserunner) using an overhand throw. Practice 10 plays before switching fielder and thrower roles. Field the ball as it is thrown to each of the directions listed below.

Success Goal = 33 out of 40 total ground ball fielding plays

9 out of 10 when the ground ball is directly at you

9 out of 10 when the ground ball is short (you must move forward to field the ball)

8 out of 10 when the ground ball is to your glove side

7 out of 10 when the ground ball is to your throwing side

Your Score =

(#) ____ plays on direct ground balls

(#) ____ plays on short ground balls

(#) ____ plays on glove-side ground balls

(#) ____ plays on throwing-side ground balls

4. Read and React to Ground Balls

When fielding in a game, you cannot predict to which side the ball will be hit. It is very unlikely that you will get 10 grounders in a row to the same side, as in the previous drill. Therefore, you need to practice fielding ground balls that come to you in random ways, the thrower increasing the difficulty of your fielding task by varying the speed and direction of the throws. You must now carefully judge the movement of each ball and quickly get into position to field it. Work on your ability to anticipate.

Set up 30 feet away from your partner. First the thrower delivers grounders directly at you, but randomly varies their speed from slow to fast. Field the ball and, in the same motion, throw the ball overhand to your partner. After 10 of these, switch roles.

Next, the thrower delivers 10 grounders to your left, again varying their speed. Switch roles. Then repeat this to your right side and switch roles.

Finally, your partner throws you 20 ground balls in which *both* the speed and the direction of the ball are mixed. These final 20 ground balls, especially, will help you practice reading and reacting to grounders. As always, focus on fielding the ball first, then make the throw to your partner. Blend the fielding and the throw into one fluid motion.

Success Goal = 38 out of 50 total successful fielding plays

8 out of 10 direct grounders, varying speeds

8 out of 10 left-side grounders, varying speeds

8 out of 10 right-side grounders, varying speeds

14 out of 20 random-direction grounders, varying speeds

Your Score =

(#) ____ out of 10 direct grounders

(#) ____ out of 10 left-side grounders

(#) ____ out of 10 right-side grounders

(#) ____ out of 20 random-direction grounders

5. React to the Bouncing Ground Ball

The fields on which you play softball games are never perfectly level. Consequently, ground balls may not roll smoothly across the ground. You need to be ready to field ground balls that bounce, sometimes quite erratically, as they approach you. Such grounders can be very difficult to field.

As in the previous drill, you and your partner set up 30 feet apart. Your partner now throws bouncing ground balls for you to field. Each throw should be directed hard at the ground and at least half of the distance to you so that the ball bounces to you.

First the thrower sends 10 bouncing grounders directly at you. Switch roles. Then the thrower sends 10 bouncing grounders to your right. Switch roles. Then you'll field 10 bouncing grounders to your left. Switch roles. Finally, the thrower sends 20 bouncing ground balls randomly to all directions.

In fielding all of these bouncing grounders, predict the location and height the ball will arrive at. Get into position and field the ball with both hands. Make the overhand throw back to your partner.

Success Goal = 36 out of 50 total bouncing grounders fielded

8 out of 10 bouncing grounders, direct

8 out of 10 bouncing grounders, right side

8 out of 10 bouncing grounders, left side

12 out of 20 bouncing grounders, random directions

Your Score =

(#) ____ out of 10, direct

(#) ____ out of 10, right side

(#) ____ out of 10, left side

(#) ____ out of 20, random directions

6. Continuous Grounder Fielding

You and your partner set up 30 feet apart. Now you both field and throw ground balls continuously. You throw a ground ball to your partner. Your partner, in one motion, fields the ball and throws a ground ball back for you to field.

Mix up the directions, the speeds, and the types of ground balls (rolling or bouncing) you throw to one another. The object is to make each other work hard so that you both will become solid, consistent fielders. However, do *not* try to throw the ball past your partner.

Repeat two sets of 30 throws (15 per partner). Count the number of successful fielding plays you and your partner make in a row. Each time one of you misses a ground ball, begin counting again at 1.

Success Goal = 24 consecutive grounders fielded out of two sets of 30 grounders

Your Score = highest (#) ____ grounders fielded consecutively

Fielding Ground Balls Keys to Success Checklist

Fielding ground balls is a major fundamental skill used in the game of softball. You have had the opportunity to work on this skill in fairly isolated practices. At this point, check your technique development by asking your teacher, your coach, or another trained observer to use the checklist below to evaluate your performance of fielding ground balls.

Preparation Phase

The fielder's

- ____ feet are shoulder width apart, with the glove-side foot ahead.
- ____ knees are well bent, with weight on the balls of the feet.
- ____ back is flat, almost parallel to the ground.
- ____ head is up, with eyes focused on the ball.

Execution Phase

The fielder's

- ____ fingertips of the glove and of the throwing hand are on the ground.
- ____ approach places the ball inside and slightly in front of the glove-side foot.
- ____ eyes focus on the ball, watching the ball go into the glove.
- ____ throwing hand covers the ball in the glove and assumes a two-finger grip (first two fingers on a seam, thumb on seam on opposite side of the ball).

Follow-Through Phase

The fielder's

- ____ weight begins to shift to the throwing-side foot.
- ____ pivot is on the back foot, turning the glove side toward the throwing target.
- ____ position is erect, keeping the knees slightly bent while bringing both hands toward the throwing-side shoulder.
- ____ throwing hand moves to the start of the throwing position, and the glove-side elbow points toward the throwing target.

Step 5 Hitting and Using the Batting Tee

Hitting is your first offensive softball skill to learn. As the *batter*, you are the person from the offensive team who is up to bat. Being at bat and attempting to hit a pitched ball is the start of every offensive play in a game.

Hitting is a very complex skill. In a very short period of time, you must make many judgments about contacting a moving object (the ball) with another moving object (the bat). When learning to hit, the factor that is easiest to eliminate is the movement of the ball. The *batting tee* is a device that allows you to practice hitting a stationary ball.

Later you practice hitting a ball tossed from a short distance. Finally, when you have developed some confidence and skill in contacting the ball with the bat, you are ready to hit a pitched ball.

WHY IS HITTING IMPORTANT?

The goal of offensive play in softball is to score runs. As explained in the section "The Game of Softball," an official game of softball is won by the team scoring the greater number of runs in 7 innings of play. A run can be scored only by a batter getting on base, progressing around the bases, and crossing home plate safely. Although there are several ways for a batter to get on base, hitting the ball is the most fun. Once there are runners on base, hitting the ball is an important technique for *advancing* the runners (causing them to move to a base closer to home plate) and eventually scoring runs. Probably the most fun of all is to hit the ball to a place that allows you (and any runners on base) to advance more than one base at a time (a double, a triple, or a home run)!

HOW TO HIT

Hitting involves moving the bat from a stationary position behind the back shoulder and into the path of the ball, making contact with the ball out in front of the front foot, and following through by completing the swing while keeping the body in a balanced position. The following description is presented at this early point so that as you work through the various progressive steps on hitting, you will have an understanding of the complete technique and a *picture in your head* of what is entailed in hitting a ball.

Stand with your feet shoulder width apart and your knees slightly bent. Take a grip on the bat so that the middle row (second row) of knuckles of both your hands line up. To achieve this grip, bend over, put the barrel end of the bat on the ground, and place your hands on the grip end of the bat so that it angles across the palms of both hands. Lift the bat up, maintaining this hand position on the bat, and check the alignment of the middle (second) knuckles of both hands (see Figure 5.1).

Hold the bat with your hands located shoulder-high, away from and to the rear of your body (away from the pitcher), as shown in Figure 5.2a. Start your swing at the ball by taking a slight step with your front foot, pointing your toes in the direction of the hit (see Figures 5.2b). Turn your hips toward the ball, pivoting on the ball of your back foot and the heel of your front foot (see Figure 5.2c). Swing the bat "through" the ball, extending your arms as you make contact. At the point of contact, your front leg is fairly straight and your back leg bent, with your hips square to the line of flight of the ball. As you contact the ball,

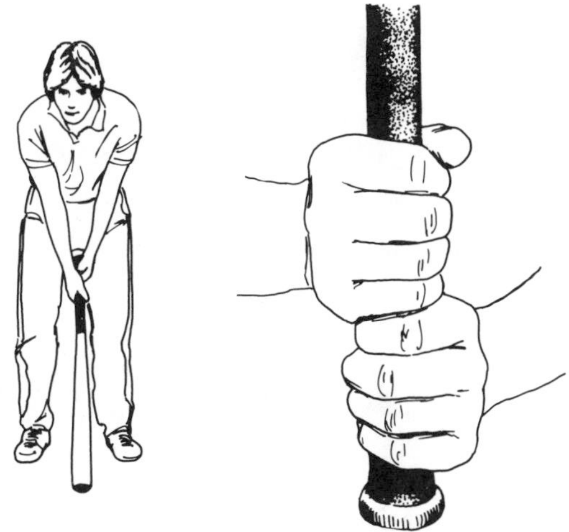

Figure 5.1 Proper grip.

your belt buckle is facing the pitcher, your head is down with your chin contacting your front shoulder, and your focus is on the ball (see Figure 5.2d). Allow your swing to continue after contact; let your hands and bat wrap around your front shoulder as shown in Figures 5.2e and f. Hold on to the bat!

HOW TO HIT THE BALL OFF THE TEE

Hitting the ball off the batting tee is one of the easiest ways to practice some of the techniques of hitting. Because the ball is stationary as you attempt to make contact, you do not have to adjust your swing to the changing positions a pitched ball would have as it approaches you.

Take a *square stance*, the toes of both feet being equidistant from the edge of home plate, or in this case the plate part of the batting tee. Because the contact point in hitting is opposite your front foot (see Figure 5.3a), take your stance with your front foot directly opposite the post of the tee. Stand at such a distance from the tee that as you swing the bat to the contact point, the center of the barrel of the bat is over the post of the tee as shown in Figure 5.3b. Adjust your position at the tee by taking a few practice swings without a ball on the tee. Then you are ready to add a ball and follow the Keys to Success in Figure 5.4.

Figure 5.2 Complete batting technique sequence.

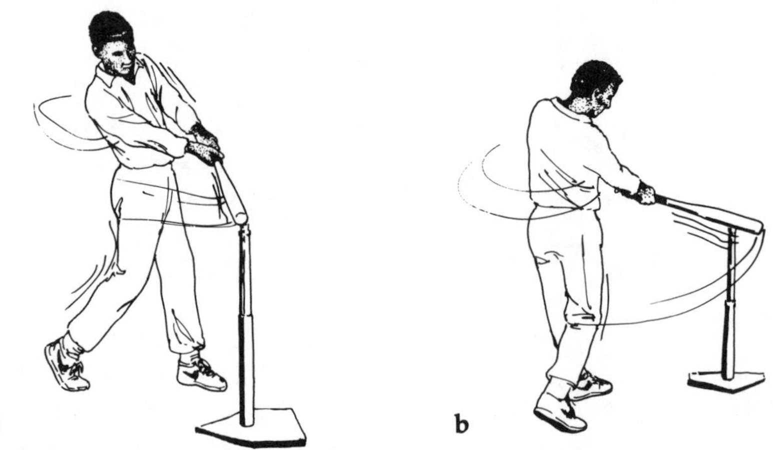

Figure 5.3 Adjust stance in relation to batting tee (a); adjust distance from batting tee (b).

Figure 5.4 Keys to Success: *Batting Tee Hitting*

Preparation Phase

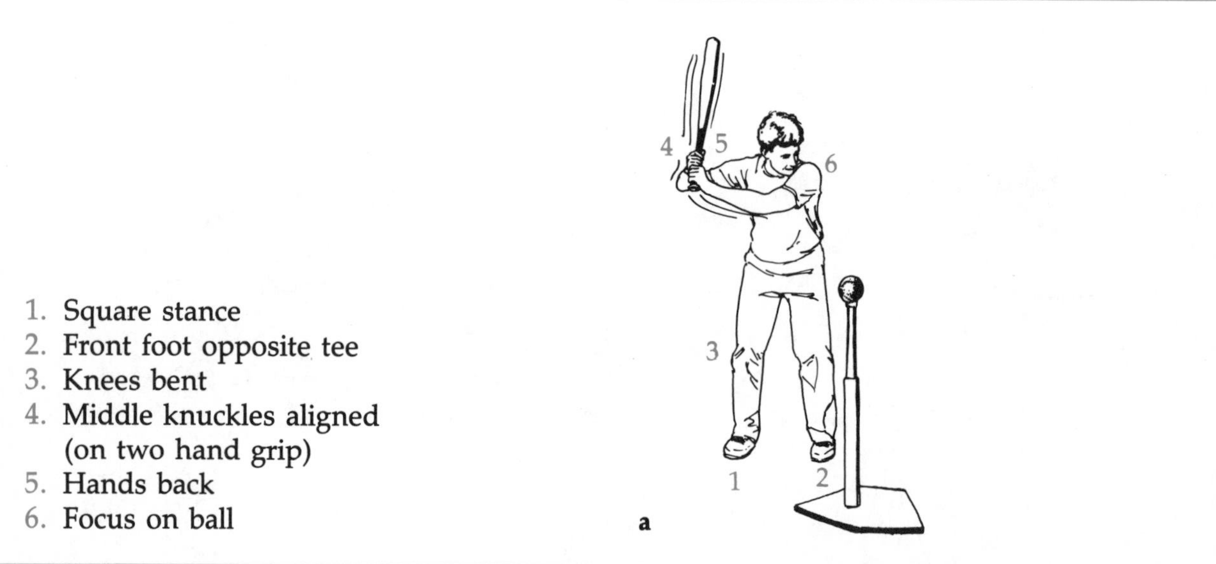

1. Square stance
2. Front foot opposite tee
3. Knees bent
4. Middle knuckles aligned (on two hand grip)
5. Hands back
6. Focus on ball
4
5
6
3
1
2
a

Execution Phase

1. Step with front foot, pointing toe toward pitcher
2. Lead swing with front elbow
3. Begin hip turn
4. Back knee bent
5. Front leg straight
6. Weight centered
7. Hips square
8. Arms extended
9. Head down

Follow-Through Phase

1. Weight centered
2. Roll wrists
3. Swing through ball
4. Back leg bent
5. Front leg straight
6. Hips square
7. Hands wrap around shoulder
8. Shoulder to chin

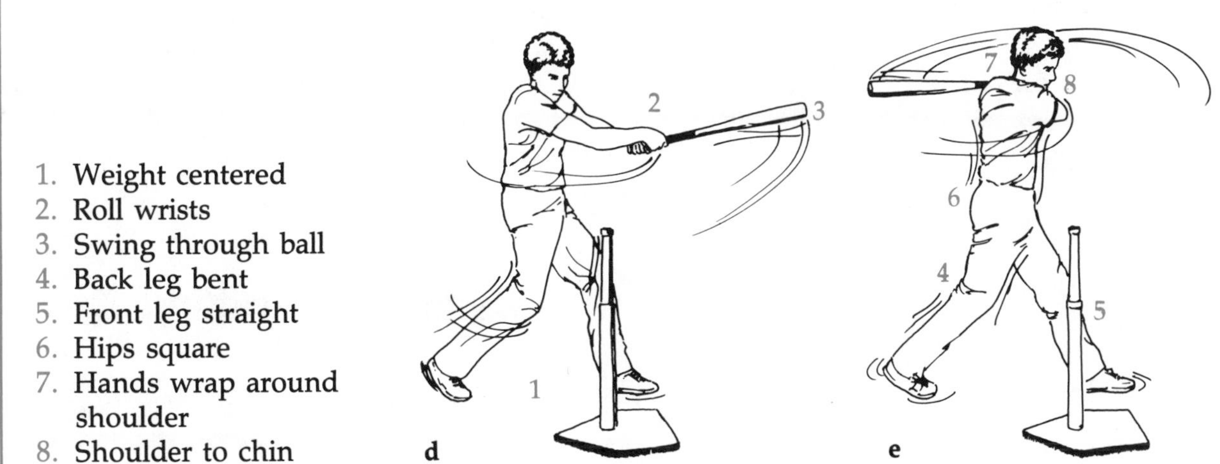

Detecting Errors in Tee Hitting

Most errors associated with using the batting tee occur because of improper body positioning at the tee. Do not stand at the tee with the midline of your body opposite the post of the tee. Remember, your contact point for hitting is ahead of your body and toward the pitcher, at the distance of your front foot. Because the teed ball is stationary, you should line up your front foot with the post of the tee. This and other common hitting problems with a batting tee are listed below, along with suggestions on how to correct them.

ERROR	CORRECTION
1. You contact the ball on the handle, close to your hands, or you have to bend your elbows for the bat to meet the ball.	1. You are probably standing with the post of the tee opposite the middle of your body. Move your stance so your front foot is opposite the post.
2. You cannot reach the ball when you swing.	2. Move closer to the tee, but in the position described in Correction 1.
3. You hit the tee instead of the ball.	3. Keep your back shoulder up; do not drop your back elbow as you swing.

Batting Tee Hitting Drills

1. Net Drill

Position the batting tee 10 feet from a hanging net. Take a stance beside the tee so that the line of flight of the batted ball is toward the net. Using a bucket of 10 regulation balls, place a ball on the tee, assume your correct batting stance beside the tee, and hit the ball off the tee and into the net.

If you are working with a partner, one person (the ball feeder) should stand with the bucket of balls on the side of the tee opposite from the hitter and place the ball on the tee. *Caution*: Hitter, wait until the feeder removes her or his hand from the ball and steps back from the tee before you swing at the ball.

Feeder, watch your partner hit the ball. Using the points from each of the three phases of the Keys to Success shown in Figures 5.4a-e, give feedback to the hitter about the technique used. Switch roles after hitting the bucket of balls. Repeat the drill so that each of you hits the ball 20 times.

Success Goal = 16 out of 20 hits with correct form

Your Score = (#) _____ of hits

2. Net Target Drill

Mark three lines on the net: one, 2 feet from the ground; another, 4 feet from the ground; and the third, 8 feet from the ground.

Using the same procedures as in the Net Drill, hit 10 balls below the 2-foot line, 10 between the 4- and the 8-foot lines, and 10 above the 8-foot line. To hit the low target (attempting to hit a ground ball), use a high-to-low swing path. To hit the middle target (attempting to hit a line drive), use a horizontal swing path. To hit the high target (attempting to hit a fly ball), use a low-to-high swing path.

Success Goal = 24 out of 30 on-target hits

8 out of 10 at low target (ground ball hits)

8 out of 10 at middle target (line drive hits)

8 out of 10 at high target (fly ball hits)

Your Score =

(#) _____ low target hits

(#) _____ middle target hits

(#) _____ high target hits

3. Bottom Hand Only Swing Drill

Put the tee at the center height setting. Stand with your regular grip on the bat. As you swing at the ball, though, let go of the bat with your top hand and make contact with the ball while holding the bat with your bottom hand only. Be sure to complete the full swing with one hand. Try to hit the ball at line drive height or lower.

Hit 10 balls to the net. Collect the balls and repeat, or switch roles with your partner if you have a partner setting balls on the tee. Count the number of balls that go line drive height or lower off of swings that do not hit the tee.

Success Goal = 16 out of 20 line drives on swings that do not hit tee

Your Score = (#) _____ line drives

4. *Low-Ball/High-Ball Drill*

Adjust the height of the batting tee to work on the low pitch, then the high pitch. Hit 20 balls at each height. Then change roles if a partner is setting balls on the tee for you, or go collect the balls if you are working alone. Strive for clean hits (no bat contact with the tee).

If the tee you are using has an adjustment to move the ball inside and outside, add those variations in combination with high and low to double your practice (high inside, high outside, low inside, and low outside). The fixed center-post tee allows you to work only on high and low balls, however.

Success Goal =

Fixed center-post tee

30 clean hits out of 40 total swings

- 15 of 20 at low setting
- 15 of 20 at high setting

or

Adjustable center-post tee

60 clean hits out of 80 total swings

- 15 at low inside setting
- 15 at low outside setting
- 15 at high inside setting
- 15 at high outside setting

Your Score =

Fixed center-post tee

(#) _____ low setting

(#) _____ high setting

or

Adjustable center-post tee

(#) _____ low inside

(#) _____ low outside

(#) _____ high inside

(#) _____ high outside

Batting Tee Hitting Keys to Success Checklist

The purpose of using the batting tee for hitting practice is to perfect your hitting stroke and to develop bat control. Your development of a smooth, coordinated swing in horizontal, high-to-low, and low-to-high swing paths is a prerequisite to your being able to successfully hit a pitched ball in a game. To check your development of these important techniques, ask your teacher, coach, or a trained observer to evaluate your hitting performance according to the checklist below.

Preparation Phase

The batter

____ stands with feet in square stance, shoulder width apart.
____ positions body so that the front foot is opposite the post of the tee.
____ bend knees slightly, with weight centered.
____ grips the bat with both hands at the base of the handle, front-side hand on bottom, hands touching.
____ grips the bat so that the middle row of knuckles of both hands line up.
____ holds bat back, hands in front of the rear shoulder, bat angled back at 45 degrees.
____ looks at ball on tee, chin contacts front shoulder.

Execution Phase

The batter

____ takes a slight step with front foot and points toes in direction of intended line of flight of the ball.
____ starts swing at ball by turning hips, pivoting on ball of back foot and heel of front foot.
____ starts bat forward by leading with front elbow, wrists staying cocked as arms begin to extend.

At the point of contact (bat to ball), the batter's

____ hips are square to the line of flight of the ball.
____ back knee is bent, heel is up, toes are pointed along intended line of flight of the ball.
____ front leg is straight, toes are pointed along intended line of flight of the ball.
____ weight is under the back knee.
____ arms are extended, wrists are extended.
____ head is down, eyes are focused on ball.

Follow-Through Phase

The batter

____ continues swing through the ball.
____ rolls wrists, top hand over bottom.
____ holds the bat securely, letting the bat wrap around the body and hands wrap around the front shoulder.
____ keeps head down and still, rear shoulder contacts chin.
____ has weight centered under back knee (since hitter is not going to run now).
____ has the back leg bent.
____ has the front leg straight.
____ has hips square.

Step 6 Fungo Hitting a Ground Ball

Fungo hitting is hitting the ball out of your hand either onto the ground (fungo hitting ground balls) or into the air (fungo hitting fly balls). It is a skill mainly used to facilitate fielding practice; it is not used in a game. However, it helps you with hand-eye and bat-eye coordination. It is a little more difficult because you now must hit a ball that is moving. Fungo hitting is easier than hitting a pitched ball because you have control over the movement of the ball.

WHY IS FUNGO HITTING GROUND BALLS IMPORTANT?

Once you have learned the basics of fielding ground balls, it is important to practice fielding balls that have actually come off a bat. If the teacher is the only one who can hit the ball toward a fielder, the practice opportunities will be severely limited. On the other hand, if everyone can fungo hit a ball, then half of you will be able to practice fielding ground balls at the same time.

HOW TO FUNGO HIT GROUND BALLS

Begin by holding the ball in the hand you place on the bottom end of the bat. Hold the bat high, back by your rear shoulder, your top hand gripping the bat as normal. Toss the ball out in front of your front foot so that it will drop to the position in space that it had on the tee. Once you toss the ball, quickly place your bottom hand on the bat to make your regular two-handed swing and hit the ball. Because you want to hit grounders, let the ball drop just below waist height (like on the tee) before contacting and swinging through the ball. Be sure to start the bat high so that your swing path is high-to-low, making the ball go down onto the ground (see Figure 6.1).

*Figure 6.1 Keys to Success: **Fungo Hitting Ground Balls***

Preparation Phase

1. Square stance
2. Knees slightly bent
3. Ball in bottom grip hand
4. Ball thigh-high over front foot
5. Bat in top grip hand
6. Bat high by back shoulder
7. Head down, focus on ball

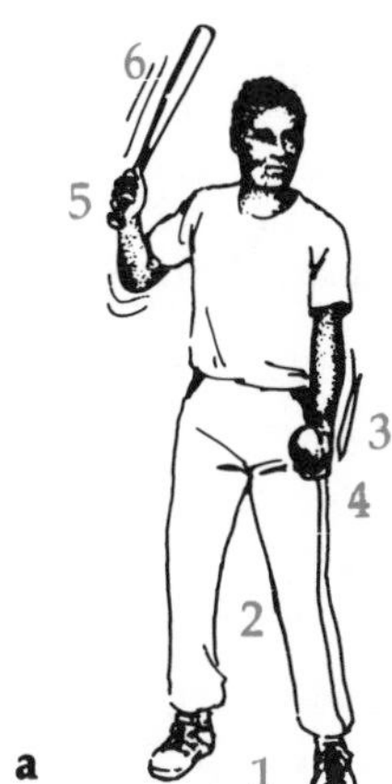

a

Execution Phase

1. Toss ball
2. Front foot step in direction of hit

3. Bottom hand to bat
4. Focus on ball

5. Pivot, swing down and through ball
6. Contact ball off front foot

Follow-Through Phase

1. Weight on front foot
2. Hips square
3. Hands wrap low around body, below shoulder

Detecting Errors in Fungo Hitting Ground Balls

Your ability to correctly and accurately hit fungo ground balls is essential to practice. Most errors in fungo hitting ground balls are caused by tossing the ball too high and by swinging too soon, thereby hitting the ball before it has a chance to drop below your waist. The most common errors in fungo hitting ground balls are listed below, along with suggestions on how to correct them.

ERROR	CORRECTION
1. The hit ball goes up in the air as a line drive or fly ball.	1. Toss the ball lower. Start your swing later. Do not contact the ball until it drops below your waist.
2. The hit ball goes off target.	2. Toss the ball directly opposite your front foot. Contact the ball off your front foot. At contact, both feet should point toward the target.
3. You miss the ball completely.	3. Watch the ball until the bat makes contact.

Fungo Hitting Ground Ball Drills

1. *Fungo Hitting Ground Ball Wall Drill*

With a bat and a ball, stand in a hitting position so that the flight of a batted ball will be toward a wall. Stand far enough away from the wall so that you have plenty of time to field the rebounding ball with your tossing hand (your other hand continuing to hold onto the bat).

Using the techniques described in Figure 6.1, hit the ball so that it contacts the floor about 5 feet from the wall. This is not an accuracy drill; aiming at the floor in front of the wall merely helps you learn to hit the ball on the ground. Remember to swing down at the ball; use a high-to-low swing path. *Note*: This same drill can be done outside against a fence. Use a bucket of 10 regulation balls (the ball will not rebound to you off the fence; thus, the bucket of balls will allow you to practice without retrieving each ball hit).

Fungo hit toward the wall (fence). Continue hitting until you get 5 consecutive fungo hits that contact the floor (ground) before they hit the wall (fence). Count the total number of fungo hits it takes you to get 5 in a row.

Success Goal = 5 consecutive fungo hits that contact the floor (ground) before the wall (fence) in less than 15 attempts

Your Score = (#) _____ fungo grounders to reach 5 in a row

2. *Accuracy Drill*

Mark three targets 3 feet by 3 feet on the floor (ground) at the base of the wall or fence. Put one target 10 feet to the right and one target 10 feet to the left of a center target. Position yourself directly in line with the center target.

Hit 10 balls to each of the three targets—center, left, and right. For ease in retrieving, use balls that do not rebound. However, if more than one hitter is nearby, use balls that rebound so you can stay out of the hitters' line of fire when retreiving. If working with a partner, use a bucket of 10 regulation balls. Hit all 10 to the center target on the wall. Have your partner field the rebounding balls and place them in an empty bucket. Switch buckets after each set of 10 hits. With your partner continuing to field, hit 10 balls to the left and then 10 to the right. Changes roles and repeat the drill.

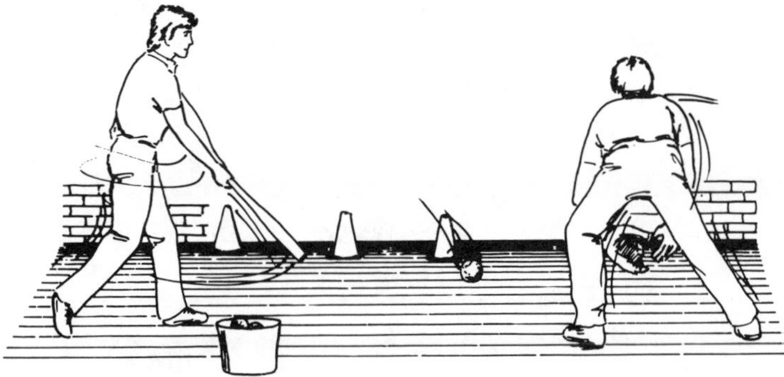

Success Goal = 24 out of 30 fungos contacting the floor or ground and then the target for each of the three targets

8 out of 10 on target hits, center target

8 out of 10 on target hits, left target

8 out of 10 on target hits, right target

Your Score =

(#) _____ on target hits, center target

(#) _____ on target hits, left target

(#) _____ on target hits, right target

3. *Partner Call Target Drill*

Using the same setup as in the previous drill, you and a partner challenge one another to a contest. Your partner names one of the three targets (left, center, right) prior to your fungo hit. Your challenge is to hit the ball into the called target. The challenge for your partner is to field the ball before it stops moving and without dropping or bobbling it. A ball that hits the target is worth 1 point. A cleanly fielded ball is worth 1 point.

Fungo hit 10 balls, then switch roles with your partner. After each of you has fungo hit 10 balls, determine the winner of that ''inning,'' the winner being the one with more points. Play three such innings and determine the winner of the game.

Success Goal = win the game

Your Score = (#) _____ innings won

Fungo Hitting Ground Balls Keys to Success Checklist

You have had a chance to test yourself on each of the Step 6 Success Goals. Next ask your teacher, your coach, or another trained observer to judge your ability by evaluating your technique subjectively according to the checklist below.

Preparation Phase

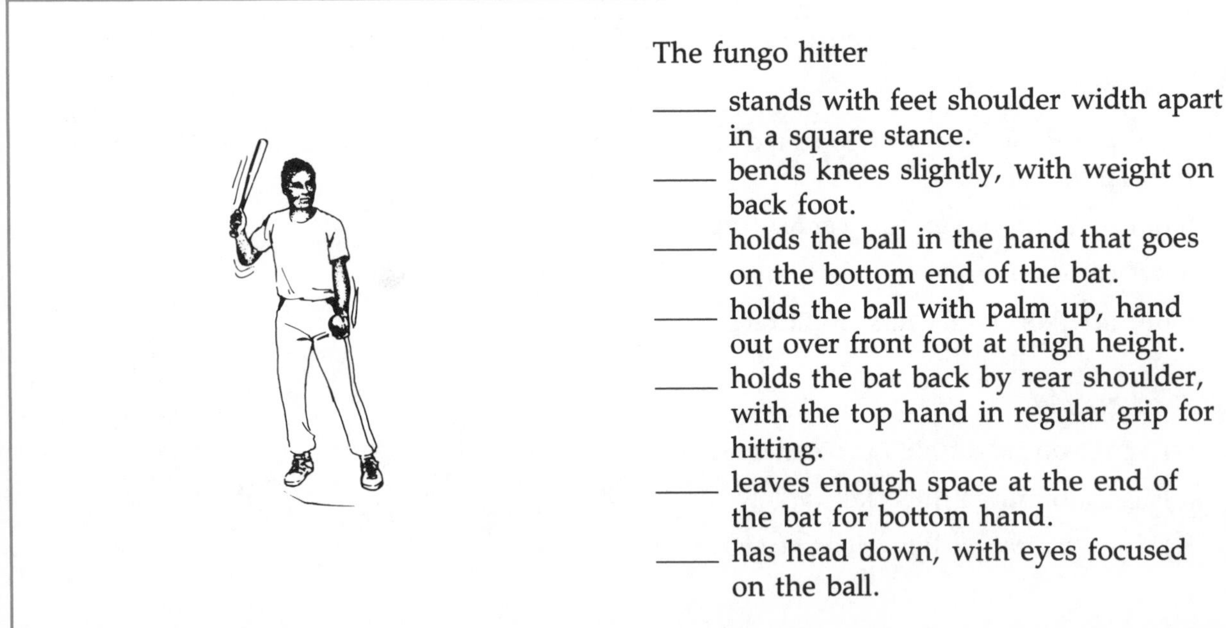

The fungo hitter

____ stands with feet shoulder width apart in a square stance.
____ bends knees slightly, with weight on back foot.
____ holds the ball in the hand that goes on the bottom end of the bat.
____ holds the ball with palm up, hand out over front foot at thigh height.
____ holds the bat back by rear shoulder, with the top hand in regular grip for hitting.
____ leaves enough space at the end of the bat for bottom hand.
____ has head down, with eyes focused on the ball.

Execution Phase

The fungo hitter

- ____ tosses the ball with a slight down-and-up motion.
- ____ starts the bat on downward swing path with top hand, as bottom hand takes its proper grip at the end of the bat.
- ____ takes a slight step with the front foot, pointing the toes in the direction of the hit.
- ____ turns hips, pivoting on the ball of the back foot and taking weight on the front foot.
- ____ swings down through the ball, contacting it just below waist height.
- ____ has head down and eyes focused on the ball at the point of contact.

Follow-Through Phase

The fungo hitter

- ____ continues swing through the ball with a high-to-low swing path.
- ____ has shifted weight to the front foot.
- ____ has hips square to the line of direction of the hit.
- ____ wraps the bat around body, hands at waist level.

Step 7 Four-Skill Combination

Your development as a softball player depends upon your ability to select and execute the appropriate skill combinations in the variety of offensive and defensive situations that occur during a game. Seldom in a team sport like softball is one skill carried out in isolation. Unlike the diver or the gymnast who does *a* dive or *a* vault, the softball player usually needs to follow fielding a ground ball with a throw to a base, or follow a hit with baserunning. In addition, often the technique used in one skill influences the performance of the follow-up skill.

You know how to execute three of the basic defensive skills used in the game of softball: catching, throwing the ball overhand, and fielding ground balls. Offensively, you know the hitting lead-ups of how to fungo hit and hit off a batting tee. Now you are ready to combine all these skills into one continuous movement, just like you have to in a game.

WHY ARE THESE SKILL COMBINATIONS IMPORTANT?

Fielding the ground ball and making an overhand throw to another player is a series of actions that occurs frequently in a regulation game. Typical gamelike situations that require these skills include the following:

- Fielding ground balls hit at varying speeds and directly at you, to your glove side, and to your throwing side.
- Fielding ground balls and making overhand throws at distances and in directions that represent the relationship of each of the infield positions to home plate.
- Fielding ground balls and making overhand throws the distance from outfield to home plate.
- Fielding ground balls and making overhand throws to the glove side.
- Fielding ground balls and making overhand throws to the throwing side (necessitates that after fielding the ball, you first make a pivot toward the target, then throw).

Fielding Ground Balls and Overhand Throwing Drills

1. Direct Grounders Drill

In groups of three, set up so that one person is the hitter, one the catcher for the hitter, and one the fielder. The fielder stands about 70 feet from the hitter and catcher (the distance from shortstop position to home). Fielder assumes the infielder's ready position (the preparation phase for fielding).

The hitter fungo hits a ground ball directly to the fielder (if the batter's fungo hitting skill is not adequate at this point, use a tee). The fielder fields the fungo hit ball and, using an overhand throw, throws the ball to the catcher. The catcher stands facing the fielder and is at least 3 feet away from the hitter across an imaginary home plate. The catcher tosses the ball to the hitter.

As the ball is hit, the fielder tracks the ball and moves into position directly in front of the ball to execute the fielding skill. The catcher gives a target with the glove.

As you do this drill focus on technique: as hitter, focus on fungo hitting the ball down; as catcher, concentrate on giving with the ball and catching with two hands; and as fielder, focus on fielding and throwing to the catcher in one motion.

After 10 repetitions, rotate roles. A successful fielding and throwing combination requires that the ball is fielded and thrown so that the catcher can catch the ball without moving more than one step.

Success Goal = 25 out of 30 total successful combinations

8 out of 10 fielding and throwing combinations

8 out of 10 hits

9 out of 10 catches at home

Your Score =

(#) ____ fielding and throwing combinations

(#) ____ hits

(#) ____ out of 10 catches

2. Moving to the Ball Drill

Use the same three-player setup as in the previous drill. The fungo hitter now directs the 10 grounders to the fielder's glove side, 5 hits at a moderate speed and 5 that are faster. The fielder fields the ball and uses an overhand throw to return the ball to the catcher. Rotate roles after 10 hits.

Next the fungo hitter directs grounders to the throwing-hand side, 5 moderate and 5 faster. The fielder uses an overhand throw to return the ball to the catcher. Rotate roles.

Finally, the fungo hitter hits 10 grounders, randomly hitting to the glove side, the throwing side, and directly at the fielder. The hitter should vary the speed of the ground balls. Rotate roles.

Success Goal = 23 out of 30 total successful fielding plays and on-target throws

8 out of 10, glove side

8 out of 10, throwing side

7 out of 10, mixed sides

Your Score =

(#) ____ glove side

(#) ____ throwing side

(#) ____ mixed sides

3. Random Hit Fielding Drill

In this drill use the same setup as in the previous drills, but when fielding you will not know in advance in what direction the ball will travel toward you, nor will you know the speed at which it will arrive. You will practice reading the ball's direction and speed as it leaves the fungo hitter's bat. Watch the ball come off the bat and move into position to field and make the overhand throw to the catcher.

The fungo hitter hits 10 ground balls to the fielder, varying the speed and the direction of each hit, but not trying to get the ball by the fielder. The hits should help the fielder develop his or her reaction to the ball's direction and speed. The fielder should have to extend to get to the ball but should be able to field and throw.

After 10 hits, rotate roles. Count the number of successful fielding plays and on-target throws. Repeat this drill for a total of 20 ground ball fielding attempts each.

Success Goal = 16 out of 20 fielding and throwing combinations

Your Score = (#) ____ out of 20 combinations

4. Triangle Drill

Two pairs of partners take part: one pair, the hitter (H) and catcher (C), position themselves as though at home plate; the other pair, fielders (F), position themselves as though at first base and third base.

The hitter hits ground balls to each of the two fielders, alternating between them. The person fielding the ball uses the overhand throw and makes the throw to the person in the other fielding position. This second fielder then throws the ball to the catcher, who tosses the ball to the hitter.

After 5 grounders to each fielder, the hitter and catcher exchange roles, and two fielders exchange positions. After the next set of 5 grounders to each fielder, the fielders exchange roles with hitter and catcher. Keep repeating the sequence.

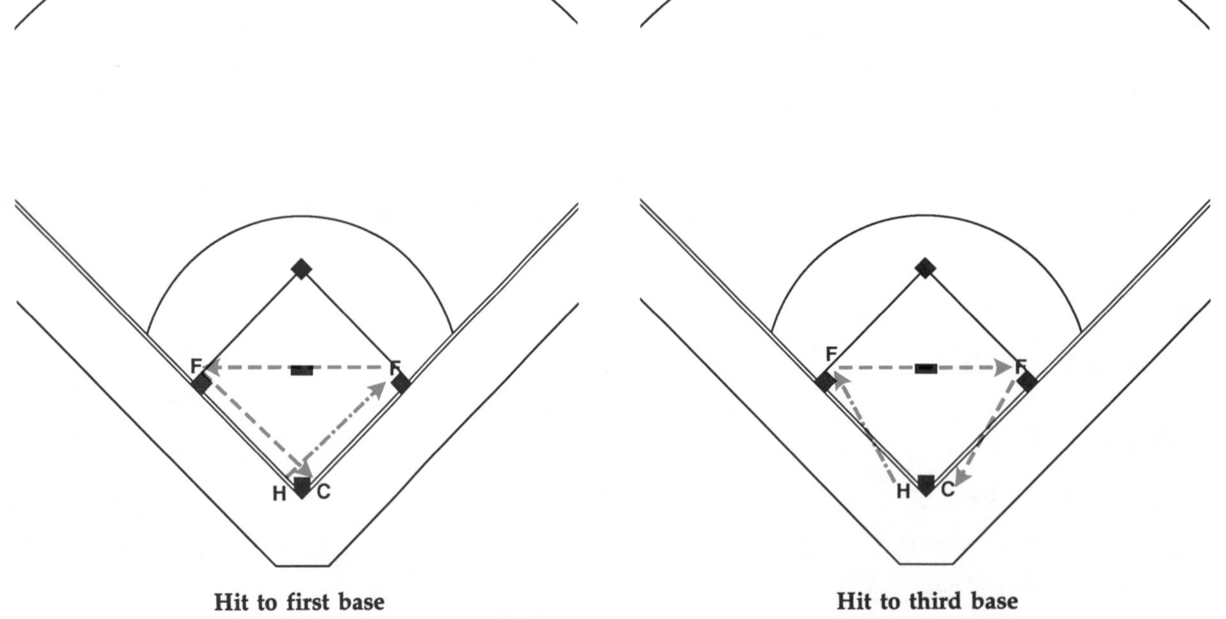

Hit to first base **Hit to third base**

Success Goal = error-free plays

8 out of 10 hits

9 out of 10 catches at home

Fielder at first base

4 out of 5 fielding plays at first-base side

3 out of 5 throws from first-base side

Fielder at third base

4 out of 5 fielding plays at third-base side

4 out of 5 throws from third-base side

Your Score =

(#) ____ hits

(#) ____ catches at home

Fielder at first base

(#) ____ fielding plays at first-base side

(#) ____ throws from first-base side

Fielder at third base

(#) ____ fielding plays at third-base side

(#) ____ throws from third-base side

5. *Get One Out Drill*

This drill is called "Get One Out" because in a real softball game, once you field the ground ball, you throw it to get a baserunner out. Outs are made one at a time in a game. Outs are made only if you cleanly field the ball and make the throw to the base accurately and on time.

You are set up in two pairs of partners as in the previous drill. The fielders are positioned at first base and shortstop. The hitter and the catcher for the hitter are positioned at home.

The hitter grounds the ball to the fielder at the shortstop position, who in one motion fields the ball and makes an overhand throw to the fielder at first base. The ball is then thrown to the catcher, who tosses it to the hitter for the next hit.

The hitter should vary the placement of the grounders, that is, to the glove side, to the throwing side, and directly at the fielder. The hitter should also vary the speed of the grounders—both hitting hard, and hitting softly so the fielder must come in on the ball to make the play. All players should work on overhand throws only.

After 10 hits, the fielders exchange positions, and the hitter and catcher exchange roles. After the next set of 10, the fielders exchange roles with the hitter and catcher. Keep repeating the sequence.

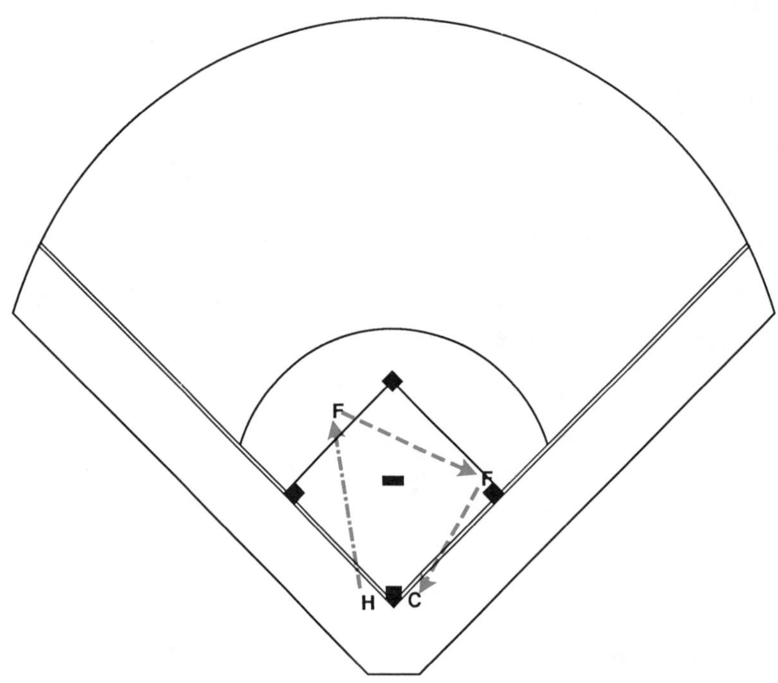

Success Goal = 41 out of 50 total error-free plays

8 out of 10 error-free fielding and overhand throwing combinations

8 out of 10 catches at first base

8 out of 10 error-free overhand throws to catcher

9 out of 10 catches at home

8 out of 10 fungo hits to intended direction

Your Score =

(#) _____ fielding and overhand throwing combinations

(#) _____ catches at first base

(#) _____ overhand throws to catcher

(#) _____ catches at home

(#) _____ on-target hits

6. *Outfielder Drill*

In this drill you will extend the distance of your overhand throw. If a ground ball should get by an infielder in a game, an outfielder would have to field the grounder and throw it to the appropriate person.

In a group of three, set up so that one person is the hitter, one the catcher for the hitter, and one the fielder. The fielder stands about 100 feet from the hitter and catcher (the distance from short fielder to catcher). The fielder assumes the outfielder's ready position—knees slightly bent, hands about waist-high, weight on the balls of the feet, and eyes focusing on the hitter.

The hitter fungo hits a ground ball to the fielder (if fungo hitting skill is not adequate at this point, use a tee). The fielder fields the ball and, using an overhand throw, throws the ball to the catcher who stands beside the hitter and faces the fielder. The catcher tosses the ball to the hitter. If you are the fielder, as the ball is hit, track the ball and move into position directly in front of the ball to execute your fielding skill. As you bring the ball to the throwing position, take a crow-hop step (hop on the glove-side foot while bringing the ball to throwing position) and make a strong overhand throw to the catcher. If you cannot throw the ball in the air all the way to the catcher, execute a one-bounce throw; aim your throw to hit the ground 10 to 15 feet in front of the catcher. The throw to that landing point should travel in a straight line.

After 10 sequences, rotate roles.

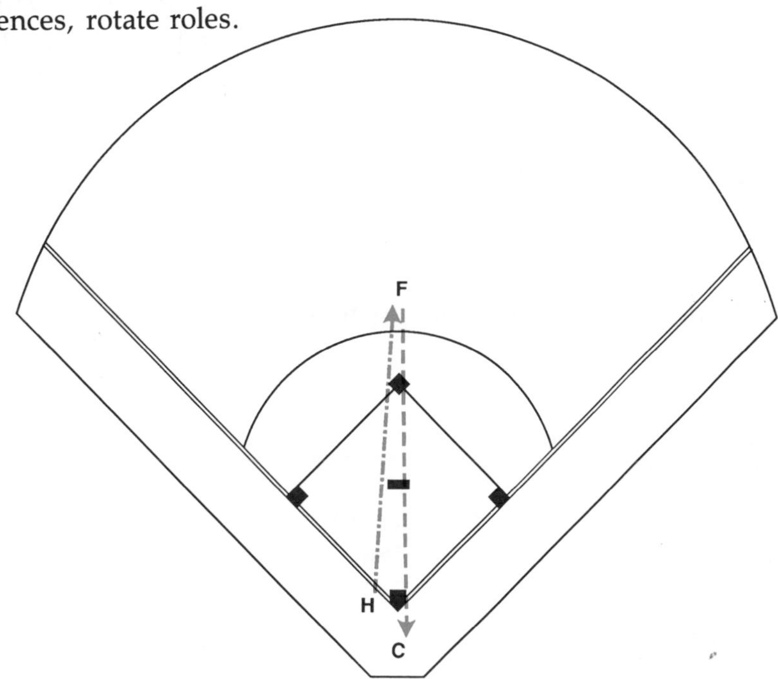

Success Goal = 24 out of 30 total error-free sequences

8 out of 10 fielding combinations

8 out of 10 on-target hits

8 out of 10 catches at home

Your Score =

(#) _____ fielding combinations

(#) _____ hits

(#) _____ catches

Step 8 Pitching

The *pitch* is the skill used by the defense to put the ball into play with every hitter in a regulation game. Pitching in slow pitch is a skill that most players can master in a relatively short time. Slow pitching is a finesse skill. It is not a power- or strength-related skill as is the fast pitch pitching style of the ''windmill'' or ''slingshot.''

WHY IS PITCHING IMPORTANT?

Can you imagine the games of tennis, volleyball, or racquetball without the serve? The serve begins the game. Likewise, in softball you cannot start or continue the game without the pitch; the batter would never get to hit without the pitch! Pitching in slow pitch softball is not as overpowering a factor in the game as it is in fast pitch; however, it is still an important skill, and the pitcher is a very important player on the team.

As pitcher, you should work on becoming as consistent as you can at getting the pitch over the plate. You want to have the hitters have a chance to hit the ball so that your teammates can catch or field the ball and make the necessary plays to get the baserunners out. Again, your job is to put the ball into play so that the game can continue.

HOW TO EXECUTE THE PITCH

Slow pitch pitching requires an underhand delivery of the ball to the batter. Swing your arm in an underhand, pendulum motion and take a step toward the batter as you release the ball. You must make the pitch with your pivot foot (throwing-side foot) in contact with the pitching rubber. Start with the ball in your throwing hand and in front of your body. Swing your arm down, back to the rear, then forward in the same path, releasing the ball so that it traverses an arc no more than 12 feet and no less than 6 feet high, while traveling a distance of 46 feet to home plate. After releasing the ball, assume your fielding position, and be ready for any ball hit back toward you (see Figure 8.1).

Figure 8.1 Keys to Success: Pitching

Preparation Phase

1. Stagger stride
2. Pivot foot on rubber
3. Weight on back foot
4. Ball held in front in three-finger grip
5. Face batter
6. Come to full stop
7. Focus on target

a

Execution Phase

1. Shift weight to pivot foot
2. Arm down and to rear
3. Step toward batter
4. Arm swings forward, fingers behind ball
5. Fingers under ball, release ball past hip

Follow-Through Phase

1. Step forward with pivot foot
2. Hand above head
3. Knees bent, square stance
4. Hands at ready position
5. Watch for hit ball

Detecting Pitching Errors

Slow pitch pitching errors usually occur because of improper timing of the release of the ball, resulting in an incorrect arc and distance. Another error is your failure to get into a good fielding position after releasing the ball. This error can be hazardous to your health! For your own safety, be sure to assume a correct fielding position after you pitch the ball. The most common pitching errors follow, along with suggestions on how to correct them.

ERROR	CORRECTION
1. The ball travels in too high an arc.	1. You are holding on to the ball too long. Release the ball as your hand passes your hip, aim for a point in space 12 feet above the ground, two thirds of the distance to home plate (about 30 feet).
2. The ball travels with no arc.	2. You are releasing too soon (see Correction 1).
3. The ball goes too long or falls short of the target.	3. The amount of force is incorrect. Try to drop the ball onto the back point of home plate.
4. You get hit with the batted ball.	4. Be sure to assume a good fielding position at the end of the pitch.

Pitching Drills

1. Pitching Wall Drill

Place marks on a wall 1, 6, and 12 feet from the floor. The 6- and 12-foot lines are the pitching height lines. The area between the floor and the 1-foot line is the target for the pitch. Place a mark on the floor 46 feet away from the wall; this mark is your pitching rubber.

When you are pitching in a regulation game, the ball must travel within a 6- to 12-foot height limitation. In order for a strike to be called, the pitch, if not hit by the batter or caught by the catcher, would land on the ground just behind the back part of the plate. Aim your pitch to traverse an arc whose peak is between the two highest lines marked on the wall (arcing higher than 6 feet, but not above 12 feet) and to hit the wall between the 1-foot mark and the base of the wall. Hitting the base of the wall in this drill is just like having the pitch come down just behind the plate.

Stand on the 46-foot floor mark with your throwing-side foot. Pitch the ball 10 times using the techniques described in Figure 8.1; remember to follow all three phases of the Keys to Success.

Success Goal = 7 out of 10 on-target pitches

Your Score = (#) ____ on-target pitches

2. Fence Drill

With a partner, position yourselves across from one another on opposite sides of the outfield fence. Each person stands about 23 feet from the fence. The distance between you (46 feet) is the regulation slow pitch pitching distance.

Pitch the ball back and forth over the fence, using it as a guide to work on the correct arc of the pitch. Pitch at least 20 balls trying to get 10 consecutive pitches that clear the fence and go to your partner. Your partner should not have to move to catch the pitch. Your partner will pitch the balls back to you. If you get 10 consecutive pitches before you reach 20 attempts, keep going and see how many times in a row you can pitch a successful pitch.

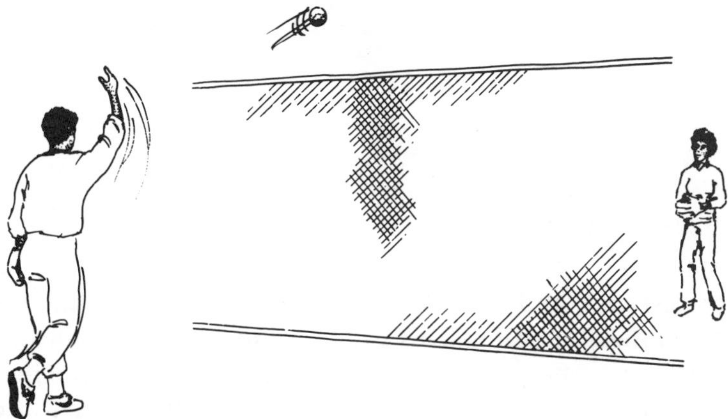

Success Goal = 10 consecutive pitches clearing the fence and going to your partner

Your Score = (#) ____ pitches

3. Bucket Drill

Place an empty bucket, milk crate, or similar container 46 feet away from your pitching rubber. Using correct technique, pitch the ball into the bucket. If you are working alone, use 10 or more balls so that you do not have to retrieve each pitch. If you are working with a partner, stand 46 feet apart, each with a bucket, and pitch one ball back and forth.

Pitch 10 pitches. Score 2 points if a pitch goes into the bucket (even if it pops out!) and 1 point if a ball hits the bucket on the fly. A pitch must traverse the regulation arc in order for the points to count. Partners should judge for one another. (If you are working without a partner, you can do this drill over the outfield fence to help you judge the regulation height of the pitch.)

Success Goal = 14 points on 10 pitches

Your Score = (#) ____ points

4. *Ball and Strike Drill*

After practicing form, distance, and accuracy of the slow-pitch pitch, you are ready to practice pitching in a more gamelike setting. This drill combines pitching with the hitter's tracking the ball (watching it from the pitcher's release until it is in the catcher's glove).

Set up in groups of three—one pitcher, one catcher, and one batter. In this drill, the pitcher and the batter score points, and the catcher is the drill facilitator. The pitcher is practicing the pitching technique, while the batter is practicing watching the pitched ball and making judgments about its position as it comes into the hitting contact zone.

As the batter stands in position at home plate, the pitcher tries to pitch strikes to the catcher crouching behind the plate. The batter doesn't swing, but calls the pitches balls or strikes, according to the official strike zone (the space over home plate between the batter's highest shoulder and knees). The catcher must verify each call. Any pitch in which the catcher disagrees with the hitter is judged a ''no pitch'' and does not count as either a ball or a strike. Two calls resulting in ''no pitch'' judgments during a single turn at bat, though, puts the batter out, resulting in 1 point for the pitcher. (One turn at bat consists of the batter's taking either 4 balls or 3 strikes.)

Scoring: 1 point for the pitcher for every strikeout (3 strikes before 4 balls on 6 or fewer consecutive pitches); 1 point for the batter for every base on balls (4 balls before 3 strikes on 6 or fewer consecutive pitches).

Rotate roles (pitcher to catcher, catcher to batter, batter to pitcher) after 3 points are scored by one person.

Success Goal = score 6 points before your opponent

Your Score =

(#) ____ points as batter

(#) ____ points as pitcher

Pitching
Keys to Success Checklist

When checking your pitching technique, the observer should watch from your pitching arm side *and* from directly behind you. The side position is best for checking your preparation position, release point, and coordination of your arm swing and your step. The rear position is best for seeing the direction of your step, the path of your arm swing, and your fielding ready position. Ask your teacher or another trained observer to evaluate your pitching technique according to the checklist below.

Preparation Phase

The pitcher's

____ feet are in staggered stride, with throwing-side foot (pivot foot) on pitching rubber.

____ weight is on back foot.

____ pitching hand holds the ball waist-high in front of the body.

____ grip is with three fingers on a seam under the ball and the thumb on a seam on top.

____ head is up, with eyes focused on the target (the ground just to the rear of the plate).

____ body is facing the batter for a full second pause.

Execution Phase

The pitcher

____ starts moving weight forward as the pitching hand starts to move down and to the rear.

____ takes weight on pivot foot and starts step with the glove-side foot in direction of batter.

____ steps onto the glove-side foot.

____ extends the pitching arm to the rear, to a horizontal position.

____ swings the pitching arm forward, keeping fingers behind and under the ball.

____ releases the ball past the hip so that it traverses a 6- to 12-foot arc.

Follow-Through Phase

The pitcher

____ continues pendulum arm action, bringing the pitching hand above head level.

____ steps with the pivot foot so that the feet are in a square stance shoulder width apart.

____ bends knees, placing body in slight crouch.

____ holds hands at waist level in ready position to field a ball hit back.

____ keeps head up, with eyes focused on the ball coming off the bat.

Step 9 Baserunning

Now that you have experienced a combination of offensive and defensive skills, let's go on to another individual offensive skill—baserunning. Once you hit the ball, you must run to first base. *Batter-baserunner* is the official rule book term to identify the batter who, one way or another, has finished a turn at bat and is moving from home plate to first base. By definition, the term *batter* refers only to the person *at bat* and does not describe that person running to first base. Technically, then, as the batter, say you hit a ground ball to the shortstop. As you run to first base, you become the batter-baserunner until you either reach first base safely or are put out there. You become a *baserunner* when you have reached first base without having been put out.

WHY IS BASERUNNING IMPORTANT?

Baserunning is the only way you can advance around the bases and score a run. Your degree of baserunning skill often makes the difference between your being safe or out on a play. Thus, baserunning technique is essential to your team's offensive success.

Running is a fundamental locomotor skill you probably learned as a young child. However, *baserunning* is a skill specific to softball and baseball that involves more than simply having the raw physical running skills: you must have the ability to use good judgment in executing those skills.

The physical skill is the development of maximum speed in short distances as you run around a 60-foot square, traveling the shortest distance possible despite making a 90-degree turn at each base. When baserunning in fast pitch softball, you may leave a base the moment that the pitcher releases the pitch. This allows you to be off the base when the ball is hit, thus shortening your distance from the next base at the moment of the hit. In slow pitch softball, though, you may not leave a base until the ball is hit or crosses home plate; thus, you must run the full distance to the next base after the ball is hit.

You must also know *when* and how to run. Will you try to advance further, or will you return to the base you just rounded? Will you slide or stand up coming into a base? These and many other questions are all related to your ability to use good judgment when baserunning.

HOW TO OVERRUN FIRST BASE

When you have hit a ground ball that is playable by an infielder, you must get out of the batter's box as quickly as possible and run at top speed *over* first base. Remember that first base and home plate are the two bases you may overrun. (The *batter's box* is the marked area at home plate in which the batter must stand when making contact with the ball. You are out if you step on home plate as you contact the ball with the bat; you are out not for stepping on home plate, but because you were out of the batter's box when you hit the ball.)

After hitting the ball, take your first step out of the batter's box with your back foot. You begin with your back foot because your swing follow-through puts your weight onto your front foot. Your run to first base *must* take you into foul territory (outside the baseline) before you reach the alley marking, which starts halfway between home and first. Otherwise, if you are running from home plate to first base in fair territory and are hit by a thrown ball, you are out.

As a left-handed batter, your first step with your back foot should be very close to being in foul territory because most of the left-handed batter's box is in foul territory on the first-base side. You would step into fair territory only if batting in the very front of the box, in which case you need to get to the foul side of the baseline as soon as possible. If you are a right-handed batter, you must make a conscious effort to start your movement to first base toward the foul line and not run directly at the base from the point at which you complete your swing follow-through.

Once you are on the foul side of the first-base foul line, run directly over the first-base bag, making contact with the front corner that is on the foul line. Overrun the base at full speed, traveling in a straight line (follow the foul line). Then start to slow down by bending your knees, taking short steps, and leaning back until you can easily come to a stop. Turn to your left, toward fair territory. If you have been called safe, return *directly* to the base (see Figure 9.1). If, as you turn toward the field of play, you see that the ball has been misplayed, make a judgment about continuing toward second base. *Remember,* once you make an *attempt* to go to second base, you are liable to be tagged out—you are no longer allowed to return to first base freely.

Figure 9.1 Keys to Success: ***Overrunning First Base***

Preparation Phase

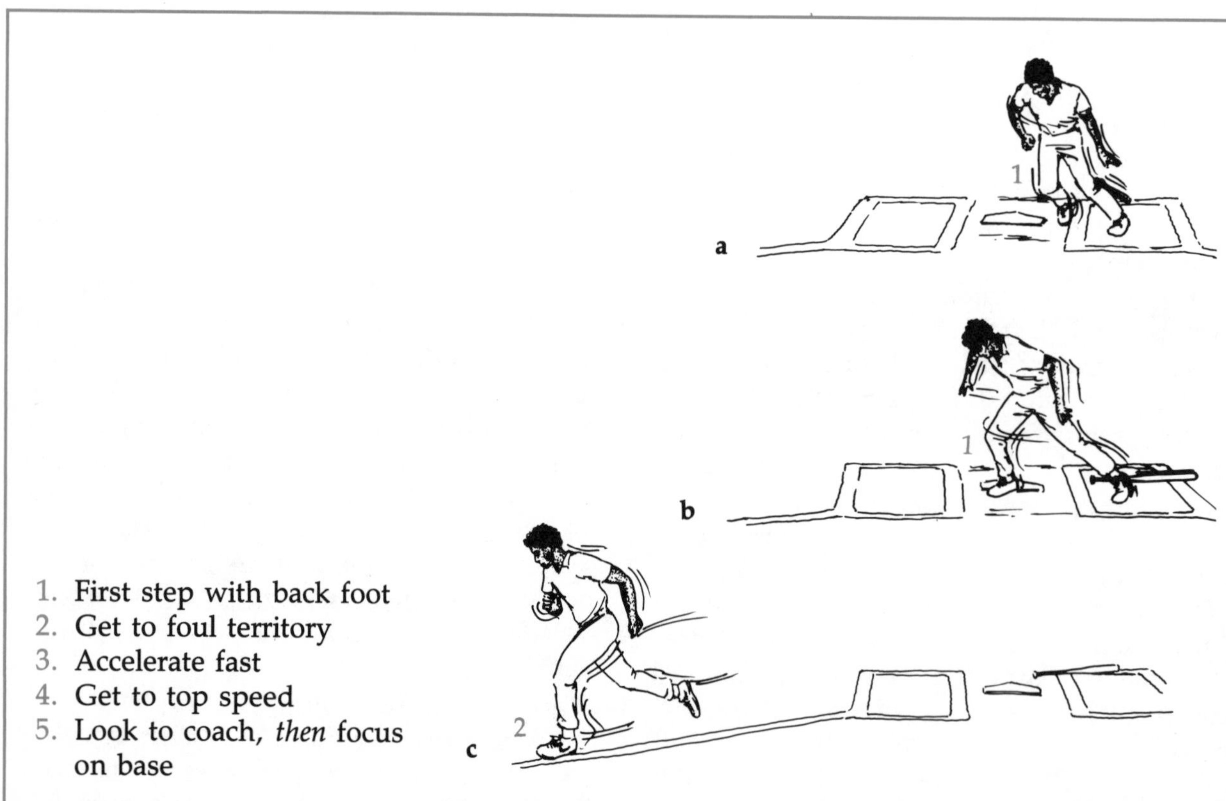

1. First step with back foot
2. Get to foul territory
3. Accelerate fast
4. Get to top speed
5. Look to coach, *then* focus on base

Execution Phase

1. Run on foul ground
2. Continue running at top speed
3. Don't break stride
4. Focus on base
5. Contact front corner of base

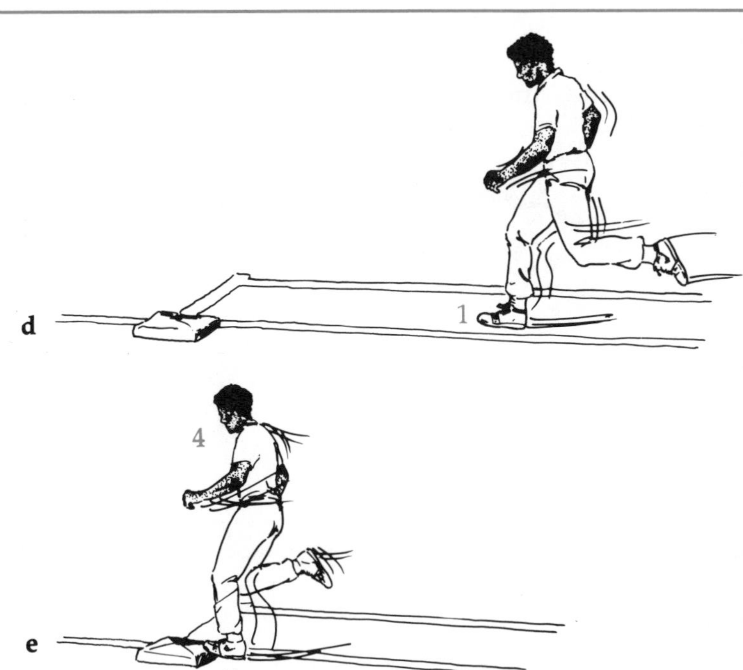

Follow-Through Phase

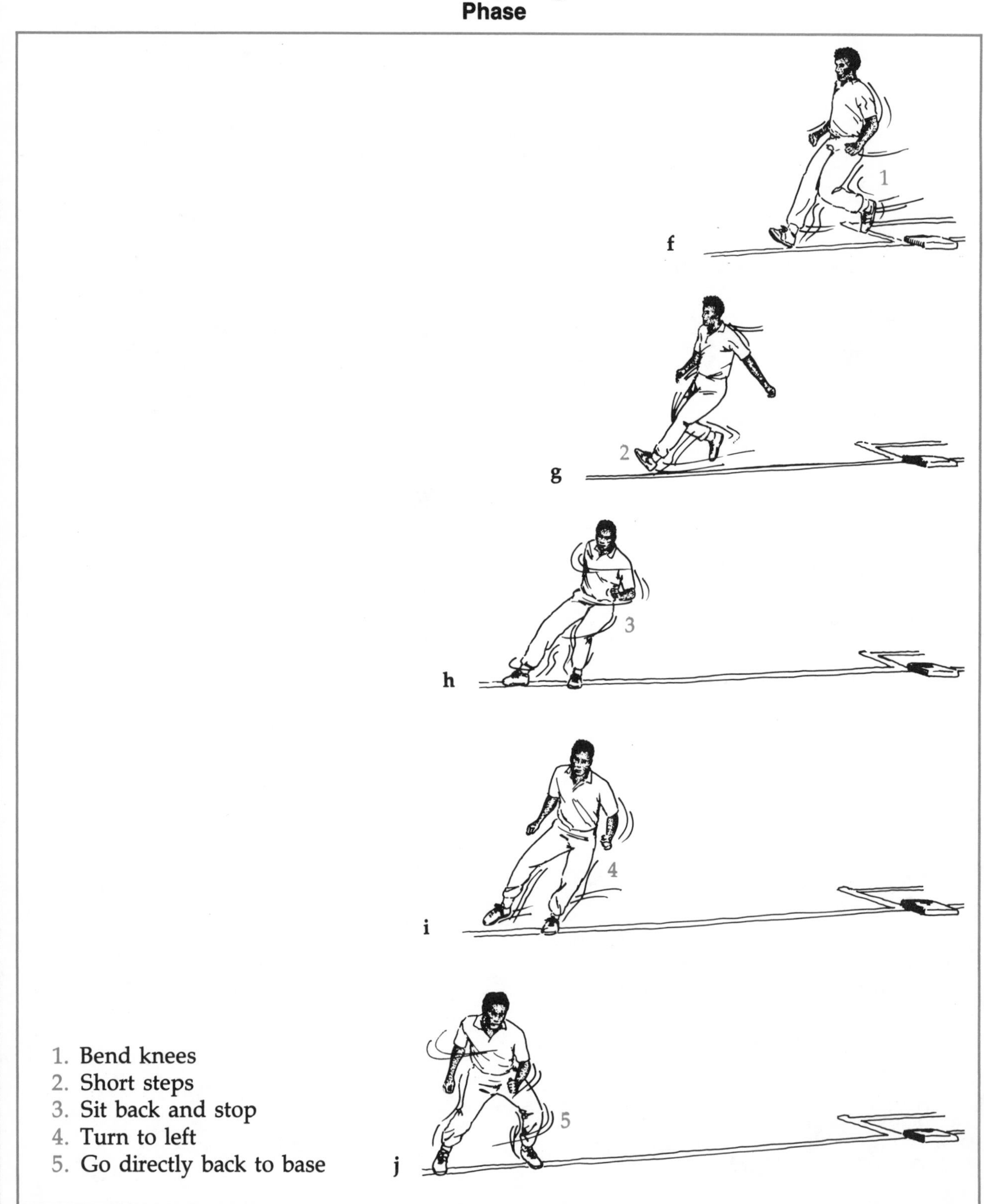

Detecting Errors in Overrunning First Base

Errors in the skill of overrunning first base usually occur at four different stages: leaving the batter's box, getting into foul territory, the speed of overrunning the base, and the turn to return to the base. Leaving the box with the back foot's taking the first step makes acceleration easier. Moving into foul territory as soon as possible eliminates the chance of your being called out for being hit by a thrown ball while in fair territory. Running at top speed over the base, slowing down, and turning to the left to return to the base makes it possible for you to see the field of play and determine whether you have the opportunity to advance another base because of a defensive misplay.

The most common errors in overrunning first base are listed below, along with suggestions on how to correct them.

ERROR	CORRECTION
1. You start to run out of the batter's box using your front foot.	1. At the end of your swing's follow-through, be sure your full weight is on your front foot, so your rear foot is free to move.
2. You run all the way to first in fair territory.	2. Get over into foul territory as soon as possible.
3. You leap for the bag or take short and long strides just before reaching first base.	3. Maintain stride and contact the front outside (foul line) corner of the base with either foot.

HOW TO ROUND A BASE

You round a base when you think you might be able to advance beyond it. When you are about 15 feet from the base, arc your path to it by swinging out to the right about three strides, then heading back in toward the base. The pattern you run looks like the outside edge of a spoon. As you cross the base, lean toward the infield and make contact with the *inside* corner of the base with your left foot (see Figures 9.2a-f).

If advancing to the next base, continue in a direct line toward that base. If you decide you cannot safely advance to the next base, take only a few steps past the base you just rounded, bend your knees, shift your weight back, and come to a stop. Retreat to the base by walking backward to it, keeping your eyes focused on the person with the ball (see Figures 9.2g-i).

When rounding a base in a game, you cannot return freely to it; instead, you are liable to be tagged out. If the ball is thrown to the person covering the base you have rounded, return to the base in a manner that best avoids the tag. Get back to the base before you are tagged; otherwise, you will be out.

Figure 9.2 Keys to Success: Rounding a Base

Preparation Phase

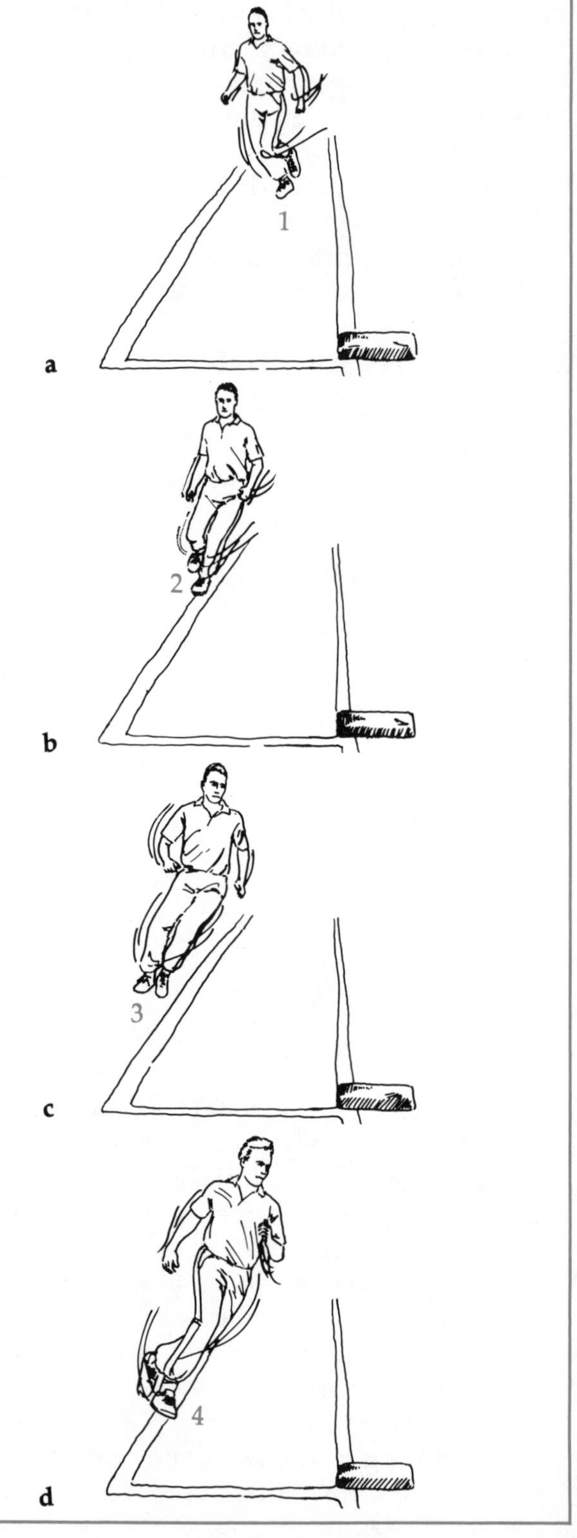

1. Run in foul territory
2. Swing out to right
3. Small arc
4. Head to base

Execution Phase

1. Lean to left
2. Contact inside corner of base
3. Contact with left foot

4. Focus on play
5. Decide whether to advance

6. If advancing, continue to run

Follow-Through Phase

g

h

i

1. If not advancing, stop
2. Bend knees, sit back
3. Push off front foot
4. Shift weight back
5. Pivot to left
6. Return to base

Detecting Errors in Rounding a Base

The primary cause of error when rounding a base is failing to swing out to the right in a small arc as you approach the base. Even though the shortest distance between two points is a straight line, it is impossible to run at full speed from home to second, making a right angle turn at first base. If you do not run in a small arc before the base, you will have to make a large, wide arc after crossing the base. That takes more time. The most common baserunning errors when using the rounding the base technique are listed below, along with suggestions on how to correct them.

ERROR	CORRECTION
1. You end up out in the right field grass when making the turn at first base.	1. You started your rounding turn too late, probably *at* the base. Start your turn 15 feet before the base.
2. You slow down at the base so you can make the turn.	2. Be sure to swing out to the right 15 feet before the base and head to the base at full speed.

HOW TO ADVANCE WHEN ON BASE

Once you have gotten on base and are stopped, you want to increase your chances of attaining the next base on a hit by a teammate. The faster you can get to the next base, the more likely it is that you will beat a throw and arrive safely.

When standing on first, second, or third base with a teammate up to bat, take a position with your left foot in contact with the center of the base (see Figure 9.3). Your right foot is either in a forward stride position (ahead of your left foot) or just behind your left foot and next to the base; you may select either foot position. In either case, turn your

Figure 9.3 Initial baserunning posture for rapid acceleration.

head to watch the batter so that you can see the ball as it goes over home plate or is hit.

At the moment the ball is hit, you leave the base by pushing off with the foot that is in contact with the base, taking your first step with your rear foot. Run directly toward the next base. Either stop at that base or round it, depending upon the circumstances.

Another situation affects the manner in which you leave a base. There is a rule in softball that applies when there is a baserunner on base with less than two outs, and a fly ball is hit. You, the baserunner, must be in contact with the base when the fly ball is caught or you are liable to be put out by a throw to the base. You may leave the base on the fly ball, but you must return to the base if the fly ball is caught. If you go too far from the base on a fly ball, the fielder who catches the fly will throw it to the base you left on the hit. Then, if the ball is in the hands of the fielder covering the base, and he or she tags you or even merely steps on the base before you get back to the base, you are out.

When leaving a base on a ball that is hit into the air, you must decide how far from the base you can venture. If you are on first base and a fly ball is hit to right field, you should go no more than two or three steps off base (8 to 10 feet). This is because the right fielder or the short fielder has only a short throw to put you out at the base if the ball is caught. On the other hand, if a fly ball is hit to left field when you are on first base, you can go nearly half the distance to second base. Then, if the ball is dropped, you will be that much closer to second base; if the ball is caught, you should be able to beat the throw back to first base.

When you are on second or third base, the same principles apply. If a fly ball is hit to an outfield position close to the base you are on, you must not leave the base and venture too far toward the next base. On the other hand, if the fly ball is hit to an outfield position far away from the base you are on, you can leave the base and venture farther toward the next base.

You must decide how far you can progress toward the next base on a fly ball and still be able to get back to your original base before the throw following a catch arrives. If the fly ball is not caught or if any ground ball is hit, your job is to run at maximum speed to the next base and either stop or round it and go to the next base. This is the judgment and decision making of baserunning—a most challenging and fun part of softball offense.

TAGGING UP AND ADVANCING ON A CAUGHT FLY BALL

Another important baserunning tactic is tagging up and advancing to the next base after the ball is caught. This opportunity usually occurs when the ball is hit deep into the outfield. As the baserunner on second base, you immediately return to second base on a deep fly ball that can be caught by the right fielder. Once the right fielder touches the ball, you may break contact with second base and advance to third base. Make sure you don't leave second *before* the fielder touches the ball. The play on you is a tag play, so you must beat the throw to the base and avoid the tag. The same tagging-up strategy applies when you are a runner on third base and a fly is hit to any deep outfield position. In such a case, you would not only be advancing a base, you would be scoring a run.

Selected Baserunning Drills

1. *Leave the Batter's Box Drill*

With a bat, assume your regular batting stance in the batter's box. Swing at 10 imaginary pitches using the technique described in Figure 5.2 for hitting. Emphasize the end of the follow-through. Make sure your weight is on your front foot, the bat is in your glove hand, your front knee is bent, and your body is leaning toward first base. Drive out of the batter's box, taking the first step with your back foot. Drop, do *not* throw, your bat on the ground to the glove side of your body as you take your second or third running step. Take several full strides toward first base (until you get into foul territory and have taken two more strides down the line for a right-handed batter, four for a left-hander).

Success Goal = 5 consecutive sequences using correct technique in the swing, in leading out of the box with the rear foot, and in getting into foul territory within 10 feet of home plate

Your Score = (#) ____ correct sequences

2. *Over-the-Base Drill*

Initiate this drill exactly as you did in the previous drill. Swing at imaginary pitches and drive out of the batter's box with the rear foot. Instead of stopping after several strides, though, continue to run to first base. Do not slow down before you get to the base. After crossing first base, turn left and return to the base. *Neither turn out to the right after crossing first base nor return directly to home plate!* Execute the skill according to the techniques outlined in Figure 9.1.

If other runners are practicing with you, quickly leave first base after you have completed your correct return to the base. Get well out of the baseline so that you will be out of the way of the next runner.

Success Goal = 10 consecutive error-free sequences

Your Score = (#) ____ sequences

3. *T-Ball and Run Drill*

In a game you have to hit a real ball and run to first base. We will simulate this situation by having you hit the ball off the batting tee and run to first base.

Set up a batting tee at home plate. Hit a ground ball off the batting tee and run to first base. If you are working alone, have a bucket of 10 balls so that you can hit all 10 (running to first base with each hit), then collect them all.

If you are working with a partner, hit the ground ball to your partner. Run to first base using correct technique. Your partner fields the ball and puts it into the bucket. When you have hit 5 balls and run to first base 5 times, change roles with your partner. Repeat the drill until you have each hit and run out 15 grounders.

Success Goal = 10 out of 15 ground balls hit and correctly run out

Your Score = (#) ____ grounds balls hit and run out

4. Overrun Base With First Baseman Drill

You need to practice overrunning first base when the first baseman is there attempting to catch the thrown ball. This is when it is critical that you contact the front foul-line corner of the base as you overrun it. You must be sure that you do not go across the center of the bag, because you may collide with the person making the catch.

This drill has two parts. You need three people for the drill. You are the batter-baserunner; one partner is the fielder, and the other partner is the first baseman. The ball is hit from the tee. Place a bucket of 10 balls beside the tee. The fielder has an empty bucket in which to place fielded ground balls.

a. No-throw portion of the drill: You hit the ball to the fielder (SS) and run to first base. The first baseman stands with one foot on the second-base side of the bag and his or her body leaning toward the shortstop. The first baseman reaches as if there were a throw coming. The fielder fields the ball, but holds it. The fielder then watches to see whether you crossed the foul-line corner of the base.

   After you have completed 5 hits and runs to first base, rotate roles. Repeat the drill. Continue the drill until each of you has made 5 hits and runs to first base.

Success Goal = 4 out of 5 runs to first base with foot contact on foul-line corner of base

Your Score = (#) ____ correct corner contacts

b. Throw portion of drill: This time the fielder (still SS) throws the ball to the player at first base after fielding the ball. The bucket is moved to the infield side of the first baseman (be sure that it is out of the way of both the first baseman and the batter-baserunner).

   Hit a ground ball to the fielder. The hit should make the fielder move no more than two steps to field it. After you hit the ball from the tee, run to first base. The shortstop fields the ball and throws it overhand to the first baseman. You, the batter-baserunner, try to beat the throw to first base. The first baseman drops the ball into the bucket.

   After 5 hits and runs to first base, rotate positions. Repeat the drill until you have each had 15 turns to hit the ball and run to first base. Count the number of times you get to first base safely (your beating the throw there or the first baseman's not catching the ball).

Success Goal = 7 out of 15 times safe at first base

Your Score = (#) ____ times safe

5. *Safe or Out Drill*

Let's play a game using the drill we just did. Do everything the same as in the previous drill. Now, however, you keep points.

The batter must hit the ground ball off the tee in the range of the shortstop, making the shortstop move no more than two or three steps to the side or forward to field the ball (otherwise, the two defensive players score a point each). If the hitter makes it to first base safely on a good grounder, he or she gets 1 point. If the shortstop and the first baseman combine to get the batter-baserunner out (the first baseman having the ball in the glove while contacting first base before the runner contacts the base), each of them gets 1 point.

Do the hit-and-run sequence 5 times, then rotate positions. Continue the game until each player has hit the ball 15 times. Keep track of your points.

Success Goal = score more points than your partners

Your Score =

(#) ____ points as hitter

(#) ____ points as shortstop

(#) ____ points as first baseman

Who won? ____________________

6. *Swing-Out Drill*

To round a base, start running directly at a base. When 15 feet from the base, swing out to the right of the baseline and run in a semicircular path to the base. Cross the inside corner of the base, heading in the direction of the next base. Continue running for several strides past the base, then stop.

If practicing on a regulation softball field, start 30 feet from first base and round the base at full speed. Slow down 10 to 15 feet past the base. Walk or jog to a position 30 feet from second base and repeat the drill. Continue in this fashion around all the bases, including home. Pay attention to the size of the semicircular path you need to take to be able to cross the base heading in the general direction of the next base. Make sure you use the keys to success shown in Figure 9.2.

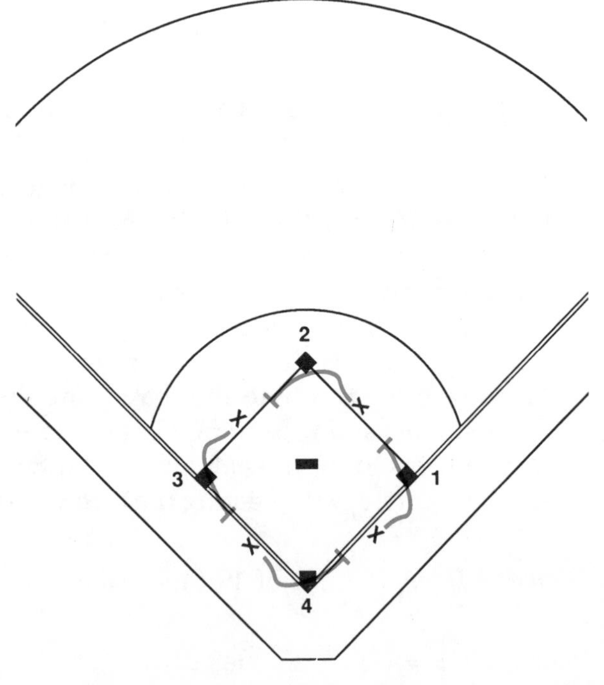

If practicing with just a single base set on the ground or floor, turn around when you have run 20 feet past the base and repeat the drill going to the same base. Swing out to the right of the base each time and cross the base from right to left. After 4 repetitions, you will be back to your original starting position.

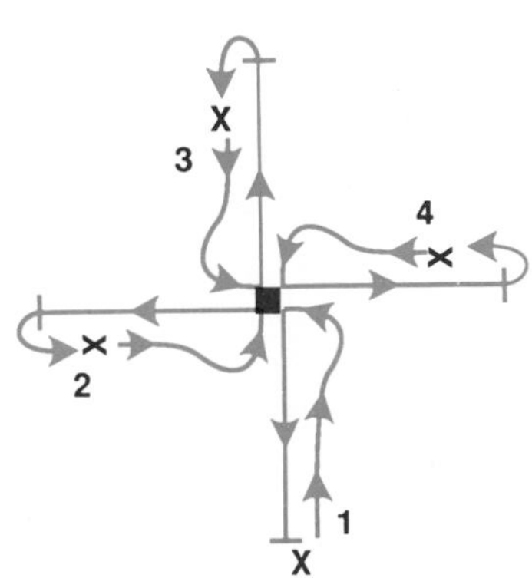

Success Goal = 8 total turning plays

Your Score = (#) ____ turning plays

7. Single, Double, Triple, Home Run Drill

You need a partner who will time you with a stopwatch for this drill. Start at home plate. Using a bat, swing at an imaginary pitch and, using the technique described in Figure 9.2, run out a single rounding first base. Retreat to first base with the correct technique. Return to home plate.

Next swing and run out a double, rounding first base and stopping at second base as you would in a game. You do this by running full speed to within 10 feet of the base, bending your knees, shifting your weight back, taking smaller steps to come to a stop on the base without overrunning it. Return to home plate. Continue the drill with a triple, rounding first and second base and stopping at third. Return to home plate.

Finish the drill sequence by running out an inside-the-park home run. When you hit the ball over the fence for a home run, you may simply jog around the bases, waving to the crowd. On an inside-the-park home run, though, the ball remains in play, and you must arrive at home plate before the defense gets the ball there to tag you out. Therefore, you must use good baserunning technique to beat the throw to the plate. Round all three bases, running at full speed all the way from home to home! Remember, you can overrun home plate so don't slow down to stop on the plate. Run full speed *over* home plate.

Success Goal = 12 total points based on the allocations below

Distance run	Time range (seconds)	Points
Home to first	More than 4.5	1
	4.0–4.5	2
	3.5–3.9	3
	Less than 3.5	4
Home to second	More than 11.0	1
	10.0–11	2
	9.0–9.9	3
	Less than 9.0	4

(Cont.)

Distance run	Time range (seconds)	Points
Home to third	More than 15	1
	14.0–15.0	2
	13.0–13.9	3
	Less than 13	4
Home to home	More than 20	1
	19.0–20.0	2
	18.0–18.9	3
	Less than 18	4

Your Score =

(#) ____ single points

(#) ____ double points

(#) ____ triple points

(#) ____ home run points

(#) ____ total points

8. React to Coach Drill

In a game situation, the batter-baserunner is assisted by a base coach in deciding whether to overrun or round first base. Usually the coach tells the batter-baserunner to round the base and look for the ball when the ball goes through the infield. The batter-baserunner must look at the coach before coming within 15 feet of the base, in order to receive the base coach's instructions in time to round the base, if necessary.

This drill is designed to help you develop your ability to react to the coach and the call he or she is making. You need a partner for this drill. You be the batter-baserunner, and your partner the coach. You swing at an imaginary pitch and run to first base. The base coach tells you either to overrun or to round the base, randomly calling either "round and look for the ball" or "overrun." You react to the base coach by doing what the call says.

You run to first base 5 times. Count the number of times you react correctly to the call of the coach. Switch roles.

Success Goal = 4 out of 5 correct actions at first base

Your Score = (#) ____ correct actions

9. Double to the Outfield

For this drill you need three people. You be the batter-baserunner. One person is an outfielder situated behind shortstop, and the other person is a second baseman situated to the outfield side of second base. You need a bucket of 5 balls and a tee at home plate, and an empty bucket near second base.

Hit 5 line drives off the tee into the outfield beyond shortstop. Run, round first base, and decide whether to run to second base. The outfielder fields the ball and throws it to the player at second base. As baserunner, try to make the correct decision: Either go back to first or get to second base before the ball is caught by the second baseman (no tag necessary in this drill). Count the number of times you make the correct decision by remaining at first base or by continuing and being safe at second.

After 5 hits, rotate roles. The hitter goes to the outfield, taking the empty bucket to second base on the way out. The outfielder goes to second base. The second base player goes to home, taking the bucket of 5 balls to hit.

Success Goal = 3 correct decisions

Your Score = (#) ____ correct decisions

10. Overrun or Round Game

Now we make the practice very similar to a game situation. For this drill six players are needed. Two players are hitters. One is the base coach. A fourth player is the shortstop, and the fifth player is the left fielder (behind the shortstop). The sixth player is the second baseman. At home plate there needs to be a bucket with 5 balls. Near second base there needs to be an empty bucket.

The batters alternate hitting 10 line drives off the batting tee, attempting to get them by the shortstop and into left field. Upon hitting the ball, the hitter runs to first base and reacts to the call of the base coach. If the ball goes by the shortstop, the base coach calls, ''Round the base and look for the ball!'' If the ball is fielded by the shortstop, the base coach calls, ''Overrun the base!'' The batter-baserunner does as the coach signals. If the call is to round the base, it is then up to the baserunner to decide whether to return to first base or to try to advance to second base.

The fielders attempt to field the ball as it comes to them. If the outfielder fields the ball, he or she throws it to the second baseman; the shortstop returns fielded balls to home.

After 5 hits by each hitter, rotate positions: one hitter to shortstop, shortstop to outfield, outfield to second base, second base to base coach, and base coach to hitter. Continue the drill until each player has hit the ball 5 times.

The goal of this game is to get as many doubles as you can. Count the number of times you make it safely to second base. See whether you can get more doubles than any of the other persons in your group.

Success Goal = 6 out of 10 hits are safe doubles

Your Score =

(#) ____ safe doubles

Who won? ________________

Baserunning Keys to Success Checklists

When you think you have acquired the appropriate techniques for overrunning first base and rounding a base, ask your teacher, your coach, or another trained observer to give you feedback on the quality of your baserunning techniques according to the checklists below.

Overrunning First Base

Preparation Phase

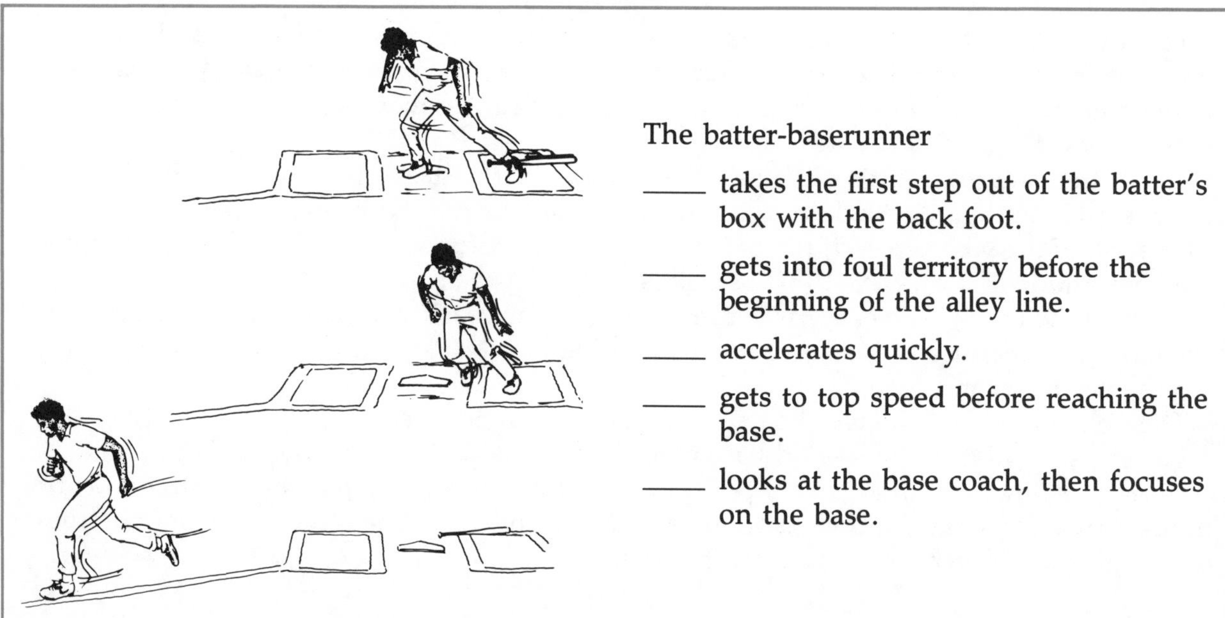

The batter-baserunner

____ takes the first step out of the batter's box with the back foot.

____ gets into foul territory before the beginning of the alley line.

____ accelerates quickly.

____ gets to top speed before reaching the base.

____ looks at the base coach, then focuses on the base.

Execution Phase

The batter-baserunner

____ runs in foul territory within the alley markings.

____ continues running at top speed until over the base.

____ doesn't break stride when contacting the base.

____ focuses on the base.

____ contacts the base's front corner that is on the foul line.

Follow-Through Phase

The batter-baserunner

____ bends the knees after crossing the base.

____ takes short steps to facilitate coming to a stop.

____ sits back and shifts the weight back to come to a stop.

____ turns to the left, toward the field of play, without making any attempt to go to second base.

____ goes directly back to the base.

Rounding a Base

Preparation Phase

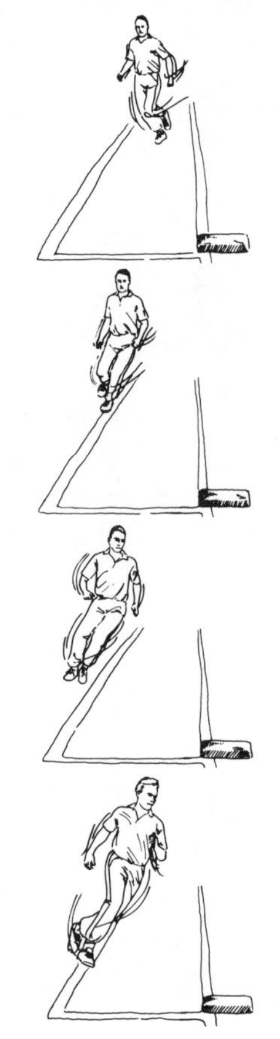

The batter-baserunner

____ runs in foul territory within the alley markings.

____ swings out to the right when within 15 feet of the base.

____ runs in a small arc.

____ leans to the left and heads toward the base.

Execution Phase

The batter-baserunner

____ leans to the left.

____ contacts the inside front corner of the base.

____ contacts the base with the left foot.

____ looks for the ball and focuses on the play.

____ decides whether to advance or return to the base.

Follow-Through Phase

The batter-baserunner

____ deciding not to advance, stops by bending the knees, sitting back, and shifting the weight back.

____ pushes off the front foot (which should be the right foot).

____ shifts the weight onto the rear foot (the left).

____ pivots to the left and returns to the base.

Step 10 Position Play

Different defensive players have different responsibilities, depending upon the positions they play and the situations to which they are called upon to react. When you play a position, you must read a developing situation and carry out the duties of your particular position.

The checklist at the end of this step includes all covering and backing-up responsibilities for each position. Only the basic concepts and some examples are described in the sections that follow. There are no drills in this step, but you will apply these concepts in all later steps.

WHY IS KNOWLEDGE OF POSITION PLAY IMPORTANT?

It is important to know the duties of all the positions so that you can interact most usefully with your teammates no matter where the ball is hit or thrown. All team members should work together smoothly, rather than be confused about responsibilities.

Good position play requires an understanding of some basic concepts, which are applied to the individual game and play situations as they arise. The two major categories of defensive duties are called *covering* and *backing up*.

COVERAGE RESPONSIBILITIES

Each of the 10 positions in slow pitch softball has a specific name, number, and coverage area. In Diagram 10.1 the sections outlined by dotted lines identify the *primary area coverage* for each starting position. The shaded areas that overlap the dotted lines identify the *interaction areas,* areas of coverage responsibility shared by adjoining positions. The diagram shows flexible approximations of these coverage areas, which may vary depending on the relative range and other skills of teammates.

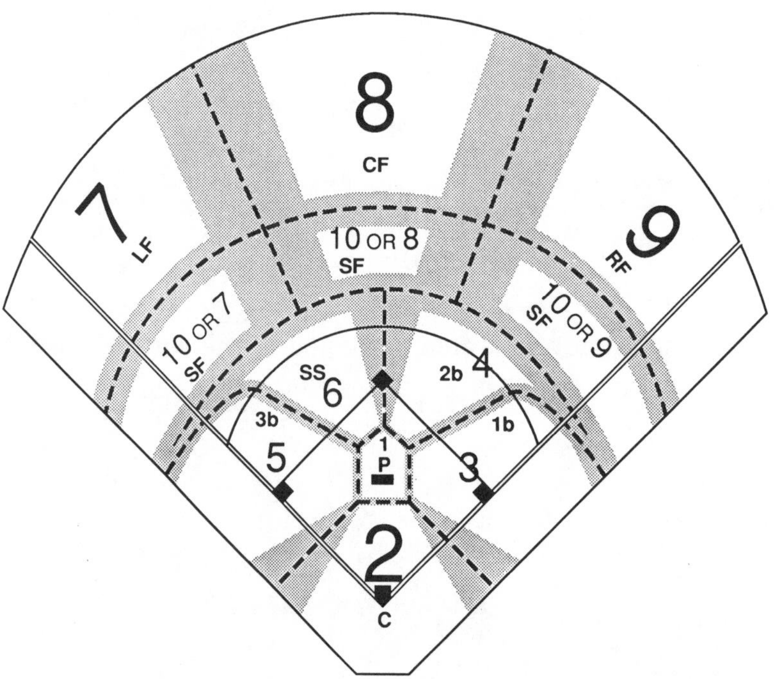

Diagram 10.1 Regular-depth starting positions and area coverage.

The term *covering* also describes the responsibility of an infielder at a base (see Diagram 10.2). For instance, on a ground ball hit to the shortstop, the first baseman covers first base to take the shortstop's throw and put out the batter-baserunner. The assigned primary coverage in Area 6, by the way, enabled the shortstop to field the ground ball without interference from another defensive player.

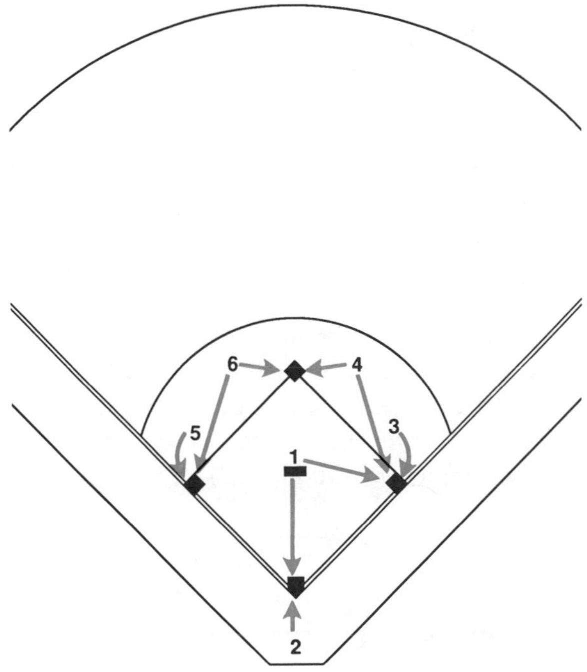

Diagram 10.2 Base coverage responsibilities.

PRIORITY SYSTEM FOR FLY BALL JURISDICTION

Any ball coming into a primary coverage area is mainly the responsibility of the defender playing the corresponding position (Diagram 10.1). A team's initial defense is based on the concept that each player has *jurisdiction* in his or her primary coverage area. Take, for example, a high pop fly hit to Area 6: The third baseman, the shortstop, the second baseman, and the pitcher would all have time to move into position to field the ball. If they all did at once, however, there would be chaos. If everyone understands that Area 6 is the responsibility of the shortstop, though, this assigned jurisdiction allows the play to be made with minimal confusion.

Verbal signs are extremely helpful for team defense. You should always *call for the ball* on a fly ball and even sometimes for ground balls playable by more than one fielder. Your call of "I have it!" or "Mine!" must be loud and clear enough for all the players in the immediate area to hear. In fact, it is helpful to call for the ball even when it is clearly in your area of responsibility. To indicate that they understood your call, other players should then call "Take it!" or call out your name.

Determining who should play a fly ball between two areas of responsibility is more of a problem, but it becomes less difficult when a priority system is established. In general, outfielders have priority over infielders. Because they are moving in on the ball, outfielders have an easier time fielding and throwing than infielders, who are running back to the ball. When the ball is between two or more outfield positions, the fielder who would be in the best throwing position after the catch has priority. When a follow-up throw is needed, the outfielder with the strongest throwing arm has priority.

In the infield, the third baseman should cut off any ground ball he or she can reach while going to the left. Because it is easier to run laterally than backward, the shortstop should field any pop-up behind the third baseman. Similarly, the second baseman should field any pop-up behind the first baseman.

Following is the priority system for most occurrences of calling for fly balls:

Fly Ball Jurisdiction Chart

Position no.	Symbol	Player position	Has fielding priority over . . .
1	P	Pitcher	No one
2	C	Catcher	Pitcher
3	1b	First baseman	Catcher, pitcher
4	2b	Second baseman	First baseman, pitcher
5	3b	Third baseman	First baseman, catcher, pitcher
6	SS	Shortstop	Third baseman, second baseman, first baseman, and pitcher
7	LF	Left fielder	All infielders, short fielder
8	CF	Center fielder	All infielders, all other outfielders
9	RF	Right fielder	All infielders, short fielder
10	SF	Short fielder	All infielders

BACKING-UP RESPONSIBILITIES

Backing up describes support or aid given to a covering player by another defensive player. The backing-up player does not make the initial play on a runner or a hit ball. For example, the short fielder or the left fielder backs up the shortstop fielding the ground ball. The catcher backs up the first baseman; the catcher's stopping a misplayed ball that has gone beyond the first baseman could prevent the runner from advancing to another base.

Backing up a play requires that you

- *know* all the possible backing-up responsibilities for the specific position you are playing,
- immediately *recognize* situations for which you have backing-up responsibilities when they begin to occur, and
- *move* into the correct backing-up positions.

Thus, backing up is two-thirds cognitive and one-third physical! Descriptions of backing-up responsibilities for each position in this step, and game opportunity experience in later steps, will help you learn this aspect of position play. Good anticipation will help you recognize situations as they unfold.

Diagram 10.3 shows the backing-up positions each fielder would likely take. An actual backing-up position is determined by the path of the ball. The backup person must get in direct line with the source of the ball (the throwing fielder or the hitter) and its receiver

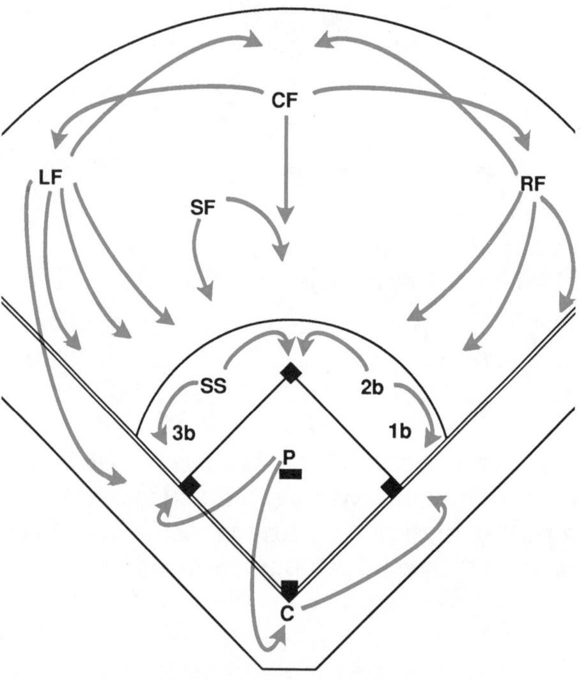

Diagram 10.3 Backing-up responsibilities.

(the fielder covering a base or the fielder making a play on the hit ball) (see Diagram 10.4). The backup player must assume the position approximately 15 to 20 feet *behind* the primary receiver. Remember, your role as the backup is to catch a misplayed ball to prevent additional advance by any baserunner; you are not trying to make the initial covering play. If you back up a play by standing too close to the covering fielder, a misplayed ball will go past you because you will not have time to react and catch the ball.

Once you catch the ball as the backup, listen for verbal assistance from your teammates as

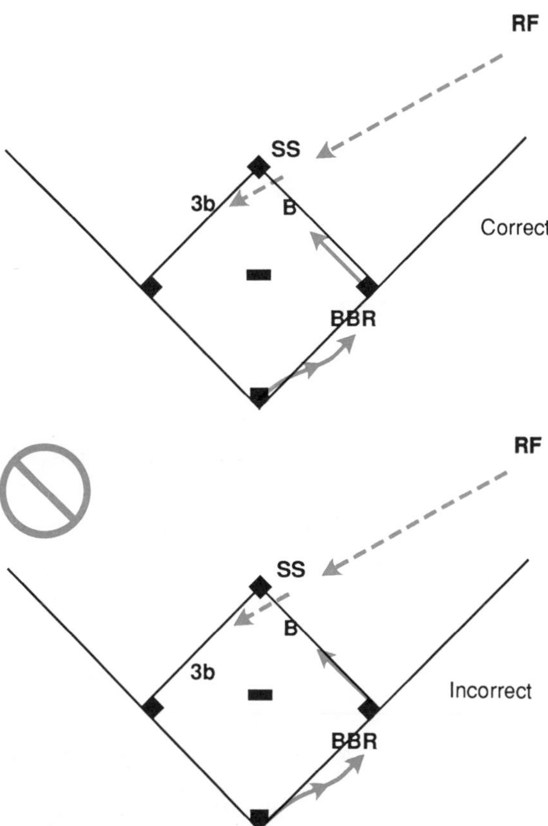

Diagram 10.4 Backing up on the overthrow line.

to what to do with the ball; also look to see what the runners, especially the lead runner, are doing. Sometimes a runner seeing an overthrow or other error will automatically dash for the next base. That is when your backing up a play really pays off, because your team then gets a second chance on the same play to get the player out.

SPECIAL SITUATIONS

Special situations may alter the starting positions and the size of the coverage areas. For example, in playing a strong right-handed pull hitter (whose swing pulls, or hits, the ball to the left field), all the infielders and outfielders would be shifted around toward the left field foul line, as shown in Diagram 10.5 (compare

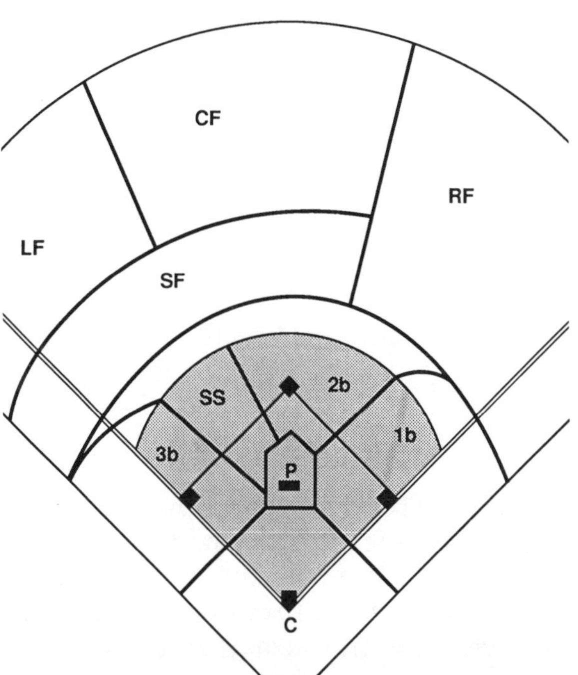

Diagram 10.5 Overshift positioning for right-handed pull hitter.

to Diagram 10.1). The areas covered by the left fielder and the third baseman become smaller. The center fielder's area remains about the same size but moves to the left; this increases the right fielder's area coverage. The second baseman's moving closer to second base increases the area coverage for the first baseman.

Another special situation comes with the bases loaded (baserunners on first, second, and third base) and less than two outs. The infielders move in a little closer to home plate (compare Diagram 10.6 to Diagram 10.1) before the ball is pitched. If a ground ball is then hit to an infielder the force play at home plate (for less experienced players) or the home-to-first double play (for those more experienced) has a greater likelihood of success. The shortened distance the ball has to travel to reach the fielder and the resulting shortened distance of the throw to home plate make a successful play more possible than if the fielders had stayed back in their regular-depth starting positions.

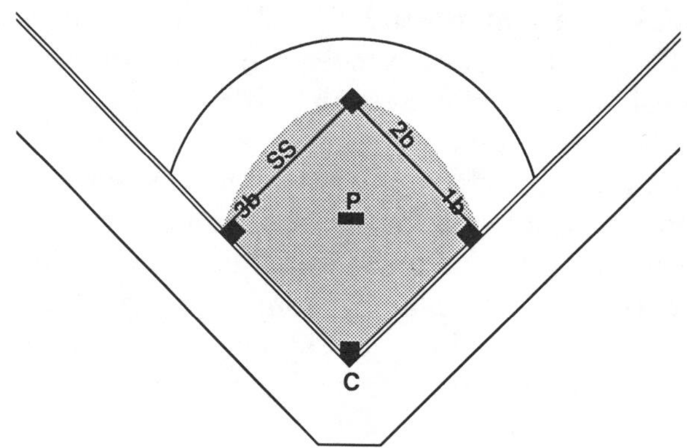

Diagram 10.6 Infield "in" positioning.

Position Play Keys to Success Checklist

Every player has individual responsibilities on every play. Defensive position play depends upon the offensive situation and the duties of the player's position. To be a successful defensive softball player, you must think about your responsibilities before each pitch.

You can also certainly become a better offensive player by knowing the defensive position play concepts. If you thoroughly understand defensive positioning and likely responses, you can make offensive moves that attack the inherent weaknesses of any defensive setup.

Use this checklist to help yourself remember what you do when playing each position. Then ask a trained observer to use the checklist to evaluate your position play during game situation drills, modified games, or official games.

Position and number	Covering responsibilities	Backing-up responsibilities
Pitcher (1)	____ Area 1 ____ First base on hit to 1b	____ Home plate on outfield (OF) throws ____ Second base on OF throws ____ Third base on OF throws
Catcher (2)	____ Area 2 ____ Home plate	____ First base on infield (IF) throws
First baseman (3)	____ Area 3 ____ First base ____ Cutoff throws from CF, RF, SF to home	____ Second base on LF throws

(Cont.)

Position and number	Covering responsibilities	Backing-up responsibilities
Second baseman (4)	____ Area 4 ____ First base when 1b out of position ____ Second base on double plays (DP) and force plays from 3b, SS ____ Second base on LF, CF, SF throws	____ Balls hit to 1b ____ Second base on P, C throws to SS
Third baseman (5)	____ Area 5 ____ Third base	____ Second base on RF throws
Shortstop (6)	____ Area 6 ____ Second base on DP and force plays from P, C, 1b, 2b ____ Second base on RF, SF throws	____ Second base on P, C throws to 2b ____ Balls hit to 3b ____ Balls hit to P
Left fielder (7)	____ Area 7	____ Balls hit to CF, SF ____ Balls hit to 3b and SS, if SF not available ____ Second base on 1b, 2b, RF, SF throws ____ Third base on C, 1b, 2b, RF, SF throws
Center fielder (8)	____ Area 8	____ Balls hit to LF, RF, SF ____ Balls hit to SS and 2b, if SF not available ____ Second base on P, C, 1b throws
Right fielder (9)	____ Area 9	____ Balls hit to CF, SF ____ Balls hit to 1b and 2b, if SF not available ____ First base on C throws ____ Second base on 3b, SS, LF, SF throws
Short fielder (10)	____ Area 10	____ Balls hit to 1b, 2b, 3b, and SS when SF is in nearby part of Area 10 ____ Second base when in overthrow line

Step 11 Force Play and Tag Play

An offensive player can be put out by the defense in a number of ways. Many of these plays fall into two categories: force plays and tag plays. A *force play* occurs whenever a baserunner must go to the next base because of the batter's becoming a baserunner. The batter must *always* go to first base after hitting the ball. Only one baserunner may be on a base at one time; therefore, when the batter goes to first base on a grounder, the runner already on first base is forced to go to second base.

The runner is put out in a force play situation by the defense's getting the ball to the base ahead of the runner. The ball itself need not actually come into contact with the base for a force-out, but the defensive player must have control of the ball and make contact with the base with some part of the body before the baserunner arrives (see Figure 11.1).

In beginning-level game play, the most basic defensive strategy is to get one out at a time. When there are less than two outs and runners are on base in a force play situation, the fundamental defensive strategy is to ''get'' the lead runner (the runner closest to home plate). The simple force play is *the* fundamental defensive concept in softball. With two outs, on the other hand, the play for the third out to end the inning is normally made at first base because that is always a force-out situation.

The *tag play* is another fundamental defensive concept. A tag play situation occurs any time a runner is not in contact with a base and is not allowed to move to any base freely. For example, a runner overrunning second base is not free to return to the base; conversely, a runner *is* free to return to the base after a foul ground ball. To put a baserunner out with a tag play, the defensive player must tag, or touch, the offensive player with the ball or with the glove holding the ball, when the runner is off base (see Figure 11.2).

Figure 11.1 Attempting to make a force-out.

Figure 11.2 Executing a tag play.

WHY IS KNOWING THE FORCE PLAY AND THE TAG PLAY IMPORTANT?

Infielders are the defensive players who normally execute force plays and tag plays because these plays occur on the infield (at or between the bases). These plays are the basic skills for getting baserunners out, preventing them from advancing and scoring runs. It is essential that you know the tag play and force play concepts and be able to execute the skills to be a successful defensive player and help your team stop the offensive play of your opponents. This is especially important if you would like to concentrate on playing an infield position.

HOW TO EXECUTE THE FORCE PLAY

The force play can be executed at all of the three bases and home plate. The basic techniques and the fundamental principles upon which the choice of techniques is based are the same regardless of the base. Make the throw for a force play about chest-high. As the defensive player covering the base, you should move to the side of the base nearest the source of the throw. This will shorten the throw's length and, thus, the time it takes for the throw to arrive in your glove. The quicker the throw arrives, the more likely it is that you will get the out.

For example, on a force play at third base with a throw coming from the catcher, the player covering the base moves to the home plate side of third base to receive the throw, as shown in Figures 11.3a-e. On the other hand, if the throw were coming to third base from the left fielder, the covering player would stand on the outfield side of third base to receive the throw (see Figures 11.4a-c).

Figure 11.3 Third base coverage on a force play, with a throw from the catcher.

Figure 11.4 Third base coverage on a force play with a throw from the left fielder.

Once the throw is on the way and you, the covering player, know exactly where the throw will arrive, place one foot on the base and stretch out to meet the throw with your glove hand and the other foot. If the throw is slightly off target to the side, step to meet the ball with the foot on the ball side, and contact the base with your other foot. For example, if the throw is off target to your left, step to the left with your left foot and contact the base with your right foot. If the play is going to be close, stretch as far as you can and catch the ball in your glove hand only. If the play is not going to be close, stretch a comfortable distance and catch the ball with both hands. Remember, you want to shorten the distance and time of the throw so the throw will get to you (and the base) before the runner does.

HOW TO EXECUTE THE TAG PLAY BETWEEN BASES

When you have the opportunity to tag out a player who is running past you between bases, you have a fairly easy play. The baserunner will be standing up, so you will have little trouble reaching him or her to make the tag. Hold the ball securely in both hands (the ball in your glove) to reduce the chance of dropping the ball. Tag the baserunner with the back of the fingers of your glove. Immediately pull both hands away so the contact with the runner does not knock the ball out of your glove. Hold on to the ball. (See Figures 11.5a-c.)

HOW TO EXECUTE THE TAG PLAY AT A BASE

First of all, remember that the tag play is required when a baserunner is not forced to go to a base. In slow pitch softball the tag play is used when a baserunner overruns second or third base, or is advancing on a hit ball and is not forced to do so. Three examples of situations that dictate that a tag play be executed when the baserunner goes into a base when not forced to:

1. With no runner on first base, a runner on second base tries to go to third base on a ground ball hit to the second baseman.
2. A runner attempts to score from second base on a base hit.
3. A runner tagging up on a fly ball tries to advance a base.

Figure 11.5 A tag play between bases.

In fast pitch softball, the tag play is also used when a baserunner attempts to steal a base.

The throw for the tag play should arrive just below the knees of the covering player. The runner will probably be sliding into the base, so the throw should be low and close to the runner. This minimizes the time it takes for the covering player to move the glove and ball into position to tag the baserunner.

There is more than one acceptable technique for covering the base for the tag play. The method recommended places you in a position at the base where you can tag the runner, but where your chances of getting knocked down are limited. As your skill increases, and if you then choose to become more aggressive in your play, you may wish to actually block the base from the runner with your body as you make the tag. For now, though, let's give the runner an open path to the base.

Your exact position at the base will depend on the path the runner is taking to the base, for instance, coming into third base from second base as opposed to coming back to third from the direction of home plate. The other factor that affects your exact position is the source and direction of the throw. Is the ball coming to you from the outfield, or from the infield side? In general, straddle the base or stand just to the side of the base facing the direction of the incoming runner. *Do not*, under any circumstance, place your leg between the base and the incoming runner! Leave the path to the base open to the runner.

Position yourself so that you can catch the ball and bring the gloved ball down to the edge of the base where the runner will arrive. As the runner slides in, let the runner tag him- or herself out by sliding into the ball held in your glove. Then sweep your glove out of the way of the runner. Even if the runner does not slide, the runner's foot must get to the base—*tag the foot*. Don't reach out to tag the runner on the chest only to find out that the feet slid into the base before you tagged the chest (see Figures 11.6a-c).

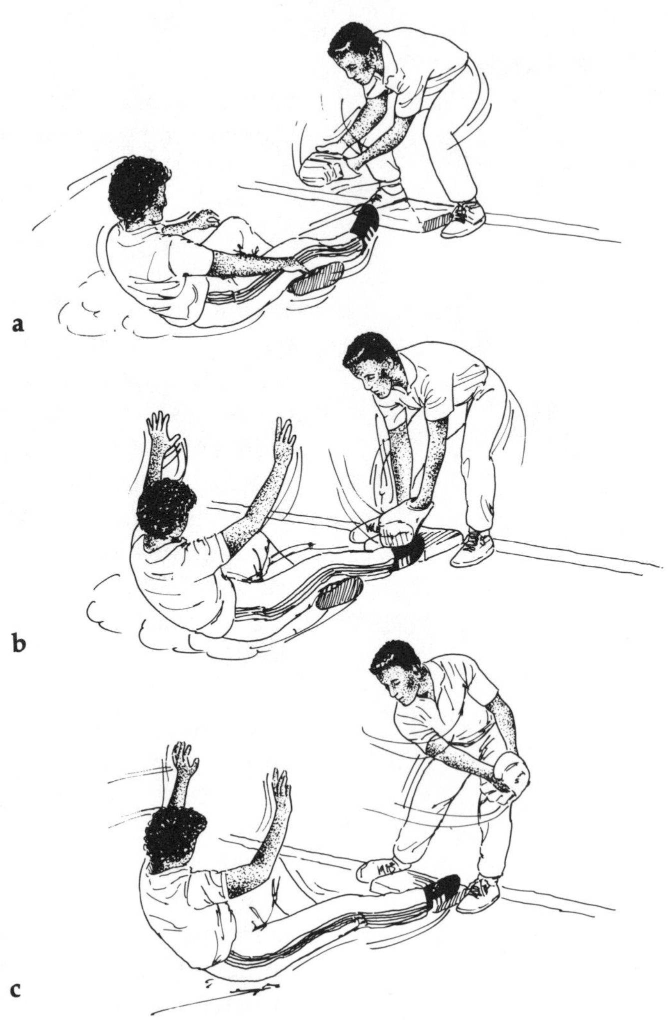

Figure 11.6 At a base, tag the foot and hold onto the ball!

Force Play and Tag Play Drills

1. Mimetic Footwork Drill

Proper footwork is required as you move from an infielder's starting position to the covering positions of force plays and tag plays. Place a loose base (home plate, if called for) on the ground or floor. Lines are helpful to use as foul lines to assist you in orienting your fielding position in relation to the base. If there are no lines available, place a cone about 15 feet from the base and use that cone to represent the direction to home plate; draw an imaginary foul line from the cone to your base. Remember that bases lie in fair territory, so the line (real or imaginary) goes on the left side of first base or the right side of third base.

When in the first baseman's role, position yourself slightly behind and about 8 feet to the right of the base. When in the third baseman's role, position yourself slightly behind and about 8 feet to the left of the base. When in the shortstop or second baseman's role, position yourself in regular fielding position. When in the catcher's role, position yourself about 2 feet behind the plate.

Without a ball, you will practice the footwork for force plays and tag plays at each base. The chart, "Force Play and Tag Play Practice Situations," lists the covering positions, the general direction, and the possible thrower that you will practice for force and tag plays.

Force Play and Tag Play Practice Situations

Covering player	General direction and possible thrower
1b (see Diagram 11.7)	RF, SF, 2b, SS, 3B, P, C
2b (see Diagram 11.8)	LF, SF, 3b, SS
SS (see Diagram 11.9)	RF, 2b, CF, SF, 1b, C, P
3b (see Diagram 11.10)	LF, SF, CF, SS, RF, 2b, 1b, P, C
C (see Diagram 11.11)	RF, 2b, 1b, CF, SF, P, LF, SS, 3b

Take your regular fielding position before each repetition in the drill, then move into coverage position at the base to receive an imaginary throw. First you move into position to make a force play. Mimic the stretch and the ball reception. Then practice the footwork needed on throws that are slightly off target to the left and to the right. Remember, on throws to your left, step with your left foot toward the throw and contact the base with your right foot; on throws to your right, step right and contact left.

Finally, do a series of repetitions in which you move to the base to make tag plays on imaginary throws from all positions listed on the chart. Mimic the actual tagging movements in each of the situations.

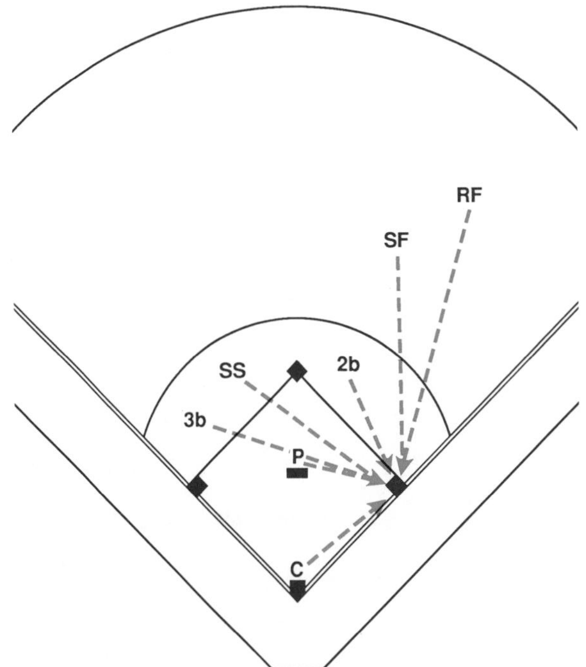

Diagram 11.7 Throws to first baseman covering first base.

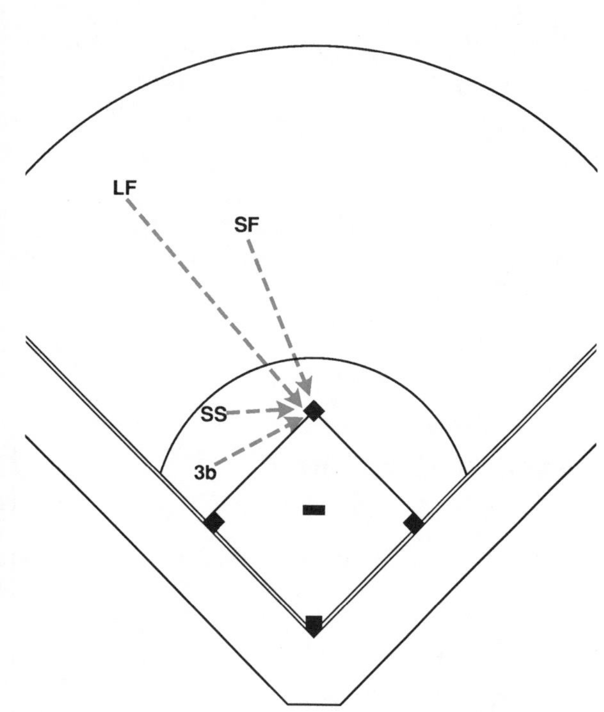

Diagram 11.8 Throws to second baseman covering second base.

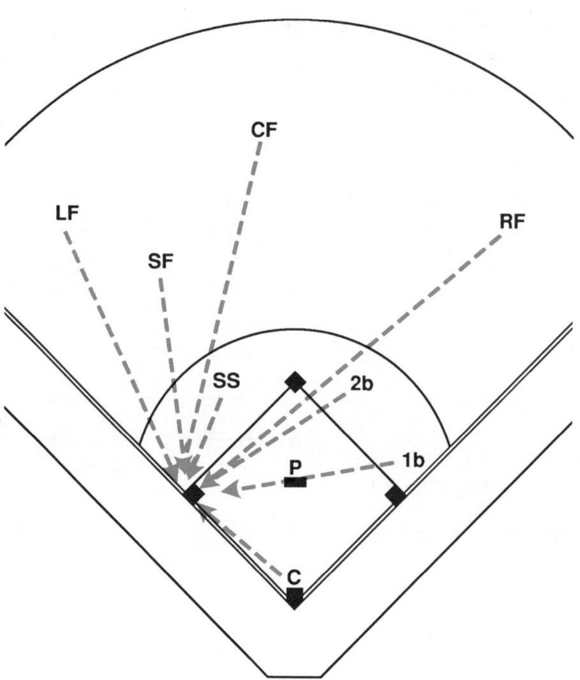

Diagram 11.10 Throws to third baseman covering third base.

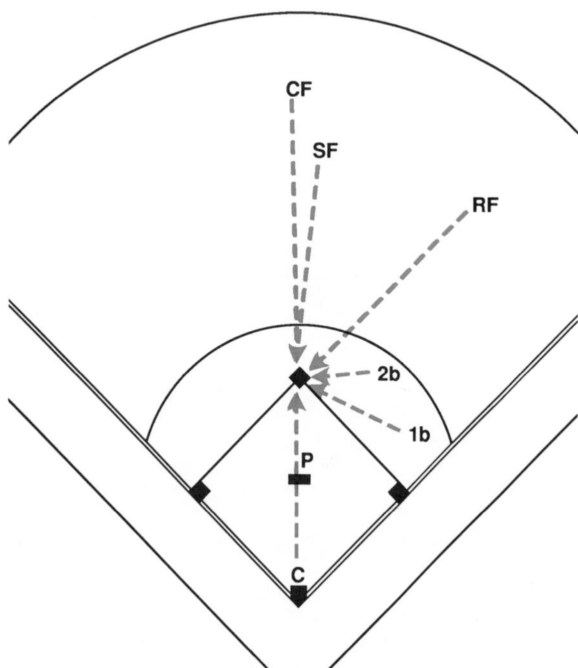

Diagram 11.9 Throws to shortstop covering second base.

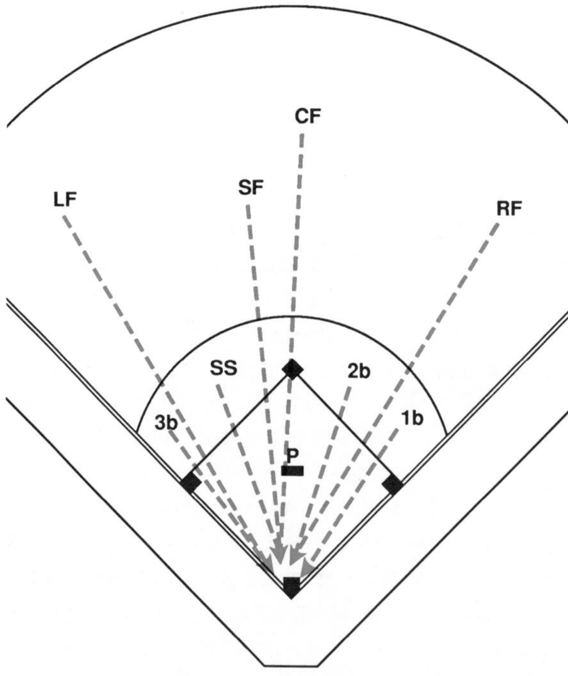

Diagram 11.11 Throws to catcher covering home plate.

Success Goal = 8 total successful repetitions (2 force plays, 2 off-target throws right, 2 off-target throws left, and 2 tag plays) at each player position

a. 1b covering first base

b. 2b covering second base

c. SS covering second base

d. 3b covering third base

e. C covering home plate

Your Score =

a. _____ completed (yes or no)?

b. _____ completed (yes or no)?

c. _____ completed (yes or no)?

d. _____ completed (yes or no)?

e. _____ completed (yes or no)?

2. *Partner Toss Footwork Drill*

You need a partner and two or three balls for this drill. You position yourself as an infielder and move to take throws for force plays and tag plays. Your partner stands about 5 feet away from the base in the general direction from which the throw is to come (see chart, ''Force Play and Tag Play Practice Situations,'' previous drill). When you have arrived at the covering position, your partner softly tosses the ball so you can practice the catching motion required by the situation.

Do 2 repetitions of each force and tag situation listed in the Success Goal. Switch roles at each base before moving on to the next base.

Success Goal = 4 repetitions (2 force plays and 2 tag plays) at each player position

a. 1b covering first base

b. 2b covering second base

c. SS covering second base

d. 3b covering third base

e. C covering home plate

Your Score =

a. _____ completed (yes or no)?

b. _____ completed (yes or no)?

c. _____ completed (yes or no)?

d. _____ completed (yes or no)?

e. _____ completed (yes or no)?

3. Force Play and Tag Play Without Runners

Two pairs of partners set up on a regulation softball field as follows: One pair of partners is a fungo hitter and a catcher at home plate; the other pair is a fielder to whom the ball will be hit and a fielder covering a base and making force and tag plays.

As the covering fielder, make two force plays and two tag plays at each of the positions listed in the "Force Play and Tag Play Practice Situation" chart in Drill 1. Switch roles within pairs after the 4 repetitions (2 force plays and 2 tag plays) have been completed. Exchange roles between pairs (fungo hitter and catcher go to two fielding positions, and vice versa) after completing all of the situations listed for a particular baseman.

Success Goal = 4 repetitions (2 force plays and 2 tag plays) at each player position

a. 1b covering first base

b. 2b covering second base

c. SS covering second base

d. 3b covering third base

e. C covering home plate

Your Score =

a. _____ completed (yes or no)?

b. _____ completed (yes or no)?

c. _____ completed (yes or no)?

d. _____ completed (yes or no)?

e. _____ completed (yes or no)?

4. Force Play and Tag Play With Runners

By now you are undoubtedly quite proficient at the footwork required to make force plays and tag plays at all the bases. In a game, however, it is not so easy to be successful because there is one factor that you have not yet practiced with—the baserunner.

As the fielder throwing the ball, you must be sure that your throw will not hit the baserunner. As the covering fielder, you must be sure that you set your glove as a target that can be hit by the thrower. Add to that the fact that the baserunner is bearing down on you, and you have to be a very cool player to react calmly and correctly under such conditions.

This drill is set up exactly like the preceding drill, with the exception that a third pair of partners are added as baserunners, who alternate doing the running. On the force play, the baserunner runs to the base at which the play is being made. On the tag play, the baserunner advances to the next base or returns to the previous base (as if having had overrun the base), as the situation calls for. The base-runner cannot leave a base to advance until the ball is hit. The baserunners must run in foul territory from home to first base. Baserunners must always wear batting helmets.

As in the previous drill, do 2 repetitions at each position, then switch roles with your partner. After both partners in a fielding pair have practiced force plays and tag plays at a particular base, rotate roles among the pairs. Baserunners rotate to fielding positions, fielders rotate to catcher and fungo hitter positions, and catcher and fungo hitter rotate to become baserunners.

The object for the fielders is to make the force play or tag play successfully, getting the baserunner out. Keep track of the number of outs you and your partner get in force play situations and in tag play situations. For each out made, you and your partner get 1 point.

The object for the baserunners is to get to the base safely. Do not slide or run into the fielder. For every base you attain safely, you and your partner get 2 points.

The object for the fungo hitter is to hit a fieldable ground ball to the fielding infielder. The catcher receives the throws from the covering fielder after the play is completed.

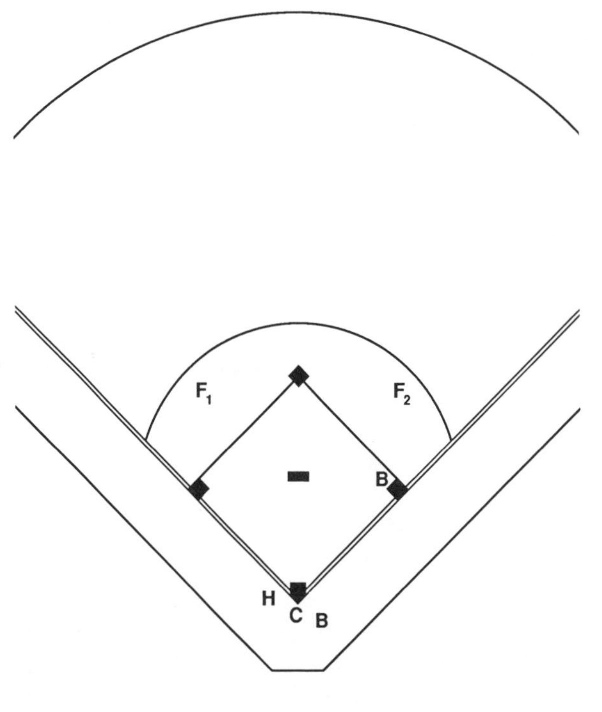

Success Goal = score more points than either of the other pairs in

a. force plays

b. tag plays

c. baserunning plays

Your Score =

a. (#) _____ force play points

b. (#) _____ tag play points

c. (#) _____ baserunning points

How did you and your partner do? Who won the force play competition? Who won the tag play competition? Were you and your partner better on force plays or tag plays? Why? Did you and your partner do better than the baserunners on your force and tag plays? Why?

Who were the most successful baserunners? Was the most successful pair simply the fastest runners, or were they smart in the way they ran the bases?

Step 12 Grounders-Only T-Ball

Modified games focus your practice on specific skills, sometimes to the exclusion of other skills. As the name might imply, Grounders-Only T-Ball, is intended to give you practice in fielding ground balls that have been hit off a batting tee. Using the tee is not gamelike; it, however, ensures that the ball is hit on every swing, thereby providing opportunities for a great number of fielding plays in a relatively short period of time. Despite the fact that this is not the same as hitting off a pitcher, this modified game gives you the opportunity to work on your hitting swing without having to wait for a pitcher to make good pitches. Also, you get a chance to practice baserunning.

The major purpose of this game, though, is to work on fielding ground balls and making throws to put runners out.

GAME SKILLS

Note that you are asked to focus on both your skill execution and your application of knowledge during the game. To be a successful softball player, you must not only have good skill technique but also know how and when to execute those skills in a game. The "good four-o'clock hitter" is a hitter who hits the ball out of the park during warm-up but doesn't get a base hit during a game. On defense, knowing how to make the force out is important, but being able to make the play with the runner coming at you in a game is the mark of a good softball player. Participation in this modified game will help you become a better game player *if* you practice the following:

1. Basic skills, including
   - hitting off the batting tee,
   - baserunning,
   - fielding ground balls,
   - overhand throws of various distances used in infield play, and
   - catching a ball.
2. Infield position-play techniques, including
   - area coverage, and
   - backing-up responsibilities.
3. Defensive and offensive tactics, including
   - force plays,
   - tag plays,
   - baserunning, and
   - place hitting.

GAME RULES

Specific rules of the game and method of play include the following:

1. The infield area, both fair and nearby foul territory, is the only area where the ball is in play. The outfield is out of play.
2. Hitting the ball off the tee, the batter *must* hit a ground ball, a hit that makes initial contact with the ground within the infield.
   - Any ball that first lands or is touched by a player in the outfield is an automatic out for the batting team. The ball is dead, and baserunners cannot advance.

- Any fly ball or pop-up is an automatic out. A pop-up in the infield area need not even be caught to be an out. The ball is in play, and baserunners advance at their own risk.
- Any line drive caught by an infielder is an out.
- Any ball (other than a fly ball or pop-up) that is misplayed by an infielder standing in the infield is a legally hit ball. If the misplayed ball goes into the outfield, the ball has become out of play. However, the batter takes first base on an *error* credited to the infielder misplaying the ball, and all other baserunners advance only one base. If the misplayed ball stays in the infield area (including nearby foul territory), the ball stays in play, and play continues.
- Any ball that is not touched by a fielder but first lands in the infield area and goes into the outfield, is a base hit. The ball is out of play, the batter takes first base with a single, and all other baserunners advance only one base.

3. There are six players per team: a pitcher, a catcher, and one player for each infield position.
4. Each team gets six outs per turn at bat. These six outs constitute a half-inning. Teams switch offensive and defensive roles after each half-inning.
5. Use a batting tee. The pitcher plays at the defensive position of pitcher but does not pitch the ball to the batter.
6. The batting order is by position number: 1 through 6, or pitcher through shortstop.
7. A baserunner cannot leave a base until the ball is hit.
8. All other situations are governed by the official slow pitch softball rules (such as those found in the *NAGWS Softball Guide*).
9. Method of scoring:
   - For every offensive and defensive play you execute correctly, 1 point is credited to your team. For every physical and mental error you commit, 1 point is subtracted from your team's score. The team with more points at the end of an equal number of half-innings of play is the winner. *Note*: The purpose of the game is to work on game concepts and the execution of skills during gamelike conditions. Therefore, performance points, not runs scored, determine which team does better.
   - Look over the Player Scoresheet (page 116) to see the kinds of actions that typically occur in this game, actions for which you can both earn and lose points. Use the Player Scoresheet to keep track of the points you contribute to your team's score. A Team Scoresheet is also provided so that each player's points can be tallied into the total team score.
   - When you are out in the field playing defense, you need to remember the plays you make and whether or not they were correctly executed. Then, when your team goes on offense, you should write down those defensive plays on the Player Scoresheet. When your team is up to bat, you can easily keep a record of your offensive tallies.

GROUNDERS-ONLY T-BALL
Player scoresheet

Name ______________________ **Position** __________

Skill/concept	*Plus performance points*	*Minus performance points*
	(Score 1 point each occurrence)	
Offensive skills		
Hitting using the tee		
ground ball	______	______
fly ball	______	______
Baserunning		
overruns 1st base	______	______
rounds a base	______	______
Defensive skills		
Fielding a ground ball	______	______
Overhand throw (not toss)	______	______
Catching a thrown ball	______	______
Game concepts		
Area coverage		
covers base	______	______
covers area	______	______
backs up play	______	______
Force play	______	______
Tag play	______	______
Baserunning	______	______

GROUNDERS-ONLY T-BALL
Team scoresheet

Position	*Player*	*Offense points*		*Defense points*		*Game concept points*		*Total team points*
		Plus	Minus	Plus	Minus	Plus	Minus	(Plus - Minus)
1.	______	___	___	___	___	___	___	______
2.	______	___	___	___	___	___	___	______
3.	______	___	___	___	___	___	___	______
4.	______	___	___	___	___	___	___	______
5.	______	___	___	___	___	___	___	______
6.	______	___	___	___	___	___	___	______

Team score ______

Step 13 Fielding a Fly Ball

When the batter hits a ball in a game, it comes to you, the fielder, as either a ground ball, a line drive, or a fly ball. You have already had practice developing your skill as a fielder for the first two types of hits. The final fielding skill that we will deal with as a separate step is catching a fly, a ball that comes to you from high in the air. Many of the principles you already know are used when you catch a fly ball.

WHY IS FIELDING FLY BALLS IMPORTANT?

The batter in slow pitch softball has a tendency to hit the ball high (and sometimes far) into the air for two main reasons. The trajectory of the slow pitch pitch is high to low; therefore, the batter tends to swing up at the ball, causing the ball to go up into the air. Second, the pitch is coming to the batter rather slowly; therefore, the batter has time to see the ball and make good contact, often trying to hit the ball out of the park on the fly for a home run. Because the hit ball in slow pitch is often a fly ball it is essential that all fielders be able to catch fly balls consistently. A caught fly ball puts the batter out and causes the baserunners to tag up before advancing once the catch is made. Runners who try to advance may get caught off base and a double play results. Thus, caught fly balls decrease the probability that baserunners will successfully advance and score runs.

Basically, all fielders need to know how to field fly balls correctly. Outfielders catch long fly balls; infielders catch pop-ups, fly balls hit in or near the infield area.

HOW TO FIELD FLY BALLS

Field the fly ball above and in front of your head. As you learned in simple catching, your fingers point up. Just as when fielding a ground ball, you want to get in front of the ball. However, when you're an outfielder who must immediately throw the caught ball a great distance, it is helpful to be moving slightly in the direction of the throw as you catch the ball. Fielding the ball in front of your throwing shoulder allows you to blend the catch and ensuing throw into one continuous motion more easily (see Figure 13.1).

Figure 13.1 Keys to Success: Fielding Fly Balls

Preparation Phase

1. Stagger stride, glove-side foot ahead
2. Knees slightly bent
3. Focus on ball
4. Hands chest-high, fingers up

Execution Phase

1. Watch ball go into glove
2. Meet ball high out front
3. Use two hands
4. Two-finger grip

Follow-Through Phase

1. Shift weight onto throwing-side foot
2. Glove side toward target
3. Ball to throwing position
4. Glove elbow toward target

Detecting Fly Ball Fielding Errors

Errors in fielding fly balls usually occur because you have difficulty tracking the ball off the bat and fail to get your body into position to deliver the throw after the catch. Another common error is fielding the ball with the glove hand only, thus delaying the throw after the catch. The most common errors committed when fielding a fly ball are listed below, along with suggestions on how to correct them.

ERROR	CORRECTION
1. You are late getting into position to field the ball.	1. Start to move to the ball as soon as it comes off the bat.
2. You have trouble judging where the ball will come down.	2. Track the ball using clouds, tree tops, and buildings to help with depth perception.
3. The ball bounces out of your glove.	3. Watch the ball go into your glove, squeeze and cover it immediately with your throwing hand.
4. You catch the ball as if in a basket below your waist.	4. *Don't*!

Fly Ball Fielding Drills

1. *Self-Toss Drill*

Toss the ball high up over your head and slightly out in front of your body. Point the fingers of your glove up. With the backs of your hands toward you, field the ball with two hands, using the points in the Keys to Success (see Figure 13.1).

Success Goal = 10 consecutive error-free tosses and catches

Your Score = (#) _____ catches

2. *Fly Ball Wall Drill*

Stand 10 feet away from a wall. Throw the ball high against the wall so that it rebounds back to you in a descending path. Move into position and catch the ball above and in front of your head. Pay particular attention to following the points described in the execution and follow-through phases of the Keys to Success (see Figure 13.1).

a. Throw 10 balls high against the wall directly in front of you. Count the number of catches you make.
b. Throw 10 balls high against the wall and to your glove side. Move into position to make the catch. Count the number of catches you make.
c. Throw 10 balls high against the wall and to your throwing side. Move into position to make the catch. Count the number of catches you make.

Success Goal = 24 out of 30 successful tosses and catches

9 of 10 catches, directly in front

8 of 10 catches, moving to glove side

7 of 10 catches, moving to throwing side

Your Score =

(#) _____ catches, directly in front

(#) _____ catches, moving to glove side

(#) _____ catches, moving to throwing side

3. *Partner Fly Ball Drill*

Stand 60 feet away from a partner, positioned so that neither of you is looking into the sun. For each option in the Success Goal, have your partner deliver 10 high throws simulating a fly ball.

Move under the ball and catch it using proper technique. In a continuous motion, throw the ball overhand back to your partner. Your partner should not have to move more than one step to catch your throw.

Success Goal = 38 out of 50 total successful catches and on-target follow-up throws

9 of 10 directly-in-front catches and throws

8 of 10 glove-side catches and throws

8 of 10 throwing-side catches and throws

6 of 10 slightly-behind catches and throws

7 of 10 random-direction catches and throws

Your Score =

(#) _____ directly in front

(#) _____ glove side

(#) _____ throwing side

(#) _____ slightly behind

(#) _____ random directions

4. *Fly Ball Target Drill*

In this drill you work both on moving to the ball to make the catch and on throwing to a target whose location varies. As the fielder, you need two other players, a thrower and a target person. The target person stands 100 feet from you, 15 feet to one side of the thrower, and near an empty bucket.

The thrower delivers fly balls of varying heights to both your glove side and your throwing side. Move into position and catch the ball. In the same motion, throw the ball to the target person, who alternates from one side of the thrower to the other. A successful throw is one that is within two steps of the target person.

Follow the sequences listed in the Success Goals. On the final 10 attempts, note that the direction and the distance of the target vary. As you move into position to catch the ball, the target person calls out the direction that you should throw. Rotate positions after you have completed all sequences.

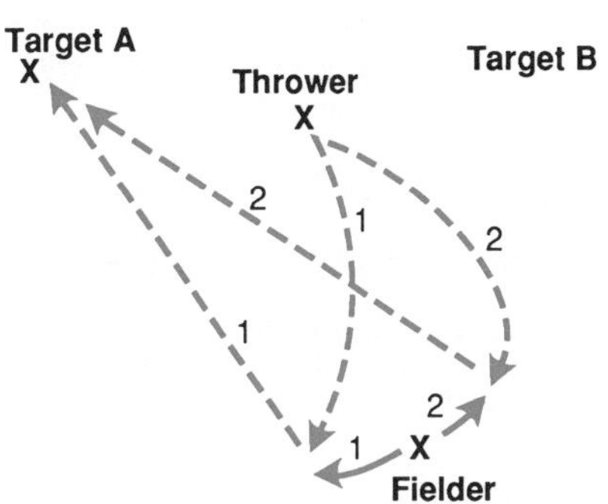

Success Goals =

a. 8 of 10 glove-side catches and on-target throws to A (100 feet away)

b. 7 of 10 throwing-side catches and on-target throws to A (100 feet away)

c. 7 of 10 glove-side catches and on-target throws to B (100 feet away)

d. 8 of 10 throwing-side catches and on-target throws to B (100 feet away)

e. 6 of 10 random side catches and on-target throws to either A or B (varying distance away)

Your Score =

a. (#) _____ glove-side catches and on-target throws to A

b. (#) _____ throwing-side catches and on-target throws to A

c. (#) _____ glove-side catches and on-target throws to B

d. (#) _____ throwing-side catches and on-target throws to B

e. (#) _____ random side catches and on-target throws to varying targets.

5. *Catcher's Pop-Up Drill*

Ask a partner to be a tosser for you. Tosser, stand slightly behind and to your throwing side. As the fielder, you assume a crouched stance like that of a catcher receiving a pitch.

The tosser throws the ball into the air well over the catcher's head and verbally signals the catcher to look up and field the pop-up. Infielders typically assist the catcher in locating pop-ups by calling ''up,'' ''up-back,'' ''up-right,'' and ''up-left.'' Using this terminology in this drill will carry over well for later game situation practice and play.

The tosser throws 10 ''pop-ups'' of varying heights and distances according to the sequences in the Success Goals. Switch roles after every set of 10 throws.

Success Goal = 26 of 40 total catches

7 of 10 directly overhead catches

7 of 10 glove-side catches

6 of 10 throwing-side catches

6 of 10 random-direction catches

Your Score =

(#) _____ overhead

(#) _____ glove side

(#) _____ throwing side

(#) _____ random direction

6. *Drop-Step Drill*

You and a partner stand 10 feet apart, facing one another. One of you is a tosser, the other a fielder. The tosser holds a ball in the throwing hand. The fielder stands in a square stance.

The tosser fakes a throw to the right or left of the fielder. On this signal the fielder starts to run back for a fly ball coming to that side, taking the first step back (a *drop step*) with the foot on the side of the indicated throw—all while maintaining visual contact with the ball, which is still in the tosser's hand.

Now the tosser throws a fly ball to the side originally signaled. It should be deep and high enough to force the fielder to continue running to catch the ball.

Change roles after each sequence listed below. The direction of the signal and ensuing throw should sometimes vary, as should the throw's distance, as follows:

a. Fake and throw right 2 times, no variance
b. Fake and throw left 2 times, no variance
c. Fake and throw right 3 times, vary distance
d. Fake and throw left 3 times, vary distance
e. Fake and throw random direction, random distance 5 times

Success Goal = 9 of 15 total catches

6 of 10 right and left catches

3 of 5 random-direction catches

Your Score =

(#) _____ right and left catches

(#) _____ random-direction catches

7. Where Is It? Drill

You and a partner stand 20 feet apart, facing the same direction, the fielder in front of the thrower. The thrower sends a fly ball into the air in any direction and at any catchable distance. The fielder, facing away from the thrower, does not know the direction of the fly ball. As the thrower releases the ball, he or she says ''Turn'' to the fielder. The fielder turns, finds the fly ball, moves under it, and makes the catch.

After each throw, switch roles. Continue until each partner has tried to catch 10 fly balls. Count the number of catches you make. If a fly ball is not catchable, the thrower must deduct 1 point from his or her catch total.

Success Goal = 5 of 10 catches, and more points than your partner

Your Score =

(#) _____ catches

(#) _____ points for you

(#) _____ points for partner

Fielding Fly Balls Keys to Success Checklist

Skill in fielding fly balls is necessary for all softball players regardless of the position played. Although a fly ball is usually considered a sure out, you know from your practice of the skill that it is easier said than done! Ask your teacher, coach, or a trained observer to evaluate your technique, according to the checklist below.

Preparation Phase

The fielder's

_____ feet are shoulder width apart, with glove-side foot ahead.
_____ knees are slightly bent, with weight going onto the throwing-side foot.
_____ head is up, with eyes focused on the ball.
_____ hands are up by chest, with fingers pointing up and thumbs together.

Execution Phase

The fielder's

_____ eyes focus on the ball, watching the ball go into the glove.
_____ catch is made in front of the body, above the head and directly in front of the throwing-side shoulder.
_____ throwing hand immediately covers the caught ball in the glove.
_____ throwing hand grabs the ball in a two finger grip, with the first two fingers on a seam and the thumb on a seam on the opposite side of the ball.

Follow-Through Phase

The fielder's

_____ pivot for the throw begins by shifting the weight onto the throwing-side foot.
_____ two hands are brought toward the throwing-side shoulder, with the throwing hand continuing to the start of the throwing position.
_____ glove side is turned toward the throwing target, with the glove-side elbow pointed toward the target.

Step 14 Hitting a Soft Toss

By now you should feel comfortable with the bat in your hands. You already know the key points about the swing itself. You have learned to hit a stationary ball off a tee and to hit a ball you tossed with your own hand.

You must now judge the path of a ball that is tossed to you by another person standing a short distance from your side. Judging your own toss was easier because you had some idea where the ball was going to go as you tossed it. Judging the short distance path of a ball tossed by someone else is a preliminary step to judging the path of a pitched ball. The toss is made from the side for the safety of the tosser. Even though the toss' direction is different than a regular pitch's, practicing hitting a moving target helps you develop your hitting technique.

WHY IS HITTING A SOFT TOSS IMPORTANT?

The most difficult part of hitting in a regular game situation is judging where the ball is going to be at the moment you will make contact. Next you must get the bat to that spot at the moment the ball arrives. Hitting a ball that is tossed softly to you, called a *soft toss* (or *ball toss* by some), allows you to work on your judgment as well as your swing. Because the soft toss travels only a short distance, contact judgments are easier to make than for hitting a ball pitched from regulation distance.

HOW TO HIT A SOFT TOSS

The soft toss really involves two skills for two people: tossing the ball and hitting the ball. Your ability to toss the ball properly is extremely important for two reasons. The batter cannot hit unless the ball is tossed correctly; the ball *must* come down ahead of the front foot of the hitter. In addition, your own safety as tosser requires proper execution; if you toss the ball too far toward the hitter's back foot, the batted ball could be hit back at you.

Figure 14.1 Correct tossing position for the soft toss.

The tosser and the hitter face each other, the hitter taking a batting stance and the tosser kneeling on the ground about 8 to 10 feet directly opposite the hitter's *back* foot. When tossing, toss the ball with a gentle down-up motion of your hand and arm. This should make the ball loop toward the batter, coming down into the hitter's contact zone and about 2 feet from the front foot. Toss the ball as if you wanted it to land on an imaginary batting tee off the batter's front foot (see Figures 14.1a, b; remember, tee is only imaginary).

When hitting a toss, stand with the same batting stance as for hitting off the tee, your feet shoulder width apart. After your partner tosses the ball, take the same swing at the ball as you did at the tee. Initiate your swing with a slight step with your front foot in the direction you intend the ball to be hit (much the same as you used for fungo hitting). Time your swing so that you contact the ball about waist-high and opposite your front foot. To protect the tosser, you must hit only balls tossed into this hitting contact zone out front and opposite your front foot. Do *not* swing if the toss does not drop to a position opposite your front foot; hitting a ball tossed toward the midline of your body, despite the fact that it might be in the strike zone, could result in the ball's hitting the tosser. After contacting the ball, complete your swing by wrapping your hands, still holding the bat, around your shoulders (see Figure 14.2).

*Figure 14.2 Keys to Success: **Hitting a Soft Toss***

Preparation Phase

For Tosser | **For Hitter**

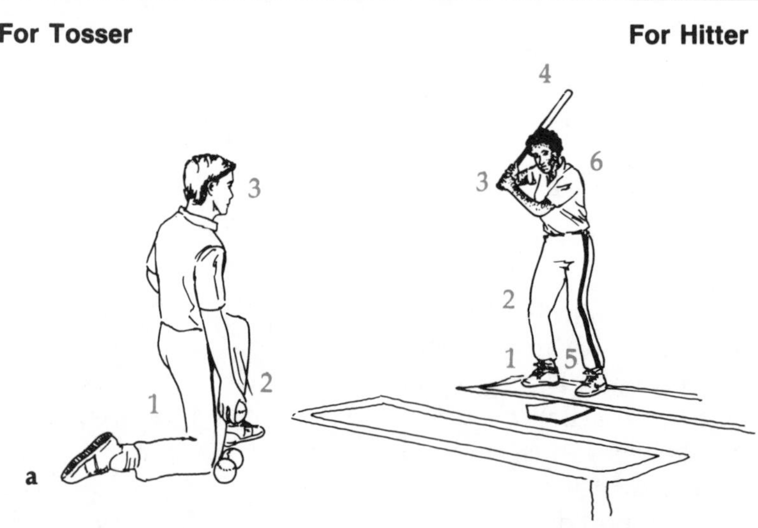

For Tosser

1. Kneel opposite hitter's back foot
2. Ball in throwing hand
3. Focus on target opposite hitter's front foot

For Hitter

1. Square stance
2. Knees bent
3. Two-hand grip, middle row of knuckles aligned
4. Hands and bat back
5. Weight on back foot
6. Head down, focus on ball

Execution Phase

For Tosser

b

c

1. Lift hand and arm
2. Release ball up and forward
3. Watch ball drop to target area
4. Watch hitter make contact

For Hitter

1. Step with front foot to open stance (point toes)
2. Hips start pivot, heel of front foot, ball of back
3. Hips square at contact
4. Back knee bent, front leg straight
5. Weight over back knee
6. Arms and wrists extended
7. Head down, focus on ball
8. Back shoulder to chin

Follow-Through Phase

For Tosser

For Hitter

d

e

f

1. Relax arm
2. Watch the hitter's technique

1. Roll wrists
2. Hands wrap around shoulder
3. Chin on shoulder
4. Weight over bent back knee, front leg straight
5. Hips square

Detecting Soft-Toss Hitting Errors

Frequently the errors that occur when performing this combined skill are caused by a poor toss. At this point, the hitter should have the basic swing under control. The hitting errors are usually caused by losing concentration and failing to watch the ball to the contact point. The most common errors for both tosser and hitter, with suggestions on how to correct them, are listed below.

ERROR	CORRECTION
For Tosser	
1. The ball goes in a straight line.	1. Make the ball traverse an arc to get to the target spot. Make your toss motion down and up, not back and forward.
2. The ball goes toward the midline of the hitter.	2. If you are directly facing the hitter, be sure your hand and arm motion is to the side and forward toward the hitter's front foot. [or] In your kneeling position, turn toward the hitter's front foot. Now make your hand and arm go down-up and straight forward toward the target spot.
For Hitter	
1. You miss the ball completely.	1. Watch the ball from the tosser's hand to your bat.
2. You just tip the ball up or to the ground.	2. Try to contact the *middle* of the ball with your bat.

Soft-Toss Hitting Drills

The drills that follow are designed to give you practice in making a consistent pattern or *groove,* in your batting stroke. As you do the drills, focus on your batting technique.

Soft-toss drills have been used by softball players for years to warm up prior to games. They enable you to get many more practice hits in a short time than hitting off a pitcher does. They can also be used between innings if you think you need some practice. These are truly some of the most useful practice hitting drills that there are.

1. Soft-Toss Net Drill

With a partner, position yourselves as hitter and tosser 10 feet from a hanging net (or blanket). Position yourselves so that the batted ball hits the net. The tosser has a bucket of 10 fleece, cloth-covered, or regulation balls. (For more experienced hitters, have 10 or 20 old tennis balls available, also. Tennis balls are smaller and, therefore, more difficult to track and hit. However, be sure that the net's hole size will restrict tennis balls before you try this drill with them.)

Conduct the drill using the techniques outlined in the Keys to Success (see Figure 14.2) for both the hitter and the tosser. The hitter switches places with the tosser after hitting 10 balls. Repeat the drill two times, for a total of 20 hits for each partner.

Success Goals =

20 on-target tosses

17 of 20 hits (on good tosses) that go directly into the net

Your Score =

(#) _____ tosses

(#) _____ hits

Fence Option for Soft-Toss Net Drill: Using the backstop or outfield fence and regulation softballs for the basic Soft-Toss Net Drill was an accepted practice for years. Recently, though, there has been litigation because of the ball's rebounding off the fence and hitting a participant. Consequently, using the bare fence for the drill is not recommended. However, it is possible to hang a large, thick gymnastic-type mat on the fence to absorb the force of the hit ball and dramatically limit the rebound. The fence option is then performed in the same manner as the Soft-Toss Net Drill.

2. *Bat-Eye Coordination Drill*

For this drill you need about 15 golf ball-size whiffle balls. You also need a wooden wand, a broom handle (cut to bat length), or a stickball bat. *Caution*: Be sure that anything you use to hit with has a taped hand grip that enables you to hold onto it. If you are making your own "bat," do not use plastic tape or any other slick or slippery material for the grip. Cloth-backed adhesive tape works well.

This drill is like the Soft-Toss Net Drill, but must be performed in an unobstructed area. It is also helpful to do the drill in groups of three; the third partner uses a container to retrieve the whiffle balls as they are hit. Rotate roles after the hitter has had a minimum of 15 swings.

The key to hitting a smaller ball is to track it carefully into the contact zone. Do not over-swing (swing too hard); just try to make solid contact with the ball. This is a difficult skill and one you can practice often if you find you are swinging and missing the ball in games.

Success Goals =

12 of 15 on-target tosses

10 of 15 full-contact hits (ball goes forward)

Your Score =

(#) _____ tosses

(#) _____ hits

3. *High-Ball and Low-Ball Hit Drill*

This drill is set up like the previous drill, except that you use 10 regular softballs and a real bat. The hitter and tosser position themselves for the ball toss. The third partner takes the empty bucket onto the field to retrieve the hit balls.

The tosser must now adjust the height of the tosses in the order given in the Success Goals. The hitter practices hitting low balls with the high-to-low swing pattern, which should produce ground balls. The batter hits high balls with the low-to-high swing pattern, which should produce fly balls.

When fielding a grounder, use proper ground ball fielding technique. Catch a fly ball with correct fly ball fielding technique. Put the fielded balls into the bucket.

After a set of 10 tosses to the hitter, rotate roles. The fielder brings the bucket in to the tosser position. The hitter takes the empty bucket out to the fielding position. The tosser moves to the hitting position.

Success Goal = 20 of 30 correct hits, according to the swing pattern

7 of 10 fly balls on high tosses (low-to-high swing)

7 of 10 ground balls on low tosses (high-to-low swing)

6 of 10 hits on random high or low tosses

Your Score =

(#) _____ fly balls

(#) _____ ground balls

(#) _____ hits on random tosses

Hitting a Soft Toss Keys to Success Checklist

Successful soft-toss hitting is dependent upon your ability, as tosser, to properly toss the ball for the hitter. You must place the ball in the contact zone, 2 feet from the hitter and in line with the front foot. If not, your safety and the hitter's opportunity to properly contact the ball are jeopardized.

As the hitter, you need to focus on two aspects of the hitting stroke: (a) timing of the swing, and (b) your position at the contact point. You have had the opportunity to evaluate your own performance based on the success goals of the soft-toss hitting drills.

Have someone use this list to check your tossing and hitting techniques.

Preparation Phase

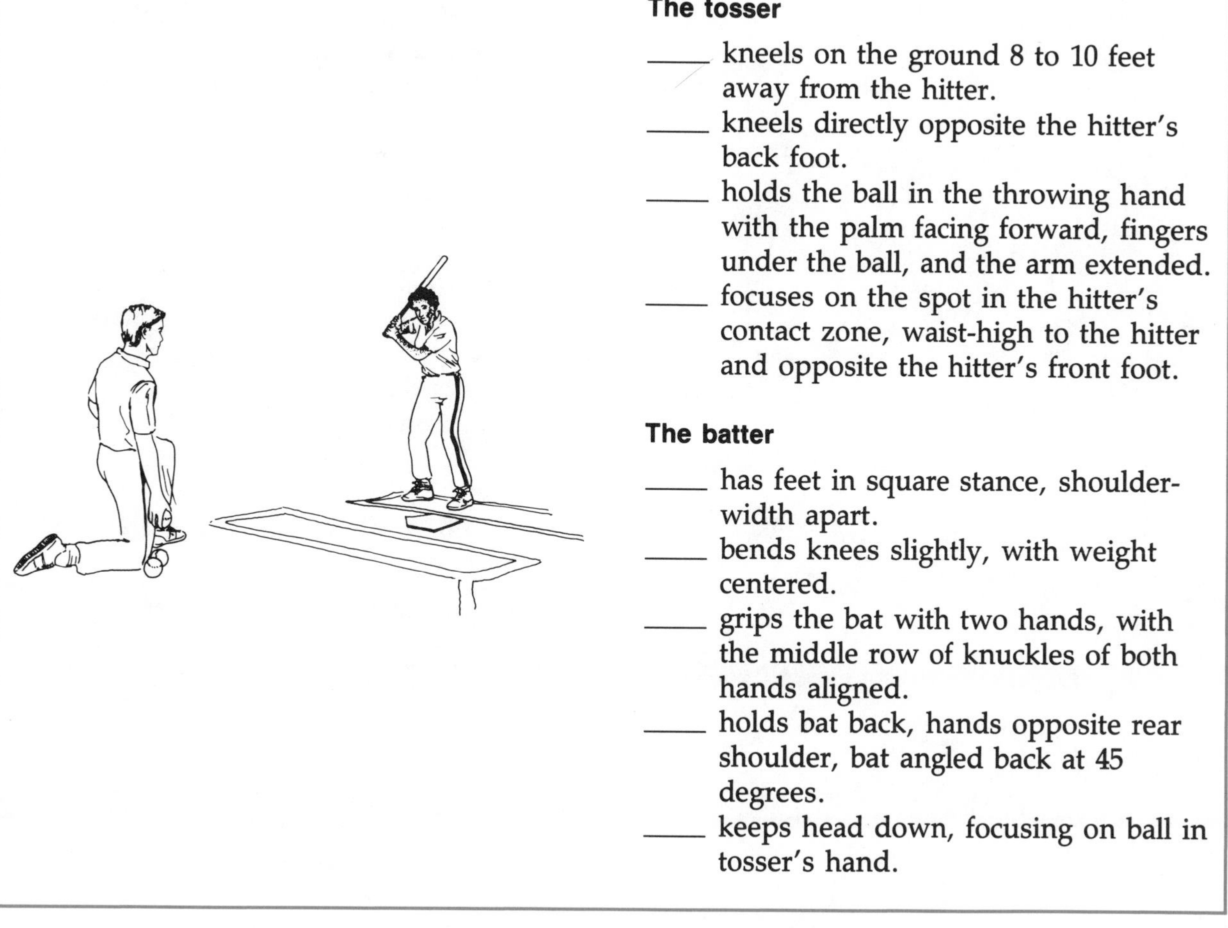

The tosser

____ kneels on the ground 8 to 10 feet away from the hitter.
____ kneels directly opposite the hitter's back foot.
____ holds the ball in the throwing hand with the palm facing forward, fingers under the ball, and the arm extended.
____ focuses on the spot in the hitter's contact zone, waist-high to the hitter and opposite the hitter's front foot.

The batter

____ has feet in square stance, shoulder-width apart.
____ bends knees slightly, with weight centered.
____ grips the bat with two hands, with the middle row of knuckles of both hands aligned.
____ holds bat back, hands opposite rear shoulder, bat angled back at 45 degrees.
____ keeps head down, focusing on ball in tosser's hand.

Execution Phase

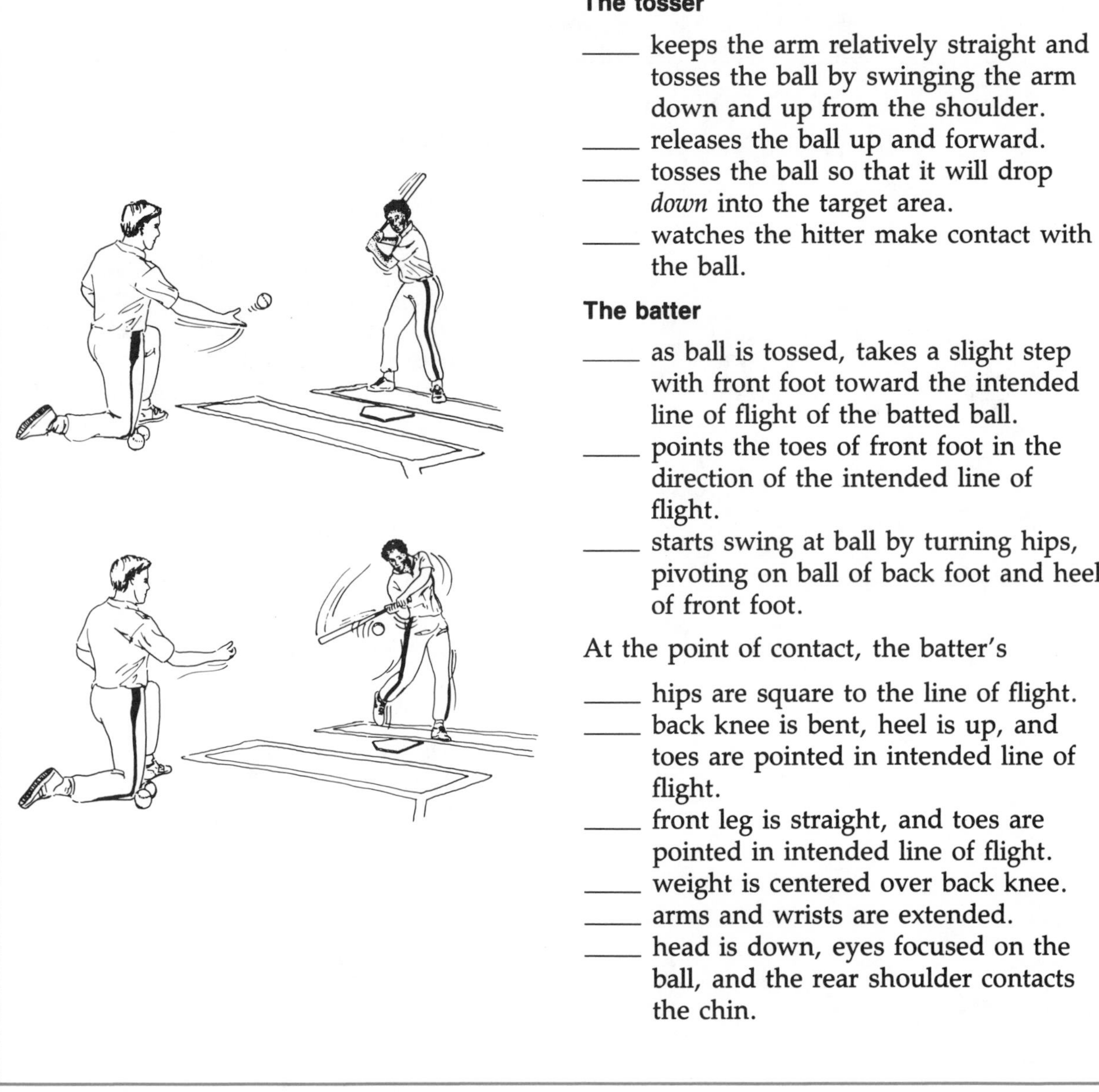

The tosser

____ keeps the arm relatively straight and tosses the ball by swinging the arm down and up from the shoulder.
____ releases the ball up and forward.
____ tosses the ball so that it will drop *down* into the target area.
____ watches the hitter make contact with the ball.

The batter

____ as ball is tossed, takes a slight step with front foot toward the intended line of flight of the batted ball.
____ points the toes of front foot in the direction of the intended line of flight.
____ starts swing at ball by turning hips, pivoting on ball of back foot and heel of front foot.

At the point of contact, the batter's

____ hips are square to the line of flight.
____ back knee is bent, heel is up, and toes are pointed in intended line of flight.
____ front leg is straight, and toes are pointed in intended line of flight.
____ weight is centered over back knee.
____ arms and wrists are extended.
____ head is down, eyes focused on the ball, and the rear shoulder contacts the chin.

Follow-Through Phase

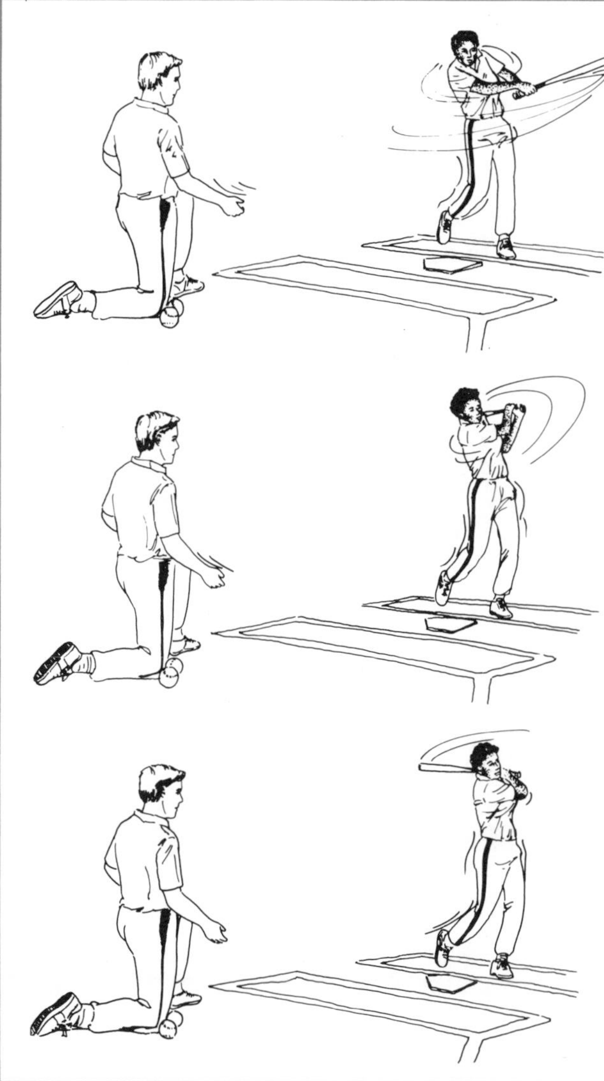

The tosser

____ gradually returns the tossing arm to the starting position.
____ watches the hitter's follow-through technique.

The batter

____ rolls wrists, top hand over bottom, and continues swing through the ball.
____ holds bat securely, lets bat wrap around back of body and hands wrap around front shoulder.
____ keeps head down and still, rear shoulder contacts chin.
____ has weight centered over back knee (batter is not going to run).
____ has back leg bent, front leg straight.
____ has hips square.

Step 15 Fungo Hitting a Fly Ball

Fungo hitting fly balls is a skill used in practice but not in games. It is another form of hitting and will give you added opportunity to work on your swing and the bat-eye coordination so necessary in all hitting. The swing path used for fungo hitting fly balls, unlike that used in hitting ground balls, is low-to-high. This path has direct carryover to hitting the ball as it descends in its slow pitch arc. A slight upward swing is needed in order to make solid contact on the descending ball.

WHY IS FUNGO HITTING FLY BALLS IMPORTANT?

The most important reason for learning this skill is so that you can provide realistic practice in fielding fly balls for others. When a fielder can react to a ball coming off a bat, it is a much more gamelike practice situation than fielding a fly ball merely thrown by a partner. You will need to be able to fungo hit fly balls in order to fully participate in the drills in the next step. Develop your fungo hitting skill with the practice opportunities provided for you in this step.

HOW TO FUNGO HIT FLY BALLS

Fungo hitting a fly ball is exactly the same as fungo hitting a ground ball except the swing path is low-to-high rather than high-to-low. You need to toss the ball a little higher than when tossing for the ground ball, and the bat must begin the swing down below your rear shoulder. Shift your weight onto your back foot and drop your rear shoulder. Swing up at the ball and make contact at about shoulder height, rather than down at waist height as is done for the ground ball. Step forward into your hit as you did for the ground ball. Your follow-through should be high, with your hands and the bat finishing well above your shoulders (see Figure 15.1).

Figure 15.1 Keys to Success: ***Fungo Hitting Fly Balls***

Preparation Phase

1. Square stance
2. Knees bent
3. Ball in bottom grip hand
4. Ball waist-high over front foot
5. Bat held low in top grip hand
6. Focus on ball

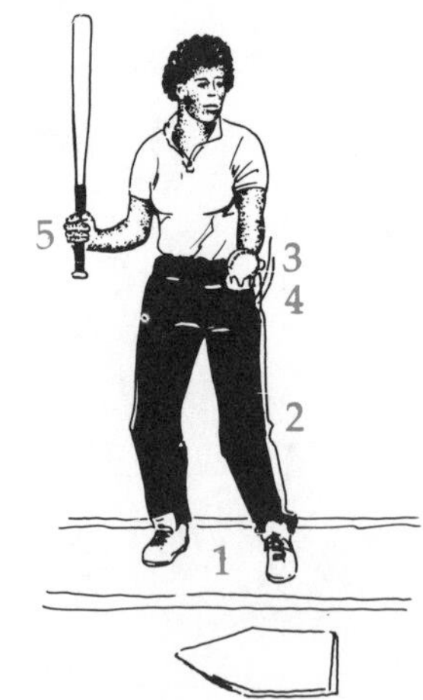

a

Execution
Phase

1. Toss ball
2. Bottom hand to bat
3. Front foot steps in direction of hit
4. Swing up and through ball
5. Contact ball at shoulder height

Follow-Through
Phase

1. Weight back
2. Back leg bent
3. Shoulder to chin
4. Hands wrap above shoulders
5. Hips square to target

Detecting Errors in Fungo Hitting Fly Balls

Most fungo hitting errors are caused by a poor toss, poor timing of the swing, and an incorrect swing path (it needs to be low-to-high). The most common errors that occur when fungo hitting fly balls are listed below, along with suggestions on how to correct them.

ERROR	CORRECTION
1. You hit a ground ball instead of a fly ball.	1. Toss the ball higher. Drop your rear shoulder and swing up at the ball.
2. The hit goes off-target.	2. Step in the direction of the intended target. Toss the ball off your front foot. At contact, feet point in line of direction to target.
3. You miss the ball completely.	3. Watch the ball until the bat makes contact.

Fungo Hitting Fly Ball Drills

1. Obstacle Drill

With a partner, select a high obstacle to hit over, such as the outfield fence, the backstop, goalposts, or even a small tree. You and your partner stand on either side of the obstacle, 50 to 80 feet away from it. Hit a ball back and forth to each other over the obstacle.

If you are practicing alone, use a bucket of balls. Hit them all over the obstacle, collect them, and hit them back. Count the number of times you try to hit the ball over the obstacle until you have 10 that have cleared it.

Success Goal = 10 hits that clear the obstacle in no more than 15 attempts

Your Score = (#) _____ clearing hits

2. *Accuracy Drill*

Mark a large target (10 feet by 10 feet) on the ground with cones or loose bases. Stand 80 to 100 feet away and hit fungo fly balls so they land within the target area.

You can do this drill alone using a bucket with 10 balls, or with a partner.

If you do the drill with a partner, your partner should stand near the target, score the number of on-target hits, collect the balls in another bucket, and then change roles with you. Repeat the drill until you and your partner have each hit 20 balls. Each ball that lands in the target is worth 1 point.

Success Goal = 15 points in 20 attempts

Your Score = (#) _____ points

3. *Increase the Distance Drill*

This is a target drill similar to the previous drill. Mark off distances of 100, 130, 160, and 190 feet from the target. Hit 5 balls from each distance. Score each on-target hit according to the scoring chart below:

Distance (feet)	On-target points		On-target hits		Points
100	1	×	(#) __________	=	__________
130	2	×	(#) __________	=	__________
160	3	×	(#) __________	=	__________
190	5	×	(#) __________	=	__________
				Total points	__________

Success Goal = 35 points on 20 hits

Your Score = (#) _____ total points

4. *Vary the Direction Drill*

This drill set up is similar to the previous drill. At the 130-foot or the 160-foot mark (select the distance at which you had more on-target hits in the last drill), make marks 15 feet to the right and 15 feet to the left of the location from which you hit the ball in the last drill.

From each of those marks, hit 5 fly balls to the target. As you are hitting the ball, you must line up your stance as if you were hitting straight ahead. Then, when you toss the ball, you must toss it more to your right when you are aiming right, or more to your left when you are aiming left. Be sure to step with the front foot in the direction of the intended hit.

Success Goal = 7 of 10 total on-target hits
4 of 5 from right side (right-handed player)
3 of 5 from left side (right-handed player)

Your Score =
(#) _____ on-target hits, right side
(#) _____ on-target hits, left side

Fungo Hitting Fly Balls Keys to Success Checklist

The skill of fungo hitting fly balls is as important for fielding practice as the skill of fielding fly balls is for the game. It is very important that you are able to fungo hit fly balls with accuracy so that others may practice fielding fly balls.

Have a skilled observer check your fungo hitting ability with the following checklist.

Preparation Phase

The fungo hitter

_____ has feet shoulder width apart in a square stance.
_____ bends the knees slightly, with weight on back foot.
_____ holds ball in hand that goes on bottom end of bat; palm is up, hand waist-high over front foot.
_____ holds bat low, well below rear shoulder, with top hand in regular grip for hitting.
_____ keeps head down, eyes focused on ball.

Execution
Phase

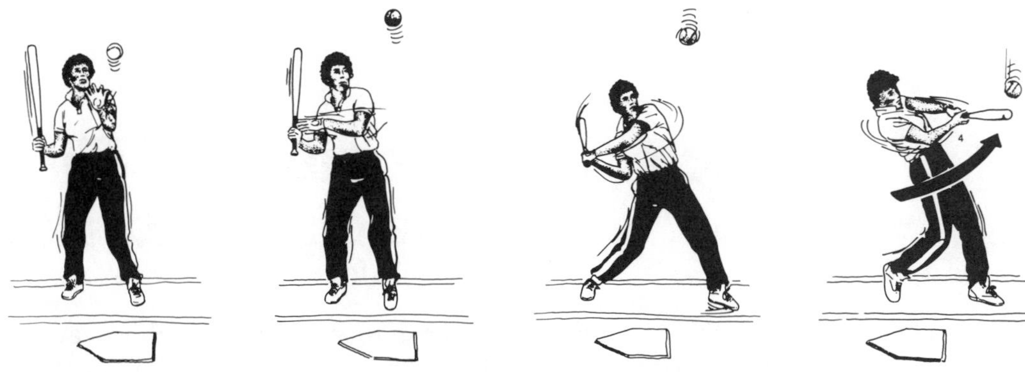

The fungo hitter

____ tosses ball with a strong down-up motion.
____ starts bat on upward swing path with top hand, as bottom hand takes proper grip at end of bat.
____ takes a slight step with front foot, pointing the toes in intended direction of the hit.
____ turns hips, pivoting on ball of back foot and taking weight on front foot.
____ swings up and through the ball.
____ contacts ball at shoulder height, with arms and wrists extended.

Follow-Through
Phase

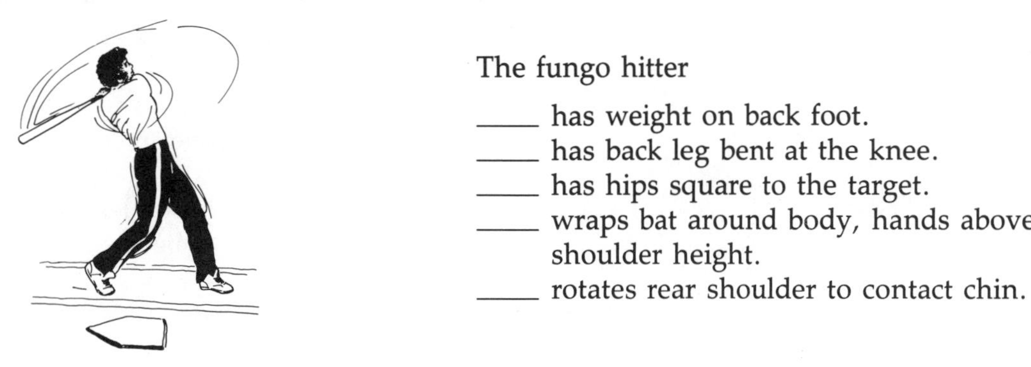

The fungo hitter

____ has weight on back foot.
____ has back leg bent at the knee.
____ has hips square to the target.
____ wraps bat around body, hands above shoulder height.
____ rotates rear shoulder to contact chin.

Step 16 Combining Fly Ball Fielding and Hitting

All defensive players must be able to field a fly ball and make a follow-up throw. You have been introduced to the new skill of fielding actual fly balls and the two practice skills of ball-toss hitting and fungo hitting fly balls, so let's put them all together.

WHY ARE THESE SKILL COMBINATIONS IMPORTANT?

In a game situation, there are two ways a ball can be hit off a pitch: into the air (either a line drive or a fly ball) or on the ground. The defense must field the ball however it is hit. Fly balls to the outfield often either advance runners into scoring position or enable them to score (*sacrifice flies*). Catching the fly ball and throwing it quickly and accurately to the proper player or base makes it much more difficult for the offensive team to move their runners ahead.

You have already worked on fielding ground balls and line drives; you need to perfect your skill in fielding fly balls to become a complete defensive player.

Fly Ball Fielding and Overhand Throwing Drills

Note: When hitting is called for in the following drills, alternate using the soft toss and the fungo techniques. The ball toss reinforces your ball-tracking skills, and fungo hitting continues your development of bat control. You could use the batting tee, but it is much more difficult to hit flies unless the post on the tee extends above waist height.

1. Basic Fly Ball Drill

In a group of three, set up so that one person is the hitter, one the catcher (and tosser, if used) for the hitter, and one the fielder. The fielder stands 130 to 160 feet from the hitter and catcher (the approximate distance from an outfield position to home plate). This practice distance can be varied according to the throwing ability of the fielder.

The fielder assumes the outfielder's ready position (preparation phase for fielding a fly ball). The hitter hits a fly ball to the fielder. As the ball is hit, the fielder tracks the ball and moves into position to field the ball. If at all possible, the fielder should be moving in the direction of the catcher while catching the ball. As the catch is made, the body movement in the direction of the throw increases force to the ensuing throw. Using an overhand throw, the fielder delivers a one-bounce throw to the catcher.

After 10 repetitions of these sequences, rotate roles. You need to keep track of the number of on-target hits (hit so that the fly ball is catchable), the number of error-free catches and throws to home, and the number of catches at home plate that you make when in each of the roles.

Success Goals =

8 of 10 on-target, error-free hits

8 of 10 error-free fielding and throwing combinations

8 of 10 error-free catches at home

Your Score =

(#) _____ hits

(#) _____ fielding and throwing combinations

(#) _____ catches at home

2. *Lateral Drill*

This drill allows you to work on moving in different directions to get under the fly balls in order to make the catches. It requires two pairs of partners. One pair is the hitter and the catcher/tosser, positioned at home plate. The other pair is fielders positioned together in center field.

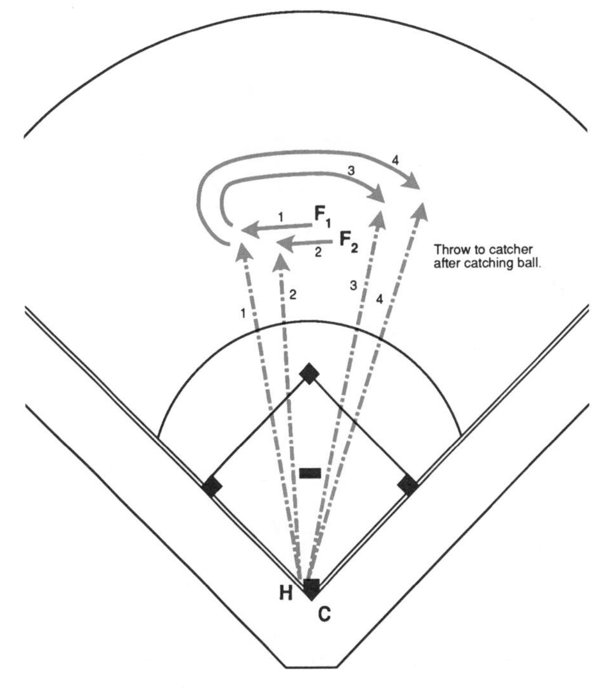

The hitter hits a fly ball so that one fielder must run laterally a minimum of 60 feet to field the ball. After the fielder makes the catch, he or she comes to a stop, steps toward the catcher, and makes an overhand throw to the catcher. The fielder should make an overhand throw to the catcher after each fielding attempt, whether a catch is made or not. When you are the fielder, be sure to concentrate on stopping your lateral movement, and getting your body moving in the direction of the throw. This will add both force and accuracy to your throw. The previous fielder moves out of the way for the next fielder to come over and make the same play. The hitter repeats the task for the next fielder. Both fielders should remain on their new location on the field.

For the next sequence, the hitter sends fly balls back toward center field, where the fielders started. The first fielder runs back in that other direction, fields the fly ball, stops, and throws overhand to the catcher. The other fielder does the same.

After 4 hits to each fielder (2 hits to the left and 2 hits to the right for each fielder), the hitter and catcher change roles; the fielders continue in their roles. After the next set of 4 hits, the fielding pair changes roles with the pair at home.

Success Goals =

5 of 8 on-target hits

10 of 16 error-free fielding and throwing combinations

6 of 8 error-free catches at home

Your Score =

(#) _____ hits

(#) _____ fielding and throwing combinations

(#) _____ catches at home

Now, just for fun, combine the numbers that you and your partner achieved in each of the previous Success Goals. Compare that score with the combined score of the other pair. If there are differences, see if you can determine whether the largest scores were achieved in hitting, fielding and throwing, or catching at home plate. Do those scores indicate to you that you need more practice on any of the skills?

3. *"I've Got It" Drill*

This drill also requires two pairs of partners and gives you the chance to practice with a teammate who is trying to make the same catch as you. One pair is the hitter and catcher/tosser, positioned at home. The other pair is fielders positioned in center field and left field (one in each field position).

The hitter hits a fly ball between the two fielders. Both fielders go for the ball. The fielder who gets under the ball first yells, "I've got it!" and catches the ball. The other fielder drops behind to back up the fielder making the play. An overhand, one-bounce throw is made to the catcher.

The fielders exchange starting positions after every sequence. After 10 hits, the hitter and the catcher exchange roles. After the next set of 10, the fielders exchange roles with the players at home.

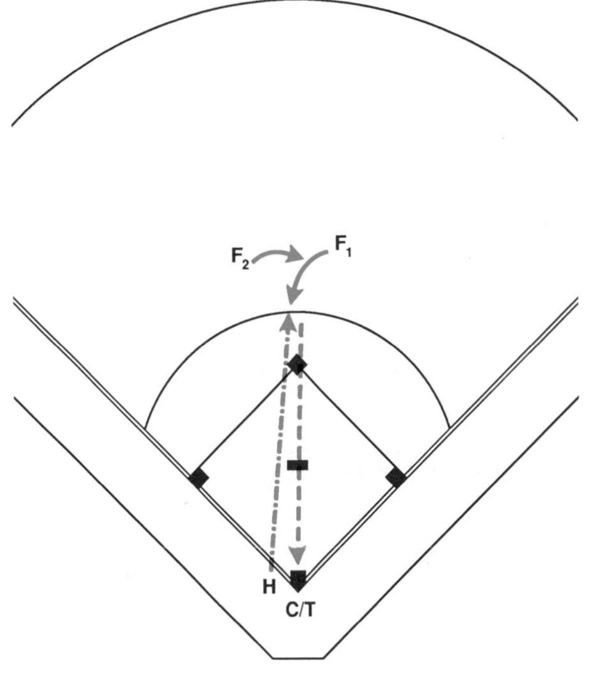

Success Goals =

8 of 10 on-target hits (between fielders)

8 of 10 error-free fielding and throwing combinations

10 well-done backing-up plays

8 of 10 error-free catches at home

Your Score =

(#) _____ hits

(#) _____ fielding and throwing combinations

(#) _____ backing-up plays

(#) _____ catches

4. In the Air Drill

Two pairs of partners are needed. One pair is a hitter and a catcher/tosser, positioned at home. The other pair is fielders, one positioned in center field and the other at second base.

A fly ball is hit to the fielder in center field. This fielder makes a throw in the air (no bounces) to the partner at second base. The ball is then thrown in the air to the catcher. These throws are relatively short and should not be bounced; they should still be thrown overhand.

After 10 sequences, one partner changes roles with the other. After the next set of 10, the fielders exchange roles with hitter and catcher.

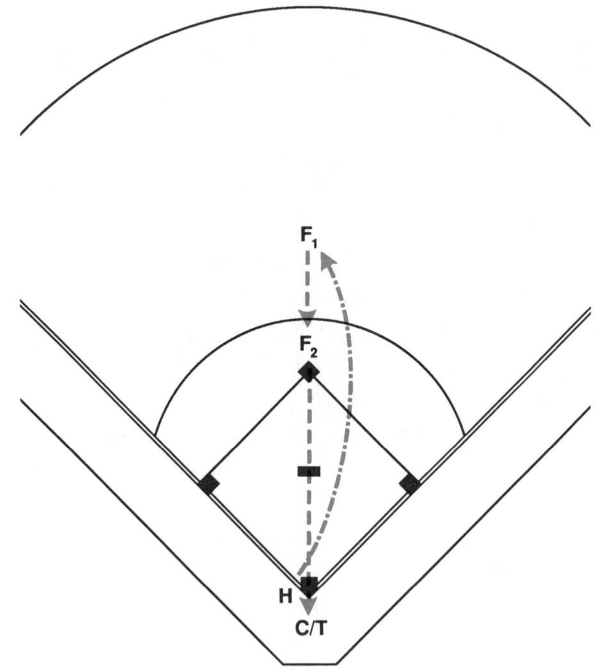

Success Goals =

8 of 10 on-target hits

8 of 10 error-free fielding and throwing combinations in the outfield

9 of 10 error-free catching and throwing combinations at second base

9 of 10 error-free catches at home

Your Score =

(#) _____ hits

(#) _____ fielding and throwing combinations in the outfield

(#) _____ catching and throwing combinations at second base

(#) _____ catches at home

5. Back to the Fence Drill

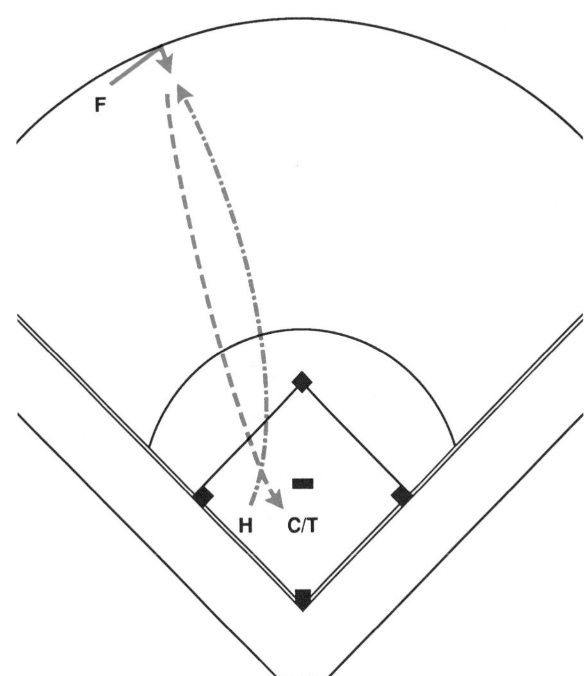

In a group of three, set up with a hitter, a catcher/tosser, and a fielder. The hitter and catcher/tosser stand about 130 feet from the outfield fence. The fielder stands 30 feet from the fence, facing the hitter.

The hitter hits a fly ball so that it comes down in the vicinity of the fence. The fielder, using the drop step, retreats toward the fence and fields the ball, then makes an overhand throw to the catcher. If the ball is very close to the fence, the fielder should locate the fence by visually sighting it, then slow his or her run in order to go right up to the fence. After locating the fence, the fielder should then come back to field the ball and make the throw.

After 10 sequences, rotate roles.

Success Goals =

7 of 10 successful hits

7 of 10 successful fielding and throwing combinations

9 of 10 successful catches back near the hitter

Your Score =

(#) _____ hits

(#) _____ fielding and throwing combinations

(#) _____ catches

Step 17 Position-Play Half-Field Games

You have enough skills now to play modified games that use position play, ground ball and fly ball fielding, fungo hitting (or soft-toss hitting), and baserunning. These games will require you to analyze each situation, anticipate the action of the other team, react as the play develops, and apply the teamwork skills you have acquired. This is where the real fun begins—putting together skills and strategy.

These are modified games for the purpose of practicing anticipation and reaction in as many different situations as possible. Play a different position every new inning of these modified games.

Try to "keep yourself in the game" for every play. Remember, no one is perfect. You can expect to make an occasional physical error, but you must forget about this kind of error and concentrate on the next play. Mental errors also occur, and these are the errors you should overcome (not just brush off) quickly. Figure out why you made it, so you won't make it again, then get ready for the next pitch.

WARM-UP BEFORE THE GAME DRILLS

Before beginning the drills, you should review the Keys to Success Checklist for Covering and Backing-Up Responsibilities, found in Step 10, "Position Play." These drills are good not only before regulation games but also between "innings" of the games that follow in this step. You will be in a different position each inning; these drills give you practice at the throws, force plays, and tag plays that occur at each new position.

You need 11 players for these drills. Players start at every defensive position, including short fielder. The eleventh player is a fungo hitter at home plate. Have a bucket of at least 6 balls near home plate (but out of the field of play).

Outfield Portion of Warm-Up

The fungo hitter begins by hitting three fly balls (unintentional grounders count) for each of the fielding situations given in the Outfield Warm-Up Chart. The outfielder fields the ball and makes an overhand throw to the designated infielder and base. The covering infielder gets into position out of the baseline but in a direct line with the throw. The infielder responsible for backing up must move into position behind the covering infielder and directly in line with the throw.

Outfield Warm-Up Chart

Fielder	Throws to	Covered by
Left fielder	Second base	Second baseman
Center fielder	Second base	Second baseman
Short fielder (in left center)	Second base	Second baseman
Right fielder	Second base	Shortstop
Short fielder (in right center)	Second base	Shortstop
Short fielder (in right center)	First base	First baseman
Right fielder	First base	First baseman
Center fielder	Third base	Third baseman
Short fielder (in left center)	Third base	Third baseman
Right fielder	Third base	Third baseman
Short fielder (in right center)	Third base	Third baseman
Left fielder	Home plate	Catcher
Short fielder	Home plate	Catcher
Center fielder	Home plate	Catcher
Right fielder	Home plate	Catcher

After completing this warm-up, only the four outfielders practice a combination drill of soft-toss hitting and fielding. Two outfielders, as tosser and hitter, position themselves on the left field foul line with a bat and a bucket of 10 balls. Two other outfielders position themselves in center field, facing the hitter and tosser, with an empty bucket for fielded balls.

The hitter hits the tosser's soft toss to the outfielders. The outfielders alternate fielding the balls, the second outfielder backing up the play.

After 5 hits the tosser and the hitter switch roles. After 10 hits the tosser and hitter exchange roles with the outfielders. The outfielders bring the bucket with the 10 balls to the hitting location; the hitter and tosser bring the empty bucket to the outfield. The outfielders repeat the drill until the infield portion of the warm-up drill (following) is completed.

Infield Portion of Warm-Up

The fungo hitter now hits 2 consecutive ground balls for each of the situations in the Infield Warm-Up Chart. The fielding infielder delivers the ball overhand to the infielder covering the base, who should be positioned on the side of the base from which the throw is coming, and out of the baseline. The baseman returns the ball to the catcher.

Before primary throws to home plate, the infielders assume the infield "in" position (see Step 10). This infield "in" position is used in real games when the bases are loaded with less than 2 outs, or when the score of the game is close and there is a runner on third base.

The fungo hitter goes through the chart two times, thus hitting a total of 4 ground balls for each situation. The first time through (2 ground balls), the covering infielder should be positioned for a force-out; the second time through (2 more), for a tag play.

Infield Warm-Up Chart

Fielder	Throws to	Covering infielder
Third baseman	First base	First baseman
Shortstop	First base	First baseman
Second baseman	First base	First baseman
Pitcher	First base	First baseman
Catcher (ball rolled not far from HP)	First base	First baseman
First baseman (ball hit close to 1b)	First base	First baseman (runs to base)
First baseman (short hit)	First base	Second baseman
Third baseman	Second base	Second baseman
Shortstop	Second base	Second baseman
Second baseman	Second base	Shortstop
First baseman	Second base	Shortstop
Pitcher	Second base	Shortstop
Catcher (rolled ball)	Second base	Shortstop
Third baseman (ball hit close to 3b)	Third base	Third baseman (runs to base)
Third baseman (short hit)	Third base	Shortstop
Shortstop	Third base	Third baseman
Second baseman	Third base	Third baseman
First baseman	Third base	Third baseman
Pitcher	Third base	Third baseman
Catcher (rolled ball)	Third base	Third baseman
Third baseman	Home plate	Catcher
Shortstop	Home plate	Catcher
Second baseman	Home plate	Catcher
First baseman	Home plate	Catcher
Pitcher	Home plate	Catcher

Infield Warm-Up Return Throws

The fungo hitter now hits 2 consecutive balls for each situation in the Infield Warm-Up Return Throws Chart. The infielder throws the ball to the first baseman, who immediately throws the ball to the catcher. The catcher then *return-throws* the ball to the infielder who originally fielded the fungo hit and who has moved into tag-play position on the base he or she is responsible for covering. The covering baseman immediately throws the ball back to the catcher for a tag play at home plate. As the covering player, be in position to receive the ball at the base without being in the baseline.

Infield Warm-Up Return Throws Chart

Fielder	Throws to	Throws to	Throws to	Throws to
Third baseman	First baseman	Catcher	Third baseman	Catcher
Shortstop	First baseman	Catcher	Shortstop (covering 2b)	Catcher
Second baseman	First baseman	Catcher	Second baseman (covering 2b)	Catcher
First baseman (playing deep)	Pitcher (covering 1b)	Catcher	First baseman (covering 1b)	Catcher
Catcher	First baseman	Catcher		

These warm-up drills probably offer the best practice of position play and throwing to the bases for both force plays and tag plays; in the future you can add double plays just prior to the return-throw portion. It will profit you greatly to alternate infield and outfield positions on successive times of doing this drill, making yourself a more versatile and proficient player.

HALF-FIELD GAMES

There are two games in this section, one using the left side of the playing field (from the second baseman's fielding position to the left field foul line), the other using the right side (from the shortstop fielding position to the right field foul line). Each game requires nine players, organized into three groups of three. In an inning, each of the three groups has a turn to score and each turn is comprised of three outs. Thus, an inning has three "half-innings" for a total of 9 outs.

There are several rules common to both games:

- A caught fly ball is an out.
- A person can be put out at a base by a force play or a tag play.
- Only the team running the bases can score. You must keep track of your group score.
- If the covering player blocks a base when a runner is approaching, the runner is safe, no matter what happens.
- A hit to the "wrong" side of the field (to right field in the left-side game, and to left field in the right-side game) is an automatic out.

Left Side of Field

Fair territory for this game is the area bounded by the left field foul line and a line extending from home plate to the outfield fence that passes through the second baseman's regular fielding position. One group of three players are a catcher, a fungo hitter, and a third baseman. A second group are a center fielder, a left fielder, and a second baseman. The third group are baserunners starting lined up at first base, with one person on the base and the others just outside the coaching box.

The game begins with one baserunner on first base. The fungo hitter, though belonging to a different team, is in the role of teammate of the baserunners. He or she hits the ball (either fly or grounder) to any place on the field of play (remember, this game uses only the left side of the field). When you are the fungo hitter, think about where to place the ball to help the lead runner advance. (Hint: Hit the ball away from the base to which the lead runner wants to advance. Also, try to hit a line drive so that an infielder cannot get to the ball.)

The runner runs to second base on a grounder or tags up on a fly ball. The runner must decide whether to advance on a fly ball; the defensive players can make a play on the baserunner.

In any case, if the baserunner is forced or tagged out, it is an out on the running team. If the baserunner makes it safely to second base (or farther), the ball goes in to the catcher at home. A second runner moves on to first base. Play continues with the fungo hitter's again hitting the ball.

The object of the game is for the baserunners to score as many runs as possible (getting all the way to home plate safely) before there are three outs. The defense (infielders, outfielders, and catcher) work together to get three outs on the baserunners as quickly as possible.

After three outs, the baserunners rotate to the outfield and second base positions. The outfielders and second baseman rotate to the hitter, catcher, and third base positions. The hitter, catcher, and third baseman become the baserunners. Every inning, rotate roles within your group so that by the end of the game you will have played every position. Play at least a three-inning game (27 outs).

Success Goals =

a. your group scores more runs than either of the other groups

b. make as few physical and mental errors as possible

Your Score =

Team	Runs, inning			Final score	Problems and noticeable improvements
	1	2	3		

Right Side of Field

Fair territory for this game is the area bounded by the right field foul line and a line extending from home plate to the outfield fence that passes through the shortstop's regular fielding position. There are three groups of three players. One group is set up as fungo hitter, third baseman, and first baseman. A second group is set up as shortstop, center fielder, and right fielder. The third group is the baserunning group, now located at home plate.

The game begins with a runner at first base and one at home plate. The fungo hitter hits the ball anywhere on the modified playing field. When the ball is hit, the runner at home plate runs to first base as a batter-baserunner. If the hit is a fly ball, the batter-baserunner is out and returns to home plate and goes to the end of the baserunner line.

The runner on first base, meanwhile, runs to second base on a grounder or tags up on a fly ball. The defensive players attempt to make a play on the lead runner. Once an out is made or play has stopped (the baserunners safe at their bases), the ball is rolled back to the fungo hitter. Play then continues with the fungo hitter's hitting another ball.

The object of this game is for the baserunners to score a "run" by advancing to third base (not home). The object of the game for the defensive players, again, is to get three outs on the baserunners as quickly as possible. Rotation is similar to the other game, and this game is also three innings long.

A new offensive strategy that can be practiced in this game is *hitting behind the runner*. With a runner on first base only and less than two outs, the hitter tries to hit a line drive to right field. The baserunner rounds second base and tries to get to third base on the hit to right field.

Success Goals =

a. your group scores more ''runs'' than either of the other groups;

b. make as few physical and mental errors as possible

Your Score =

Team	Runs, inning			Final score	Problems and noticeable improvements
	1	2	3		

Step 18 Hitting a Pitched Ball

You have already covered the fundamentals of hitting and developed your hitting expertise through a variety of hitting drills, including hitting off a batting tee, hitting fungo ground balls and fly balls, and hitting the soft toss. Now it is time to *polish* your technique by practicing situations in which you must deal with the judgment that goes into hitting a pitched ball.

WHY IS HITTING A PITCHED BALL IMPORTANT?

Although you have thus far used several modified forms to practice the basic techniques of hitting a softball, official game rules require you to hit a pitched ball. *Situational hitting* is a special aspect of hitting a pitched ball in a game. Certain game situations call for you to be able to hit the ball to a particular part of the field. This is called *place hitting*, and this skill is very important if you are to become an outstanding hitter.

For example, if you want to advance a runner from first base to second, it is important to hit the ball to a location that gives the baserunner the best chance of making it to second base safely. Thus, a hit deep to the outfield between the outfielders or a hit that goes between the persons playing first and second base will give the baserunner the best opportunity to safely advance. Similarly with a runner on third base and less than two outs, you would want to hit the ball deep into the outfield. Whether the ball is caught on the fly or fielded on the ground, the throw to home plate is of considerable distance and will take longer to arrive at home than will the baserunner.

HOW TO HIT A PITCHED BALL

By now you should have techniques well established for the grip with the middle knuckles aligned; square, closed, and open stances; and horizontal, high-to-low, and low-to-high swing paths. All of the techniques for hitting you have learned so far are applicable to hitting a pitched ball. Your challenge in hitting a pitched ball is to predict the flight of the pitched ball and time your swing to make contact with the ball in your hitting contact zone (see Figure 18.1a-e).

Figure 18.1 Hitting a pitched ball in the hitting contact zone.

Pitched Ball Hitting Drills

Note: Any of the following drills that involve actually hitting the pitched ball can be done as combination drills, with fielders practicing the skills of fielding ground balls and fly balls, throwing, and catching. If drills are done in partners or groups of three, you must make provision for collecting the hit balls. Using a large plastic bucket with a handle for ease in carrying is recommended. You can also use a plastic milk crate, but it is more difficult for one person to carry. If throwing is not one of the practice tasks when fielders are used, use two buckets. Start with a full one by the pitcher and an empty one in the field. Collect the fielded balls in the bucket in the field. Have the players who are rotating into and out of the hitting role ferry the buckets in and out.

1. Bleacher Ball Drop Drill

You need a partner for this drill. The hitter places a glove on the ground (to simulate home plate) approximately 2 feet away from the end of a set of bleachers. The hitter stands at "home plate" across from the bleachers with the front foot opposite the plate. The hitter takes a couple of practice swings to make sure the bat covers home plate but will not contact the bleachers during the swing. The partner stands at the end of the bleacher, slightly in front of, and above the head of the hitter. There needs to be a bucket of 20 balls (regulation balls if outside; whiffle or fleece if inside) beside the partner in the bleachers.

This partner drops the ball so that it angles down toward the back edge of the plate simulating the slow pitch pitch. The hitter swings, hitting the ball as it comes into the contact zone. Hit 10 balls then change roles. Collect the balls after both partners hit. Repeat the sequence.

Success Goal = 8 of 10 balls hit beyond a target distance of 100 feet

Your Score = (#) ____ successful hits

2. Call Ball or Strike Drill

Game play requires that hitters be able to read the pitch that is being delivered. It is critical that a hitter not stand and watch three strikes go by or swing at pitches that are not in the strike zone. This wastes a turn at bat. Hitters need practice at watching the ball come from the pitcher so that they can tell a strike from a ball.

This drill requires a group of three—a hitter, a pitcher, and a catcher. The pitcher stands at the regulation, 46-foot pitching distance from home plate. The catcher is in catcher's position with a mask on. The hitter takes a regular batting stance at the plate.

The pitcher should be focusing on pitching technique and attempting to throw strikes during this drill. The pitcher pitches a regulation softball. The batter takes a stride and initiates the swing by starting the hip and shoulder turn only—hands and arms keeping still, *not swinging at the ball!* The batter watches the ball pass over (or by) the plate and calls a ball if the pitch was outside the strike zone or a strike if the pitch went through the strike zone. The catcher verifies the call of the pitch. Remember, the slow pitch strike zone is the area over home plate between the batter's high shoulder and the top of the knees.

Rotate roles after 10 pitches.

Success Goals =

6 out of 10 strikes (pitcher)

8 out of 10 agreed-upon calls (hitter and catcher)

Your Score =

(#) ____ strikes

(#) ____ calls in agreement

3. Contact Drill

This drill needs a group of three, with additional fielders as an option. Set up as in the previous drill. Now, though, the hitter takes a full swing at each pitch and attempts to make contact with the ball. When hitting, you should focus on contacting the ball out front, opposite your front foot. Watch the ball all the way to the bat. Take a smooth, fluid swing at the ball. Just make contact—do not try to hit the ball hard. Hit the middle of the ball.

Score each hit according to the ball's path as follows:

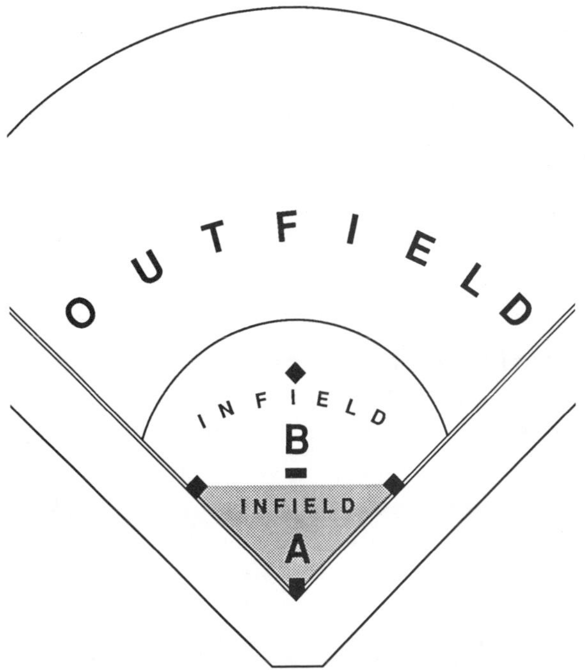

Landing zone	Settle zone	Contact points		Hits		Points scored
Foul	Foul	1	×	____	=	____
Infield A	Infield A or B	2	×	____	=	____
Infield B	Infield B	3	×	____	=	____
Infield A or B	Outfield	4	×	____	=	____
Outfield	Outfield	5	×	____	=	____
				Total points	=	____

Success Goal = 25 points on 10 pitches

Your Score = (#) ____ points

4. *Between the White Lines Drill*

This drill is set up as in the two previous drills. The hitter attempts to hit a fair ball into the outfield. Distance is not the focus. Making contact and keeping the ball in play is the intent of this drill. It is not often that the home run wins the game. The home run is difficult to hit and occurs very seldom. Consequently, it is more important to be able to hit the ball and keep it in play. In other words, when the ball is hit into the field of play, baserunners have the chance to advance and either score or become threats to score.

Success Goal = 8 out of 10 balls landing in fair territory and ending up in the outfield

Your Score = (#) ____ successful hits

5. *Line Drives Drill*

Mark the outfield with two arcing lines going from foul line to foul line, one 100 feet and the other 130 feet from home plate. The area between the lines is the target landing area for line drives.

The hitter attempts to hit a line drive that would clear the infield but land in front of the outfielders. In addition to landing in the target area, a successful hit must be judged to be a line drive (it must traverse a relatively horizontal path) and not a fly ball. If available, a trained observer can make the judgment, or participants in the drill can judge.

Timing of the swing is critical for success in this drill. The ball must be contacted above the waist as it drops through the strike zone. The swing must be horizontal, and the middle of the ball should be contacted.

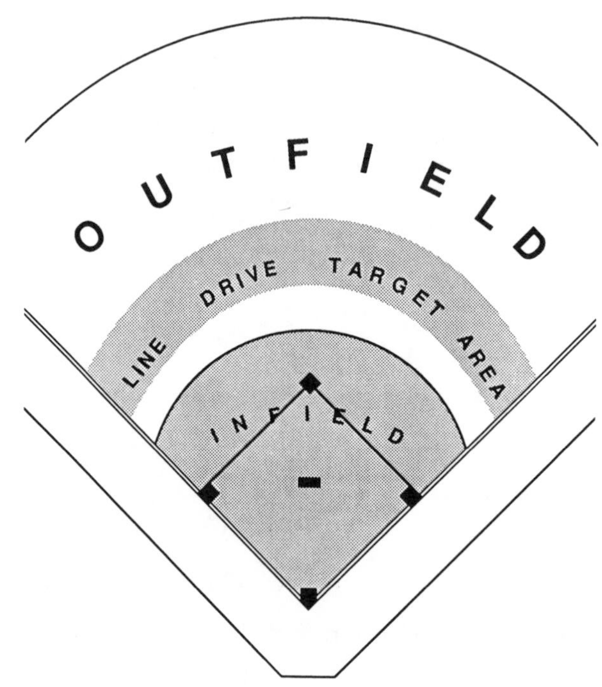

Success Goal = 6 out of 10 hit balls landing in the target area

Your Score = (#) ____ line drives

6. *Hit the Gap Drill*

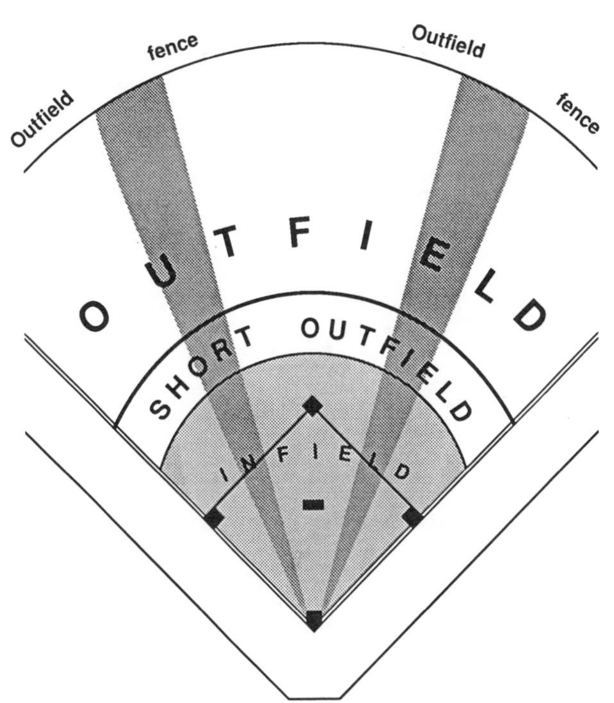

From home plate, mark two fan-shaped alleys (a narrow strip of the field). One alley goes through the shortstop position to the left-center field fence, and the other through the second baseman's position to the right-center field fence.

The hitter attempts to hit the ball into these target areas and all the way to the fence. A hit into these alleys (gaps in coverage by the defensive setup) will most often result in a safe trip to first base and in advancing the base-runner one or more bases.

Score each on-target hit according to the ball's path as follows:

Landing zone	Settle zone	Points		Hits		Points scored
Infield alley	Short outfield	1	×	____	=	____
Outfield alley	Outfield	3	×	____	=	____
Outfield alley	Outfield fence	5	×	____	=	____
				Total points	=	____

Success Goal = 20 points on 10 hits

Your Score = (#) ____ points

7. *Hit Behind the Runner Drill*

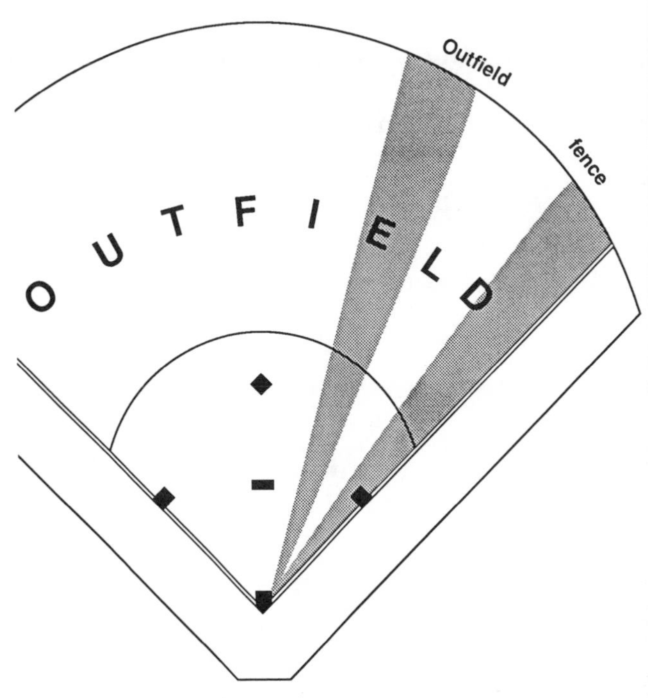

Use the alley marking to right-center field from the previous drill as one target area. For the second target area, start with the right field foul line as one of its boundary lines; from the foul line at the outfield fence, go into fair territory 30 feet and mark the target's inside boundary line from the fence to home plate.

The hitter attempts to hit the ball to these right-side targets. In a game situation, a runner on first base could advance to third or beyond on your hit into these alleys.

Success Goal = 19 out of 30 on-target hits

8 out of 10 hits landing in any part of either target area

6 out of 10 hits landing in an outfield target area

5 out of 10 hits landing in an outfield target area and going all the way to the fence

Your Score =

(#) ____ hits in either target area

(#) ____ hits landing in outfield target

(#) ____ hits landing in outfield target and rolling to fence

8. Situation Drill

Now only two players are needed. This is a "let's pretend" drill. A pitcher and a hitter alternate setting hypothetical game situations for the hitter. For example, one player says, "There's a runner on third base, no one out in the last of the seventh inning, with the score tied." The batter attempts to execute a hit (even an out) that would produce positive offensive results for the hypothesized situation.

The pitcher attempts to get the batter out. Using the regulation three strikes (an out) and four balls (a base on balls), the batter has one turn at bat to accomplish a given task. Judgments on hits must be made as to whether they would be base hits or outs of various kinds, and whether catchable fly balls would be deep enough to advance a runner, and so on.

The batter scores 1 point for each turn at bat that produces a positive offensive result for the specified situation. A base on balls is always 1 point for the batter. The pitcher scores 1 point when the batter fails to produce a positive result, that is, makes an out that doesn't advance a runner. Players change roles after 6 situations have been completed (each player has set 3 situations).

Success Goal = 6 points scored in 12 situations

Your Score = (#) ____ points

Step 19 Situation-Ball Game

The modified game, Situation Ball, requires you to make judgments about *how* and *when* to use the various skill responses and game concepts you have developed to date. Because the game play will be more controlled than in a regulation game, you will have more time to prepare yourself both physically and mentally for each of the situations.

Be confident on both defense and offense. As a defensive player, always concentrate on each situation, focus on the roles you may play, and anticipate what your responses should be, especially regarding covering and backing up. The Keys to Success Checklist for Reading a Defensive Situation (later in this Step) suggests factors that you should quickly run through in your mind before each pitch in every situation. Your answers to these questions will provide the basis for you to decide which responses you could use in a given situation. The response you actually use when the ball is hit will be a matter of instantaneously selecting the best option, based upon your on-the-spot judgment about the exact nature of the hit ball and the unfolding of the baserunning.

You also need to concentrate and plan ahead as a batter. Before you step into the batter's box, you must evaluate the situation and anticipate what you are going to try to do as the hitter. The Keys to Success Checklist for Reading an Offensive Situation lists factors to think about to make informed decisions in any situation. Your initial decision is to determine the major objective of the situation: to get on base, to advance a runner, or to score a runner? Your next step is just like the defensive player's: Determine the options you have for accomplishing the major objective. Next read the positioning of the defense to determine your best possible option.

Although the batter figures out what to do even before stepping up to the plate, both baserunners and defensive players must wait until the ball is hit to make their *final* decisions on plans of action. This is why anticipation is so important for these players. As a baserunner, you usually do not have too many options. If you are in a force situation, you must run on a grounder; if you cannot be forced, you need to make judgments based on all the things listed in the Keys to Success Checklist for Reading an Offensive Situation. Be aware of applicable rules, anticipate, read the defense, and think of your potential responses. Once the ball is hit, there is only enough time to select the *best* action to take. Hopefully, your prior analysis of the situation and of your options provide you a successful response.

WHAT TO PRACTICE

In a regulation softball game, it may be very difficult to get experience with a specific situation because that situation may never present itself in an entire game! For example, say you and the other outfielders on your team feel that you all need work on throwing out a runner tagging up and trying to score on a fly ball. However, if your opponents never have a runner on third base with less than two outs, or, if they do, the batter never hits a fly ball, you won't get a chance to practice that type of throw. With Situation Ball, though, you can set most any situation to practice. If you'd like, the offensive team must try to hit a sacrifice fly.

Look over the following list of game components that you can select to work on in Situation Ball.

1. Basic skills
    - pitching
    - hitting the pitched ball
    - fielding ground balls
    - fielding fly balls
    - throwing overhand
    - catching the ball
    - baserunning

2. Game concepts
   - position play area coverage
   - position play covering and backing up responsibilities
   - the force play
   - the tag play
   - baserunning
   - advancing a runner via a(n): base hit, place hitting, error, fielder's choice, base on balls (walk), sacrifice fly, or a tag after a caught fly ball (fair and foul).
3. Knowledge of rules
   - fair ball and foul ball
   - pitching regulations
   - batting regulations
   - ways a batter may be put out
   - ways a baserunner may be put out
   - in play, out of play
   - infield fly
   - regulation game, innings
4. Safety concerns
   - use of equipment
   - the playing area

HOW TO PLAY

The following rules and method of playing the game have been established to both facilitate play and ensure that opportunities are presented to work on specific offense and defensive situations under gamelike conditions. Prior to the start of the game, members of each team should get together and decide on a specified number of offensive and defensive situations the team wants to work on. (The number of situations specified will depend on the anticipated number of innings to be played.) Determine the order for selection of situations to be practiced (for example, Team A selects the situation for the odd number innings and Team B selects for the even number innings). Teams should check with each other to see that similar situations are not repeated.

Until you get familiar with the Keys to Success Checklist for reading offensive and defensive situations, it would be helpful to put them on a three-by-five card to keep in your pocket and to review from time to time to become more proficient in their application. During Situation Ball, new game situations will undoubtedly arise as plays unfold. Keep a list of those different game situations so that you will remember to practice them another day. It also helps you develop your anticipation and reaction skills to note the situations that continue to give you trouble. Mentally practice these situations to acquire an edge the next time you play Situation Ball or a regulation game.

Specific rules of the game and method of playing the game are:

1. Use the regulation field of play. You must establish ground rules for out-of-play areas.
2. Teams are regulation slow pitch teams with 10 players, one at each official position.
3. The batting order will be by order of position number, 1 through 10. In other words, starting with the pitcher (position 1) as the first batter, progress to the short fielder (position 10) as last batter. The player(s) who made the last out(s) in an inning serve as base runner(s) for the situation set up in the next inning.
4. Both teams' half-innings of a single inning start with the same specified situation set up. For example, the first inning for each team starts with the #10 batter as a runner on first base. The team that is first up in the inning stays at bat until three outs have been made. The team that started the inning on defense then comes to bat and sets up the same situation and plays until making three outs. At the start of the second inning, a new situation is established. Because each team has the opportunity to play through the same situation as the other, it is important that complete innings are played.
5. All play is governed by the official slow pitch rules once the situation to start the half-inning has been established.

Reading a Situation Keys to Success Checklists

The Keys to Success Checklists outline aspects of both the defensive and offensive situations that you as a player need to mentally check off before each pitch. At first glance, the checklists look long. However, as you work with this mental exercise in a modified game, you will find the task easier than it looks. For example, when you take the field as a defensive player you know the inning and the score of the game before you get out to your position. You answer the questions about the batter before the batter gets into the box. Determine the number of outs, position of runners on base, and the ball and strike count on the batter during the inning. The answers to these questions *must* run through your mind before every pitch. As the batter, you answer most of the check items before you step into the batter's box. The same is true if you are a baserunner.

Reading a Defensive Situation

1. What inning is it?	1__2__3__4__5__6__7 __ __ __
2. What is the score of the game?	Visitor ____ Home ____
3. How many outs are there?	0 outs ____ 1 out ____ 2 outs ____
4. Runners on what bases?	First ____ Second ____ Third ____
5. Speed of the runners on base?	Fast ____ Slow ____ at first
	Fast ____ Slow ____ at second
	Fast ____ Slow ____ at third
6. Who is the batter?	
a. Know batter!	Yes ____ No____
b. Right- or left-handed hitter?	Right ____ Left____
c. Running speed?	Fast ____ Slow____
d. Strong pull hitter?	Yes ____ No____
e. Line drive hitter?	Yes ____ No____
f. Long-ball hitter?	Yes ____ No____
g. Contact hitter?	Yes ____ No____
h. Can hit to opposite field?	Yes ____ No____
i. Can hit with power to opposite field?	Yes ____ No____
7. What is the ball and strike count?	____ Balls and ____ Strikes
8. How does pitcher like to pitch to the batter?	Inside ____ Outside____
	High ____ Low____

Reading an Offensive Situation

1. What inning is it?	1__2__3__4__5__6__7 __ __ __
2. What is the score of the game?	Visitor ____ Home ____
3. How many outs are there?	0 outs ____ 1 out ____ 2 outs ____
4. Runners on what bases?	First ____ Second ____ Third ____
5. Speed of the runners on base?	Fast ____ Slow ____ at first Fast ____ Slow ____ at second Fast ____ Slow ____ at third
6. Major objective for batter?	Get on base ____ Advance the runner(s) ____ Score runner(s) ____
7. What is the ball and strike count?	____ Balls and ____ Strikes
8. Positioning of the outfield defense?	SF aligned with LF, CF, and RF SF in front of LF, CF, and RF in L ____ C____ R ____ area Playing straightaway ____ Shifted to left ____ Shifted to right ____ Largest gap(s) at LF foul line ____ left center ____ right center ____ RF foul line ____ Strongest arm at LF _____ CF _____ RF _____ SF _____ Weakest arm at LF _____ CF _____ RF _____ SF _____

(Cont.)

Reading an Offensive Situation (Cont.)

9. Positioning of the infield defense?

3b shifted toward SS ____ third ____
deep ____ shallow____

SS shifted toward second ____ third ____
deep ____ shallow____

2b shifted toward second ____ third ____
deep ____ shallow____

1b shifted toward second ____ third ____
deep ____ shallow____

Strongest arm at
P ___ C ___ 1b ___ 2b ___ 3b ___ SS ___

Weakest arm at
P ___ C ___ 1b ___ 2b ___ 3b ___ SS ___

Step 20 Throwing Sidearm

The sidearm throw should be used sparingly at a beginning level of softball. However, there are instances when a less forceful throw than the full overhand throw is needed. Use the sidearm throw for short-distance throws, such as the second baseman's throw to first base after fielding a ground ball, and the close feeds at any base for a force play or first out of a double play.

WHY IS THE SIDEARM THROW IMPORTANT?

The trajectory of the sidearm throw is horizontal or slightly low to high. Most throws should arrive to the fielder about chest-high. A short-distance overhand throw tends to go high to low, arriving at the fielder's feet, making for a very difficult catch. The sidearm throw in this situation stays horizontal or rises slightly to the fielder, arriving at the desired chest height. Although the sidearm throw is less accurate than the overhand, the shorter distance in which the throw is used reduces the likelihood of error. The sidearm throw also gets to the intended base more quickly because the thrower releases the ball from a crouched position, rather than coming to the erect overhand throwing posture.

HOW TO THROW SIDEARM

As you field a ground ball, bring the ball to the throwing position, keeping your back flat, knees bent, and torso parallel to the ground. If you are a right-handed second baseman throwing to first, or a right-handed shortstop throwing to second, your glove side is already pointed toward your target so you don't need to pivot. Simply bring the ball across the front of your body, keeping your throwing arm parallel to the ground (see Figure 20.1). All other aspects of the throw are exactly the same as for the overhand throw.

When using the sidearm throw to your throwing-arm side, though, pivot your feet after fielding the ball so your glove side is toward the throwing target and then proceed as described for throwing to the glove side. A left-handed player in the example above would be using the sidearm throw to the throwing side and therefore would need to use the pivot prior to throwing.

Figure 20.1 Keys to Success: Throwing Sidearm

Preparation Phase

Glove-Side Target

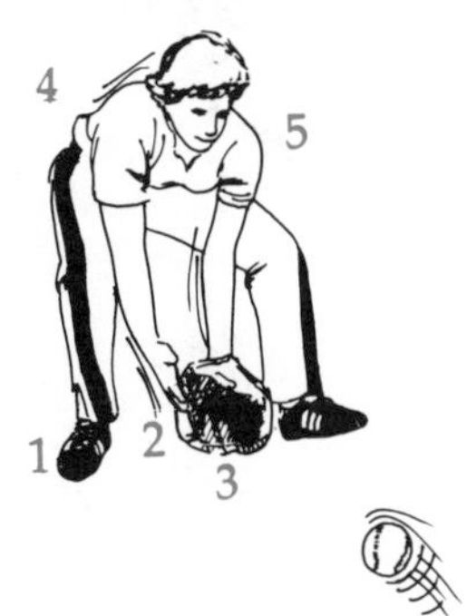

1. Feet more than shoulder width apart
2. Hands low, fingers pointed down
3. Glove open to ball
4. Back flat
5. Focus on ball

Throwing-Side Target

Execution Phase

Glove-Side Target

1. Shift weight to throwing side
2. Start glove and hand to throwing shoulder, two finger grip

Throwing-Side Target

1. Weight on throwing-side foot
2. Start pivot to throwing-side target

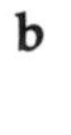

3. Stay low, flex at waist
4. Step toward target
5. Ball to throwing position
6. Glove elbow toward target
7. Weight to glove side
8. Hips square to target
9. Throwing arm parallel to ground
10. Snap wrist on release of ball

Follow-Through Phase

Glove-Side Target

1. Weight on glove-side foot
2. Knees bent
3. Throwing arm moves horizontally toward target
4. Glove elbow back

Throwing-Side Target

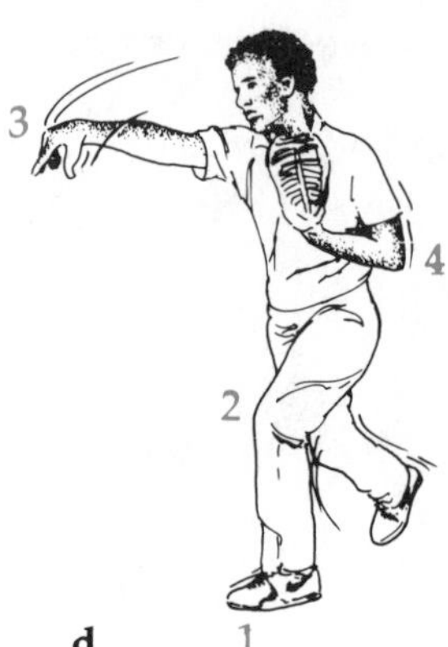

Detecting Sidearm Throwing Errors

Sidearm throwing errors are usually caused by an improper point of release and failure to keep the throwing arm parallel to the ground through the throwing movement. The most common sidearm throwing errors are listed below, along with suggestions on how to correct them.

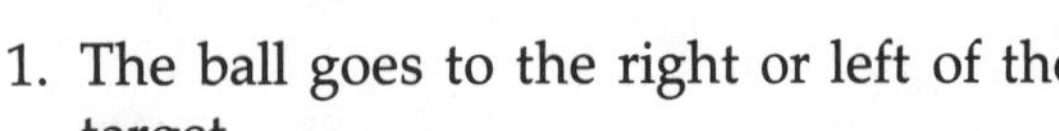

ERROR	CORRECTION
1. The ball goes to the right or left of the target.	1. Snap your wrist directly at the target, not before or after.
2. The trajectory of the ball is high or low.	2. Do not "stand up" to throw—stay low. Bend and keep your torso parallel to the ground.
3. You "push" the ball rather than throw it.	3. Lead with your elbow as you bring the ball across your body.
4. When you throw to a target on your throwing side, the ball is off-line.	4. Stay low, but pivot your torso so your glove side is toward the target.

Throwing Sidearm Drills

Note: The first three drills could be done outside, using a fence for the throwing target. Because the ball will not rebound off the fence, though, use a partner with a bucket of balls to roll a ball to you from a short distance.

If no partner is available, take 10 balls and place them 1 foot apart in a straight line in front of you, going away from you. Move up to field each ball and make the sidearm throw to the fence. Remember, these are not accuracy drills; your target is simply the wall or the fence.

If your skill in throwing is good and you *want* to work on accuracy, though, make small targets on the wall or fence. Throw to hit those targets.

1. Wall Drill to Glove Side

This drill is best done in a gymnasium. Stand in a corner of a gym, facing one wall, with the other wall to your glove side. Position yourself 30 feet from each wall.

Throw the ball against the front wall so that a ground ball rebounds to you. Field the ball and make a sidearm throw to the wall on your glove side, using the techniques described in the Keys to Success (see Figure 20.1) for the sidearm throw. Turn to the wall on your glove side, field the rebounding ball, and move it directly into throwing position. Do not immediately throw the ball. Assume your starting position. Repeat the sequence.

The major purpose of this drill is for you to practice the sidearm throw. However, take the opportunity to work on your ground ball fielding skill, also!

Success Goal = 10 consecutive error-free sequences

Your Score = (#) ____ sequences

2. Wall Drill to Throwing Side

Stand as in the previous drill, except what was the side wall is now the front wall, and the side wall is now on your throwing-arm side.

Throw the ball to the front wall so that the rebound is a ground ball. Field the ball, pivot to your throwing side, and throw the ball sidearm to that wall. Field the rebound and return to your starting position.

Success Goal = 6 consecutive error-free sequences

Your Score = (#) ____ sequences

3. *Combined Wall Drill*

As in the previous drills, stand in a corner of a gym, 30 feet from each wall.

Face one wall, throw the ball, field the ground ball, and make a sidearm throw to the other wall. Field the rebounding throw and make a sidearm throw to the first wall. Be sure to pivot your feet on throws to the throwing-arm side. Continue with no break in the sequence. Work on making the fielding and the ensuing throw all one continuous movement.

Continue this drill until you have completed 30 sequences (throw to wall, field rebound, throw to other wall, field rebound). Count the number of error-free sequences you make in a row. If you make an error, begin the count again from 1.

Success Goal = 10 consecutive error-free sequences

Your Score = (#) ____ sequences

4. *Partner Drill*

You will now practice sidearm throws after fielding a partner's fungo hits. Again, position yourself in the corner of a gym as if on an infield. Place small targets on the wall.

a. Your partner has a bucket of 20 balls and hits fungo ground balls to you. You field them and make sidearm throws to the target on the wall. You field the rebound and roll the ball back to your partner.

   You first field 10 balls hit to your glove side, then 10 hit to your throwing side. These balls should be hit at moderate speed and should not make you take more than one step in either direction. Count the number of good fielding plays and on-target throws. Reverse roles with your partner and repeat.

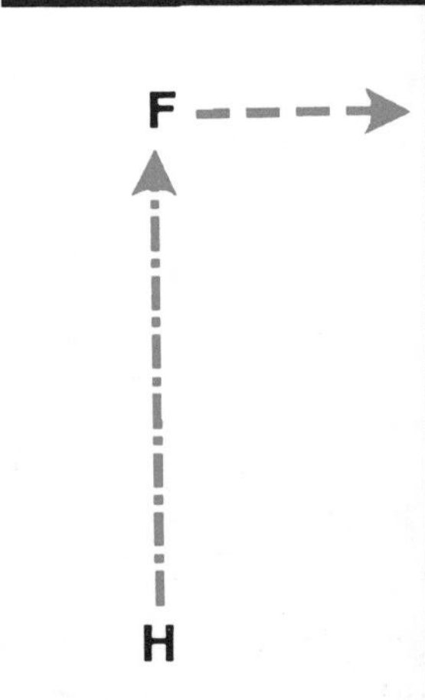

b. Next, ask your partner to vary the speed and the range of the hits, and alternatively hit 10 balls to your glove side, then to your throwing side. Reverse roles and repeat.

c. In order to get you to react to the ball coming off the bat, the fungo hitter now randomly varies the speed, the range, and the side to which the hit is made. Field 10 balls and throw to the wall target. Reverse roles and repeat.

Success Goal = 32 out of 50 good fielding plays and on-target throws

a. 7 of 10, glove side
   6 of 10, throwing side
b. 7 of 10, glove side, various speeds
   6 of 10, throwing side, various speeds
c. 6 of 10, random sides, random speeds

Your Score =

a. (#) ____ glove side
   (#) ____ throwing side
b. (#) ____ glove side, varied speeds
   (#) ____ throwing side, varied speeds
c. (#) ____ random sides, random speeds

5. Four-Player Drill

Now you will practice fielding and throwing the ball to a partner. One important consideration in this drill is that your short sidearm throw be ''sympathetic'' in force so it can be caught.

This drill requires four people. Set up this drill as on a field. You, the primary fielder, play in the position of shortstop. A partner with a bucket of 10 balls fungo hits ground balls to you from home plate. Another partner plays second base and the last plays third base.

The hitter hits 10 ground balls of various speeds and directions to you. In one smooth motion, you field each ball and throw sidearm to the base on the side the ground ball came. Focus first on clean fielding, then on the throw. Be sure that your throw does not overpower the base player. The base player catches the ball and rolls it back to the hitting station.

Count the number of good fielding plays and sidearm throws. After your 10 practice attempts, rotate positions and repeat the drill. Were you more successful going to your glove or your throwing side? Were your throws at a force that was catchable? Work on those areas that gave you problems so that you can do them consistently. Continue until each partner has completed 10 attempts at fielding and throwing sidearm to a base player.

Success Goal = 6 of 10 good fielding plays and sidearm throws

Your Score = (#) ____ fielding plays and throws

Throwing Sidearm Keys to Success Checklist

If you want to play shortstop or second base, you must be able to execute the sidearm throw with accuracy. Outfielders, on the other hand, seldom have occasion to use this skill. You may not have a preference of positions at this stage in your development, so you should learn all the skills.

Ask your teacher or another trained observer to check your sidearm throwing ability to both glove and throwing sides. Except for the pivot shown in Figure 20.1b, the techniques are the same for both sides.

Preparation Phase

Glove-Side Target

Throwing-Side Target

The thrower

____ has the feet more than shoulder width apart, with the glove-side foot slightly ahead as in the execution phase of fielding a ground ball.

____ has hands low, fingers down, and the glove open to the ball.

____ stays low, with the back horizontal to the ground.

____ focuses on the ball.

Execution Phase

Glove-Side Target | **Throwing-Side Target**

The thrower

____ has weight shifted to the throwing-side foot.

____ begins the pivot if the target is to the throwing side.

____ has the hands separated, the glove elbow pointed at the target, and the ball hand away from the target in throwing position with a two-finger grip on the ball.

____ shifts weight or steps in the direction of the throwing target with the glove-side foot (depends on distance of target).

____ stays low, keeping the back as close to horizontal as possible.

____ forcefully extends the throwing-side forearm toward the target, keeping it horizontal, and snaps the wrist, releasing the ball toward the target.

____ has hips square to the target.

Follow-Through Phase

Glove-Side Target | **Throwing-Side Target**

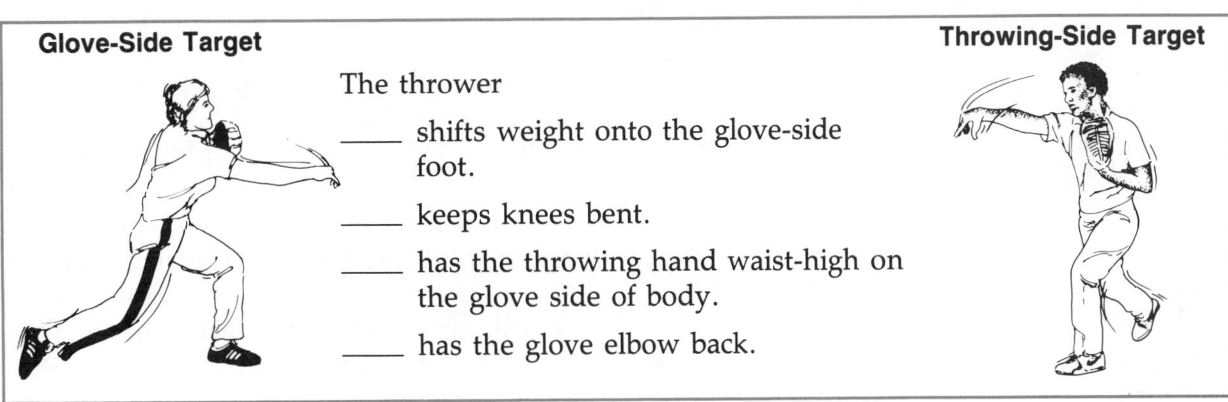

The thrower

____ shifts weight onto the glove-side foot.

____ keeps knees bent.

____ has the throwing hand waist-high on the glove side of body.

____ has the glove elbow back.

Step 21 Double Plays

As you have learned, your beginning-level defensive strategy was to get one out at a time; with less than two outs and runners in a force situation, the play should be on the lead runner. In this step, you focus on the double play, an advanced play that calls upon the defense to get two outs from continuous action.

The most common double play occurs with one or more runners on base in a force situation with less than two outs. The batter hits a ground ball to an infielder, who throws to a base to put the lead runner out, immediately followed by that covering player's throw to first base to put the batter-baserunner out. Infield double plays are made second to first when there is a lone runner on first base, third to first with runners on first and second, and home to first when the bases are loaded.

A double play occurs anytime two players are put out during continuous action: for example, a fly or line drive out, and the baserunner is put out after failing to tag up; a fly out, and the baserunner is thrown out attempting to advance; in fast pitch a strikeout on the batter, and a baserunner is thrown out trying to steal a base. However, in this step your focus is on the second-to-first double play on an infield grounder. The skills that you will work on for this double play are the shortstop's drag step and inside pivot techniques, and the second baseman's crossover and rocker pivot techniques.

WHY IS THE DOUBLE PLAY IMPORTANT?

The double play is important because the defense can get two players out on one pitch. Not only is the batter put out, but a baserunner already on base is also put out as part of the continuous play.

In game play, the successful execution of a double play can deal a psychological blow to the offensive team and at the same time can serve as a significant motivator to the team that has successfully executed it. Thus, the momentum in a game often changes hands as a result of a successful double play.

HOW TO EXECUTE THE DOUBLE PLAY

Double plays on ground balls are typically executed with a throw from an infielder to second base, or to third base, or to home plate. The baseman playing the base to which the infielder's throw was made then throws the ball to first base. The initial out is made on the lead runner, who is forced; the second out is made on the batter at first base, which is also technically a force-out. In fast pitch softball, where runners can lead off the base on the pitch, the only hope of getting the second out rests with the batter. Even in slow pitch, the batter will be at least a step behind the other baserunners because of completing the swing and shifting gears to get out of the batter's box. The right-handed batter in particular has a greater distance to run than any other runner because he or she is on the side of home plate farthest away from first base. The recommendation for both fast and slow pitch play is to get the second out of the double play at first base.

You have already had experience executing portions of the double play, namely, throwing, catching, and footwork for a force play. The sidearm throw you worked on in the previous step is used by the shortstop when executing the *feed* (throw) to the second or the third baseman for the first out of the double play. If the second baseman initially fields the ball, he or she uses the sidearm throw to feed the ball to the shortstop who is at second base for the *front end* (first out) of the double play. You use the overhand throw, which you can execute well by this point, when the throw is fairly long for either of the outs in the double play.

When executing the double play, you need to make only slight additions to the footwork technique you have already practiced for force-outs at home and at third. To make a strong throw for the second out of the double play, you need to take a step toward first base as you throw the ball. Thus, for the force-out at third or home for the first out of the double play, do not stretch out as far as you would to catch the ball on a regular force-out. You need to be in balance to take the step toward first base on your throw for the second out. Stretching to catch the ball for the first out of the double play puts your body in an unbalanced position. You will consequently be slower and less powerful in your throw to first base. On the other hand, if you're the first baseman, you must stretch as far as possible in order to shorten the length of the throw for the second out. You try to cut down the time it takes for the ball to arrive at first base because the play there is usually very close.

You have not had experience to date with the footwork involved in making the first out of the double play at second base. This footwork is much different from the footwork done at third or home when that is the location of the first out. In your practice as both shortstop and as second baseman, you have learned how to make a force play at second base, but not a force play that must be followed by a throw to first base. The footwork for the double play at second base is called a *pivot*. Basically, your pivot must position you to make the force-out on the runner at the base, let you move into position to make a strong throw to first base for the second out, and get you out of the way of the runner.

In order to be in a position for a strong throw to first base, you must be out of the path of the oncoming runner, and you must also be able to step in the direction of first base on the throw. In softball, unlike baseball, the first out of the double play at second base is almost always made with the runner close to the base. Therefore, the need to get out of the path of the runner is ever present for you, the fielder, making the pivot at second base. You do not want to be knocked down by the runner, because then you cannot make the throw to first. In addition, you must have a clear path for the ball to travel to first base; the path must not be blocked by the runner. The runner has no option but to go to the base. You, therefore, must be the one to move away from the base in order to find the clear path needed for your throw. The following sections provide the techniques used by the shortstop and the second baseman when making the pivot on the double play at second base.

Note: Descriptions of techniques for executing the first out of the double play at second base, as used by the shortstop (and, later, the second baseman), are for a right-handed player. A left-handed player would have great difficulty getting into position to receive the throw at the base for the force-out, and then to get out of the way of the runner, still leaving the left side of second base unobstructed for the throw to first base. For this and other reasons, left-handers do not usually play shortstop or second base.

HOW TO EXECUTE THE SHORTSTOP'S DRAG STEP

The *drag step* is used by the shortstop to make contact with second base while simultaneously catching the ball for the force-out and positioning the body for the throw to first base for the second out. This technique gets the shortstop out of the baseline so that the runner cannot in any way be in the line of the throw. The drag step is used when the feed for the play is thrown by the second or first baseman from the outfield side of the baseline between first base and second base.

When the ground ball is hit to the first-base side of the infield in a double play situation, you, the shortstop, move from your initial fielding position to one just behind second base. You straddle the back corner of the base (the corner pointing to center field) with the inside of your right foot contacting the very corner of the base and your shoulders parallel with an imaginary line between first and third base. If there is not time to get to the base and stop in the straddle position over the back corner, move through that position without stopping. As you catch the ball to make the force-out, step toward the right outfield grass with your left foot (see Figure 21.1a), dragging the toes of your right foot across the back corner of the base (see Figure 21.1b). Still moving, close your right foot to your left (see Figure 21.1c), step out with your left foot again (see Figure 21.1d), and throw to first base for the second out (see Figure 21.1e,f). To accomplish these objectives, without the runner's interference, the direction of your movement past the base must be toward the right outfield area *not* down the baseline toward first base!

Figure 21.1 Keys to Success: ***Shortstop's Drag Step***

Preparation Phase

1. Straddle back corner
2. Weight on right foot
3. Face thrower
4. Focus on ball
a
1
2
3
4

Execution Phase

1. Catch ball
2. Step past base with left foot
3. Drag right foot across back corner of base
4. Close right foot to left
5. Step with left foot
6. Ball to throwing position
7. Throw to first base

Follow-Through Phase

1. Weight on glove-side foot
2. Knees bent
3. Throwing arm horizontal
4. Glove arm back

f

HOW TO EXECUTE THE SHORTSTOP'S INSIDE PIVOT

In Step 10, "Position Play," you learned that the shortstop is responsible for covering second base on a double play when the ball is hit to the catcher or to the pitcher. The reason for this coverage is that the shortstop has easier skills to execute than the second baseman does when making the play at second base. The shortstop, when moving toward second base from the regular fielding position, is already going in the general direction of first base. Thus, the whole flow of the play is in the direction of the throw for the second out of the double play. The movement of the second baseman going to cover second base, on the other hand, is away from first base and, therefore, away from the direction of the throw for the second out.

Similar principles apply to covering a base for a double play force-out, as for any force play. As you know, when the feed for the force play is coming from the infield side of the baselines, the covering position is at the inside corner of the base. In a double play situation, when the ball is hit up the middle to either the pitcher or the catcher, the shortstop covers second base on the inside corner and uses an *inside pivot* to make the force-out and complete the throw to first base for the double play. On the other hand, if the ball is clearly hit to the first-base or third-base side and is played from the infield side of the baselines, the regular guidelines for base coverage apply (grounder to first-base side, shortstop covers second base; third-base side, second baseman covers).

When the ball is hit up the middle, you, the shortstop, come to the inside corner of second base, step on the inside corner of the base with your left foot, and face the player making the feed. As you make the catch for the force-out, bend your knees, take the weight fully on your left foot, and spring away from the base (off your left foot) toward the pitching rubber, landing on your right foot well clear of the baserunner. Step toward first base with your left foot, and make the throw for the second out of the double play (see Figures 21.2a-g).

*Figure 21.2 Keys to Success: **Shortstop's Inside Pivot***

Preparation Phase

1. Step on inside corner with left foot
2. Hips square to thrower
3. Focus on ball

a

Execution Phase

1. Catch ball
2. Take full weight on left foot
3. Bend knees
4. Spring from base off left foot
5. Land on right foot
6. Step onto left foot
7. Ball to throwing position
8. Throw to first base

Follow-Through Phase

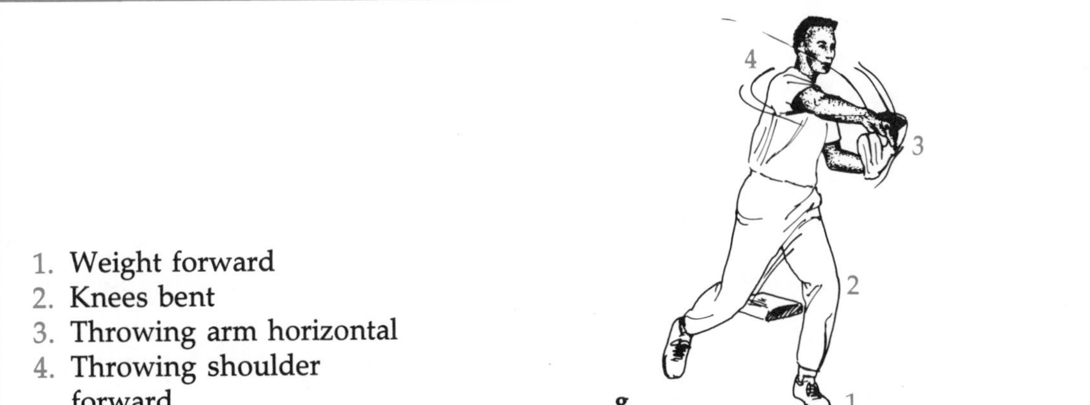

1. Weight forward
2. Knees bent
3. Throwing arm horizontal
4. Throwing shoulder forward

HOW TO EXECUTE THE SECOND BASEMAN'S CROSSOVER PIVOT

For the second baseman, turning the double play at second base involves a true pivot, or change of direction. The second baseman must clear the base after making the force-out, stop the momentum of the body going away from first base, and step back toward first on the throw. The second baseman uses the crossover step pivot when the feed is coming from the infield side of the baseline between second and third, or when timing demands shortening the distance the feed has to travel (that is, on any throw from the third baseman and a long throw from the shortstop). As the second baseman, you should move to cover second base from your regular fielding position so that you can cross the base in a direct line toward the person feeding you the ball. If time allows, move to a position just short of the base, facing the direction of the incoming throw. As the thrown ball approaches, step on the base with your left foot, move over the base, and catch the ball on the far side of the base while still in contact with the base with your left foot for the force-out.

Immediately block your forward momentum by landing on your right foot and bending that knee. This step onto your right foot should carry you well clear of the base in order to get out of the path of the runner. Shift your weight to the left side. Step left and throw the ball to first base (see Figures 21.3a-e).

The major-league style of play calling for the second baseman to leap straight up into the air over second base to avoid the incoming runner, spin in the air, and—while suspended in midair—make the throw to first base is technique above and beyond our expectations of you in this book!

Figure 21.3 ***Keys to Success: Second Baseman's Crossover Pivot Step***

Preparation Phase

1. Weight on right foot
2. Face thrower
3. Focus on ball

Execution
Phase

b

c

d

1. Step on base with left foot
2. Hands to ball height
3. Body moves forward over base
4. Catch ball
5. Block with right foot
6. Step left to target
7. Shift weight to left and throw

Follow-Through
Phase

1. Weight on glove-side foot
2. Knees bent
3. Hips square to target
4. Throwing shoulder forward

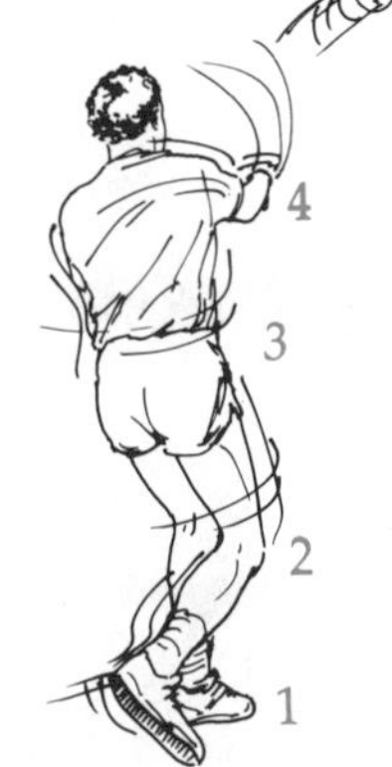

e

HOW TO EXECUTE THE SECOND BASEMAN'S ROCKER PIVOT

The *rocker pivot* is used by the second baseman when the feed from the shortstop is initiated very close to second base. As second baseman, you move to the base and place the toes of your right foot in contact with the outfield side of the base. With your weight on your left foot, catch the ball (making the force-out), step back onto your right foot, step left toward first base, and throw the ball. Stepping back with the right foot in this way, called the *deep drop*, gets you out of the path of the runner. Another rocker technique involves standing with your weight on your *right* foot, kicking the base with your *left* foot for the force-out, and stepping left to throw. This move, the *short drop*, is quicker by one step, but leaves you in the baseline. Base your choice of footwork technique on the position of the runner at the time of the play at the base. Use the deep drop if the runner is close and the short drop if the runner is farther away. Figures 21.4a-e show how to execute both the deep and short drop rocker pivots.

Figure 21.4 Keys to Success: ***Second Baseman's Rocker Pivots***

Preparation Phase

Deep Drop

1. Approach base, weight on left foot
2. Hands to ball height
3. Focus on ball
4. Right foot contacts base
5. Catch ball

Short Drop

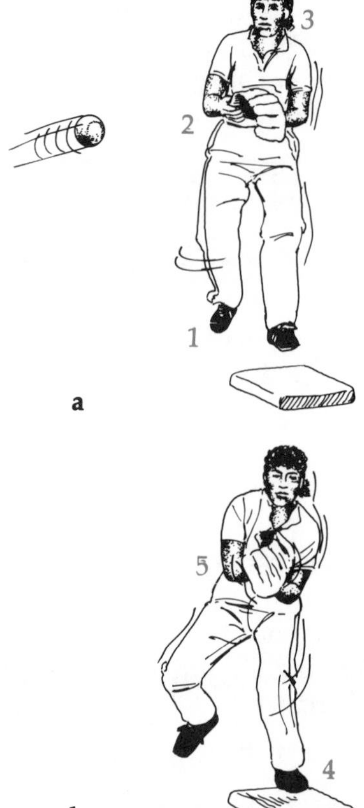

1. Approach base, weight on right foot
2. Hands to ball height
3. Focus on ball
4. Left foot contacts base
5. Catch ball

Execution Phase

Deep Drop

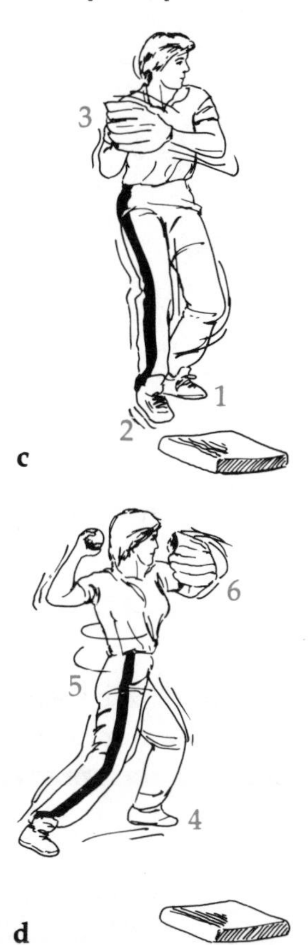

1. Weight on left foot
2. Step back on right foot
3. Ball to throwing shoulder
4. Step left toward target
5. Right hip drives forward
6. Left elbow points toward target, throw to first base

Short Drop

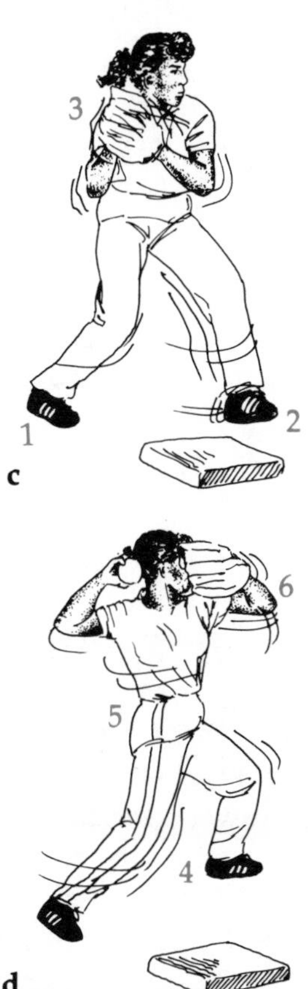

1. Weight on right foot
2. Step left toward target
3. Ball to throwing shoulder
4. Weight to left side
5. Right hip drives forward
6. Left elbow points toward target, throw to first base

Follow-Through Phase

Deep Drop

Short Drop

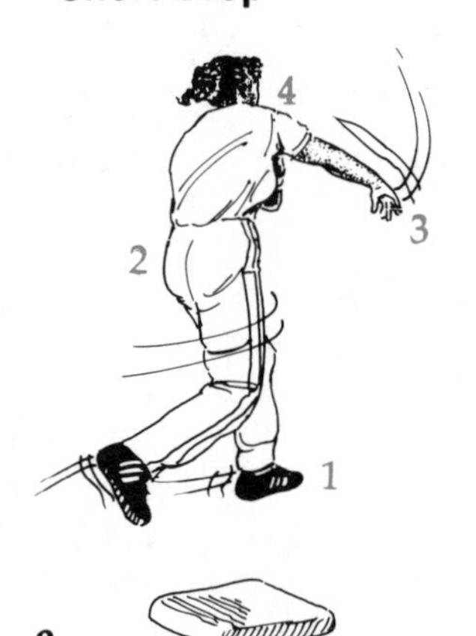

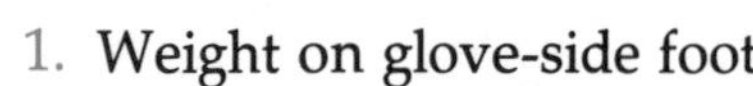

1. Weight on glove-side foot
2. Hips square to target
3. Throwing hand low
4. Throwing shoulder forward

Double Play Drills

1. Mimetic Footwork Drill

You and a partner are positioned at the regular shortstop and second base fielding positions. Without using a ball, you and your partner practice the footwork for double plays. First the shortstop moves to second base and executes the correct footwork for the drag step. That player goes back to the starting shortstop position and repeats the drill for a total of 5 times. Then the shortstop practices the inside pivot.

The partner in the second baseman's position then executes the crossover step pivot. Be sure to go back to the regular fielding position between repetitions. After completing 5 crossover step pivots, the second base player practices the rocker step pivot.

After each of you practices both pivots in the initial fielding position, switch positions and practice the other pivots.

Success Goal = 20 correctly executed pivots

5 shortstop drag steps

5 shortstop inside pivots

5 second baseman crossover step pivots

5 second baseman rocker step pivots

Your Score =

(#) _____ drag steps

(#) _____ inside pivots

(#) _____ crossover pivots

(#) _____ rocker pivots

2. *Full Double Play Mimetic Drill*

Position yourselves as in the previous drill. Repeat the drill but add a mimetic catch of the ball for the force-out each time. The second baseman should practice the feed mimetically when the shortstop is executing the double play footwork. The shortstop should practice the feed mimetically when the second baseman is practicing the footwork. After the mimetic catch, the player continues the action by mimetically throwing the ball to first base for completion of the double play.

Be sure to go back to your starting fielding position between mimetic practices. Do 5 repetitions of each of the four types of footwork.

Success Goal = 20 correctly executed catch-pivot-throw combinations

5 drag step combinations

5 inside pivot combinations

5 crossover pivot combinations

5 rocker pivot combinations

Your Score =

(#) ____ drag step combinations

(#) ____ inside pivot combinations

(#) ____ crossover pivot combinations

(#) ____ rocker pivot combinations

3. *Simulated Hit Drill*

This drill requires three pairs of partners. One pair consists of a feeder and a pivot player. A second pair is the first base cover player and a "hitter." The third pair is a catcher and a pitcher/third baseman (this person plays either position, according to the chart below).

All fielders take regular fielding positions. The hitter stands only 20 feet from the feed player on a direct line to home plate. There should be a bucket of 5 balls beside the hitter.

The hitter merely rolls a ground ball to the feed player, who throws to the pivot player covering second base. This player, after correct footwork to tag the base, throws to first base for the completion of the double play. The first baseman throws the ball to the catcher.

Repeat the sequence 5 times for each of the variations on the chart. After completing the 6 variations, exchange positions within pairs. When both partners have completed the double play variations, the pairs rotate. The hitter and the first baseman become the shortstop and the second baseman. The catcher and the third baseman/pitcher become the hitter and the first baseman. The shortstop and the second baseman become the catcher and the third baseman/pitcher.

If you and the others are more experienced players, you can add complexity to the drill by having a baserunner run from first to second base on each play. If you do, be sure the baserunner wears a batting helmet to protect against errant throws.

Feeder, pivot player, and first base coverage variations:

Feeder	Pivot player	First base cover
Second baseman	Shortstop	First baseman
First baseman	Shortstop	Second or first baseman
Catcher	Shortstop	First baseman
Pitcher	Shortstop	First baseman
Shortstop	Second baseman	First baseman
Third baseman	Second baseman	First baseman

Success Goal = 4 of 5 successful executions of each variation (24 of 30 total) required as feeder and pivot player

Your Score = total (#) ____ successful variation executions as feeder and pivot player

4. Fungo Double Play Drill

This drill is set up and practiced like the previous drill, except the hitter fungo hits ground balls from home plate to the feed players. The hitter starts with a bucket of balls at home plate. The first base cover doesn't throw the balls to the catcher, but deposits them in an empty bucket. In addition, there is a fourth pair of players who serve as runners.

This drill is more gamelike because you execute the double play off of a hit ball. Furthermore, on the last 3 of the 5 practice sequences of each play variation (see Drill 3), the baserunners are added. One runner goes from home to first on the ground ball while the other runs from first to second. Baserunners, *be sure you wear batting helmets.*

Success Goal = 4 of 5 successful executions of each variation (24 of 30 total) required as feeder and pivot player

Your Score = total (#) ____ successful variation executions as feeder and pivot player

5. Home-to-First Double Plays

By now you should be proficient as both pivot player and feeder for all of the double play combinations from second to first base. The following variations are possible combinations for home-to-first double plays:

Feeder	Pivot player	First base cover
Third baseman	Catcher	First baseman
Shortstop	Catcher	First baseman
Second baseman	Catcher	First baseman
Pitcher	Catcher	First baseman
First baseman	Catcher	Second baseman

All of the previous drills can be set up for practicing the double play when the first out occurs at third or home. The coverage at third is by the third baseman, except when the ball is hit to the third baseman, in which case the shortstop or pitcher covers third for the first out of the double play. The coverage at home is by the catcher, and the coverage at first is by the first baseman.

Set up with a catcher, a first baseman, a runner at third, a runner at home, a hitter, and a fielder who moves around to play each of the various feed positions. Five repetitions at each different infield position give the fielder good practice making feeds (overhand throws) to the catcher. The catcher practices positioning to receive throws from the various infield positions while at the same time getting out of the way of the runner and making the throw to first base.

Success Goals =

a. 20 of 25 double plays as feeder

b. 20 of 25 double plays as pivot player

c. 20 of 25 covers at first base

Your Score =

a. (#) ____ out of 25 double plays as feeder

b. (#) ____ out of 25 double plays as pivot player

c. (#) ____ out of 25 covers at first base

Double Play Keys to Success Checklists

The double play is an extremely important defensive play. Double plays in fast pitch softball are very difficult to accomplish because a baserunner can lead off when the ball is pitched and is consequently that much closer to the next base when the ball is hit. Making the double play in slow pitch softball is easier because the baserunner cannot leave the base until the ball is hit. Still, success in completing the second-to-first double play depends largely upon your correct and timely execution of the techniques you have worked on in this step.

Ask your teacher or another trained observer to use the following checklists to evaluate your performance in executing the various double play pivots as a shortstop and as a second baseman.

Shortstop's Drag Step

Preparation Phase

The shortstop

____ straddles the back corner of second base.

____ has weight on the right foot.

____ faces the thrower with hips and shoulders square.

____ focuses on the ball.

Execution
Phase

The shortstop

____ catches the ball with both hands.

____ steps past the base with the left foot.

____ drags the toes of the right foot across the back corner of the base.

____ closes the right foot to the left, bringing the ball to the throwing shoulder.

____ brings the ball to the throwing position.

____ steps toward first base with the left foot.

____ throws the ball to first base.

Follow-Through
Phase

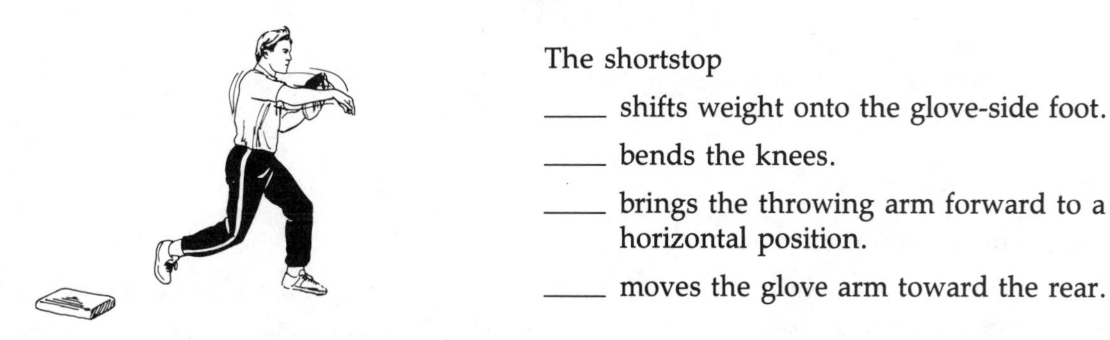

The shortstop

____ shifts weight onto the glove-side foot.

____ bends the knees.

____ brings the throwing arm forward to a horizontal position.

____ moves the glove arm toward the rear.

Shortstop's Inside Pivot

Preparation Phase

The shortstop

____ steps on the inside corner of second base with the left foot.

____ has hips square to the thrower.

____ focuses on the ball.

Execution Phase

The shortstop

____ catches the ball in two hands.

____ takes the full weight on the left foot, freeing the right foot.

____ bends the knees.

____ springs off the base toward the pitching rubber.

____ lands on the right foot.

____ brings the ball to the throwing position.

____ steps toward first base with the left foot.

____ throws the ball to first base.

Follow-Through Phase

The shortstop

____ shifts weight forward.

____ bends the knees.

____ brings the throwing arm forward to a horizontal position.

____ has the throwing shoulder forward.

Second Baseman's Crossover Pivot

Preparation Phase

The second baseman

____ stands behind second base with weight on the right foot.

____ faces the thrower.

____ focuses on the ball.

Execution Phase

The second baseman

- ____ steps on second base with the left foot.
- ____ keeps the hands at ball height.
- ____ moves the body forward over the base.
- ____ catches the ball in both hands.
- ____ blocks the forward momentum by stepping with the right foot, bending the knee, and shifting weight back.
- ____ steps toward first base with the left foot.
- ____ shifts weight forward and throws the ball to first base.

Follow-Through Phase

The second baseman

- ____ has weight on the glove-side foot.
- ____ bends the knees, has hips square to the target.
- ____ has the throwing shoulder forward.

Second Baseman's Rocker Pivots

Preparation Phase

The second baseman

Deep Drop	*Short Drop*	
____		approaches second base shifting the weight onto the left foot.
	____	approaches second base shifting the weight onto the right foot.
____	____	brings hands to ball height.
____	____	focuses on the ball.
____		places the right foot in contact with the base.
	____	places the left foot in contact with the base.
____	____	catches the ball.

Execution Phase

The second baseman

Deep Drop	*Short Drop*	
____		with weight on the left foot, steps back toward right field with the right foot.
	____	with weight on the right foot, steps toward first base with left foot.
____	____	brings the ball and glove to the throwing shoulder.
____		steps left toward first base.
	____	shifts weight to the left side.
____	____	drives the right hip forward.
____	____	points the left elbow toward the target and throws the ball to first base.

Follow-Through Phase

Deep Drop	*Short Drop*	The second baseman
____	____	shifts weight onto the glove-side foot.
____	____	has hips square to the target.
____	____	has the throwing hand low.
____	____	thrusts the throwing shoulder forward.

Deep Drop

Short Drop

Step 22 Relays and Cutoffs

Most recently, you have had opportunities to work on some skills and game concepts used in intermediate-level game play. Two new defensive skills, the relay and the cutoff, are not necessarily intermediate-level skills in terms of difficulty. However, the *relay* (especially) and the *cutoff* (in some instances) require a strong, accurate overhand throwing ability, and good catching ability. By now you should have had enough practice time to develop sufficient throwing and catching skills so that you will have success in executing both the relay and cutoff.

WHY ARE THE RELAY AND CUTOFF PLAYS IMPORTANT?

Both plays increase the defense's chance of making an out on a hit ball. The *relay*, as its name might imply, uses more than one player to get the ball to its destination. It is used when the throwing distance is too great for one player to execute a fast, accurate throw. The *cutoff*, interrupting the flight of a throw, is used when the baserunner being played on cannot be thrown out but another baserunner is *in jeopardy*, in a position to be put out. Without the use of these two plays in the situations described, all runners might be safe.

HOW TO EXECUTE THE RELAY

The relay is typically executed by the shortstop or the second baseman. When a hit ball goes past the outfielders in left and center fields, or when these fielders retrieve a hit ball and the throwing distance is beyond their capabilities, the shortstop goes out to receive the throw and relay it to its ultimate destination. Similarly, when a ball is hit to right field in like situations, the second baseman is the relay person. Another option in the slow pitch game is to use the short fielder as a relay person if he or she is covering the area between the other three outfielders and the infield and is in a good position to take the outfielder's throw. When the short fielder is the fourth outfielder and all are equidistant from home plate, the shortstop and second baseman must assume the relay responsibility.

The player executing the relay faces the outfielder with the ball and raises his or her arms to make a big target. The throw from the outfielder should arrive on the fly chest-high to the relay person. The relayer catches the ball, pivots by turning to the glove side, and throws the ball to the intended destination (see Figures 22.1a-f).

Figure 22.1 Keys to Success: Relay

Preparation Phase

a

b

1. Face outfielder with ball
2. Identify yourself as relay person
3. Raise arms to increase your size as target
4. Focus on ball
5. Extend hands to prepare to catch
6. Begin pivot by stepping toward target with glove-side foot.

Execution Phase

c

d

e

1. Catch ball
2. Complete pivot
3. Begin crow-hop step
4. Weight on throwing-side foot
5. Glove side toward target
6. Step toward target with glove-side foot
7. Throw ball using two-finger grip

Follow-Through Phase

1. Weight on glove-side foot
2. Throwing-side shoulder forward
3. Throwing hand pointed at target

HOW TO EXECUTE THE CUTOFF PLAY

In slow pitch softball, the cutoff play on a throw to home plate is executed by the first baseman (see Diagram 22.1) when the throw is from right or center field, and by the third baseman (see Diagram 22.2) when the throw is from left field. Because in fast pitch the first baseman plays in front of the baseline, much closer to home, the first baseman is the cutoff player on *all* throws from the outfield to home. On throws to third base from right and center field (see Diagram 22.3), the shortstop is the cutoff person in both slow and fast pitch softball.

As the cutoff person, you assume a position about 35 feet from the original target base, facing the source of the throw. Stand with the

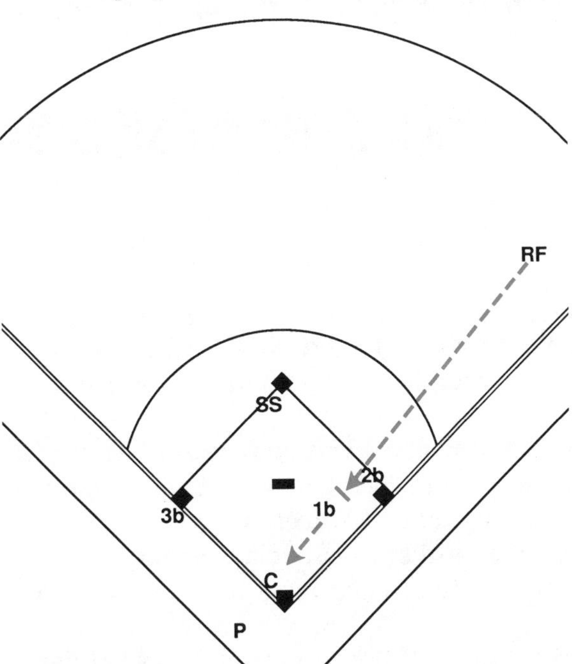

Diagram 22.1 Cutoff play on throw from right or center field to home.

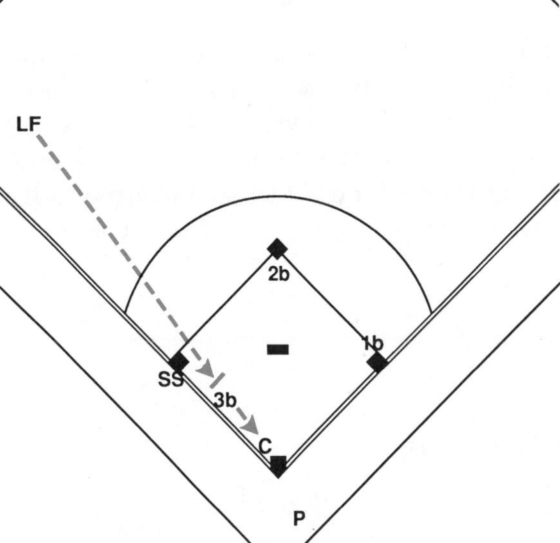

Diagram 22.2 Cutoff play on throw from left field to home (slow pitch softball).

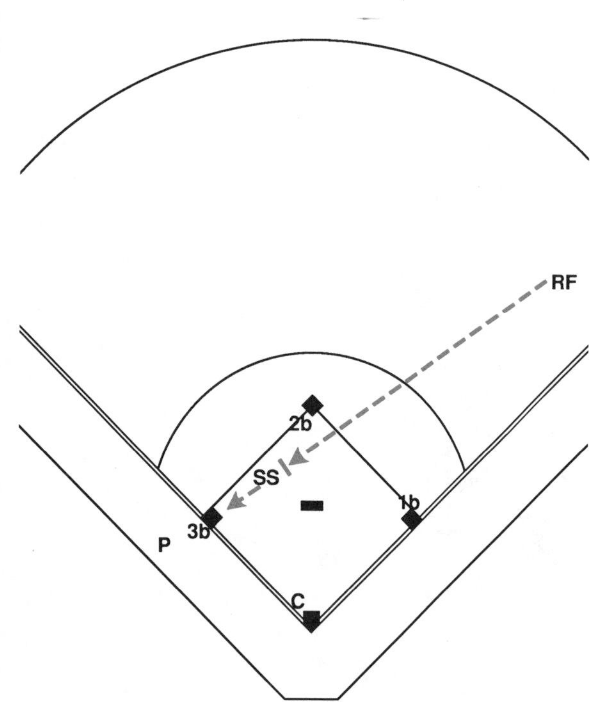

Diagram 22.3 Cutoff play on throw from right or center field to third base (slow or fast pitch softball).

glove side of your body adjacent to the line of flight of the ball so that you do not block the covering player's view of the ball. Because the play on the baserunner will be a tag play, the throw from the fielder should arrive at the covering player about knee-high. The throw should thus be passing you, the cutoff person, at about head or shoulder height.

Cutoff plays require strong communication between the intended fielder and the person who cuts off the throw. The covering player directs the cutoff play. If there is still a play to be made on the incoming baserunner and the throw is on target, the covering player says nothing, and you allow the throw to go through to him or her. On the other hand, if the incoming runner is already safe or the throw is off target, the covering player calls ''cut'' and further indicates to you by a verbal signal where (if at all) to play the ball. ''Cut second'' or ''cut two'' indicates that you should cut off the incoming throw and then make a throw to second base, making a play on the batter-baserunner trying to advance. On an off-target throw to home, ''cut home'' or ''cut four'' tells you to cut off the throw and relay the ball to home plate because there is still a play on the incoming runner. ''Cut'' with no further direction tells you to cut off the throw and simply hold it.

As with the relay person, any pivot toward home by you as the cutoff player is done to your glove side. Pivoting to your glove side puts your body in direct throwing position with only a quarter-turn change of direction.

Relay and Cutoff Drills

1. Mimetic Pivot Drill

You and a partner are in a ''boss-worker'' setup with the boss calling directions and the worker executing the pivots and mimetically throwing to the intended target. Both of you should use mental imagery as you participate in this drill.

At the command, ''hit the relay,'' the worker as an outfielder turns and runs a few steps after an imaginary ball. He or she picks it up from the ground, pivots to the glove side, and makes an overhand throw to an imaginary relay person about 100 feet away.

At the command, ''relay home'' or ''relay four,'' the worker as the relay person catches an imaginary throw from an outfielder at chest height. He or she pivots to the glove side and makes an overhand throw to home plate.

At ''cut second, third or home'' or ''cut two, three, or four,'' the worker as the cutoff person assumes a starting position as the first or the third baseman. From this position, she or he moves to the cutoff position for an imaginary throw to home. The worker follows the call of the boss by mimetically cutting the throw, pivoting, and making the throw to the base called.

Do 5 combinations (mimetically get or catch the ball and throw) in each role. The boss and worker change roles after practice at each position.

Success Goal = 4 out of 5 performances executed with correct form and technique at each position

Your Score =

(#) _____ combinations as outfielder

(#) _____ combinations as relay person

(#) _____ combinations as cutoff person

2. *Wall Drills for Relay Pivot and Accuracy*

On a wall, mark a 2-foot-square target five feet from the floor. Take a position 80 to 100 feet from the wall.

a. With your back to the target, slowly roll a ball away from you, run to retrieve it, pivot to your glove side, and make an overhand throw to the target. Field the rebounding ball and repeat the sequence. If a wall is not available, use a fence or a net and a bucket of 10 balls.

b. Now repeat the drill, but this time roll the ball away from you toward your glove side. Run to retrieve it, pivot to your glove side and make an overhand throw to the target. Field the rebounding ball and repeat the sequence for a total of 10 throws. Next, do the same thing except roll the ball to your throwing side.

Success Goals =

a. 8 out of 10 on-target throws

b. 8 of 10 on-target throws, glove side roll

8 of 10 on-target throws, throwing side roll

Your Score =

a. (#) _____ on-target throws

b. (#) _____ on-target throws, glove side roll

(#) _____ on-target throws, throwing side roll

3. *Three-Person Relay Drill*

Three of you stand 100 feet apart from each other in a straight line. One end person acts as an outfielder and the other as the catcher; the middle position is the relay person. The relayer and the outfielder are facing the catcher.

The catcher throws a ball past the outfielder (using a fence or wall to stop the ball facilitates the drill). The outfielder turns, retrieves the ball, pivots to the glove side, and throws to the relay person, who has turned to face the outfielder. The relay person at this point has his or her back to the catcher and cannot see the straight line relationship needed for an efficient relay play, that being the shortest distance between the outfielder and the catcher. Therefore, the catcher must be lining up the relay person between the moving outfielder and the catcher by giving verbal directions of ''right'' or ''left.'' The relay person catches the ball, pivots to the glove side, and throws to the catcher.

After 10 sequences, rotate positions. Score a successful catch and throw combination as 1 point. A point is scored when the outfielder fields the ball and makes an on-target throw to the relayer and when the relayer catches the outfielder's throw and makes an on-target throw to the catcher. There are 2 points possible on each relay sequence. Accumulate points for *consecutive* error-free throws by the outfielder and the relay person. Because your goal is consistency in performance, you must begin scoring again at 1 after any error in execution.

You will be attaining points when you are in the outfielder's position and when you are in the relayer's position. Add the points you get in each position to establish a total score. The object of this drill is for each of the three of you to come within 2 points of each other in total score.

Success Goal = 30 total points in two sets of 10 sequences, one set at outfielder and one set at relayer positions

15 points at outfielder position

15 points at relayer position

Your Score =

(#) _____ __________ __________ points as outfielder

(#) _____ __________ __________ points as relayer

_____ __________ __________ Total scores

(you) (partner 1) (partner 2)

Step 23 Scrub and One-Pitch Games

Modified game play in preceding steps provided you with opportunities to respond to predetermined game situations in controlled settings. They were designed to give you specific offensive and defensive situations to work on in gamelike conditions. You had the advantage of knowing in advance the particular tactic or technique that you would be called upon to use. The rules were designed to increase the number of opportunities you had to execute a particular technique.

WHY PLAY SCRUB AND ONE-PITCH?

The modified games, Scrub and One-Pitch, are less controlled and more like a regulation game. The situations that come up in each of these games result naturally from your play and the play of the others participating in the game. Thus, if no one gets on first base with fewer than two outs, you will not be able to work on the second-to-first double play. You now must recognize situations as they spontaneously occur and be able to effectively respond both mentally and physically.

The games in this step are modified to only a limited extent, primarily to increase participation. In the game Scrub, the rules of regulation softball are modified to allow you to play different defensive positions. The game One-Pitch is played like a regulation game with the exception of some new pitching rules that speed up the game so that you can be more active. The official rules govern the remainder of the play situations in both games. Thus, you can continue to develop your knowledge and understanding of the rules of the game. Also, review and use the defensive and offensive checklists from Step 19 to enhance your softball game sense as you participate in these games.

You may play a game for a variety of reasons. Having fun, enjoying the company of others, and soaking up the rays are all legitimate reasons for playing softball. However, playing a game can also be looked upon as a test of sorts. Although we may not always think of tests in a positive light, a softball game can be a fun test. It is a chance for you to see what you can do with the skills and knowledge you have learned. The game skills of hitting a ball, running the bases, fielding ground balls, making a force-out on a baserunner, and so forth can provide challenges.

As you play the modified games, list the purposes you know and execute with ease and with success. Also list the areas with which you have difficulty. The problem areas should become your focal points as you play additional innings in a game and as you play future games. Your weak areas can become your strengths; to make this happen, however, you need to recognize your problem skills and concepts, then work hard to master them. A complete softball player executes the skills proficiently and knows what every game situation calls for (review the game components outlined in Step 19).

SCRUB

Scrub is a good game to play when you do not have enough players for two full teams. It is also a good game for increasing your understanding of the total game because you must play all of the defensive positions.

Rules of Scrub and Method of Play

The following rules govern the method of play.

1. Use the regulation field of play. You must establish ground rules (that apply only to a given field and game) for out-of-play areas.
2. The participating group should number between 14 (minimum) and 16 (maximum). Ten players start in the field (one in each slow pitch position), and the remainder start as the ''team'' at bat.
3. A player remains in the group at bat until he or she makes an out. A player hitting a ball that is caught in the air by a fielder (a fly ball, a pop-up, or a line drive) exchanges places with *that* fielder before play continues. The fielder comes in to join the group at bat and hits at the end of the batting order. A player (either batter or baserunner) making an out on any other kind of play goes out into the field in Position 10 (short fielder). All fielders then rotate to the next lower numbered position, 10 to 9, 9 to 8, and so on. Remember, shortstop is Position 6, so the left fielder (7) rotates into the infield there and not to third base (5). The shortstop (6) rotates to third base (5), not around the infield to second base (4). The pitcher (1) moves to the end of the batting order of the group at bat.
4. After three outs have been made, clear the bases of runners and start a new half-inning. Remember, players remain in the group at bat until making an out.

Scoring

There can actually be no team score in the game of Scrub. Scoring, if desired, must be oriented toward the individual player. For example, keep track of the offensive statistics of each person: the number of times at bat; the number of base hits (singles, doubles, triples, and home runs); the number of bases on balls (walks); and the number of runs scored. At the end of a predetermined amount of time, winners are declared in each category. Use the Offensive Player Scorecard to record your score for this game.

OFFENSIVE PLAYER SCORECARD

Name ______________________________

Times at bat	_____	× 1 point	= _____	points
Walks	_____	× 1 point	= _____	points
Runs scored	_____	× 2 points	= _____	points
Base hits	_____			
Singles	_____	× 2 points	= _____	points
Doubles	_____	× 3 points	= _____	points
Triples	_____	× 4 points	= _____	points
Home runs	_____	× 5 points	= _____	points
		Total	= _____	points

DEFENSIVE PLAYER SCORECARD

Name ______________________________

Number of fielding attempts

ground balls (#) ______ attempts (#) ______ successes × 2 = ______ points

fly balls (#) ______ attempts (#) ______ successes × 1 = ______ points

Number of throwing plays

(#) ______ attempts

(#) ______ on-target × 1 = ______ points

(#) ______ to proper position × 2 = ______ points

Number of covering and backing-up plays

(#) ______ correct covers × 2 = ______ points

(#) ______ correct backups × 2 = ______ points

(#) ______ incorrect covers or backups × −1 = ______ points

Number of pitches thrown (if 50 percent pitches are strikes add 2 points)

(#) ______ balls (#) ______ strikes × 1 = ______ points

Total = ______ points

If you wish to keep track of your individual defensive play, you can make a scorecard of defensive skills and concepts, such as the Defensive Player Scorecard. You know best just what you want to work on, so adapt your scorecard to reflect this.

You can use these scorecards in Scrub and in all softball games. If you analyze the scorecards after you play, they can indicate your physical and mental levels of play. If you see areas where scores are not satisfactory, challenge yourself to work harder on those areas next time you play.

ONE-PITCH

One-Pitch is played exactly like a regulation game except that the batter is allowed only one pitch per time at bat: If the pitch is a ball, the batter walks; if the pitch is a strike, the batter must swing. This pitching-rule modification speeds up the game.

One-Pitch provides you with opportunities for practicing all softball skills and game concepts. You might review the scorecards you used in Scrub. If there are any particular skills or concepts in which you are weak, concentrate on them when relevant situations occur in the game. Anticipate the actions that might be called for by each situation; then your reaction is likely to be appropriate.

It will help you become more versatile if you play different positions during the game. Do not be afraid to change positions. In fact, you can become very skillful and knowledgeable by practicing *all* the infield and outfield positions, and don't forget the pitcher and catcher positions. You know the old saying, "Try it, you'll like it!"

Rules of One-Pitch and Method of Play

The following rules govern the method of play.

1. Use the regulation field of play. You must establish ground rules for out-of-play areas.
2. Use official teams, in terms of the number of people on a team and the positions played.
3. Official rules govern play except that the pitcher delivers one pitch per batter, with the following results:
   - If the pitch is a ball and the batter does not swing at it, the batter gets a walk (base on balls).
   - If the pitch is a strike and the batter does not swing, the batter is out on a strikeout.
   - If the batter fouls off the pitch, the batter is out.
   - If the batter hits the pitched ball fair, the hit ball is played out.

Success Goal = your team scores more runs than the opposing team

Your Score =

(#) ____ your team's score

(#) ____ opposing team's score

Step 24 Rundowns

Rundown situations occur when a runner is caught between bases by the defense. Remember, you may freely overrun first base and home plate, but not second or third base; also, if you go toward second base after overrunning first, the defense can make a play on you. Anytime you are off a base and not free to return to that base, and the defense tries to pick you off (catch you off base) by throwing the ball to a covering fielder, you are in a rundown situation.

WHY IS THE RUNDOWN IMPORTANT?

On defense, you want to take advantage of any mistake made by a baserunner and get an out. The challenge is always there for the defensive players and for the baserunner. As a baserunner, however, you do not want to be caught in a run-down situation, unless a run can score on the play. If there is no opportunity for a run to score, it means that you were not concentrating and got caught between bases. As defensive players, you want to be alert to the fact that a runner caught in a rundown when there is another runner on third base may be trying to distract you and allow the run to score. Be sure that you keep an eye on the runner at third base.

Planned and practiced techniques are necessary to ensure that a runner caught off base will be tagged out. If no set play were used, the runner could more easily escape the situation and reach a base safely.

HOW THE DEFENSE EXECUTES THE RUNDOWN

The rundown is most effectively executed when the defensive players have backup help in case the runner gets past one of the original chasers. The defensive players closest to the bases between which the runner is caught, the original chasers, are called the *primary fielders* for the play. The players next closest to the two bases involved back up those original chasers and are called the *backup fielders* for the play. A backup fielder takes an initial position in front of the base he or she is responsible for. As the rundown play develops, the backup fielder should stay at least 10 feet behind the primary fielder, unless the play has moved close to the base (see Figure 24.1).

Figure 24.1 Rundown between second base and third base.

The following chart identifies the fielders and their responsibilities in rundown plays.

Runner caught between	Primary fielders	Backup fielders
First and second	First baseman Second baseman	Pitcher Shortstop
Second and third	Shortstop Third baseman	Second baseman Pitcher
Third base and home	Third baseman Catcher	Shortstop Pitcher

BASIC DEFENSIVE RUNDOWN PLAY

The primary fielders can execute the rundown alone unless one is passed by the baserunner. When passed, the primary fielder steps aside to allow the backup fielder to move up to receive the throw. The original primary fielder rotates back behind the now-active backup fielder and assumes the backup role. The original backup fielder has now moved up to assume the new primary role. Rotation in this manner continues each time a primary fielder is passed by the baserunner. Of course, this rotation should not be necessary if the runner is put out with no throws or one throw! All persons involved in the rundown must maintain positions in front of the bases. The backup fielder, particularly, must not line up behind the bases, because a runner passing a primary fielder could then get to a base safely before encountering the backup fielder.

There are several possible methods of executing the run-down play in addition to the one described above. However, certain principles apply to any run-down situation, regardless of the method used to get the runner out.

1. Try to get the runner out with the least possible number of throws. Zero throws are ideal, one is good, two are all right, and more than two are too many.
2. The person with the ball (not necessarily one of the primary or backup fielders) initiates the play by holding the ball and running at (directly toward) the

runner until she or he commits to moving toward one base or the other. (This person can actually make the tag if he or she is close enough before the baserunner gets to the base.) The ball is then thrown to the fielder toward whom the runner is going.

3. Keep the runner away from the bases, but closer to the base last touched rather than the one not yet reached.
4. Throw the ball back and forth beside the runner, not over the head of the runner.
5. When making the tag, grip the ball with your bare hand and hold the ball and hand securely in your glove. Tag the runner with the back of your glove.

HOW THE OFFENSE EXECUTES THE RUNDOWN

As the runner, you must watch the person with the ball. Once the initial throw is made, you must decide whether to turn and run at full speed to the base away from the throw or stay in the run-down situation. If the throw is a long throw and you are fairly close to the person throwing the ball, your chances of making it to the base are good. If a preceding runner has a chance to score while the defense plays on you in the run-down situation, stay in the rundown and try to avoid being tagged out until the run scores.

Rundown Play Drills

1. Rotation Drill

This drill requires four persons positioned between two bases, two in primary roles and two in backup roles. No runner is used.

One primary fielder with a ball starts the drill by throwing the ball to the other primary fielder. After each throw, the assumption is made that the imaginary runner passes the thrower. Therefore, the thrower in the primary role rotates back, and the person in the backup role moves up to receive the next throw.

Success Goal = 30 seconds of continuous error-free run-down play (throwing, catching, and rotation movement)

Your Score = (#) _____ seconds

2. Full Rundown Play Number 1

Use the same setup as in the previous drill except a runner is used and rotation occurs only when the runner actually passes a primary fielder. The baserunner must wear a batting helmet in this drill.

Success Goals =

a. as runner, reach base safely or stay in the rundown for 30 seconds without being tagged

b. as defense, tag the runner out in less than 30 seconds, with no more than two throws

Your Score =

a. as runner, reach base safely YES ____ NO ____

(#) ____ seconds in rundown

b. as defense, (#) ____ throws

(#) ____ seconds before out

3. *Full Rundown Play Number 2*

Three pairs of partners are set up with both a backup fielder and primary fielder in front of two bases, and the third pair acting as the runner and the "initiator."

The initiator, with the ball, stands 20 feet from the runner (remember to wear a helmet), who is in the baseline midway between the two bases. The initiator starts the drill by running toward the runner. Once the runner commits to run to a base, the initiator throws the ball to the primary fielder at that base. From that point the drill proceeds as in the previous drill.

After three 20-second run-down bouts, rotate roles within each set of partners. After the next set of three 20-second bouts, pairs of partners rotate roles: The base 1 pair become runner and initiator; runner and initiator become base 2 fielders; and base 2 fielders move to base 1. Continue the drill until the rotation has gone full circle.

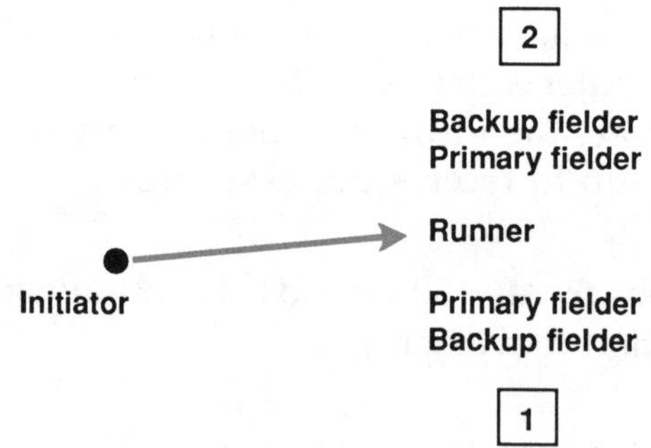

Success Goals =

a. as runner, reach base safely or stay in rundown for 20 seconds

b. as initiator, tag the runner out yourself or make a throw so that the runner stays in the run-down situation for the time being

c. as defense, tag the runner out in less than 20 seconds, with no more than two throws

Your Score =

a. as runner, reach base safely YES ____ NO ____

b. as initiator, tag the runner out yourself YES ____ NO ____

runner remains in run-down situation YES ____ NO ____

c. as defense, (#) ____ throws

(#) ____ seconds

Step 25 Coed Slow Pitch Game

Because this book has been designed both for your individual use and for use in instructional settings, this final step presents, for your participation and enjoyment, one of the official games that is particularly appropriate for class use. Many classes today are coeducational, females and males learning and participating together. To facilitate play in coed settings, specific rules have been developed for coed slow pitch softball. However, only the basic rules are presented here. You will need to refer to an official rule book for a description of the complete set of rules.

WHY IS COED SLOW PITCH SOFTBALL IMPORTANT?

Participation in coed softball takes place in a setting that is reflective of society. Males and females working together toward a common goal is an everyday occurrence in the workplace and in the family, but usually not so in sport, especially team sports. Softball is one of the few team sports (volleyball being another) that has an official game designed specifically for coed play. Coed softball thus provides class and recreational play opportunities in which women and men can learn to work together and develop respect for one another's abilities.

COED SLOW PITCH SOFTBALL

All the skills and knowledge that you have worked so hard to develop over the past 24 steps are now put to use in an official game between two teams. There are no special rules to ensure that you have ample opportunity to practice a particular skill or utilize a specific game concept. This game is for real!

As you undoubtedly noticed before you were far along on your climb up the steps to success staircase, softball is not a game that can be played alone. It isn't even very easy to practice the individual skills alone. You can play catch with yourself using a wall, hit a ball off a tee, or run around the bases all by yourself, but only for a short time. Others had to join you on your climb up the staircase if, indeed, you were to make much progress in your quest to become a skillful softball player. Softball is a team sport, and the ultimate in enjoyment occurs when two teams take the field together and challenge one another's skill and knowledge in game situations.

RULES AND METHOD OF PLAY

Uniforms and a freshly lined field are not a prerequisite for your enjoyment of a class version of coed softball, but teams made up of five females and five males are a must. Most of the rules that you have learned for slow pitch softball are applicable to the coed game. Some of the rules that specifically apply to the coed game follow.

1. Use a regulation field of play. You must establish ground rules for out-of-play areas.
   - Baseline distances: 65 feet for players 13 years old and older; 60 feet for youths 11 and 12 years old; 55 feet for youths 10 years old and younger

- Pitching distances: 46 feet for players 13 years old and older; 40 feet for youths 11 and 12 years old; 35 feet for youths 10 years old and younger

2. Defensive positioning (5 females and 5 males)
   - Outfield: 2 females and 2 males in any of the four positions
   - Infield: 2 females and 2 males in any of the four positions
   - Pitcher and catcher: 1 female and 1 male in either position
3. Batting rules
   - Batting order alternates between the sexes
   - Walk (base on balls) to a male batter: The next batter (a female) after a male has walked can, before stepping into the batter's box, choose between an automatic walk or hitting. *Note:* This is the official rule, which could be modified (or even not used) in an unofficial class game, if desired. The official rule's purpose is to prevent the pitcher from intentionally walking the male batters in order to have to pitch only to female batters.
4. Scoring
   - Official rules concerning scoring runs and deciding the winner apply. The team with more runs wins the game.

Enjoy the game!

Rating Your Total Progress

Throughout this book you have been working on developing both the physical skills needed for softball and the psychological preparation for play. The following self-rating inventory is provided so that you can rate your overall progress. Read the statements carefully and respond to them thoughtfully.

PHYSICAL SKILLS

The first general success goal in a softball course is to acquire the physical skills needed to practice and to play the game. How would you rate yourself on these skills?

	Very good	Good	Okay	Poor
Catching	_____	_____	_____	_____
Throwing overhand	_____	_____	_____	_____
Fielding ground balls	_____	_____	_____	_____
Hitting using the batting tee	_____	_____	_____	_____
Fungo hitting ground balls	_____	_____	_____	_____
Pitching	_____	_____	_____	_____
Baserunning				
Overrunning the base	_____	_____	_____	_____
Rounding the base	_____	_____	_____	_____
Fielding fly balls	_____	_____	_____	_____
Hitting a soft toss	_____	_____	_____	_____
Tossing for soft toss	_____	_____	_____	_____
Fungo hitting fly balls	_____	_____	_____	_____
Hitting a pitched ball	_____	_____	_____	_____
Sidearm throwing	_____	_____	_____	_____
Detecting your errors	_____	_____	_____	_____
Correcting your errors	_____	_____	_____	_____
Detecting your partner's errors	_____	_____	_____	_____
Helping your partner correct errors	_____	_____	_____	_____

GAME CONCEPT SKILLS

The second general success goal in a softball course is utilizing your game-concept skills to improve your game play. How would you rate your physical and mental abilities to utilize the following to your advantage?

	Very good		Good		Okay		Poor	
	Physical	Mental	Physical	Mental	Physical	Mental	Physical	Mental
Position play as an infielder								
Covering	____	____	____	____	____	____	____	____
Backing up	____	____	____	____	____	____	____	____
Position play as an outfielder								
Covering	____	____	____	____	____	____	____	____
Backing up	____	____	____	____	____	____	____	____
Force play	____	____	____	____	____	____	____	____
Tag play	____	____	____	____	____	____	____	____
Double plays								
Shortstop drag step	____	____	____	____	____	____	____	____
Shortstop inside pivot	____	____	____	____	____	____	____	____
Second baseman crossover	____	____	____	____	____	____	____	____
Second baseman rocker	____	____	____	____	____	____	____	____
Third to first	____	____	____	____	____	____	____	____
Home to first	____	____	____	____	____	____	____	____
Relays	____	____	____	____	____	____	____	____
Cutoffs	____	____	____	____	____	____	____	____
Rundowns (defensive)	____	____	____	____	____	____	____	____
Rundowns (offensive)	____	____	____	____	____	____	____	____
Rules of play								
Do you know the rules?	____	____	____	____	____	____	____	____
Do you play by them?	____	____	____	____	____	____	____	____

OVERALL SOFTBALL PROGRESS

Considering all of the factors you rated above, how would you rate your softball progress?

____ Very successful

____ Successful

____ Barely successful

____ Unsuccessful

Are you pleased with your progress?

____ Very pleased

____ Pleased

____ Somewhat pleased

____ Not pleased

ADDITIONAL COMMENTS AND QUESTIONS

Look back over your self-ratings. What are your strengths and weaknesses? Are you willing to spend time to improve your game? Where do you want to go from here?

If you have made some improvement and feel comfortable in at least some aspects of the game, you should feel good about those accomplishments. The real rewards for becoming skillful do not always come in the form of trophies or plaques, but rather in the way you feel about yourself. That is what this game (or any game) is really all about. Helping you develop some skill and knowledge—so that you can have fun, meet some new and interesting people, and enjoy yourself when playing softball—is the goal of this book.

Individual Program

INDIVIDUAL COURSE IN ____________ GRADE/COURSE SECTION ____________

STUDENT'S NAME ____________ STUDENT ID # ____________

	SKILLS/CONCEPTS	TECHNIQUE AND PERFORMANCE OBJECTIVES	WT* ×	POINT PROGRESS**				= FINAL SCORE***
				1	2	3	4	

Note. From "The Role of Expert Knowledge Structures in an Instructional Design Model for Physical Education" by J.N. Vickers, 1983, *Journal of Teaching in Physical Education*, **2**(3), p. 17. Copyright 1983 by Joan N. Vickers. Adapted by permission.

*WT = Weighting of an objective's degree of difficulty.

**PROGRESS = Ongoing success, which may be expressed in terms of (a) accumulated points (1, 2, 3, 4); (b) grades (D, C, B, A); (c) symbols (merit, bronze, silver, gold); (d) unsatisfactory/satisfactory; and others as desired.

***FINAL SCORE equals WT times PROGRESS.

Glossary

ball A pitch that does not enter the strike zone.

base on balls A batter gains first base when four pitches judged to be balls are delivered to the batter during a turn at bat; also called a *walk*.

base path An area 6 feet wide running between the bases, the center of which is a direct line from base to base.

baserunner A player of the offensive team who has reached first base safely.

batter-baserunner A player who has finished his or her turn at bat and has not yet reached first base.

catch A ball caught by a fielder in the bare hand or the glove.

chopped ball An illegal hitting technique in slow pitch softball, the batter's hitting the ball with a short downward swing.

coed softball An official game of softball played by teams made up of 5 men and 5 women positioned so that 2 men and 2 women are in the outfield, 2 men and 2 women are in the infield, and 1 man and 1 woman pitch and catch.

defensive team The team in the field.

double A two-base hit.

double play A play by the defense in which two offensive players are put out by continuous action.

fair ball A batted ball that is touched or comes to rest in fair territory in the infield; that touches first, second, or third base; that is touched or first lands on or over fair territory in the outfield (beyond first, second, or third base).

fair territory That part of the playing field between the first and third base foul lines, including home plate.

fast pitch softball An official game of softball, played by teams of 9 players, in which the underhand pitch is delivered to the batter with considerable speed.

fly ball A batted ball that goes into the air.

foul ball A batted ball that is touched or comes to rest in foul territory in the infield; that is touched or first lands on or over foul territory in the outfield (beyond first or third base).

foul territory That part of the playing field between the first and third base foul lines and the out-of-play area surrounding the field.

foul tip A batted ball that goes, not higher than the batter's head, directly back to and is caught by the catcher.

ground ball A batted ball that is hit directly onto the ground; a *grounder*.

ground rule A rule of play established, typically, to identify the boundaries of a field, especially out-of-play areas, for a playing field that is not an enclosed ballpark.

ground rule double A base hit that, because of going into an out-of-play area, limits the batter to two bases.

home team The team upon whose ground the game is played.

infield That area of the playing field in fair territory typically covered by infielders; for most softball fields, it is the area of the playing field that is dirt, or "skinned."

inning That portion of a game in which each team has three outs while on offense and three on defense; a softball game is seven innings long.

line drive A fly ball that travels into the playing field relatively parallel to the ground.

middle line of knuckles The batting-grip knuckle alignment in which the second knuckles of the fingers of both hands are lined up one over the other in a straight line.

offensive team The team up at bat.

outfield That area of the playing field in fair territory beyond the infield.

overthrow A thrown ball that goes beyond the intended receiver; in rules terminology, it is a thrown ball that goes into foul territory or out of play.

pop fly A fly ball hit in the infield area; a *pop-up*.

runner A term used synonymously with *baserunner* and *batter-baserunner*.

slow pitch softball An official softball game played by teams of 10 players in which the pitch must be delivered underhand and must traverse an arc of between 6 and 12 feet.

strikeout Occurs when the batter swings at and misses, or fails to swing at, a third strike.

strike zone In fast pitch, that area over home plate between the batter's armpits and the top of the knees; in slow pitch that area over home plate between the batter's highest shoulder and the knees.

About the Authors

Diane L. Potter, EdD, is professor of physical education at Springfield College in Springfield, Massachusetts. A teacher with over 27 years of experience in physical education professional preparation with responsibility for softball skills and coaching classes, Dr. Potter also coached the Springfield College softball team for 21 years. In addition she was a player for 15 years in Amateur Softball Association (ASA) Class A Fast Pitch. Dr. Potter has been an international clinician in softball, conducting clinics in Italy and The Netherlands and taking teams to The Netherlands in 1971, 1975, and 1982. In 1982 she was awarded the Silver Medaillen by the Koninklijke Nederlandse Baseball en Softball Bond (the Royal Dutch Baseball and Softball Association); she is the only woman so honored.

Dr. Potter is an outstanding leader in women's sport; she has served as a member of the AIAW Ethics and Eligibility Committee and was inducted into the National Association of Collegiate Directors of Athletics Hall of Fame in 1986. She is an outstanding teacher and coach and has served as a role model for many fine women teacher-coaches who have been her students and her players. Her recreational activities include horseback riding, fishing, camping, and listening to classical music.

Gretchen A. Brockmeyer, EdD, is an associate professor of physical education at Springfield College, in Springfield, Massachusetts. She has been a teacher educator for over 20 years, with primary responsibility for secondary physical education methods and supervision of field-based teaching experience. A coach of many different sports, she served as assistant coach for the Springfield College softball team for 7 years. Dr. Brockmeyer is an exceptional teacher educator and role model. She is committed to helping her students become physical educators who possess the skills and the professional commitment necessary to provide meaningful learning opportunities for those they teach. Her students have consistently confirmed her faith in their abilities by becoming successful teachers who are caring human beings. Dr. Brockmeyer's leisure time is devoted to golfing, reading, music, and nurturing Seth and Alyssa.